Bioactive Phytochemicals

Perspectives for Modern Medicine

— Volume 2 —

The Editor

Dr. Vijay Kumar Gupta, Ph.D., FLS, London (born 1953-) Chief Scientist, CSIR- Indian Institute of Integrative Medicine, Jammu, India. He did his M.Sc. (1975) and Ph.D. (1979) in Zoology both from University of Jammu, Jammu-India. His research capabilities are substantiated by his excellent work on histopathology, ecology and reproductive biology of fishes, turtles, birds and mammals, which has already got recognition in India and abroad.

Dr. Gupta has to his credit more than 100 scientific publications and review articles which have appeared in internationally recognized Indian and foreign journals. Founder fellow, life member and office bearer of many national societies, academies and associations. He has successfully completed a number of research/consultancy projects funded by government, private and multinational agencies. His current areas of interest are histopathology, toxicology, pre-clinical safety pharmacology, reproductive efficacy studies of laboratory animals and biodiversity.

He is the Series Editor of the recently published multi-volume set of books, "**Comprehensive Bioactive Natural Products (Vols. 1-9)**", published by M/S Studium Press, LLC, USA. He is also Editor-in-Chief of the books, "**Utilisation and Management of Medicinal Plants**", "**Medicinal Plants: Phytochemistry, Pharmacology and Therapeutics (Vols. 1 and 2)**", "**Traditional and Folk Herbal Medicine**", "**Natural Products: Research Reviews**", "**Bioactive Phytochemicals: Perspectives for Modern Medicine**", "**Perspectives in Animal Ecology and Reproduction (Vols. 1-9)**" and "**Animal Diversity, Natural History and Conservation**". The Editor-in-Chief of the American Biographical Institute, USA, has appointed him as *Consulting Editor* of *The Contemporary Who's Who*. Dr. Gupta also appointed as Nominee for the *Committee for the Purpose of Control and Supervision of Experiments on Animals* (CPCSEA, Govt. of India). The *Linnaean Society of London, U.K.* has awarded fellowship to him in November 2009 in recognition of his contribution towards the cultivation of knowledge in Science of Natural History. Recently, Modern Scientific Press, USA has nominated Dr. Gupta as the Editor of the *International Journal of Traditional and Natural Medicine*.

Bioactive Phytochemicals

Perspectives for Modern Medicine

— Volume 2 —

Editor

V.K. Gupta

Chief Scientist

CSIR-Indian Institute of Integrative Medicine

Canal Road, Jammu – 180 001,

India

2014

Daya Publishing House®

A Division of

Astral International Pvt. Ltd.

New Delhi – 110 002

Cataloging in Publication Data—DK
Courtesy: D.K. Agencies (P) Ltd. <docinfo@dkagencies.com>

Bioactive phytochemicals : perspectives for modern medicine / editor, V.K. Gupta.
v. 2 cm.
Includes bibliographical references and index.
ISBN 9788170357780 (series)
ISBN 9789351300915 (series) (International edition)
ISBN 9788170359111 (v. 2) (Hardbound)
ISBN 9789351300922 (v. 2) (International edition)

1. Pharmacognosy. 2. Phytochemicals. 3. Antineoplastic agents. I. Gupta, V. K. (Vijay Kumar), 1953-

DDC 615.32 23

Published by : **Daya Publishing House®**
A Division of
Astral International Pvt. Ltd.
– ISO 9001:2008 Certified Company –
81, Darya Ganj, Near Hindi Park,
Delhi Medical Association Road,
New Delhi - 110 002
Phone: 011-4354 9197, 2327 8134
Fax: +91-11-2324 3060
E-mail: info@astralint.com
Website: www.astralint.com

Laser Typesetting : **Classic Computer Services**, Delhi - 110 035

Printed at : **Salasar Imaging Systems**, Delhi - 110 035

PRINTED IN INDIA

Editorial Board

Prof. Ian Fraser Pryme

Department of Biomedicine,
University of Bergen,
Jonas Lies vei 91, N-5009 Bergen, NORWAY
Tel: +47-55-586438
Fax: +47-55-586360
E-mail: ian.pryme@biomed.uib.no

Dr. George Qian Li

Herbal Medicines Research and Education Centre,
Faculty of Pharmacy, The University of Sydney,
NSW 2006, AUSTRALIA
Tel: 612 9351
Fax: 9351 8638
E-mail: george.li@sydney.edu.au

Dr. Julia Serkedjieva

Institute of Microbiology,
Bulgarian Academy of Sciences,
26, Academician Georgy Bonchev St.,
1113 Sofia, BULGARIA
Tel: +359 2 979 31 85
Fax: +359 2 870 01 09
E-mail: jserkedjieva@microbio.bas.bg

NIIGATA UNIVERSITY 新潟大学
Faculty of Engineering
Department of Chemistry and Chemical Engineering
Ikarashi 2-chome, Nishiku, Niigata 950-21, Japan

Foreword

Bioactive phytochemicals have served as an important source of drugs since ancient times and about half of the useful drugs today are derived from natural sources. Recently, a renewed interest in obtaining bioactive secondary metabolites from plants and microorganisms including fungal endophytes has been observed, notwithstanding the impressive progress competing methodologies such as combinatorial chemistry and high-throughput screening or genetic engineering. This world-wide attention on bioactive phytochemicals is based on their low or absent toxicity, their complete biodegradability, their availability from renewable sources, and their low cost compared with that of corresponding synthetic drugs. The development of separation techniques as well as that of spectroscopic methods allows the isolation of bioactive phytochemicals and their characterization for a short period. The development of efficient screening methods as well as that of biological studies in cell level makes it possible to find bioactive phytochemicals and their reaction mechanisms using a small amount of sample for a short period. In developing countries, the search for biologically active compounds obtainable from locally available plants is also important to reduce public health costs there.

This book contains contributions of experts in multidisciplinary fields related to bioactive phytochemicals, covering alkaloids, flavonoids, neem leaf glycoprotein, betulinic acid, iridoids, iridoid glycosides, bibenzyls and bis(bibenzyls)s, naphthoquinones, diterpenoids, steroids, tritepenoids, capsaicin, imidazole alkaloids, podophyllotoxin-A and its related compounds, coumarins, L-DOPA, anticancer agents from microbes and plants, indol alkaloids, sesquiterpene lactones, phenols, picroside-1, kutkoside, ginkglides, ochratoxin-A, and embelin. The plant sources of the researches are neem leaf, the bark of the white birch tree, fruits of *Vitex* plants, bryophytes, *Heterophragma quadriloculare* and *Heterophragma adenophyllum*, *Pilocarpus jaborandi*, *Podophyllum peltatum*, *Calophyllum inophyllum*, *Mucuna pruriens*, *Catharanthus roseus*, endophytic fungi, *Cichorium* species, *Picrorhiza kurroa* (Picroliv), *Aspilia africana*, *Elsholtzia communis*, amazon fruits, *Coriandrum sativum*, *Bauhinia variegata*, and *Embelia*. The biological activities of the medicinal chemicals discussed in this book are antiphlogistic and antiarthritic activities of alkaloids and flavonoids, promotion of therapeutic anti-tumor immunity by neem leaf glycoprotein, anticancer activities of betulinic acid (triterpene) and its derivatives, antibacterial, antioxidant, cytotoxic activities toward human cancer cell lines, and hepatoprotective activities of fruits of *Vietex* plants, antifungal activity of bibenzyl and bis(bibenzyls)s, antitumor, antimicrobial, antifungal, antiviral, and anti-inflammatory activities of lapacol and its analogs, anticancer activity of capsaicin, parasympathominetic activity of imidazole alkaloids, anticancer activity of podophyllotoxin, anti-HIV, antitumor, antiulcer, antiinflammatory, and antibacteria activities of chemical constituents of leaves of *Calophyllum inophyllum*, anti-Parkinson activity of L-DOPA, anticancer agents of natural products from biodiverse organisms, anticancer activity of indole alkaloids, antidiabetic and hypolipidemic activities of compounds from fungal endophytes, antibacterial, antihepatotoxic, antiulcerogenic, and antiinflammatory activities of root extracts of *Chichorium* species, hepatoprotective cell-protective, and gene expression regulatory activities of picroside I and kutokoside, antiinflammatory, antimicrobial, contraceptive, and antimalarial activities of constituents of *Aspilia africana*, anti-oxidant activity of extracts of *Elsholtzia communis*, antishitosomal, antiviral, and antihepatotonic activities of lignoids from Burseraceae, antioxidant and cytoprotective activities of lignans in Amazon fruits, antiinflammatory activities of flavonoids from *Bauhinia variegate*, antibacterial, antiulcer, and antidiabetic activities of extracts of bark of *Bauhinia variegate*, and antinociceptive, antiinflammatory, antioxidant activities, and X-linked inhibitor of apotosis of embelin.

Continuing first volume, the present volume ***Bioactive Phytochemicals: Perspectives for Modern Medicine, Vol. 2*** deals with recent progress on bioactive phytochemicals. I am sure these findings will provide lead material for further investigations to researchers interested in the discovery of new drugs of natural origin. I hope, this volume will be widely read and used by all interested in promoting sustainable health security.

I feel delight to prepare this foreword and applaud the great effort of editor, Dr. V.K. Gupta in editing this excellent volume and thanks the contributors in contributing their valuable and informative research/review chapters.

Masayoshi Ando

Emeritus Professor

Preface

Medicinal plants are a rich source of bioactive phytochemicals or bionutrients and have been used as drugs for millennia. The phytochemicals present in plants are responsible for preventing disease and promoting health have been studied extensively to establish their efficacy and to understand the underlying mechanism of their action. Such studies have included identification and isolation of the chemical components, establishment of their biological potency both by *in vitro* and *in vivo* studies in experimental animals and through epidemiological and clinical-case control studies in man.

Nature has continuously provided mankind with a broad and structurally diverse array of pharmacologically active compounds that have proved to be indispensable for the cure of deadly diseases or as lead structures for novel pharmaceuticals. Their dominant role is evident in the approximately 60% of anticancer compounds and 75% of drugs for infectious diseases that are either natural products or natural product derivatives. At present 25% of the modern medicines are developed from plants that were first used traditionally, and many synthetic drugs have also been obtained from natural precursors. Almost, 70% modern medicines in India are derived from natural products. Studies carried out during the past have shown that these phytochemicals have an important role in preventing chronic diseases like cancer, diabetes and coronary heart disease. For example, Hippocrates may have prescribed willow tree leaves to abate fever. Salicin, having anti-inflammatory and pain-relieving properties, was originally extracted from the bark of the white willow tree and later synthetically produced became the staple over-the-counter drug called Aspirin. There is evidence from laboratory studies that phytochemicals in fruits and vegetables may reduce the risk of cancer, possibly due to dietary fibers, polyphenol antioxidants and anti-inflammatory effects. An important cancer drug, Taxol (paclitaxel), is a phytochemical initially extracted and purified from the Pacific yew tree. Among phytochemicals from edible plants with promise for deterring disease, diindolylmethane, from Brassica vegetables is being tested against recurring respiratory papillomatosis tumors, is in

Phase III clinical trials for cervical dysplasia and is in several clinical trials for prostate cancer. Some phytochemicals with physiological properties may be elements rather than complex organic molecules. Abundant in many fruits and vegetables, selenium, for example, is involved with major metabolic pathways, including thyroid hormone metabolism and immune function. Particularly, it is an essential nutrient and cofactor for the enzymatic synthesis of glutathione, an endogenous antioxidant.

With the advances in synthetic methodology and the development of more sophisticated isolation and analytical techniques, many more of these phytochemicals should be identified as effective as synthetic medicines with fewer or no side effects and contraindications. There is, however, much scope for further systematic research in screening medicinal plants for these phytochemicals and assessing their potential in protecting against different types of diseases. It has been proved that although the effects of natural remedies may seem slower, the results are sometimes better on the long run especially in chronic diseases.

The chapters included in the present volume "***Bioactive Phytochemicals: Perspectives for Modern Medicine, Vol.** 2*" are likely to lead further researches in this direction and it is hoped that this publication would attract world wide audience of researchers and the academicians of allied disciplines engaged in research of new drug from plant resources.

V.K. Gupta

Contents

Bioactive Phytochemicals: Perspectives for Modern Medicine Vol. 2 (2014)
Editor: **V.K. Gupta**
Published by: **DAYA PUBLISHING HOUSE, NEW DELHI**

Pages **1–49**

1

Alkaloids and Flavonoids: Versatile and Novel Bioactive Natural Products, a New Avenue for the Development of New Generation Anti-Phlogistic and Antiarthritic Agents

Suresh Ramnath Naik[1*], Santosh Manohar Wala[1]
and Priyank Dilipkumar Shah[1]

ABSTRACT

Inflammation is an endogenous repair mechanism(s) or protective responses by activating both local and systemic responses after injury due to physical, chemical, thermal insults and/or micro-organisms infection. The use of traditional medicine is expanding to new horizons and plants still remained as the novel source of structurally important compounds that lead to the development of novel drugs. The traditional system of medicine remained as a major guidelines for the use of plants in the treatment of various inflammatory diseases. Many substances which interfere with the inflammatory responses have been isolated from plants/herbs. The present review describes the various alkaloids and flavonoids that were isolated and identified from the plants, which have been practiced and used for the treatment of inflammation and inflammation associated disorders. Keeping such traditional practices, folk medicine of diverse countries of the world, an attempt was made to unfold and define such therapeutic activities and importance with versatile experimental results including the underlying mechanism(s) of action. In addition, the application of the advanced knowledge as well as the state-of-the-art- of-methods and instrumentation coupled with novel approaches

1 Department of Pharmacology, Sinhgad Institute of Pharmaceutical Sciences, Off Mumbai-Pune Expressway, Kusgaon (Bk.), Lonavala, Pune – 410 401, Mh., India

* *Corresponding author*: E-mail: srnaik5@rediffmail.com

for the development of new generation phytopharmaceuticals for the treatment as well as the interventions of inflammatory disorders. Flavonoids are polyphenolic compounds that are distributed ubiquitously in different parts of the plant. Similarly, alkaloids are also a group of natural products with low-molecular-weight, nitrogen-containing compounds present in various plants and herbs. Flavonoids and alkaloids exhibited antiinflammatory activity in in-vitro and in-vivo models. The associated diverse mechanisms of flavonoids and alkaloids are: (I) inhibition of eicosanoid generating enzymes phospholipase A_2, cyclooxygenases, and lipoxygenases, (II) the modulation of proinflammatory gene expression such as cyclooxygenase-2, inducible nitric oxide synthase (iNOS), and several pro-inflammatory mediators (cytokines). With such distinct cellular and molecular mechanisms, flavonoids and alkaloids are considered to be novel promising antiinflammatory phytopharmaceuticals.

Keywords: Alkaloids, Flavonoids, Antiinflammatory, Antioxidant, Prostaglandin, Cytokines.

Introduction

Inflammation is defined as the local response or body defense mechanism(s) to eliminate foreign matters, pathogens and noxious substances such as chemical or physical injury followed by removal of necrosed cells and tissues (Medzhitov, 2008). The immediate response to noxious stimulus would result in increased local blood flow, capillary permeability, production of soluble cellular and intra cellular mediators (Vasoactive amines, cytokines, chemokines, lipid mediators) (Germolec *et al.*, 2010) by the resident cells (*i.e.*, tissue macrophages, dendritic cells, lymphocytes, endothelial cells, fibroblasts and mast cells) at the inflammation site which promote the development of efflux of proteins and influx of granulocytes from the blood stream (Gordon *et al.*, 2005). The cardinal signs of inflammation are: pain, oedema, (due to increase vascular permeability) redness, heat and loss of function. Chronic inflammation is defined as a progressive shift in the type of cells present at the site of the inflammation and is characterised by simultaneous destruction and repairing of the tissue from the inflammatory process. The acute inflammatory response encompasses a complex and well coordinated cascade of events involving a large number of pharmacological, physiological, biochemical, and molecular changes. Acute inflammation has therapeutic consequence whereas chronic inflammation leads to degeneration of tissues as well as structural cum organ alterations (arthritis, spondylitis, pancreatitis etc.).

Bioactive Flavonoids as Antiinflammatory Agents

Flavonoids are a group of chemical entities of benzopyrone derivatives or a polyphenolic compounds widely distributed in the plant kingdom and are found in fruits, vegetables, grains, bark, roots, stems, flowers, tea, and wine (Kuhnau, 1976; Middleton, 1998). They are largely classified as chalcones, flavan-3-ols, flavanones, flavones, flavonols, isoflavones, and biflavonoids on the basis of their chemical structure (Table 1.1) (Rice-Evans *et al.*, 1996; Beecher, 2003). They possess simple chemical structures, and exists more than 4,000 derivatives, indicating their chemical diversities and most of them are responsible for the attractive colours of flowers, fruits, and leaves (De Groot and Rauen, 1998). Flavonoids, also known as nature's tender drugs, possesses wide spectrum of chemotherapeutical and pharmacological activities including antineoplastic, antibiotic, antiviral, antiulcer, hepatoprotective, antiinflammatory, immunomodulatory, and antithrombotic activities (Havsteen, 1983).

Table 1.1: Main classes of flavonoids, the individual phytoconstituents, and their sources.

Class/Groups	*Phytoconstituents*	*Sources of Phytoconstituents*
Flavones	Apigenin, luteolin, Chrysin, kaempferol, Myricetin, Rutin, Sibelin, quercetin.	Apple skins, Berries, Broccoli, Luteolin Celery, Fruit peels, Grapes, Lettuce, Olives, Onions, Parsley, Cranberries, Green leafy spices.
Flavanones	Fisetin, naringenin, hesperetin, eriodictyol, Narigin, Taxifolin.	Citrus fruit, Citrus peel.
Flavanols	Catechin, gallocatechin, epicatechin.	Tea leaves, red grapes and red wines.
Isoflavones	Daidzein, genistein, glycitein, biochanin A.	Soybeans, soy foods, and legumes.
Flavonols	Kaempferol, myricetin, quercetin, isorhamnetin.	Found abundantly in many foods.
Catechins	Catechin, Epicatechin, Epigallocatechin gallate.	Red wine, Tea
Anthocyanins	Cyanidin, Delphinidin, Malvidin, Pelargonidin, Peonidin, Petunidin.	Berries, Cherries, Grapes, Tea, Fruit peels with dark pigments, Raspberries, Red grapes, Red wine.

Flavonoids elicit antiinflammatory activity on various *in-vivo* and *in-vitro* animal models of inflammation. They are found to regulate/modulate cellular activities of the inflammation associated cells: mast cells, macrophages, lymphocytes, and neutrophils. Some flavonoids are known to inhibit histamine release from mast cells and others inhibit T-cell proliferation (Middleton *et al.*, 2000). Flavonoids (Quercetin, luteolin, 3,4-dihyroxyflavone, galangin, and morin) (Lanni *et al.*, 1985; Bauman *et al.*, 1980) affect the cascade of enzyme *viz.* phospholipase A_2 (PL A_2) (Kim *et al.*, 2001), cyclooxygenase (COX) (Nakadate *et al.*, 1985), and lipoxygenase (LOX) (Laughton *et al.*, 1991) and the nitric oxide (NO) (Krol *et al.*, 1995) producing enzyme, nitric oxide synthase (NOS), which are actively participating in arachidonic acid (AA) and eicosanoids metabolism. Inhibition of such enzymes by flavonoids reduces the production of AA, prostaglandins (PG), leukotrienes (LT), and NO, thereby preventing the synthesis of such potent inflammatory mediators. Also, flavonoids inhibit the various pro-inflammatory mediators like cytokines (IL-1, IL-6, IL-13, TNF-α, etc.) in *in-vitro* models. The recent studies demonstrated that certain flavonoids are found to be responsible for the modulation of gene expression, partly, via suppression of transcription factor activation by inhibition of protein kinases involved in signal transduction including protein kinase C (PKC) and mitogen activated protein kinase (MAPK) (Ferriola *et al.*, 1989; Weinstein *et al.*, 1992). Through the inhibition of such enzymes, DNA-binding capacity of transcription factors such as NF-8B or activator protein-1 (AP-1) is regulated. Therefore, it is likely that such genetic propagated events seem to control the expression rate of the target gene, and may be responsible for the attenuation of the inflammation processes. In addition, there are few reports

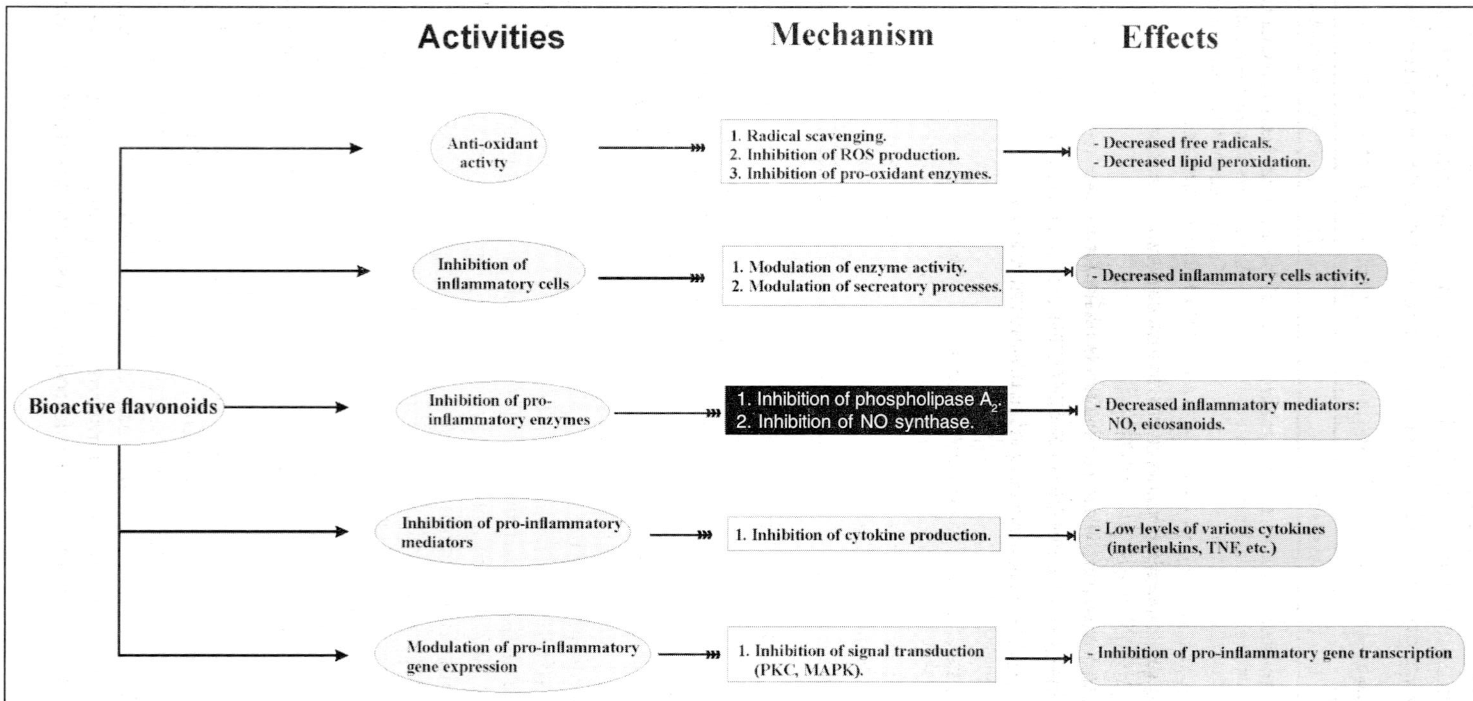

Figure 1.1: The biochemical and pharmacological effects of flavonoids on inflammation and inflammation associated diseases.

suggesting that flavonoids might be able to inhibit iNOS and COX-2 expression by activating peroxisome proliferator-activated receptor- γ (Banerjee *et al.*, 2002; Liang *et al.*, 2001) and likely to act as inhibitors of proteasome activity (Kazi *et al.*, 2003).

Some Important Plants Species Containing Flavonoids Eliciting Promising Antiinflammatory Activity and Further Avenue for Development of Plant Derived Products

Grindelia robusta Nutt.

G. Robusta Nutt. (Compositae), extracts contained important constituents (large amount of resin rich in grindelan-type diterpenes, flavonoids, acetylenes, essential oil, saponins, tannins and phenolic acids) in Italian traditional medicinal preparation due to their antiinflammatory, antioxidant, antimicrobial and expectorant activities. The antiinflammatory activity (Inhibition of human neutrophil elastase activity) was observed with the extract of *G. robusta,* due to the presence of active components, quercetin-3-methylether and 6-OH-kaempferol-3, 6-dimethylether (flavonoids). Subsequently, it was confirmed that quercetin-3-methylether (Figure 1.2) was 10 fold more potent than 6-OH-kaempferol-3, 6-dimethylether in its antiinflammatory activity. Therefore, Quercetin-3-methylether is considered to be a marker compound for standardisation of herbal preparation of *G. Robusta* (Liselotte *et al.*, 2009).

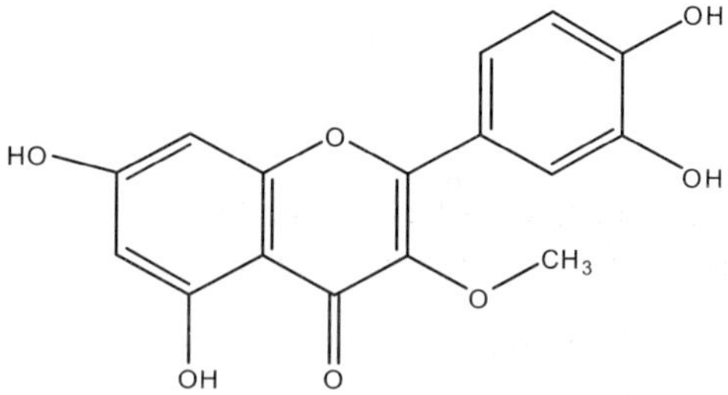

HO OH OH O OH O CH3 O

Quercetin 3-methylether

Eupatorium arnottianum Griseb.

Eupatorium arnottianum Griseb. (Compositae) is herb that grows in the North-east and centre of Argentina and South of Bolivia. The rural populations of Santa Victoria e Iruya used for the treatment of gastric pains and by "kallawaya" healers from the bolivian altiplano against respiratory disorders (asthma, bronchitis, colds) and topically for the treatment of bone fractures and dislocations by incorporating into the plasters. *E. arnottianum* extracts obtained from Dichloromethane and ethanol showed antiinflammatory activity in TPA mouse ear oedema and carrageenan rat paw oedema models. The flavonoids (nepetin, jaceosidin and hispidulin) isolated from dichloromethane extract of *E. arnottianum* found to be responsible for antiinflammatory activity in TPA-induced ear oedema in mice and carrageenan rat paw oedema by

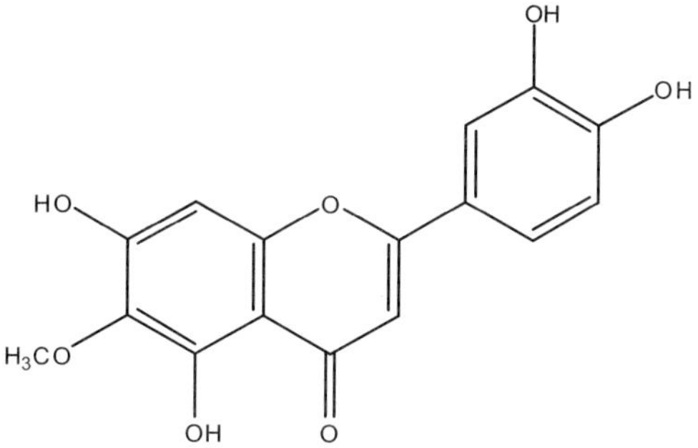

OH OH HO O H_3CO OH O

Nepetin

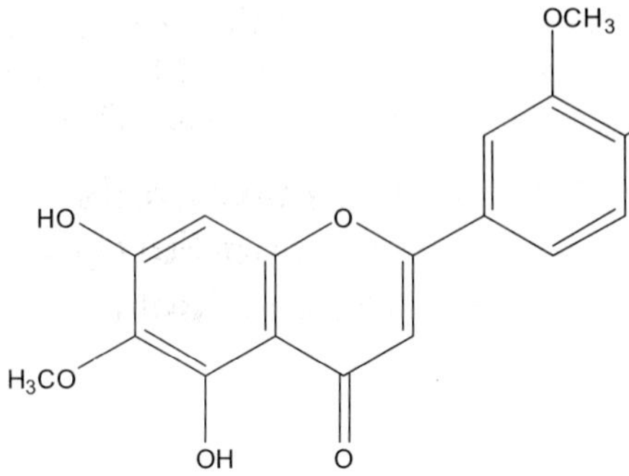

Jaceosidin

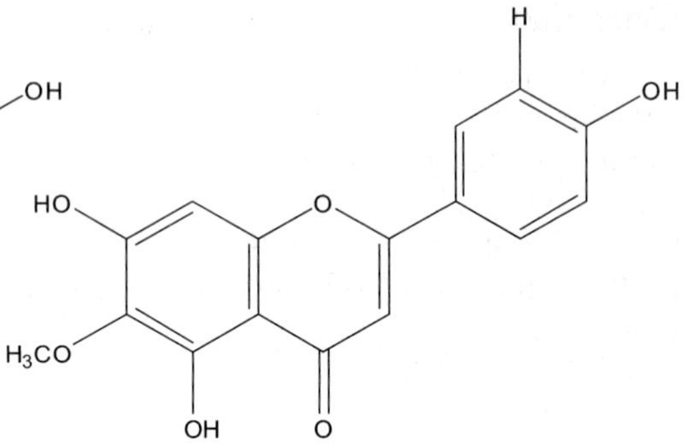

Hispidulin

inhibiting NF-8B induction, a biomolecule required for the expression of intracellular pro-inflammatory mediators (Clavin *et al.*, 2007).

Boldoa purpurascens Cav.

Boldoa purpurascens Cav. (Nyctaginaceae), a plant, commonly known as 'nitro', is a bush distributed in Latin America, including Cuba, where its leaves are used as a potent diuretic. *B. purpurascens* exhibited the antiinflammatory activity in dextran 1 per cent induced rat paw oedema model (acute inflammation) and the cotton pellet induced granuloma in rodents (chronic inflammation). Subsequently, isolated

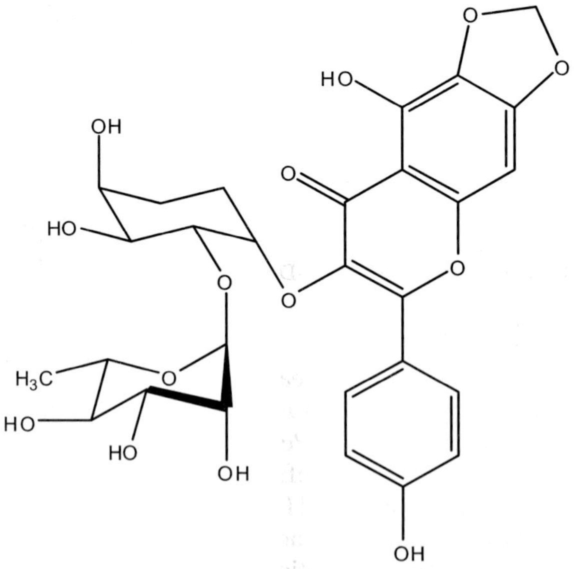

4, 5-dihydroxy-6, 7 methylenedioxyflavonol 3-O-α-L-rhamnopyranosyl-(1-2)-β-D-xylopyranoside

4, 5-dihydroxy-6, 7 methylenedioxyflavonol 3-O-α-L-rhamnopyranosyl-(1-2)-β-D-xylopyranoside, (flavonoid glycoside) from the leaves extract of *B. purpurascens* and found to be responsible for antiinflammatory activity by the inhibition of the release of inflammatory mediators (Mosquera *et al.*, 2011).

Caralluma attenuata Wight.

Caralluma attenuata Wight. (Asclepiadaceae) is a thick, succulent perennial herb growing wild in and around Warangal and several other districts of Andhra Pradesh, India. The cocktail of *C. attenuata* and black pepper is recommended for the treatment of migraine [2]. The flavone glycoside chemically, luteolin-4-O-[α-(L-rhamnopyranosyl (1.2)-β-D-glucopyranoside)] is a major constituents of plant *C. attenuata* is responsible for the antiinflammatory activity observed in carrageenan induced rat paw oedema, which was comparable to that of ibuprofen, a potent non-steroidal anti inflammatory drug (Ramesh *et al.*, 1998; Kumar *et al.*, 2011).

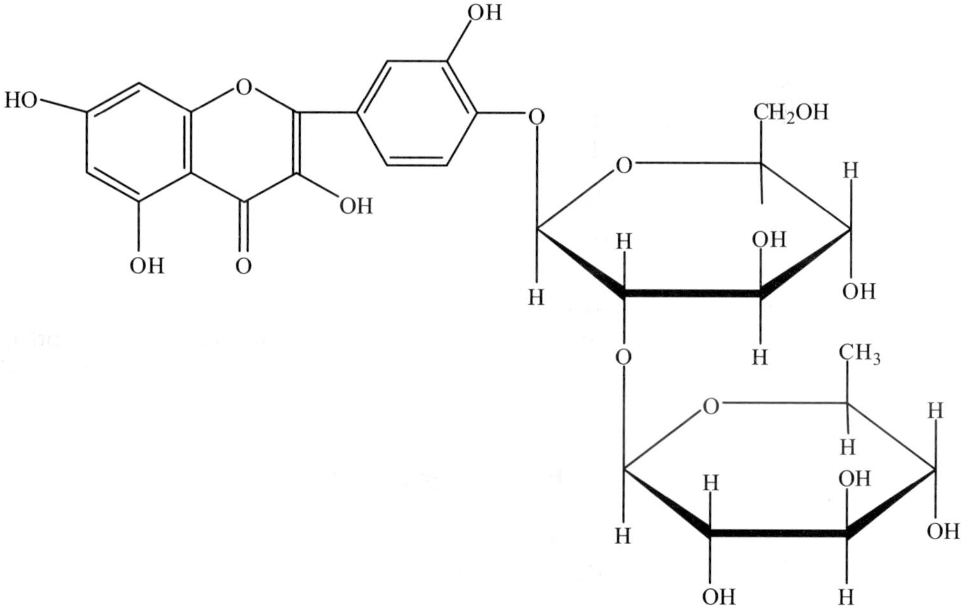

HO, O, OH, OH, O, O, CH_2OH, O, H, H, OH, H, OH, H, O, H, CH_3, O, H, H, H, OH, H, OH, OH, H

luteolin-4-O-[α-(L-rhamnopyranosyl (1.2)-α-D-glucopyranoside)]

Cayaponia tayuya (Well.) Cogn

Cayaponia tayuya cogn or tayuya (Cucurbitaceae) is a climbing, lignified plant with a large swollen root that has been used traditionally for the treatment of inflammatory and rheumatic disease in Brazil, Peru, and Colombia. It was first recorded in the Brazilian Pharmacopoeia as an official herbal drug in 1929. This official compendium describes the syrup prepared from the *C. tayuya* roots "Caroba composto" obtained along with other natural extracts. The cocktail elicited various pharmacological activities, such as analgesic, diuretic, antiinflammatory, tonic, blood purifier and detoxifier, and has been used in popular medicine for treatment of various diseases, like skin disorders, arthritis, and rheumatism. Furthermore, syrup preparation also induced antiinflammatory activity in various acute and chronic animal models of inflammation. The butanol extract of *C. tayuya* roots contain flavonoids namely vicenin-2, acacetin, apigenin, demethylnobiletin, nobiletin,

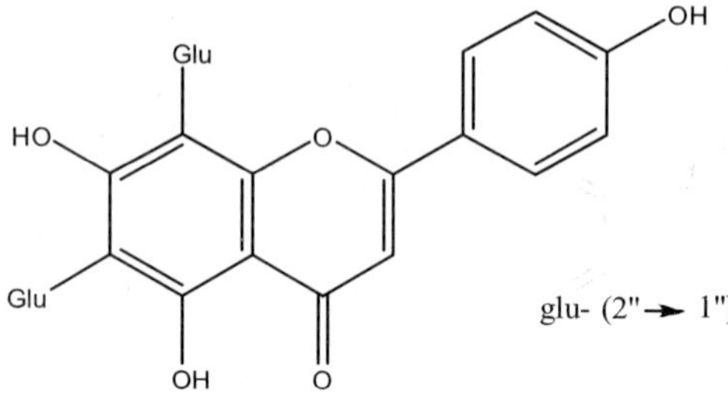

Vicenin-2

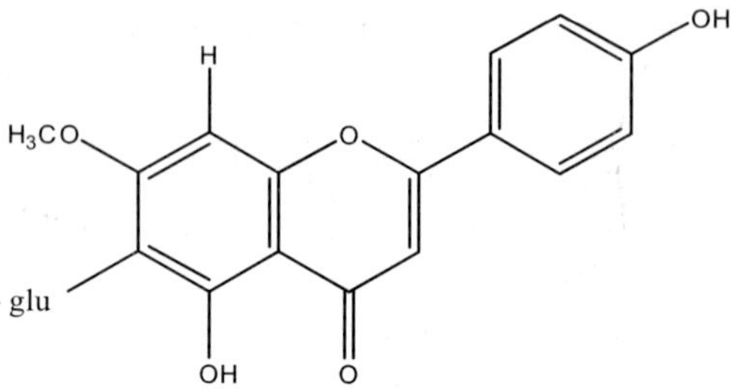

Spinosin

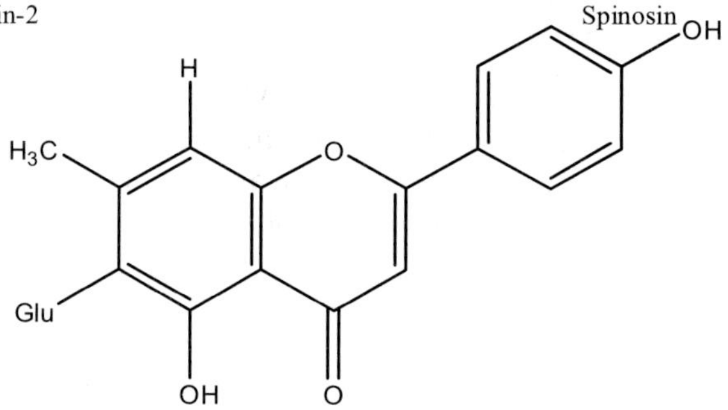

Swertisin

swertisin and spinosin showing antiinflammatory activity in TPA-induced mouse ear oedema (acutely and sub chronically). These flavonoids inhibit arachidonic acid metabolism, induction of NOS-2 at the level of NOS-2 gene transcription, nitric oxide production, suppress the expression of iNOS and COX-2. It is likely that the flavonoids from *C. tayuya* elicit antiinflammatory activity through the inhibition of COX and NOS induction (Aquila *et al.*, 2009).

Cistus laurifolius

In Turkish folk medicine, 5 *Cistus* species and *Cistus laurifolius* are reported to be effective against a broad range of diseases. As leaves decoction is used to treat diarrhoea, rheumatic pain, high fever, urinary tract infection and hypoglycaemia, while flower decoction is used to treat hyperacidity and gastric ulcer. The extracts obtained from the leaves and branches of various *Cistus* species are being used in Turkish traditional medicine for the treatment of rheumatism and renal inflammations. Antiinflammatory and analgesic activity of the *Cistus laurifolius* L. (Cistaceae) extracts and fractions from the leaves extracts with non woody branches were observed in carrageenan-induced hind paw oedema, acetic acid-induced vascular permeability and p-benzoquinone-induced writhing episodes, respectively in rodents. The

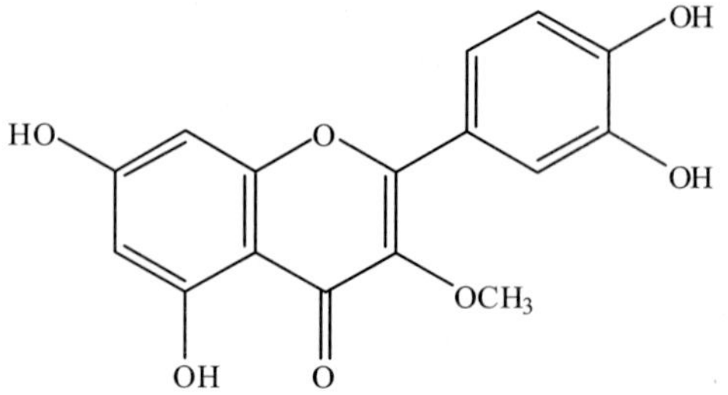

3-O- methyl quercetin

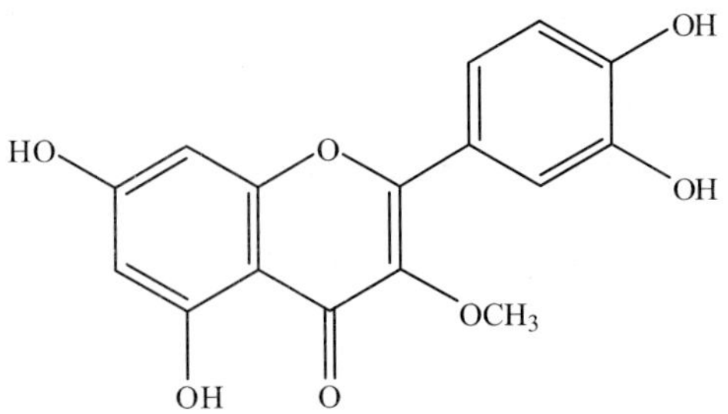

3,7-O- Dimethyl quercetin

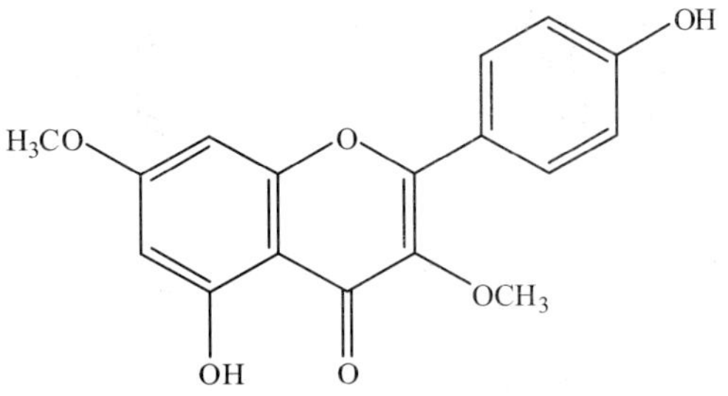

3,7-O- dimethyl kaempferol

extract of *C. laurifolius* contains flavonoids *viz.* (i) 3-O-methylquercetin, (ii) 3,7-O-dimethylquercetin and (iii) 3,7-O-dimethylkaempferol and responsible for antiinflammatory activity (affecting both early and late phase of inflammatory response) and analgesic activity (Kupeli and Yesilada, 2007).

Sophora flavescens

Sophora flavescens. (Leguminosae) is belongs to species of plant in the genus *Sophora*. The roots of *S. flavescens* (*Sophorae radix*) were used as in traditional Chinese medicine for the treatment of fever, inflammatory disorders, ulcers and skin burns. The formulation containing *Sophorae radix* was shown to inhibit angiogenesis of collagen-induced arthritic rats (Li *et al.*, 2003). *S. flavescens* also inhibited allergic reactions *in vitro* and *in vivo*, as well as passive cutaneous anaphylaxis, and

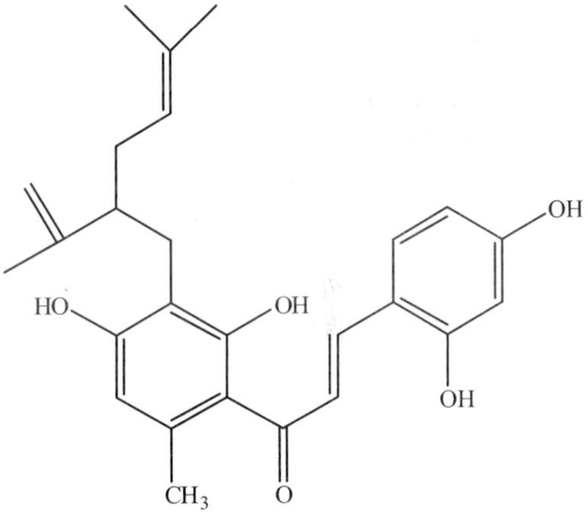

Kurarinone

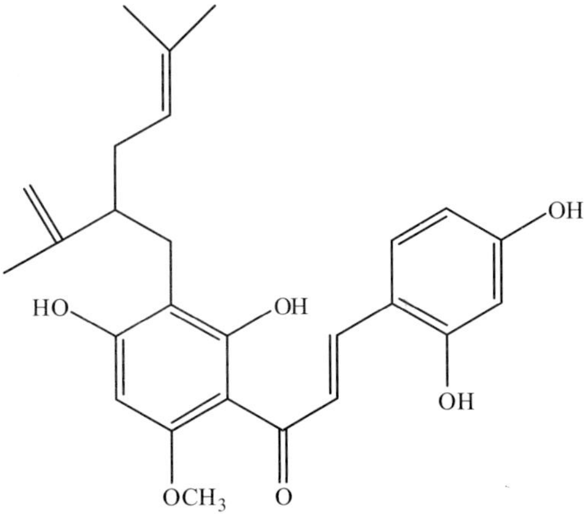

Kuraridin

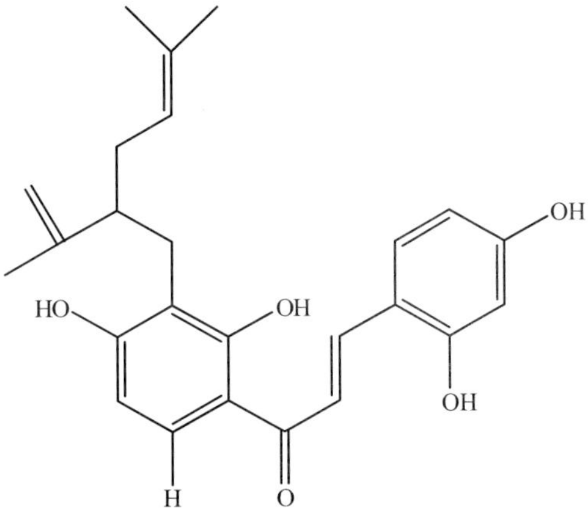

Sophora flavonone G

exhibited antipruritic activity (Miyamoto *et al.*, 2003). Furthermore, the extract inhibited histamine release and cytokine production in the human mast cell line, probably by the inhibition of nuclear factor-8B (NF-8B) activation pathway. The major active constituents of *S. flavescens*: prenylated flavonoids, sophoraflavanone G, kuraridin and kurarinone, are found to be responsible for the inhibition of COX-1, COX-2, 5- lipoxygenase (5-LOX) and 12-LOX, in adjuvant-induced arthritis rat model (Jin *et al.*, 2010; Kim *et al.*, 2002).

Citrus aurantium

Citrus aurantium L. (Rutaceae) is a spiny evergreen tree native to Southern Vietnam and been used as a traditional herbal medicine in Korea and China in inflammatory disorders. It is cultivated for the essential oil extraction from the fruit, and for neroli oil and orange flower water, which are obtained via distillation of flower's extract of *C. aurantium* L. The Korea *C. aurantium* L. contains flavonoids (nobiletin, naringin and hesperidin) which exert inhibition of the pro-inflammatory mediators including cytokines by blocking nuclear factor-kappa B (NF-8B) and mitogen-activated protein kinase (MAPK) signaling in lipopolysaccharide (LPS)-stimulated RAW 264.7 macrophages cell line. These flavonoids also exhibit the suppression of mRNA and protein

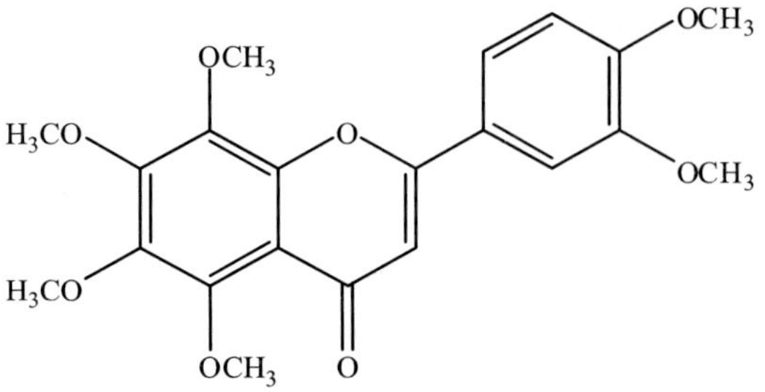

Nobiletin

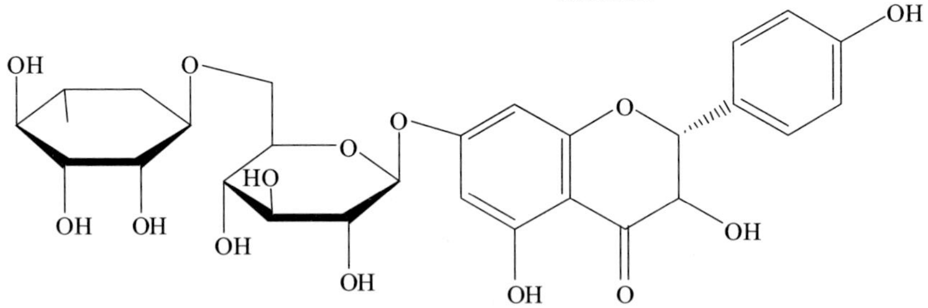

Naringin

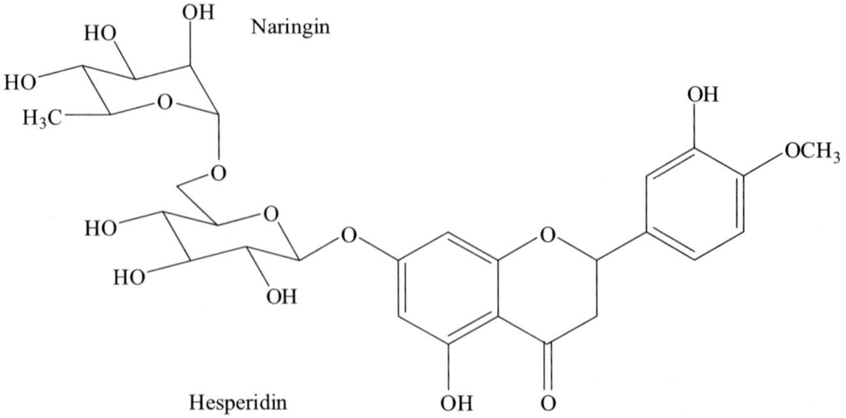

Hesperidin

expression of COX-2 and iNOS in LPS-induced macrophages (Kanno *et al.*, 2006). These findings further suggest that flavonoids are responsible for antiinflammatory activity by suppressing the expression of COX-2, iNOS and cytokines via blocking the NF-8B and MAPK signalling pathways. Hesperidin inhibited inflammatory cytokine production and suppressed infection-induced endotoxin shock in mice (Choi *et al.*, 2007). Nobiletin treatment reduced the inflammation in (I) human synovial fibroblasts, (II) mouse macrophages (III) phorbol ester-induced skin inflammation and (IV) amelioration of oxidative stress in animal experiments (Murakami *et al.*, 2000). Further, it was observed that the antiinflammatory activity of Naringin is due to inhibition of COX-2, iNOS and cytokines (Kang *et al.*, 2011).

Cirsium maackii

Cirsium maackii (Compositae) is herbaceous perennial native to Korea, and grow abundantly in Japan, China and used in oriental folk medicine because of its antiinflammatory properties. It is also popularly known as Korean thistle *C. maackii*. Two hundred and fifty species of *C. maackii* have been identified worldwide, and

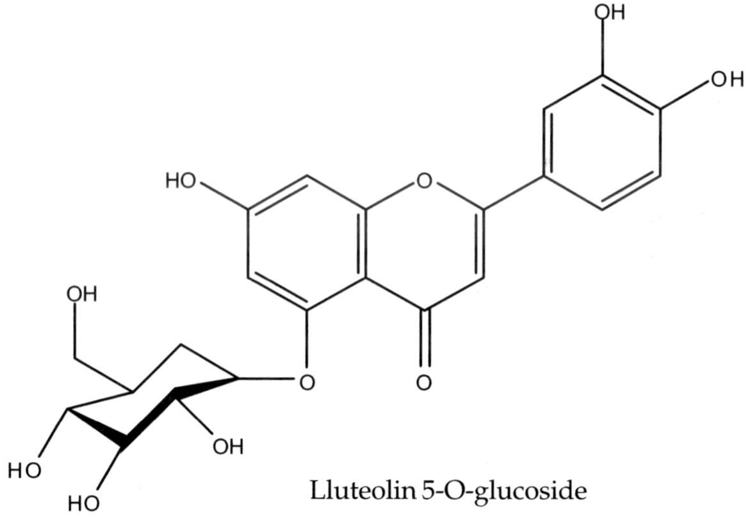

Lluteolin 5-O-glucoside

milk thistle is one of the most popular herbal remedies used in treatment of liver disease. The luteolin 5-O-glucoside, a major flavonoid isolated from the ethyl acetate fraction of *C. maackii*, was studied its effects on RAW 264.7 murine macrophage cells and demonstrated that: (I) inhibition of lipopolysaccharide induced nitric oxide production, iNOS, COX-2 protein expression, and (II) tert-butyl hydroperoxide (t-BHP)-induced reactive oxygen species (ROS) generation (Chen *et al.*, 2007). These findings suggest that the luteolin 5-O-glucoside, active constituent of *C. maackii* found to be useful therapeutically as well as prophylactically in the treatment of inflammatory and oxidative stress-related diseases (Jung *et al.*, 2012).

***Scutellaria baicalensis* Georgi**

Scutellaria baicalensis Georgi (Lamiaceae) is a species of flowering plant, skullcap native to North America. It is one of the 50 fundamental herbs used in traditional

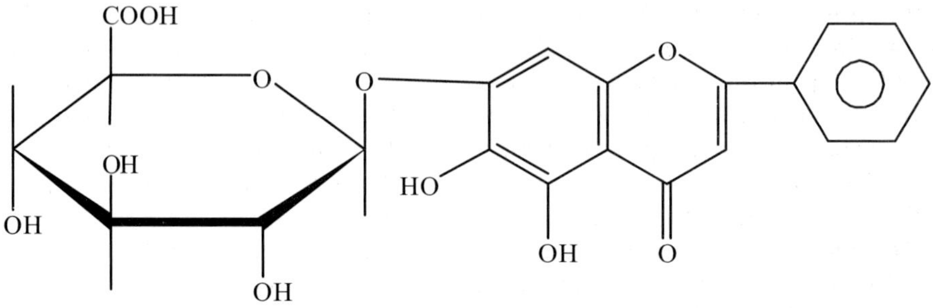

Baicalin

Chinese medicine. Several phytochemical components have been isolated from the root of *S. baicalensis viz.* baicalein, baicalin, wogonin, and β-sitosterol. Among these Baicalin (BA) (7-glucuronic acid,5,6-dihydroxy-flavone) is a flavonoid compound isolated from *S. baicalensis* Georgi and used in the treatment of a variety of inflammatory diseases such as bronchitis, nephritis, hepatitis, asthma, and atopic dermatitis. Baicalin showed antiinflammatory and antiallergic activities mainly by inhibiting IL-8-induced neutrophil infiltration (*in-vivo*). Antiinflammatory activity of baicalin was further evaluated using *in-vitro* models and found that BA inhibited chemokine's binding to human cells, reduced the ability of chemokines ligands to bind its receptors, and inhibited chemotactic activity of chemokines ligands (Li *et al.*, 2000).

Dioclea grandiflora

Dioclea grandiflora Mart. ex Benth (Leguminosae) known as Mucuna, is a species of legume native to the American continent and used as sedative, tonic, and antiepileptics. Dioclein is a flavonoid isolated from the roots of *Dioclea grandiflora* Mart. ex Benth showed antiinflammatory activity by (I) significant suppression of the pro-inflammatory expression of IL-6, TNF-α, CXCL1/KC, CCL2/JE and NO, by LPS-stimulated macrophages *in-vitro*; (II) inhibition of the function of isolated PDE_4 isoenzyme (III) has significant reactive oxygen species scavenging activity both in macrophages and cell free systems and (IV) inhibition of PDE_4 and reduced of the concentration of reactive oxygen species in the medium that contributed to the antiinflammatory activity of Dioclein (Guabiraba *et al.*, 2010).

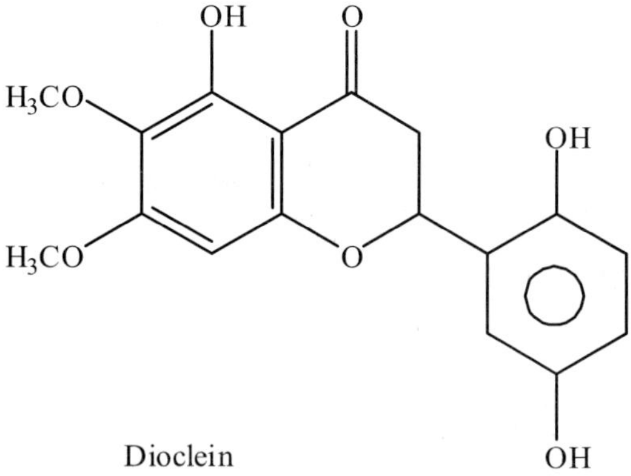

Dioclein

Butea Monosperma (Lam)

Butea monosperma (Fabaceae) is a medicinal plant found largely in Myanmar, India and Sri Lanka. The flowers of Butea monosperma are used as tonic, astringent, aphrodiasic and diuretic. The decoction of the bark is used in cold, cough, fever, haemorrhages, in menstrual disorders, and in the preparation of tonics and elixirs. The stem bark of *Butea monosperma* is reported to elicit antitumour, antiulcer, antifungal and anti-diarrhoeal activities. In addition, the powder of the stem bark is used to apply on mechanical induced injury, the juice used for the treatment of goiter and the paste of the stem bark are applied on body swelling. The roots are used in the treatment of filariasis, night blindness, helmenthiasis, piles, ulcers, and tumors. The methanolic extract of stem bark of *Butea monosperma* showed antiinflammatory and analgesic activity due to presence of flavonoid fraction, which contains two isoflavones, genistein and prunetine. The antiinflammatory activity of *Butea monosperma* was evaluated by using acute inflammation model (carrageenan induced paw oedema) and chronic model (cotton-pellet induced granuloma). The flavonoid fraction inhibited the COX-2 and 5-LOX enzyme which is responsible for the inhibition of the carrageenan rat paw oedema and cotton pellet granuloma in rodents (Muralidhar *et al.*, 2010).

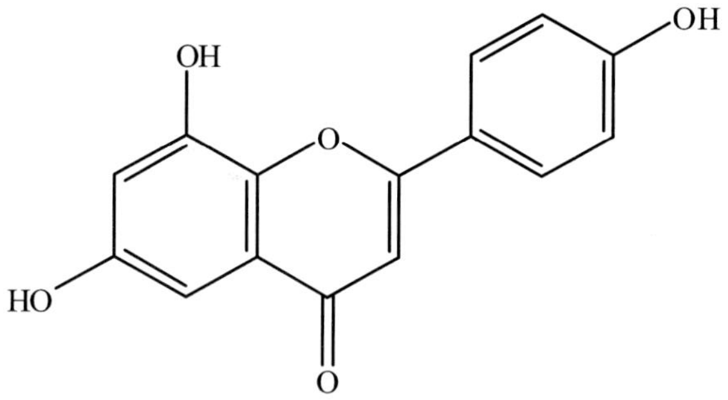

OH, O, OH, HO, O

Genistein

H_3C, O, O, OH, O, OH

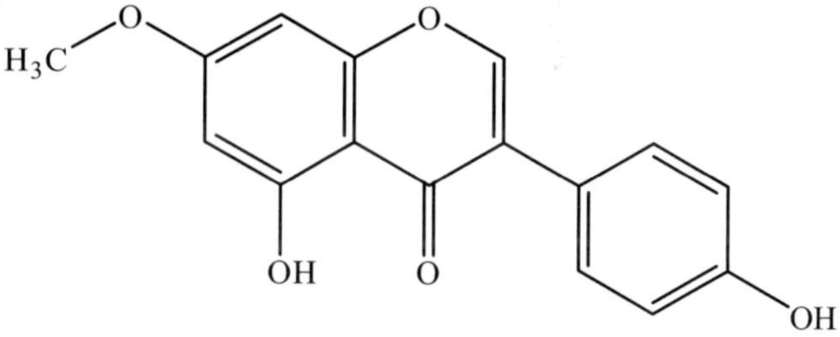

Prunetine

Viscum album

The genus *Viscum* L. comprises semi-parasitic plant which grows in Europe, western and southern Asia, on various host trees and shrubs. *Viscum album* L. (Loranthaceae) is a species of mistletoe, also known as European Mistletoe or Common Mistletoe. *V. album* L. has long been used in the European and Asian, popular medicine for treatment of various disorders likely: headache, hypertension, atherosclerosis, bone and joint disorders including periarthritis, spondylitis, hyperalgesia, arthritis, and epilepsy. The flavonoid components isolated from the ethyl acetate fractions of the extract of *V. album* L. includes (1) 5,7-dimethoxy-flavanone-4-O-β-D-glucopyranoside, (2) 2-hydroxy-4,6-dimethoxy-chalcone-4-O-β-D-glucopyranoside, (3) 5,7-dimethoxy-flavanone-4-O-[2-O-(5-O-trans cinnamoyl)-β-D apiofuranosyl]-β-D-glucopyranoside, (4) 2-hydroxy-4,6 dimethoxy-chalcone-4-O-[2-O-(5-O-trans-cinnamoyl)-β-D-apiofuranosyl]-β-D-glucopyranoside, and (5) 5,7-dimethoxy-flavanone-4-O-[β-D-apiofuranosyl- (1, 2)]-β-D-glucopyranoside, and were studied for antinociceptive and antiinflammatory activities. The compounds (2) and (5)

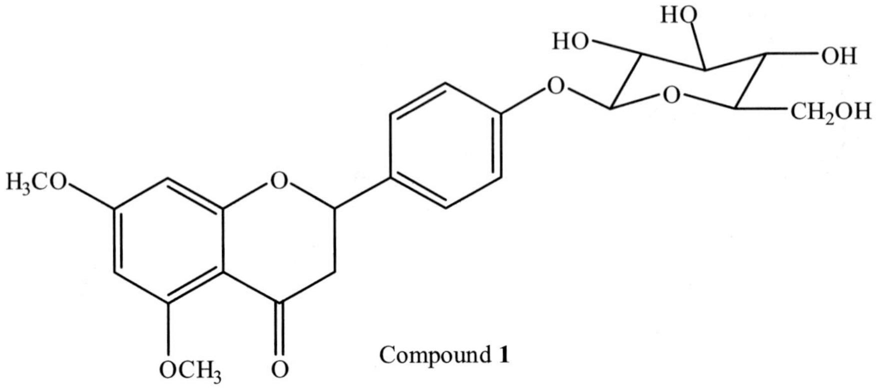

Compound **1**

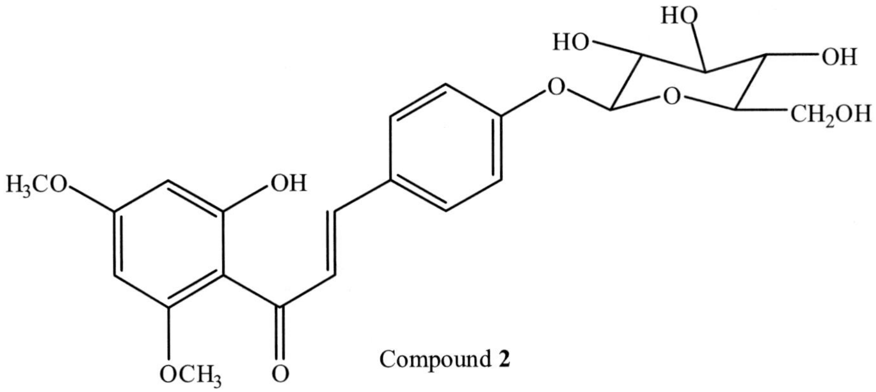

Compound **2**

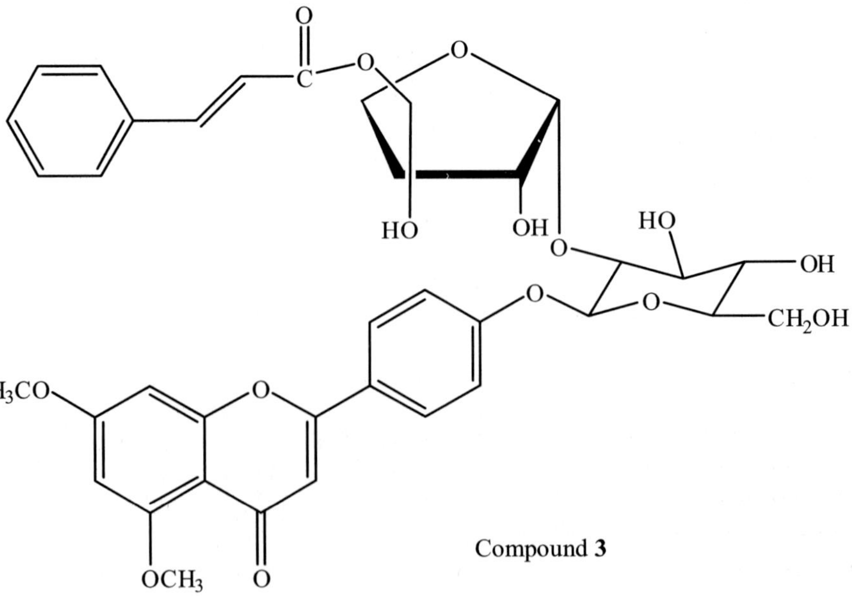

Compound **3**

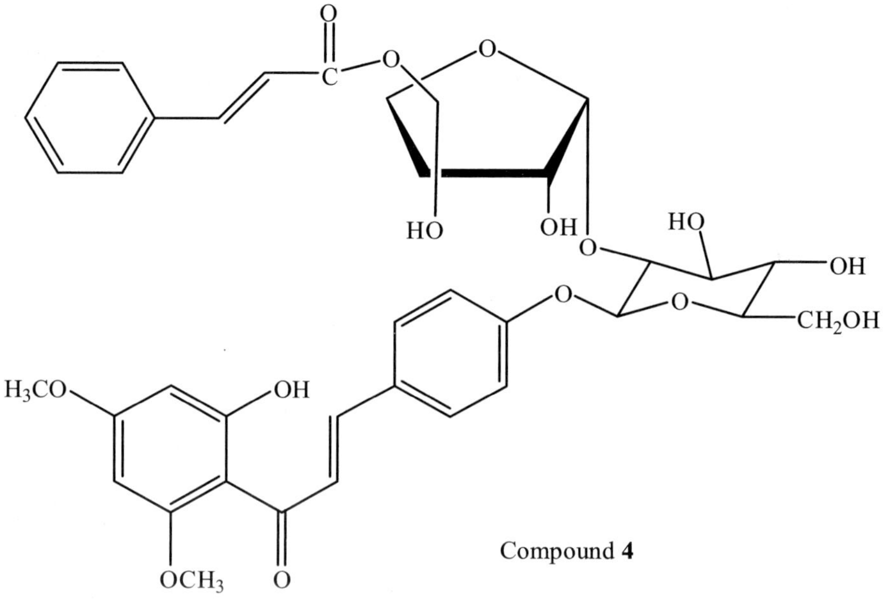

Compound **4**

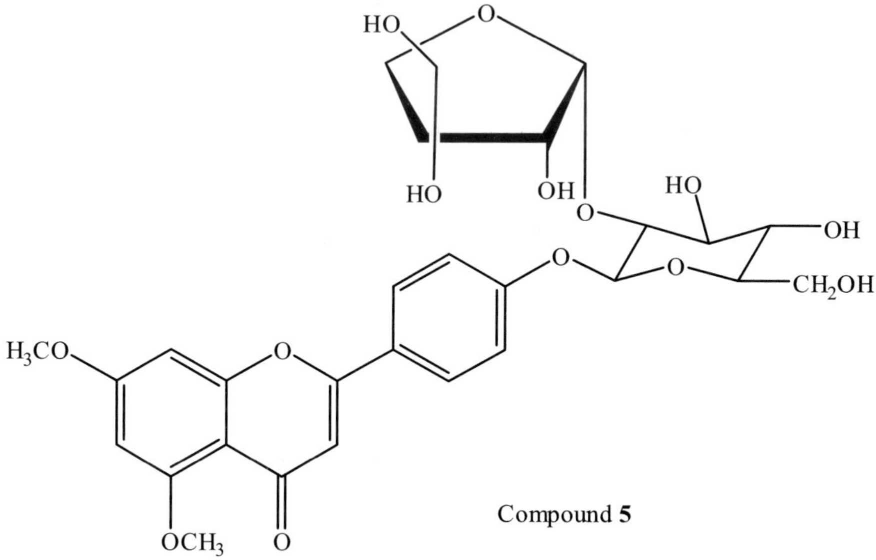

Compound **5**

showed significant analgesic (p-benzoquinone-induced writhing model) and antiinflammatory (carrageenan induced rat paw oedema) activity dose dependently and devoid of epigastric distress in rodents. The antiinflammatory activity of the compound (2) was comparable to that of indomethacin (Orhan *et al.*, 2006).

Santolina insularis (Genn. ex Fiori) Arrigoni

Santolina species are widely used in the Italian traditional medicine for their antiinflammatory properties. *Santolina insularis* Arrigoni (Asteraceae) is an endemic shrub growing in the mountainous area of Sardinia (Italy). The plant has been used in folk medicine as an intestinal vermifuge against horse strongy loidiasis and as a parasite repellent. The essential oil and chemical components of the acetone extract of the aerial parts of *S. insularis* showed significant anti HSV-1 and anti HSV-2 activity (*in-vitro*).

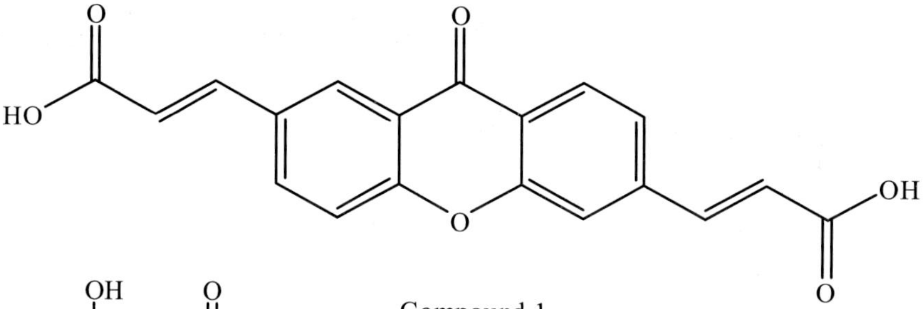

Compound 1

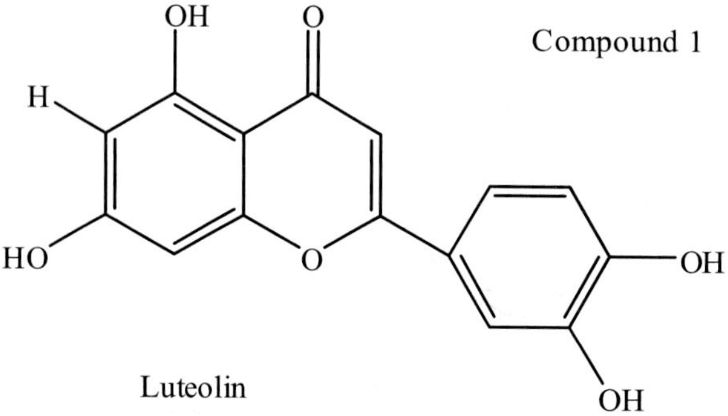

Luteolin

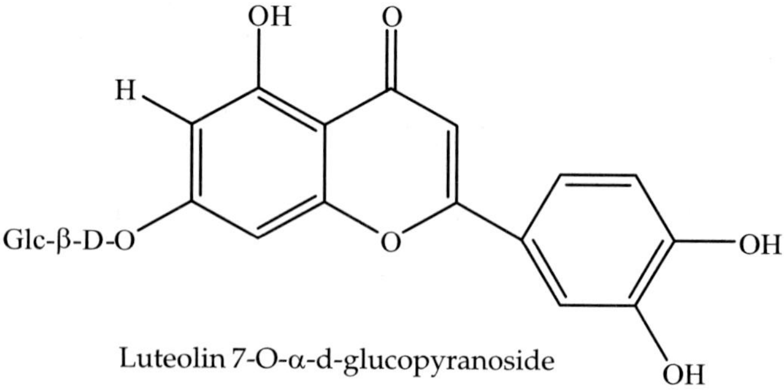

Luteolin 7-O-α-d-glucopyranoside

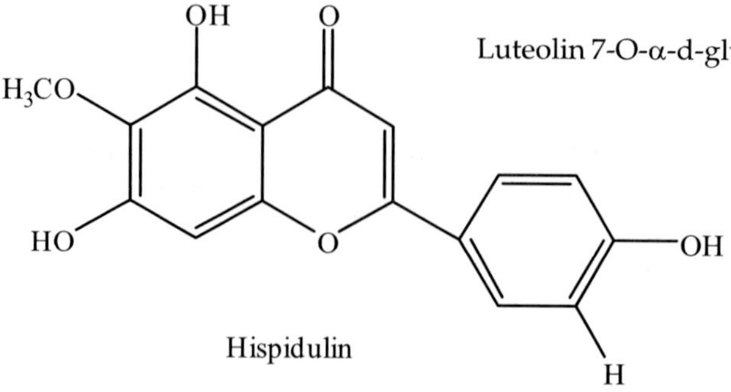

Hispidulin

A new xanthone compound (1), [(E)-3-{6-[(E)-3-hydroxy-3-oxo-1-propenyl]-9- oxo-9H-xanthen-2-yl}-2-propenoic acid] and six known flavonoids, (hispidulin, nepetin, cirsimaritin, rhamnocitrin, luteolin and luteolin 7-O-α-d-glucopyranoside) were isolated from the methanolic extract of the leaves of *S. insularis*. Of these isolated flavonoids, luteolin showed promising antiinflammatory activity against croton oil induced ear oedema in mice by topical application and the antiinflammatory activity was greater than indomethacin (Cottigliaa *et al.*, 2005).

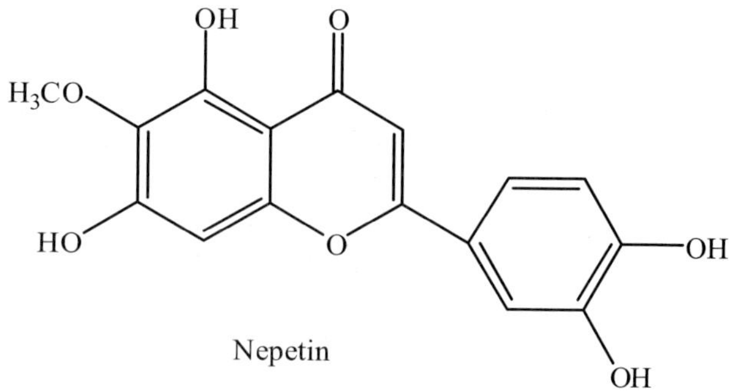

Nepetin

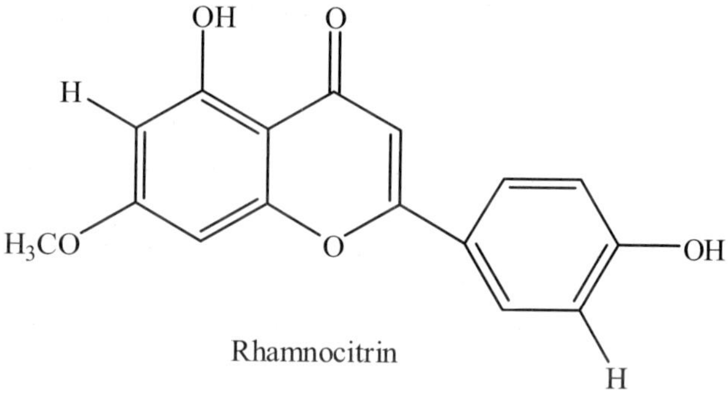

Rhamnocitrin

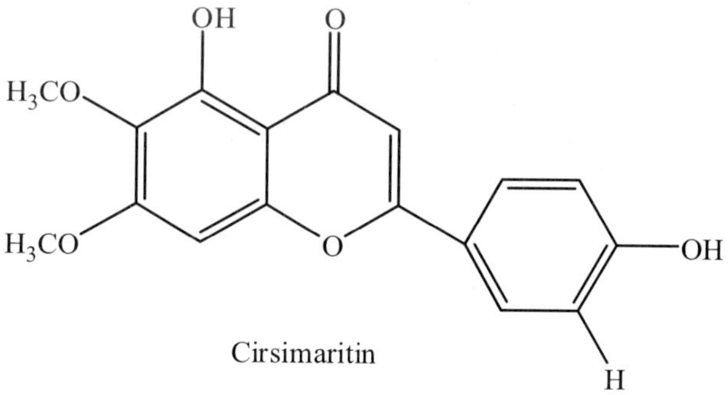

Cirsimaritin

Tanacetum microphyllum DC

In Spanish traditional medicine, *Tanacetum microphyllum* DC (Compositae), an endemic species of the Iberian Peninsula, is used for the treatment of inflammation, rheumatism, and also as cytoprotective agent. It has been experimentally demonstrated that extracts of Spanish herb showed antiinflammatory and gastroprotective activity in the rats. The dichloromethane

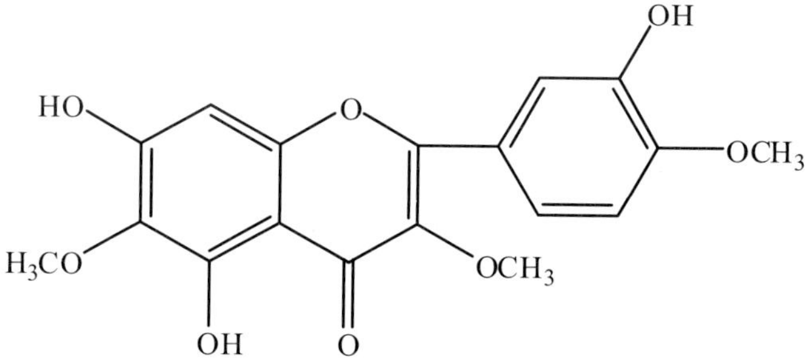

Centaureidin

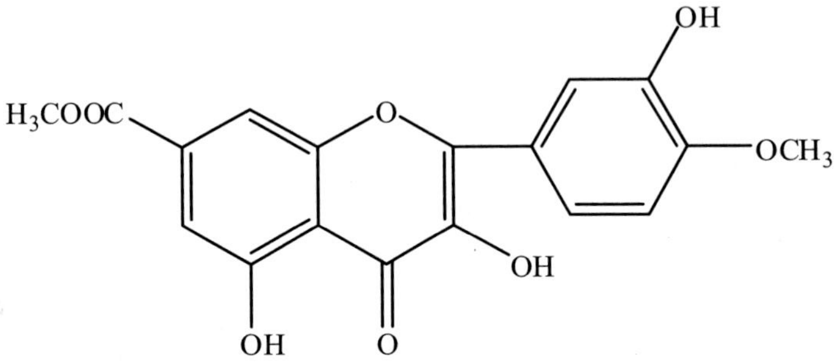

5,3-dihydroxy-4-methoxy-7-carbomethoxyflavonol

extract of *T. microphyllum*, contains two flavonoids: centaureidin and 5, 3-dihydroxy-4'- methoxy-7-carbomethoxyflavonol, and inhibited the synthesis of soybean lipooxygenase as well as cyclooxygenase products (*in-vitro*). Among these two flavonoids, 5, 3-dihydroxy-4-methoxy-7-carbomethoxyflavonol, found to be more potent prostaglandin inhibitor. These experimental findings indicate that *T. microphyllum* may be useful in the treatment of inflammatory disorders (Abad *et al.*, 1995).

Microtea debilis

Microtea debilis Sw (Phytolaccaceae), commonly known as weak jumby pepper, is a small creeping annual weed originating from South America. It has long been used in the traditional medicine to treat cough, stomach ache, acute renal failure, and proteinuria. The ethanolic extract of the aerial parts of *M. debilis* contains two, 5-O-glucosylflavones, (1) 5-O-β-D-glucopyranosyl cirsimaritin and (2) 5, 42-O-β-D-diglucopyranosyl cirsimaritin, and flavonoids (cirsimarin, cirsimaritin, salvigenin, 4,5-dihydroxy-7-methoxyflavone, and vomifoliol) were isolated. Of these flavonoids, 42 , 5-dihydroxy-7-methoxyflavone and cirsimarin were found to be potent inhibitor of nitrite production in LPS induced macrophages as well as antioxidant by lipid peroxidation inhibition in adipose tissue (Bai *et al.*, 2011).

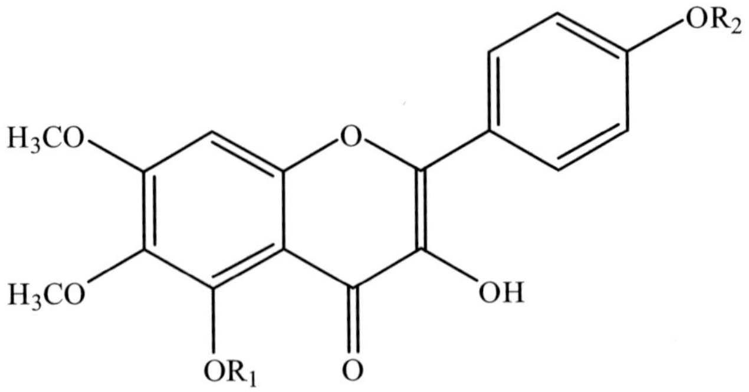

	Flavonoids	R_1	R_2
(1)	5-O-β-D-glucopyranosyl cirsimaritin	β-D-glucopyranosyl	H
(2)	5,42-O-β-D-diglucopyranosyl cirsimaritin	β-D-glucopyranosyl	β-D-glucopyranosyl

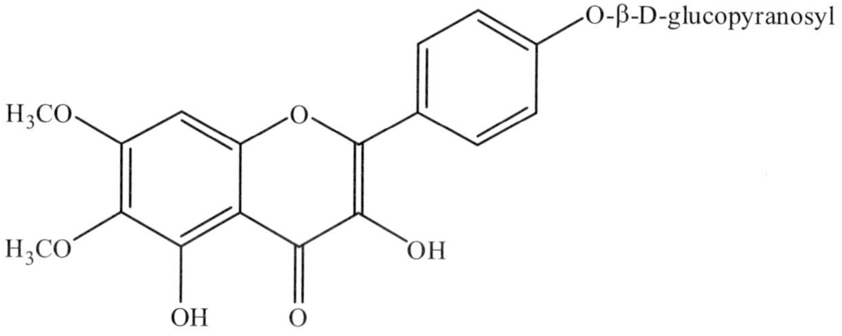

Cirsimarin

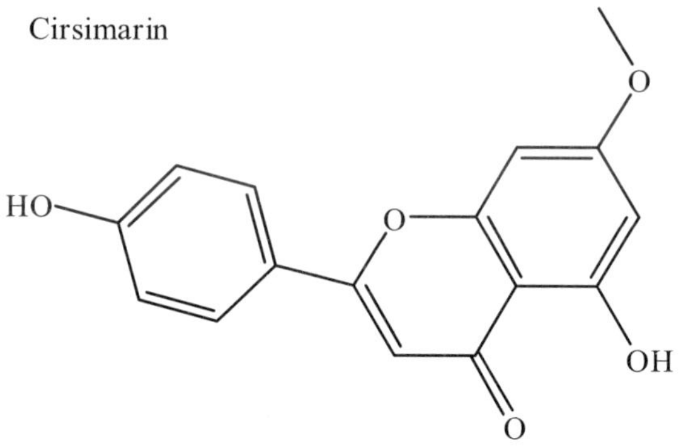

4,5-Dihydroxy-7-methoxyflavone

Rhus verniciflua

Rhus verniciflua (Anacardiaceae) is a deciduous tree native to Korea, Japan and China and been used traditionally as a food supplement and herbal medicine, to improve blood circulation and prevent blood stasis. The documented report indicate that the extract of *R. verniciflua* exert antioxidant, antiinflammatory, antiapoptotic, antiproliferative, antimutagenic and antiobesity activities. The antiinflammatory activities of the methanolic extract of *R. verniciflua* were investigated with microglial cells. The various flavonoids were isolated from the methanolic extract of *R. verniciflua* (fustin, fisetin, sulfuretin, butein, butin, eriodictyol, morin hydrate, quercetin, kaempferol and isoliquiritigenin). Among these flavonoids fisetin, sulfuretin, butein, and butin showed antioxidant as well as antiinflammatory activity (*in-vitro* macrophage RAW 264.7 cell line). In addition, butein showed a dose dependent suppression of the expression of both iNOS and COX-2 in cell line experiments.

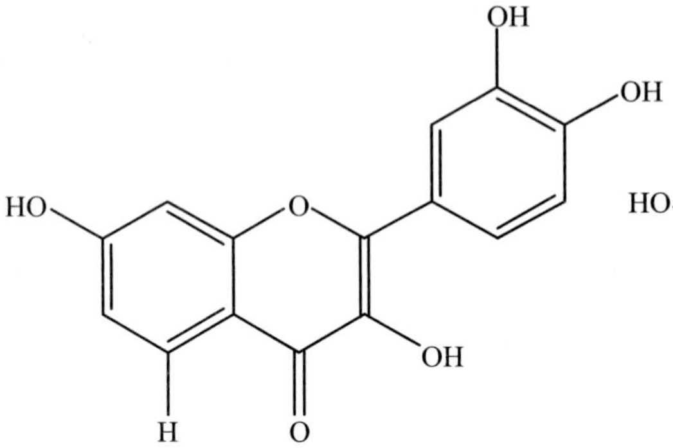

Fisetin

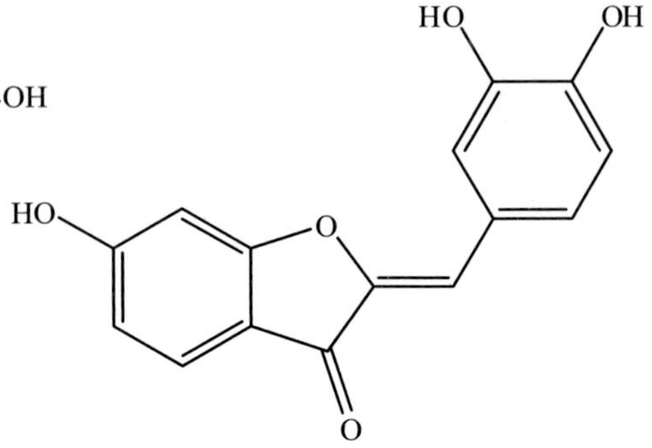

Sulfuretin

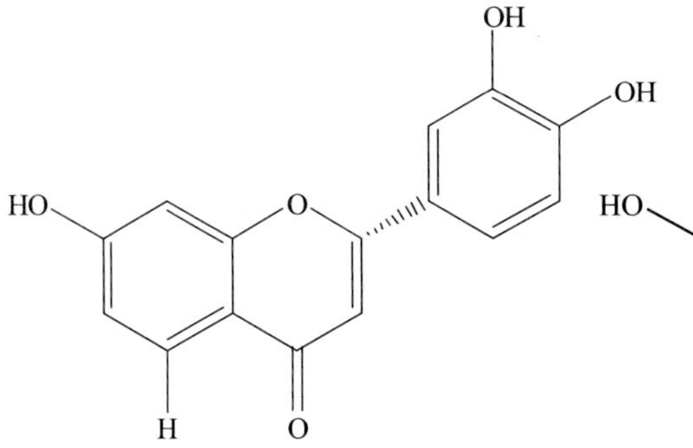

Butin

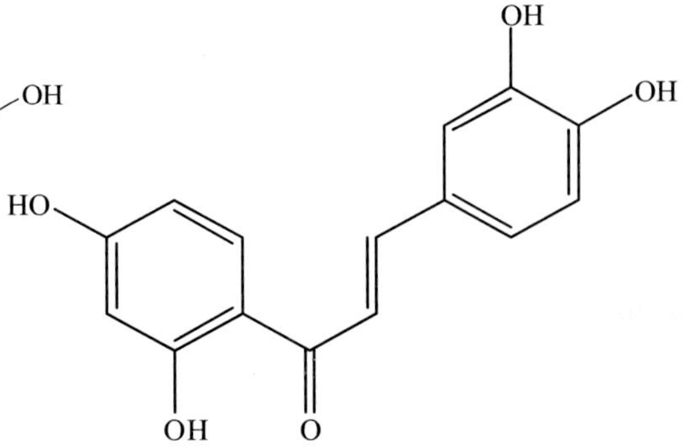

Butein

There by indicating therapeutic potential of *R. verniciflua* in the treatment of inflammatory diseases (Cho *et al.*, 2012).

Tilia argentea

Tilia argentea Desf. ex DC (Tiliaceae) flower is a popular remedy worldwide (Britain, North America, and in Asia) against the symptoms of common cold, bronchitis and cough. Moreover, it is also used as diaphoretic for feverish colds and infectious diseases, occasionally used as stomachic, antispasmodic, diuretic and sedative. The two main flavonoid

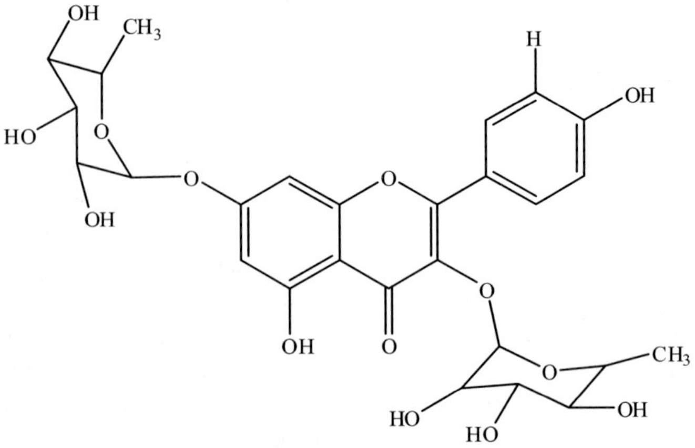

Kaempferol 3,7-O-α-L-dirhamnoside

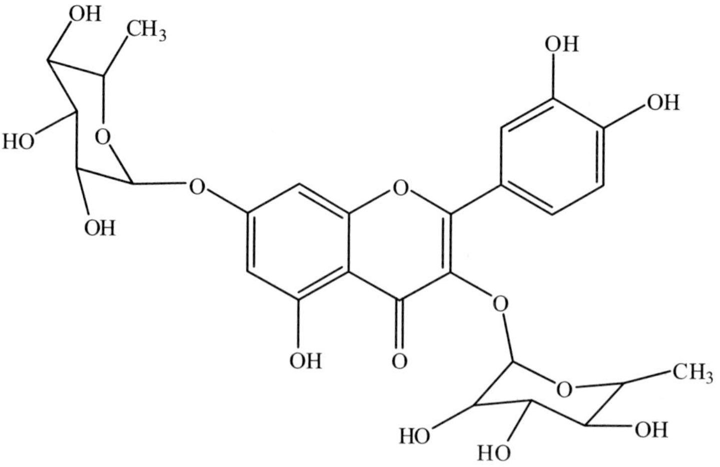

Quercetin 3,7-O-α-L-dirhamnoside

glycosides: kaempferol-3, 7-O-α-dirhamnoside and quercetin-3,7-O-α-dirhamnoside, were isolated from the leaves of *T. argentea*, and found to exhibit potent analgesic (active against p-benzoquinone-induced writhing episodes) and antiinflammatory (carrageenan-induced hind paw oedema) activity. Furthermore, both flavonoid glycosides inhibited cyclooxygenase and lipoxygenase pathways of the arachidonic acid (*in-vitro*) (Toker *et al.*, 2004)

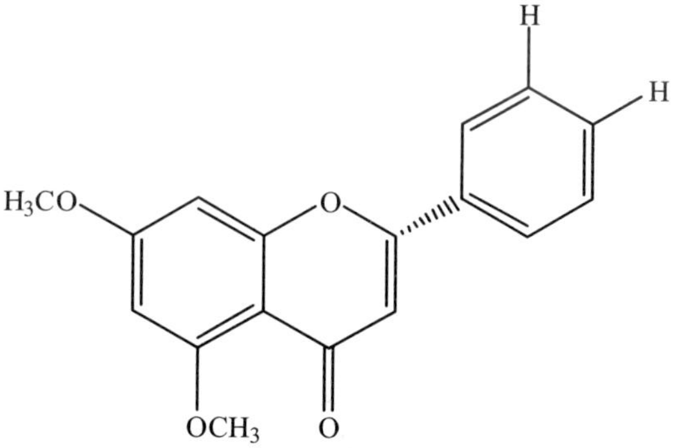

5,7-dimethoxyflavanone

Caesalpinia pulcherrima

Caesalpinia pulcherrima L. Swartz (Leguminosae) is an ornamental flower plant with medicinal importance in India, Taiwan and South-East Asian countries. In traditional practice, different parts of these plants are used for the treatment of inflammation, bronchitis, malarial infection, diarrhoea and dysentery. The five flavonoids, (1) 5, 7-dimethoxyflavanone, (2) 5, 7-dimethoxy-3, 4-methylenedioxy-flavanone, (3) isobonducellin, (4) 2-hydroxy-2, 3, 4, 6-tetramethoxychalcone and (5) Bondu-

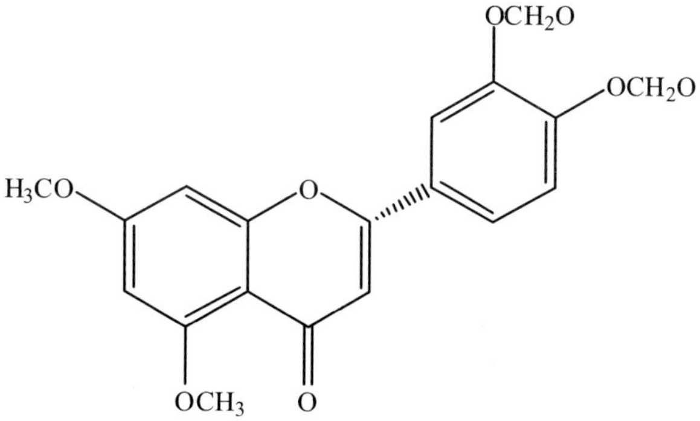

5,7-dimethoxy-3,4-methylenedioxyflavanone

cellin, were isolated from the acetone extract of aerial parts of *C. pulcherrima*. The antiinflammatory activity of *C. pulcherrima* has been demonstrated by its dose dependent inhibitory effect on inflammatory mediators like NO, TNF-α, IL-12 in LPS and IFN-γ activated murine peritoneal macrophages culture. Thus, the anti-inflammatory activity of *C. pulcherrima* in traditional medicine has been attributed to its flavonoid compounds (Rao *et al.*, 2005).

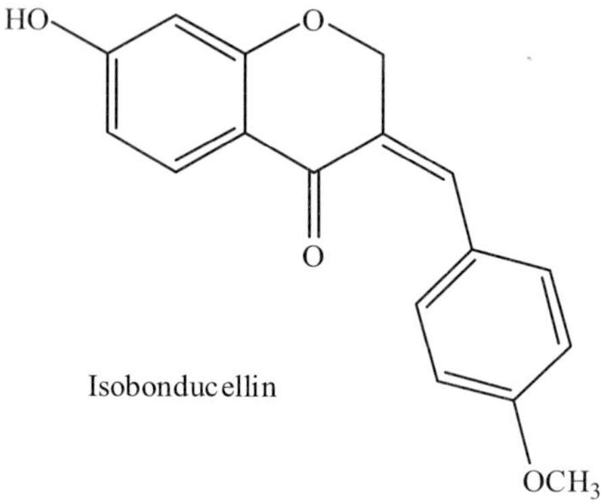

Isobonducellin

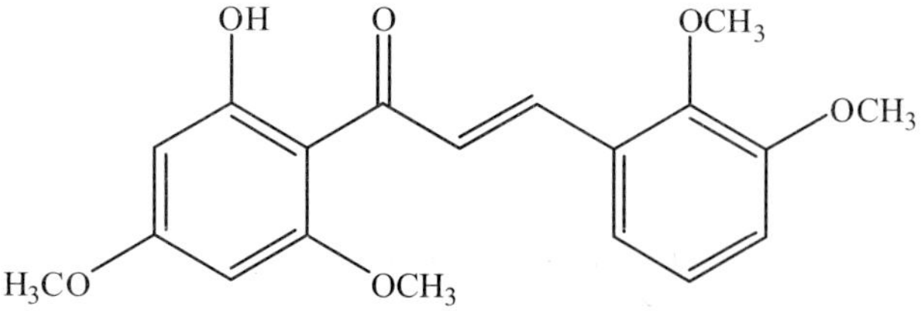

2-Hydroxy-2,3,4,6-tetramethoxychalcone

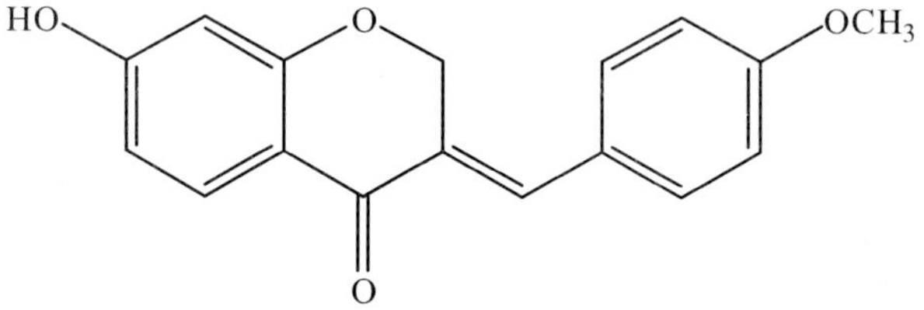

Bonducellin

Oxytropis falcata

Oxytropis falcata Bunge (Leguminosae), locally known as one of the "three antiinflammatory drugs" and the "king of herb" in Tibetan traditional medicine and was first to be included in the Chinese Pharmacopoeia as an official herbal drug in 1977. It has been largely prescribed in Tibetan and Mongolian medicine with multiple traditional therapy in influenza, hyperpyrexia, pain, wounds, haemorrhage, inflammation and anthrax. The four flavonoid compounds (1) 2,4-dihydroxy chalcone, (2) 7-hydroxy flavonone, (3) 5,7-dihydroxy-4-methoxy flavonol and (4) 5,7-dihydroxy-4-methoxy flavone-3-O-β-D-galactoside, were isolated and purified from

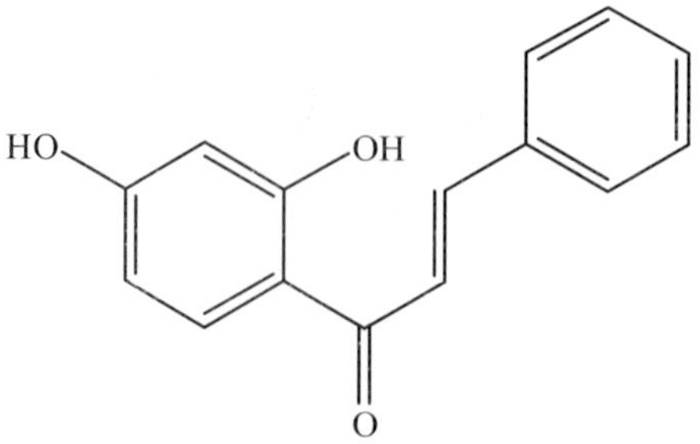

2,4-dihydroxychalcone

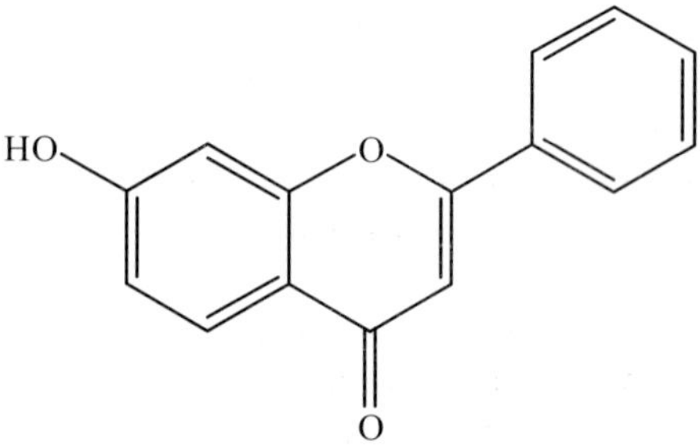

7-hydroxyflavonone

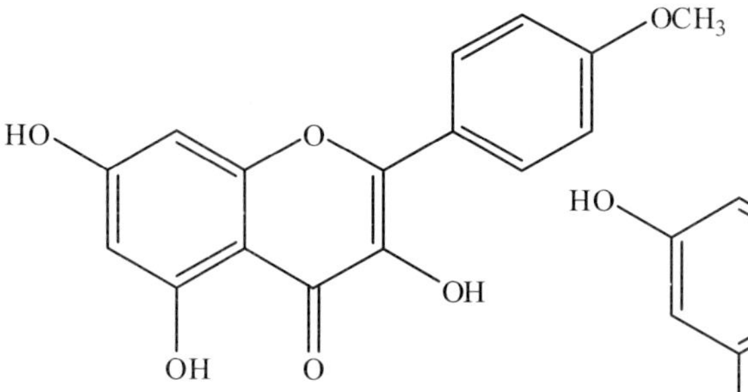

5, 7-dihydroxy-4-methoxy flavonol

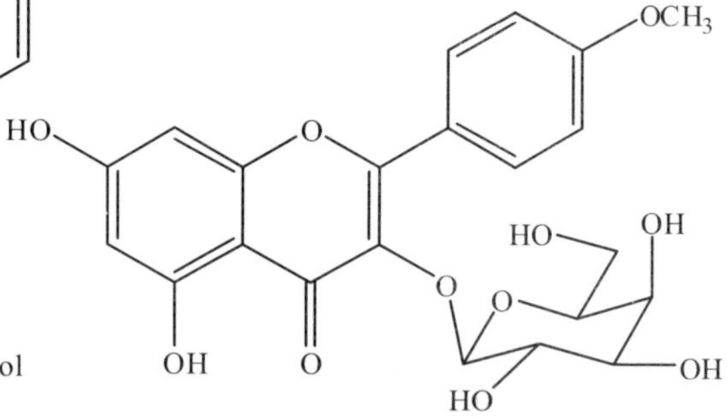

5, 7-dihydroxy-4-methoxy flavone-3-O-β-D-galactoside

the ethyl acetate extracts of *O. falcata*. Of these flavonoids, (1) 2, 4-dihydroxy chalcone, (2) 7-hydroxy flavonone, inhibited NO in LPS-induced RAW 264.7 cells and (3) 5, 7-dihydroxy-4-methoxy flavonol, marked free radical-scavenging activities (*in-vitro*). The flavonoid fractions of *O. falcata* inhibited PGE2 levels, elevated enzyme antioxidants and antiulcer activity in animal experiments (Dong *et al.*, 2010).

Bioactive Alkaloids as Antiinflammatory Agents

Alkaloids are a group of low-molecular-weight, nitrogen-containing compounds found in various natural products including plants and herbs (Facchini, 2001). Alkaloids are commonly found in flowering plants of the family *viz.* Papaveraceae, Papilonaceae, Ranunculaceae, and Solanaceae. They are also found in lower plants, insects, marine organisms, microorganisms and animals. A wide range of biological activity has been reported for alkaloids, including emetic, anticholinergic, antitumor, diuretic, sympathomimetic, antiviral, antihypertensive, analgesic, antidepressant, muscle relaxant, antitussive, antimicrobial and antiinflammatory activities (Souto *et al.*, 2011). Alkaloids showed great diversity in their botanical and biochemical origin, in chemical structure and pharmacological activity. The alkaloids are broadly categorised into 2 groups on the basis of chemical structures present in it. (I) Heterocyclic or typical alkaloids, (II) Non-heterocyclic or atypical alkaloids, sometimes called 'protoalkaloids' or biological amines (Table 1.2). Table 1.2 describes the different types of alkaloids and their occurrence in various plants along with basic chemical moiety. Alkaloids are also classified according to biosynthetic pathways and their precursors which are involved in the synthesis of various alkaloids (Table 1.3) (Gomes *et al.*, 1978; Geoffrey and Cordell, 2012; Hegnauer 1988).

Alkaloids showed antiinflammatory activity (carrageenan paw oedema, cotton pellet granuloma/pouch, polyarthritis, skin inflammation etc.) and analgesic activity (acetic acid induced pain syndrome, heat induced pain threshold) in various animal models. In addition, they also inhibited inflammatory mediators (histamine, kinin-kallikrein, prostaglandins, leukotrienes etc.), capillary permeability, vasodilation,

Table 1.2: The different types of alkaloids.

Type	*Chemical Ring Moiety*	*Examples*
	(I) True alkaloids (heterocyclic alkaloids)	
Pyrrole and pyrolidine	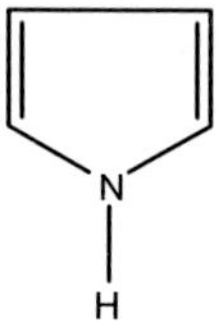 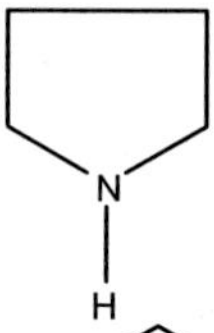 N–H; N–H	Coca species
Pyridine and piperidine	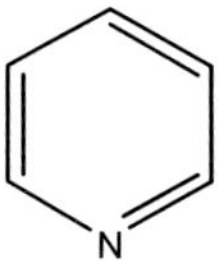 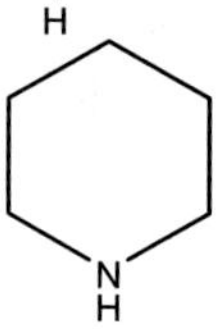 N; N H	Lobeline, arecoline, anabasine, coniine etc.
Pyrrolizidine	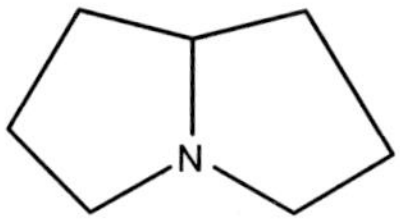 N	Senecionine, seneciphylline, Echimidine, Retronecine
Tropane	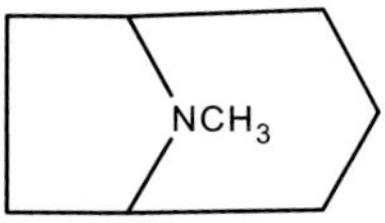 NCH_3	Atropine, hyoscine, hyoscyamine, cocaine, pseudopelletierine
Quinoline	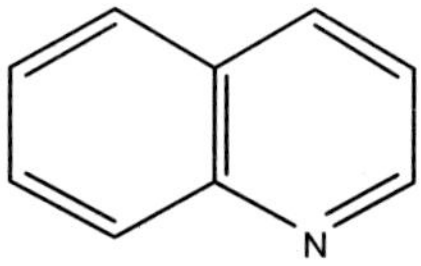 N	Quinine, quinidine, cinchonine, cinchonidine, cupreine, camptothecin
Isoquinoline	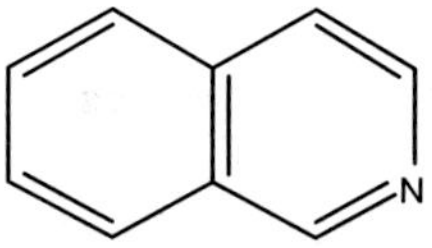 N	Morphine, codeine, berberine, emetine, cephaeline, papaverine, nicotine
Indole	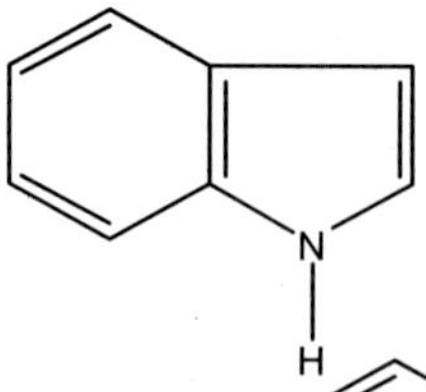 N–H	Ergotamine, ergometrine, reserpine, vincristine, vinblastine, strychnine, brucine, physostigmine
Aporphine	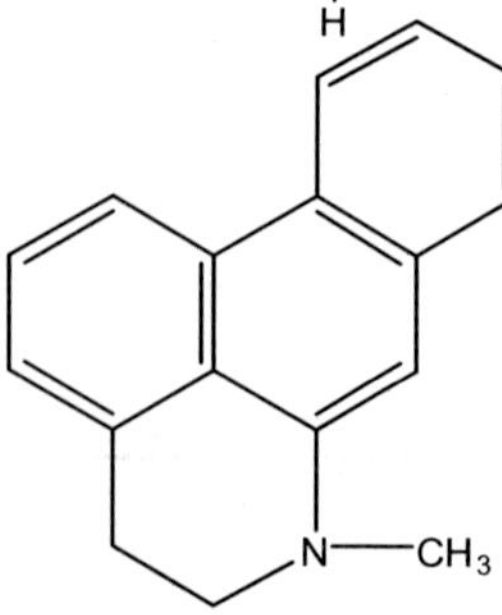 N—CH_3	Boldine

Contd...

Table 1.2–*Contd...*

Type	*Chemical Ring Moiety*	*Examples*
Imidazole	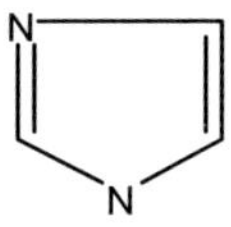	Pilocarpine, isopilocarpine, pilosine
Purine	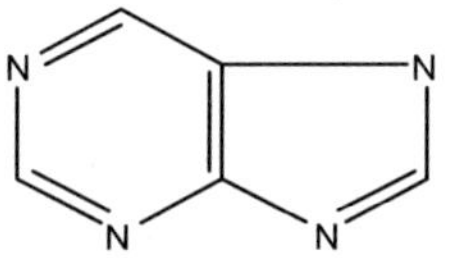	Caffeine, theobromine, theophylline
Norlupinane	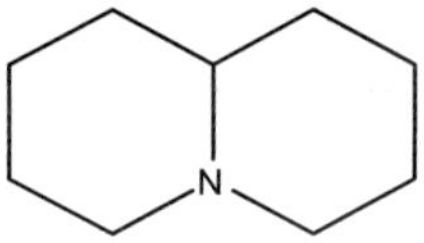	Lupanine, laburnine, cytosine
Steroidal	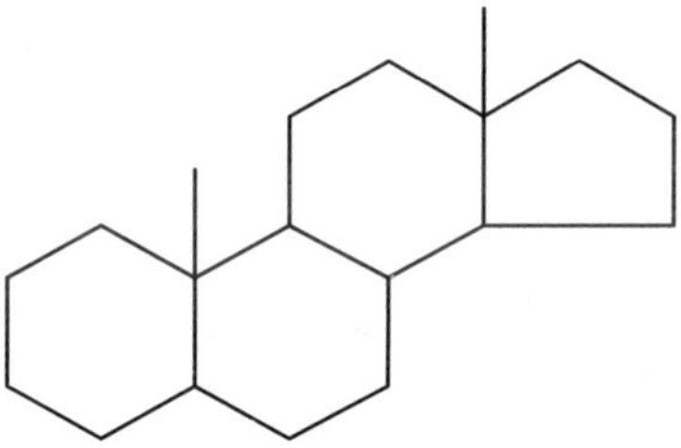	Solanidine, conessine
	(II) Proto alkaloids	
Amino or alkylamine	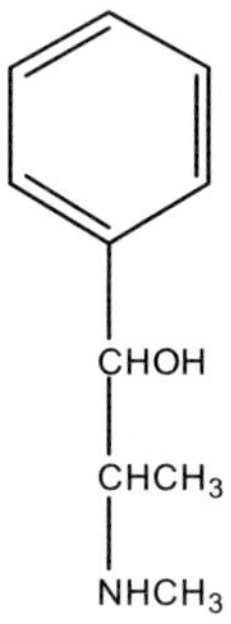 CHOH $CHCH_3$ $NHCH_3$	Ephedrine, pseudoephedrine, colchicines, mescaline
Diterpenes	$C_{20}H_{32}$	Aconitine, aconine, hypoaconitine.

Table 1.3: Classification of alkaloids according to biosynthetic pathways (precursor) and their basic structural moiety.

Biosynthetic Precursors	*Structural Moieties of Derived Alkaloids*			
Ornithine	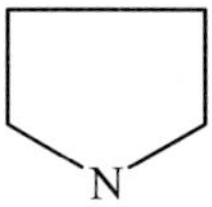 Pyrrolidine 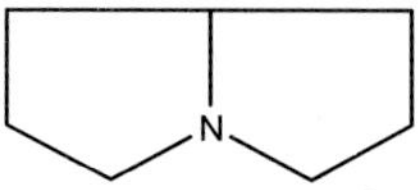	Pyrrolizidine 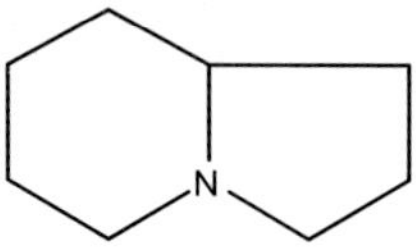	Indolizidine 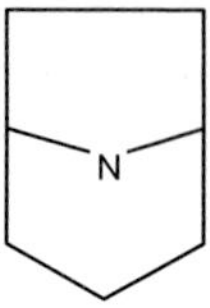	Tropane
Lysine	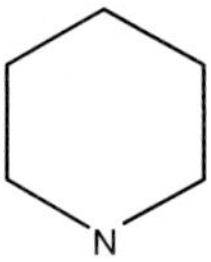 Piperidine 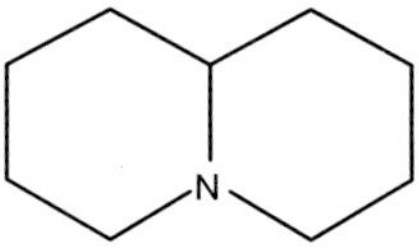	Quinolizidine		
Histidine	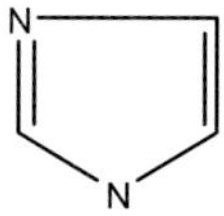 Imidazole			
Phenylalanine	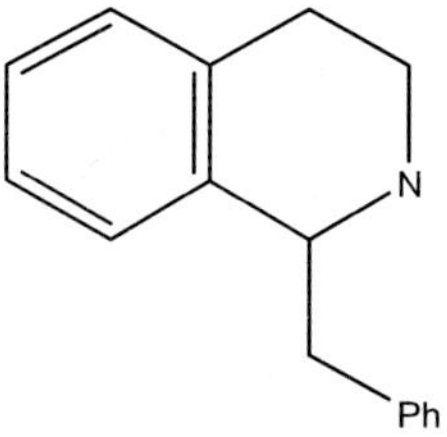 Benzylisoquinoline 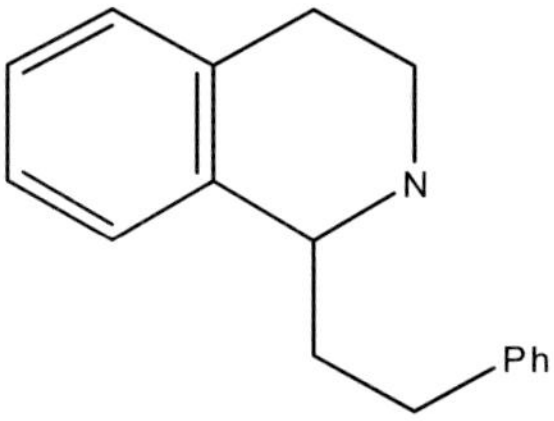	Phenylethylisoquinoline 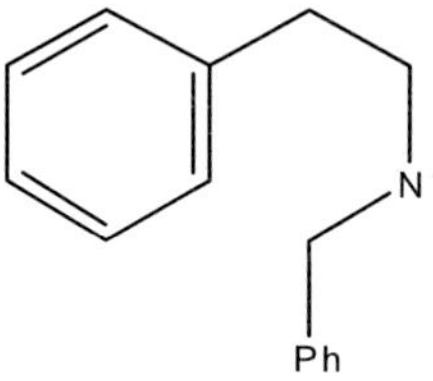	benzyl-phenylethylamine	
Anthranilic acid	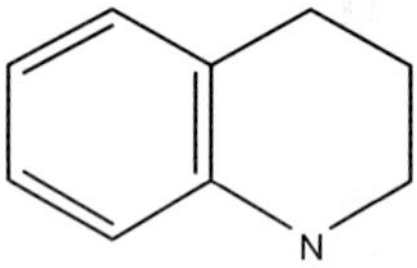 Quinoline 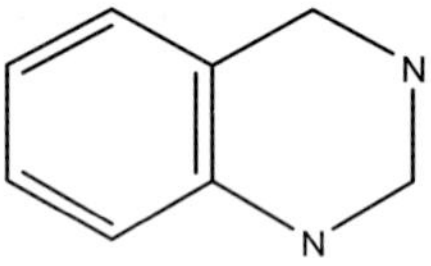	Quinazoline 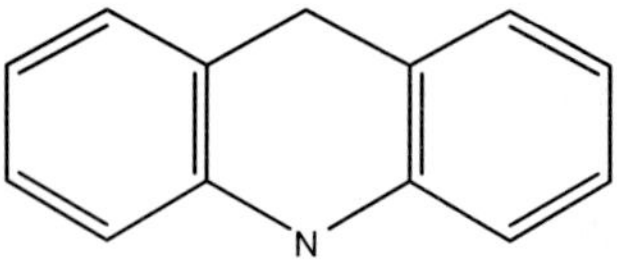	Acridine	
Mevalonic acid	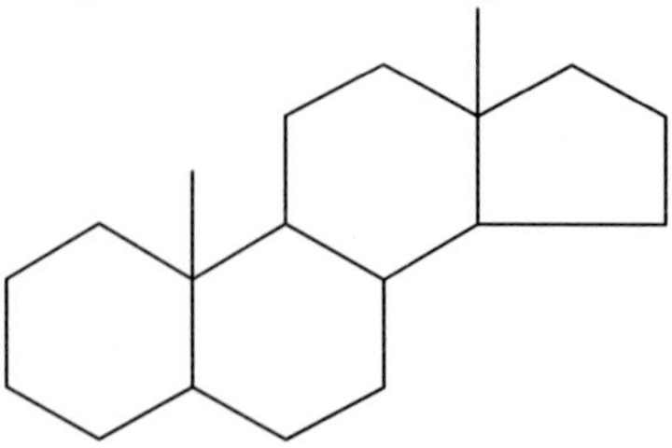 Steroid			

Contd...

Table 1.3–*Contd...*

Biosynthetic Precursors	*Structural Moieties of Derived Alkaloids*
Tryptophan + mevalonic acid	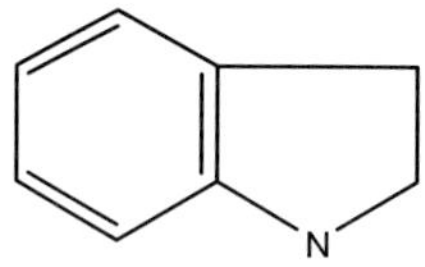 Indole
Nicotinic acid	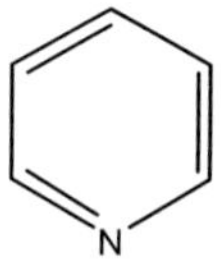 Pyridine

free radicals formation, connective tissue metabolism, catheptic enzymes, platelet aggregation, endothelial derived factors and intracellular pro-inflammatory mediators. In addition, alkaloids able to restore body defense mechanisms by augmenting antioxidants and other associated biomolecules. All these salient biological and biochemical effects of alkaloids are responsible for anti-phlogistic as well as antiarthritic activity (Figure 1.2). Furthermore, such phytochemicals may further be developed into a new generation antiinflammatory drugs with the application of the state-of-the-art of techniques and methods.

Some Important Natural Products (Plants/Herbs) Containing Alkaloids Showed Promising Antiinflammatory Activity and Further Scope for the Development of Plant Derived Products

Sida cordifolia Linn.

Sida cordifolia Linn (Malvaceae) is herb used as herbal drug in the Indian subcontinent. The water extract of the leaves was reported to elicit analgesic and

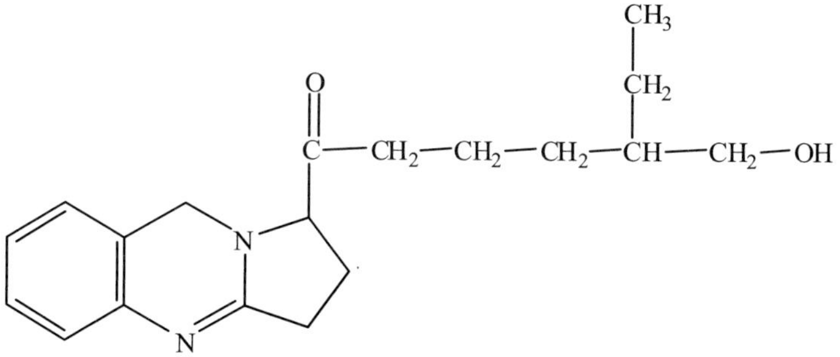

5'-Hydroxymethyl-1'-(1, 2, 3, 9-tetrahydro-pyrrolo [2, 1-β] quinazolin-1-yl)-heptan-1-one)

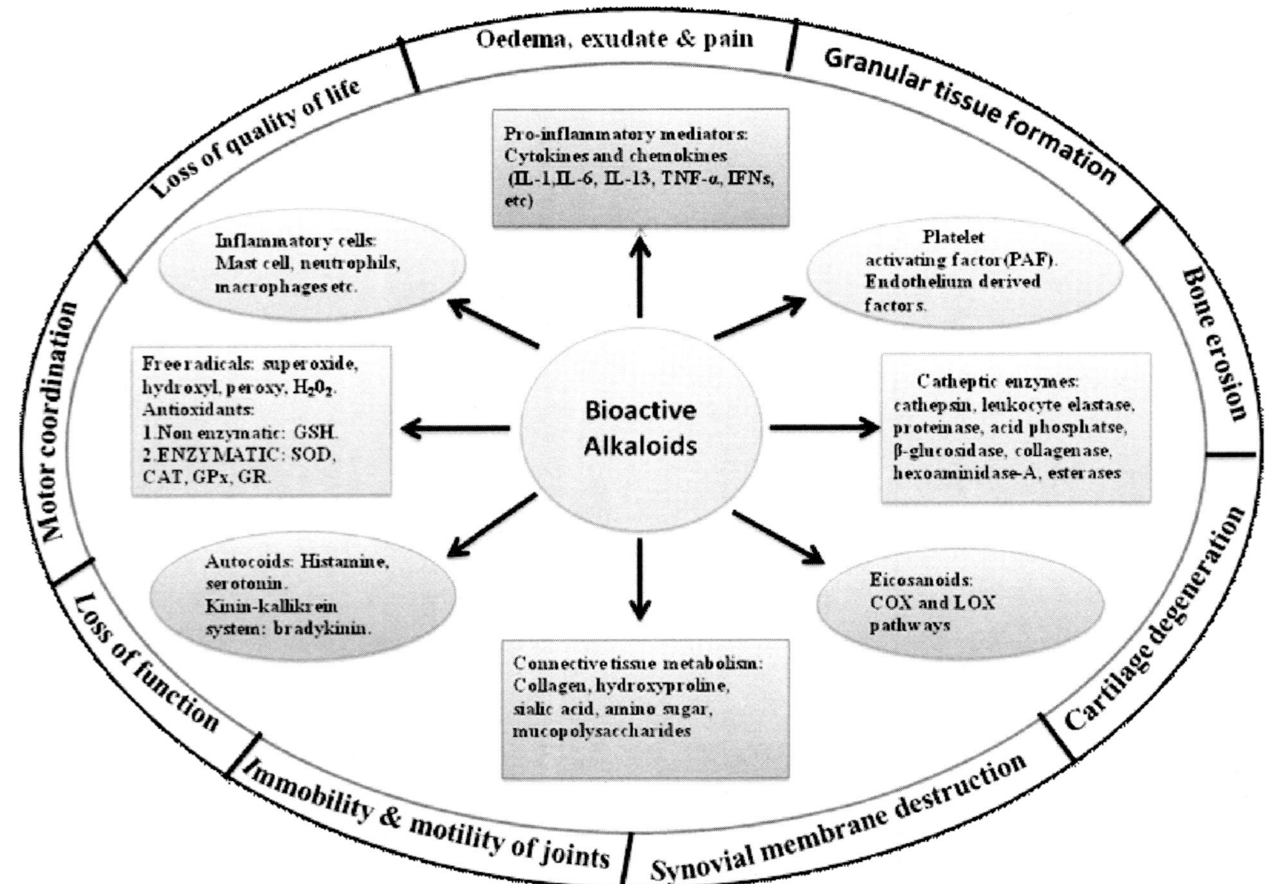

Figure 1.2: The biochemical and pharmacological effects of alkaloids on inflammation and its derived products.

antiinflammatory activities in animal models and used in the treatment of rheumatism. *S. cordifolia* Linn contains phytochemical constituent *viz.* ephedrine, vasicinol, vasicinone and N-methyl tryptophan. The active alkaloid: (5'-Hydroxymethyl-1'-(1, 2, 3, 9-tetrahydro-pyrrolo [2, 1-β] quinazolin-1-yl)-heptan-1-one), isolated from aerial part of *S. cordifolia* Linn using different solvent systems. The crude alkaloid was further purified by preparative TLC. The compound was characterised and identified by spectroscopic methods (UV, IR, 13C NMR, GC-MS etc). The pure compound exhibited a dose dependent antiinflammatory activity in carrageenan induced rat hind paw oedema. In addition, the alkaloid also showed analgesic activity in acetic acid induced writhing, and radiant heat tail flick methods. The activity has been attributed to the inhibition of cyclooxygenase (Sutradhar *et al.*, 2006).

Mahonia swaseyi

Mahonia swaseyi Fedde (Berberidaceae) is an evergreen shrub growing to 2.5 m. The various isoquinoline alkaloids, berbamine, berberine and tetradrine were isolated from *Mahonia swaseyi* Fedde and studied for their antiinflammatory activity. Of these, alkaloids berbamine and tetradrine were found to be promising hence further

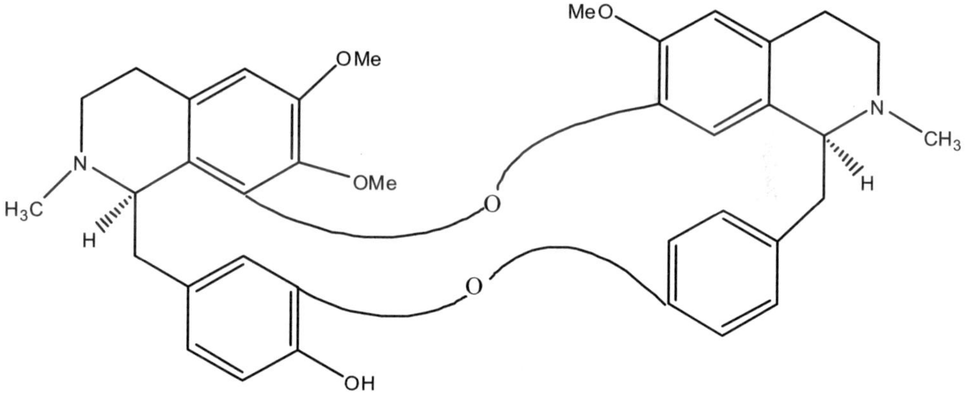

Berbamine

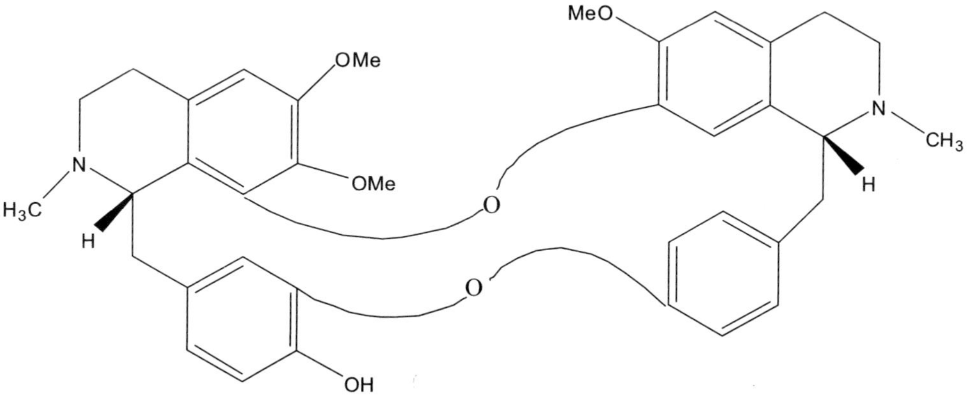

Tetradrine

evaluated for antiinflammatory activity using various *in-vivo* and *in-vitro* models and described below. The bisbenzylisoquinoline alkaloids, berbamine and tetradrine were studied for their effect on IL-1, TNF and PAF in the rat granuloma pouch model of inflammation and on PMN infiltration induced by PAF (*in-vitro*) (Wong *et al.*, 1992). Tetradrine found to be more effective than berbamine in suppressing of leukocyte infiltration in the exudates of granuloma pouch induced by IL-1, TNF and PAF induced PMN infiltration. Berbamine and tetradrine elicited significant antiinflammatory activity in arachidonic acid and TPA induced inflammation in rodents (Yasukawa *et al.*, 1993; Ono, 1994) as well as suppressed adherence, locomotion and 3H deoxyglucose uptake by neutrophils (*in-vitro*) and significantly inhibited NK cell cytotoxicity. All these experimental findings suggest that both tetradrine and berbamine may be useful in the treatment of chronic inflammatory diseases in which IL-1, TNF and PAF are actively participating in the pathogenesis (Ying *et al.*, 1989).

Berberis vulgeris

Berberis vulgaris/European barberry is a shrub in the family Berberidaceae, native to central and Southern Europe, Northwest Africa and Western Asia; it is also naturalised in northern Europe, including the British Isles and Scandinavia, and North America. Although not naturalised, in rural New Zealand it has been widely cultivated as a hedge on farms. It is cultivated for its fruits, primarily in Iran. In traditional Eastern medicine, several Chinese remedies that have been used in the treatment of rheumatic diseases are based on the extracts of the *Berberis* species. In Bulgarian folk medicine extracts from *Berberis vulgaris* roots and stem bark are topically applied in chronic inflammations of liver, gall bladder and kidney, arthritis and podagra. The analysis on *B. vulgaris* chemical content proves the presence of protoberberines and bisbenzylisoquinoline alkaloids. The protoberberines are prevalent in the roots and from which berberine alkaloid was isolated. Berberine is an another isoquinoline alkaloid isolated from roots and stem-bark of *B. vulgeris* showed promising antiinflammatory activity in acute and chronic inflammatory models such as carrageenan induced paw oedema, adjuvant induced arthritis in rats, cotton pellet granuloma pouch in rodents, granulation tissue using fertile eggs, croton oil granuloma pouch and granulation tissue on chick embryo chorio-allantoic membrane (Ivanovska and Philipov, 1996).

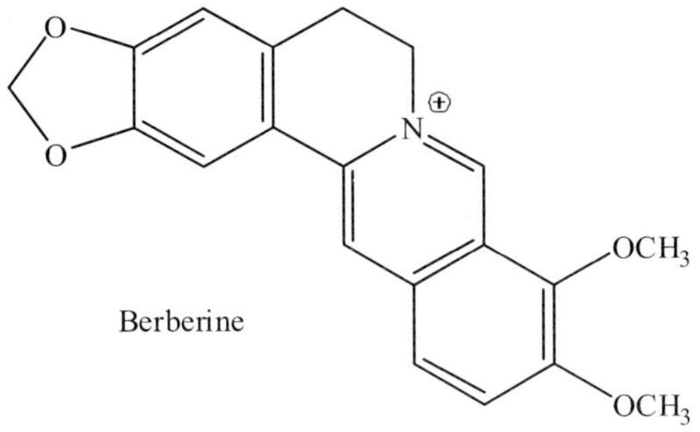

O O N $^{\oplus}$ OCH_3 OCH_3

Berberine

Gloriosa superb

Gloriosa superb L. (Liliaceae) is a perennial climber and is used as an Ayurvedic medicinal herb to cure diseases in various parts of Africa and Southeast Asia. The plant is used in the treatment of arthritis, gout, rheumatism, inflammation, ulcers, bleeding piles, skin diseases, leprosy, impotency, snakebites, etc.(Jana and Shekhawat, 2011). Various alkaloids have been isolated from the plant parts mainly tubers and

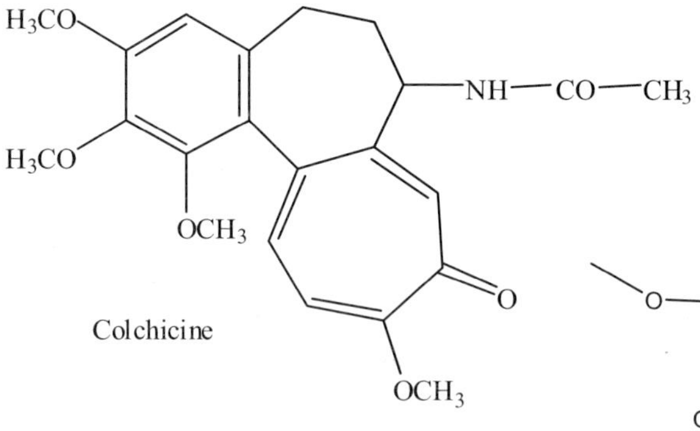

Colchicine

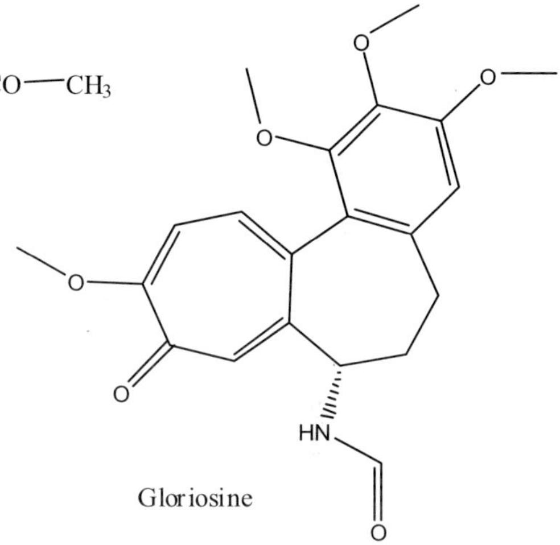

Gloriosine

seeds, *viz.*, colchicine, colchicoside, superbine, gloriosine, lumicolchicine, 3-demethyl-N-deformyl-N-deacetylcolchicine, 3-demethylcolchicine, N-formyldeacetyl-colchicine. The alkaloids from plants (Colchicine and Gloriosine) are used in treatment of gout and rheumatism. The hydroalcoholic extract of aerial part of *G. Superb* possesses an antiinflammatory effect in animal models like carrageenan induced paw oedema and cotton pellet induced granuloma suggesting its effects mainly on the proliferative phase of inflammation. They have reported significant antiinflammatory activity largely due to the presence of colchicines and its derivatives (John *et al.*, 2010).

Peumus boldus

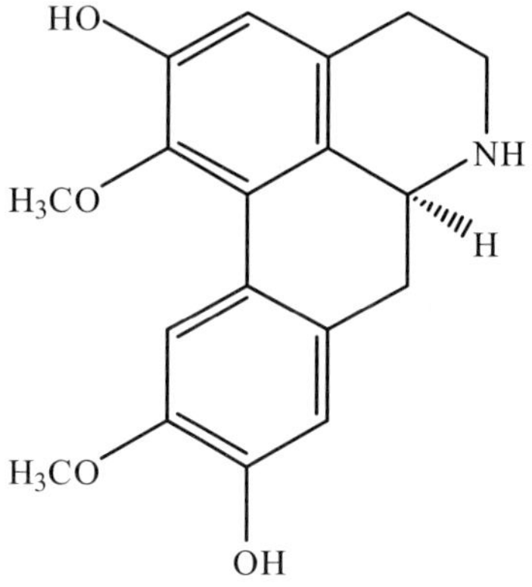

Peumus boldus (Monimiaceae) is natively endemic to the central region of Chile, and also been introduced to Europe and North Africa, though it is not often seen outside botanical gardens. It is a characteristic component of the sclerophyllous forest endemic to central Chile and leaves have a strong, woody and slightly bitter flavour and camphor-like aroma are used for culinary purposes, primarily in Latin America. The leaves are used in a similar manner to bay leaves and also used as an herbal tea, primarily in Chile, Bolivia, Argentina, Paraguay, Peru, Uruguay, Brazil and bordering countries in South America. Boldine is an antioxidant alkaloid isolated from *P. boldus* elicited a dose-dependent antiinflammatory activity in the carrageenan-induced guinea pig paw oedema model. ED50 of boldine was 34 mg/kg by oral route. Boldine (60 mg/kg p.o) treatment inhibited (51- 98 per cent) bacterial-LPS induced hyperthermia in rabbits. Further findings in *in-vitro* studies on rat aortal rings clearly indicated that antiinflammatory activity of boldine is due to the inhibition of prostaglandin biosynthesis (Backhouse *et al.*, 1994). Boldine also showed protection

Boldine

against acid-induced oedema of the colon in rats reflected by the decreased total colon weight (Gotteland *et al.*, 1997).

Aconitum flavum

3-acetyl aconitine, an isoprenoid pathway-derived alkaloid, isolated from *Aconitum flavum* Hand-Mazz and *A. pendulum* Busch, (Ranunculaceae) grown in china, found to inhibit the increased vascular permeability induced by acetic acid/ histamine, carrageenan and egg albumin induced hind pedal oedema (Tang *et al.*, 1984). The curative effect of rheumatism is one of the reputed therapeutic effects of extracts prepared from *Aconitum* roots in traditional Chinese and Japanese medicine.

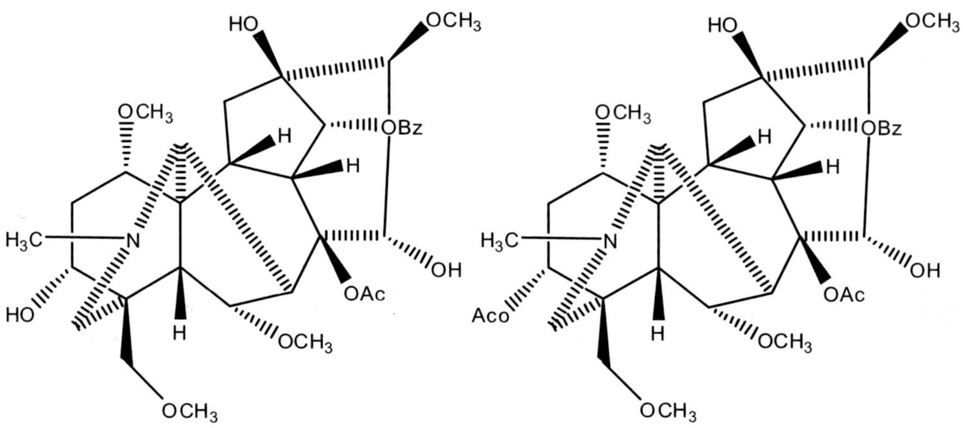

Mesaconitine 3-Acetyl aconitine

The *Aconitum* alkaloids mesaconitine (Hikino *et al.*, 1982) and 3-acetyl aconitine (Tang *et al.*, 1984) were experimentally demonstrated to possess antiinflammatory activity in addition to their above described antinociceptive effect. Both mesaconitine and 3-acetyl aconitine were shown to inhibit the increased vascular permeability induced by acetic acid in mouse peritoneal cavity and by histamine in rat intra dermal sites. The antiinflammatory effect of both alkaloids were observed in sham-operated mice as well as adrenalectomized mice, conforming that the antiinflammatory activity was not mediated by glucocorticoids secreted as a result of stimulation of the hypophysis-adrenal system. Mesaconitine prevented hind-paw oedema induced by carrageenan or by subplantar injection of histamine, serotonin and prostaglandin E1. These experimental findings indicate that mesaconitine is effective at an early, exudative stage of inflammation.

Alchornea cordifolia

A. cordifolia Schum and Thonn (Euphorbiaceae) is a straggling shrub or small tree of swampy, found along the coastal areas of West Africa. The leaves are simple, ovate, basically cordite, acuminate with sessile glands at the base close to the petiole. The flowers are greenish white. The plant is widely used in traditional medicine, in

Diisopentenyl guanidine

Triisopentenyl guanidine

the Niger Delta Region of Nigeria to treat wound, dermatitis, asthma, hepatitis, rheumatism, arthritis, piles, toothache and inflammatory diseases. The ethanol fraction as well as methanol fraction from hexane extracts of *A. cordifolia* leaves and bark exhibited potent antiinflammatory activity in rat models *viz.* egg albumin induced hind paw oedema, topical ear oedema by croton oil better than indomethacin. The ethanol fraction of hexane extract contains two active alkaloids, diisopentenyl guanidine and triisopentenyl guanidine are found to be responsible for antiinflammatory activity (Osadebe *et al.*, 2003; Mavar-Mangaa *et al.*, 2004; 2008).

Sapium baccatum Roxb

Sapium baccatum Roxb (Euphorbiaceae) is a tropical tree found in South Asia, Eastern Himalaya, India (Sikkim), Bangladesh and China. The bukittinggine was isolated by petroleum ether from the leaves of the *S. baccatum* Roxb. The antiinflammatory activity of extract was evaluated by carrageenan-induced hind paw oedema, adjuvant-induced arthritis, cotton pellet induced granuloma, and carrageenan induced pleurisy in rodents species. In all these animal models bukittinggine showed promising antiinflammatory activity comparable to that of aspirin. Furthermore, it was observed that bukittinggine exert marked inhibition on exudative formation, accumulation of leukocytes and on PGE_2-like activity in the exudates. In addition, bukittinggine elicit marked antipyretic and analgesic activity in different tests. It was presumed that the mechanism of antiinflammatory, analgesic and antipyretic activity may be similar to that of aspirin (Panthong *et al.*, 1998).

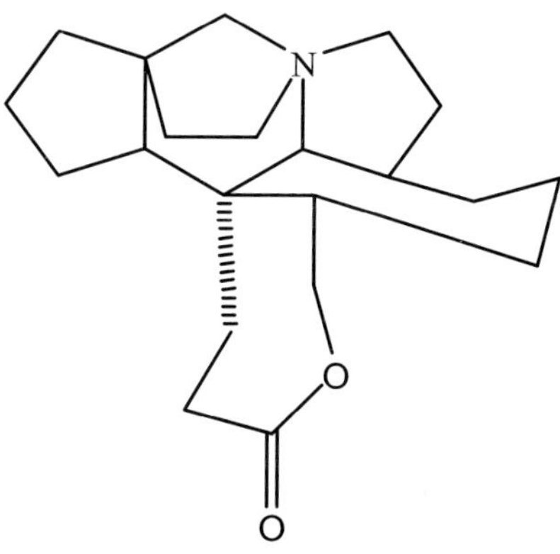

Bukittinggine

Camellia kucha

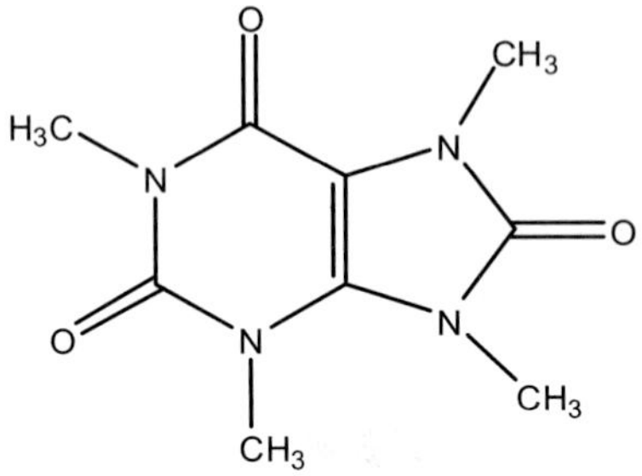

Camellia kucha Hung T. Chang (Theaceae) is a tea plant endemic to the wild woods of Yunnan and has been consumed as a tea for a long time. *C. kucha* found to contain a purine alkaloid theacrine structurally different from caffeine, the predominant purine alkaloid found in traditionally cultivated tea plants (Johnson, 1937; Zheng *et al.*, 2002). The antiinflammatory activity of theacrine (1, 3, 7, 9-tetramethyluric acid), were investigated on animal models *viz.* xylene induced ear oedema, acetic acid-induced vascular permeability and λ-carrageenan-induced paw oedema. Acetic acid-induced writhing and hot-plate tests were used to determine analgesic effect. Oral administration of theacrine (8–32 mg/kg) induced dose-related antiinflammatory and analgesic effects. Theacrine significantly inhibited the formation of xylene induced ear oedema by preventing the mediators such as histamine, serotonin, and bradykinin to produce the antiinflammatory effect. The results of λ-carrageenan experiment indicated that theacrine mainly functions on the first phase of inflammation presumably by interacting with histamine and serotonins and other similar mediators. In addition, the antiinflammatory activity of theacrine was also demonstrated by inhibition of increased vascular permeability induced by acetic acid. Theacrine showed significant antiinflammatory activity by blocking vascular permeability, and thereby conforming its membrane stabilizing activity (Wang *et al.*, 2010).

Theacrine

Bulbus fritillariae

Bulbus fritillariae cirrhosae is the bulb of fritillaria genus from the Liliaceae family, called *fritillaria* cirrhosae is widely used to treat cough and asthma in traditional Chinese Medicine. The likely alkaloids present are: imperialine, chuanbeinone,

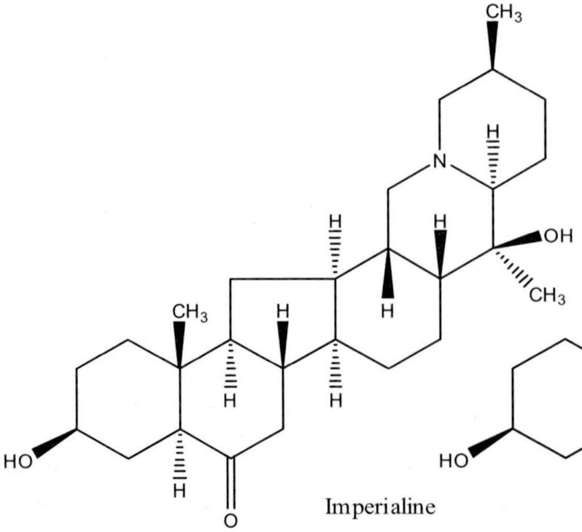

Imperialine

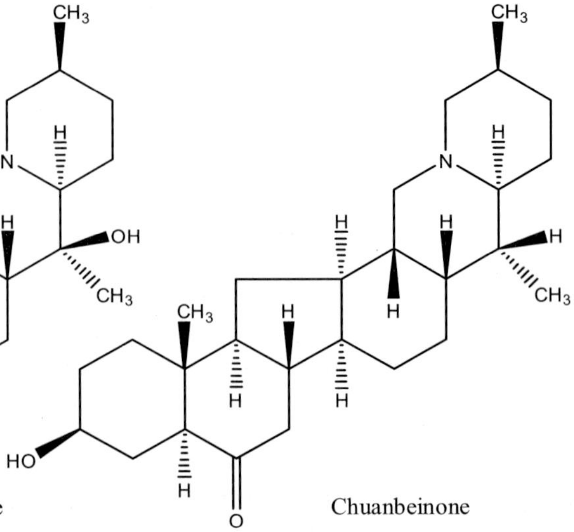

Chuanbeinone

verticinone and verticine from the *B. fritillariae* bulb. The antiinflammatory activity of these alkaloids, were investigated on animal models *viz.* xylene induced ear oedema. At doses of 3.0 mg/kg body weight, imperialine and chuanbeinone elicited significant inhibition of xylene-induced mice ear oedema, the inhibitory effect was found to be greater than dexamethasone. Such antiinflammatory activity of imperialine and chuanbeinone may be useful in the treatment of inflammatory diseases of the respiratory tract such as cough, asthma and chronic bronchitis (Wang *et al.*, 2011).

Corydalis turtschaninovii

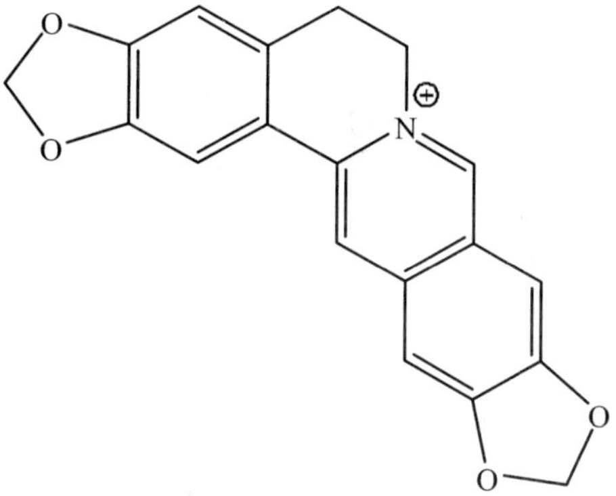

Corydalis is a genus of about 470 species of annual and perennial herbaceous plants in the Papaveraceae family, native to the temperate northern hemisphere and the high mountains of tropical eastern Africa. They are abundantly found in China and the Himalayas, with at least 357 species in China. The pseudocoptisine, a quaternary alkaloid with a benzylisoquinoline skeleton, was isolated from the tubers of *C. turtschaninovii*. Pseudocoptisine showed inhibitory effect on pro-inflammatory mediators in lipopolysaccharide (LPS)-stimulated murine macrophage RAW 264.7 cells. Pseudocoptisine caused dose-dependent reductions in the levels of inducible nitric oxide (iNOS) and cyclooxygenase-2 (COX-2) at both protein and mRNA levels and concomitant decrease in PGE-2 and NO production. In addition, it was found that pseudocoptisine suppressed the production and mRNA expressions of inflammatory cytokines, such as, TNF-α and IL-6.

Pseudocoptisine

Furthermore, the effect at molecular level revealed that pseudocoptisine inhibited the LPS-stimulated DNA binding activity and the transcription activity of nuclear factor-kappa B (NF-κB). Moreover, this effect was accompanied by decreased phosphorylation of inhibitory κB (κB)-α and in the subsequent blocking of p65 subunit of NF-κB translocation to the nucleus. In addition, pseudocoptisine dose-dependently inhibited the phosphorylations of ERK and p38. Taken together these results indicate that pseudocoptisine has the ability to reduce levels of the pro-inflammatory mediators, such as, iNOS, COX-2, TNF-α and IL-6 through the inhibition of NF-κB activation via the suppression of ERK and p38 phosphorylation (RAW 264.7 cells). These findings partly explain the molecular basis of the antiinflammatory activity of pseudocoptisine (Yun *et al.*, 2009).

Picralima nitida

The dried seeds of *Picralima nitida* (Apocyanaceae) are used in traditional medicine throughout West Africa, particularly in Ghana as well as in the Ivory Coast and Nigeria.

The five indole alkaloids (akuammine, akuammicine, akuammidine, akuammigine and pseudo-akuammigine) were isolated from *P. nitida* seeds. Pseudo-akuammigine, an alkaloid from *P. nitida* seed extract, was investigated for antiinflammatory and analgesic activity using the carrageenan-induced rat paw oedema and tail flick method.

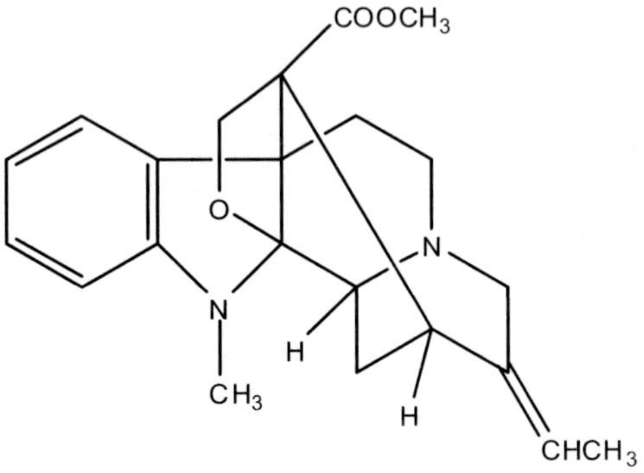

COOCH3 O N CH3 H N H CHCH3

Pseudo-akummigine

On carrageenan induced rat paw oedema, a dose-dependent inhibitory effect was revealed over a dose range of 1-50 mg/kg p.o. Carrageenan induced oedema is primarily a vascular event initiated by dilatation of arterioles and an ultimately increase in permeability of post-capillary venules causing exudation of inflammatory cells and accumulation of fluid at the site of injury. Inhibition of this acute phase of inflammation, will therefore ultimately abort the inflammatory process. Pseudo-akuammigine may therefore be responsible for the observed antiinflammatory activity in the aqueous ethanolic extracts of *P. nitida* (Duwiejua *et al.*, 1995; Duwiejua *et al.*, 2002).

Caulerpa racemosa

Caulerpa racemosa is a species of green alga, seaweed in the family Caulerpaceae. It is commonly known as sea grapes and is found in many areas of shallow sea around the world. A plant of *C. racemosa* consists of a number of branches linked to stolons which are anchored to the sandy substrate by rhizoids. *C. racemosa* is widely distributed in shallow temperate and tropical seas. This bisindole alkaloid was isolated from the lipoid extract of *C. Racemosa* collected in the Northeast of Brazil. Caulerpin comes from a family of bisindole natural products, and has an extra eight-member ring between the two indole rings which are incorporated directly with the carbonyl group. This alkaloid elicited a wide variety of important biological activities like antitumor, growth regulator and the plant root growth stimulant properties. The antiinflammatory activity of caulerpin was evaluated on capsaicin induced ear oedema mice model. Topical application of capsaicin was observed to cause a significant rise in plasma extravasations when compared to control ear in the same animal treated with a vehicle. The exudate formation was significantly reduced (55.8 per cent of the inhibition) with the pre-treatment of caulerpin. Caulerpin pre-treatment also reduced carrageenan induced peritonitis in rodents (De Souza *et al.*, 2009).

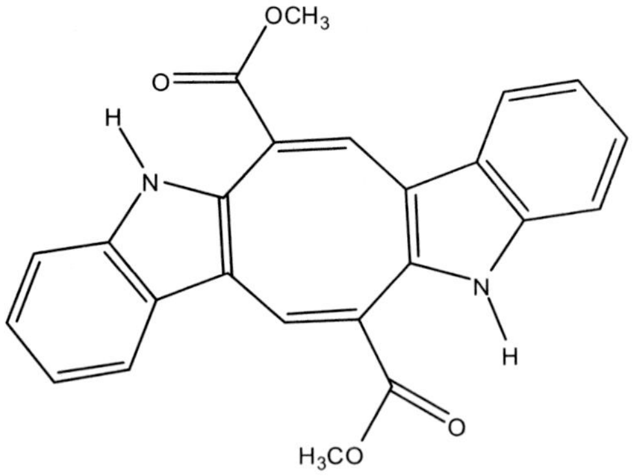

OCH3 O H N N H O H3CO

Caulerpin

Rauwolfia serpentina

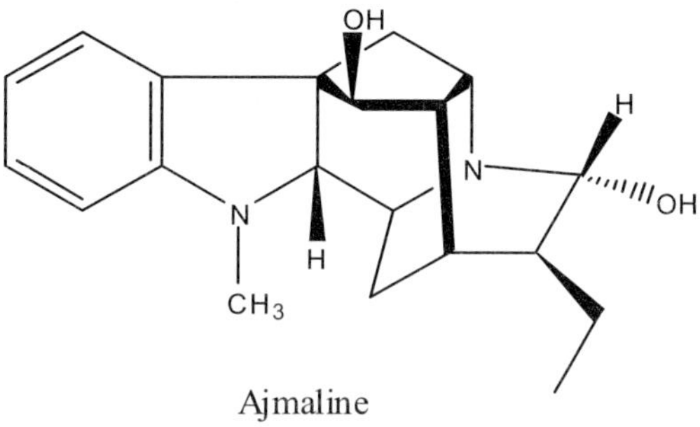

Rauwolfia serpentina, or 'snake-root' or 'sarpagandha' (Apocyanaceae) is a species of flowering plant in the family Apocynaceae. It is native to South and East Asia (from India to China and Indonesia). *R. serpentina* contains a number of bioactive phytocompounds, which includes yohimbine, reserpine, ajmaline, deserpidine, rescinnamine, serpentinine. The effects of ajmaline on human platelet aggregation, arachidonic acid metabolism and platelet activating factor (PAF)-induced lethality in rabbits were studied. Platelet aggregation induced by several stimuli (ADP, collagen, and PAF) was inhibited in a concentrations dependent manner by ajmaline. The potency of ajmaline was higher when PAF was employed as stimulating agent in comparison with other agonists, whereas ajmaline had no effect against arachidonic acid-induced aggregation. In contrast, ajmaline inhibited arachidonic acid metabolism in platelet homogenates. The formation of both thromboxane A_2 and 12-hydroxy-eicosatetraenoic acid was inhibited by ajmaline and the effect was comparable to that of aspirin. Pre-treatment of rabbits with ajmaline (50 mg/kg) completely abolished the lethal effects of PAF (11 micrograms/kg) given intravenously ($P < 0.001$). In addition, ajmaline at doses ranging from 50 to 100 mg/kg inhibited carrageenan induced rat paw oedema ($P < 0.001$). In this test ajmaline was three fold potent than aspirin.

Ajmaline

In the light of these experimental findings it was concluded that ajmaline; a known antiarrhythmic agent is a PAF antagonist and a dual inhibitor of platelet cyclooxygenase and lipoxygenase enzymes with potent antiinflammatory properties (saeed *et al.*, 1993).

Sanguinaria canadensis

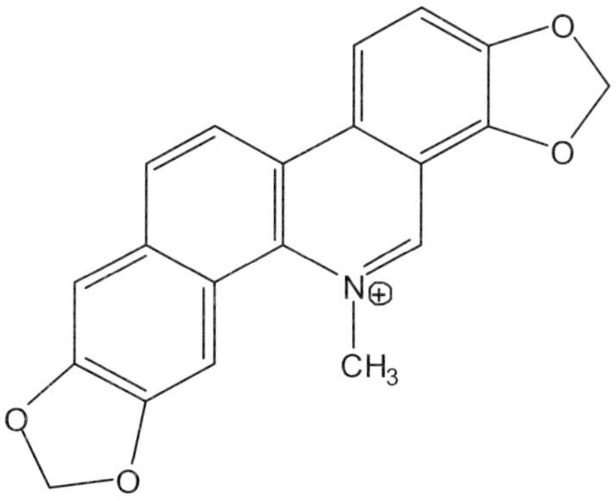

Sanguinaria canadensis (Papaveraceae) is a perennial, herbaceous flowering plant native to eastern North America from Nova Scotia, Canada southward to Florida, United States, and west to Great Lakes and down the Mississippi embayment. It is the only species in the genus *Sanguinaria*, included in the family Papaveraceae, and most closely related to Eomecon of Eastern Asia. Sanguinarine (SANG,13-methyl-[1,3] benzodioxolo [5,6-c]-1,3- dioxolo [4,5-I] phenanthrdinium) is an alkaloid present in the roots of *Sanguinaria canadensis*. This alkaloid is used as a naturopathic therapy for the

Sanguinarine

treatment of infections, and functions as a pain reliever, expectorant, sedative, and emetic. Epidemiological studies have suggested that the use of sanguinarine as an oral rinse and in toothpaste effectively suppresses dental plaque formation and reduces gingival inflammation via its antioxidant, antimicrobial, and antiinflammatory activity. The antiinflammatory activity of a sanguinarine was evaluated by xylene induced ear oedema, formaldehyde or carrageenan induced paw oedema, and the lipopolysaccharide (LPS) induced endotoxic shock in rodents. Sanguinarine elicited significant antiinflammatory activity in LPS stimulated peritoneal macrophages (*in-vitro*) as well as in acute (carrageenan induced paw oedema) and chronic inflammatory (formaldehyde induced paw oedema) *in-vivo* rodents models. These findings further demonstrated that sanguinarine potently inhibits the expression of inflammatory mediators both *in-vivo* and *in-vitro* experiments. Additionally, *in-vitro* experimental results showed that sanguinarine suppressed the activation of mitogen activated protein kinase (MAPK), which altered the synthesis and release inflammatory mediator (Niu *et al.*, 2012).

Alstonia scholaris

Alstonia scholaris (Apocynaceae) is evergreen, tropical tree native to the Indian subcontinent and Southeast Asia. The extract of leaves of *A. scholaris* was developed commercially and used for the treatment of chronic respiratory diseases like tracheitis, bronchitis and cold in the good olden days in China. From the ethyl acetate fraction of leaf extract of *A. scholaris*; alkaloids (picrinine, vallesamine and scholaricine), were isolated, which showed antiinflammatory activity in granuloma pouch as well as xylene induced ear oedema models. Furthermore, the alkaloids fraction also showed antioxidant activity (increasing SOD and decreased MDA) and reduced inflammatory mediators (NO, PGE_2) in the exudate of granuloma pouch. Such antiinflammatory activity of alkaloid fraction was subsequently confirmed in the inhibition of COX-1, COX-2 and 5-LOX in *in-vitro* system. Thus, the antiinflammatory activity of

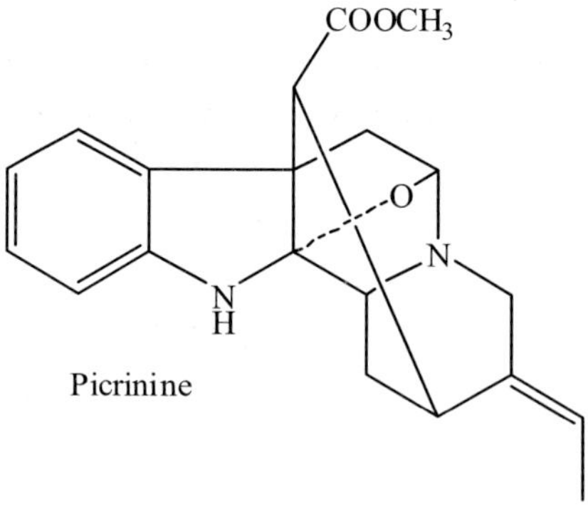

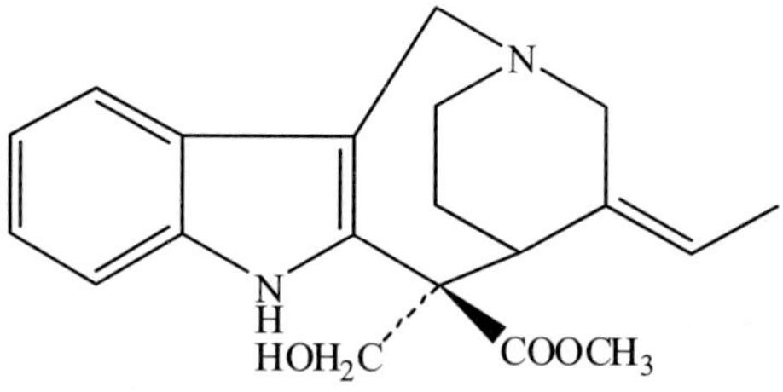

Picrinine

Vallesamine

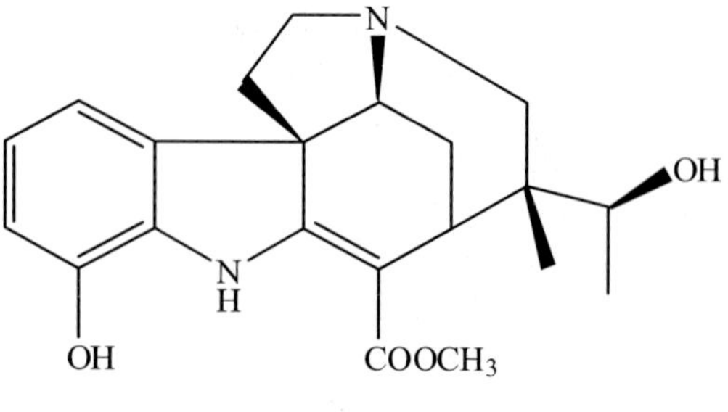

Scholaricine

alkaloid fraction observed was found to be mediated through its effect on prostaglandins and leukotriene pathways with significant antioxidant activity. In addition, alkaloids fraction also showed promising analgesic activity in acetic acid writhing mouse model (Shang *et al.*, 2010).

Withania somnifera

Withania somnifera is a small, erect, evergreen woody under shrub belongs to Solanaceae family. *W. somnifera* is a popular in the Ayurvedic and traditional medicinal system for over 3000 years and cultivated in many of the drier regions of India, such as mandsaur district of Madhya Pradesh, Punjab, Sindh, Rajasthan and Nepal. *W. somnifera* is also popularly known as ashwagandha, Indian ginseng, and winter cherry. In-depth studies carried out with the extracts of *W. somnifera* showed multi therapeutic activity like antioxidant, antitumor, antistress, antiinflammatory, immunomodulatory, haematopoetic, antiageing, anxiolytic, rejuvenating properties, and also influences various neurotransmitter receptors in the central nervous system leading to antidepressant as well as adaptogenic activity (Sharma *et al.*, 2011). Furthermore, the *W. somnifera* extract showed anticancer activity on different cancer cell lines. The active constituents, Withaferin A and Sitoindosider VIII-X of methanolic and ethanolic leave extract of *W. somnifera* showed significant antiinflammatory (carrageenan- induced paw oedema) and antiarthritic activity which has been correlated to its antioxidant activity, inhibition of cyclooxygenase 1 and 2 (Jayaprakasam and Nair, 2003). Similarly, the aqueous extracts of root powder of *W. somnifera* showed promising chondroprotective activities on human osteoarthritic cartilage matrix damage, and inhibition of the gelatinase and collagenase-2 enzyme in *in-vitro* (Sumantrana *et al.*, 2007). The root extract of *W. somnifera* inhibited monosodium urate crystal-induced gouty arthritis and the antiarthritic effect was correlated to the inhibition of the release of lysosomal enzymes (β-glucuronidase and lactate dehydrogenase) from Polymorphonuclear leukocytes (Rasool and Varalakshmi, 2006). The other beneficial effects derived from *W. somnifera* root extracts are: analgesic, antipyretic and antiulcer activity. The root extract of *W. somnifera* reported to enhance the levels of IFN-γ, IL-2, IL-4, IL-6, IL-10, IL-12, and GM-CSF in variety of inflammation (using Balb/c mice) models. Such effects of *W. somnifera* may help for antiinflammatory, immunomodulatory and adaptogenic activity (Davis and Kuttan, 1999). Furthermore, such biochemical derived effects of *W. somnifera* can be

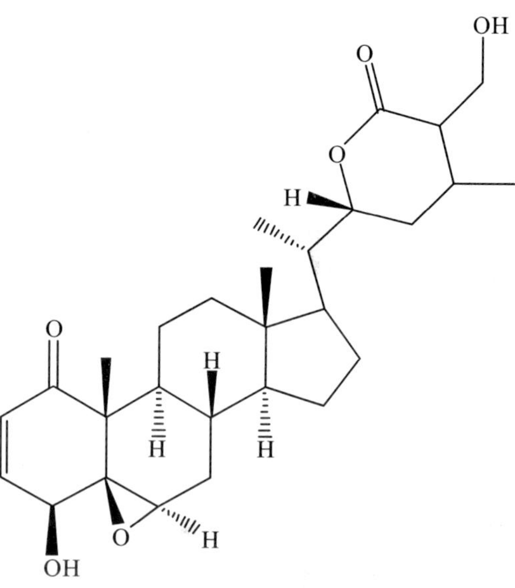

Withaferin A

combined with other herbs like *Grifola frondosa*, to induce effective antiinflammatory and immunomodulatory activity (Vetvicka and Vetvickova, 2011).

Plumeria acutifolia

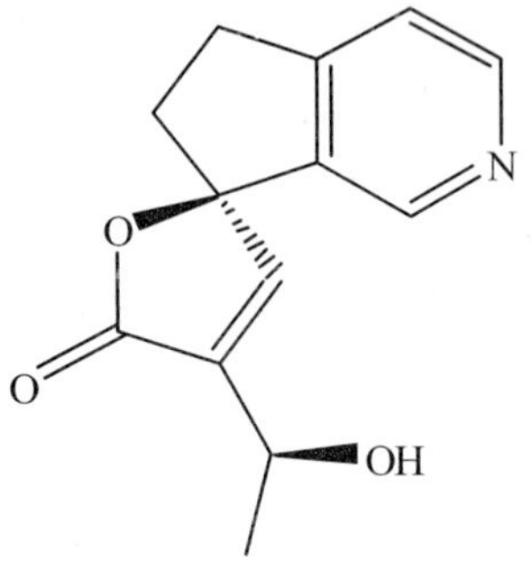

Plumeria is a genus of flowering plants belonging to family: Apocynaceae, includes, 7-8 species of mainly deciduous shrubs and small trees. *Plumeria acutifolia* are native to Central America, Mexico, the Caribbean, Brazil and many other parts of the world. In many parts, of the world's traditional medicine *P. acutifolia* used for the treatment of inflammation, allergy, diuresis, diarrhoea, ulcer, leprosy, liver cirrhosis and infections. The major active phytoconstituents, plumerianine isolated from the root bark of *Plumeria acutifolia* was found to be responsible for antianaphylactic (passive cutaneous anaphylaxis, passive paw anaphylaxis), antiinflammatory (carrageenan induced paw oedema and cotton pellet granuloma) and membrane stabilizing activity (*i.e.*, protection against haemolysis of RBC's by hypotonic solution). The antiinflammatory activity has been attributed to inhibition of the early phase and late phase of inflammatory mediators (Alakshmi *et al.*, 2011).

Plumerianine

Sternbergia fisheriana

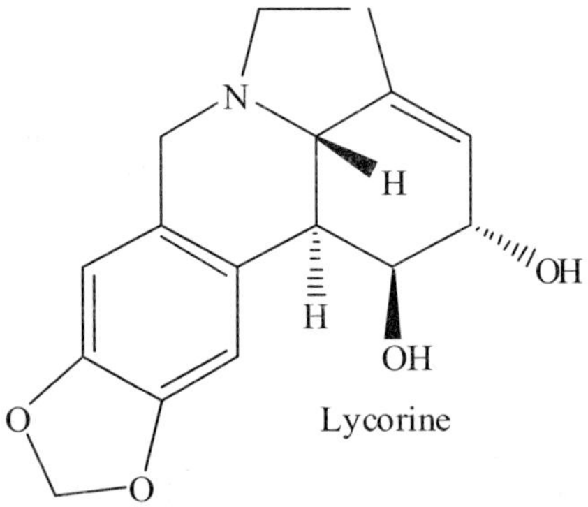

Sternbergia fisheriana (Herbert) Rupr. (Amaryllidaceae) is a yellow flowering spring blooming species that abundantly found in the stony places in Turkey, Iran, and Iraq. The alkaloid Lycorine containing pyrrolol-phenantridine ring was isolated from *Sternbergia fisheriana*. The lycorine was formulated as expectorant to treat wide variety of ailments like antitumor, antiviral, antifungal, antimalarial, antiinflammatory, antioxidant, antiparasitic, hepatoprotective and antifeedant in Russia. The Lycorine exhibited antiinflammatory activity in carrageenan induced hind paw oedema and analgesic activity in acetic acid induced writhing and tail-flick tests. The antiinflammatory activity of lycorine was comparable to that of indomethacin, a reference drug (Saltan Citoglu *et al.*, 2012).

Lycorine

Chelidonium majus

Chelidonium majus (Papaveraceae) is a herbaceous perennial plant, abundantly grows in Europe and Western Asia and North America. The *C. Majus*, herbal preparation has been traditionally used for the treatment of liver disorders. From the ethyl acetate extract of *C. Majus*, a new alkaloid, methyl 20-(7,8-dihydrosanguinarine-8-yl) acetate (Compound 1), was isolated together with other six known alkaloids *viz.* stylopine, protopine, norchelidonine, chelidonine, berberine, and

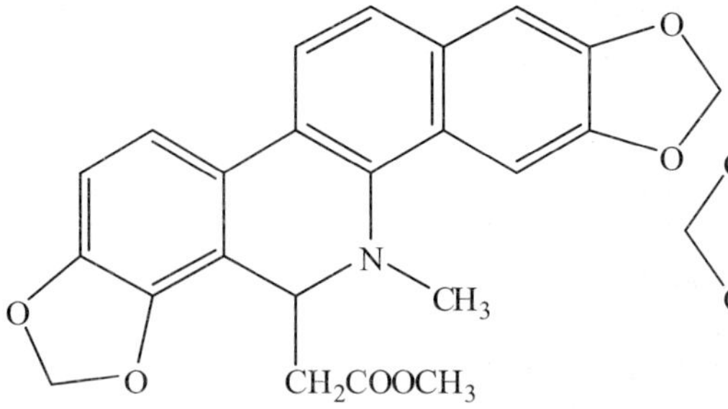

Compound 1

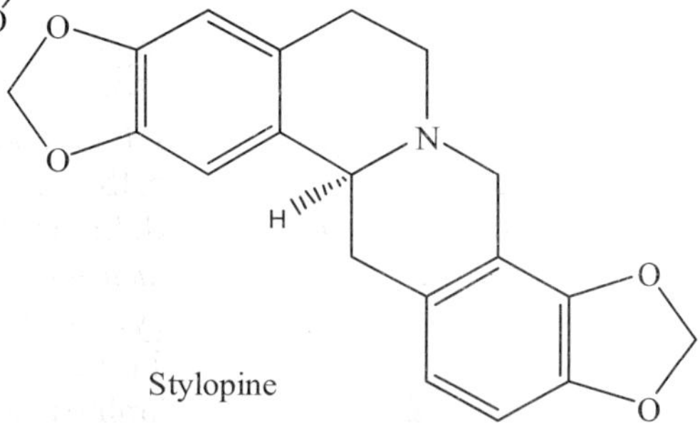

Stylopine

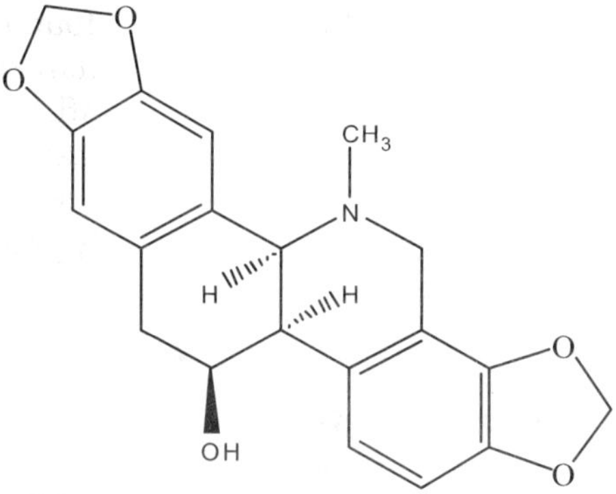

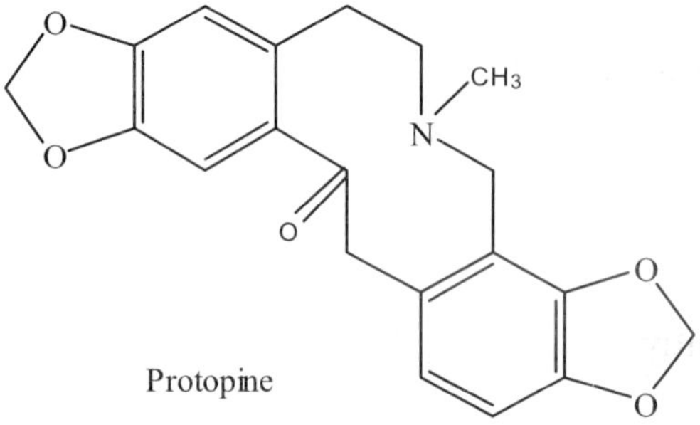

Protopine

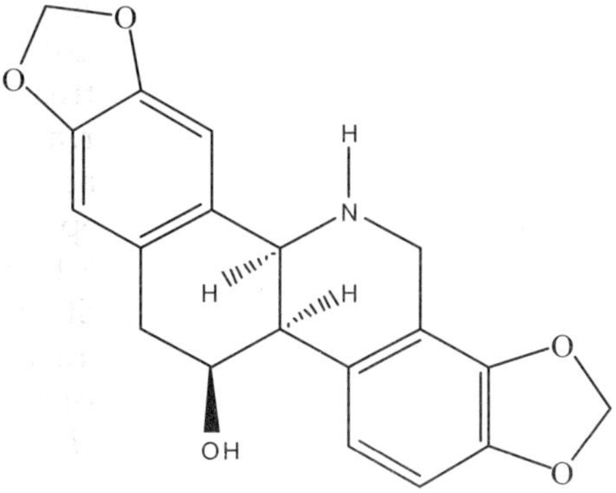

chelidonine

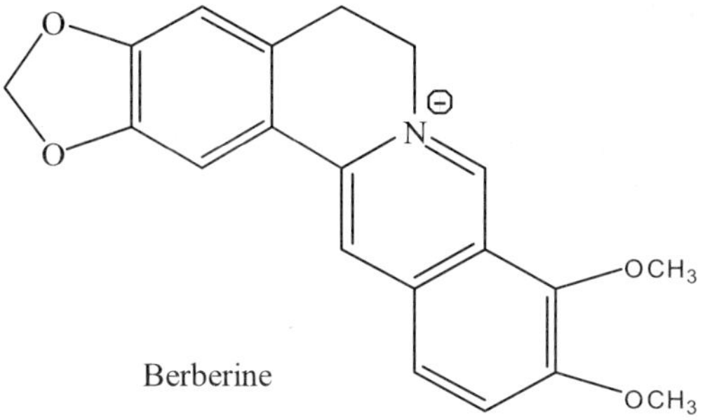

Berberine

Norchelidonine

8-hydroxydihydrosanguinarine. The *in-vitro* studies on these alkaloids showed antiinflammatory activity by the inhibition of LPS induced NO production in macrophage RAW264.7 cells, among these chelidonine and 8-hydroxydihydrosanguinarine showed greater inhibitory activity on the LPS induced NO production in macrophage RAW264.7 cells. Furthermore, chelidonine and 8-hydroxydihydrosanguinarine inhibited dose dependently the induction of COX-2 and iNOS mRNA suggesting its inhibitory effect on both synthesis as well as transcriptional level (Park *et al.*, 2011).

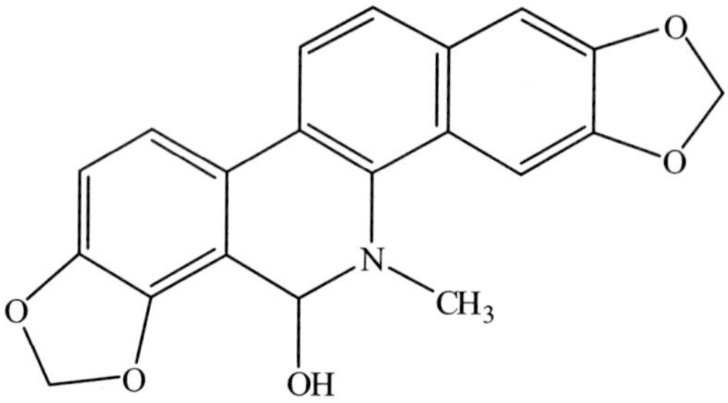

8-hydroxydihydrosanguinarine

Conclusions

The flavonoids and alkaloids are important phytoconstituents of large number of herbs and plants, and long been used in traditional medicine for the treatment of various diseases in number of countries. The flavonoids and alkaloids of various plant extracts showed significant antiinflammatory, antiarthritic activity including other inflammation related disorders in different type of *in-vivo* and *in-vitro* animal models. In addition, other therapeutic beneficial activities like antioxidant, membrane stabilizing, analgesic, and antipyretic activity have also been reported. The cellular (mast cells, leukocytes, macrophages, catheptic enzymes, etc.) and molecular level (cytokines, prostaglandins, leukotrienes, iNOS, NF-8B, PKC, MAPK etc.) effects of flavonoids and alkaloids during inflammatory conditions (acute and chronic) have helped to understand their probable underlying mechanism(s) of action of antiinflammatory and antiarthritic activity. Some of the alkaloids and flavonoids showed greater antiinflammatory activity than the standard drugs. Some of the alkaloids and flavonoids mechanism(s) of action may help to understand patho-physiological process or aetiology of inflammatory disorders.

The review covers large number of herbs containing flavonoids and alkaloids having antiinflammatory activity which can be further explored with the advanced techniques and methods to develop therapeutically effective phytochemicals. Since the review includes large number of diverse herbs and plants used in different traditional system of medicine, it should be beneficial for researchers, the research organisations and teaching professionals.

References

Abad, M.J., Bermejo, P., and Villar, A. (1995).The activity of flavonoids extracted from *Tanacetum microphyllum* DC. (Compositae) on soybean lipoxygenase and prostaglandin synthetase. *Gen. Pharmac.*, **26(4):** 815-819.

Alakshmi V., chandiran, R., Velraj M., Hemalatha., and Jayakumari. (2011). Anti-Anaphylactic and anti-inflammatory activities of a bioactive alkaloid from the root bark of *Plumeria acutifolia* Poir. *Iranian Journal of Pharmaceutical Research*, **10 (3):** 525-533.

Aquila, S., Giner, R.M., Recio, M.C., Spegazzini, E.D., and Ríos, J.L. Anti-inflammatory activity of flavonoids from *Cayaponia tayuya* roots. *Journal of Ethnopharmacology*, **121**: 333–337.

Backhouse, N., Delporte, C., Givernau, M., Cassels, B. K., Valenzuela, A., and Speisky, H. (1994). Anti-inflammatory and antipyretic effects of boldine. *Agents Actions*, **42**: 114-117.

Bai, N., He, K., Roller, M., Lai, C.S., Shao, X., Pan, M.H., Bily, A., and Ho, C.T. (2011). Flavonoid glycosides from *Microtea debilis* and their cytotoxic and anti-inflammatory effects. *Fitoterapia*, **82:** 168–172.

Banerjee, T., Valacchi, G., Ziboh, V.A., and van der Vliet, A. (2002). Inhibition of TNF-α induced cyclooxygenase-2 expression by amento-flavone through suppression of NF-8B activation in A549 cells. *Mol Cell Biochem.*, **238**: 105–110.

Bauman, J., Bruchhausen, F.V., and Wurm, G. (1980). Flavonoids and related compounds as inhibitors of arachidonic acid peroxidation. *Prostaglandins*, **20**: 627–639.

Beecher, G.R. (2003). Overview of dietary flavonoids: nomenclature, occurrence and intake. *J Nutr.*, **133**: 3248S–3254S.

Chen, C.Y., Peng, W.H., Tsai, K.D., and Hsu, S.L. (2007). Luteolin suppress inflammation associated gene expression by blocking NF-8B and AP-1 activation pathway in mouse alveolar macrophages. *Life Sci.*, **81**: 602–1614.

Cho, N., Choi, J.H., Yang, H., Jeong, E.J., Lee, K.Y., Kim, Y.C., and Sung, S.H. (2012). Neuroprotective and anti-inflammatory effects of flavonoids isolated from *Rhus verniciflua* in neuronal HT22 and microglial BV2 cell lines. *Food and Chemical Toxicology*, **50:** 1940–1945.

Choi, I.Y., Kim, S.J., Jeong, H.J., Park, S.H., Song, Y.S., and Lee, J.H., *et al.* (2007). Hesperidin inhibits expression of hypoxia inducible factor-1 alpha and inflammatory cytokine production from mast cells. *Molecular and CellularBiochemistry*, **305:** 153–161.

Clavin, M., Gorzalczany, S., Macho, A., Munoz, E., Ferraro, G., Acevedo, C., and Martino, V. (2007). Anti-inflammatory activity of flavonoids from *Eupatorium arnottianum. Journal of Ethnopharmacology*, **112**: 585–589.

Cottigliaa, F., Casua, L., Bonsignorea, L., Casub, M., Florisb, C., Sosac, S., Altinierc, G., and Loggia, Z.R.D. (2005). Topical anti-inflammatory activity of flavonoids and a new xanthone from *Santolina insularis. Naturforsch.*, **60c**: 63-66.

Davis, L. and Kuttan, G. (1999). Effect of *Withania somnifera* on cytokine production in normal and cyclophosphamide treated mice. *Immunopharmocol.Immunotoxicol.*, **21**: 695-704.

De Groot, H., and Rauen, U. (1998). Tissue injury by reactive oxygen species and the protective effects of flavonoids. *Fundam Clin Pharmacol.*, **12**: 249–55.

De Souza, E.T., De Lira, D.P., De Queiroz, A.C., Da Silva, D.J., De Aquino, A.B., and Campessato Mella, E.A. *et al.* (2009). The antinociceptive and anti-inflammatory activities of caulerpin, a bisindole alkaloid isolated from seaweeds of the genus *Caulerpa. Mar. Drugs*, **7:** 689-704.

Dong, W., Wei, T., Ming, Y.G., and Chang, C.B. (2010). Anti-inflammatory, antioxidant and cytotoxic activities of flavonoids from *Oxytropis falcata* Bunge. *Chinese Journal of Natural Medicines*, **8(6):** 0461-0465.

Duwiejua, M., Obiri, D. D., Zeitlin, I. J., and Waterman, P. G. (1995). Anti-inflammatory activity in extracts from *Picralima nitida* (Fam. Apocynaceae). *J. Pharmacol.*, **116:** 360.

Duwiejua, M., Woode, E., and Obiri, D.D. (2002). Pseudo-akuammigine, an alkaloid from *Picralima nitida* seeds, has anti-inflammatory and analgesic actions in rats. *Journal of Ethnopharmacology*, **81**: 73-79.

Facchini, P.J. (2001). Alkaloid biosynthesis in plants: Biochemistry, cell biology, molecular regulation, and metabolic engineering applications. *Annu. Rev. Plant Physiol. Plant Mol. Biol.*, **52:** 29–66.

Ferriola, P.C., Cody, V., and Middleton, E. (1989). Protein kinase C inhibition by plant flavonoids. Kinetic mechanisms and structural-activity relationships. *Biochem Pharmacol.*, **38**: 1617–1624.

Geoffrey, A., and Cordell, J. (2012). Fifty years of alkaloid biosynthesis in Phytochemistry. *Phytochem.*, **05:** 012.

Germolec, D.R., Frawley, R.P., and Evans, E. (2010). Markers of inflammation. *Methods Mol Biol.*, **598**: 53–73.

Gomes, C. M. R., and Gottlieb, O. R. (1978). Alkaloid evolution and angiosperm systematics. *Biochemical Systematics and Ecology*, **8:** 81-87.

Gordon, S., and Taylor, P.R. (2005). Monocyte and macrophage heterogeneity. *Nat Rev Immunol.*, **5(12)**: 953–964.

Gotteland, M., Jimenez, I., Brunser, O., Guzman, L., Romero, S., Cassels, B.K., and Speisky, H. (1997). Protective effect of boldine in experimental colitis. *Planta Med.*, **63**: 311-315.

Guabiraba, R., Campanha-Rodrigues, L.A., Souza, A.L.S., Santiago, H.C., Lugnier, C., and Alvarez-Leite, J. *et al.* (2010). The flavonoid Dioclein reduces the production of pro-inflammatory mediators *in vitro* by inhibiting PDE4 activity and scavenging reactive oxygen species. *European Journal of Pharmacology*, **633:** 85–92.

Havsteen, B. (1983). Flavonoids, a class of natural products of high pharmacological potency. *Biochem Pharmacol.*, **32**: 1141– 1148.

Hegnauer, R. (1988). Biochemistry, distribution and taxonomic relevance of higher plant alkaloids. *Phytochemistry*, **21(8):** 2423-2427.

Hikino, H., Takata, H., Fujiwara, M., Konno, C., and Ohuchi, K. (1982). Mechanism of inhibitory action of mesaconitine in acute inflammations. *Eur. J. Pharmac.*, **82:** 65-71.

Ivanovska, N., and Philipov, S. (1996). Study on the anti-inflammatory action of *Berberis vulgaris* root extract, alkaloid fractions and pure alkaloids. *Int J Immunopharmacol*, **18**: 553-561.

Jana, S., and Shekhawat, G. S. (2011). Critical review on medicinally potent plant species: *Gloriosa superb*. *Fitoterapia*, **82:** 293–301.

Jayaprakasam, B., and Nair, M. G. (2003). Cyclooxygenase-2 enzyme inhibitory withanolides from *Withania somnifera* leaves. *Tetrahedron*, **59:** 841–849.

Jin, J.H., Kim, J.S., Kang, S.S., Son, K.H., Chang, H.W., and Kim, H.P. (2010). Anti-inflammatory and anti-arthritic activity of total flavonoids of the roots of *Sophora flavescens*. *Journal of Ethnopharmacology*, **127:** 589–595.

John, J.C., Fernandes, J., Nandguda, T., Niphade, S.R., Savla, A., and Deshmukh, P.T. (2010). Analgesic and anti-inflammatory activities of the hydrochloric extract from *G. superba* Linn. *Int J Green Pharma.*, **3**: 215–9.

Johnson, T.B. (1937). Purines in the plant kingdom: the discovery of a new purinein tea. *J Am Chem Soc.*, **59**: 1261–1263.

Jung, H.A., Jin, S.E., Min B.S., Kim, B.W., and Choi, J.S. (2012). Anti-inflammatory activity of Korean thistle *Cirsium maackii* and its major flavonoid, luteolin 5-O-glucoside. *Food and Chemical Toxicology*, **50:** 2171–2179.

Kang, S.R., Park, K.I., Park, H.S., Lee, D.H., Kim, J. A., and Nagappan, A. *et al.* (2011). Anti-inflammatory effect of flavonoids isolated from Korea *Citrus aurantium L.* on lipopolysaccharide-induced mouse macrophage RAW 264.7 cells by blocking of nuclear factor-kappa B (NF-κB) and mitogen-activated protein kinase (MAPK) signalling pathways. *Food Chemistry*, **129:** 1721–1728.

Kanno, S., Shouji, A., Tomizawa, A., Hiura, T., Osanai, Y., and Ujibe, M., *et al.* (2006). Inhibitory effect of naringin on lipopolysaccharide (LPS)-induced endotoxin shock in mice and nitric oxide production in RAW 264.7 macrophages. *Life Sciences*, **78**: 673–681.

Kazi, A., Daniel, K.G., Smith, D.M., Kumar, N.B., and Dou, Q.P. (2003). Inhibition of the proteasome activity, a novel mechanism associated with the tumor cell apoptosis-inducing ability of genistein. *Biochem Pharmacol.*, **66**: 956–976.

Kim, D.W., Chi, Y.S., Son, K.H., Chang, H.W., Kim, J.S., Kang, S.S., and Kim, H.P. (2002). Effects of sophoraflavanone G, a prenylated flavonoid from *Sophora flavescens*, on cyclooxygenase-2 and *in vitro* inflammatory response. *Archives of Pharmacal Research*, **25:** 329–335.

Kim, H.P., Pham, H.T., and Ziboh, V.A. (2001). Flavonoids differentially inhibit guinea pig epidermal cytosolic phospholipase A2. *Prostaglandins Leukot Essent Fatty Acids*, **65**: 281–286.

Krenn, L., Wollenweber, E., Steyrleuthner, K., Görick, C., and Melzig, M. F. (2009). Contribution of methylated exudate flavonoids to the anti-inflammatory activity of *Grindelia robusta*. *Fitoterapia*, **80:** 267–269.

Krol, W., Czuba, Z.P., Threadgill, M.D., Cunningham, B.D., and Pietse, G. (1995). Inhibition of nitric oxide (NO) production in murine macrophages by flavones. *Biochem Pharmacol.*, **50**: 1031–1035.

Kuhnau, J. (1976). The flavonoids: A class of semi-essential food components: their role in human nutrition. *World Res Nut Diet.*, **24:** 117-191.

Kumar, A.S., Kavimani, S., and Jayaveera, K.N. (2011). A review on medicinal plants with potential antidiabetic activity. *International Journal of Phytopharmacology*, **2(2)**: 53-60.

Kupeli, E., and Yesilada, E. (2007). Flavonoids with anti-inflammatory and antinociceptive activity from *Cistus laurifolius* L. leaves through bioassay-guided procedures. *Journal of Ethnopharmacology*, **112**: 524–530.

Lanni, C., and Becker, E.L. (1985). Inhibition of neutrophil phospholipase A2 by p-bromophenylacyl bromide, nordihydroguaiaretic acid, 5,8,11,14-eicosatetrayenoic acid and quercetin. *IntArchs Allergy Appl Immun.*, **76**: 214–217.

Laughton, M.J., Evans, P.J., Moroney, M.A., Hoult, J.R.S., and Halliwell, B. (1991). Inhibition of mammalian 5-lipoxygenase and cyclooxygenase by flavonoids and phenolic dietary additives. *Biochem Pharmacol.*, **42**: 1673–1681.

Li, B.Q., Fu, T., Gong, W.H., Dunlop, N., Kung, H.F., and Yan, Y.D. *et al.* (2000). The flavonoid baicalin exhibits anti-inflammatory activity by binding to chemokines. *Immunopharmacology*, **49:** 295–306.

Li, S., Lu, A.P., Wang, Y.Y., and Li, Y.D. (2003). Suppressive effects of a Chinese herbal medicine *qing-luoylin* extract on the angiogenesis of collagen-induced arthritis in rats. *American Journal of Chinese Medicine*, **31**: 713–720.

Liang, Y.C., Tsai, S.H., Tsai, D.C., Lin-Shiau, S.Y., and Lin, J.K. (2001). Suppression of inducible cyclooxygenase and nitric oxide synthase through activation of peroxisome proliferator-activated receptor-γ by flavonoids in mouse macrophages. *FEBS Lett.*, **496**: 12–18.

Mavar Manga, H., Brkic, D., Marie, D.E.P., and Quetin-Leclercq, J. (2004). *In vivo* anti-inflammatory activity of *Alchornea cordifolia* (Schumach. and Thonn.) Mull. Arg. (Euphorbiaceae). *Journal of Ethnopharmacology*, **92**: 209– 214.

Mavar-Mangaa, H., Haddada, M., Pieters, L., Baccelli, C., Penge, A., and Quetin-Leclercq, J. (2008). Anti-inflammatory compounds from leaves and root bark of *Alchornea cordifolia* (Schumach. and Thonn.) Mull.Arg. *Journal of Ethnopharmacology*, **115:** 25–29.

Medzhitov, R. (2008). Origin and physiological roles of inflammation. *Nature*, **454**: 428–435.

Middleton, E., Kandaswami, C., and Theoharides, T.C. (2000). The effects of plant flavonoids on mammalian cells: implications for inflammation, heart disease, and cancer. *Pharmacol Rev.*, **52**: 673– 751.

Middleton, E.J. (1998). Effect of plant flavonoids on immune and inflammatory cell function. *AdvExp Med Biol.*, **439**: 175–182.

Miyamoto, Y., Kawasuji, T., Kuraishi, Y., and Suzuki, H. (2003). Antipruritic effects of *Sophora flavescens* on acute and chronic itch-related responses in mice. *Biological and Pharmaceutical Bulletin*, **26:** 722–724.

Mosquera., D.M.G., Ortega, Y.H., Kilonda, A., Dehaen, W., Pieters, L., and Apers, S. (2011). Evaluation of the *in vivo* anti-inflammatory activity of a flavonoids glycoside from *Boldon purpurascens*. *Phytochemistry Letters*, **4**: 231–234.

Murakami, A., Nakamura, Y., Torikai, K., Tanaka, T., Koshiba, T., and Koshimizu, K., *et al*. (2000). Inhibitory effect of *citrus nobiletin* on phorbol ester-induced skin inflammation, oxidative stress, and tumor promotion in mice. *Cancer Research*, **60:** 5059–5066.

Muralidhar, A., Sudhakar Babu, K., Ravisankar, T., Reddanna, P., Reddy, G.V., and Latha, J. (2010). Anti-inflammatory activity of flavonoid fraction isolated from the stem bark of *Butea monosperma* (Lam): a mechanism based study. *International Journal of Phytopharmacology*, **1(2)**: 124-132.

Nakadate, T., Aizu, E., Yamamoto, S., and Kato, R. (1985). Effects of chalcones derivatives on lipoxygenase and cyclooxygenase activities of mouse epidermis. *Prostaglandins*, **3**: 357–368.

Niu, X., Fan, T., Li, W., Xing, W., and Huang, H. (2012). The anti-inflammatory effects of sanguinarine and its modulation of inflammatory mediators from peritoneal macrophages. *European Journal of Pharmacology*, **689:** 262–26.

Ono, M. (1994). Inflammation inhibitors containing cepharanoline or berbamine. *Patent-Japan* Kokai Tokkyo Koho., **6**: 211-661.

Orhan, D.D., Küpeli, E., Yesilada, E., and Ergun, F.Z. (2006). Anti-Inflammatory and antinociceptive activity of flavonoids isolated from *Viscum album* ssp. *Album. Naturforsch*, **61c**: 26-30.

Osadebe, P. O., and Okoye, E.C. (2003). Anti-inflammatory effects of crude methanolic extract and fractions of *Alchornea cordifolia* leaves. *Journal of Ethnopharmacology*, **89:** 19–24.

Panthong, A., Kanjanapothi, D., Thitiponpunt, Y., Taesotikul, T., and Arbain, D. (1998). Anti-inflammatory activity of the alkaloid bukittinggine from *Sapium baccatum*. *Planta Med.*, **64(6)**: 530-5.

Park, J.E., Cuong, T.D., Hung, T.M., Lee, I.S., Na, M.K., Kim, J.N., Ryoo, S.W., Lee, J.H., Choi, J.S., Woo, M.H., and Min, B.S. (2011). Alkaloids from *Chelidonium majus* and their inhibitory effects on LPS-induced NO production in RAW264.7 cells. *Bioorganic and Medicinal Chemistry Letters*, **21:** 6960–6963.

Ramesh, M., Nageshwar Rao, Y., Appa Rao, A.V.N., Prabhakar, M.C., SeshagiriRao, C., Muralidhar, N., and Reddy, M.B. (1998). Antinociceptive and anti-inflammatory activity of a flavonoid isolated from *Caralluma attenuate*. *Journal of Ethnopharmacology*, **62**: 63–66.

Rao, K., Fang, S.H., and Tzeng, Y.M. (2005). Anti-inflammatory activities of flavonoids isolated from *Caesalpinia pulcherrima* Yerra. *Journal of Ethnopharmacology*, **100:** 249–253.

Rasool, M., and Varalakshmi, P. (2006). Suppressive effect of *Withania somnifera* root powder on experimental gouty arthritis: An *in vivo* and *in vitro* study. *Chemico-Biological Interactions*, **164:** 174–180.

Rice-Evans, C.A., Miller, N.J., and Paganga, G. (1996). Structure-antioxidant activity relationships of flavonoids and phenolic acids. *Free RadicBiol Med.*, **20**: 933–956.

Saeed, S.A., Simjee, R.U., Mahmood, F., and Rahman, N.N. (1993). Dual inhibition of platelet-activating factor and arachidonic acid metabolism by ajmaline and effect on carrageenan-induced rat paw oedema. *J Pharm Pharmacol.*, **45(8)**: 715-9.

Saltan Çitoglu, G., Bahadýr Acýkara, O., Sever Yýlmaz, B., and Ozbek, H. (2012). Evaluation of analgesic, anti-inflammatory and hepatoprotective effects of lycorine from *Sternbergia fisheriana* (Herbert) Rupr. *Fitoterapia*, **83:** 81–87.

Shang, J.H., Cai, X.H., Fenga, T., Zhao, Y.L., Wang, J.K., Zhang, L.Y., Yan, M., and Luo, X.D. (2010). Pharmacological evaluation of *Alstonia scholaris*: Anti-inflammatory and analgesic effects. *Journal of Ethnopharmacology*, **129:** 174–181.

Sharma, V., Sharma, S., Pracheta, and Paliwal, R. (2011). *Withania somnifera*: A rejuvenating ayurvedic medicinal herb for the treatment of various human ailments. *International Journal of PharmTech Research*, **3(1):** 187-192.

Souto, A.L., Tavares, J.F., Da Silva, M.S., Melo Diniz, M.F., Athayde-Filho, P.F., and Barbosa Filho, J.M. (2011). Review anti-inflammatory activity of alkaloids: An update from 2000 to 2010. *Molecules*, **16:** 8515-8534.

Sumantran, V. N., Kulkarni, A., Boddul, S., Chinchwade, T., Koppikar, S. J., Harsulkar, A., Patwardhan, B., Chopra, A., and Wagh, U.V. (2007). Chondroprotective potential of root extracts of *Withania somnifera* in osteoarthritis. *J. Biosci.*, **32(2):** 299–307.

Sutradhar, R., Rahman, Ahmad, M., Bachar, S., Saha, A., and Guha, S. (2006). Bioactive alkaloid from *sida cordifolia* linn. With analgesic and anti-inflammatory activities. *Iranian journal of pharmacology and therapeutics*, **5:** 175-178.

Tang, X.C., Lin, Z.G., Cai, W., Chen, N., and Shen, L. (1984). Anti-inflammatory effect of 3-acetylaconitine. *ActaPharmacol. Sin.*, **5:** 85–89.

Toker, G., Kupeli, E., Memisoglu, M., and Yesilada, E. (2004). Flavonoids with antinociceptive and anti-inflammatory activities from the leaves of *Tilia argentea* (silver linden). *Journal of Ethnopharmacology*, **95:** 393–397.

Vetvicka, V., and Vetvickova, J. (2011). Immune enhancing effects of WB365, a novel combination of Ashwagandha (*Withania somnifera*) and Maitake (*Grifola frondosa*) extracts. *North American Journal of Medical Sciences*, **3:** 7.

Wang, D., Zhu, J., Wang, S., Wang, X., Ou, Y., Wei, D., and Li, X. (2011). Antitussive, expectorant and anti-inflammatory alkaloids from *Bulbus Fritillariae* Cirrhosae. *Fitoterapia*, **82:** 1290–1294.

Wang, Y., Yang, X., Zheng, X., Li, J., Ye, C., and Song, X. (2010). Theacrine, a purine alkaloid with anti-inflammatory and analgesic activities. *Fitoterapia*, **81:** 627–631.

Weinstein, S.L., Sanghera, J.S., Lemke, K., DeFranco, A.L., and Pelech, S.L. (1992). Bacterial lipopolysaccharide induces tyrosine phosphorylation and activation of mitogen-activated protein kinases in macrophages. *J Biol Chem.*, **267**: 14955–14962.

Wong, C.W., Seow, W. K., O'Callaghan, J.W., and Thong, Y. H. (1992). Comparative effects of tetradrine and berbamine on subcutaneous air pouch inflammation induced by interleukin-1, tumour necrosis factor and platelet-activating factor. *Inflammation Research*, **36(1-2)**: 112-118.

Yasukawa, K., Akasu, M., Takeuchi, M., and Takido, M. (1993). Bisbenzylisoquinoline alkaloids inhibit tumor promotion by 12-*O*-tetradecanoylphorbol-13-acetate in two-stage carcinogenesis in mouse skin. *Oncology*, **50**: 137-140.

Ying, L.S., Hua, L.L., Teh, B.S., Seow, W.K., and Thong, Y.H. (1989). Anti-inflammatory and immunosuppressive properties of the bis-benzylisoquinolines: *In vitro* comparisons of tetrandrine and berbamine. *International Journal of Immunopharmacology*, **11(4):** 395–401.

Yun, K.J., Shin, J.S., Choi, J.H., Back, N., Chung, H.G., and Lee, K.T. (2009). Quaternary alkaloid, pseudocoptisine isolated from tubers of *Corydalis turtschaninovi* inhibits LPS-induced nitric oxide, PGE_2, and pro-inflammatory cytokines production via the down-regulation of NF-κB in RAW 264.7 murine macrophage cells. *International Immunopharmacology*, **9:** 1323–1331.

Zheng, X.Q., Ye, C.X., Kato, M., Crozier, A., and Ashihara, H. (2002). Teacrine (1, 3, 7, 9-tetramethyluric acid) synthesis in leaves of a Chinese tea, kucha (*Camellia assamica* var. *kucha*). *Phytochemistry*, **60**: 129–34.

Bioactive Phytochemicals: Perspectives for Modern Medicine Vol. 2 (2014) *Pages* **51–82**
Editor: **V.K. Gupta**
Published by: **DAYA PUBLISHING HOUSE, NEW DELHI**

2

Tumor and its Environment: Effort by Neem Leaf Glycoprotein to keep it Green

Subhasis Barik[1], Kuntal Kanti Goswami[1], Atanu Mallick[1], Saptak Banerjee[1], Anamika Bose[2] and Rathindranath Baral[1*]

ABSTRACT

Tumor is not only the accumulation of tumor cells; it is a heterogeneous mass consisting of tumor cells, along with numerous other cells. This includes fibroblasts, inflammatory immune cells (such as, macrophages, smooth muscle cells), blood vessel constituting cells (endothelial cells and pericytes), embedded in an extracellular matrix. In fact, just like seeds, tumor cells are impregnated in soil, i.e., its environment. Environment as well as its secretory factors (VEGF, TGFβ, IL-10 etc.) provide all necessary extrinsic signals to proceed for further malignant growth and metastasis. Tumor microenvironment (TME) takes the critical decision in determining tumor progression versus dormancy or destruction. Accordingly, TME demands research attention to find out way to inhibit the tumor growth. Unfortunately, up to mid 1990s, cancer researchers paid little attention to TME, because of little understanding on the alliance between tumor and its environment causes tumor proliferation, angiogenesis, drug resistance, metastasis, immune paralysis and escape. Available therapeutic measures inhibit tumor growth at the cost of severe pollution in TME in terms of induction of hypoxia, T cell anergy/exhaustion/apoptosis, protumor bias of cytokine/chemokine network and several other tumor promoting activities. Thus, in true sense, targeting of tumor only give transient response and requires normalization of TME to obtain durable tumor restriction. As complicated network of signaling is operational within TME between the tumor and several other tumor associated

---

1 Department of Immunoregulation and Immunodiagnostics, Chittaranjan National Cancer Institute (CNCI), 37, S.P. Mukherjee Road, Kolkata – 700 026, W.B., India

2 Department of Molecular Medicine, Bose Institute, P1/12, C.I.T. Road, Scheme-VIIM, Kolkata – 700 054, W.B., India

* *Corresponding author*: E-mail: baralrathin@hotmail.com; baral.rathindranath@gmail.com

cellular components, normalization of TME requires synchronization of multiple events. In order to do this important job, the first task is to find out the candidate drug(s) for this particular purpose. Obtained experimental and clinical data confirmed that chemotherapeutic drugs and radiotherapeutic measures inhibit tumor proliferation in one hand, but creates protumor microenvironment in other hand. Different medicinal plants have proved worthy for cancer treatment from very early times, and, evaluated in clinical studies or currently being investigated to understand their tumoricidal actions against various cancers. Although, extensive research is waiting, it is expected that plant derived anti-tumor drugs cause limited pollution to TME.

Neem leaf Glycoprotein (NLGP), a glycoprotein from aqueous preparation of neem (Azadirachta indica) leaf is efficient to restrict murine tumor growth. NLGP has no direct tumoristatic or tumorilytic activities, but it offers tumor growth restriction chiefly by immunomodulation as evident from several studies from our group. Such success may not be possible only by activating immunocompetent cells, like, T cells or NK cells, but also requires normalization of TME. TME obtained from NLGP treated sarcoma and melanoma bearing mice (NLGP-TME), is completely different (normalized) from untreated tumor bearing mice (PBS-TME), immunologically as well as biochemically. Proportion of $CD8^+$ T cells increased within NLGP-TME with downregulation of $CD4^+CD25^+Foxp3^+$ Tregs and $CD11b^+Gr1^+$ MDSCs. Analysis of cytokine milieu revealed IL-10, TGF-β, IL-6 rich TME with type-2 characters that switched to create type-1 microenvironment with dominance of IFN-γ secretion from cellular components of NLGP-TME. NLGP defends T cells from TME-induced anergy, as indicated by higher level of pNFAT and related downstream signaling. Moreover, low expression of $FasR^+$ cells within $CD8^+$ T cell population denotes prevention of activation induced cell death. Using CFSE as a probe, better migration of T cells was noted within TME from NLGP treated mice than PBS cohort. $CD8^+$ T cells isolated from NLGP-TME exhibited greater cytotoxicity to sarcoma/melanoma cells in vitro and these cells show higher expression of cytotoxicity related molecules like perforin and granzyme B. Adoptive transfer of NLGP-TME exposed T cells inhibited the growth of different tumors. Accumulated evidences strongly suggest that NLGP is able to maintain green of cellular environment by reducing toxic nature of TME that allows T cells to perform optimally to kill the tumor cells. The proposed review would discuss the efforts of NLGP to maintain the normal functions of several components of TME, thereby, to keep the TME green.

Keywords: Neem leaf glycoprotein, Tumor microenvironment, Cytotoxic T cells, Tumor associated macrophages, Dendritic cells, Regulatory cells.

Introduction

Genomic instability and high proliferation rates impose a unique mutational signature to each cancer cell that gives rise to genetic heterogeneity of a tumor (Pedraza-Farina *et al.*, 2006). For the majority of tumors, genetic heterogeneity helps cancer cells to grow and to acquire drug resistance in cancer therapy (Turner *et al.*, 2012). In addition, cellular heterogeneity provides a growth and survival advantage to the cancer cells within a growing tumor. As the cancer cells grow, they are stressed by environmental factors, like, lack of oxygen, nutrients and pressure of the nearby tissues (Höckel *et al.*, 2001). Without a steady supply of nutrition and oxygen, microscopic colonies of cancer cells ("carcinoma *in situ*", CIS) cannot grow and transform into invasive carcinoma, thus, tumors require their environmental support for progression. Tumor cells send out chemical signals to modulate the different populations of the host's healthy cells (Whiteside *et al.*, 2006) and to recruit blood-derived immune cells simultaneously (Penn *et al.*, 1973). As tumor develops, these surrounding environment and recruited normal cells co-evolve into an alternatively-activated state through a continuous paracrine as well as autocrine communication, thereby creating a microenvironment that promotes cancer growth and ultimately leads to metastatic disease (Wels *et al.*, 2008). Accordingly, malignant cells create a tolerant neighborhood in which they can function with limited interference. Collectively this neighborhood consisting of connective tissue, extracellular matrix, and the resident non-cancerous cells surrounding the cancer cells are known as tumor stroma. Many researchers prefer the broader term "tumor microenvironment (TME)," instead of "stroma," as it includes infiltrating cells of the immune system (innate and adaptive), and cell-free molecules (proteases and growth factors) in addition to the residential stromal components (Pietras *et al.*, 2010) Although the same basic building blocks constitute the stroma of all tumors, the actual composition of the TME is quite variable, with differences seen between individuals, between different tumor types, and in different areas of the same tumor. Furthermore, the composition of the TME is also altered as the disease progresses (Diaz-Cano, 2012).

Tumor and its Environment

From the beginning of the discovery of cancer to late 1990s, scientists have focused on the cancer cell itself; more recently they begun to think; how the rest of our body or surrounding tissues influence the cancer cells. The microenvironment: the normal cells and molecules that surround a tumor cells- are becoming extremely important in the process of tumorigenesis. Tumor is an assorted mass of different cell types, which includes cancer cells and several other cells, including fibroblasts, inflammatory immune cells (*e.g.*, macrophages), smooth muscle and endothelial cells, all embedded in an extracellular matrix. It also contains special kinds of growth factors, chemical messenger molecules, called chemokines and cytokines, and also chemicals, like oxygen. The communication between the tumor cells and the surrounding cells of the microenvironment helps to drive the process of tumor progression (Figure 2.1). So transformation from normal to benign, benign to malignant, malignant to metastatic is driven not just by what's happening inside the tumor cell itself but also depends on what's happening around it. Recently, it is becoming more and more clear that TME plays a very important role in assisting tumor cells for metastasis (Chantrain *et al.*, 2008). Interestingly, the association between TME and metastasis is named as "Seed and soil" hypothesis, first time suggested by Stephen Paget over a century ago in 1889 to explain organ preferred metastasis (Langley and Fidler, 2011). This hypothesis suggested that tumor cells (seed) can colonize only to those distant organs (soil), where favourable environment for tumor growth exists. Stephen Paget concluded that metastasis results only when the seed and soil were compatible (Paget 1989). However, only recently Paget's idea received

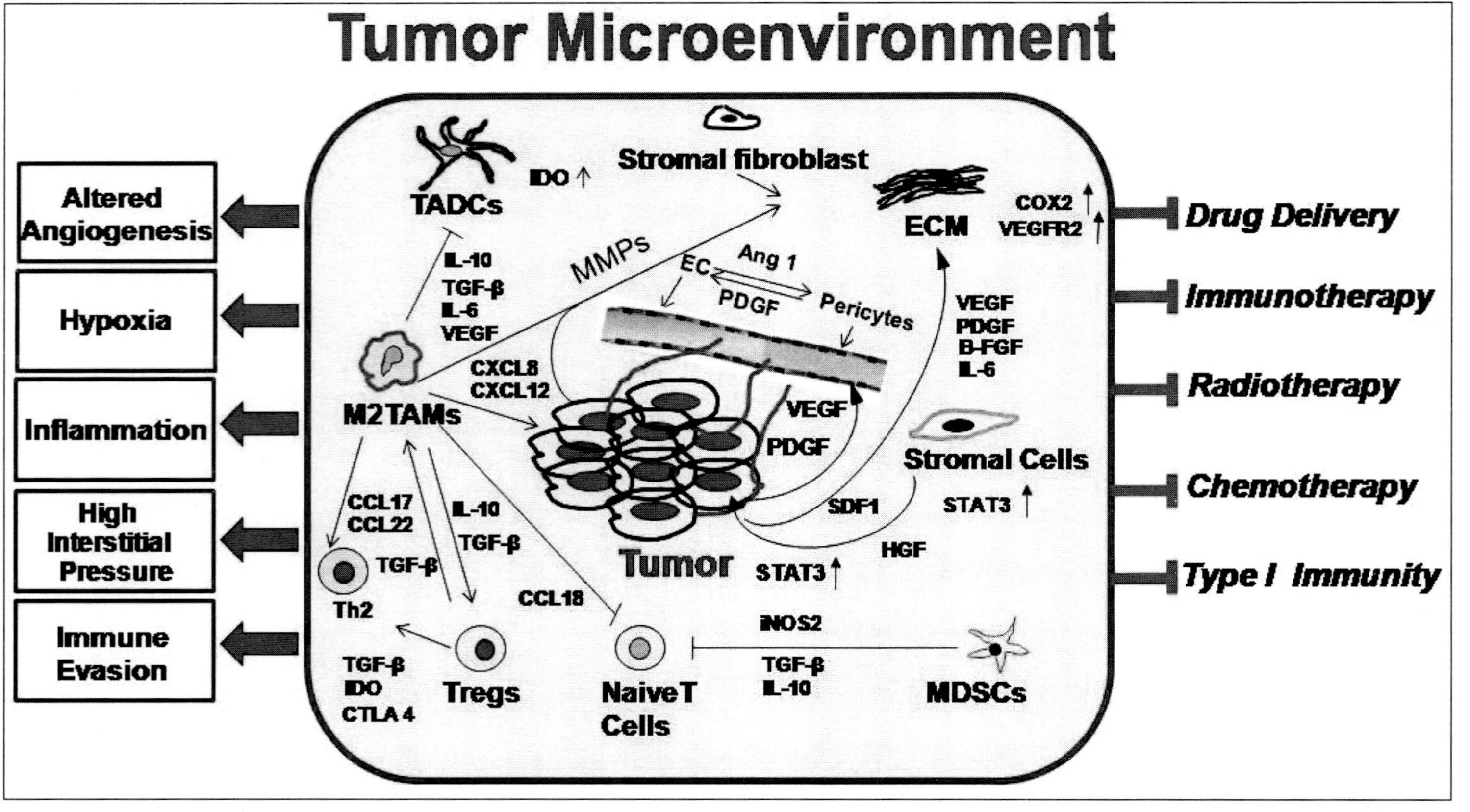

Figure 2.1: Immune and non-immune components of Tumor Microenvironment (TME). *Left panel* describes hallmarks of TME. *Right panel* exhibits obstacles in anti-tumor immunity and therapy due to TME. *Middle panel* represents interactions between tumor cells and other immune and non-immune cells through cytokines, chemokines etc. TADC, Tumor associated dendritic cells; M2TAM, M2 Tumor associated macrophages; ECM, Extracellular matrix; Tregs, regulatory T cells; MDSCs, Myeloid derived suppressor cells (Barik *et al.*).

a lot of attention, and TME in cancer progression, metastasis and drug-response became a very actively investigated topic.

The tumor cells are highly heterogeneous with diverse differentiation grades. The success of tumor cells to grow seems to be dependent on their ability to control and shape the surrounding microenvironment to favour their own survival (de Visser *et al.*, 2006). Eventually the fact is established that tumor progression or metastasis is governed on a systemic level involving various non-malignant (stroma) cells, soluble factors and ECM components (Deryugina and Quigley, 2006). *This research field is still at the very early phase, and the direct contribution of each element of TME in tumor progression or metastasis remains to be elucidated in detail. It has been proposed that targeting tumor cells together with the targeting of implicated TME factors can improve the efficacy of anticancer therapy.*

Tumor Microenvironment (TME) and Associated Factors

A hallmark of chronic inflammatory tissues and adenocarcinomas is the presence of an abundant fibrotic component, *i.e.*, desmoplasia (Korc., 2007). The interlinked relationship between tumor cells and its stroma promotes tumor growth and metastasis by supporting activation of fibroblasts (Liao and Luo, 2009). Stromal components are important for creation and homeostasis of the inflammatory TME, consists of infiltrating immune cells, including, neutrophils, dendritic cells (DCs), macrophages, eosinophils, mast cells, and T cells. These cells will also recruit additional immune cells to the tumor milieu by secretion of chemotactic factors. Infiltrating immune cells support tumor progression by the release of growth and survival factors, matrix remodeling factors, and reactive oxygen species etc. (Erez *et al.*, 2010). The composition and characteristics of the TME vary widely and are important in determining the anti-tumor immune responses. For example, certain cells of the immune system, including NK cells, DCs and effector T cells, are capable of driving potent anti-tumor responses. However, several subtypes of immune cells [such as myeloid-derived suppressor cells (MDSCs) and regulatory T cells (Tregs)] that commonly infiltrate solid tumor harbor immunosuppressive activity and foster tumor development by expression of potent pro-tumor mediators. These cells undoubtedly restrict the effectiveness of anticancer strategies. Formation of new blood vessels, *i.e.*, angiogenesis is essential for the growth of solid tumors beyond 1-2 mm^3 (Bhowmick *et al.*, 2004). The cancer cells could promote angiogenesis directly by secreting VEGF, bFGF, CXCL8, and placenta growth factor (PIGF), and also indirectly by taking over the role of the thrombocytes by attracting immune cells and by the transformation of stellate cells to highly proliferating CAFs (Figure 2.1).

Inflammation and Cancer

Inflammation is an essential event in the development and progression of tumors and is suggested to be the seventh hallmark of cancer (Colotta *et al.*, 2009). The inflammatory environment is created by the release of proinflammatory cytokines, defined as "alarm cytokines" present early in the carcinogenesis, such as TNFα, IL-1α, and IL-1β. These cytokines initiate inflammatory responses and are secreted by infiltrating leukocytes and malignant cells. Besides being inflammatory initiators, these cytokines also induce expression of other proinflammatory genes, such as COX-2, inducible nitric oxide synthase (iNOS), chemokines, cytokines, and matrix metalloproteinases (MMPs) (Apte and Voronov, 2008). MMPs are zinc-dependent endopeptidases that function to degrade all kinds of extracellular matrix proteins. The MMPs have been shown to play important roles in tissue remodeling associated with various physiological and pathological processes such as morphogenesis, angiogenesis, tissue repair, cirrhosis, arthritis, and metastasis.

Immune Components of TME

The composition and characteristics of the TME vary widely and are important in determining the anti-tumour immune response. Certain cells of the immune system, including macrophages, NK cells, DCs and effector T cells are capable of driving potent anti-tumour responses. However, tumor cells often induce an immunosuppressive microenvironment, which favours the development of immunosuppressive populations of immune cells, such as MDSCs and Treg cells. Understanding the complexity of immunomodulation by tumors is important for the development of immunotherapy.

Effector Cells

The immune system contains several types of cytotoxic cells, which are able to lyse host and foreign cells. Cytotoxic T-lymphocytes (CTL) appear to play the most important role among the killer cells but other lymphatic cells, such as NK cells and lymphokine-activated killer (LAK) cells as well as macrophages are also highly effective in the lyses of appropriate targets. The various cytotoxic effector cells differ distinctly concerning origin, phenotype, morphology and target cell specificity, but they bear the common feature that they destroy the target cells in a contact-dependent process. CTL recognize specific antigens which are presented in context with molecules of class I major histocompatibility complex (MHC). NK cells, on the other hand, kill the appropriate targets without prior immunization and without requiring recognition of MHC molecules at the target cells. They also bear a typical pattern of surface markers which differ in several aspects from that of CTL. CTL, NK and LAK cells appear to possess similar mechanisms for cytolysis including secretion of pore-forming proteins, serine proteases and other proteins. Macrophages differ substantially from other cytotoxic effector cells concerning morphology, phenotype, kinetics of activation and target cell spectrum. They perform a variety of functions whereby contact-dependent target cell lysis represents only one of their properties. After binding with target cell they release over 20 different molecules such as IL-1 and TNFα as mediators for cytolysis. Within TME, effector cells become paralyzed due to interactions between tumor cell and altered stromal cells, along with suppressive activity of regulatory cells. The tumor cells cannot be targeted by CTL as an efficient APC within TME is unable to deliver a co-stimulatory signal required for productive interactions with T lymphocytes. These CTLs become anergized and undergo activation induced cell death.

Regulatory Cells

Regulatory cells (Tregs, MDSCs) of immune system down regulate the activity of other immune cells, thus preventing the development of autoimmune diseases or allergies. Treg cells may secrete suppressor cytokines that can directly inhibit the function of responder T cells and myeloid cells. Treg cells express high CD25, the IL-2 receptor alpha chain, and have the capacity to compete with effector T cells for IL-2 resulting in cytokine-mediated deprivation of the effector cells and Bim-mediated apoptosis. Treg cells also express CTLA4 that mimics CD28 on T cells and promotes crosstalk with APCs by means of B7 molecules and inhibits antigen presentation (André *et al.*, 2009). Activated Foxp3$^+$ Treg cells may function as cytotoxic cells and directly kill effector cells in a manner similar to CD8$^+$ cytotoxic cells. Thus, Tregs within TME play a role in promoting tumor growth and metastasis by inhibiting the immune response against cancer by means of suppression of activation of antigen specific effector immune cells (both CD4$^+$ and CD8$^+$ T cells).

Immunosuppressive functions of other suppressor cells, MDSCs, require direct cell-cell contact (Wang *et al.*, 2012), which suggests that they act either through cell-surface receptors and/or through the release of short-lived soluble mediators like arginase, iNOS, ROS and peroxynitrite (Gabrilovich and Nagaraj, 2009). It also induces development of Treg cells. Some studies also observed that MDSCs abrogated NK cell cytotoxicity against tumor cells and inhibited IFNγ production by NK cells (Jewett and Tseng, 2011).

Non-Immune Component of TME

Besides immune components, TME contains resident non-cancerous cells (fibroblasts, endothelial cells), connective tissue, and extracellular matrix (ECM; components of tissues that provide structural support, such as proteins like collagen). These components are equally important both in tumor initiation and progression. Collectively, these components are known as stroma, required to maintain epithelial tissues. Changes in epithelium cause inevitable alteration in stroma. *In cancer, changes in the stroma drive invasion and metastasis, hallmarks of malignancy.*

Vascular Endothelial Cells

Endothelial cells in direct contact with blood are called vascular endothelial cells (VEC) whereas those in direct contact with lymph are known as lymphatic endothelial cells (Alitalo and Detmar, 2011). VECs use cell adhesion molecules, such as integrins and cadherins to attach themselves to each other and to the vascular extracellular matrix. Tumor cells secrete angiogenic factors (like, VEGF) to stimulate VEC proliferation and induce angiogenesis. These tumor-associated growth factors stimulate the direct interaction between tumor cells and VEC via MAPK and Notch signaling pathways, thereby promoting tumor angiogenesis and tumor growth (Veikkola *et al.*, 2000).

Pericytes

Blood vessels are composed of two interacting cell types. VECs from the inner lining of the vessel wall, and perivascular cells-referred to as pericytes, vascular smooth muscle cells or mural cells-envelop the surface of the vascular tube (Bergers and Song, 2005). VECs in blood vessels under formation are stabilized by the recruitment of pericytes, both in normal tissues and during angiogenesis in pathological situations, including neoplasia (Franco *et al.*, 2011). In the tumor vasculature, besides supporting the functionality of blood flow, pericytes protect VECs from anti-angiogenic therapies, and have thus been implicated in clinical resistance to vascular targeting drugs (Franco *et al.*, 2011). Recently, another function of pericytes within TME is documented by us, *i.e.*, tumor associated pericytes induce tolerance to $CD4^+$ and $CD8^+$ T cells, as tumors and tumor derived factors alter immunophenotypes and functionality of tumor resident pericytes *(unpublished observation)*.

Fibroblasts

A fibroblast is a type of cell that synthesizes the extracellular matrix and collagen, the structural framework (stroma) for animal tissues, and plays a critical role in wound healing. Fibroblasts are the most common cells of connective tissue in animals. The main function of fibroblasts is to maintain the structural integrity of connective tissues by continuously secreting precursors of the extracellular matrix (Ge *et al.*, 2004). Fibroblasts are the most prominent cell type within the tumor stroma of many cancers (Cirri and Chiarugi, 2011); these populations share some properties collectively leading to their "activation state". The main activation markers, fibroblast specific protein (FSP), PDGFRα and fibroblast activation protein (FAP) have been found over expressed in stromal fibroblasts of solid tumors (Pietras and Ostman, 2010).

Extracellular Matrix (ECM)

ECM constitutes a major component of microenvironment of a cancer cell, plays important roles in cancer development. It is a complex network of macromolecules with distinctive physical and biochemical properties (Lu *et al.*, 2012). ECM has many important roles, collectively measured by its production, degradation, and remodeling. Abnormal ECM affects cancer progression directly by promoting cellular transformation and metastasis. Notably, this abnormal ECM affects character of stromal cells, assist tumor-associated angiogenesis and inflammation, and lead to creation of a tumorigenic microenvironment (Hanahan and Weinberg, 2011).

Cytokines

During carcinoma formation, cancer cells release various cytokines and growth factors into their surroundings and recruit and reprogram many other types of cells in order to establish a TME (Sheu *et al.*, 2008). Consequently, the tumor tissues almost always contain a large number of VECs, fibroblasts, and infiltrating inflammatory cells that in turn fabricate a variety of cytokines, like, IL-10, IL-6, TGFβ (Ma *et al.*, 2012). The cytokines produced by these cells have been positioned as key factors in modulating immune response either against or in favour of tumorigenesis in the microenvironment. The interactions that take place between immune and cancer cells are complex, involving multiple cascades of cytokines, chemokines, and/or growth factors (Coussens and Werb, 2002).

Chemokines

In the immune system, chemokines exert vital homeostatic functions by regulating the organization of lymphoid organs, lymphocyte differentiation, and lymphocyte homing, thus inuencing inammation and adaptive immune responses (Stein and Nombela-Arrieta, 2005). It has also been realized that tumor cell migration and growth is dependent on direct chemokine signals to tumor cells (Kulbe *et al.*, 2044). During tumor progression chemokines play a paramount role. Both tumor cells and stromal cells elaborate chemokines and cytokines. These act either by autocrine or paracrine mechanisms to sustain tumor cell growth, induce angiogenesis and facilitate evasion of immune surveillance through immunoediting (Gouwy *et al.*, 2005).

Importance of Normalization of TME for Anti-Tumor Immunity

It has been widely recognized, though recently, that a tumor grows towards a malignant phenotype by altering its microenvironment in a way that accelerates its malevolent potential (Lorusso and Rüegg, 2008). Consequently, current anti cancer therapeutic strategies target not only the intracellular distortions of a malignant cell but also cause deformation in its extracellular microenvironment (Prager *et al.*, 2012). The remedial strategies are designed with a major aim to optimize the antitumor immunity in the tumor vicinity, as the alterations in the TME are dominantly manifested as an impaired immune response. As tumor is seeded in the soil, *i.e.*, TME, its progression, and, in some cases regression solely depends on the constituents of TME (Witz, 2009). Components of TME interact with each other and contribute significantly in tumor progression. Cells (TAM, MDSCs, tumor associated DCs and tumor cell itself) from TME produce angiogenic factors, like, VEGF and PDGF (Rehman *et al.*, 2003). Moreover, TGFβ, HGF secreted from fibroblasts stimulate tumor cell proliferation (Bhowmick *et al.*, 2004). Tumor conditioned immune cells with type 2 characters, like, Tregs, MDSCs and TAMs synthesize and secrete VEGF, PDGF, HGF, TGFβ etc, to promote tumor progression. As tissue becomes cancerous, pathological interactions between cancer cells and host immune cells in the TME create an immunosuppressive network that protects the tumor from immune attack (Whiteside, 2006) leading to tumor growth, progression, invasion and metastasis (Mbeunkui and Johann, 2009). Thus, normalization of TME is the principle task of cancer immunotherapy.

Plant Products in Normalization of Immune System and Cancer Therapy

Medicinal plants impart significant roles in preventing human being from various pathogenic microorganisms and the diseases like cancer (Singh *et al.*, 2011). In nature there are various medicinal plants, used as immunomodulatory agents (Kannan and Singh, 2010). In Indian medicinal literature a large number of plants are included, that promote defense mechanisms of the body. In other hand a large number of medicinal plants included in Rasayanas have been claimed to possess immunomodulatory activities (Kumar *et al.*, 2011). Medicinal plants are used as an immunomodulator to provide alternative potential to conventional chemotherapy for cancerous diseases, especially in relation to host defense mechanism. The use of plant products like polysaccharides, lectins, peptides, flavonoids and tannins has been established in the modification of immune response in various *in vitro* models (Mazid *et al.*, 2011).

Natural Products Modulates Tumor Microenvironment

Several plant products, like, Combretastatin A4, Cinnamon extract, Curcumin have been identified for their anti cancer activity and they have utilized a variety of mechanisms. However, few of these compounds were tested in context of modulation or normalization of components of TME.

Combretastatin A4

Isolated from the stem wood of the South African tree *Combretum caffreum*, is a vascular disruptive agent, thus, playing an important role in normalization of angiogenesis in cancer. It inhibits tumor blood vessel growth, causing tumor cell death and necrosis. Phase I trials have shown some clinical activity of combretastatin A4 and a favourable toxicological profile (Rustin *et al.*, 2003).

Cinnamon Extract

Cinnamon is one of the most widely used herbal medicines with diverse bioactive effects (Yu *et al.*, 2010). However, little evidence has been reported about the potential anti-tumor effects of cinnamon. Cinnamon may act by normalization of TME by means of reduction of excessive blood flow within tumors. *In vitro* (in melanoma cells) and *in vivo* (in experimental melanoma model) studies showed that cinnamon treatment strongly inhibited the expression of pro-angiogenic factors and master regulators of tumor progression (Kwon *et al.*, 2009). In addition, cinnamon treatment increased the anti-tumor activities of $CD8^+$ T cells by increasing the levels of cytolytic molecules (Kwon *et al.*, 2009). Cinnamon extract has the potential to be an alternative medicine for tumor treatment. Anti-tumor effect of cinnamon extracts is directly linked with enhanced pro-apoptotic activity and inhibition of NFκB and AP1 activities and their target genes (Kwon *et al.*, 2010). However, no report is available regarding their effects on other components of TME.

Curcumin

Curcumin is one of the most studied chemopreventive agents (Duvoix *et al.*, 2005). It is a natural compound extracted from the rhizome of *Curcuma longa* L. that allows suppression, retardation or inversion of carcinogenesis (Singh and Singh, 2011). Recent evidences suggest that curcumin exhibits strong anti-inflammatory and antioxidant activities and modulates the expression of transcription factors, cell cycle proteins, and signal transducing kinases etc (Surh *et al.*, 2001). All of these functions are major challenges within TME, thus, curcumin may work in the direction of TME normalization. Curcumin has different immunomodulatory effects on the development of immune responses, which include its effect on lymphoid cell populations, antigen presentation, humoral and cell-mediated immunity, and cytokine production (Bill *et al.*, 2009). Although, studies are coming up to measure the modulation of intratumoral immunity by curcumin, it may be hypothesized that anti tumor activity of curcumin may be mediated by TME normalization.

Resveratrol

Resveratrol, a natural compound found in red wine, some fruits and nuts, has shown anti-tumor activity in concentrations and regimens that appear to show minimal toxicity to normal cells (Van Ginkel *et al.*, 2007). It slows tumor growth and, depending on the dose and route of administration, can also lead to the regression of local tumors (Van Ginkel *et al.*, 2007, Chen *et al.*, 2004). The anti-tumor action of Resveratrol has been tested in mice bearing different types of malignant tumors such as breast, lung, liver, melanoma and neuroblastoma (Busquets *et al.*, 2007). Indirect evidences suggest that Resveratrol has some effect on TME or TME components. Positive effects of Resveratrol in ameliorating inflammatory conditions such as colonic inflammation, (Martín *et al.*, 2004) lipopolysaccharide-induced airway inflammation and osteoarthritis, and even in preventing allograft rejection (Wu *et al.*, 2005) have been reported. Resveratrol has been shown to inhibit activation of the transcription factor NF-κB induced by IL-1β (Estrov *et al.*, 2003) and by TNFα, and also to inhibit activation of AP-1, MAPK and c-Jun N-terminal kinase (Manna *et al.*, 2000). PBMC exposure to Resveratrol produced a biphasic effect on the anti-CD3/anti-CD28-induced development of both IFNγ/IL-2 and IL-4 producing $CD8^+$ and $CD4^+$ T cells, with stimulation at low Resveratrol concentrations and suppression at high concentrations. Similarly, the compound was found to induce a significant enhancement at low concentrations and suppression at high concentrations of both CTL and NK cell cytotoxic activity.

We have attempted to present an overview on TME in the earlier discussion and it is understandable that TME has great influence on cancer progression and metastasis. Thus, targeting of only tumor cells may not offer maximum clinical benefit. Effort to normalization of TME would be crucial to design future medicine. So far, very little attempt has been done in this direction of research and we have found few reports using natural products to normalize TME or its components. We are extensively exploring the role of neem leaf glycoprotein (NLGP) in cancer immunotherapy. Now, we are describing the effect of NLGP on various tumor components to predict the possible ways to restrict the tumor growth by restoring the environment of tumor towards normal as much as possible.

Neem Traditional View

The neem tree has been found as a traditional medicinal plant from ancient Harappa and Mohenjo-Daro civilizations in Indian subcontinent. The medical practitioners of that age studied a variety of plants and trees having therapeutic values and neem tree was one of them. The earliest indication of neem tree being used for its medicinal properties in households began nearly 5000 years ago. Neem is said to have been widely used in traditional systems of medicine like Ayurveda and this is also mentioned in earliest Indian scriptures of medicine- the Charak Samhita and Sushruta Samhita. Hence neem received the key interest in traditional medicine as SARBOROGANIBARANI (can cure all forms of diseases) and eight parts of this plant have wonderful biological actions (Biswas *et al.*, 2002). In 1992, US National Academy of Science designated this tree, as 'A tree sloving global problem' (National Research Council, 1992). Our group identified the tumor restricting property of an aqueous preparation from neem leaf, termed Neem Leaf Preparation (NLP) (Baral and Chattopadhyay, 2004) and a glycoprotein, named Neem Leaf Glycoprotein (NLGP), which appeared as an active principal for tumor growth restriction and associated immunomodulation (Chakraborty K *et al.*, 2008). NLGP is totally nontoxic in nature and have potential efficacy to regress tumor in different cancer models (Haque *et al.*, 2006; Mallick *et al.*, 2012).

NLGP from Chemical Viewpoint

NLGP is a purified glycoprotein isolated from aqueous preparation of neem leaf. Purified NLGP appeared in nondenatured PAGE as a single band and a single peak in HPLC, whereas, three bands having molecular weight of 15 kd, 23 kd and 47 kd in SDS-PAGE. This data indicates its hetero-trimeric nature. This glycoprotein constitutes of the carbohydrate moiety of about 33 per cent sugar, consisting of arabinose, galactose and glucose. Protein moiety of glycoprotein consists of sixteen amino acids, except arginine. Tumor restricting activity of NLGP is dependent on both glycol- part as well as protein nature of the NLGP. This hypothesis was proved by exposing NLGP in an array of temperature (100°C), pH and enzymes. Exposure to the adverse temperature (100°C), pH (5.7) and enzymes (papain, neuraminidase), resulted complete disappearance of the tumor growth restricting activity of NLGP (Baral *et al.*, 2010). Further works to characterize the protein are in progress.

NLP, NLGP and Restriction of Murine Tumors

We have started our work with NLP to see its effect on tumor growth restriction. In that effort, mice were grouped into two. Mice (Swiss and C57BL/6) of first group were subcutaneously injected with NLP weekly for four weeks in total and other group received PBS or extract obtained from leaf of the tree belongs to same Melliesse family (Mehagani). Following immunization carcinoma and melanoma cells were inoculated in Swiss and C57BL/6 mice respectively of both groups. In NLP immunized group, tumor growth was restricted significantly (Haque and Baral, 2006). Repetition of such experiments several times yielded identical results. Our *in vitro* studies with several murine and human cancer cells, along with normal cells, revealed that NLP has no cytotoxic effect on neither cancer cells nor normal cells. However, inoculation of spleen cells of NLP treated mice mixed with B16Mel cells in a group of mice, revealed significant reduction of tumor growth and increased survivability of the tumor hosts in comparison to mice inoculated with tumor along with normal spleen cells. This observation compelled us to think about a strong immune-boosting ability of NLP. Similar studies were conducted with NLGP and identical results were obtained. Most recently, we have observed that use of NLGP in therapeutic settings effectively restricts the growth of murine sarcoma (Mallick *et al.*, 2013), carcinoma and melanoma (*unpublished observation*) *in vivo*. Pattern of immunomodulation by NLGP on different immune cells, mostly in periphery, was elucidated in a series of investigations. These observations help to investigate their behaviour within TME in close vicinity of other TME components, like, vascular endothelial cells, pericytes, fibroblast etc.

Immunomodulation by NLP/NLGP

Immunomodulatory effect of NLP and NLGP on macrophages, DCs, NK cells, NK-T cells, T cells, regulatory T cells, B cells, etc. and secretion of various cytokines and chemokines was discussed in a series of studies. Principal observations are discussed below.

Regulation of Macrophages, NK cells, NK-T cells by NLP/NLGP in Periphery

In the initial phase of our research on NLP, we observed that NLP was found to activate PBMC from either healthy subjects or HNSCC patients, demonstrating significant enhancement in the cytotoxicity of tumor cells. We found $CD56^+$ and $CD16^+$ population increases within the PBMC fraction obtained from healthy and HNSCC patients upon NLP stimulation suggesting the participation of NK cells in cytotoxic function. NLP efficiently upregulated cytotoxic efficacy of NK cells ($CD16^+$ and $CD56^{dim}$) by means of enhanced killing of NK-sensitive K562 cells. Such killing is dependent on IL-12. Moreover, NLP upregulated the expression of cytotoxic molecules, perforin-granzyme B and FasL. Studies using specific inhibitors suggested that NLP-mediated induction of cytotoxicity is chiefly dependent on perforin-granzyme B system (Bose and Baral, 2007).

In addition, NLP unregulated the expression of CD40 on $CD14^+$ monocytes and CD40L in $CD56^+$ lymphocytes. Neutralization of CD40 and CD40L in NLP-stimulated PBMC culture resulted in significant downregulation of IL-12 release and cytotoxicity of NK cells; this clearly suggested that IL-12 is the key regulator of the NLP-mediated NK cell cytotoxicity, demonstrating the role of CD40-CD40L interaction in the observed functions (Bose and Baral, 2007). After identification of NLGP as an active component, similar experiments with NK cell functions were performed and identical results were obtained. Within TME NLGP may activate suppressed NK cell functions, but experiments in context of TME are yet to be done. After identification and purification of active principle of NLP *i.e.*, NLGP, we have demonstrated augmentation of the NK ($CD3^-CD56^+$) and NK-T ($CD3^+CD56^+$) cells mediated tumor cell cytotoxicity. These NK cells were isolated from the peripheral blood of HNSCC patients with a state of immunosuppression (Bose *et al.*, 2009). In addition to human system, we have also observed that mononuclear cells from blood and spleen of NLP-activated Swiss and C57BL/6 mice cause enhanced cytotoxicity to murine tumor cells *in vitro*. Flow cytometric analysis revealed in both blood and spleen, NK cells ($DX5^+$ or $NK1.1^+$) and NK-T cells ($CD3^+/DX5^+$ or $CD3^+/NK1.1^+$) were increased in number in Swiss, C57BL/6 and athymic nude mice those were NLP pretreated and tumor bearing. NLP-stimulated spleen cells showed greater secretion of TNFα and IFNγ. Thus, NLP-activated NK and NK-T cells in mice may regulate tumor cell cytotoxicity by enhancing the secretion of different cytotoxic cytokines (Haque and Baral, 2006).

Regulation of Macrophages, NK cells, NK-T cells by NLP/NLGP in TME

Within solid tumors marked myeloid cell infiltrates are seen and activation of these, chiefly, tumor associated macrophages (TAMs), is now known to play a key role in tumor progression. Monocytes enter tumors through blood vessels during the life span of tumors, from early-stage tumor nodules that are beginning to vascularize to late-stage tumors that are invasive and metastatic. A number of tumor-derived

Figure 2.2: Representative demonstration of modulation of immune components in periphery (I) and TME (II) of tumor host.

Periphery (I). *T cells.* Alterations in blood $CD4^+$ and $CD8^+$ T cells of sarcoma bearing mice having NLGP treatment (A.1). Upregulation of cytotoxic molecules, perforin and granzyme B (A.2), IFNg (A.3) and type 1 specific transcription factor, T-bet (A.4) in $CD8^+$ T cells from HNSCC patients after *in vitro* NLGP treatment. *Treg cells.* Downregulation of $CD4^+CD25^+Foxp3^+$ Treg cells from HNSCC patients after *in vitro* NLGP treatment (B.1). Inhibition of tumor supernatant mediated migration of Treg cells by NLGP (B.2). Reduced phosphorylation of STAT3 in tumor conditioned T cells under influence of NLGP (B.3). *Dendritic cells.* Morphology of NLGP matured DCs (C.1). Expression of MHC-I, MHC-II, CD80, CD86 and CD83 on NLGP matured DCs (C.2). IL-12 and IL-10 gene expression in NLGP matured DCs (C.3). *Macrophages.* Increased number of phagocytic cells after NLGP treatment (D.1). Secretion of IL-12 and IL-10 from peritoneal macrophages of colorectal carcinoma bearing mice (D.2). Increased nitric oxide release from peritoneal macrophages of NLGP treated melanoma bearing mice (D.3).

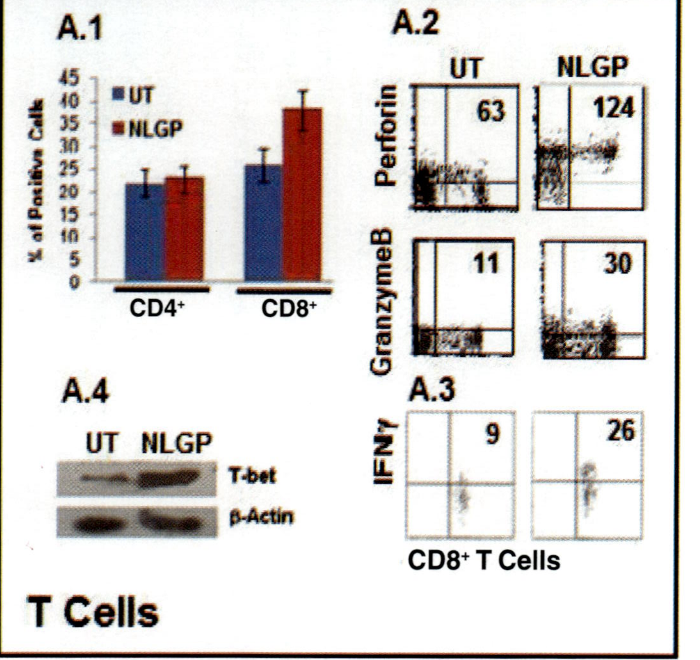

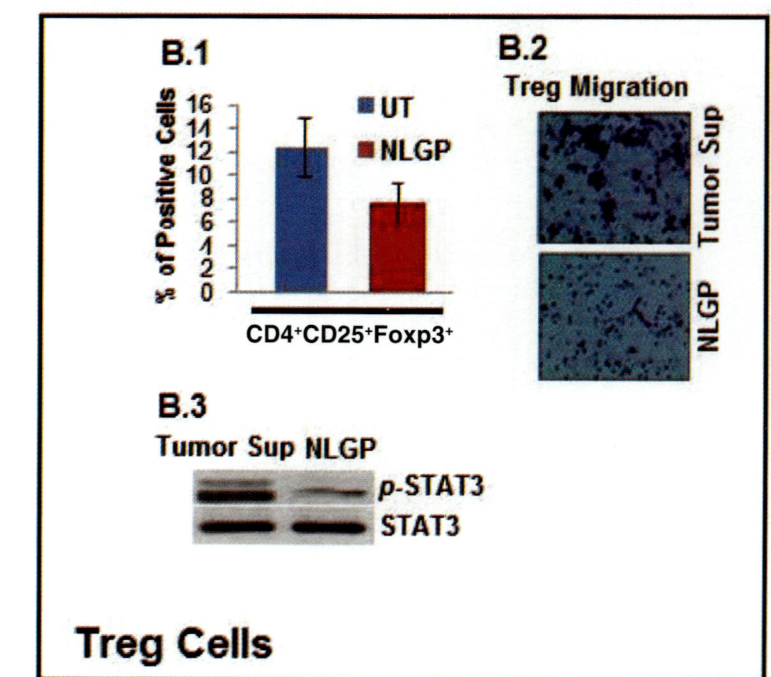

Contd...

Figure 2.2 Periphery (I)—*Contd...*

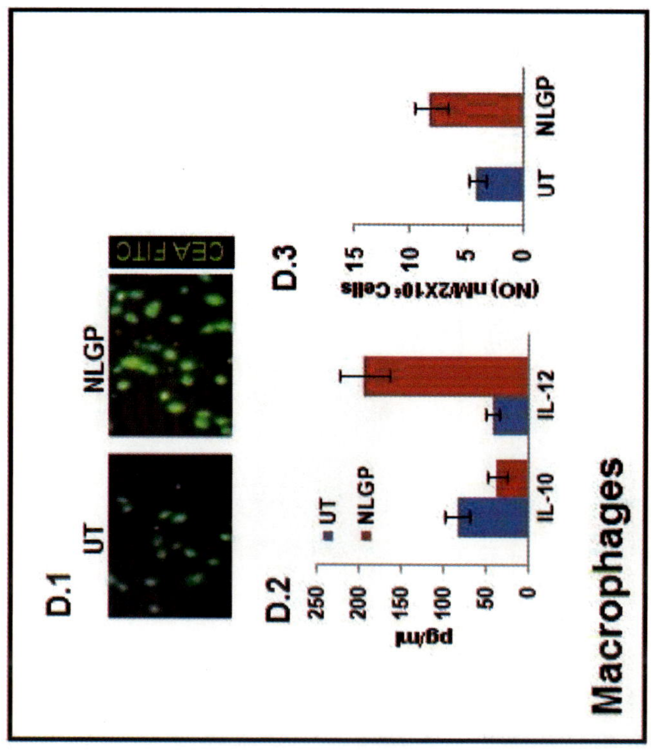

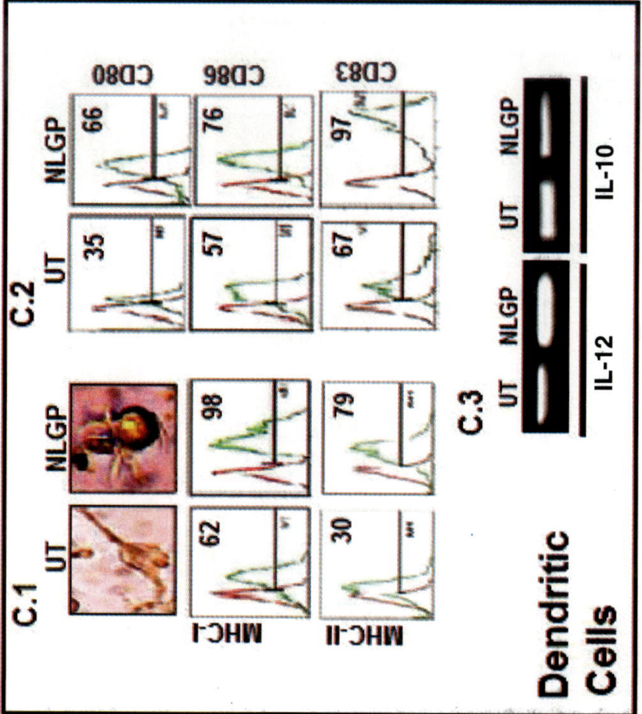

Contd...

Figure 2.2–*Contd...*

TME (II). *T cells.* Increased number of CD8⁺ T cells within sarcoma tumor from NLGP treated mice (E.1). CD8⁺ T cells expressed greater number of early activation marker CD69 in tumor draining lymph node cells and tumor infiltrating lymphocytes from NLGP treated sarcoma bearing mice (E.2). NLGP mediated therapeutic tumor growth restriction is withdrawn after CD8⁺ T cell depletion (E.3). Exhaustion of T cells with TME is significantly less in NLGP treated mice, as shown by reduced expression of exhaustion markers, lag3, pd1 and tim3 (E.4). *Treg cells.* Decreased number of CD4⁺Foxp3⁺ Treg cells within tumor from NLGP treated mice (F.1). Treg influenced growth of mouse tumor is inhibited by NLGP treatment (F.2). Reduced expression of transcription factor, Foxp3, and, CCR4 ligand, CCL22 within tumor from NLGP treated mice (F.3). *Dendritic cells.* Increased frequency of CD11c⁺ DCs within NLGP treated mouse melanoma (G.1). Increased expression of MHC-I, CD80 and CD86 within CD11c⁺ DCs after NLGP treatment of melanoma bearing mice (G.2). *Macrophages.* Reduced frequency of F4/80 positivity (M2) within CD11b⁺ cells from NLGP treated murine melanoma (H.1). Increased expression of IL-12 and reduced expression of IL-10 in same tumor from NLGP treated mice (H.2). *Angiogenesis.* Decreased number of CD31⁺ cells within mouse melanoma after NLGP treatment (I.1). Downregulated VEGF gene (I.2) and protein (I.3) expression in NLGP treated B16 melanoma.

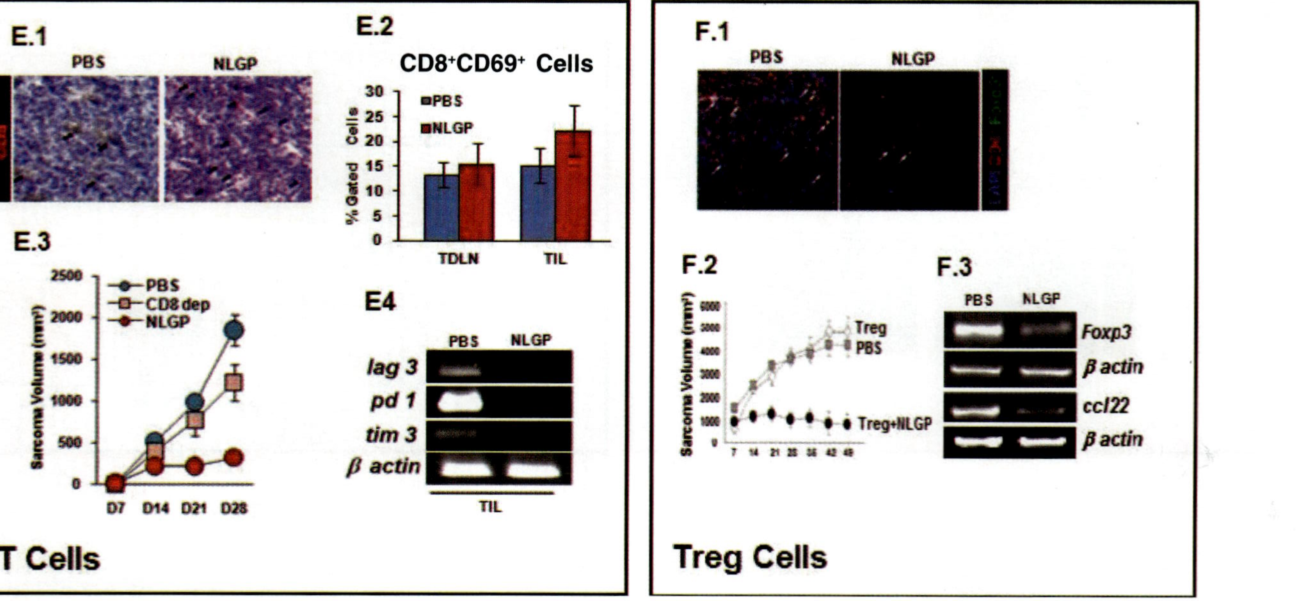

Contd...

Figure 2.2 TME (II)—*Contd...*

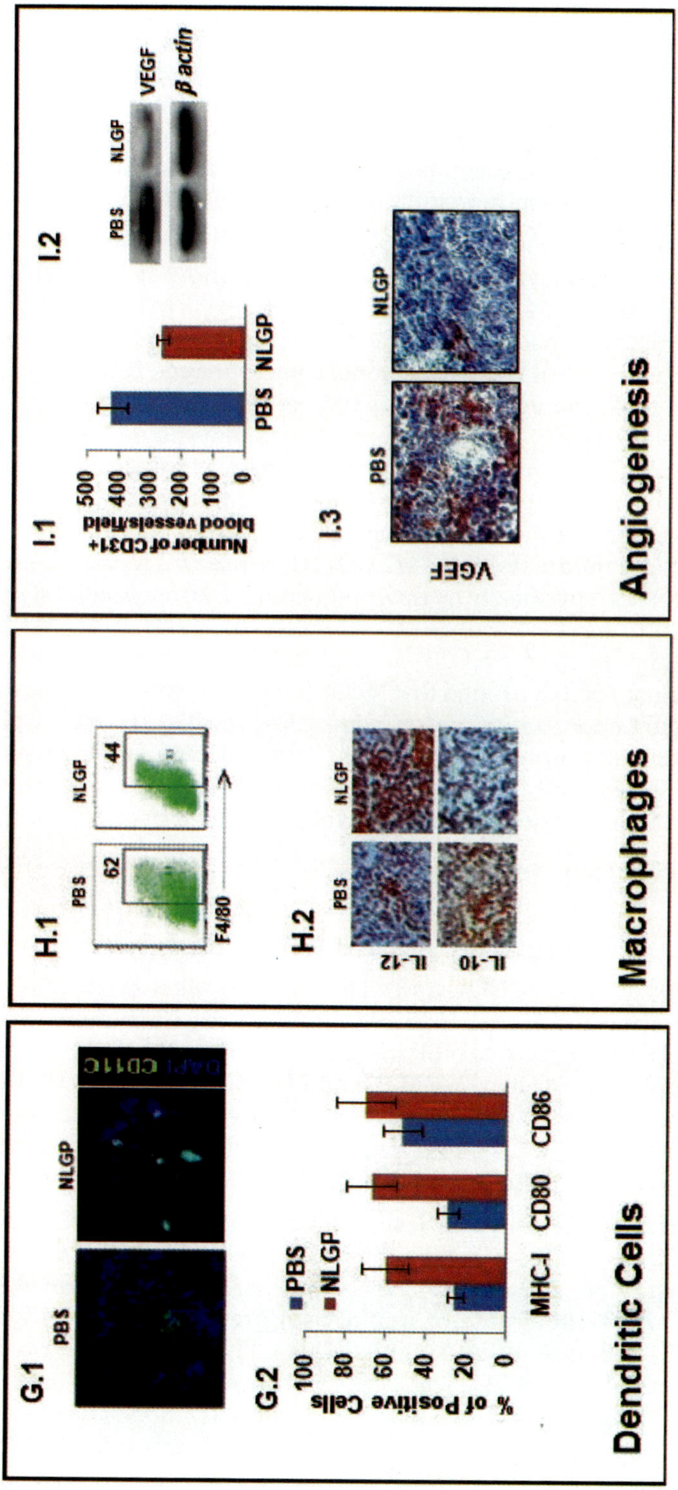

chemoattractants are thought to ensure this ongoing recruitment, including colony-stimulating factor-1 (CSF-1 also known as M-CSF), the CC chemokines, CCL2, CCL3, CCL4, CCL5, CCL8 and VEGF. After recruitment according to the local tissue microenvironment these monocytes are either polarized into proinflammatory anti-tumorous M1 type or anti-inflammatory protumorous M2 form. We studied the effect of TME on the generation of alternate M2 type macrophages in B16 melanoma system, in association with the immunoregulatory function of NLGP.

Our *in vitro* and *in vivo* studies confirmed that B16 melanoma TME polarizes macrophages in alternative M2 form where they show IL-12^{low}, IL-10^{high}, Nitric Oxide^{low}, Arginase^{high}, F4/80^{high}, ManR^{high}, Fizz1^{high}, YM1^{high}, CCR5^{low}, CCR7^{low} phenotypic profile and poor antigen presenting ability that are in contrary to that of the classical phenotype of normal peritoneal macrophages. Results further confirm that macrophage alteration does not only stay restricted to genotypic and phenotypic level as these alternatively activated macrophages show altered functional property and subvert normal immunological functioning in a way by paralyzing T effector cell functions, secrete immunosuppressive cytokines, augment Treg population, slows and/or stops DC maturation, negatively affect activated NK cell status. NLGP was found to partially modulate these M2 characteristics of TAMs and polarize them towards M1 version specifically by downregulating the upregulated status of STAT3 in M2 macrophages (Figures 2.2H1-H2). In particular, to assess the immunomodulatory role of NLGP, we isolated TAMs from both tumor core and peripheral regions and have found that NLGP has prominent effect in downregulating the M2 status in tumor core by means of impeding the IL-10 signaling pathway that upon activation causes upregulation of STAT3 responsible for giving type 2 genotypic expression. Therefore, NLGP can be used as a natural contrivance to reduce M2 load from tumor that in majority of the cases is responsible for poor prognosis.

Studies on NLGP regulation of NK and NK-T cell functions within TME is awaiting.

Regulation of T Cells by NLP/NLGP in Periphery

NLP helps to increase CD4^{+} and CD8^{+} T cell population upon its treatment in both tumor bearing and normal mice. Significant increase in lymphocyte count both in blood as well as in spleen of NLP treated mice suggested that NLP is effective in inducing lymphocytosis. Increase in the number of CD4^{+} and CD8^{+} T lymphocyte subpopulation in spleen and blood of the NLP treated mice also suggested NLP induced proliferation of T cell subpopulation, which may play a major role in restricting tumor growth in the NLP treated mice (Baral and Chattopadhyay, 2004).

After identification and purification of active principle of NLP *i.e.*, NLGP, we have demonstrated increase in the number of T cells (CD3^{+}CD8^{+}CD56^{-}) mediated tumor (KB) cell cytotoxicity. These T cells were isolated from the peripheral blood of HNSCC patients with a state of immunosuppression. NLGP induces TCRαβ-associated CTL reaction to kill KB cells. This CTL reaction is assisted by NLGP-mediated up-regulation of CD28 on T cells and HLA-ABC, CD80/86 on monocytes. CTL-mediated killing of KB cells is associated with activation of these cells by NLGP. This activation is evidenced by increased expression of early activation marker CD69 with altered expression of CD45RO/CD45RA. NLGP is a strong inducer of IFNγ from T cells (Figures 2.2A1-A4). This NLGP-induced cytotoxicity is dependent on up-regulated perforin/granzyme B expression in killer cells, which is again IFNγ dependent in T cells. Although, FasL expression is increased by NLGP, it may not be truly linked with the cytotoxic functions, as brefeldin A could not block such NLGP-mediated cytotoxicity, like, concanamycin A, a perforin inhibitor. On the basis of these results, we conclude that NLGP might be effective to recover the suppressed cytotoxic functions of T cells from HNSCC patients.

Regulation of T Cells by NLP/NLGP in TME

In order to evaluate the mechanism of tumor growth restriction in therapeutic settings, we have analyzed TME from sarcoma and melanoma bearing mice with NLGP therapy. A significant increase of $CD8^+$ T cells was observed within TME from NLGP treated mice (NLGP-TME), in comparison with control tumor (Figures 2.2E1-E2). $CD8^+$ T cells isolated from NLGP-TME exhibited greater cytotoxicity to sarcoma/melanoma cells *in vitro* and *in vivo* (Figure 2.2E3) and these cells show higher expression of cytotoxicity related molecules, perforin and granzyme B. Anergy, exhaustion and AICD are the terminal fates of impaired T cells; thus, we studied these phenomena within NLGP-TME and PBS-TME. Anergy is a hyporesponsive state of T cells commonly encountered at tumor vicinity that is mediated by poor co-stimulation in absence of IL-2, a condition that upregulates various anergy related genes *cbl-b*, *egr2*, *egr3*, itch, GRAIL and DGKα in T cells. Negligible expression of these anergy related genes along with higher level of phosphorylated NFAT in T cells incubated within NLGP-TME as well as those recovered from *in vivo* NLGP treated tumors in comparison to PBS-TME and ionomycin treated controls is a noteworthy observation. T cell exhaustion is also reduced by NLGP, as demonstrated by down regulated expression of related markers, tim3, pd1 and Lag3 (Figure 2.2E4). AICD causes premature turnover of T cells at tumor site and is marked by enhanced expression of FasR, caspase 3 and 8. T cells incubated with NLGP-TME showed a diminished expression of these markers and decreased expression of FasR on $CD8^+$ T cells isolated from NLGP treated tumors compared to control. Accumulated evidences strongly suggest that NLGP mediated normalization of TME allows T cells to perform optimally to inhibit the tumor growth.

Regulation of Treg Cells by NLP/NLGP in Periphery

Downregulation of Tregs is considered as a promising cancer immunotherapeutic approach due to its immense immune-suppressive nature. As NLGP restricts tumor growth in mice by immune activation, we were interested to see whether NLGP can modulate Tregs in association with tumor growth restriction. We have observed that NLGP down regulates $CD4^+CD25^+Foxp3^+$ Tregs within tumors. NLGP-mediated down regulation of CCR4 along with its ligand CCL22 restricts Treg migration at the tumor site. NLGP is not apoptotic to Tregs but significantly downregulates the expression of Foxp3, CTLA4 and GITR *in vitro*. It also reverses the functional impairment of T effector cells by Tregs, in terms of IFNγ secretion, cellular proliferation and tumor cell cytotoxicity. NLGP also facilitates reconditioning of TME by increasing IFNγ and IL-12 but decreasing IL-10, TGFβ, VEGF and IDO, creating an antitumor niche. Interaction between Foxp3, *p*NFATc3 and *p*Smad2/3, needed for successful Treg function, is also inhibited by NLGP (Chakraborty *et al.*, 2011) (Figure 2.2B1-B3).

In separate observation we have shown that NLGP inhibits the Treg induced suppression of tumoricidal functions of $CD14^+CD68^+$ monocyte/macrophages (MO/Mö) from human peripheral blood. Cytotoxic efficacy of MO/Mö toward macrophage sensitive cells, U937, is decreased in presence of Tregs (induced), however, it was increased further by treatment of NLGP *in vitro*. Associated Treg mediated inhibition of perforin/granzyme B expression and nitric oxide release from MO/Mö was normalized by NLGP. Altered status of signature cytokines, like, IL-12, IL-10, IL-6, TNFα from MO/Mö under influence of Tregs is also rectified by NLGP. Tregs significantly enhanced the expression of altered marker, mannose receptor (CD206) on $CD68^+$ cells that was downregulated upon NLGP exposure. Treg mediated inhibition of MO/Mö chemotaxis in contact dependent manner was also normalized partially by NLGP, where participation of CCR5 was documented. Overall results suggest that Treg influenced pro-tumor MO/Mö functions are rectified in a significant extent by NLGP to create an anti-tumor immune environment. All of these coordinated events might result in inhibition of Treg associated-tumor growth and therefore increased survivability of mice having NLGP treatment before or/and after tumor inoculation.

Regulation of Treg Cells by NLP/NLGP in TME

In cancer, enhanced frequency of suppressor cells, like, Tregs make the situation more favorable for tumorigenesis. These cells negatively interfere in optimum anti-tumor T cell functions. Previously we have observed that NLGP treatment efficiently down regulates Treg generation, but *in vivo* data suggest that NLGP does not reduce $CD4^+CD25^+$ T cells within TME of NLGP treated mice. NLGP may reduce the Treg functionality within TME. Investigation in this direction is under progress. Western blot analysis showed the down regulation of Foxp3 within NLGP-TME. In addition, it was observed that TME from NLGP treated mice showed lower mRNA expression of IDO1 and CTLA-4 in comparison to PBS-TME. Minimization of the suppressor functions within TME by NLGP would be translated into inhibition of tumors in therapeutic settings. As optimum therapeutic outcome may depend on the balance between effector to regulatory T cell ratios, we determined the ratio between $CD8^+$ effector T cells and $CD4^+CD25^+$ regulatory T cells within TME, the obtained data revealed increment in such ratio after NLGP treatment in tumor bearing mice, favoring the T cell mediated anti-tumor immunity (Figures 2.2F1-F3).

Regulation of B Cells by NLP/NLGP in Periphery

The conditional tumor growth retardation, observed in mice treated with NLP before tumor inoculation, may be regulated by NLP mediated immune activation, having prominent role in the humoral and cellular immune functions of the tumor host. The first evidence to regulate B- cells by NLP was found when we observed mice (Swiss and Balb/c) and rats (Sprague Dawley) immunized with BTAA (identified and partially characterized from human breast tumors) along with NLP have a higher IgG antibody response and a lower IgM response than mice immunized with poorly immunogenic Breast Tumor Associated Antigen (BTAA) alone. Antibody generated by immunization with BTAA and NLP can induce antibody-dependent cellular cytotoxicity (ADCC) and CTL response towards BTAA-expressing MCF-7 cells whereas a poor cytotoxic response was found, when serum from BTAA control mice was used. The occurrence of ADCC and CTL response induced by BTAA plus NLP vaccination was possibly assisted by the induction of a Th1 response, as evidenced by the enhanced secretion of IFNγ and decreased release of IL-10 from spleen cells and the greater production of IgG2a antibody in immunized mice (Mandal *et al.*, 2007). Similar observation was found when a significant higher antibody (IgG2a) titre was found in mice immunized with B16MelSAg + NLP, in comparison to B16MelSAg alone (Baral *et al.*, 2005). After identification of NLGP from NLP, adjuvant function of it was studied in context to carcinoembryonic antigen (CEA) expressing tumors. CEA was presented using macrophages, in combination with NLGP and such vaccination generates significantly higher antibody (IgG2a) and T cell response than the immunization protocol without NLGP. NLGP controls the function of both B cells and macrophages by altering the expressions of various regulatory molecules, like, CD19, CD11b, etc. NLGP also directs CEA vaccination towards Th1 bias, by modulating cytokine secretion. This NLGP-generated anti-CEA immune response would be effective as a vaccine to lyse CEA^+ tumors *in vitro* and *in vivo* (Sarkar *et al.*, 2008).

Regulation of B Cells by NLP/NLGP in TME

B cell components within TME might be in some dysregulated state. Role of NLGP in rectification of B cell functions within TME is not studied. However, enormous scope is lying in the research in this direction.

Regulation of Dendritic Cells by NLP/NLGP in Periphery

Previously we observed NLGP regulates the function of both B cells and macrophages by altering the expression of various regulatory molecules, like, CD19, CD11b, etc. as well as directs type 1 skewness, by modulating cytokine secretion. These results provided an indication that it might have a role on generation and functioning of DCs, a major antigen presenting cell population. Interestingly, we observed that GMCSF/IL-4 differentiated $CD14^+$ monocytes were matured with NLGP. NLGP matured DCs (NLGP-DCs) show upregulated expression of CD83, CD80, CD86, CD40 and MHCs, in a comparable extent of control, LPS (Figures 2.2C1-C2). NLGP stimulated macrophages and NLGP matured DCs secrete high amount of IL-12p70 with low IL-10 (Figures 2.2C3, D2-D3). NLGP also stimulated the uptake of antigen in macrophages (Figure 2.2D1). NLGP upregulates the expression of crucial transcription factor, ikaros, indicating maturation towards DC1 phenotype. Increased expression of CD28 and CD40L on T cells following co-culture with NLGP-DCs was noticed to promote DC-T cell interactions. As a result, T cells secrete high amount of IFNγ with low IL-4 and generates anti-tumor type 1 immune microenvironment. Such NLGP-DCs present CEA effectively to T cells to increase CTL mediated cytotoxicity of CEA^+ tumor cells *in vitro* and *in vivo* (Goswami *et al.*, 2010). In other experimental settings, we have observed that vaccination with NLGP matured CEA^+ pulsed DCs enhances antigen-specific humoral and cellular immunity against CEA and restricts the growth of murine CEA^+ tumors as it helps better CEA uptake, processing and presentation to T/B cells. This vaccination (NLGP matured CEA pulsed DCs (DCNLGPCEA)) elicits mitogen induced and CEA specific T cell proliferation, IFNγ secretion and induces specific cytotoxic reactions to CEA^+ colon tumor cells. In addition to T cell response, DCNLGPCEA vaccine generates anti-CEA antibody response, which is principally IgG2a in nature. This antibody participates in cytotoxicity of CEA^+ cells in antibody-dependent manner, which helped to protect mice from tumor development and these mice remained tumor free following second tumor inoculation suggesting the generation of memory response (Sarkar *et al.*, 2010).

In cancer, dysfunctional maturation of DCs is a common cause of dysregulated anti tumor T-cell functions. We have observed that in myeloid-derived DCs generated from monocytes obtained from stage IIIB cervical cancer (CaCx IIIB) patients instruct poor T cell functions. Accordingly, we were interested to see whether NLGP could rectify DC maturation at least partially in CaCxIIIB patients. *In vitro* studies showed that NLGP treatment of immature DCs (iDCs) obtained from CaCx IIIB patients results in upregulated expression of various cell surface markers (CD40, CD83, CD80, CD86, and HLA-ABC), which indicates DC maturation. Consequently, NLGP-matured DCs displayed balanced cytokine secretions, with type 1 bias with remarkable functional properties. These DCs displayed substantial T cell allostimulatory capacity and promoted the generation of CTLs. Although NLGP-matured DCs derived from CaCx monocytes are generally subdued compared to those with a healthy monocyte origin, considerable revival of the suppressed DC-based immune functions is noted *in vitro* at a fairly advanced stage of CaCx, and thus, further exploration of *ex vivo* and *in vivo* DC-based vaccines is proposed. Moreover, the DC maturating efficacy of NLGP might be much more effective in the earlier stages of CaCx, where the extent of immune dysregulation is less and, thus, the scope of further investigation may be explored (Roy *et al.*, 2011).

Regulation of Dendritic Cells by NLP/NLGP in TME

From our observations on biology of DCs in periphery of healthy individuals and cancer patients, it was proved that NLGP optimizes its maturation and establishes DC1 phenotypes (Figure 2.2G2). At the same time, we observed that Th1 bias within TME from NLGP treated sarcoma and melanoma bearing mice. Higher level of secretory IL-12 within TME suggests the presence of mature DCs in a greater percentage. Our preliminary immuno-histochemical observation supported the presence of significant $CD11c^+$ cells in TME from NLGP treated melanoma bearing mice (Figures 2.2G12) *(unpublished observation)*. This observation indicates that NLGP not only rectifies DC's property but also helps in migration of DCs within TME. Further research is required in this direction.

Regulation of Cytokine/Chemokine Signaling by NLGP in Periphery

Obtained supernatant from NLP stimulated human PBMC, inhibits tumor cell proliferation and induces apoptosis. Presence of cytotoxic cytokines, IFNγ and TNFα in such supernatant was detected by ELISA and inhibition of secretion of IFNγ and TNFα caused a significant decrease in tumor cell apoptosis (Bose *et al.*, 2007). In other study, NLP was found to activate NK cells to enhance their tumor cell cytotoxic ability and to stimulate the release of IL-12 from macrophages from healthy individuals and HNSCC patients. In addition, NLP upregulated the expression of CD40 and CD40L on CD14$^+$ monocytes and CD56$^+$ lymphocytes respectively, which is directly related to IL-12 release (Bose and Baral, 2007).

We have analyzed immune activation and associated immune polarization upon NLGP treatment. NLGP-activated T cells secrete greater amount of signature Th1 cytokines IFNγ and a lower amount of the Th2 cytokine, IL-4. Similar type-1 directiveness is also observed in antigen-presenting monocytes and DCs by upregulation of IL-12, TNFα and downregulation of IL-10. This type-1 biasness is also assisted by NLGP-induced downregulation of Foxp3$^+$ Treg cells and upregulation of type 1-specific transcription factor, T-bet. In order to find out the mechanism behind it, we have found increased phosphorylation of STAT1 and STAT4 with decreased phosphorylation of STAT3. Overall, it was observed from our studies on human and mouse systems that NLGP is effective to release type-1 cytokines, like, IFNγ, IL-12, IL-2, TNFα and inhibited secretion of IL-10, IL-6 and TGFβ (Bose *et al.*, 2009).

Effect of NLGP in rectification of the dysregulated IFNγ dependent chemokine and its receptor CXCR3 splice variants was investigated. Upregulated expression of CXCR3B in HNSCC-PBMC was downregulated following *in vitro* NLGP treatment. Unchanged expression of CXCR3A+B by NLGP with downregulation of the CXCR3B indirectly suggests the upregulation of the CXCR3A, responsible for cellular migration. However, stimulation of healthy-PBMC with NLGP maintains physiological homeostasis of CXCL10 and increases IFNγ secretion. The suppressed chemotaxis of HNSCC-PBMC could be restored either by *in vitro* treatment with NLGP or during use of NLGP stimulated PBMC supernatant as a chemoattractant. Ligands (RANTES, MIP1α and MIP1β) of this chemokine receptor were also secreted in lesser quantity from MO/Mw of HNSCC patients in comparison with healthy individuals (Chakraborty *et al.*, 2008). In a separate experiment, we found that NLGP upregulated CCR5 expression, as evidenced from studies on MO/Mw of peripheral blood from HNSCC patients as well as healthy individuals. Expression of RANTES, MIP1α and MIP1β was also upregulated following NLGP treatment of these cells *in vitro*. Interestingly, NLGP has little effect on the expression of CCR5 and the ligand RANTES in oral cancer cells. This restored CCR5 receptor-ligand signaling seen in MO/Mw was reflected in improved CCR5-dependent, p38MAPK mediated migration of MO/Mw after NLGP treatment to a standard chemoattractant. In addition, NLGP-treated MO/Mϕ-primed T cells can effectively lyse tumor cells *in vitro* (Chakraborty *et al.*, 2010). These results suggest a new approach in cancer immunotherapy by modulating dysregulated CCR5 signals from MO/Mϕ.

Regulation of Cytokine/Chemokine Signaling by NLGP within TME

As TME is dominated with immunosuppressive cytokines, like, IL-10, TGFβ etc., along with low IFNγ and IL-12, normalization of TME is desired to initiate antitumor immune response, especially T cell response. To investigate the role of NLGP in TME normalization, two groups of mice with established sarcoma (average tumor volume, 100 mm^3) were treated with NLGP and PBS respectively. Tumors were harvested in three different days (n=3, in each time point) following initiation of the NLGP treatment and cytokine secretory status was assessed from cellular contents by ELISA and immunoblotting. Tumors of NLGP treated mice exhibited significant upregulated level of IFNγ, IL-2 and IL-12 in day dependent manner and reduced release of IL-6, IL-10 and TGFβ within TME, in comparison to mice having PBS injection.

Chemokine-mediated T cell migration is essential for optimal anti-tumor immune response (Lu *et al.*, 2012). Similar to the cytokine network, chemokine milieu is also dysregulated within TME, thus, hampering movement of effector $CD8^+$ T cells in tumors. In an objective to know the status of chemokine receptor-ligand profile, we have analyzed different genes related to chemokine ligands (*ccl3, ccl4, ccl5, ccl8, cxcl9, cxcl10* and *cxcl12*) and chemokine receptors (*ccr5, cxcr3* and *cxcr4*) at transcriptional level. We observed chemokine receptors, *cxcr3* and *ccr5*, were upregulated in NLGP treated tumors. At the same time it was found that ligands for *ccr5* like *ccl3, ccl4, ccl5* and *ccl8* were upregulated within NLGP-TME, although extent of upregulation is varied. Same trend of upregulation was observed in case of *cxcr3* ligands, *cxcl9*, and *cxcl10*. We also observed a marked downregulation of *cxcr4* and *cxcl12* in case of NLGP treated tumor than PBS controls *(unpublished observation)*.

Earlier discussion demonstrated the recruitment of enhanced number of type 1 polarized $CD8^+$ T cells within TME. Such recruitment to tumor sites occur via CXCR3 mediated chemotaxis in response to the CXCL9-11 chemokines, produced within the TME. Accordingly in order to determine whether Tumor draining lymph nodes (TDLN)/Tumor derived T cells in NLGP treated animals are expressing more chemotactic components over PBS mice, we assessed $CD8^+$ T cells for their expression of CXCR3 and corresponding ligands on day 21 sarcoma mice. We found that $CXCR3^+$ subpopulation of $CD8^+$ T cells as well as *cxcr3* transcripts were significantly increased in TDLN of NLGP treated mice. Expression of mRNAs corresponding to *cxcl9* and *cxcl10* also increased within TILs from tumors (Mallick *et al.*, 2013). We also found upregulation of $CCR5^+$ cells and enhanced *ccr5* expression at mRNA levels. Though CCR5 is expressed generally on monocytes, it is also involved in T cell migration (Hanahan *et al.*, 2011). Consistently, we found higher expression of its corresponding ligands, *ccl3, ccl4, ccl5* and *ccl8* within TME. Similar result was obtained in case of TDLN for *ccl4* and *ccl5*, without noticeable upregulation in *ccl3* and *ccl8*. Interestingly, CXCR3 and CCR5 expression was unchanged on sarcoma cells upon *in vitro* NLGP treatment. These data suggest that NLGP therapy not only stimulates T cell expansion, but also licenses these cells for trafficking to peripheral tissues in which CXCR3/CCR5 ligands are expressed, such as TME.

Regulation of Altered Angiogenesis within TME by NLGP

The growth of primary and metastatic tumors to larger than a 2-3 millimeters requires the recruitment of neighboring blood vessels and vascular endothelial cells to support their metabolic requirements. Tumors utilize a number of mechanisms to promote their vascularization, and in each case they subvert normal angiogenic processes to suit this purpose. Tumor angiogenesis is a complex process involving many different cell types that must proliferate, migrate, invade, and differentiate in response to signals from the TME. VECs sprout from host vessels in response to VEGF, bFGF, Ang2 and other proangiogenic stimuli within TME. Sprouting is stimulated by VEGF/VEGFR2, Ang2/Tie-2, and integrin/extracellular matrix interactions. Tumor cells themselves may directly form parts of vascular channels within TME. The pattern of vessel formation is haphazard: vessels are tortuous, dilated, leaky, and branched in random ways. This leads to uneven blood ow within the tumor, with areas of acidosis and hypoxia (which stimulate release of angiogenic factors) and high intratumoral pressures that inhibit delivery of therapeutic agents, leading to tumor growth. Accordingly, normalization of angiogenic vasculature as well as tumor microenvironment is the essential focus in current cancer therapeutic research. We have observed more normalized blood vasculature within and surrounding tumors of carcinoma and melanoma bearing mice having prophylactic or therapeutic NLGP treatment (Figures 2.2I.1). NLGP may act directly on VECs of blood vessels as evidenced by downregulation of CD31, VEGFR2, whereas, expression of VEGFR1 is unaltered. NLGP also stabilized blood vessels by normalizing pericytes, as evidenced by downregulation of NG2. At the same time, NLGP make significant downregulation of VEGF, PDGF, IL-10, TGFβ, IDO as detected by Western blotting, RT-PCR and immunohistochemical analysis (Figures 2.2I.21-I.3). Downregulation of HIF 1α by NLGP also indicates reduction of hypoxia and normalization of the vascular interstitial pressure. All of these synchronized events may promote infiltration of antitumor immune competent cells and therapeutic regimens within tumors that ultimately correlated with tumor growth inhibition.

Concluding Remarks

In view of the **'Triple E'** (**E**limination, **E**quilibrium and **E**scape) theory of cancer, malignant tumor can be formed when cancer cells escaped from the immune surveillance, otherwise, they can be eliminated by immune system. In the process of escape, TME helps tumor enormously. Again, in the creation of TME tumors play an active role. Immune and non-immune components of TME help tumors to grow in different ways, thus, normalization of TME would be an effective choice for anti-tumor therapy. TME induces type-2 immunity by active contribution of different TME components, like, TAM, that promotes tumor. We have reported earlier that NLGP is able to maintain type-1 immunity in tumor host (Bose *et al.*, 2009) and it also can increase the type 1 cytokines within TME. This is a big step towards keeping the TME green by NLGP. TME derived factors also establish a chemokine gradient that can pool suppressor cells within TME, in spite of effector cells. We have specially accumulated data on the efficacy and migration of $CD8^+$ T cells under the influence of NLGP within TME (Mallick *et al.*, 2013). It is clear that NLGP activated T cells can be more cytotoxic and secretes type 1 cytokines (Bose *et al.*, 2009), even within TME. Abnormal functioning of antigen presenting cells, like, macrophages, DCs, are nicely tackled by NLGP (Sarkar *et al.*, 2008; Goswami *et al.*, 2010; Roy *et al.*, 2011), so that T cells obtain required antigenic activation (Sarkar *et al.*, 2009). Suppressor cells and their secreted cytokines, like, IL-10, TGFβ make immune system dysfunctional and it is clearly demonstrated that NLGP downregulates IL-10, TGFβ, thereby helping other immune cells to become more functional (Chakraborty *et al.*, 2011). Metastasis promoting alteration of TME also corroborated with dysregulated degradation of extracellular matrix by MMPs, thus, expression of those proteins increased significantly. NLGP may inhibit metastasis by downregulating MMPs. Effect of NLGP on extracellular matrix is yet to be elucidated. Hypoxic environment of TME promotes angiogenesis and vasculature becomes chaotic, tortuous, leaky, thus, creating increased interstitial pressure and uneven blood flow. This angiogenic character prevents the normal flow of immune cells within tumor. It also prevents drug delivery within tumor, if patients are under chemotherapeutic treatment. Evidences are already obtained that NLGP downregulates angiogenesis related molecules, like, VEGF, HIF1α and make the blood flow normalized *(unpublished observation).* We also initiated studies on mesenchymal stem cells that has significant suppressive role on immune functions. NLGP also shows good promise in this regard. In addition to these, several other complicated interactions are always ongoing within TME and role of NLGP on those aspects should be investigated to understand the mechanism by which NLGP attempts to keep the TME green. Maintenance of pollution free green environment is essential as TME not only attenuates the efficacy of endogenous antitumor immune responses, but also represents a barrier to therapeutic modalities, including chemotherapy and immunotherapy.

References

Alitalo, A., and Detmar, M. (2011). Interaction of tumor cells and lymphatic vessels in cancer progression. *Oncogene.* Epub ahead of print, Dec 19, 2011.

André, S., Tough, D.F., Lacroix-Desmazes, S., Kaveri, S.V., and Bayry, J. (2009). Surveillance of antigenpresenting cells by CD4+CD25+regulatory T cells in autoimmunity: immunopathogenesis and therapeutic implications. *Am J Pathol.,* **174:** 1575-1587.

Apte, R. N., and Voronov, E. (2008). Is interleukin-1 a good or bad 'guy' in tumor immunobiology and immunotherapy? *Immunol Rev.,* **222:** 222-241.

Baral, R.N., Sarkar, K., Mandak-Ghosh, I., and Bose, A. (2010). Neem leaf glycoprotein as a new vaccine adjuvant for cancer immunotherapy. *In:* Comprehensive Bioactive Natural Products: Immune-modulation and Vaccine Adjuvants, Vol. 5, Ed. By Gupta, V.K., Studium Press LLC, USA, pp. 21-45.

Baral, R., Mandal, I., and Chattopadhyay, U. (2005). Immunostimulatory neem leaf preparation acts as an adjuvant to enhance the efficacy of poorly immunogenic B16 melanoma surface antigen vaccine. *Int Immunopharmacol.,* **5:** 1343-1352.

Baral, R., and Chattopadhyay, U. (2004). Neem (*Azadirachta indica*) leaf mediated immune activation causes prophylactic growth inhibition of murine Ehrlich carcinoma and B16 melanoma. *Int Immunopharmacol.,* **4:** 355-366.

Bergers, G., and Song, S. (2005). The role of pericytes in blood-vessel formation and maintenance. *Neuro Oncol.,* **7:** 452-464.

Bhowmick, N. A., Neilson, E. G., and Moses, H.L. (2004). Stromal fibroblasts in cancer initiation and progression. *Nature,* **432:** 332-337.

Bill, M.A., Bakan, C., Benson, D.M J.r, Fuchs, J., Young, G., and Lesinski, G.B. (2009). Curcumin induces proapoptotic effects against human melanoma cells and modulates the cellular response to immunotherapeutic cytokines. *Mol Cancer Ther.,* **8:** 2726-2735.

Biswas, K., Chattopadhyay, I., Banerjee, R.K., and Bandopadhyay, U. (2002). Biological activities and medicinal properties of neem (*Azadirachta indica*). *Curr Sci.,* **82:** 1336-1345.

Bose, A., and Baral, R. (2007). Natural killer cell mediated cytotoxicity of tumor cells initiated by neem leaf preparation is associated with CD40-CD40L-mediated endogenous production of interleukin-12. *Hum Immunol.,* **68:** 823-831.

Bose, A., Chakraborty, K., Sarkar, K., Goswami, S., Chakraborty, T., Pal, S., and Baral, R. (2009). Neem leaf glycoprotein induces perforin-mediated tumor cell killing by T and NK cells through differential regulation of IFNγ signaling. *J Immunother.,* **32:** 42-53.

Bose, A., Haque, E., and Baral, R. (2007). Neem leaf preparation induces apoptosis of tumor cells by releasing cytotoxic cytokines from human peripheral blood mononuclear cells. *Phytother Res.,* **21:** 914-920.

Bose, A., Chakraborty, K., Sarkar, K., Goswami, S., Haque, E., Chakraborty, T., Ghosh, D., Roy, S., Laskar, S., and Baral, R. (2009). Neem leaf glycoprotein directs T-bet-associated type 1 immune commitment. *Human Immunol.,* **70:** 6-15.

Busquets, S., Ametller, E., Fuster, G., Olivan, M., Raab, V., Argilés, J.M., and López-Soriano,.FJ. (2007). Resveratrol, a natural diphenol, reduces metastatic growth in an experimental cancer model. *Cancer Lett.,* **245:** 144-148.

Chakraborty, K., Bose, A., Pal, S., Sarkar, K., Goswami, S., Ghosh, D., Laskar, S., Chattopadhyay, U., and Baral R. (2008). Neem leaf glycoprotein restores the impaired chemotactic activity of peripheral blood mononuclear cells from head and neck squamous cell carcinoma patients by maintaining CXCR3/CXCL10 balance. *Int Immunopharmacol.,* **8:** 330-340.

Chakraborty, K., Bose, A., Chakraborty, T., Sarkar, K., Goswami, S., Pal, S., and Baral, R. (2010). Restoration of dysregulated CC chemokine signaling for monocyte/ macrophage chemotaxis in head and neck squamous cell carcinoma patients by neem leaf glycoprotein maximizes tumor cell cytotoxicity. *Cell Mol Immunol.,* **7:** 396-408.

Chakraborty, T., Bose, A., Barik, S., Goswami, K.K., Banerjee, S., Goswami, S., Ghosh, D., Roy, S., Chakraborty, K., Sarkar, K., and Baral, R. (2011). Neem leaf glycoprotein inhibits CD4+CD25+Foxp3+ Tregs to restrict murine tumor growth. *Immunotherapy,* **3:** 949-969.

Chantrain, C.F., Feron, O., Marbaix, E., and DeClerck, Y.A. (2008). Bone marrow microenvironment and tumor progression. *Cancer Microenviron.,* **1:** 23-35.

Chen, Y., Tseng, S.H., Lai, H.S., and Chen, W.J. (2004). Resveratrol-induced cellular apoptosis and cell cycle arrest in neuroblastoma cells and antitumor effects on neuroblastoma in mice. *Surgery,* **136:** 57-66.

Cirri, P., and Chiarugi, P. (2011). Cancer associated fibroblasts: the dark side of the coin. *Am J Cancer Res.,* **1:** 482-497.

Colotta, F., Allavena, P., Sica, A., Garlanda, C., and Mantovani, A. (2009). Cancer-related inflammation, the seventh hallmark of cancer: links to genetic instability. *Carcinogenesis,* **30:** 1073-1081.

Coussens, L. M. and Werb, Z. (2002). Inflammation and cancer. *Nature.* 420: 860–867. Diaz-Cano, S.J. (2012). tumor heterogeneity: mechanisms and bases for a reliable application of molecular marker design. *Int J Mol Sci.,* **13:** 1951-2011.

Deryugina, E.I., and Quigley, J.P. (2006) Matrix metalloproteinases and tumor metastasis. *Cancer Metastasis Rev.,* **25:** 9-34.

de Visser, K.E., Eichten, A., and Coussens, L.M. (2006). Paradoxical roles of the immune system during cancer development. *Nat Rev Cancer,* **6:** 24-37.

Duvoix, A., Blasius, R., Delhalle, S., Schnekenburger, M., Morceau, F., Henry, E., Dicato, M., and Diederich, M. (2005). Chemopreventive and therapeutic effects of curcumin. *Cancer Lett.,* **223:** 181-190.

Erez, N., Truitt, M., Olson, P., Arron, S.T., and Hanahan, D. (2010). Cancer-Associated Fibroblasts are Activated in Incipient Neoplasia to Orchestrate Tumor-Promoting Inflammation in an NF-κB-Dependent Manner. *Cancer Cell.,* **17:** 135-147.

Estrov, Z., Shishodia, S., Faderl, S., Harris, D., Van, Q., Kantarjian, H.M., Talpaz, M., Aggarwal, B.B. (2003). Resveratrol blocks interleukin-1β-induced activation of the nuclear transcription factor NF-κB, inhibits proliferation, causes S-phase arrest, and induces apoptosis of acute myeloid leukemia cells. *Blood,* **102:** 987-995.

Franco, M., Roswall, P., Cortez, E., Hanahan, D., and Pietras, K. (2011). Pericytes promote endothelial cell survival through induction of autocrine VEGF-A signaling and Bcl-w expression. *Blood,* **118:** 2906-2917.

Gabrilovich, D.I., and Nagaraj, S. (2009). Myeloid derived suppressor cells as regulators of the immune system. *Nat Rev Immunol.,* **9:** 162-174.

Ge, G., Seo, N.S., Liang, X., Hopkins, D.R., Höök, M., and Greenspan, D.S. (2004). Bone morphogenetic protein-1/tolloid-related metalloproteinases process osteoglycin and enhance its ability to regulate collagen fibrillogenesis. *J Biol Chem.,* **279:** 41626-41633.

Goswami, S., Bose, A., Sarkar, K., Roy, S., Chakraborty, T., Sanyal, U., and Baral, R. (2010). Neem leaf glycoprotein matures myeloid derived dendritic cells and optimizes anti-tumor T cell functions. *Vaccine,* **28**: 1241-1252.

Gouwy, M., Struyf, S., Proost, P. and VanDamme, J. (2005). Synergy in cytokine and chemokine networks amplifies the inflammatory response. *Cytokine Growth Factor Rev.,* **16:** 561-580.

Hanahan, D., and Weinberg, R.A. (2011). Hallmarks of cancer: the next generation. *Cell,* **144:** 646-674.

Haque, E., and Baral, R. (2006). Neem (*Azadirachta indica*) leaf preparation induces prophylactic growth inhibition of murine Ehrlich carcinoma in Swiss and C57BL/6 mice by activation of NK cells and NK-T cells. *Immunobiology,* **211:** 721-731.

Haque, E., Mandal, I., Pal, S., and Baral, R. (2006). Prophylactic dose of neem (*Azadirachta indica*) leaf preparation restricting murine tumor growth is nontoxic, hematostimulatory and immunostimulatory. *Immunopharmacol Immunotoxicol.*, **28:** 33-50.

Höckel, M. and Vaupel, P. (2001). Tumor hypoxia: definitions and current clinical, biologic, and molecular aspects. *J Natl Cancer Inst.*, **93**: 266-276.

Jewett, A., and Tseng,H.C. (2011). Tumor induced inactivation of natural killer cell cytotoxic function; implication in growth, expansion and differentiation of cancer stem cells. *J Cancer*, **2:** 443-457.

Kannan, M., and Singh, R.A.J.A. (2010). An immuno-pharmacological investigation of Indian medicinal plant *Nyctanthes arbor-tristis* Linn. *World Applied Sciences Journal*, **11:** 495-503.

Korc, M. (2007). Pancreatic cancer-associated stroma production. *Am J Surg.*, **194:** 84-86.

Kulbe, H., Levinson, N.R., Balkwill, F. and Wilson JL., 2004. The chemokine network in cancer much more than directing cell movement. *Int J Dev Biol.*, **48:** 489-496.

Kumar, S., Gupta, P., Sharma, S., and Kumar, D. (2011). A review on immunostimulatory plants. *Zhong Xi Yi Jie He Xue Bao.*, **9:** 117-128.

Kwon, H.K., Jeon, W.K., Hwang, J.S., Lee, C.G., So, J.S., Park, J.A., Ko, B.S., and Im, S,H. (2009). Cinnamon extract suppresses tumor progression by modulating angiogenesis and the effector function of $CD8^+$ T cells. *Cancer Lett.*, **278:** 174-82.

Kwon, H.K., Hwang, J.S., So, J.S., Lee, C.G., Sahoo, A., Ryu, J.H., Jeon, W.K., Ko, B.S., Im, C.R., Lee, S.H., Park, Z.Y., and Im, S.H. (2010). Cinnamon extract induces tumor cell death through inhibition of NFκB and AP1. *BMC Cancer*, **10:** 392.

Langley, R.R., and Fidler, I.J. (2011). The seed and soil hypothesis revisited—the role of tumor-stroma interactions in metastasis to different organs. *Int J Cancer*, **128:** 2527-2535.

Liao, D., and Y. Luo, *et al.* (2009). Cancer associated fibroblasts promote tumor growth and metastasis by modulating the tumor immune microenvironment in a 4T1 murine breast cancer model. *PLoS One*, **4:** 7965.

Lorusso, G., and Rüegg, C. (2008). The tumor microenvironment and its contribution to tumor evolution toward metastasis. *Histochem. Cell. Biol.*, **130:** 1091-1103.

Lu, P., Weaver, V.M., and Werb, Z. (2012). The extracellular matrix: a dynamic niche in cancer progression. *J Cell Biol.*, **196:** 395-406.

Ma, F., Li, Y., Jia, L., Han, Y., Cheng, J., Li, H., Qi, Y., and Du, J. (2012). Macrophage-stimulated cardiac fibroblast production of IL6 is essential for TGFβ/ Smad activation and cardiac fibrosis induced by angiotensin II. *PLoS One*, **7:** 35144.

Mandal-Ghosh, I., Chattopadhyay, U., and Baral, R. (2007). Neem leaf preparation enhances Th1 type immune response and anti-tumor immunity against breast tumor associated antigen. *Cancer Immunity*, **7:** 8.

Mallick, A., Barik, S., Goswami,K.K., Banerjee, S., Ghosh, S. Sarkar, K., Bose, A., and Baral, R. (2013). Neem leaf glycoprotein activates CD8+ T cells to promote therapeutic anti-tumor immunity inhibiting the growth of mouse sarcoma. PLOS One, *(in press)*.

Manna, S.K., Mukhopadhyay, A. and Aggarwal, B.B. (2000). Resveratrol suppresses TNF-induced activation of nuclear transcription factors NF-κB, activator protein-1, and apoptosis: potential role of reactive oxygen intermediates and lipid peroxidation. *J Immunol.*, **164:** 6509-6519.

Martín, A.R., Villegas, I., La Casa, C., and de la Lastra, C.A. (2004). Resveratrol, a polyphenol found in grapes, suppresses oxidative damage and stimulates apoptosis during early colonic inflammation in rats. *Biochem Pharmacol.*, **67:** 1399-1410.

Mazid, M., Khan, T.A., and Mohammad, F. (2011). Role of secondary metabolites in defense mechanisms of plants. *Biology and Medicine,* **3:** 232-249.

Mbeunkui, F., and Johann Jr, D.J. (2009). Cancer and the tumor microenvironment: a review of an essential relationship. *Cancer Chemother. Pharmacol.*, **63:** 571-582.

Paget, S. (1989). The distribution of secondary growths in cancer of the breast. *Lancet,* **133:** 571-573.

Pedraza-Farina, L.G. (2006). Mechanisms of oncogenic cooporation in cancer initiation and metastasis. *Yale j Biol Med.*, **79:** 95-103.

Penn, I., and Starzl, T.E. (1973). Immunosuppression and cancer. *Transplant Proc.*, **5:** 943-947.

Pietras, K., and Ostman, A. (2010). Hallmarks of cancer: interactions with the tumor stroma. *Exp Cell Res.*, **316:** 1324-1331.

Prager, G.W., Poettler, M., Unseld, M. and Zielinski, C.C. (2012). Angiogenesis in cancer: Anti-VEGF escapes mechanisms. *Transl. Lung Cancer Res.*, **1:** 14-25.

Rehman, J., J. Li, C. Orscell, and K. March. (2003). Peripheral blood endothelial cells are derived from monocyte/macrophages and secrete angiogenic growth factors. *Circulation,* **107:** 1164-1169.

Roy, S., Goswami, S., Bose, A., Chakraborty, K., Pal, S., Haldar, A., Basu, P., Biswas, J., and Baral, R. (2011). Neem leaf glycoprotein partially rectifies suppressed dendritic cell functions and associated T cell efficacy in patients with stage IIIB cervical cancer. *Clin Vaccine Immunol.*, **18:** 571-579.

Rustin, G.J.S., Galbraith, S.M., Anderson, H., Stratford, M., Folkes, L.K., Sena, L., Gumbrell, L., and Price, P.M. (2003). Phase I clinical trial of weekly combretastatin A4 phosphate: Clinical and pharmacokinetic results. *J.Clin. Oncol.*, **3:** 2815-2822.

Sarkar, K., Goswami, S., Roy, S., Mallick, A., Chakraborty, K., Bose, A., and Baral, R. (2010). Neem leaf glycoprotein enhances carcinoembryonic antigen presentation of dendritic cells to T and B cells for induction of anti-tumor immunity by allowing generation of immune effector/memory response. *Int Immunopharmacol.*, **10:** 865-874.

Sarkar, K., Bose, A., Chakraborty, K., Haque, E., Ghosh, D., Goswami, S., Chakraborty, T., Laskar, S., and Baral R. (2008). Neem leaf glycoprotein helps to generate carcinoembryonic antigen specific anti-tumor immune responses utilizing macrophage-mediated antigen presentation. **26:** 4352-4362.

Schmitt, M.W., Prindle, M.J, and Loeb, L.A. (2012). Implications of genetic heterogeneity in cancer. *Ann N Y Acad Sci.,* **1267:** 110-116.

Sheu, B.C., Chang, W.C., Cheng, C.Y., Lin, H.H., Chang, D.Y., and Huang, S.C. (2008). Cytokine regulation networks in the cancer microenvironment. *Front Biosci.,* **13:** 6255-6268.

Singh,M., and Singh, N. (2011). Curcumin counteracts the proliferative effect of estradiol and induces apoptosis in cervical cancer cells. *Mol Cell Biochem.,* **347:** 1-11.

Singh V.K., Sharma, P.K., Dudhe, R., and Kumar, N. (2011). Immunomodulatory effects of some traditional medicinal plants. *J. Chem. Pharm. Res.,* **3:** 675-684.

Stein, J.V., and Nombela-Arrieta,C. (2005) Chemokine control of lymphocyte trafficking: A general overview. *Immunology,* **116:** 1-12.

Surh, Y.J., Chun, K.S., Cha, H.H., Han, S.S., Keum, Y.S., Park, K.K., and Lee, S.S. (2001).Molecular mechanisms underlying chemopreventive activities of anti-inflammatory phytochemicals: down-regulation of COX and iNOS through suppression of NF-κB activation. *Mutat Res.,* 480-481: 243-268.

Turner, N.C., and Reis-Filho, J.S. (2012). Genetic heterogeneity and cancer drug resistance. *Lancet Oncol.,* **13:** 178-185.

Van Ginkel, P.R., Sareen, D., Subramanian, L., Walker, Q., Darjatmoko, S.R., Lindstrom, M.J., Kulkarni, A., Albert, D.M., and Polans, A.S. (2007). Resveratrol inhibits tumor growth of human neuroblastoma and mediates apoptosis by directly targeting mitochondria. *Clin Cancer Res.,* **13:** 5162-5169.

Veikkola, T., Karkkainen, M., and Claesson-Welsh, R. (2000). Regulation of angiogenesis via vascular endothelial growth factor receptors. *Cancer Res.,* **60:** 203-212.

Wang, Y., Gu, X., Xiang, J., and Chen, Z. (2012). Myeloid derived suppressor cells participate in preventing graft rejection. *Clin Dev Immunol.,* **2012:** 731486.

Wels, j., Kaplan, R.N., Rafii, S., and Lyden, D. (2008). Migratory neighbors and distant invaders: tumor-associated niche cells. *Genes Dev.,* **22:** 559-574.

Whiteside, T. L. (2006). Immune suppression in cancer: Effects on immune cells, mechanisms and future therapeutic intervention. *Semin. Cancer Biol.,* **16:** 3-15.

Witz, I.P. (2009). The tumor–microenvironment interactions: the making of a paradigm. *Cancer Microenviron.,* **2:** 9-17.

Wu, S.L., Yu, L., Meng, K.W., Ma, Z.H., and Pan, C.E. (2005). Resveratrol prolongs allograft survival after liver transplantation in rats. *World J Gastroenterol.,* **11:** 4745-4759.

Yu, Y.B., Dosanjh, L., Lao, L., Tan, M., Shim, B.S., and Luo, Y. (2010). *Cinnamomum cassia* bark in two herbal formulas increases life span in Caenorhabditis elegans via insulin signaling and stress response pathways. *PLoS One,* **5:** 9339.

Bioactive Phytochemicals: Perspectives for Modern Medicine Vol. 2 (2014)
Editor: **V.K. Gupta**
Published by: **DAYA PUBLISHING HOUSE, NEW DELHI**

Pages **83–92**

3

Betulinic Acid and its Derivatives: The Potential Anticancer Agents

Swee Keong Yeap[1], Wan Yong Ho[1], Nur Hana Faujan[2], Abdul Manaf Ali [3] and Noorjahan Banu Alitheen[1]*

ABSTRACT

Pentacyclic triterpenes have a wide spectrum of biological activities and significant pharmacological properties that may be useful in medicine. Betulinic acid (3β-hydroxy-lup-20(29)-ene-28-oic acid), an example of a pentacyclic triterpene that can be chemically derived from betulin, a substance found in the outer bark of white birch tree Betula alba. Some biological activities have been ascribed to betulinic acid which include anti-inflammatory, anti-tumor, anti-viral, anti-neoplastic and anti-plasmodial. Betulinic acid exerts a selective anti-tumor activity on cultured human melanoma, neuroblastoma, malignant brain tumor, leukemia cells, lung carcinomas, ovarian and cervical carcinomas without significant side-effects. Moreover, its activity against neuroectodermal has been reported and related to the induction of apoptosis via direct mitochondrial alterations. The favorable therapeutic index from the lack of toxicity towards normal cells suggested betulinic acid to be an attractive and promising anti-tumor agent. Moreover, ranges of derivatives have been synthesized from betulinic acid and all these potential anti-tumor compounds possessed various degree of toxicity against different types of cancerous cells. This review covers the potential cytotoxicity of betulinic acid and its potent derivatives.

***Keywords**:* Betulin, Betulinic acid, Cancer, Cytotoxic, Pentacyclic triterpenes, Derivaties.

1 Department of Cell and Molecular Biology, Faculty of Biotechnology and Biomolecular Sciences, Universiti Putra Malaysia, 43400, Serdang, Selangor, Malaysia

2 Faculty of Sciences, Universiti Putra Malaysia, 43400, Serdang, Selangor, Malaysia

3 Faculty of Agriculture and Biotechnology, Universiti Sultan Zainal Abidin, Kampus Kota, Jalan Sultan Mahmud, 20400 Kuala Terengganu, Malaysia

* *Corresponding author*: E-mail: noorjahan@biotech.upm.edu.my

Introduction

Betulinic acid, 3β-hydroxy-lup-20(29)-en-28-oic acid (Figure 3.1) with $C_{30}H_{48}O_3$ formula is a pentacyclic lupane type of triterpene that is widely distributed in the plant kingdom (Cichewicz and Kouzi, 2004). Betulinic acid was first extracted in 1948 from bark of the London plane tree (*Platanus acerifolia*) (Jung and Duclos, 2006). It can also be isolated from tropical African tree *Ziziphus mauritiama* Lam (Rhamnaceae) (Fujioka *et al.*, 1994), *Tryphyllum peltatum, Ancistrocladus heyneaus, Tetracera boliviana, Zizyphus joazeiro, Syzygium formosanum* and from the leaves of *Syzigium claviflorum* and *Aerva javanica* (Amaranthaceae) which grow wild in the eastern desert of Egypt (Kim *et al.*, 1997).

Other than isolation, betulinic acid can also be synthesized from betulin, a pentacyclin triterpene found in abundance in the outer bark of white birth trees (*Betula alba*) through a two steps synthesis process that give higher yield of betulinic acid as compared to isolation (Figure 3.1). In this process, betulin was converted to betulinic acid in over 75 per cent yield by Jones' oxidation (CrO_3/H_2SO_4/acetone/0°C) followed by reduction ($NaBH_4$/THF) to afford betulinic acid in quantitative yield (Kim *et al.*, 1998; Rene *et al.*, 2006). These betulin and betulinic acid have been reported to possess a wide spectrum of biological and pharmacological activities (Sami *et al.*, 2006).

Betulinic Acid and its Anticancer Property

Betulinic acid possesses selective apoptotic activity towards various tumor cells including neuroectodermal origin (Fulda *et al.*, 1998a,b), B16 (Liu *et al.*, 2004), medullablastoma (Daoy), glioblastoma (A172) and melanoma (Mel-Juso) cancer cells (Fulda and Debatin, 2005). Both betulinic acid and betulin had contributed to the major cytotoxicity of birch tree bark extract on skin epidermoid carcinoma, breast carcinoma and cervix adenocarcinoma cells (Soica *et al.*, 2012)

Betulinic acid exerts a selective anti-tumor activity towards cultured human melanoma cell lines *in vitro* and *in vivo* (Pisha *et al.*, 1995). The cytotoxic data indicated that betulinic acid is melanoma specific cytotoxic agent which is active against human melanoma cell lines including MEL-1 (lymph node), MEL-2 (pleural fluid), MEL-3 (liver), MEL-7 (metastatic melanoma from growing area) (Chatterjee *et al.*, 2000), drug-sensitive melanoma (MeWo), drug-resistant melanoma MeWo CIS (cisplatin), MeWo ETO (etoposide), MeWo VIN (vinblastin), MeWo FOTE (fotemusine) cell lines and normal melanocyte NHEM-neo cell line (Surowiak *et al.*, 2009) as opposed to several other cell types including human epidermoid carcinoma of the mouth (KB) (Chatterjee *et al.*, 2000). Moreover, it has also inhibited the growth of melanoma cells implanted in athymic mice (Pisha *et al.*, 1995). When betulinic acid was used in combination with anticancer drug vincristine, betulinic acid has shown synergistic cytotoxic effect on the growth and metastases of melanoma cells (B16F10) *in vivo*. These two drugs induced cell cycle arrest at different points and caused apoptosis of melanoma cells (Sawada *et al.*, 2004). The selective toxicity towards malignant melanoma as compared to normal melanocytes was underscored by a study reporting a lack of toxicity towards melanocytes in an *in vivo* model (Fulda *et al.*, 1997; Pisha *et al.*, 1995). Moreover, betulinic acid does not show any toxicity on normal human fibroblast *in vitro* (Fulda

Betulin

Jone's oxidation
(CrO_3/H_2SO_4/acetone/0°C)

Betulonic acid

$NaBH_4$/THF

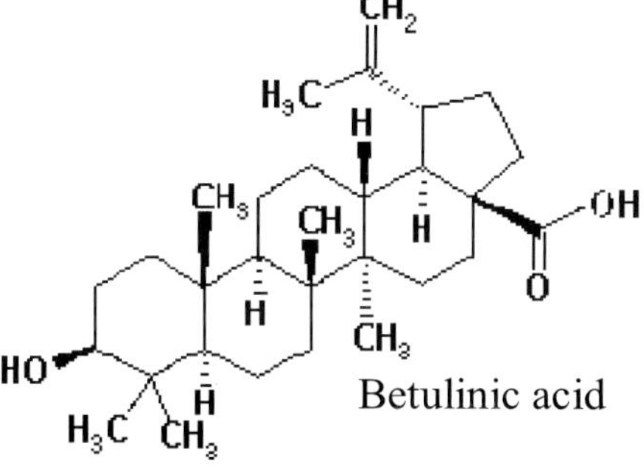

Figure 3.1: A schematic conversion of betulin to betulinic acid (Kim *et al.,* 1998).

and Debatin, 2005) and mice *in vivo* up to 500 mg/kg (Pisha *et al.*, 1995). The antitumor activity carried by this compound is mediated by the induction of apoptosis, as demonstrated by a variety of cellular responses (Pisha *et al.*, 1995).

Betulinic acid and its derivatives showed cytotoxicity against variety of tumor and cancer cell lines comparable to some clinically used drug. Jeremias *et al.* (2004) found that *in vitro* betulinic acid induced specific cell death of more than 75 per cent in primary glioblastoma multiforme cells at substantially higher rate than established cytotoxic drugs vincristine. Betulinic acid triggers an apoptotic cascade in all five human glioma cell lines as well. The concentration values of a compound where 50 per cent of its maximal effect is observed (EC_{50}) were in the range of 20 mM for LN-229, 25 mM for U87 MG and T98G, 70 mM for LN-18, and 100 mM for LN-308. BA-induced apoptosis requires new protein, but not RNA synthesis, indicating that translation of preexisting mRNA mediates BA cytotoxicity. BA-induced apoptosis is also found to be independent of *p53*, and results in *p21* protein accumulation in the absence of a cell cycle arrest. However, BA-induced apoptosis involves the activation of caspases that cleave poly-(ADP ribose) polymerase. Interactions of death ligand/receptor pairs of the CD95/CD95 ligand family did not mediate BA-induced caspase activation (Wick *et al.*, 1999). In another study, 9 human neuroblastoma cell lines, SKNSH, SHSY5Y, IMR-5, LAN-5, NBL-S, NBAS5, NBAS6, NBL-W and NBL-WR were treated *in vitro* with betulinic acid at the concentration of 0 to 20 µg/mL for 0 to 6 days. At the highest concentration tested (20 µg/mL), all neuroblastoma cells were killed within 72 hours. Morphological and biochemical analysis noted that betulinic acid induced apoptosis in these neuroblastoma cells. DNA fragmentation analysis showed ladder formation in the 100 to 1200 bp region in neuroblastoma cell lines treated with BA for 24 to 72 hours (Schmidt *et al.*, 1997). Furthermore, betulinic acid was able to enhance the radiotherapy effect on human malignant glioma cells especially under the hypoxic situation (Bache *et al.*, 2011).

When betulinic acid was compared for *in vitro* efficiency against several leukemia cell lines with conventionally used cytotoxic drugs, betulinic acid was more active than 9 out of standard therapeutic such as L-asparaginase, dexamethasone, prednisolone and 6-thioguanine (Ehrhardt *et al.*, 2004). It was also found to induce cell death on HL-60, CEM, Wehi-3B (Alitheen *et al.*, 2010) and Jurkat cells (Chen *et al.*, 2009) through cell cycle arrest (Chen *et al.*, 2009) and apoptosis (Faujan *et al.*, 2010).

Other than melanoma, glioblastoma and leukemia, betulinic acid has shown low endothelial cell specificity (ECS) against human tumor cell lines DU145 (prostate), LI32 (lung), A549 (lung) and PA-I (ovary). It posses both anti-tumor and anti-angiogenic potential which could be envisioned to act effectively with acceptable toxicity profiles (Mukherjee *et al.*, 2004). Moreover, it was also able to induce apoptosis and inhibit hedgehog signaling on soft tissue rhabdomyosarcoma (Eichenmuller *et al.*, 2010) and hepatoblastoma (Eichenmuller *et al.*, 2009). Betulinic acid induced apoptosis is not exerted through a ligand/receptor system but is mediated by a decrease in mitochondrial permeability, the release of mitochondrial cytochrome c into the cytosol (Fulda and Debatin, 2005), the formation of reactive oxidative species and the activation of crm-A-insentive caspase activity (Wick *et al.*, 1999). Betulinic acid also enhances the levels of *bax* and *bcl-2* proteins. Ectopic expression of *bcl-2* prevents betulinic acid

-induced caspase activation, DNA fragmentation, and cell death. The generation of reactive oxygen species is blocked by *bcl-2* and requires new protein synthesis but is unaffected by caspase inhibitors, suggesting that betulinic acid toxicity sequentially involves new protein synthesis, formation of reactive oxygen species, and activation of caspases (Wick, *et al.*, 1999). Recently, betulinic acid was found to inhibit *in vitro* and *in vivo* growth of estrogen receptor negative MDA-MB-231 cell line via downregulation of the expression of specificity protein (Sp) transcriptions factors Sp 1, Sp3 and Sp4 which are overexpressed in tumors. The BA-induced Sp1, Sp3 and Sp4 downregulation had promoted the activation of zinc finger ZBTB10 expression which is a putative Sp-repressor and the inhibition of microRNA-27a (regulator of ZBTB10) (Mertens-Talcott *et al.*, 2012).

Poor solubility of betulinic acid was also resolved by liposomes delivery system that was able to inhibit proliferation of human lung and colon cancer *in vivo* (Mullauer *et al.*, 2011). Amino asid conjugate of betulinic acid are of interest as inhibitors of human melanoma (Jeong *et al.*, 1999). Added to this, transformation of betulinic acid to mono- and di-substituted esters of 1-amino acids can also increase the water solubility of betulinic acid associated with earlier cytotoxic effect (Drag-Zalesinska *et al.*, 2009).

Betulinic Acid Derivatives and its Cytotoxicity

Betulinic acid has three positions (C-3, C-20 and C-28) where chemical modifications can be performed to yield derivatives for structure-activity relationship (SAR) study (Kim *et al.*, 2000). Simple modifications of the parent structure of betulinic acid at the position C-3 and C-28 can produce potentially important derivatives, which may be developed as anti-tumor drugs (Kim *et al.*, 1997). The toxicity of betulinic acid derivatives at C-3, C-20 and C-28 positions has been evaluated against cultured human melanoma (MEL-2) and human epidermoid carcinoma of the mouth (KB) cell lines (Kim *et al.*, 1998).

Chemical modifications were performed on C-20 position of betulinic acid for a structure–activity relationship study. The evaluation of the compounds using human colon carcinoma HCT-116, human prostate adenocarcinoma PC3, and human melanoma cell lines M14-MEL, SK-MEL-2, and UACC-257 did not show any selective cytotoxicity towards melanoma cells. The results from both MTT reduction assay and SRB staining assay were comparable that no remarkable differences in cytotoxicity profile of the compounds were noticed. The C-20 position was found to be sensitive to the size and the electron density of the substituents in retaining the cytotoxicity of betulinic acid and was found to be an undesirable position to derivative (Kim *et al.*, 2004).

The compound 23-hydroxybetulinic acid derived from the roots of *Pulsatilla chinensis* (Bunge) Regel, is a newly isolated derivative of betulinic acid. 23-hydroxybetulinic acid acts as an effective cytotoxic agent against HL-60 cells by inducing growth arrest and apoptosis in concentration- and duration-dependent manners. The concurrent inhibition of *bcl-2* expression and telomerase activity was observed 3 hours after 23-hydroxybetulinic acid treatment, while the apoptotic cells only became detectable 6 hours after drug treatment, suggesting that the downregulation of *bcl-2* and telomerase may play a causative role in 23- hydroxybetulinic acid-induced apoptosis (Ji *et al.*, 2002).

Betulinic acid and dihydrobetulinic acid were treated with 3,3-dimethylglutaric, diglycolic, glutaric or succinic anhydrides in pyridine in the presence of 4-dimethylaminopyridine to furnish the corresponding 3-O-acyl derivatives. In contrast, similar treatment of betulinic acid and dihydrobetulinic acid with 2,2-dimethylsuccinic anhydride afforded a mixture of 3-O-(2′,2'-dimethylsuccinyl)- and 3-0-(3',3'-dimethylsuccinyl)-betulinic acid and -dihydrobetulinic acid, respectively. 3-O-(3',3'-dimethylsuccinyl)-betulinic acid and 3-O-(3',3'-dimethyl-succinyl)-dihydrobetulinic acid were reported to show extremely potent inhibitory activity with EC_{50} values of <3.5 × 10^{-4} μM, and remarkable *in vitro* therapeutic index (TI) values of 20,000 and 14,000, respectively (Hashimoto *et al.*, 1997).

3-O-(3',3'-dimethylglutaryl)-betulinic acid, 3-O-(3',3'-dimethylglutaryl)-dihydrobetulinic acid, 3-O-diglycolyl-betulinic acid, 3-O-diglycolyl-dihydrobetulinic acid and 3-O-glutaryl-betulinic acid were also potent inhibitors of HIV replication with EC_{50} values ranging from 0.04 to 2.3 × 10^{-3} μM and TI values from 292 to 2344. In addition, compounds 3-O- (3',3'-dimethylsuccinyl)-dihydrobetulinic acid and 3-O-(3',3'-dimethylglutaryl)-dihydrobetulinic acid were also active against HIV replication in a monocyte cell line and in peripheral blood mononuclear cells. However, 3-O-(2',2'-dimethylsuccinyl)-betulinic acid was also found to be an inhibitor of HIV-induced membrane fusion with an IC_{100} value of 20 μg/ml, though it displayed significantly lower anti-HIV activity than foregoing compounds with an EC_{50} value of 2.7 μM and TI of 6.7 (Hashimoto *et al.*, 1997).

The natural triterpene betulinic acid and its analogues (betulinic aldehyde, lupeol, betulin, methyl betulinate and betulinic acid amide) are potent membrane-active agents exerting their stomatocytogenic or echinocytogenic action in concentration region, where the *in vitro* antiplasmodial effect is observed. Thus, the analogues with a functional group ac C-28 having a capacity of donating a hydrogen bond (COOH, CH_2OH, $CONH_2$) caused formation of echinocytes, whereas those lacking this ability (CH_3, CHO, $COOCH_3$) induced formation of stomatocytes. Both kinds of erythrocyte alterations were prohibitive with respect to *Plasmodium falciparum* invasion and growth. All compounds were inhibitory with IC_{50} values in the range 7–28 μM, and the growth inhibition correlated well with the extent of membrane curvature changes assessed by transmission electron microscopy (Ziegler *et al.*, 2004).

Betulinic acid, isolated from the leaves of *Bacopa monniera* was subjected to hydrogenation to create dihydrobetulinic acid, which is a novel lead compound for antileishmanial therapy. It acts by targeting DNA topoisomerases which have been identified as promising target for anti-parasitic therapeutics. In this study, pulsed field gradient gel electrophoresis, confocal microscopy, and transmission electron microscopy were performed to assess cytotoxicity of the compound and ultrastructural damage of the parasite. Apoptosis was studied by the isolation of DNA from dihydrobetulinic acid -treated parasites and subsequent electrophoresis in 1 per cent agarose gel. Dihydrobetulinic acid inhibits the growth of *Leishmania donovani* promastigotes and amastigotes with an IC_{50} of 2.6 and 4.1 mμ, respectively. Treatment of infected golden hamsters with the compound markedly reduces (> 92 per cent) parasitic burden, both in spleen and liver (Chowdhury *et al.*, 2003).

Betulinic acid acetate appears as a pale yellow solid crystal. It also can be prepared by enzymatic reaction of betulinic acid and acetic anhydride in the present of lipase (Ahmad *et al.*, 2005). Structure modification at C-3 hydroxy group of betulinic acid produced betulinic acid acetate which is further studied for the cytotoxic activities. The relationships between the structure and activities of betulinic acid and betulinic acid acetate on the induction of HL-60 cell differentiation and apoptosis were studied. The activity of these triterpenes on the induction of HL-60 cell differentiation and apoptosis demonstrated that the hydroxy function at C-3 enhanced their differentiation inducing activities and the carbonyl group at C-17 may played an important role on determining the strength of their toxicities (Faujan *et al.*, 2010).

Dihydrobetulonic acid, 2-Bromo dihydrobetulonic acid, 2,2-Dibromo dihydrobetulonic acid, 2-Hydroxymethylene-betulonic acid derived from betulin and betulinic acid were found to be more sensitive against CEM cell line (IC_{50} = 4–1 lmol/ L) as compared to betulinic acid (Urban *et al.*, 2012).

28-(1*H*-imidazol-1-yl)-lup-20(29)-en-3,28-dione (RS01), 28-(1*H*-imidazol-1-yl)-lup-1,20(29)-dien-3,28-dione (RS02) and 28-hydroxy-lup-20(29)-en-3β-yl-1*H*-imidazole-1-carboxylate (RS03) synthesized from betulin and betulinic acid were able to inhibit the proliferation of HepG2 (hepatocellular carcinoma), HeLa (cervical adenocarcinoma) and Jurkat (leukemia) cells with IC_{50} values ranging from 0.8 to 11.1 μM which was 18-45 times greater than betulinic acid. Introduction of imidazolyl moiety was suggested as the major contributor that arrested the cell at S phase and further mediated activation of caspasese 9 and 3 that contributed to the DNA fragmentation (Santos *et al.*, 2011). http: //www.sciencedirect.com/science/article/ pii/S0300908411000708–bib34#bib34

In summary, betulinic acid and its derivatives showed selective cytotoxic towards various cell line and not toxic to normal cell. This review demonstrated that those compounds have the potential to be exploited as suitable anti-cancer drug since they are non-toxic toward normal cells. Nevertheless, a better understanding of the cytotoxic mechanism of betulinic acid and its derivatives is of great importance in view of their potential clinical application.

References

Ahmad, F.B.H., Anisah Issak, M.B., Nur Hana, F, Yamin, Y., and Ali, A. M. (2005). Synthesis of 3β-acetoxy-lup-20(29)-ene-28-decanoate using lipase as the Catalyse. *J. Chem. Environ. Res.*, **14:** 207-213.

Alitheen, N.B., Mashitoh, A.R., Yeap, S.K., Shuhaimi, M., Abdul Manaf, A. and Nordin, L. (2010). Cytotoxic effect of damnacanthal, nordamnacanthal, zerumbone and betulinic acid isolated from Malaysian plant sources. *International Food Research Journal*, **17:** 711-719.

Bache, M., Zschornak, M.P., Passin, S., Kessler, J., Wichmann, H., Kappler, M., Paschke, R., Kaluderovic, G.N., Kommera, H., Taubert, H., and Vordermark, D. (2011).

Increased betulinic acid induced cytotoxicity and radiosensitivity in glioma cells under hypoxic conditions. *Radiation Oncology,* **6:** 111.

Chatterjee, P., Kouzi, S.A., Pezzuto, J.M., and Hamann, M.T. (2000). Biotransformation of the antimelanoma agent betulinic acid by *Bacillus megaterium* ATCC 13368. *Applied and Environment Microbiology,* **66:** 3850-3855.

Chen, Z., Wu, Q.L., Chen, Y., and He, J. (2008). Effect of betulinic acid on proliferation and apoptosis in Jurkat cells and its mechanism. *Chinese Journal of Oncology,* **30:** 588-592.

Chowdhury, A.R., Mandal, S., Goswami, A., Ghosh, M., Mandal, L., Chakraborty, D., Ganguly,A., Tripathi,G., Mukhopadhyay, S., Bandyopadhyay, S., and Majumderi, H. (2003). Dihydrobetulinic acid induces apoptosis in *Leishmania donovani* by targeting DNA topoisomerase I and II: Implications in antileishmanial therapy. *Molecular Medicine,* **9:** 26-36.

Cichewicz, R.H., and Kouzi, S.A. (2004). Chemistry, biological activity, and chemotherapeutic potential of betulinic acid for the prevent and treatment of cancer and HIV infection. *Medicinal Research Reviews,* **24:** 90-114.

Drag-Zalesinska, M., Kulbacka, J., Saczko, J., Wysocka, T., Zabel, M., Surowiak, P., and Drag, M. (2009). Esters of betulin and betulinic acid with amino acids have improved water solubility and are selectively cytotoxic toward cancer cells. *Bioorganic and Medicinal Chemistry Letters,* **19:** 4814-4817.

Ehrhardt, H., Fulda, S., Fuhrer, M., Debatin, K., and Jeremias, I. (2004). Betulinic acid induced apoptosis in leukemia cells. *Leukemia,* 18: 1406-1412.

Eichenmuller, M., Hemmerlein, B., Von Schweinitz, D., and Kappler, R. (2010). Betulinic acid induces apoptosis and inhibits hedgehog signialling in rhabdomyosarcoma. *British Journal of Cancer,* **103:** 43-51.

Eichenmuller, M., Von Schweinitz, D., and Kappler, R. (2009). Betulinic acid treatment promotes apoptosis in hepatoblastoma cells. *International Journal of Oncology,* **35:** 873-879.

Faujan, N.H., Alitheen, N.B., Yeap, S.K., Ali, A.M., Muhajir, A.H., and Ahmad, F.B.H. (2010). Cytotoxic effect of betulinic acid and betulinic acid acetate isolated from *Melaleuca cajuput* on human myeloid leukemia (HL60) cell line. *African Journal of Biotechnology,* **9:** 6387-6396.

Fujioka, T., Kashiwada, Y., Kilkuskie, R.E., Cosentino, L.M., Ballas, L.M, Jiang, J.B., Janzen, W.P., Janzen, W.P., Chen, I-S., and Lee, K-H. (1994). Anti-Aids agents in betulinic acid and platonic acid as anti-HIV principles from *Syzigh claviflorum* and the anti-HIV activity of structurally related triterpenoids. *Journal of Natural Products,* **52:** 243-247.

Fulda, S., and Debatin, K.M. (2005). Sensitization for anticancer drug-induced apoptosis by betulinic acid. *Neoplasia,* 7: 162-170.

Fulda, S., Friesen, C., Los, M., Scaffidi, C., Mier, W., Benedict, M., Nunez, G., Krammer, P. H., Peter, M. E., and Debatin, K.-M. (1997). Betulinic acid triggers CD95 (APO-

1/Fas)-and p53-independent apoptosis via activation of caspases in neuroectodermal tumors. *Cancer Research,* **57:** 4956-4964.

Fulda, S., Susin, S.A., Kroemer, G., and Debatin, K.M. (1998a). Molecular ordering of apoptosis induced by anticancer drugs in neuroblastoma cells. *Cancer Research,* **58:** 4453-4460.

Fulda, S., Scaffidi, C., Susin, S.A., Krammer, P.H., Kroemer, G., Peter, M.E., and Debatin, K.M. (1998b). Activation of mitochondria and release of mitochondrial apoptogenic factors by betulinic acid. *The journal of biological chemistry,* **273:** 33942–33948.

Hashimoto, F., Kashiwada, Y., Cosentino, L.M., Chen, C.H., Garrett, P.E., and Lee, K.H. (1997). Anti-AIDS agents—XXVII. Synthesis and anti-HIV activity of betulinic acid and dihydrobetulinic acid derivatives. *Bioorganic and Medicinal Chemistry,* **5:** 2133-2143.

Jeong, H.J., Chai, H.B., Park, S.Y., and Kim, D.S. (1999). Preparation of amino acid conjugates of betulinic acid with activity against human melanoma. *Bioorg. Med. Chem. Lett.,* **9:** 1201-1204.

Jeremias, I., Steiner, H.H., Benner, A., Debatin, K.M., and Henro, H.M. (2004). Cell death induction by betulinic acid, ceramide and TRAIL in primary glioblastoma multoforme cells. *Acta Neurochir.,* 146: 721-729.

Ji, Z.N., Ye, W.C., Liu, G.G., Hsiao, W.L. (2002). 23-hydroxybetulinic acid-mediated apoptosis is accompanied by decreases in *bcl-2* expression and telomerase activity in HL-60 cells. *Life Sciences,* **72:** 1-9.

Jung, M.E., and Duclos, B.A. (2006). Synthetic approach to analogues of betulinic acid. *Tetrahedron,* **62:** 9321-9334.

Kim Y.K., Yoon, S.K., and Ryu, S.Y. (2000). Cytotoxic triterpenes from stem bark of *Physocarpus intermedius. Planta Medica,* **66:** 485–486.

Kim, D.S., Chen, Z., Nguyen, v.T., Pezzuto, J.M., Qiu, S., and Lu, Z.Z. (1997). A concise semi-synthetic approach to betulinic acid from betulin. *Synthetic Communication,* **27:** 1607.

Kim, D.S. H. L., Pezzuto, J. M., and Pisha, E. (1998). Synthesis of betulinic acid derivatives with activity against human melanoma. *Bioorganic and Medicinal Chemistry Letters,* **8:** 1707-1712.

Kim, J.Y., Koo, H.M., and Kim, D.S. (2001). Development of C-20 modified betulinic acid derivatives as antitumor agents. *Bioorganic and Medicinal Chemistry Letters,* **11:** 2405-2408.

Metens-Talcott, S.U., Noratto, G.D., Li, X., Angel-Morales, G., Bertoldi, M.C., and Safe, S. 2012. Betulinic acid decreases ER-negative breast cancer cell growth *in vitro* and *in vivo*: Role of Sp transcription factors and microRNA-27a: ZBTB10. *Molecular Carcinogenesis,* doi: 10.1002/mc.21893.

Mukherjee, R., Jaggi, M., Rajendran, P., Siddiqui, M.J.A., Srivastava, S.K., Vardhan, A., and Burmana, A.B. (2004). Betulinic acid and its derivatives as anti-angiogenic agents. *Bioorganic and Medicinal Chemistry Letters,* **14:** 2181-2184.

Mullauer, F.B., Van Bloois, L., Daalhuisen, J.B., Ten Brink, M.S., Storm, G., Medema, J.P., Schiffelers, R.M., and Kessler, J.H. (2011). Betulinic acid delivered in liposomes reduces growth of human lung and colon cancers in mice without causing systemic toxicity. *Anti-Cancer Drugs,* **22:** 223-233.

Pisha. E., Chai, H., Lee, I.S., Chagwedera, T.E., Farnsworth, N.R., Cordell, G.A., Beecher, C.W., Fong, H.H., Kinghorn, A.D., and Brown, D.M. (1995). Discovery of betulinic acid as a selective inhibitor of melanoma that functions by induction of apoptosis. *Nature Medicine,* **10:** 1046-1050.

Sami, A., Taru, M., Salme, K., and Jari, Y.K. (2006). Pharmacological properties of the ubiquitous natural product betulin. *European Journal of Pharmaceutical Science,* **29**: 1-13.

Santos, R.C., Salvador, J.A., Cortes, R., Pachon, G., Marin, S., and Cascante, M. (2011). New betulinic acid derivatives induce potent and selective antiproliferative activity through cell cycle arrest at the S phase and caspase dependent apoptosis in human cancer cells. *Biochimie,* **93:** 1065-1075.

Sawada, N., Katoaka, K., Kondo, K., Arimochi, H., Fujino, H., Monden, Y., and Ohnishi, Y. (2004). Betulinic acid augments the inhibitory effects of vincristine on growth and lung metastasis of B16F10 melanoma cells in mice. *British Journal of Cancer,* **90**: 1672-1678.

Schmidt, M.L., Kuzmanoff, K.L., Ling-Indeck, L., and Pezzuto, J.M. (1997). Betulinic acid induces apoptosis in human neuroblastoma cell lines. *European Journal of Cancer,* **33:** 2007-2010.

Soica, C.M., Dehelean, C.A., Peev, C., Aluas, M., Zupko, I., Kasa, P.Jr., and Alexa, E. (2012). Physico-chemical comparison of betulinic acid, betulin and birch bark extract and *in vitro* investigation of their cytotoxic effects towards skin epidermoid carcinoma (A431), breast carcinoma (MCF-7) and cervix adenocarcinoma (HeLa) cell lines. *Natural Product Research,* **26:** 968-974.

Surowiak, P., Drag, M., Materna, V., Dietel, M., and Lage, H. (2009). Betulinic acid exhibites stronger cytotoxic activity on the normal melanocyte NHEM-neo cell line than on drug-resistant and drug-sensitive MeWo melanoma cell lines. *Molecular Medicine Reports,* **2:** 543-548.

Urban, M., Vlk, M., Dzubak, P., Hajduch, M., and Sarek, J. (2012). Cytotoxic heterocyclic triterpenoids derived from betulin and betulinic acid. *Bioorganic and Medicinal Chemistry,* **20:** 3666-3674.

Wick, W., Grimmel, C., Wagenkknecht, B., Dichgans, J., and Weller, M. (1999). Betulinic acid-induced apoptosis in glioma cells: A sequential requirement for new protein synthesis, formation of reactive oxygen species, and caspase processing. *The Journal of Pharmacology and Experimental Therapeutics,* **289:** 1306-1312.

Ziegler, H. L., Franzyk, H., Sairafianpour, M., Tabatabai, M., Tehrani, M. D., Bagherzadeh, K., Hagerstrand, H., Stærka, D., and Jaroszewskia, J. W. (2004). Erythrocyte membrane modifying agents and the inhibition of *Plasmodium falciparum* growth: structure–activity relationships for betulinic acid analogues. *Bioorganic and Medicinal Chemistry,* **12:** 119–127.

Bioactive Phytochemicals: Perspectives for Modern Medicine Vol. 2 (2014) *Pages* **93–119**
Editor: **V.K. Gupta**
Published by: **DAYA PUBLISHING HOUSE, NEW DELHI**

4

Biological Activities and Phytochemicals of the Fruits of *Vitex* Plants

J.F. Pío-León[1], J. Montes-Avila[1], S.P. Díaz-Camacho[1] and F. Delgado-Vargas[1*]

ABSTRACT

The Vitex genus (Lamiaceae) is composed of 270 species, trees or shrubs; that are distributed in tropics worldwide. Vitex has been the source of a large number of compounds with pharmacological activities, many of them associated with their fruits. In spite of the high number of Vitex spp.*, fruit studies have been mainly focused in a reduced group; Vitex agnus-castus is the best studied, and several fruit-derived drugs, used to treat premenstrual syndrome, are commercialized in the United States of America and Europe. Other species with a good number of studies are V. rotundifolia, V. trifolia, V. negundo and V. doniana. Interesting activities associated with Vitex fruits are antitumoral (e.g., against breast, gastric, colon, leukemia and cervix cells), antioxidant, antimicrobial and anti-diarrhoea. Identified fruit compounds belong to different phytochemical families such as flavonoids (e.g., casticin, orientin, and isoorientin), terpenes (e.g., rotundifaran, ferruginol), iridoids (e.g., agnoside, aucubin) and essential oils (e.g., 1,8-cineolene, sabinene). Interestingly, these fruits showed dark colours that have been poorly studied; these pigments could be associated with important biological activities based on the actual knowledge about natural pigments. In this review, we are showing the main studies about the chemicals and biological activities of Vitex* spp. *fruits, and it will be a reference material for future studies.*

***Keywords*:** *Vitex* fruits, Biological activities, Chemical composition, Nutritional and nutraceutical properties, Essential oils, Chocolate berries, Gynecologic disorders, Anti-diarrheal.

1 Facultad de Ciencias Químico Biológicas de la Universidad Autónoma de Sinaloa. Ciudad Universitaria s/n, Culiacán, Sinaloa, México, CP 80010

* *Corresponding author*: E-mail: fdelgado@uas.edu.mx

Introduction

Linnaeus described the genus *Vitex* (Lamiaceae before Verbenaceae) in 1753 and four species were registered (*V. agnus-castus, V. negundo, V. pinnata* and *V. trifolia*); nowadays, this genus includes between 250 and 300 species of trees and bushes (rarely lianas). These plants are mainly distributed in tropical or sub-tropical areas over the world (Chantaranothai, 2011; de Kok, 2008; Munir, 1987). *Vitex agnus-castus* shows a wide distribution in Mediterranean Europe and Central Asia and it is the best known and studied plant of this genus (Padmalatha *et al.,* 2009).

The leaves of *Vitex* spp. are opposite, decussate with three to five folioles, rarely two or one as in *V. rotundifolia;* flowers are zygomorphic, calyx with five fused sepals, persistent, tubular-like or bell-shaped like, sometimes dentate; corolla with five fused petals, deciduous of tubular-like form with internal villous and sometimes pubescent (Munir, 1987). Historically, *Vitex* spp. have been widely used in traditional medicine; and up to date, they have a large number of ethnopharmacological applications such as: treatment of premenstrual and gynecologic affections, bacterial infections, gastrointestinal problems and inflammation; as well as insect repellent and against stings of venomous animals (Padmalatha *et al.,* 2009).

A large number of the medicinal properties of *Vitex* spp. are associated with their fruits; some of them are important as foods in highly marginated and low income areas. *Vitex* spp. fruits are small succulent drupes, globular or ovoid of 0.2-2 cm size (usually smaller than 2 cm); they possess a hardened endocarp that is divided in four pyrenes, each one with a seed (Chantaranothai, 2011; de Kok, 2008; Munir, 1987). Fruits of different *Vitex* spp. have similar morphology and sometimes they are difficult to differentiate. The ripe pulp of these fruits is characterized by dark-purple to brown colours (Figure 4.1) (Chantaranothai, 2011; Munir, 1987; NRC, 2008; Webster, 2008). In this review, we are showing the main characteristics of the *Vitex* spp. fruits, emphasizing their phytochemicals (more than 100 identified compounds) and biological activities. The current knowledge of these fruits is discussed as well as the future research trends.

Nutritional Characteristics of *Vitex* Fruits

Fruits are important foods in human diet; they mainly provide vitamins, minerals and fiber. Recently, functional and nutraceutical foods, which provide benefits for the prevention/treatment of diseases in addition to their nutritional components, have acquired an increased relevance; within these food groups, fruits are highly represented (*e.g.,* grapes, cranberries, blueberries, pomegranate) and their consumption have been associated with a healthy condition and longevity (Kaliora and Dedoussis, 2007; Kalra, 2003; Kaur and Kapoor, 2001; Rissanen *et al.,* 2003).

Many of the *Vitex* fruits are edible; about 70 species have been registered just in Africa, where they are known as "chocolate berries". These wild berries have a significant contribution to food security and sustainable rural development for some African communities and many of them are valuable commercial products in those regions (NRC, 2008). The most representative chocolate berries are *V. doniana, V. payos, V. fischeri, V. grandifolia, V. simplicifolia, V. madiensis, V. mombassae,*

Figure 4.1: Fruits of different *Vitex* spp.: *V. mollis* **(A)**, *V. agnus-castus* **(B)**, *V. rotundifolia* **(C)**, *V. vestita* (**D**), *V. cymosa* (**E**), *V. schliebenii* (**F**), *V. polygama* (**G**), *V. payos* (**H**), *V. doniana* (**I**), *V. fischeri* (**J**) and *V. cochinchinensis* (**K**).

Sources of the images: **(A)** Authors of this work, **(B)** http: //www.biodiversidadvirtual.org/herbarium/Vitex-agnus-castus-L-img57586.html, **(C)** http: //eol.org/data_objects/1990800, **(D)** Chantaranothai (2011) **(E)** http: //eol.org/data_objects/13508823, **(F)** http: //eol.org/data_objects/19246219, **(G)** http: //www.bananasraras.org/frutasrarasingles/vitex2.htm, **(H)** Mandu (1999), **(I)** http: //www.prota4u.org/searchresults.asp, **(J)** Emery-Thompson *et al.* (2008), **(K)** Chantaranothai (2011).

V. ferruginea and *V. pooara* (Maundu, 1999; NRC, 2008). *Vitex* fruits are consumed fresh, dried or in boiled preparations with sugar (*e.g.* jelly, marmalade) (NRC, 2008). Research groups have developed new products based on the properties of chocolate berries, for example, the *V. doniana* pulp was used to prepare a syrup with sensorial characteristics and acceptability similar to those of honey (Egbekun *et al.*, 1996); a pasteurized juice of *V. mombassae* showed good acceptability and its physicochemical characteristics were preserved up to 3 months of storage (Ndabikunze *et al.*, 2010). In Mexico, *V. mollis* is used for sweet preparations that can be consumed alone or mixed with milk; however, the properties of such products have not been established (Montiel-Herrera *et al.*, 2004).

Table 4.1: Physico-chemical and nutritional parameters of *Vitex* spp. fruits.

Parameter	*V. mollis*[1]	*V. doniana*[2]	*V. keniensis*[3]	*V. fischeri*[3]	*V. mombassae*[5]
Proximate (per cent f.w.)					
Moisture	87	39.42	40.56	37.74	
Protein	0.60	0.85	0.87	0.98	–
Lipids	0.38	2.44	2.35	2.66	–
Ash	0.66	3.41	3.40	3.66	–
Fiber	1.92	11.48	10.42	12.38	–
Carbohydrates	9.48	43.12	42.49	42.67	–
Minerals (mg/100 g d.w.)					
Ca	45	320	–	–	55
Mg	–	72	–	–	156
Na	300	100	–	–	43
K	1610	880	–	–	1757
P	–	200	–	–	309
Fe	4	–	–	–	2.69
Zn	4.4	–	–	–	1.53
Cu	0.38	–	–	–	0.27
Ascorbic acid (mg/100g f.w.)	5.8	81.6-100	–	–	40.4
pH	4.86	4.38	–	–	3.56
Acidity (per cent)	0.13	0.36	–	–	0.14
Soluble solids (°Brix)	14	5.2	–	–	12.3

–: Non reported.

Cited papers–1: Montiel-Herrera *et al.* (2004); 2: Egbekum *et al.* (1996); and 2,3: Ochieng and Nanwa (2010).

In spite of the use of chocolate berries as foods, their physico-chemical, phytochemical and nutritional/nutraceutical characteristics have been poorly studied (NRC, 2008). Nevertheless, it has been clearly demonstrated that *Vitex* fruits are good sources of fiber, vitamin C and some minerals; they showed an uncommon low content of water (40 per cent) (Table 4.1). After moisture and carbohydrates, the most important component of *Vitex* spp. is the fiber. *Vitex doniana* fruits are rich in calcium and vitamin C; *V. mombassae* in potassium; and *V. mollis* in copper and zinc (Table 4.1). Consumption of 100 g fresh weight (f.w.) of *V. doniana* provides 100 and 19.4 per cent of the recommended daily intakes for vitamin C and calcium (90 and 1000 mg, respectively) (FNB/FNIC, 2010).

Vitex fruits show characteristic odor and colour. Odor is influenced by the composition of fruit essential oils which will be discussed in section 7 of this review. With respect to colour, the dark hue of *Vitex* fruits has not been associated with specific components. The dark intense colourations of common fruits such as blackberry, chokeberry or elderberry, are produced by anthocyanins, soluble pigments of low to intermediate molecular weight which are characterized for colour changes induced by pH variation, property used to quantitate anthocyanins (Delgado-Vargas and Paredes-López, 2003; Wrolstad, 2002). Our research group is studying the dark pigment of *V. mollis*. This pigment is very stable to pH and thermal treatment; it presents high antioxidant activity and a relative high molecular weight (personal communication). Thus, *V. mollis* pigment is not an anthocyanin and we have hypothesized that melanin pigments are responsible of the dark colours of *Vitex* fruits. Melanins are amorphous polymers, of dark colours, low solubility in common solvents and high molecular weight; in plants, these pigments are produced by the oxidation-polymerization of phenolic compounds, as it has been demonstrated for the black tea (Sava *et al.*, 2001). On the other hand, some fungi have edible melanins which are associated with sugars and they are water soluble (Seniuk *et al.*, 2010). The chemical and biological characterization of the *Vitex* pigments is an unexplored and promising research area.

Characterized Phytochemicals of *Vitex* Fruits

Vitex fruits have been the source of many new compounds; considering the most studied fruits (*i.e. V. agnus-castus*, V. *rotundifolia*, *V. trifolia*, *V. negundo* and *V. cannabifolia*), 100 chemicals are presented in Table 4.1 and Figure 4.2. The main components of *Vitex* spp. are terpenoids, mainly diterpenes, and flavonoids. Casticin (also known as vitexicarpin) is the *Vitex* spp. compound with the highest number of demonstrated biological activities (*e.g.*, estrogenic, opioid, antitumoral, antioxidant, anti-inflammatory) (Lee *et al.*, 2011; Xu *et al.*, 2012; Ye *et al.*, 2010). The properties of other interesting *Vitex* compounds, such as agnoside and rotundifaran, are discussed in following sections.

Gynecologic Properties of *Vitex* Fruits

The main traditional uses of *Vitex* fruits have been for gynecologic and hormonal problems such as premenstrual syndrome (PMS), post-menopause disorders, and *libido* modulation. As a matter of fact, *V. agnus-castus* has been used since before Christ by Greek and Roman cultures; fruit preparations have been applied as anti-inflammatory; to diminish the *libido*, the dysmenorrhoea and menopause symptoms, PMS and acne; to treat infertility and to stop lactation (AMR, 2009; Halaska *et al.*, 1999; Webster, 2008). Nowadays, many of the traditional uses of *V. agnus-castus* have

Figure 4.2: Chemical structures of compounds obtained from *Vitex* spp. fruits.

Terpenoids

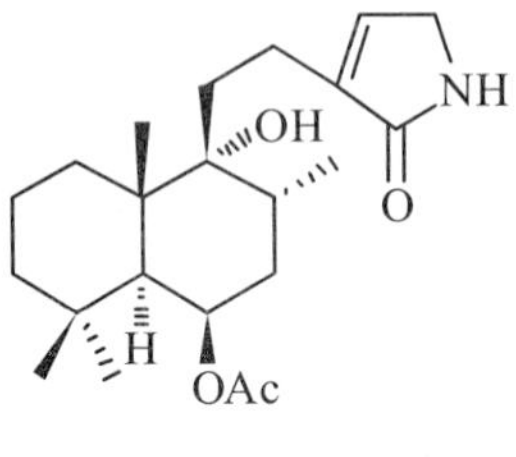

(1)

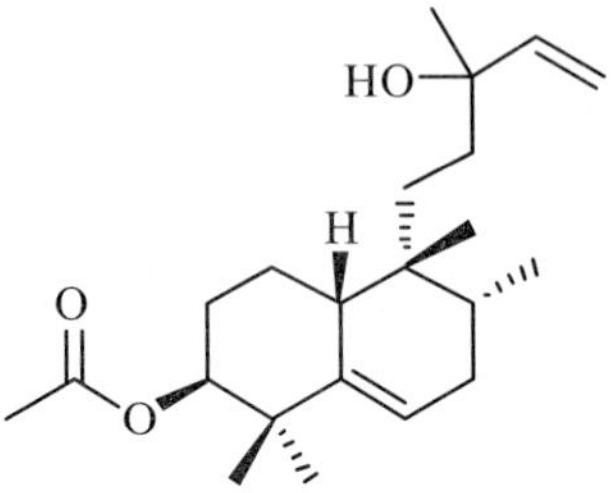

(2)

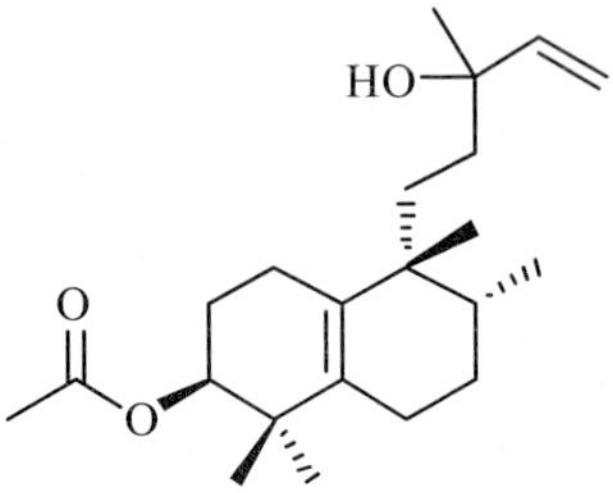

(3)

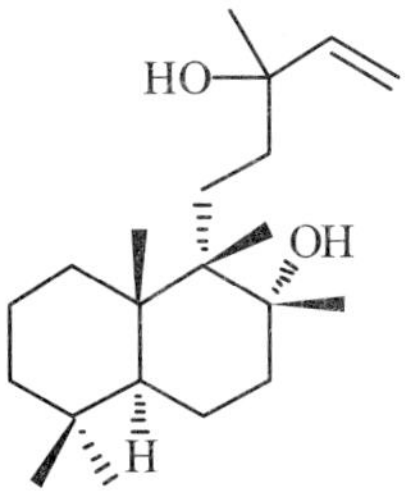

(4)

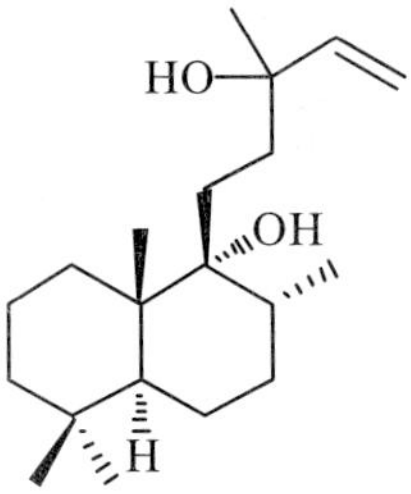

(5)

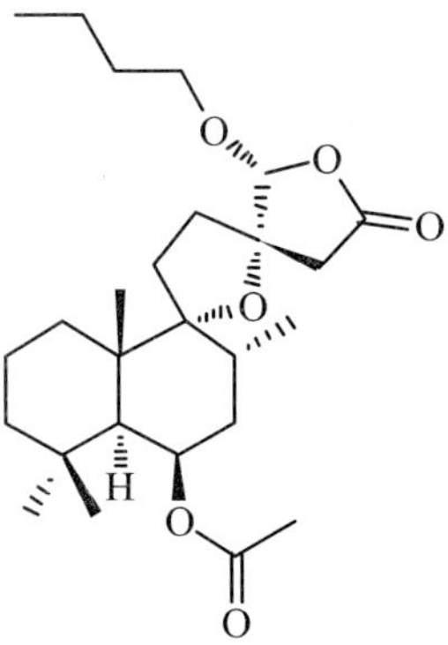

(6)

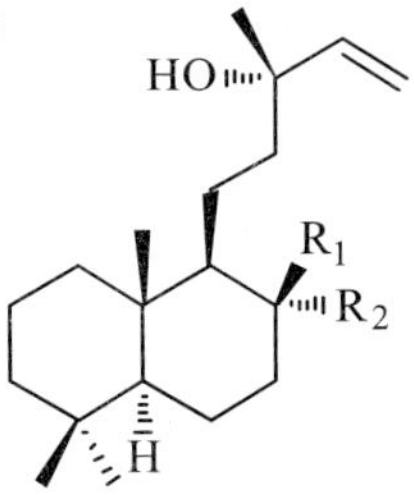

(7); R_1 = OH, R_2 = CH_3

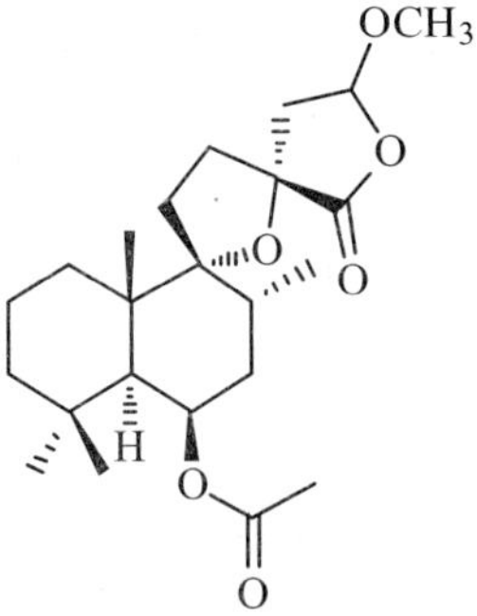

(8)

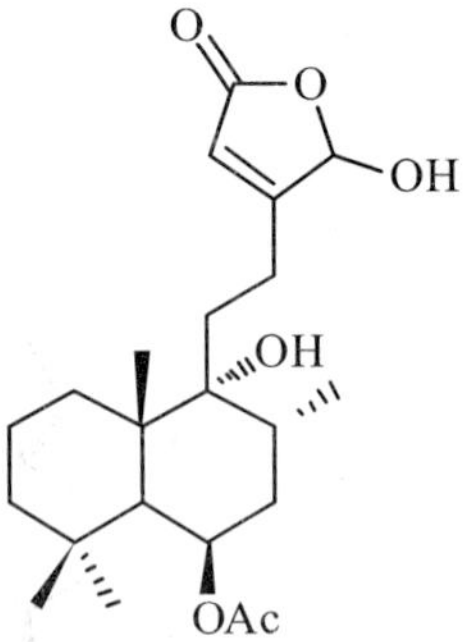

(9)

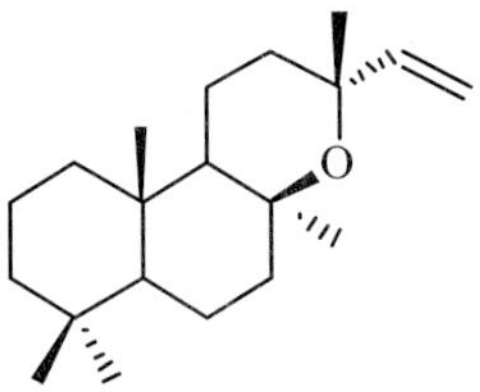

(10)

Contd...

Figure 4.2–Terpenoids–*Contd...*

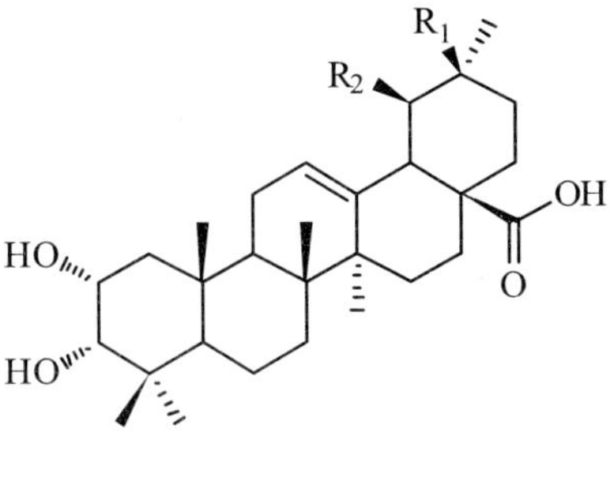

(**11**) R_1 = Me, R_2 = H
(**12**) R_1 = H, R_2 = Me

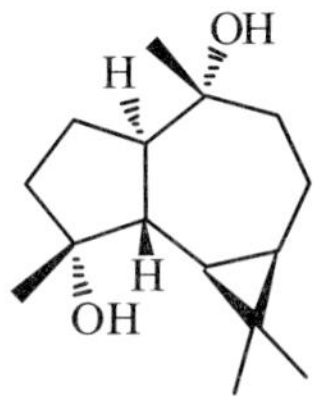

(**13**)

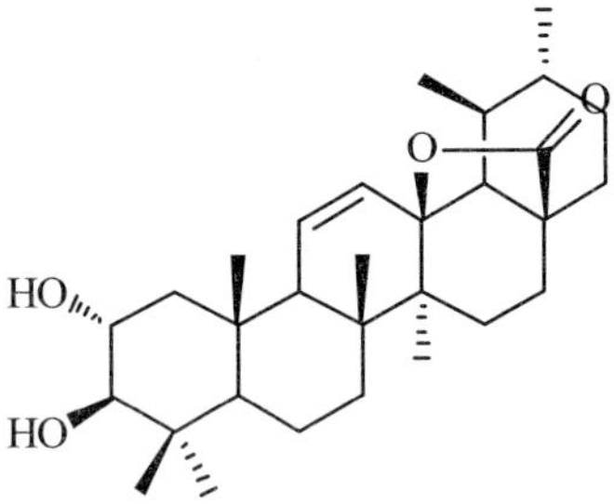

(**14**)

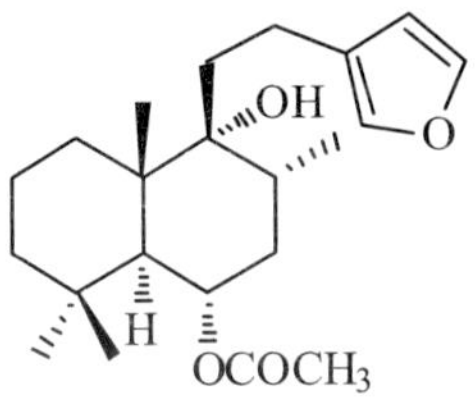

(**15**)

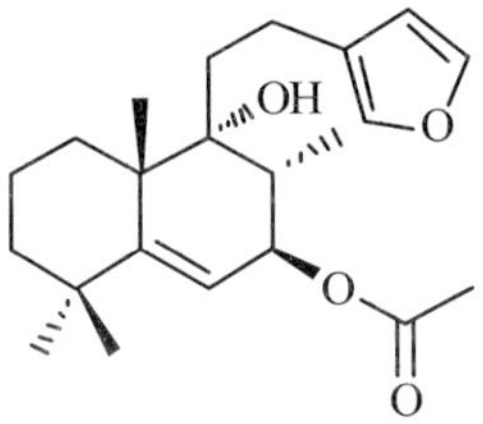

(**16**)

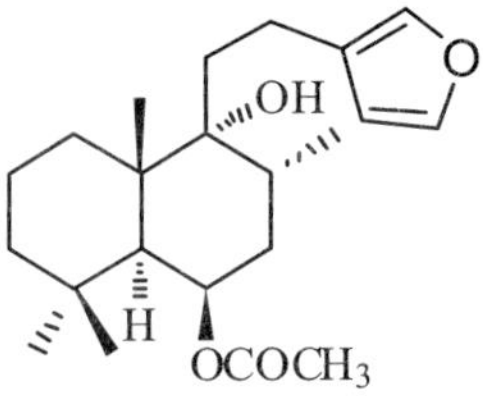

(**17**)

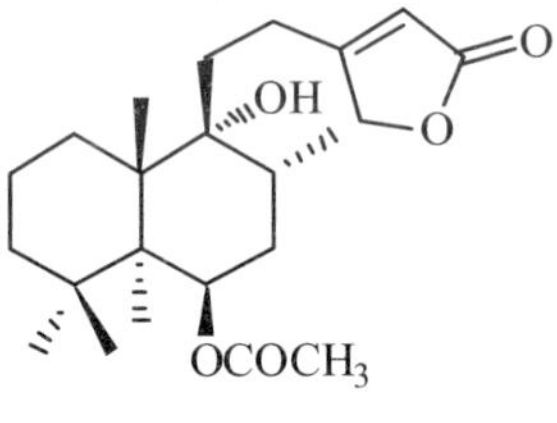

(**18**)

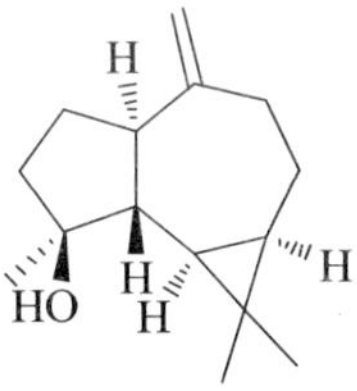

(**19**)

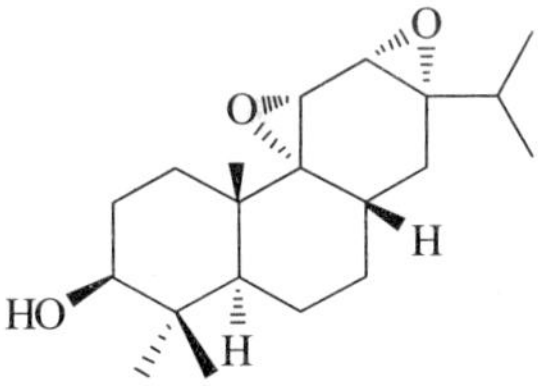

(**44**)

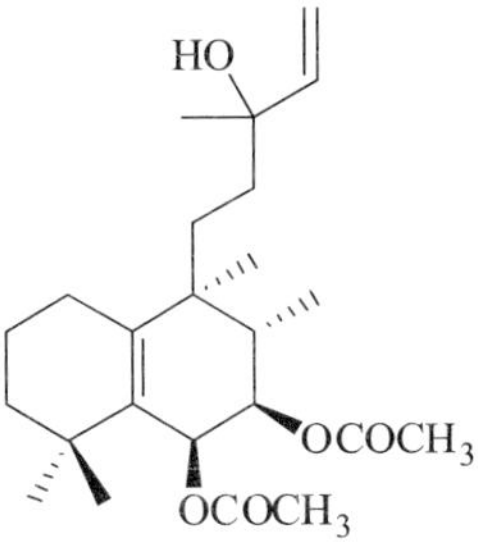

(**45**)

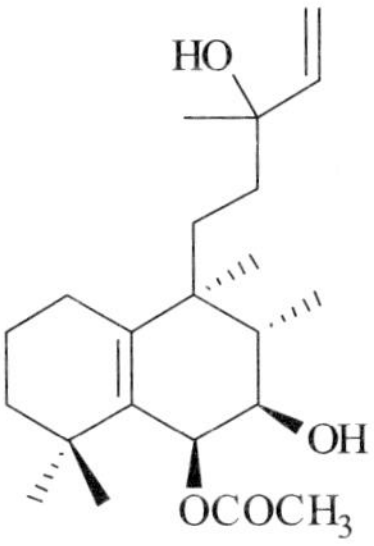

(**46**)

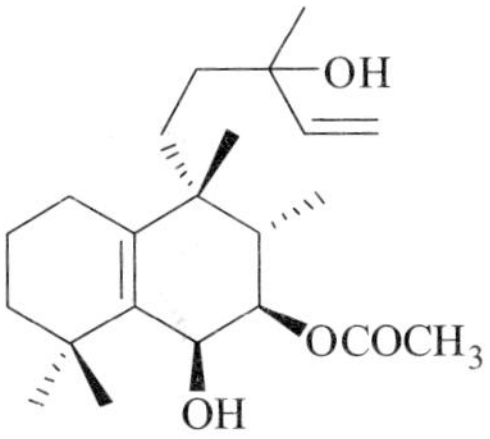

(**47**)

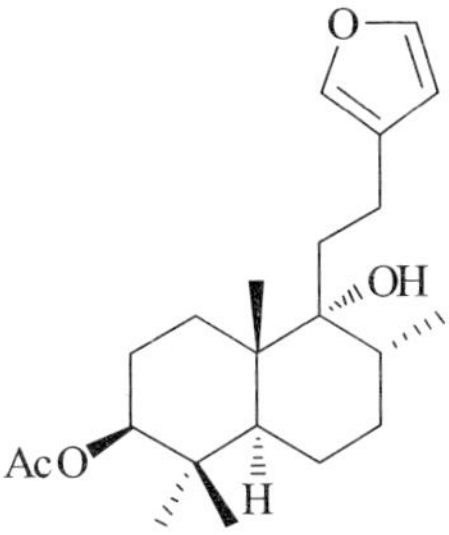

(**48**)

Contd...

Figure 4.2–Terpenoids–*Contd...*

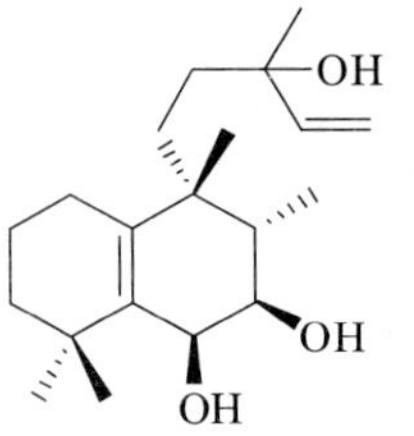

OH
OH
OH

(49)

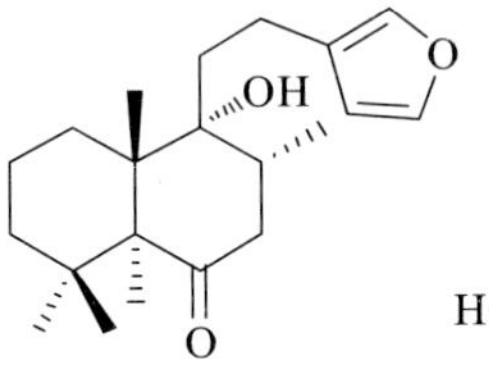

OH
O
O

(50)

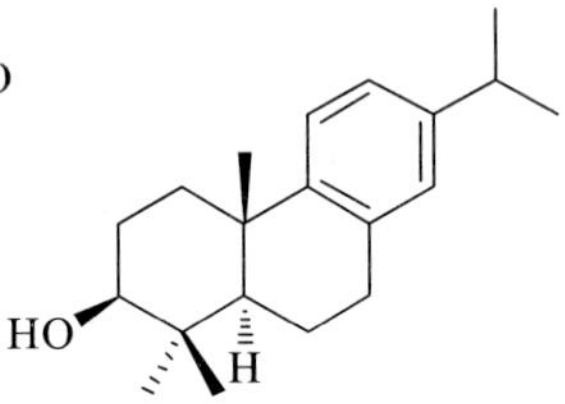

HO
H

(51)

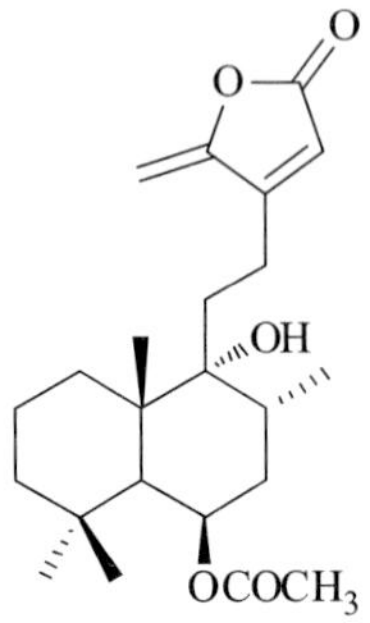

O
O
OH
OCOCH$_3$

(52)

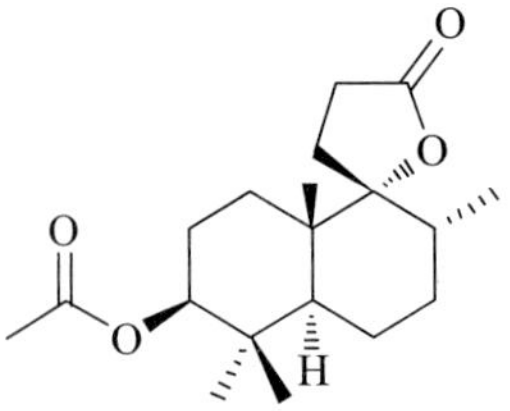

O
O
O
O
H

(75)

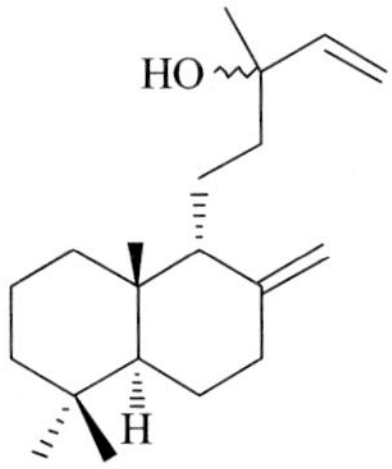

HO
H

(89)

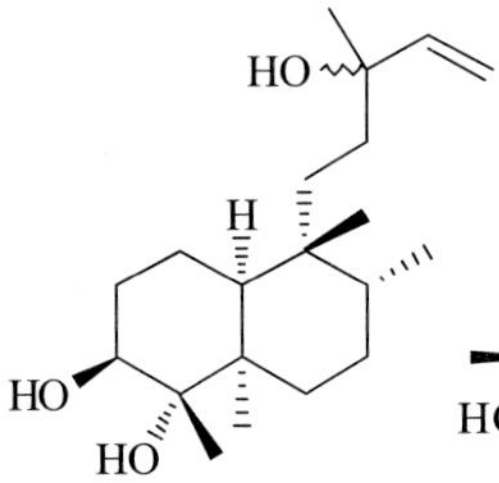

HO
H
HO
HO

(90)

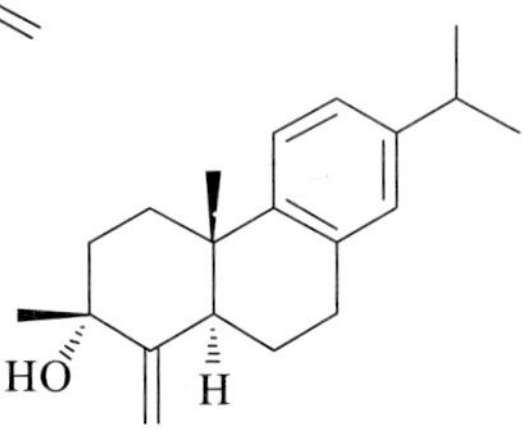

HO
H

(91)

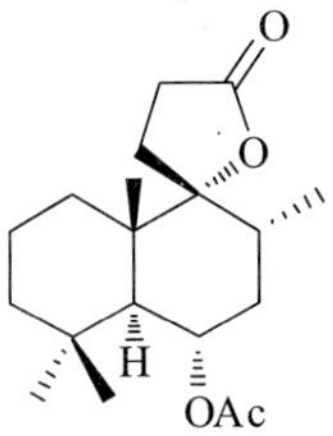

O
O
H
OAc

(92)

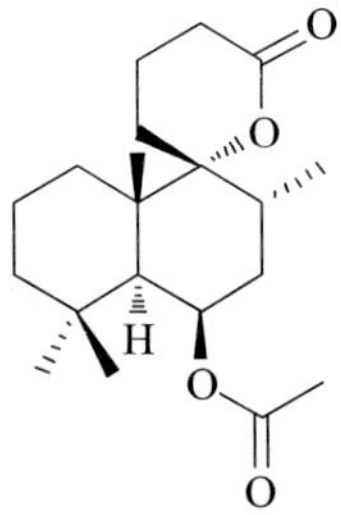

O
O
H
O
O

(93)

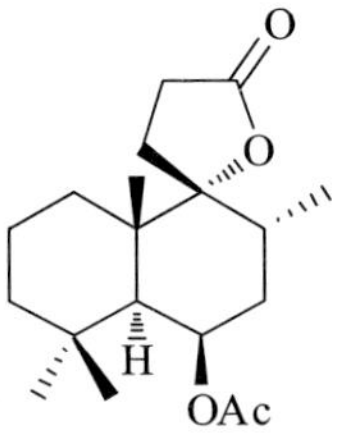

O
O
H
OAc

(94)

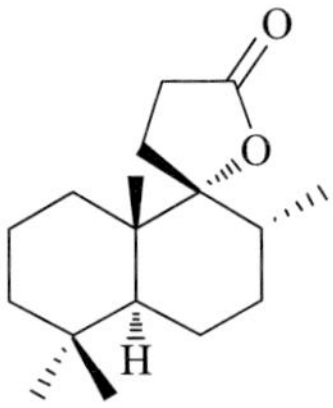

O
O
H

(95)

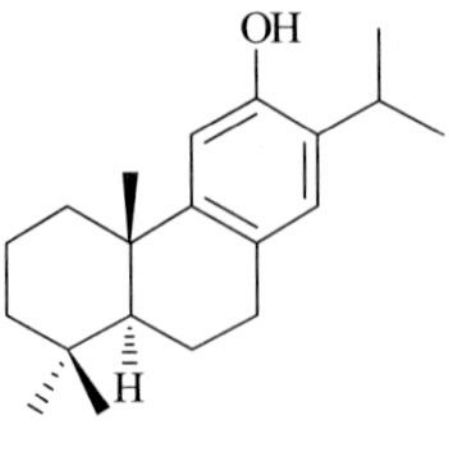

OH
H

(96)

O
O
OH
OH
O
O
OH
HO
OH
H
OCOCH$_3$

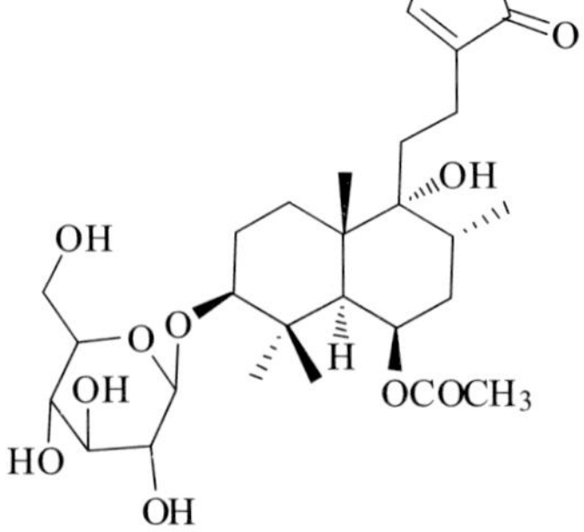

(97)

Contd...

Figure 4.2–*Contd...*

Flavonoids

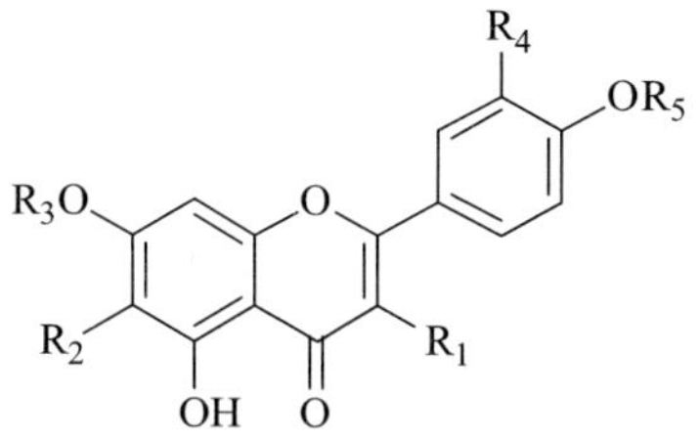

(**20**) R_1 = R_2 OMe, R_3 = R_5 Me, R_4 = OH
(**21**) R_1 = R_2 = OMe, R_3 = Me, R_4 = R_5 = H
(**22**) R_1 = R_2 = R_3 = R_4 = R_5 = H
(**23**) R_1 = OMe, R_2 = R_3 = R_4 = R_5 = H
(**24**) R_1 = R_2 = R_3 = R_5 = H, R_4 = OH
(**25**) R_1 = OH, R_2 = R_3 = R_4 = R_5 = H
(**26**) R_1 = OMe, R_2 = R_3 = R_5 = H, R_4 = OH
(**27**) R_1 = OMe, R_2 = R_5 = H, R_3 = Me, R_4 = OH

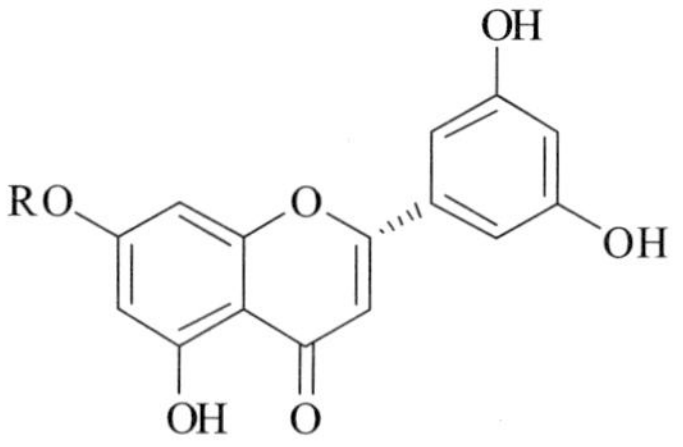

(**28**) R = H
(**29**) R = Me

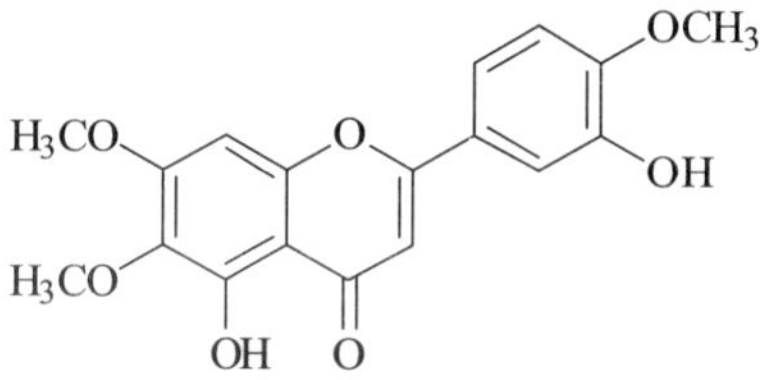

(**30**)

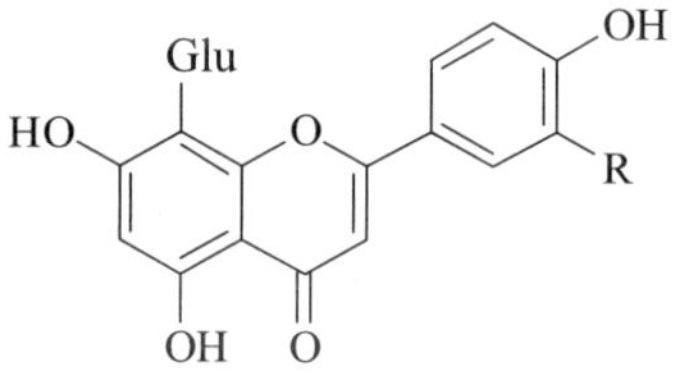

(**31**) H
(**32**) OH

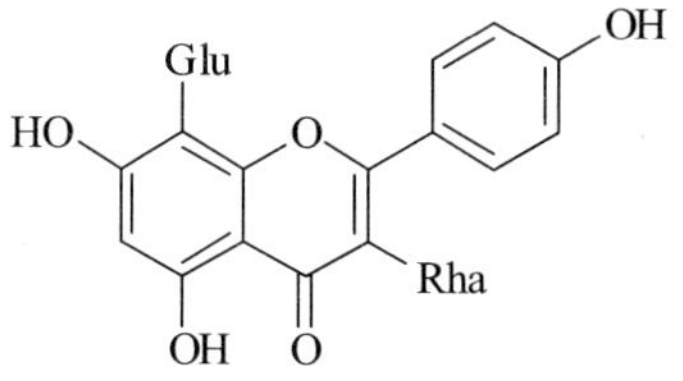

(**33**)

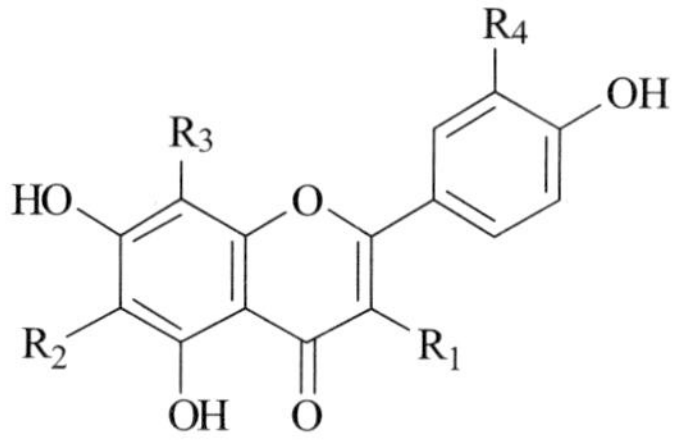

(**61**) R_1 = R_3 = H, R_2 = Glc, R_4 = OH
(**62**) R_1 = R_2 = H, R_3 = Glc, R_4 = OH

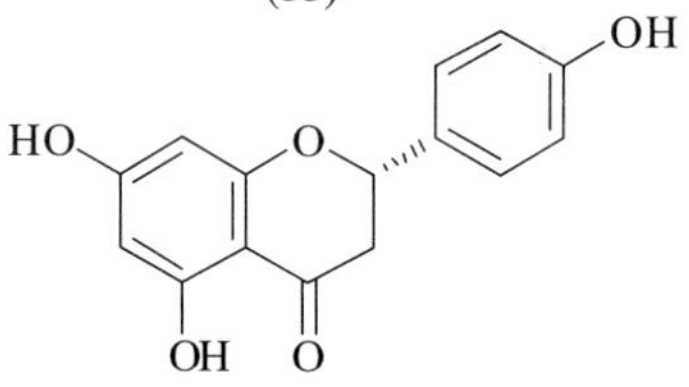

(**78**)

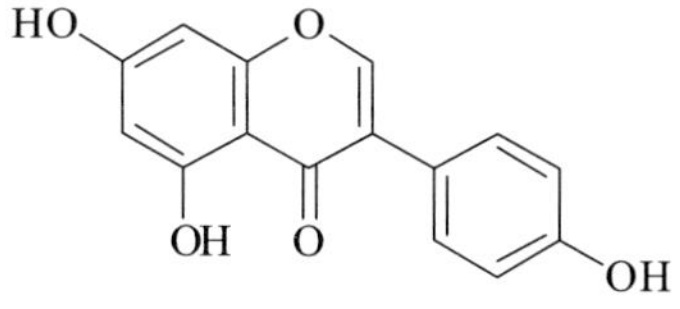

(**79**)

Contd...

Figure 4.2–Flavonoids–*Contd...*

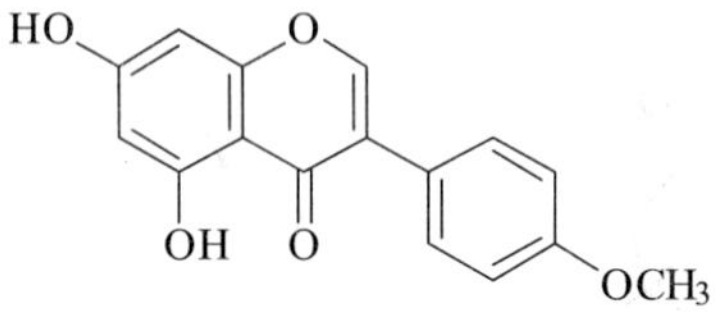

(**80**)

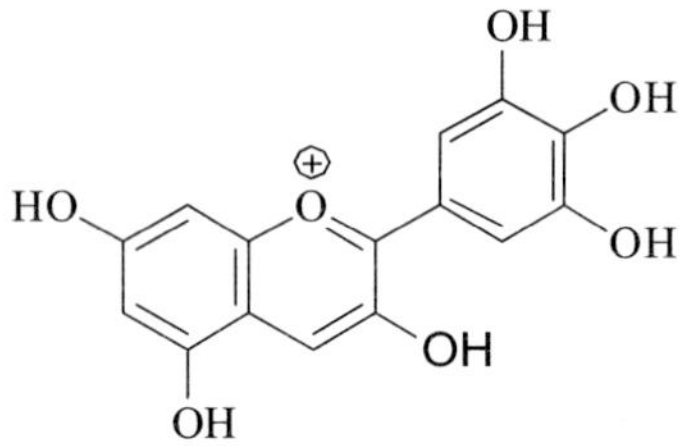

(**81**)

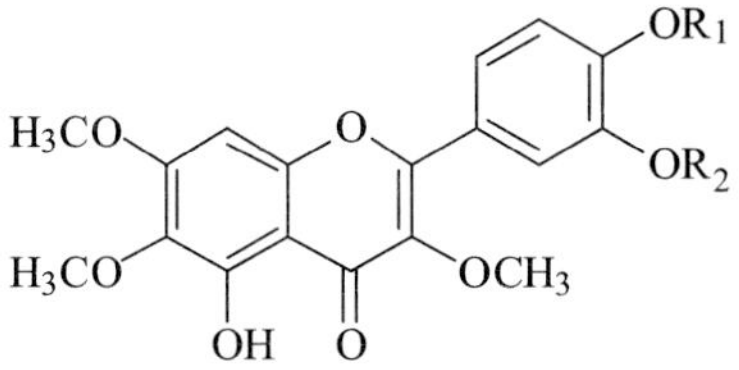

(**98**) $R_1 = R_2 = CH_3$
(**99**) $R_1 = R_2 = H$

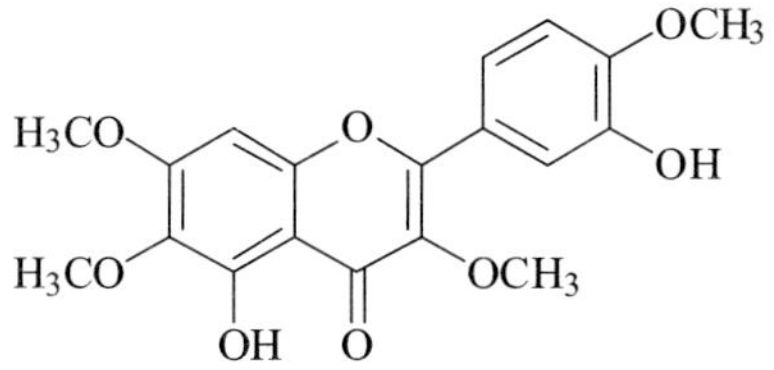

(**100**)

Simple phenolics

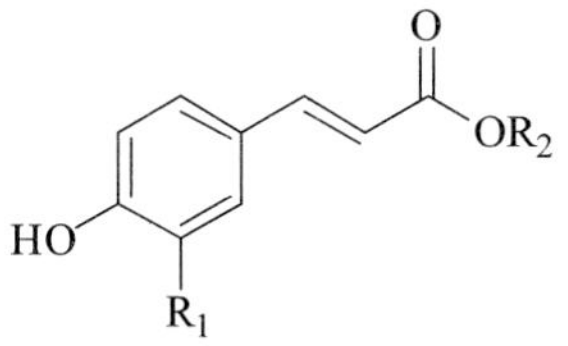

(**34**) R_1 = OMe, R_2 = H
(**35**) R_1 = H, R_2 = *p*-hydroxyphenylethanol
(**36**) $R_1 = R_2 = H$

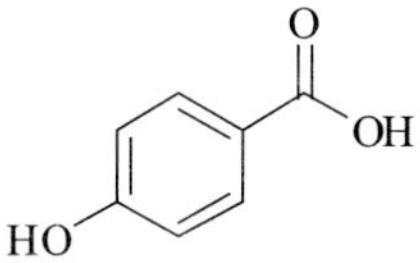

(**37**)

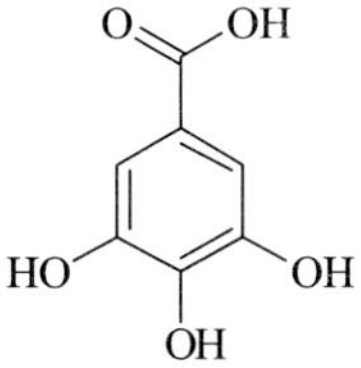

(**82**)

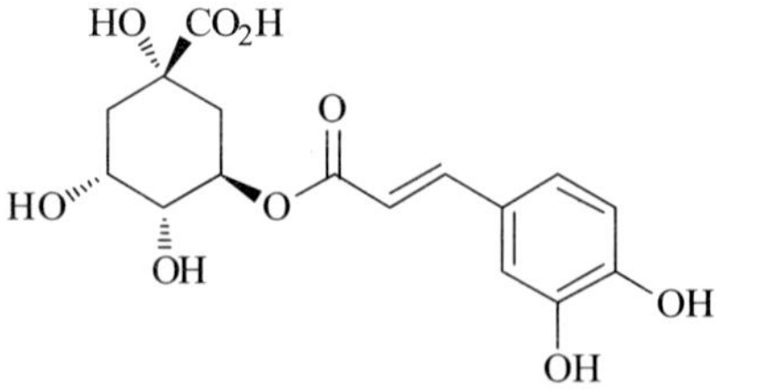

(**83**)

Contd...

Figure 4.2–*Contd...*

Iridoids

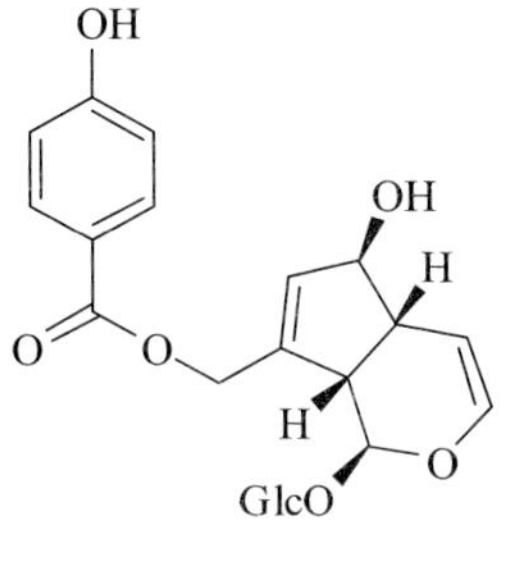

(**42**) 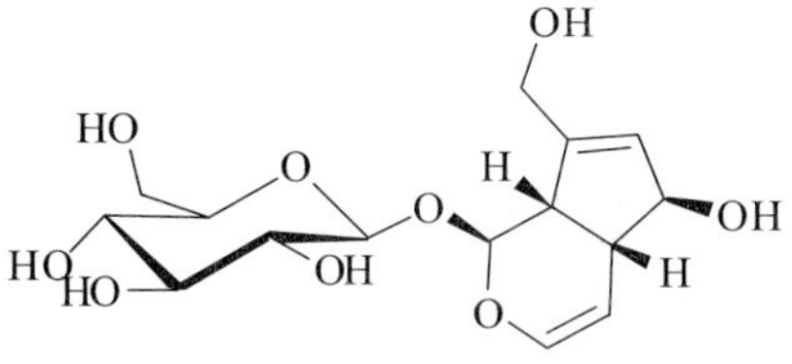 (**43**) 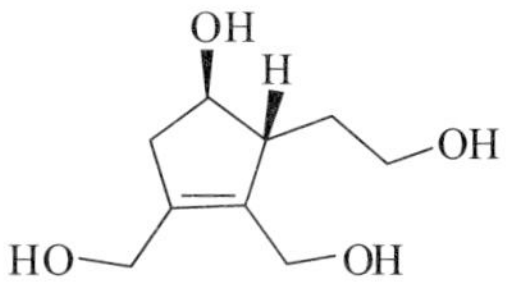 (**60**)

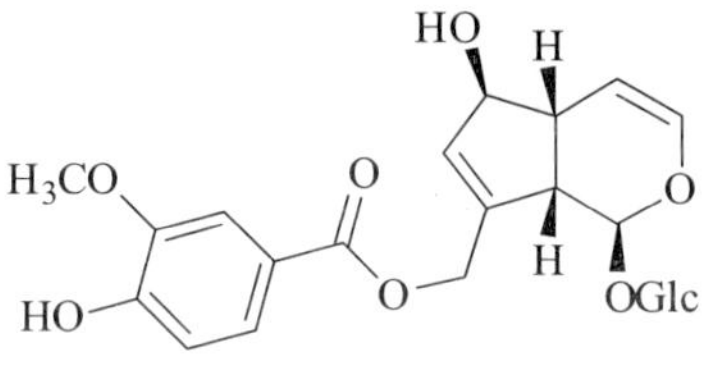

(**71**) 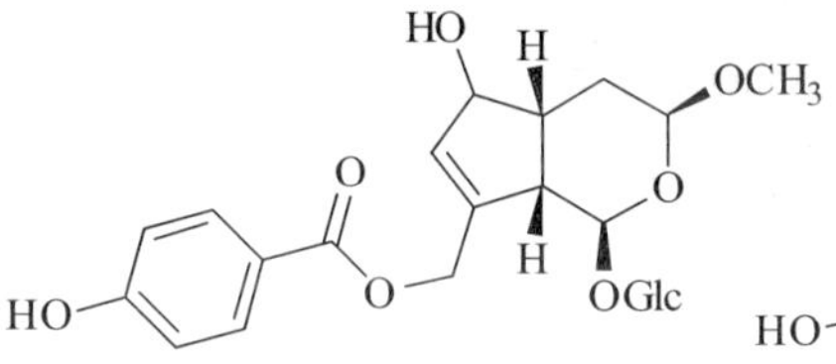 (**72**) 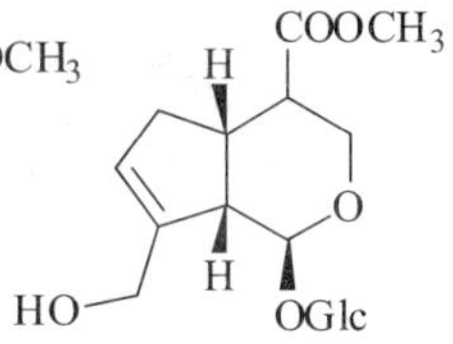 (**73**)

Lignan derivatives

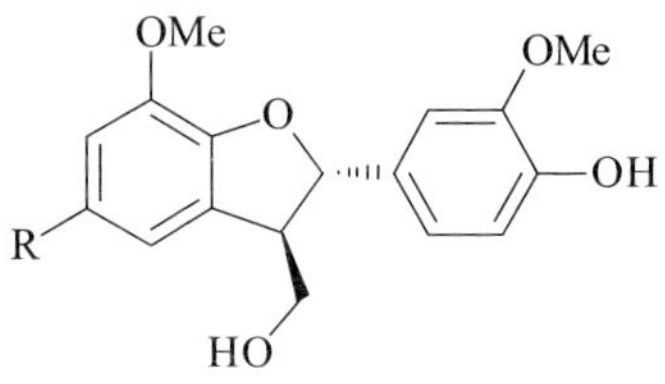

(**38**) R = -CHO
(**39**) R = $-CH_2CH_2CH_2OH$
(**40**) R = -(*E*)-CH=CH-CHO

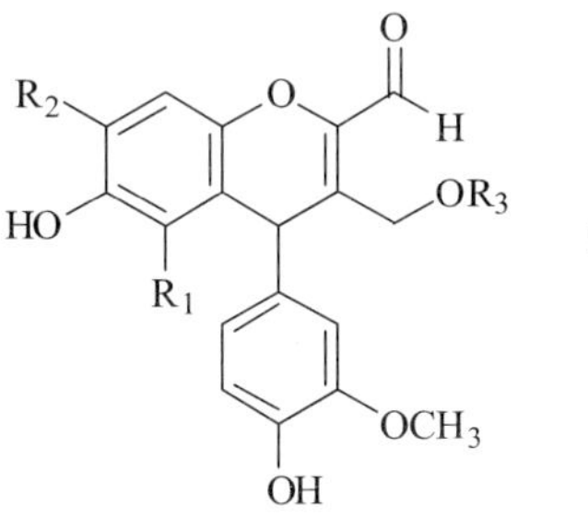

(**63**) $R_1 = OCH_3$, $R_2 = H$, $R_3 = Glc$
(**64**) $R_1 = H$, $R_2 = OCH_3$, $R_3 = Glc$
(**65**) $R_1 = H$, $R_2 = OCH_3$, $R_3 = H$
(**66**) $R_1 = H$, $R_2 = OCH_3$, $R_3 = OH$

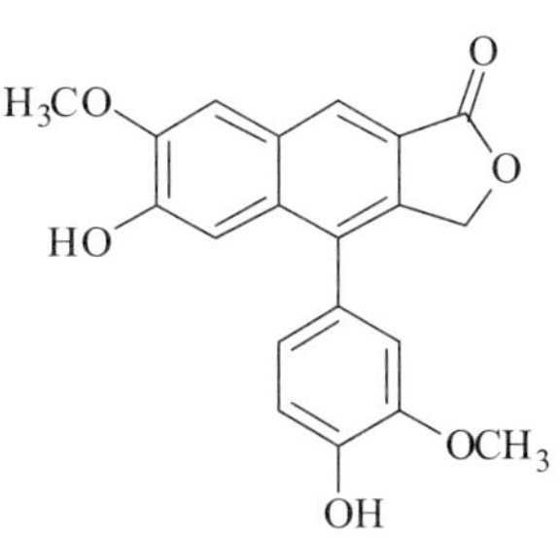

(**67**)

Contd...

Figure 4.2–Lignan derivatives–*Contd...*

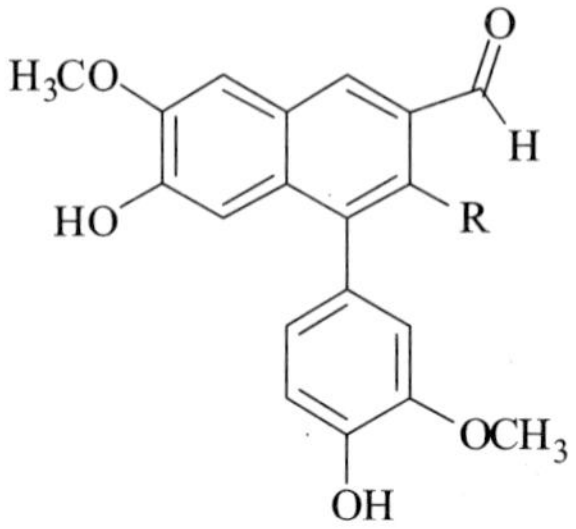

(**68**) R = H
(**69**) R = OH

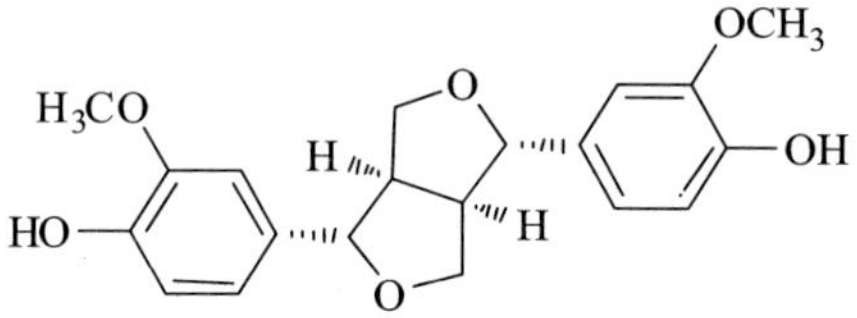

(**70**)

Catechins

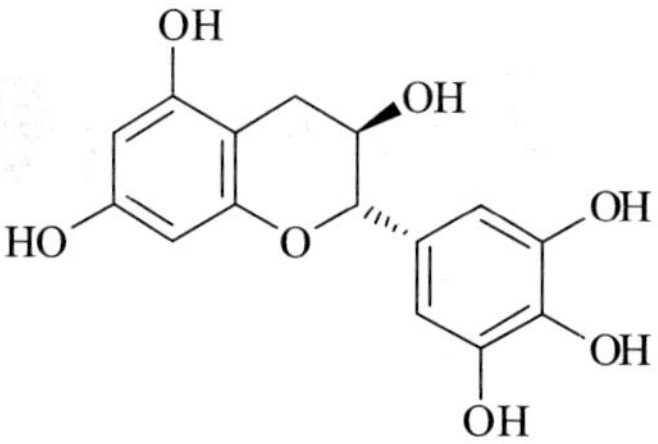

(**84**)

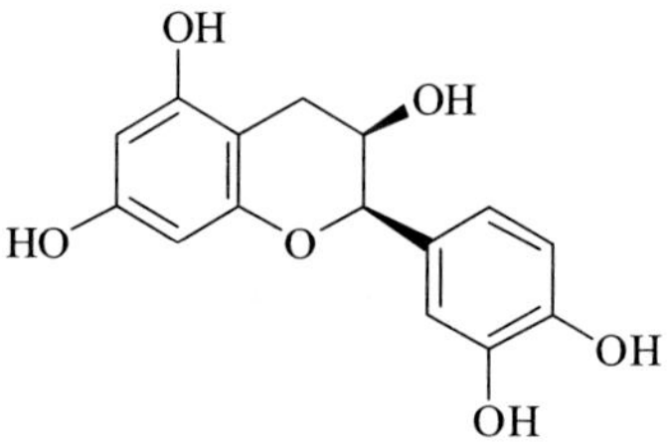

(**85**)

Anthraquinones

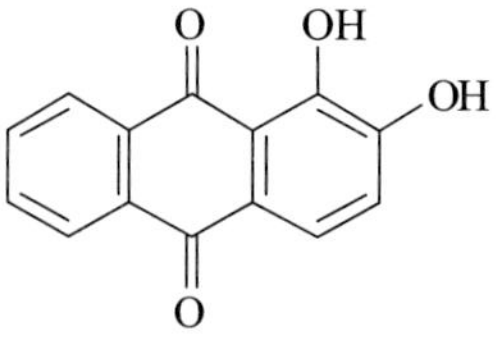

(**86**)

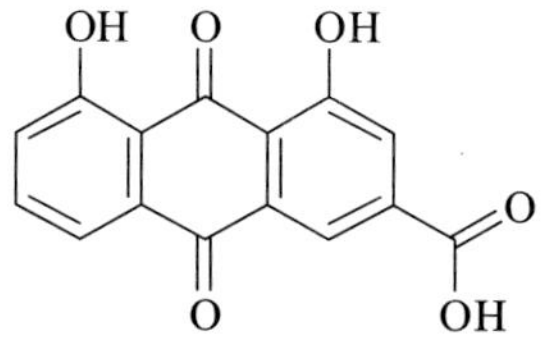

(**87**)

Others

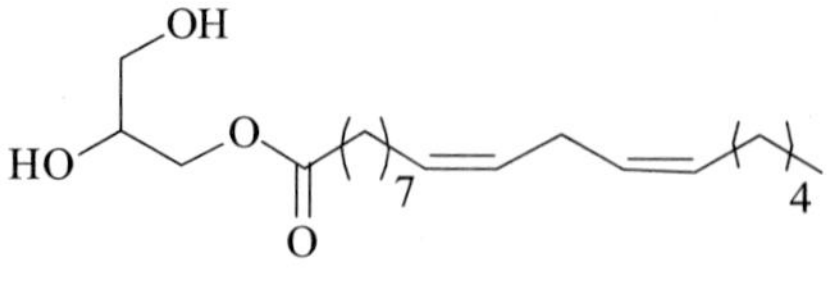

(**41**)

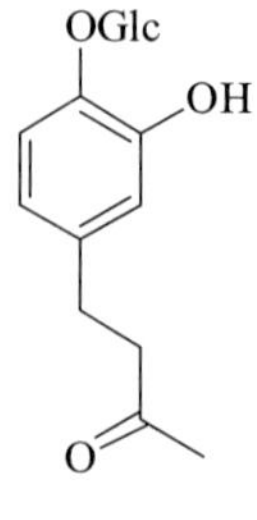

(**74**)

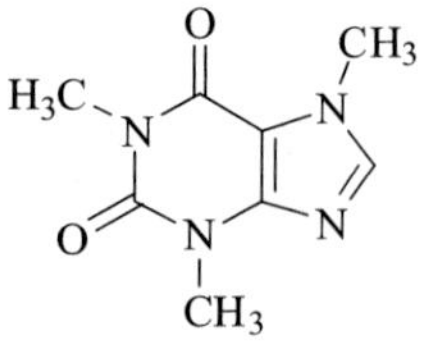

(**88**)

Table 4.2: Characterized phytochemicals obtained from *Vitex* spp. fruits.*

Vitex spp.	*Family of Compounds: Identified Compounds*	*References*
Vitex agnus-castus	**Terpenoids**: Vitexlactam A **(1)**	Li *et al.* (2002)
	Viteagnusin A **(2)**, viteagnusin B **(3)**, viteagnusin C **(4)**, viteagnusin D **(5)**, viteagnusin E **(6)**, 8-*epi*-sclareol **(7)**, (*rel* 5*S*,6*R*,8*R*,9*R*,10*S*,13*R*)-6-acetoxy-9,13-epoxy-15-methoxy-labdan-16, 15-olide **(8)**	Ono *et al.* (2008)
	Viteagnusin I **(9)**, 8-epi-manoyl oxide **(10)**, 3-epi-maslinic acid **(11)**, 3-epi-corosolic acid **(12)**, aromadendrane-4α,10α-diol **(13)**, ilelatifol D **(14)**	Chen *et al.* (2011)
	Vitetrifolin B **(15)**, vitetrifolin C **(16)**, rotundifaran **(17)**, vitexilactone **(18)**, spathulenol **(19)**	Hajdu *et al.* (2007), Webster *et al.* (2011)
	Flavonoids: Casticin **(20)**, penduletin **(21)**, apigenin **(22)**, 3-methylquercetin **(23)**, luteolin **(24)**, kaempferol **(25)**, 3,7-dimethylquercetin **(26)**, 3-*O*-methylkaempferol **(27)**, 5,7,32,52-tetrahydroxyflavanon **(28)**, 5,32,52-trihydroxymethoxylflavanone **(29)**, eupatorin **(30)**, vitexin **(31)**, orientin **(32)**, apigenin 3, 8-di-C-glycosides **(33)**	Chen *et al.* (2011), Hajdu *et al.* (2007), Webster *et al.* (2011), Ibrahim *et al.* (2008)
	Phenolics: Ferulic acid **(34)**, Phydroxyphenylethanol-*p*-coumarate **(35)**, p-Coumaric acid **(36)**, 4-hydroxybenzoic acid **(37)**. **Neolignans:** Ficusal **(38)**, vladirol F **(39)**, balanophonin **(40)**, **Glycerides**: Glyceryl linoleate **(41)**.	Chen *et al.* (2011)
	Iridoids: Agnuside **(42)**	Webster *et al.* (2011)
	Aucubin **(43)**	Ibrahim *et al.* (2008)
Vitex trifolia	**Terpenoids**: Vitetrifolin A **(44)**, vitetrifolin B **(15)**, vitetrifolin C **(16)**, vitetrofilin D **(45)**, vitetrofolin E **(46)**, vitetrifolin F **(47)**, vitetrifolin H **(48)**, vitetrifolin I **(49)**, rotundifaran **(17)**, dihydrosolidagenone **(50)**, abietatriene 3β-ol **(51)**, vitexilactone **(18)**, (rel 5S,6R,8R,9R,10S)-6-acetoxy-9-hydroxy-13(14)-labden-16,15-olide **(52)**.	Ono *et al.* (2000), Li *et al.* (2005), Wu *et al.* (2009)
	Vitrifol A **(53)**, dihydrodehydrodiconifenyl alcohol **(54)**, stigmast-4-en-6β-ol-3-one **(55)**	Gu *et al.* (2008)
	Vitexoid **(56)**, 6-acetoxy-9-hydroxy-13(14)-labdane-16,15-olide **(57)**, previtexilactone **(58)**, 6-acetoxy-9,13;15,16-diepoxy-15-methoxylabdane **(59)**	Wu *et al.* (2009)
	Flavonoids: Casticin **(20)**	Xu *et al.* (2012)
	Iridoids: Eucommiol **(60)**	Gu *et al.* (2008)

Contd...

Table 4.2–*Contd...*

Vitex spp.	*Family of Compounds: Identified Compounds*	*References*
Vitex cannabifolia	**Flavonoids**: 5,4'-dihydroxy-3,6,7,8,3'-pentamethoxyflavone **(61)**, isoorientin **(62)**, orientin **(32)**	
	Lignans: vitecannaside A **(63)**, vitecannaside B **(64)**, vitedoin A **(65)**, 6-hydroxy-4-(4-hydroxy-3-methoxyphenyl)-3-hydroxymethyl-7-methoxy-3,4-dihydro-2-naphthaldehyde **(66)**, detetrahydro-conidendrin **(67)**, vitrofolal E **(68)**, vitrofolal F **(69)**, pinoresinol **(70)**.	Yamasaki *et al.* (2008)
	Iridoids: agnuside **(42)**, 10-*O*-vanilloyl aucubin **(71)**, nisindaside **(72)**, geniposide **(73)**.	
	Phenylbutanone glucoside: 4-(3,4-dihydroxyphenyl)-butan-2-one-4-*O*-β-D-glucoside **(74)**.	
Vitex negundo	**Terpenoids:** vitedoin B **(75)**. **Lignans**: Vitedoin A **(65)**, vitedoamine A **(76)**, 6-hydroxy-4-(4-hydroxy-3-methoxyphenyl)-3-hydroxymethyl-7-methoxy-3,4-dihydro-2-naphthaldehyde **(66)**, detetrahydro-conidendrin **(67)**, vitrofolal E **(68)**, vitrofolal F **(69)**, 2α,3β-7-*O*-methylcedrusin **(77)**.	Ono *et al.* (2004) Guha *et al.* (2010)
	Flavonoids: naringenin **(78)**, genistein **(79)**, biochanin A **(80)**, delphinidin **(81)**. **Phenolics**: 4-Hydroxybenzoic acid **(37)**, gallic acid **(82)**, chlorogenic acid **(83)**. **Catechins**: (-)-gallocatechin **(84)**, epicatechin **(85)**. **Anthraquinones**: alizarin **(86)**, rhein **(87)**. **Others:** Caffeine **(88)**.	
Vitex rotundifolia	**Diterpenoids**: Vitexifolin A **(89)**, vitexifolin B **(90)**, vitexifolin C **(91)**, vitexifolin D **(92)**, vitexifolin E **(93)**, vitetrifolin D **(45)**, trisnor-γ-lactone **(94)**, *iso*-ambreinolide **(95)**, ferruginol **(96)**, abietatrien-3β-ol **(51)**, viteoside A **(97)**	Ono *et al.* (1998; 1999; 2002)
	Rotundifaran **(17)**	Hu *et al.* (2007)
	Flavonoids: Casticin **(20)**, luteolin **(26)**, artemetin **(98)**, quercetagetin **(99)**, 5,3'-dihydroxy-6,7,4'–trimethoxyflavanone **(100)**.	Kobayakawa *et al.* (2004), Ko *et al.* (2000), Ono *et al.* (2002), Hu *et al.* (2007)
	Iridoid: agnuside **(42)**	Hu *et al.* (2007)

* Bold type numbers within parenthesis correspond to the structure presented in **Figure 4.2**.

been demonstrated and in particular its use for PMS (Roemheld-Hamm, 2005; Webster, 2008; Wuttke *et al.*, 2003). In herbal medicine, a tincture was employed for PMS and prepared by mixing 1/5 of dried *V. agnus-castus* fruits with alcohol (50 per cent or 70 per cent), with a recommended dose of 20 drops/day (about 1 mL/day); alternatively, the German Commission E has suggested the consumption of 30–40 mg of dried fruit/day to treat not only PMS but also mastalgia and irregular menstruation (Webster, 2008). Now, diverse standardized extracts are commercialized: Agnolyt (caps and solution), Agnucaston (tablets), Mastonynon (tablets and solution), Femarpim and Ze 440 (tablets). These preparations were obtained by extraction with ethanol (50–70 per cent) and standardized for the content of casticin or agnoside (0.3–0.6 per cent). To date, the effect of these extracts in *libido* modulation has not been demonstrated (Roemheld-Hamm, 2005; Webster, 2008). In general, the commercial preparations of *V. agnus-castus* are well tolerated and secondary effects (*i.e.*, headache, seasickness, tiredness or thirst) are non-frequent or slight (Daniele *et al.*, 2005; Roemheld-Hamm, 2005). It has been proposed that *V. agnus-castus* extracts are good for the symptoms but not for curing, since symptoms return within some days of leaving the treatment (Berger *et al.*, 2000).

The mechanism of action of the *V. agnus-castus* on PMS has not been established but multiple targets have been proposed; one of them involves the inhibition of hyperprolactinemy during the premenstrual stage. High prolactin in serum stimulates breast milk production and induces mastalgia. The *V. agnus-castus* compounds bind dopamine receptors and inhibit the pituitary liberation of prolactin; the proposed chemicals involved are diterpenes (Wuttke, *et al.*, 2003) and the flavonoid casticin (Ye, *et al.*, 2010); remarkably, these compounds have also been isolated from other *Vitex* spp. (Table 4.2). Other proposed mechanisms are the estrogenic (Hu *et al.*, 2007; Ibrahim *et al.*, 2008; Liu *et al.*, 2001), contributing to the hormonal homeostasis, and opioid activities (Webster *et al.*, 2011; Webster *et al.*, 2006) of *Vitex* fruits. Extracts and flavonoids of *V. agnus-castus* (*e.g.*, casticin, apigenin, 3-methylkaempferol, and luteolin) show affinity for the μ and δ opioid receptors, acting analogously to β-endorphin, but not for the κ receptor. During PMS, opioid activity decrease and the appearance of pain and mood changes are facilitated; thus, *V. agnus-castus* is acting in PMS by reactivation of opioid activity (Webster, *et al.*, 2011; Webster, *et al.*, 2006). Among the evaluated flavonoids, casticin shows the highest affinity for the opioid receptors (Chen *et al.*, 2011). Similar effects have been obtained with other opiaceous such as morphine; the advantage of the *Vitex* extracts over morphine is that after hundred years of use, they do not induce secondary effects (Webster *et al.*, 2011).

In some places, *V. agnus-castus* is not available and people use the fruits of *V. rotundifolia* or *V. trifolia*. These fruits share similar ethnopharmacological applications. Some commercial preparations of *V. agnus-castus* are adulterated with *V. rotundifolia* or *V. trifolia* and up to date, the use of these mixtures have not induced secondary effects but it is suggested that the desirable pharmacological effect is reduced (Webster, 2008).

Essential oils of *V. agnus-castus* have been used for treatment of post-menopausal symptoms. The oils obtained from fruits or leaves diminished the hot flashes, night sweets, dry vagina with pain during sex and insomnia. Moreover, treatment with the oil of leaves reduces depression symptoms, changes in personality, prolonged bleedings, loss of memory, confusion and *libido*. Within the adverse effects of essential oil application are queasiness, headache, depression and nightmares. It has been suggested that these preparations must not be used in combination with anticonceptives, progesterone or anti-neuroleptics (Lucks, 2003; Lucks *et al.*, 2002).

Studies with chimpanzees of the Gombe National Park, Tanzania, showed that consumption of *V. fischeri* fruits by females increased the urinary progesterone in contrast with seasons of non-consumption or the levels of estrogens or male testosterone; these last two parameters did not change (Emery-Thompson *et al.*, 2008).

Antitumoral Properties of *Vitex* Fruits

Vitex fruits have shown antitumoral activity (Table 4.3). *Vitex agnus-castus, V. trifolia* and *V. rotundifolia* preparations induce apoptosis in tumoral cell lines; some of the identified compounds do not affect the growth of normal cells. The mechanism of action is still unknown but it has been proposed that apoptosis induction is mediated by inhibition of the cell cycle (G0/G1, G2-M), of the dynamics microtubule formation/degradation or by inducing cell oxidative stress. Amongst the *Vitex* compounds, casticin has shown the highest antitumoral activity. It has been suggested that substituent hydroxyls (3′ and 5) and methoxyls (3 and 4′) are important structural characteristics associated with its activity. Ono *et al.* (2002) showed higher activity of casticin against tumoral colon cells (HCT116) (IC_{50} = 119 ng/mL) than cisplatin (IC_{50} = 794 ng/mL). On the other hand, rotundifaran was also better against leukemia cells K562 and tsFT210 than cisplatin (IC_{50}, 33.3 and higher than 100 µg/mL, respectively) (Table 4.3). In rats, the availability of oral casticin was good (45.5 per cent) with a half life of 36.48 min; it was suggested that casticin has high potential for future clinical applications (Xu, *et al.*, 2012).

Antioxidant Properties of *Vitex* Fruits

A large body of evidence has established a correlation between fruit and vegetable consumption and a decreased risk of chronic degenerative diseases (Liu *et al.*, 2000; Rissanen, *et al.*, 2003). Nevertheless, the main causes of morbidity and mortality over the world are associated with an inadequate alimentation (WHO, 2008). Fruits and vegetables not only provide essential nutrients (*e.g.*, vitamins and minerals) but also useful compounds for prevention/treatment of different diseases and those materials are catalogued as "functional foods" or "nutraceuticals" (Kaur and Kapoor, 2001). Oxidative stress is common to a large number of diseases (*e.g.*, atherosclerosis, hypertension, cardiac failure, Alzheimer, diabetes); thus, antioxidant activity is one of the most evaluated properties of plant foods to designate them as nutraceutical (Kaliora and Dedoussis, 2007; Mladênka *et al.*, 2010).

Some *Vitex* fruits are characterized by their high antioxidant capacity. The antioxidant activity of methanol-acetone-water (6: 3: 1 v/v) extracts of three *Vitex* spp. (*V. doniana, V. keniensis* and *V. fischeri*) from Kenya were analyzed by three methods (DPPH, FRAP y NO scavenging) (Ochieng and Nanwa, 2010); ripe fruits showed higher activity than green-ripe fruits and similar to that of the standards used (*i.e.*, vitamin C, rutin and gallic acid). Moreover, total phenolics contents of the three *Vitex*

Table 4.3: Cytotoxic activity and suggested action mechanism of compounds isolated of *Vitex* spp. fruits.*

Fruit	*Cell Line Evaluated*	*Responsible Compounds*	*Proposed Mechanism of Action*	*References*
Vitex rotundifolia	HL-60 (myeloid leukemia)	**(99)** IC_{50} = 4.03 μM; **(20)** $\mathbf{IC_{50}}$ **= 0.12 μM;** **(98)** IC_{50} = 30.98 μM; **(17)** IC_{50} = 22.5 μM.	Apoptosis induction (DNA fragmentation and releasing of nucleosomes to cytoplasm)	Ko *et al.* (2000; 2001)
	PC-12 (lung cancer), HCT116 (colon cancer).	**(20)** IC_{50} PC-12 = 114 ng/mL, IC_{50} = HCT116 = 119 ng/mL; **(98)** IC_{50} PC-12= 2270 ng/mL, IC_{50} HCT116 = 2200 ng/mL.	Inhibition of tubulin polymerization. Substituent groups are important for activity: 3',5-hydroxyl and 3,4'-methoxyl	Ono *et al.* (2002)
	KB (nasopharynx cancer)	**(20)** $\mathbf{IC_{50}}$ **= 0.23 μM;** **(98)** IC_{50} = 16 μM; **(99)** IC_{50} = 15.3 μM; **(100)** IC_{50} = 18.6 μM	Casticin stop the cell cycle at G2/M by inhibition of the processing of microtubules. Groups 3',5-hydroxyl and 3,4'-methoxyl are important for the activity	Kobayakawa *et al.* (2004)
Vitex trifolia	tsFT210 y K562 (myeloid leukemia)	**(17)** IC_{50} tsFT210 = 36.2 μg/mL, IC_{50} K562 = 26.5 μg/mL; **(45)** IC_{50} tsFT210 = 41.3 μg/mL, IC_{50} K562 = 35.2 μg/mL; **(18, 52, 46)** IC_{50} tsFT210 >80 μg/mL, IC_{50} K562 >50 μg/mL.	Apoptosis and inhibition of the cell cycle progression in the G0/G1 phase	Li *et al.* (2005)
	HeLa (cervical cancer)	**(49)** IC_{50} = 4.9 μMol/L; **(18)** IC_{50} = 9.5 μMol/L; **(45, 46, 48)** IC_{50} = 15 μMol/L	Inhibition of the cell cycle in the stage G0/G1	Wu *et al.* (2009)
Vitex agnus-castus	KATO-III (gastricsignet-ring carcinoma), SKOV-3 (breast cancer), COLO 201 (colon cancer), Lu-134-A-H (small cell lung carcinoma), MCF-7 (ovarian cancer), SKG-3[a] (cervical cancer)	Ethanolic extract	Apoptosis in the cell lines: KATO-III, SCOV-3, COLO 201 and Lu-134-A-H. In the KATO-III is induced by cell oxidative stress	Ohyama *et al.* (2003; 2005)

*IC_{50} = Half maximal (50 per cent) inhibitory concentration. Bold type numbers within parenthesis corresponds with structures of **Figure 4.2**.

fruits were high and similar among them; phenolics were also higher in ripe fruit (572–719 mg of gallic acid equivalents/100 g f.w.) than green-ripe fruit (290–371 mg of gallic acid equivalents/100g f.w.). Fruit ripening of *Vitex* fruits is accompanied by colour changes from green to dark intense hue, which could be correlated with the phenolics content.

Aqueous extracts of *V. agnus-castus* show higher antioxidant activity than low polarity preparations; the activities of aqueous extracts from fruits and leaves (2.27 and 2.50 mM Trolox equivalents, respectively) were better than those of ethanolic and hexanic extracts (0.097-1.57 mM Trolox equivalents) evaluated by the ABTS method (Saglam *et al.*, 2007). Similar results were reported by Sirikurkcy *et al.* (2009), who found that the aqueous extract of *V. agnus-castus* fruits was more active than the low polarity extracts and essential oils by using the methods of DPPH, β-carotene bleaching and iron reducing power; in addition, the antioxidant results by the β-carotene bleaching method with the aqueous extract were similar to those registered for the food approved antioxidants butylated hydroxytoluene (BHT) and butylated hydroxyanisole (BHA) (93 – 99 per cent). Total phenolics of the aqueous extracts (112.5 mg of gallic acid equivalents/g of extract) were higher than those of the other preparations (21.7 – 46.6 mg/g of extract); however, dichloromethane extract had a higher concentration of flavonoids (43.5 mg of quercetin equivalents/g of extract). Hajdu *et al.* (2007) identified casticin as the antioxidant molecule in an ethyl acetate fraction obtained from a methanolic extract of *V. agnus-castus*; they also demonstrated that casticin shows higher inhibition of lipid peroxidation (IC_{50} = 0.049 mM) than vitamin C (IC_{50} = 0.703 mM) in a rat brain homogenate.

The Masateru Ono group from Japan analyzed the antioxidant activity of phytochemicals obtained from the methanolic extracts of different *Vitex* spp. fruits; they identified ferruginol (abietane-type diterpene) from *V. rotundifolia* which showed higher antioxidant activity than BHT by the ferric thiocyanate method, but lower by the DPPH method (Ono *et al.*, 1999); isoorientin and orientin were obtained from *V. cannabifolia* and identified as better antioxidants than L-cysteine and α-tocopherol by the DPPH method (Yamasaki *et al.*, 2008); this group also characterized the two new lignans viteodin A and vitedoamine A (structures 65 and 76, Figure 4.2) from *V. negundo* seeds, which were more potent than L-cysteine by the DPPH method (Ono *et al.*, 2004).

Essential Oils of *Vitex* spp. Fruits

Vitex fruits are an interesting source of essential oils and the best studied species is *Vitex agnus-castus* (Table 4.4). The main components of the *Vitex agnus-castus* essential oil are 1,8-cineole, sabinene and α-pinene and its composition is almost the same at different maturity stages (Novak *et al.*, 2005; Stojkovic *et al.*, 2011). However, a clear variation between species has been registered and essential oil composition can be used for species identification; based on this consideration, pharmacological applications associated with essential oil composition must change with the *Vitex* sp. On the other hand, contrasting the essential oil composition of fruits and leaves from *Vitex agnus-castus*, they were similar (Novak, *et al.*, 2005; Stojkovic, *et al.*, 2011). It has been registered that secondary effects of treatment with *V. agnus-castus* essential oil depends on the source (leaf or fruit), but both have reduced the symptoms associated with gynecologic problems (see section 4) (Lucks, 2003; Lucks, *et al.*, 2002).

Table 4.4: Composition (per cent) of essential oils of *Vitex* spp. fruits.

Compound	*V. agnus-castus (Ripe fruit)*	*V. agnus-castus (Unripe fruits)*	*V. negundo*	*V. pseudonegundo*
1,8-Cineole	14-30.2	13.7-28.2	–	–
Sabinene	11.6-48.2	11.8-48.2	–	2.9
α-Pinene	1.1-25.5	1.1-19.4	–	31.7
Limonene	tr-6.8	tr-5.3	–	11.5
trans-β-Farnesene	0-9.3	0-7.5	–	–
α-Terpinyl acetate	0-6.6	0-6.5	–	1.2
Caryophyllene oxide	0-4.6	0-3.8	3	–
trans-β-Caryophyllene	0.6-6.7	1-6.4	5	1.8
τ-Cadinol	3	0.7	–	–
Terpinen-4-ol	0.3-3.6	0.4-3	–	–
α-Terpineol	0.6-3.6	0.6-4.5	–	–
Sclarene	0-2	0-1.7	–	–
Geranyl linalool	2.25-2.3	0.7	–	–
Myrcene	1.4-2.9	1.1-3	–	1.7
Abietatriene	1.2	1	–	–
trans-Sabinene hydrate	0-0.9	0-0.9	–	–
β-Pinene	0-2.2	1.0-1.8	–	–
cis-β-Farnesene	0-0.7	0-0.5	–	4.6
Ledol	0-1.8	0.3-1.6	–	–
β-Selinene	–	–	22	–
α-Cedrene	–	–	14	–
Germacrene D	0-2.1	0-1.4	8	4.3
Germacrene B	–	–	–	2.0
Aristolene	–	–	8	–
Hexadecanoic acid	–	–	8	–
α-Copaene	–	–	5.4	–
α-Humulene	–	–	4	–
Guaia-3,7-diene	–	–	2	–
Bicyclogermacrene	1.9-9.9	1.2-9.4	–	14.5
Citronellyl acetate	–		–	1.4
allo-Aromadendrene	–		–	2.7
α-Phellandrene	–		–	1.2
β-Sesquiphellandrene	–		–	1.2
α-Gurjunene	–		–	1.1
trans-α-Bergamotene	–		–	1.1

–: Non reported.

The essential oil of *V. agnus-castus* showed low antioxidant activity when evaluated in aqueous media (DPPH and ferric ion reducing power), but high activity was obtained in an emulsion system (β-carotene bleaching) (Sarikurkcu, *et al.*, 2009). The essential oils of *V. agnus-castus* and *V. doniana* were potent antimicrobials (Khokra *et al.*, 2008; Stojkovic, *et al.*, 2011; Suleiman and Yusuf, 2008). The *V. agnus-castus* oil was effective against *Staphylococcus aureus* ATCC 6538, *Micrococcus flavus* ATCC 9341, *Bacillus subtilis* ATCC 10907, *Salmonella typhimurium* ATCC 13311 and *Escherichia coli* ATCC 35210 (minimal inhibitory concentration, MIC = 44.5–445 μg/mL), and also against the fungi *Alternaria alternata* DSM 2006, *Aspergillus flavus* ATCC 9643, *Aspergillus niger* ATCC 6275, *Aspergillus ochraceus* ATCC 12066, *Fusarium tricinctum* CBS 514478, *Penicillium ochrochloron* ATCC 9112, *Penicillium funiculosum* ATCC 36839 and *Trichoderma viride* JCM 22452 (MIC = 44.5–219 μg/mL). The active constituents of this essential oil were 1,8-cineole and α-pinene and their MIC values for bacteria and fungi (4–8 μg/mL) were better than those of streptomycin (MIC = 50–100 μg/mL) and bifonazole (MIC = 100–150 μg/mL). The *V. doniana* essential oil inhibited the growth of *S. aureus, B. subtilis* and *E. coli*, in the disc diffusion assay; diameters were slightly smaller than those obtained with ciprofloxacin (Khokra, *et al.*, 2008; Stojkovic, *et al.*, 2011; Suleiman and Yusuf, 2008).

Other Biological Activities of *Vitex* spp. Fruits

Vitex mollis and *V. doniana* are commonly employed for diarrhea and gastrointestinal disorders. Fractions of the methanolic extract of *V. mollis* were active against enteropathogenic bacteria; the ethyl acetate fraction was enriched in phenolics (tannins and flavonoids) and showed the highest activity, being specially effective *vs. Shigella dysenteriae* (causal agent of dysentery) (MIC = 2 mg/mL) (Delgado-Vargas *et al.*, 2010). On the other hand, the aqueous extract of *V. doniana* (0.3 – 2.4 mg/mL) inhibited the pig intestine contractions induced by treatment with acetylcholine; in addition, it showed a dose-response effect (150 – 650 mg/kg) in the reduction of the gastrointestinal transit time and in the diarrhea index produced by treating mice with castor oil (Suleiman and Yusuf, 2008). As it happened with other *Vitex* spp., *V. doniana* was also rich in flavonoids and tannins. It has been proposed that phenolics are useful for diarrhea treatment because of their antibacterial activity and effect in the reduction of the gastrointestinal transit time (Palombo, 2006; Ríos and Recio, 2005), supporting the observations registered for the *Vitex* spp. fruits and their use as antidiarrheic agents.

A flavonoid rich extract obtained from *V. negundo* seeds induced infertility in rats (15 and 30 mg/rat/day/15 days). The treated rats showed reduced weight of sexual organs and lower density, motility and morphological abnormalities of spermatozoids (double headed and separation of head and tail); in addition, they had reduced the α-glucosidase activity in the epididymis, the citric acid in prostate and the fructose in the seminal vesicles.

Some arthropods bite humans and animals and can be vectors of a range of diseases such as borreliosis, malaria, yellow fever and dengue; natural repellents have been used to prevent the attack of these insects. At this respect, a supercritical CO_2 extract of *V. agnus-castus* seeds was repellent, up to 8 h, against ticks

(*Ixodes ricinus* and *Rhipicephalus sanguineus*), fleas (*Ctenocephalides felis*), mosquitoes (*Culex quinquefasciatus, Anopheles stephensi* and *Aedes aegypti*) and flies (*Stomoxys calcitrans*), evaluated in human and animal models. Remarkably, the extract (1–3 per cent) did not produce allergic responses (Mehlhorn *et al.*, 2005).

The aqueous extract of *V. rotundifolia* shows antiallergenic effects. The polymer 48/80 is produced by condensation of *N*-methyl-*p*-methoxyphenethylamine with formaldehyde and induces mast cells to release histamine; then, an anaphylactic allergic reaction is produced that can end with the death of the affected animal. All rats injected (i.p., 8 mg/kg) with the 48/80 compound died; but mortality was avoided when rats were treated with 1 g/kg of the aqueous extract of *V. rotundifolia*, 1 h before or 5 to 10 min after the injection of 48/80. The analysis of the rat plasma showed that the aqueous extract decreased by 70 per cent the histamine release, related with the peak of histamine produced when only the allergen was applied (Shin *et al.*, 2000). These results support the traditional use in Korea of *V. rotundifolia* for treatment of diverse allergenic diseases.

Concluding Remarks

Vitex spp. fruits show a variety of biological properties that can be used for humanity benefits. These species are widely distributed around the world and people could benefit from those available in every region. Many species share ethnopharmacological uses and compounds, suggesting that *Vitex* spp. share biological activities, but this needs to be scientifically tested. For example, some of the *Vitex* spp. are used to treat pre-menstrual syndrome but it has been validated only for *V. agnus-castus*. Other research area involves the chemical and biological characterization of the less studied *Vitex* fruits; casticin, agnoside and rotundifaran are markers for these species and they are associated with specific biological activities. Moreover, essential oils of the *Vitex* fruits share components and some of them have shown high antibacterial and antifungal activity. It is particularly interesting that *Vitex* pigments have not been studied and preliminary results have showed their high antioxidant activity. Outstandingly, aqueous extracts of *V. agnus-castus* showed higher antioxidant activity than low polarity extracts, while those of *V. doniana* and *V. rotundifolia* were effective against diarrhea in mice and as antiallergenic, respectively. It is possible that such biological properties could be related to the dark soluble pigments. This hypothesis remains to be tested as most of the chemical-biological studies of *Vitex* fruits have been carried out with low to intermediate polarity extracts without analyzing the aqueous extract.

Briefly, we are pointing out the following research areas with the *Vitex* spp. fruits:

- ☆ Scientific validation of the ethnopharmacological properties of *Vitex* spp. with similar uses.
- ☆ Isolation and characterization of representative compounds of the genus *Vitex* and their association with pharmacological relevant activities (*e.g.*, casticin and agnoside).

- ☆ New product development of foods and standardized extracts based on *Vitex* spp. and demonstration of their nutritional and nutraceutical properties.
- ☆ Chemical and biological characterization of the dark pigments of *Vitex* fruits.

Acknowledgments

Authors acknowledge to López-Valenzuela J.A. by the critical reading of manuscript.

References

AMR. (2009). *Vitex agnus-castus*. Monograph. *Alt. Med. Rev.*, **14**: 67-71.

Berger, D., Schaffner, W., Schrader, E., Meier, B., and Brattstrom, A. (2000). Efficacy of *Vitex agnus castus* L. extract Ze 440 in patients with pre-menstrual syndrome (PMS). *Arch. Gynecol. Obstet.*, **264**: 150-153.

Chantaranothai, P. (2011). A revision of the genus *Vitex* (Lamiaceae) in Thailand. *Trop. Nat. Hist.*, **11**: 91-118.

Chen, S. N., Friesen, J. B., Webster, D., Nikolic, D., van Breemen, R. B., Wang, Z. J., Fong, H. H., Farnsworth, N. R., and Pauli, G. F. (2011). Phytoconstituents from *Vitex agnus-castus* fruits. *Fitoterapia*, **82**: 528-533.

Daniele, C., Thompson Coon, J., Pittler, M. H., and Ernst, E. (2005). *Vitex agnus castus*: a systematic review of adverse events. *Drug Safety*, **28**: 319-332.

de Kok, R. (2008). The genus *Vitex* (Labiatae) in the flora Malesiana region, excluding New Guinea. *Kew Bull.*, **63**: 17-40.

Delgado-Vargas, F., Félix-Favela, F., Pío-León, J. F., López-Angulo, G., López-Valenzuela, J. A., Díaz-Camacho, S. P., and Uribe-Beltrán, M. J. (2010). Antibacterial activity and qualitative phytochemical analysis of *Vitex mollis* fruit. *Int. J. Green Pharm.*, **4**: 288-291.

Delgado-Vargas, F., and Paredes-López, O. (2003). Natural colorants for food and nutraceutical uses. CRC Press. Boca Raton, FL, USA.

Egbekun, M. K., Akowe, J. I., and Ede, R. J. (1996). Physico-chemical and sensory properties of formulated syrup from black plum (*Vitex doniana*) fruit. *Plant Food Hum. Nutr.*, **49**: 301-306.

Emery-Thompson, M., Wilson, M. L., Gobbo, G., Muller, M. N., and Pusey, A. E. (2008). Hyperprogesteronemia in response to *Vitex fischeri* consumption in wild chimpanzees (*Pan troglodytes schweinfurthii*). *Am. J. Primatol.*, **70**: 1064-1071.

FNB/FNIC. (2010). Food and Nutrition Board, Food and Nutrition Information Center: Dietary reference intakes. Retrived from http: //fnic.nal.usda.gov/nal_display/index.php?info_center=4&tax_level=3&tax_subject=256&topic_id=1342&level3_id=5140.

Gu, Q., Zhang, X. M., Zhou, J., Qiu, S. X., and Chen, J. J. (2008). One new dihydrobenzofuran lignan from *Vitex trifolia*. *J. Asian Nat. Prod. Res.,* **10**: 499-502.

Guha, G., Rajkumar, V., and Ashok Kumar, R. (2010). Polyphenolic constituents of methanolic and aqueous extracts of *Vitex negundo* render protection to Hep3B cells against oxidative cytotoxicity. *Food Chem. Toxicol.,* **48**: 2133-2138.

Hajdu, Z., Hohmann, J., Forgo, P., Martinek, T., Dervarics, M., Zupko, I., Falkay, G., Cossuta, D., and Mathe, I. (2007). Diterpenoids and flavonoids from the fruits of *Vitex agnus-castus* and antioxidant activity of the fruit extracts and their constituents. *Phytother. Res.,* **21**: 391-394.

Halaska, M., Beles, P., Gorkow, C., and Sieder, C. (1999). Treatment of cyclical mastalgia with a solution containing a *Vitex agnus castus* extract: results of a placebo-controlled double-blind study. *Breast,* **8**: 175-181.

Hu, Y., Hou, T., Zhang, Q., Xin, H., Zheng, H., Rahman, K., and Qin, L. (2007). Evaluation of the estrogenic activity of the constituents in the fruits of *Vitex rotundifolia* L. for the potential treatment of premenstrual syndrome. *J. Pharm. Pharmacol.,* **59**: 1307-1312.

Ibrahim, N. A., Shalaby, A. S., Farag, R. S., Elbaroty, G. S., Nofal, S. M., and Hassan, E. M. (2008). Gynecological efficacy and chemical investigation of *Vitex agnus-castus L.* fruits growing in Egypt. *Nat. Prod. Res.,* **22**: 537-546.

Kaliora, A. C., and Dedoussis, G. V. Z. (2007). Natural antioxidant compounds in risk factors for CVD. *Pharmacol. Res.,* **56**: 99-109.

Kalra, E. K. (2003). Nutraceutical–definition and Introduction. *AAPS PharmSci.,* **5**: 1-2.

Kaur, C., and Kapoor, H. C. (2001). Antioxidants in fruits and vegetables–the millennium's health. *Int. J. Food Sci. Technol.,* **36**: 703-725.

Khokra, S. L., Prakash, O., Jain, S., Aneja, K. R., and Dhingra, Y. (2008). Essential oil composition and antibacterial studies of *Vitex negundo* linn. extracts. *Indian J. Pharm. Sci.,* **70**: 522-526.

Ko, W. G., Kang, T. H., Lee, S. J., Kim, N. Y., Kim, Y. C., Sohn, D. H., and Lee, B. H. (2000). Polymethoxyflavonoids from *Vitex rotundifolia* inhibit proliferation by inducing apoptosis in human myeloid leukemia cells. *Food Chem. Toxicol.,* **38**: 862-865.

Ko, W. G., Kang, T. H., Lee, S. J., Kim, Y. C., and Lee, B. H. (2001). Rotundifuran, a labdane type diterpene from *Vitex rotundifolia,* induces apoptosis in human myeloid leukaemia cells. *Phytother. Res.,* **15**: 535-537.

Kobayakawa, J., Sato-Nishimoria, F., Moriyasub, M., and Matsukawa, Y. (2004). G2-M arrest and antimitotic activity mediated by casticin, a flavonoid isolated from Viticis Fructus (*Vitex rotundifolia* Linne fil.). *Cancer Lett.,* **208**.

Lee, S. M., Lee, Y. J., Kim, Y. C., Kim, J. S., Kang, D. G., and Lee, H. S. (2011). Vascular protective role of vitexicarpin isolated from *Vitex rotundifolia* in human umbilical vein endothelial cells. *Inflammation,* **35**: 584-593.

Li, S. H., Zhang, H. J., Qiu, S. X., Niu, X. M., Santarsiero, B. D., Mesecar, A. D., Fong, H. H. S., Farnsworth, N. R., and Suna, H. D. (2002). Vitexlactam A, a novel labdane diterpene lactam from the fruits of *Vitex agnus-castus*. *Tetrahedron Letters,* **73**: 5131–5134.

Li, W., Cui, C., Cai, B., and Yao, X. (2005). Labdane-type diterpenes as new cell cycle inhibitors and apoptosis inducers from *Vitex trifolia* L. *J. Asian Nat. Prod. Res.,* **7**: 95-105.

Liu, J., Burdette, J. E., Xu, H., Gu, C., van Breemen, R. B., Bhat, K. P., Booth, N., Constantinou, A. I., Pezzuto, J. M., Fong, H. H., Farnsworth, N. R., and Bolton, J. L. (2001). Evaluation of estrogenic activity of plant extracts for the potential treatment of menopausal symptoms. *J. Agric. Food Chem.,* **49**: 2472-2479.

Liu, S., Manson, J. E., Lee, I. M., Cole, S. R., Hennekens, C. H., Willett, W. C., and Buring, J. E. (2000). Fruit and vegetable intake and risk of cardiovascular disease: the Women's Health Study. *Am. J. Clin. Nutr.,* **72**: 922-928.

Lucks, B. C. (2003). *Vitex agnus castus* essential oil and menopausal balance: a research update. *Complement Ther. Nurs. Midwifery,* **9**: 157-160.

Lucks, B. C., Sorensen, J., and Veal, L. (2002). *Vitex agnus-castus* essential oil and menopausal balance: a self-care survey. *Complement Ther. Nurs. Midwifery,* **8**: 148-154.

Maundu, P. M. (1999). Traditional Food Plants of Kenya Retrieved from http: // www.fastonline.org/CD3WD_40/CD3WD/AGRIC/H1093E/EN/ B567_1.HTM

Mehlhorn, H., Schmahl, G., and Schmidt, J. (2005). Extract of the seeds of the plant *Vitex agnus* castus proven to be highly efficacious as a repellent against ticks, fleas, mosquitoes and biting flies. *Parasitol. Res.,* **95**: 363-365.

Mladênka, P., Zatloukalová, L., Filipský, T., and Hrdina, R. (2010). Cardiovascular effects of flavonoids are not caused only by direct antioxidant activity. *Free Radical Biol. Med.,* **49**: 963–975.

Montiel-Herrera, M., Camacho-Hernández, I. L., Ríos-Morgan, A., and Delgado-Vargas, F. (2004). Partial physicochemical and nutritional characterization of the fruit of *Vitex mollis* (Verbenaceae). *J. Food Comp. Anal.,* **17**: 205-215.

Munir, A. A. (1987). A taxonomic revision of the genus *Vitex* L. (Verbenaceae) in Australia. *J. Adelaide Bot. Gard.,* **10**: 31-79.

Ndabikunze, B. K., Masambu, B. N., and Tiisekwa, B. M. (2010). Vitamin C and mineral contents, acceptability and shelf life of juice prepared from four indigenous fruits of the Miombo woodlands of Tanzania. *J. Food Agric. Environm.,* **8**: 91-96.

Novak, J., Draxler, L., Gohler, I., and Franz, C. M. (2005). Essential oil composition of *Vitex agnus-castus*–comparison of accessions and different plant organs. *Flavour Frag. J.*, **20**: 186-192.

NRC. (2008). Lost Crops of Africa: Volume III: Fruits. The National Academies Press. Washington, DC.

Ochieng, C. O., and Nanwa, B. O. (2010). Proximate composition, phenolic content and antioxidant activities of three black plum (*Vitex* sp.) fruits: preliminary results. *J. Food Technol.*, **8**: 118-125.

Ohyama, K., Akaike, T., Hirobe, C., and Yamakawa, T. (2003). Cytotoxicity and apoptotic inducibility of *Vitex agnus-castus* fruit extract in cultured human normal and cancer cells and effect on growth. *Biol. Pharm. Bull.*, **26**: 10-18.

Ohyama, K., Akaike, T., Imai, M., Toyoda, H., Hirobe, C., and Bessho, T. (2005). Human gastric signet ring carcinoma (KATO-III) cell apoptosis induced by *Vitex agnus-castus* fruit extract through intracellular oxidative stress. *Int. J. Biochem. Cell Biol.*, **37**: 1496-1510.

Ono, M., Ito, Y., and Nohara, T. (1998). A labdane diterpente glycoside from fruit of *Vitex rotundifolia*. *Phytochemistry*, **48**: 207-209.

Ono, M., Nishida, Y., Masuoka, C., Li, J. C., Okawa, M., Ikeda, T., and Nohara, T. (2004). Lignan derivatives and a norditerpene from the seeds of *Vitex negundo*. *J. Nat. Prod.*, **67**: 2073-2075.

Ono, M., Sawamura, H., Ito, Y., Mizuki, K., and Nohara, T. (2000). Diterpenoids from the fruits of *Vitex trifolia*. *Phytochemistry*, **55**: 873-877.

Ono, M., Yamamoto, M., Ito, Y., Yamashita, M., and Nohara, T. (1999). Diterpenes from the fruits of *Vitex rotundifolia*. *J. Nat. Prod.*, **62**: 1532-1537.

Ono, M., Yamasaki, T., Konoshita, M., Ikeda, T., Okawa, M., Kinjo, J., Yoshimitsu, H., and Nohara, T. (2008). Five new diterpenoids, viteagnusins A–E, from the fruit of *Vitex agnus-castus*. *Chem. Pharm. Bull.*, **56**: 1621-1624.

Ono, M., Yanaka, T., Yamamoto, M., Ito, Y., and Nohara, T. (2002). New diterpenes and norditerpenes from the fruits of *Vitex rotundifolia*. *J. Nat. Prod.*, **65**: 357-541.

Padmalatha, K., Jayaram, K., Raju, N. L., Prasad, M. N. V., and Arora, R. (2009). Ethnopharmacological and biotechnological significance of *Vitex*. *Biorem. Biod. Bioavailaility*, **3**: 6-14.

Palombo, E. A. (2006). Phytochemicals from traditional medicinal plants used in the treatment of diarrhoea: modes of action and effects on intestinal function. *Phytother. Res.*, **20**: 717-724.

Ríos, J. L., and Recio, M. C. (2005). Medicinal plants and antimicrobial activity. *J. Ethnopharmacol.*, **100**: 80-84.

Rissanen, T. H., Voutilainen, S., Virtanen, J. K., Venho, B., Vanharanta, M., Mursu, J., and Salonen, J. K. (2003). Low intake of fruits, berries and vegetables is associated with excess mortality in men: the kuopio ischaemic heart disease risk factor (KIHD) study. *J. Nutr.*, **133**: 199-204.

Roemheld-Hamm, B. (2005). Chasteberry. *Am. Fam. Physician*, **72**: 821-824.

Saglam, H., Pabuccuoglu, A., and Kivcak, B. (2007). Antioxidant activity of *Vitex agnus-castus* L. extracts. *Phytother. Res.*, **21**: 1059-1060.

Sarikurkcu, C., Arisoy, K., Tepe, B., Cakir, A., Abali, G., and Mete, E. (2009). Studies on the antioxidant activity of essential oil and different solvent extracts of *Vitex agnus castus* L. fruits from Turkey. *Food Chem. Toxicol.*, **47**: 2479-2483.

Sava, V. M., Yang, S. M., Hong, M. Y., Yang, P. C., and Huang, G. S. (2001). Isolation and characterization of melanic pigments derived from tea and tea polyphenols. *Food Chem.*, **73**: 177-184.

Seniuk, O. F., Gorovoj, L. F., Beketova, G. V., Savichuk, H. O., Rytik, P. G., Kucherov, I. I., Alla B. Prilutskay, A. B., and Prilutsky, A. I. (2010). Anti-infective properties of the melanin-glucan complex obtained from medicinal tinder bracket mushroom, *Fomes fomentarius* (L.: Fr.) (Aphyllophoromycetideae). *Int. J. Med. Mushrooms*, **13**: 7-8.

Shin, T. Y., Kim, S. H., Lim, J. P., Suh, E. S., Jeong, H. J., Kim, B. D., Park, E. J., Hwang, W. J., Rye, D. G., Baek, S. H., An, N. H., and Kim, H. M. (2000). Effect of *Vitex rotundifolia* on immediate-type allergic reaction. *J. Ethnopharmacol.*, **72**: 443–450.

Stojkovic, D., Sokovic, M., Glamoclija, J., Dzamic, A., Ciric, A., Ristic, M., and Grubisic, D. (2011). Chemical composition and antimicrobial activity of *Vitex agnus-castus* L. fruits and leaves essential oils. *Food Chem.*, **128**: 1017-1022.

Suleiman, M. M., and Yusuf, S. (2008). Antidiarrheal activity of the fruits of *Vitex doniana* in laboratory animals. *Pharm. Biol.*, **46**: 387-392.

Webster, D. E. (2008). Botanical, chemical, genetic, and pharmacological studies of *Vitex agnus-castus*. ProQuest. Ann Arbor, Michigan.

Webster, D. E., He, Y., Chen, S. N., Pauli, G. F., Farnsworth, N. R., and Wang, Z. J. (2011). Opioidergic mechanisms underlying the actions of *Vitex agnus-castus L. Biochem. Pharmacol.*, **81**: 170-177.

Webster, D. E., Lu, J., Chen, S. N., Farnsworth, N. R., and Wang, Z. J. (2006). Activation of the mu-opiate receptor by *Vitex agnus-castus* methanol extracts: implication for its use in PMS. *J. Ethnopharmacol.*, **106**: 216-221.

WHO. (2008). The global burden of disease: 2004 update. World Health Organization. Switzerland.

Wrolstad, R. E. (2002). Current Protocols in Food Analytical Chemistry. John Wiley and Sons, Inc.

Wu, J., Zhou, T., Zhang, S. W., Zhang, X. H., and Xuan, L. J. (2009). Cytotoxic terpenoids from the fruits of *Vitex trifolia* L. *Planta Medica*, **75**: 367-370.

Wuttke, W., Jarry, H., Christoffel, V., Spengler, B., and Seidlova-Wuttke, D. (2003). Chaste tree (*Vitex agnus-castus*)-pharmacology and clinical indications. *Phytomedicine*, **10**: 348-357.

Xu, J., Zhang, Q., Zhao, L., Wang, Y., Xue, L., Han, T., Zheng, C., and Qin, L. (2012). Quantitative determination and pharmacokinetic study of casticin in rat plasma by liquid chromatography-mass spectrometry. *J. Pharm. Biomed. Anal.*, **61**: 242-246.

Yamasaki, T., Kawabata, T., Masuoka, C., Kinjo, J., Ikeda, T., Nohara, T., and Ono, M. (2008). Two new lignan glucosides from the fruit of *Vitex cannabifolia*. *J. Nat. Med.*, **62**: 47-51.

Ye, Q., Zhang, Q., Zheng, C., Wang, Y., and Qin, L. (2010). Casticin, a flavonoid isolated from *Vitex rotundifolia*, inhibits prolactin release *in vivo* and *in vitro*. *Acta Pharmacol. Sin.*, **31**: 1564–1568.

Bioactive Phytochemicals: Perspectives for Modern Medicine Vol. 2 (2014) *Pages* **121–136**
Editor: **V.K. Gupta**
Published by: **DAYA PUBLISHING HOUSE, NEW DELHI**

5

Bioactive Bibenzyls and Bis(bibenzyls)s from Bryophytes: Therapeutic Potentials

Abhijit Dey[1] and Jitendra Nath De[2*]

ABSTRACT

Bryophytes represent a group of non vascular plants with diverse phtyto-constituents with pharmacological activity. These small-sized plats are classified into three groups such as liverworts, hornworts and mosses. They are among the pioneer colonizers on the barren rock surface and are subjected to a number of diverse environmental urgencies such as biotic (predation, microbial pathogenecity) and abiotic (heavy metals and other pollutants, temperature, ultraviolet radiation) stresses. Bryophytes, especially the liverworts possess a class of secondary metabolites termed as bibenzyls and bis(bibenzyls) known for their diverse pharmacological efficacy such as antibacterial, antifungal, anti-trypanosomal, leishmanicidal, anti-feedant, cytotoxic/anti-cancer/anti-tumor/anti-proliferative/anti-neoplastic/pro-apoptotic, vasorelaxant, antioxidant and free radical scavenging and nitric oxide inhibitory activities. Although, the higher plants are also able to synthesize such compounds, the authors, in the present review only concentrate on the bibliographical compilation of the available literature related to bioactivity of these compounds found in bryophytes with a note on their therapeutic potential.

Keywords: Bibenzyls, Bis(bibenzyls)s, Bryophytes, Antifungal, Cytotoxic, Pharmacology.

1 Assistant Professor, Department of Botany, Presidency University (Formerly Presidency College), 86/1, College Street, Kolkata – 700 073, West Bengal, India
E-mail: abhijitbio25@yahoo.com

2 Retired Reader and Head of the Department of Botany, Charuchandra College, 22, Lake Road, Kolkata – 700 029, West Bengal, India.

Present address: Flat No. C-103, Sunrise Estate, 14C Radhanath Chaudhuri Road, Kolkata – 700 015, West Bengal, India.

* *Corresponding author*: E-mail: jnde@rocketmail.com

Introduction

Bryophytes, evolutionary placed between algae and pteridophyte, are classified into three groups such as liverworts or hepatics, hornworts and mosses. Bryophytes are considered as one of the pioneer colonizers on barren rock surface during plant succession. They are also subjected to a number of biotic and abiotic stresses due to which a number of secondary metabolites are produced in them (Xie and Lou, 2009). Bryophytes are able to cope up with stresses such as oxidative, water, heavy metal, ultraviolet radiation, bacterial, fungal, insect attack etc. (Wu *et al.*, 2007; Xie and Lou, 2009; Charron and Quatrano, 2009; Dey and De, 2012a). Members of this small sized plant group has also been reported to possess antioxidative (Pejin *et al.*, 2012), nitric oxide scavenging (Chobot *et al.*, 2008; Dey and De, 2012b), cytotoxic (Jensen *et al.*, 2012), antibacterial (Singh *et al.*, 2011; Mukherjee *et al.*, 2012), antifungal (Dey and De, 2011) activities. Several investigations have been carried out to elucidate the phytochemistry, biological efficacy and potential medicinal importance of bryophytes, their crude extracts and isolated compounds (Asakawa, 2001, 2008; Nagashima and Asakawa, 2011; Asakawa *et al.*, 2012).

Bibenzyls and bis(bibenzyls) are two of the major stilbenoids formed by a branch of the flavonoid biosyntheitic pathway. Bibenzyls are classified into simple and substituted (prenylated, geranylated, hydroxybenzyl substituted etc.) types (Xiao *et al.*, 2008). Bibenzyls and bisbibenzyls, being predominant in liverworts, also occur in different plant groups such as mosses (Zheng *et al.*, 1994), pteridophytes (Oiso *et al.*, 1999), monocots (Li *et al.*, 2009a) and dicots (Kosenkova *et al.*, 2007). Bibenzyls are known to possess anti-HIV (Manfredi *et al.*, 2001), antioxidant (Hsiao *et al.*, 1996), nitric oxide (NO) inhibitory (Zhang *et al.*, 2007), cytotoxic (Zheng *et al.*, 1994), anticancer (Sun *et al.*, 2011), antimicrobial (Takagi *et al.*, 1983), antifungal (Zhang *et al.*, 2008), anti-protozoal (Roldos *et al.*, 2008), anti-feedant (Labbé *et al.*, 2007), vasorelaxant (Morita *et al.*, 2011), anti-angiogenic and anti-inflammatory (Liu *et al.*, 2012), anti *P*-glycoprotein and multi drug resistance (MDR) reversal (Xi *et al.*, 2010) activities.

Enumeration

The present review systemically compiles the various bioactivity of bibenzyls and bis(bibenzyls)s from bryophytes. The authors have not only incorporated the bio-efficacy of bryophyte derived bibenzyls and bis(bibenzyls)s, but have also tried to consider the activity of bibenzyls and bis(bibenzyls)s derived and related synthetic compounds. Extensive literature searches using the common scientific databases have generated the following tabulated representation of our present review denoting the source (alphabetically), name and type of the compound, their bioactivity and relevant reference. Bioactivity of some chemically synthesized bibenzyls and bis(bibenzyls)s related to bryophyta has also been compiled. In cases, where the exact name of the bryophyte/liverwort is missing, the source was denoted as bryophyte/liverwort.

Discussion

Bioassays involving bryophyte related bibenzyls and bis(bibenzyls)s have revealed various pharmacological activities. Cytotoxicity of these types of compounds

Table 5.1: A list of bryophyte derived and related bibenzyls and bis(bibenzyls)s and their bioactivity.

Source	*Compound Name*	*Compound Type*	*Bioactivity*	*Reference*
Asterella angusta	Asterelin A, asterelin B, 11-*O*-demethyl marchantin I, Dihydroptychantol A	Bis(bibenzyl)s	Antifungal	Qu *et al.*, 2007
Asterella angusta	Marchantin C	Macrocyclic bisbibenzyl	Pro-apoptotic	Shi *et al.*, 2008a
Asterella angusta	Dihydroptychantol A (chemically synthesized) and derivatives	Macrocyclic bisbibenzyl	Multidrug resistance reversal activity	Sun *et al.*, 2009
Asterella angusta	Dihydroptychantol A (chemically synthesized)	Macrocyclic bisbibenzyl derivative	Reversal of *P*-glycoprotein-mediated multidrug resistance	Li *et al.*, 2009b
Asterella angusta	Dihydroptychantol A	Macrocyclic bisbibenzyl	Pro-apoptotic	Li *et al.*, 2011
Bazzania trilobata	Bazzanin B, bazzanin S, isoplagiochin D	Bisbibenzyls	Antifungal	Scher *et al.*, 2004
Blasia pusilla	Riccardin C	Macrocyclic bisbibenzyl	Liver X receptor (LXR) alpha agonist and an LXRbeta antagonist	Tamehiro *et al.*, 2005
Conocephalum japonicum	Marchantin A, marchantin E, marchantin C and others	Bibenzyls	Cytotoxicity	Liu *et al.*, 2011
Dumortiera angusta	Marchantin C	Macrocyclic bisbibenzyl	Microtubule inhibitor, nti-tumor	Shi *et al.*, 2009
Dumortiera angusta	Marchantin C	Macrocyclic bisbibenzyl	Anti-angiogenic	Lv *et al.*, 2012
Dumortiera hirsuta	Lunularin	Bibenzyl	Cytotoxicity, antimicrobial	Lu *et al.*, 2006
Dumortiera hirsuta	Riccardin D	Macrocyclic bisbibenzyl	Antifungal	Cheng *et al.*, 2009
Dumortiera hirsuta	Riccardin D	Macrocyclic bisbibenzyl	Anti-angiogenic, anticancer	Sun *et al.*, 2011
Dumortiera hirsuta	Riccardin D	Macrocyclic bisbibenzyl	Anti-intestinal adenoma formation, pro-apoptotic, anti-inflammatory	Liu *et al.*, 2012
Dumortiera hirsuta	Riccardin D	Macrocyclic bisbibenzyl	Antifungal	Li *et al.*, 2012

Contd...

Table 5.1–*Contd...*

Source	*Compound Name*	*Compound Type*	*Bioactivity*	*Reference*
Dumortiera hirsuta	Riccardin D	Macrocyclic bisbibenzyl	Antiproliferative, DNA topoisomerase II inhibitor, pro-apoptotic	Xue *et al.*, 2012a
Dumortiera hirsuta	Riccardin D	Macrocyclic bisbibenzyl	Pro-apoptotic, anticancer	Xue *et al.*, 2012b
Frullania inouei	3,3',4,4'-tetramethoxybibenzyl, chrysotobibenzyl, brittonin A, brittonin B	Methoxylated bibenzyls Resistance (MDR)	Cytotoxicity, reversal of multidrug	Guo *et al.*, 2010
Frullania muscicola	3-hydroxy-4'-methoxybibenzyl	Bibenzyl	Antifungal	Lou *et al.*, 2002
Marchantia diptera	Marchantin H	Macrocyclic bisbibenzyl	Antioxidant, free radical scavenger	Hsiao *et al.*, 1996
Marchantia emarginata subsp. *tosana*	Marchantin A	Cyclic bis(bibenzyl ether)	Pro-apoptotic, anticancer, free radical-scavenging activity	Huang *et al.*, 2010
Marchantia polymorpha	Marchantin A	Cyclic bis(bibenzyl)	Antibacterial	Kámory *et al.*, 1995
Marchantia polymorpha	Plagiochin E, 13, neomarchantin A, 13'-*O*-isoproylidenericcardin D	Bis(bibenzyls)	Antifungal	Niu *et al.*, 2006
Marchantia polymorpha	Plagiochin E (obtained in powdered form)	Macrocyclic bisbibenzyl	Fungal resistance reversal	Guo *et al.*, 2008
Marchantia polymorpha	Plagiochin E	Macrocyclic bis(bibenzyl)	Antifungal	Wu *et al.*, 2008
Marchantia polymorpha	Plagiochin E	Macrocyclic bisbibenzyl	P-glycoprotein inhibitory activity, reversal of multidrug resistance in adriamycin-Resistant K562/A02 cells	Shi *et al.*, 2008b
Marchantia polymorpha	Marchantin A	Macrocyclic bisbibenzyl	Marchantin A+Aurora-A kinase inhibitor MLN8237 show cytoxic and anticancer activity	Jensen *et al.*, 2012
Marchantia polymorpha, *Marchantia paleacea*	Several compounds	Bis(bibenzyls)	Vasorelaxant	Morita *et al.*, 2011
Marchantia species	Marchantin A trimethyl ether	Macrocyclic bis(bibenzyl) derivative	Skeletal muscle relaxation activity	Taira *et al.*, 1994

Contd...

Table 5.1–*Contd...*

Source	*Compound Name*	*Compound Type*	*Bioactivity*	*Reference*
Plagiochasm intermedium	Riccardin F, pakyonol	Macrocyclic bisbibenzyls	Reversal of *P*-glycoprotein-mediated multidrug resistance	Ji *et al.*, 2011
Plagiochila fruticosa	Isoplagiochins A and B	Macrocyclic bis(bibenzyls)	Antimitotic	Morita *et al.*, 2009
Plagiochila stephensoniana	4-hydroxy-3'-methoxybibenzyl	Bibenzyl	Antifungal, antibacterial	Lorimer *et al.*, 1993
Polytrichum pallidisetum	Pallidisetin A, pallidisetin B	Cinnamoyl bibenzyls	Cytotoxic	Zheng *et al.*, 1994
Ptogiochasma intermedlum	Riccardin F, riccardin C	Macrocyclic bis(bibenzyls)	Antifungal, modification of fungal resistance (riccardin C with fluconazole)	Xie *et al.*, 2010
Radula appressa	2-geranyl-3,5-dihydroxybi-benzyl	Prenyl bibenzyl	Nitric oxide inhibitory	Harinantenaina *et al.*, 2006
Radula kojana	3,5-dihydroxy-2-(3-methyl-2-butenyl)- and 2-geranyl-3,5-dihydroxybibenzyls	Bibenzyls	5-lipoxygenase and calmodulin inhibitory activity	Asakawa *et al.*, 1991a
Radula perrottetii	Perrottetin A–D	Prenyl bibenzyls	Antimicrobial	Asakawa *et al.*, 1982
Radula sp.	Some prenyl-containing bibenzyls	Prenyl bibenzyls	5-lipoxygenase and calmodulin inhibitory activity, vasopressin antagonist activity	Asakawa *et al.*, 1991b
Riccardia multifida	Riccardin A, riccardin B	Cyclic bis(bibenzyls)	Cytotoxicity	Asakawa *et al.*, 1983
Riccardia polyclada	Several compounds	Polychlorinated bibenzyls	Antifeedant, antifungal, cytotoxicity	Labbé *et al.*, 2007
Ricciocarpos natans	14-hydroxylunularin	Bibenzyl	Anti-protozoal	Roldos *et al.*, 2008
Marchantia polymorpha, *M. paleacea* var. *diptera*, *Plagiochila sciophila*, *Radula kojana*	Marchantin A, B, E, plagiochin A, perrottetin F	Macrocyclic bis(bibenzyls)	Anti-influenza	Iwai *et al.*, 2011
Riccardia marginata	2,4,6-trichloro-3-hydroxybi-benzyl, 2,4-dichloro-3-hydroxy-bibenzyl, 2-chloro-3-hydroxy-bibenzyl	Chlorinated bibenzyls	Antimicrobial	Baek *et al.*, 2004

Contd...

Table 5.1–*Contd...*

Source	*Compound Name*	*Compound Type*	*Bioactivity*	*Reference*
Schistochila glaucescens	Neomarchantins A, and B, marchantin C	Bisbibenzyls	Cytotoxic	Scher *et al.*, 2002
Asterella angusta, Plagiochasm intermedium, Asterella angusta, Marchantia polymorpha	Riccardin C, pakyonol, marchantin M, plagiochin E	Cyclic bisbibenzyls	Cytotoxic, pro-apoptotic	Xu *et al.*, 2010
Marchantia polymorpha and other liverworts, chemically synthesized	Marchantin A, plagiochin A, 2(R)-2-isopropenyl-6,7-dihydroxy-4-(2-phenylethyl) dihydrobenzofuran	Bis(bibenzyls)s and bibenzyls	Antitrypanosomal	Otoguro *et al.*, 2012
Liverwort	Marchantin C	Macrocyclic bisbibenzyl	Anti-invasion and anti-migration, anti-angiogenic activity, anti-tumor	Shen *et al.*, 2010
Liverworts	Marchantin A, marchantin B riccardin A, riccardin F etc.	Bis(bibenzyls)	Nitric oxide inhibitory	Harinantenaina *et al.*, 2005
Liverworts, chemically synthesized	Several compounds	Bisbibenzyls and bibenzyls	Antifungal	Zhang *et al.*, 2011
Chemically synthesized	4,4'-dihydroxybibenzyls	Bibenzyls	Anti-oestrogenic and antifertility	Emmens *et al.*, 1968
Chemically synthesized	Marchantin C, riccardin D and plagiochin E	Macrocyclic bisbibenzyls	Anti-tubulin, anticancer	Jiang *et al.*, 2012

Figure 5.1: Chemical structure of some bioactive bibenzyls and bis(bibenzyls)s.

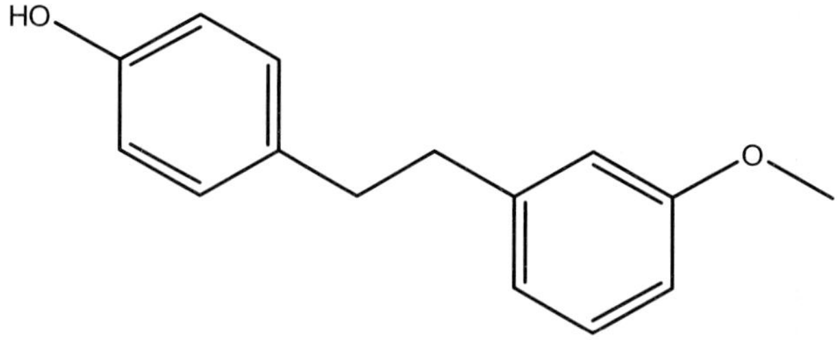

HO
O

4-hydroxy-3'-methoxybibenzyl

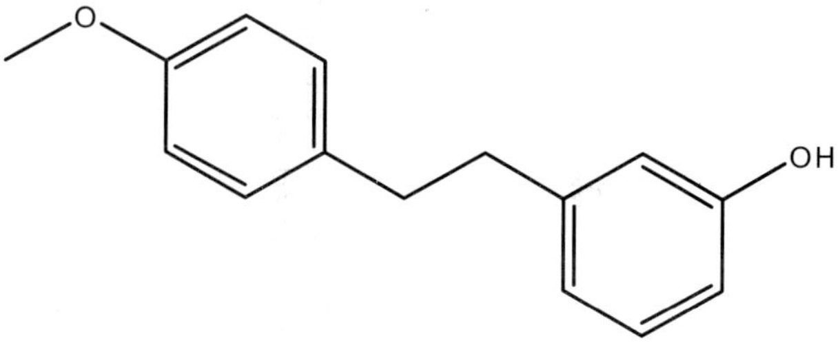

O
OH

3-hydroxy-4'-methoxybibenzyl

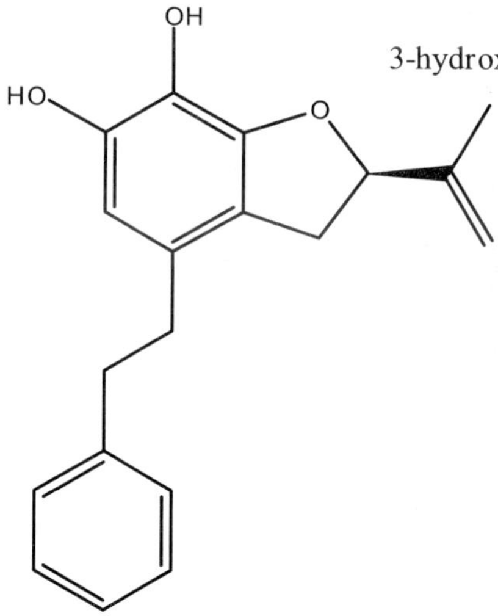

OH
HO
O

2(R)-2-isopropenyl-6,7-dihydroxy-4-(2-phenylethyl)dihydrobenzofuran

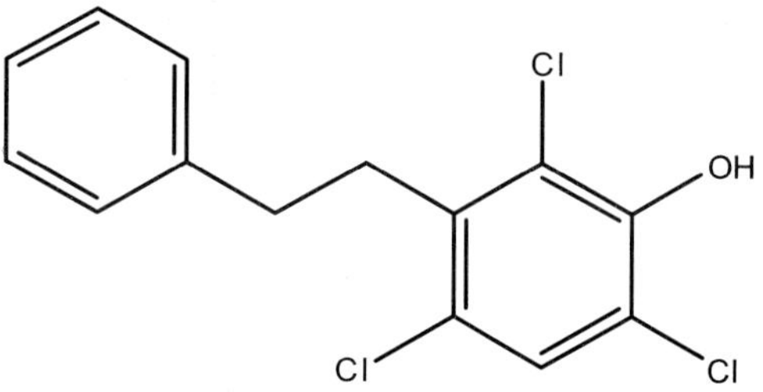

Cl
OH
Cl
Cl

2,4,6-Trichloro-3-hydroxybibenzyl

Contd...

Figure 5.1–*Contd...*

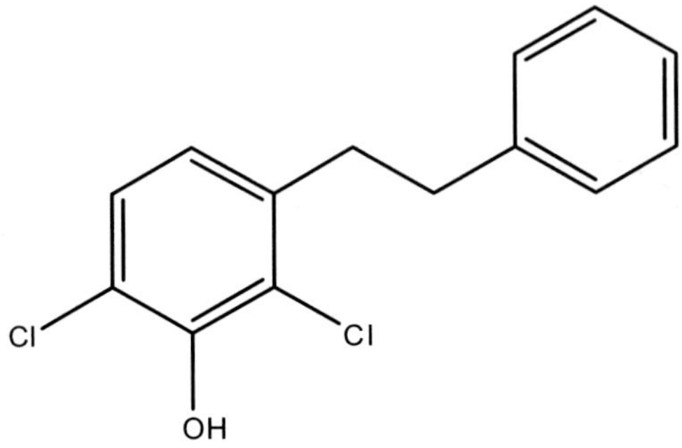

2,4-Dichloro-3-hydroxybibenzyl

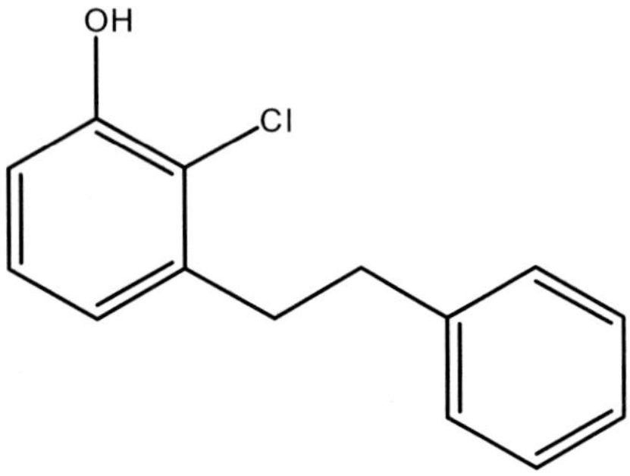

2-Chloro-3-hydroxybibenzyl

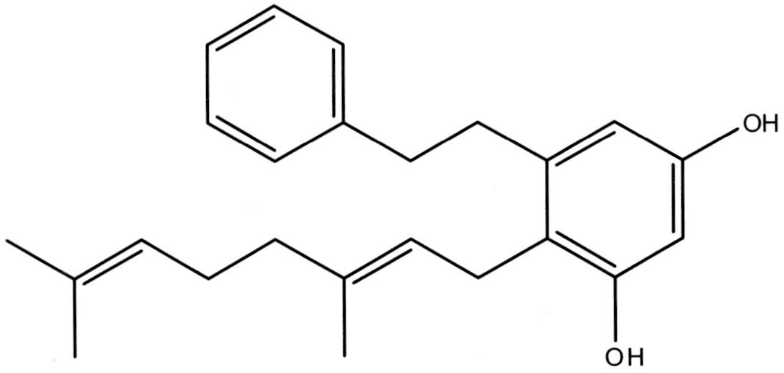

2-geranyl-3,5-dihydroxybibenzyl

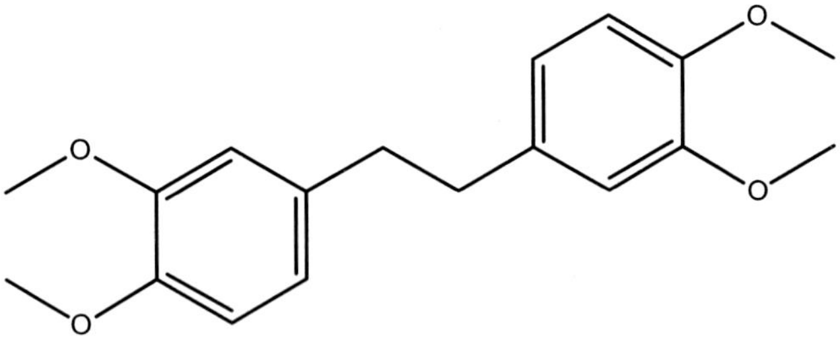

3,3',4,4'-tetramethoxybibenzyl

Contd...

was noted mostly on human tumor/cancer cell lines (Lou *et al.*, 2002) such as pharyngeal squamous carcinoma (KB) (Liu *et al.*, 2011), osteosarcoma U2OS cells (Li *et al.*, 2011), prostate cancer PC3 cells (Xu *et al.*, 2010), umbilical vascular endothelial cells (HUVEC) (Sun *et al.*, 2011), MCF-7 breast cancer cells (Huang *et al.*, 2010), glioma A172 cells (Shi *et al.*, 2008a), tumor cell lines RPMI-7951 melanoma (Zheng *et al.*, 1994) etc. Cytotoxic potential was also tested by brine shrimp lethality bioassay (*Artemia salina*) (Labbé *et al.*, 2007). Cytotoxicity and anticancerous properties of these compounds were attributed to inhibition of tubulin polymerization (Shi *et al.*, 2009;

Figure 5.1–*Contd...*

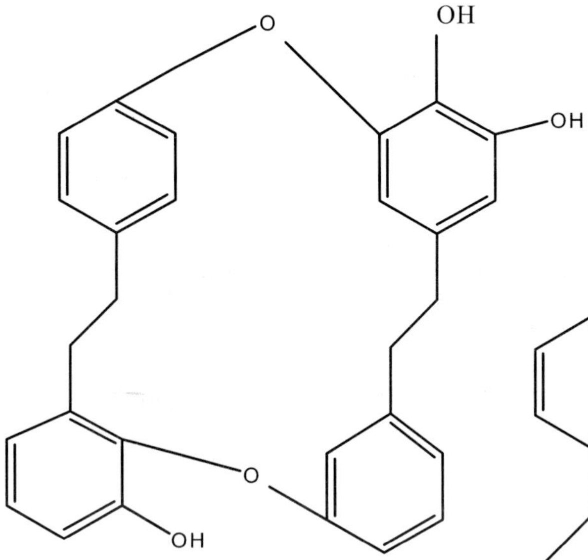

Marchantin A

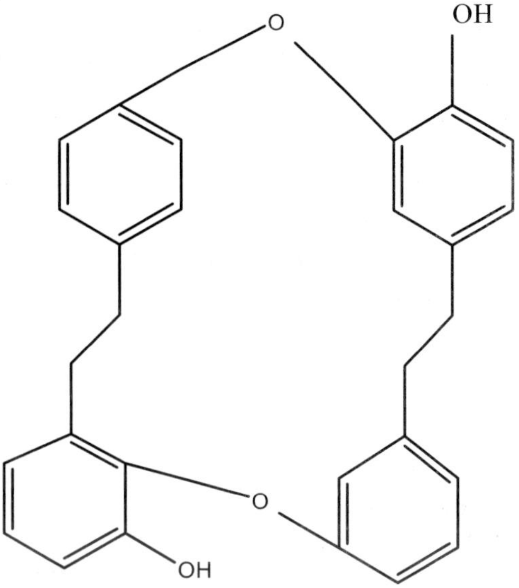

Marchantin C

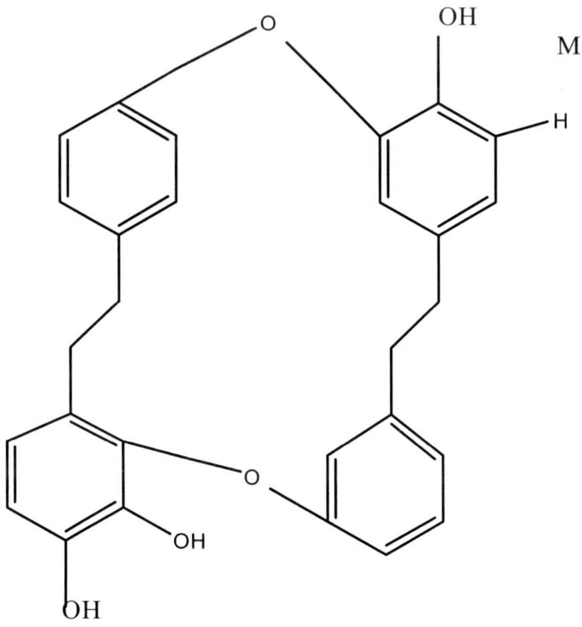

Marchantin H

Morita *et al.*, 2009), enhancement of expression of pro-apoptotic Bax (Xu *et al.*, 2010), reduction in expression of cyclin B1 and D1 genes (Huang *et al.*, 2010), cell cycle arrest at G/M-phase (Li *et al.*, 2011), inhibition of DNA topoisomerase II (Xue *et al.*, 2012a), controlling the spread of cancer cell migration (Shen *et al.*, 2010), anti-inflammatory activity (Liu *et al.*, 2012) and pro-apoptotic activity (Xue *et al.*, 2012b). Anti-angiogenic activity of these compounds against glioma cells and monocytes was also noted (Lv *et al.*, 2012). Anticancer activity of a macrocyclic bisbibenzyl was ascribed to the anti-angiogenic property of the compound (Sun *et al.*, 2011). Furthermore, reversal of multidrug resistance (MDR) vincristine-resistant KB/VCR was noted following the application of certain bibenzyls and bisbibenzyl (Sun *et al.*, 2009; Guo *et al.*, 2010). Reversal of P-glycoprotein (P-gp)-mediated MDR in K562/A02 cells by macrocyclic bisbibenzyl and bisbibenzyl derivatives was also reported (Shi *et al.*, 2008b; Li *et al.*, 2009b; Ji *et al.*, 2011). Synergistic cytotoxicity of marchantin A together with the aurora kinase inhibitor MLN8237 was noted against a human cancer cell line which was attributed to their different inhibitory potentials of various molecular aspects associated with cancer (Jensen *et al.*, 2012). In this connection, the chemopreventive potentials of bibenzyls (Orsini and Verotta, 1999) have also been to be noted.

Antifungal properties of such compounds were mostly tested against the clinical pathogen *Candida albicans* (Niu *et al.*, 2006; Qu *et al.*, 2007). Antifungal experimentation was performed by measuring the minimum inhibitory concentrations (MIC). Reversal or modification of fungal resistance and positive synergistic activity of such compounds in combination with popular antibiotics were also noted (Guo *et al.*, 2008; Xie *et al.*, 2010; Li *et al.*, 2012). Antifungal activity of bibenzyls was assessed against *Trichophyton mentagrophytes* (Baek *et al.*, 2004), causal organism of ringworm in animals especially in rodents and against *Cladosporium herbarum*, a fungal plant pathogen (Labbé *et al.*, 2007). Bisbibenzyls are known to exert antifungal efficacy by modulating a phosphatase encoding gene leading to affected biofilm formation and thus inhibiting the fungus *C. albicans* (Zhang *et al.*, 2011). Other biofilm inhibition related antifungal activity of bibenzyls was also demonstrated (Cheng *et al.*, 2009; Li *et al.*, 2012).

Liverworts derived macrocyclic bisbibenzyls were represented as potential candidature for anti-influenza drugs. *In-vitro* inhibition of influenza PA endonuclease activity by several such compounds was noted and a particular common moiety was thought to be responsible for the bioactivity (Iwai *et al.*, 2011). 14-hydroxylunularin, a hydroxybibenzyl derivative was noted for antiprotozoal efficacy in terms of anti-trypanosomal and leishmanicidal activities (Roldos *et al.*, 2008). In another experiment, *in-vitro* trypanocidal activity of liverworts derived bis(bibenzyls)s and bibenzyls was found (Otoguro *et al.*, 2012). Antifeedant activity of bibenzyls was also noted by relevant bioassays (Labbé *et al.*, 2007).

Hydroxybibenzyls are also known to exhibit anti-oestrogenic and antifertility effects, some of which were found to be pro-oestrogen also (Emmens *et al.*, 1968). Furthermore, liverworts derived bis(bibenzyls) are also known to exhibit nitric oxide inhibitory potential by inhibiting lipopolysaccharide (LPS)-induced inducible nitric oxide synthase (iNOS) in RAW 264.7 macrophages (Harinantenaina *et al.*, 2005,

2006). Among the other bioactivities, skeletal muscle relaxation activity (Taira *et al.*, 1994), antioxidant and free radical-scavenging activity (Hsiao *et al.*, 1996; Huang *et al.*, 2010) and vasorelaxant effects (Morita *et al.*, 2011) were also ascribed to these compounds.

Conclusions

Most of the pharmacological investigations have shown that the activity of bibenzyls and bis(bibenzyls)s acts in a dose dependent manner. Structure-activity analyses were found to be another aspect of relevant research and some active moieties were found for pharmacological efficacy of certain compounds. Diverse pharmacological activities of these natural bio-active compounds provide an opportunity for clinical and commercial applications since the synthetic drugs face certain limitations. Expensiveness, development of drug resistance and side effects of commercially available conventional medicine can be dealt with the natural products having medicinal potentials based upon complementary and alternative medications.

References

Asakawa, Y. (2001). Recent advances in phytochemistry of bryophytes-acetogenins, terpenoids and bis(bibenzyl)s from selected Japanese, Taiwanese, New Zealand, Argentinean and European liverworts. *Phytochemistry,* **56(3):** 297-312.

Asakawa, Y. (2008). Liverworts-potential source of medicinal compounds. *Curr Pharm Des.,* **14(29):** 3067-3088.

Asakawa, Y., Hashimoto, T., Takikawa, K., Tori, M., and Ogawa, S. (1991b). Prenyl bibenzyls from the liverworts *Radula perrottetii* and *Radula complanata. Phytochemistry,* **30(1):** 235-251.

Asakawa, Y., Kondo, K., Takikawa, N.K., Tori, M., Hashimoto, T., and Ogawa, S. (1991a). Prenyl bibenzyls from the liverwort *Radula kojana. Phytochemistry,* **30(1):** 219-234.

Asakawa, Y., Ludwiczuk, A., and Nagashima, F. (2012). Phytochemical and biological studies of bryophytes. *Phytochemistry,* [Epub ahead of print].

Asakawa, Y., Takikawa, K., Toyota, M., and Takemoto, T. (1982). Novel bibenzyl derivatives and ent-cuparene-type sesquiterpenoids from *Radula* sp. *Phytochemistry,* **21(10):** 2481–2490.

Asakawa, Y., Toyota, M., Taira, Z., Takemoto, T., and Kido, M. (1983). Riccardin A and riccardin B, two novel cyclic bis(bibenzyls) possessing cytotoxicity from the liverwort *Riccardia multifida* (L.) S. Gray. *J. Org. Chem.,* **48(13):** 2164-2167.

Baek, S.H., Phipps, R.K., and Perry, N.B. (2004). Antimicrobial chlorinated bibenzyls from the liverwort *Riccardia marginata. J Nat Prod.,* **67(4):** 718-720.

Charron, A.J., and Quatrano, R.S. (2009). Between a rock and a dry place: the water-stressed moss. *Mol Plant,* **2(3):** 478-486.

Cheng, A., Sun, L., Wu, X., and Lou, H. (2009). The inhibitory effect of a macrocyclic bisbibenzyl riccardin D on the biofilms of *Candida albicans. Biol Pharm Bull.,* **32(8):** 1417-1421.

Chobot, V., Kubicová, L., Nabbout, S., Jahodár, L., and Hadacek, F. (2008). Evaluation of antioxidant activity of some common mosses. *Z Naturforsch C.*, **63(7-8):** 476-482.

Dey, A., and De, J.N. (2011). Antifungal bryophytes: a possible role against human pathogens and in crop protection. *Research Journal of Botany*, **6(4):** 129-140.

Dey, A., and De, J.N. (2012a). Antioxidative potential of Bryophytes: Stress tolerance and commercial perspectives: A Review. *Pharmacologia*, **3(6)**: 151-159.

Dey, A., and De, J.N. (2012b). Nitric oxide inhibitory potential of bryophytes. *Insight Ethnopharmacol.*, **2(1):** 3-4.

Dey, A., De, A., Ghosh, P. and Mukherjee, S. (2013). Altitude and tissue type influence the antioxidative potential of *Pellia endiviifolia* (Dicks.) Dumort. from Darjeeling Himalaya. *Pak J. Bio. Sci.* (In Press).

Emmens, G.W., Collins, D.J., Hobbs, J.J., Miller, B.G., and Owen, W.H. (1968). Anti-oestrogenic and antifertility properties of some 4,4'-dihydroxybibenzyls. *J Reprod Fertil.*, **16(1):** 1-6.

Guo, D.X., Xiang, F., Wang, X.N., Yuan, H.Q., Xi, G.M., Wang, Y.Y., Yu, W.T., and Lou, H.X. (2010). Labdane diterpenoids and highly methoxylated bibenzyls from the liverwort *Frullania inouei*. *Phytochemistry*, **71(13):** 1573-1578.

Guo, X.L., Leng, P., Yang, Y., Yu, L.G., and Lou, H.X. (2008). Plagiochin E, a botanic-derived phenolic compound, reverses fungal resistance to fluconazole relating to the efflux pump. *J Appl Microbiol.*, **104(3):** 831-838.

Harinantenaina, L., Quang, D.N., Takeshi, N., Hashimoto, T., Kohchi, C., Soma, G., and Asakawa, Y. (2005). Bis(bibenzyls) from liverworts inhibit lipopolysaccharide-induced inducible NOS in RAW 264.7 cells: a study of structure-activity relationships and molecular mechanism. *J Nat Prod.*, **68(12):** 1779-1781.

Harinantenaina, L., Takahara, Y., Nishizawa, T., Kohchi, C., Soma, G., and Asakawa, Y. (2006). Chemical constituents of malagasy liverworts, part V: prenyl bibenzyls and clerodane diterpenoids with nitric oxide inhibitory activity from *Radula appressa* and *Thysananthus spathulistipus*. *Chem Pharm Bull (Tokyo).*, **54(7):** 1046-1049.

Hsiao, G., Teng, C.M., Wu, C.L., and Ko, F.N. (1996). Marchantin H as a natural antioxidant and free radical scavenger. *Arch Biochem Biophys.*, **334(1):** 18-26.

Huang, W.J., Wu, C.L., Lin, C.W., Chi, L.L., Chen, P.Y., Chiu, C.J., Huang, C.Y., and Chen, C.N. (2010). Marchantin A, a cyclic bis(bibenzyl ether), isolated from the liverwort *Marchantia emarginata* subsp. tosana induces apoptosis in human MCF-7 breast cancer cells. *Cancer Lett.*, **291(1):** 108-119.

Iwai, Y., Murakami, K., Gomi, Y., Hashimoto, T., Asakawa, Y., Okuno, Y., Ishikawa, T., Hatakeyama, D., Echigo, N., and Kuzuhara, T. (2011). Anti-influenza activity of marchantins, macrocyclic bisbibenzyls contained in liverworts. *PLoS One.*, **6(5):** e19825.

Jensen, J.S., Omarsdottir, S., Thorsteinsdottir, J.B., Ogmundsdottir, H.M., and Olafsdottir, E.S. (2012). Synergistic cytotoxic effect of the microtubule inhibitor marchantin A from *Marchantia polymorpha* and the Aurora kinase inhibitor MLN8237 on breast cancer cells *in vitro*. *Planta Med.*, **78(5):** 448-454.

Ji, M., Shi, Y., and Lou, H. (2011). Overcoming of P-glycoprotein-mediated multidrug resistance in K562/A02 cells using riccardin F and pakyonol, bisbibenzyl derivatives from liverworts. *Biosci Trends*, **5(5):** 192-197.

Jiang, J., Sun, B., Wang, Y.Y., Cui, M., Zhang, L., Cui, C.Z., Wang, Y.F., Liu, X.G., and Lou, H.X. (2012). Synthesis of macrocyclic bisbibenzyl derivatives and their anticancer effects as anti-tubulin agents. *Bioorg Med Chem.*, **20(7):** 2382-2391.

Kámory, E., Keserü, G.M., and Papp, B. (1995). Isolation and antibacterial activity of marchantin A, a cyclic bis(bibenzyl) constituent of Hungarian *Marchantia polymorpha*. *Planta Med.*, **61(4):** 387-388.

Kosenkova, Y.S., Polovinka, M.P., Komarova, N.I., Korchagina, D.V., Kurochkina, N.Y., Cheremushkina, V.A., and Salakhutdinov, N.F. (2007). Riccardin C, a bisbibenzyl compound from *Primula macrocalyx*. *Chemistry of natural compounds*, **43(6):** 712-713.

Labbé, C., Faini, F., Villagrán, C., Coll, J., and Rycroft, D.S. (2007). Bioactive polychlorinated bibenzyls from the liverwort *Riccardia polyclada*. *J Nat Prod.*, **70(12):** 2019-2021.

Li, X., Sun, B., Zhu, C.J., Yuan, H.Q., Shi, Y.Q., Gao, J., Li, S.J., and Lou, H.X. (2009b). Reversal of *p*-glycoprotein-mediated multidrug resistance by macrocyclic bisbibenzyl derivatives in adriamycin-resistant human myelogenous leukemia (K562/A02) cells. *Toxicol In Vitro*, **23(1):** 29-36.

Li, X., Wu, W.K., Sun, B., Cui, M., Liu, S., Gao, J., and Lou, H. (2011). Dihydroptychantol A, a macrocyclic bisbibenzyl derivative, induces autophagy and following apoptosis associated with p53 pathway in human osteosarcoma U2OS cells. *Toxicol Appl Pharmacol.*, **251(2):** 146-154.

Li, Y., Ma, Y., Zhang, L., Guo, F., Ren, L., Yang, R., Li, Y., and Lou, H. (2012). *In vivo* inhibitory effect on the biofilm formation of *Candida albicans* by liverwort derived Riccardin D. *PLoS One.*, **7(4):** e35543.

Li, Y., Wang, C.L., Wang, Y.J., Guo, S.X., Yang, J.S., Chen, X.M., and Xiao, P.G. (2009a). Three new bibenzyl derivatives from *Dendrobium candidum*. *Chem Pharm Bull.*, **57(2):** 218-219.

Liu, H.P., Gao, Z.H., Cui, S.X., Sun, D.F., Wang, Y., Zhao, C.R., Lou, H.X., and Qu, X.J. (2012). Inhibition of intestinal adenoma formation in APC(Min/+) mice by Riccardin D, a natural product derived from liverwort plant *Dumortiera hirsuta*. *PLoS One.*, **7(3):** e33243.

Liu, N., Guo, D.X., Wang, Y.Y., Wang, L.N., Ji, M., and Lou, H.X. (2011). Aromatic compounds from the liverwort *Conocephalum japonicum*. *Nat Prod Commun.*, **6(1):** 49-52.

Lorimer, S.D., Perry, N.B., and Tangney, R.S. (1993). An antifungal bibenzyl from the New Zealand liverwort, *Plagiochila stephensoniana*. Bioactivity-directed isolation, synthesis, and analysis. *J. Nat. Prod.,* **56(9):** 1444–1450.

Lou, H.X., Li, G.Y., and Wang, F.Q. (2002). A cytotoxic diterpenoid and antifungal phenolic compounds from *Frullania muscicola* Steph. *J Asian Nat Prod Res.,* **4(2):** 87-94.

Lu, Z.Q., Fan, P.H., Ji, M., and Lou, H.X. (2006). Terpenoids and bisbibenzyls from Chinese liverworts *Conocephalum conicum* and *Dumortiera hirsuta*. *J Asian Nat Prod Res.,* **8(1-2):** 187-192.

Lv, Y., Song, Q., Shao, Q., Gao, W., Mao, H., Lou, H., Qu, X., and Li, X. (2012). Comparison of the effects of marchantin C and fucoidan on sFlt-1 and angiogenesis in glioma microenvironment. *J Pharm Pharmacol.,* **64(4):** 604-609.

Manfredi, K.P., Vallurupalli, V., Demidova, M., Kindscher, K., and Pannell, L.K. (2001). Isolation of an anti-HIV diprenylated bibenzyl from *Glycyrrhiza lepidota*. *Phytochemistry,* **58(1):** 153-157.

Morita, H., Tomizawa, Y., Tsuchiya, T., Hirasawa, Y., Hashimoto, T., and Asakawa, Y. (2009). Antimitotic activity of two macrocyclic bis(bibenzyls), isoplagiochins A and B from the Liverwort *Plagiochila fruticosa*. *Bioorg Med Chem Lett.,* **19(2):** 493-496.

Morita, H., Zaima, K., Koga, I., Saito, A., Tamamoto, H., Okazaki, H., Kaneda, T., Hashimoto, T., and Asakawa, Y. (2011). Vasorelaxant effects of macrocyclic bis(bibenzyls) from liverworts. *Bioorg Med Chem.,* **19(13):** 4051-4056.

Mukherjee, S., Ghosh, P., De, A, and Dey, A. (2012). *In vitro* antibacterial activty of different tissue types of *Dumortiera hirsuta* from different altitudes of eastern Himalaya. *Asian Pacific J. Tro. Dis.,* **2(Suppl. 1):** S285-S290.

Nagashima, F., and Asakawa, Y. (2011). Terpenoids and bibenzyls from three Argentine liverworts. *Molecules,* **16(12):** 10471-10478.

Niu, C., Qu, J.B., and Lou, H.X. (2006). Antifungal bis[bibenzyls] from the Chinese liverwort *Marchantia polymorpha* L. *Chem Biodivers.,* **3(1):** 34-40.

Oiso, Y., Toyota, M., and Asakawa, Y. (1999). Occurrence of a bis-bibenzyl derivative in the Japanese fern *Hymenophyllum barbatum*: First isolation and identification of Perrottetin H from the pteridophytes. *Chem. Pharm. Bull.,* **47(2):** 297-298.

Otoguro, K., Ishiyama, A., Iwatsuki, M., Namatame, M., Nishihara-Tukashima, A., Kiyohara, H., Hashimoto, T., Asakawa, Y., Omura, S., and Yamada, H. (2012). *In vitro* antitrypanosomal activity of bis(bibenzyls)s and bibenzyls from liverworts against *Trypanosoma brucei*. *J Nat Med.,* **66(2):** 377-382.

Pejin, B., Bogdanovic-Pristov, J., Pejin, I., and Sabovljevic, M. (2012). Potential antioxidant activity of the moss *Bryum moravicum*. *Nat Prod Res.,* [Epub ahead of print].

Qu, J., Xie, C., Guo, H., Yu, W., and Lou, H. (2007). Antifungal dibenzofuran bis(bibenzyl)s from the liverwort *Asterella angusta*. *Phytochemistry,* **68(13):** 1767-1774.

Roldos, V., Nakayama, H., Rolón, M., Montero-Torres, A., Trucco, F., Torres, S., Vega, C., Marrero-Ponce, Y., Heguaburu, V., Yaluff, G., Gómez-Barrio, A., Sanabria, L., Ferreira, M.E., Rojas de Arias, A., and Pandolfi, E. (2008). Activity of a hydroxybibenzyl bryophyte constituent against *Leishmania* spp. and *Trypanosoma cruzi*: in silico, *in vitro* and *in vivo* activity studies. *Eur J Med Chem.*, **43(9):** 1797-1807.

Scher, J.M., Burgess, E.J., Lorimer, S.D., and Perry, N.B. (2002). A cytotoxic sesquiterpene and unprecedented sesquiterpene-bisbibenzyl compounds from the Liverwort *Schistochila glaucescens*. *Tetrahedron*, **58(39):** 7875–7882.

Scher, J.M., Speakman, J.B., Zapp, J., and Becker, H. (2004). Bioactivity guided isolation of antifungal compounds from the liverwort *Bazzania trilobata* (L.) S.F. Gray. *Phytochemistry*, **65(18):** 2583-2588.

Shen, J., Li, G., Liu, Q., He, Q., Gu, J., Shi, Y., and Lou, H. (2010). Marchantin C: a potential anti-invasion agent in glioma cells. *Cancer Biol Ther.*, **9(1):** 33-39.

Shi, Y.Q., Liao, Y.X., Qu, X.J., Yuan, H.Q., Li, S., Qu, J.B., and Lou, H.X. (2008a). Marchantin C, a macrocyclic bisbibenzyl, induces apoptosis of human glioma A172 cells. *Cancer Lett.*, **262(2):** 173-182.

Shi, Y.Q., Qu, X.J., Liao, Y.X., Xie, C.F., Cheng, Y.N., Li, S., and Lou, H.X. (2008b). Reversal effect of a macrocyclic bisbibenzyl plagiochin E on multidrug resistance in adriamycin-resistant K562/A02 cells. *Eur J Pharmacol.*, **584(1):** 66-71.

Shi, Y.Q., Zhu, C.J., Yuan, H.Q., Li, B.Q., Gao, J., Qu, X.J., Sun, B., Cheng, Y.N., Li, S., Li, X., and Lou, H.X. (2009). Marchantin C, a novel microtubule inhibitor from liverwort with anti-tumor activity both *in vivo* and *in vitro*. *Cancer Lett.*, **276(2):** 160-170.

Singh, M., Singh, S., Nath, V., Sahu, V., and Rawat, A.K. (2011). Antibacterial activity of some bryophytes used traditionally for the treatment of burn infections. *Pharm Biol.*, **49(5):** 526-530.

Sun, B., Yuan, H.Q., Xi, G.M., Ma, Y.D., and Lou, H.X. (2009). Synthesis and multidrug resistance reversal activity of dihydroptychantol A and its novel derivatives. *Bioorg Med Chem.*, **17(14):** 4981-4989.

Sun, C.C., Zhang, Y.S., Xue, X., Cheng, Y.N., Liu, H.P., Zhao, C.R., Lou, H.X., and Qu, X.J. (2011). Inhibition of angiogenesis involves in anticancer activity of riccardin D, a macrocyclic bisbibenzyl, in human lung carcinoma. *Eur J Pharmacol.*, **667(1-3):** 136-143.

Taira, Z., Takei, M., Endo, K., Hashimoto, T., Sakiya, Y., and Asakawa, Y. (1994). Marchantin A trimethyl ether: its molecular structure and tubocurarine-like skeletal muscle relaxation activity. *Chem Pharm Bull.*, **42(1):** 52-56.

Takagi, S., Yamaki, M., and Inoue, K. (1983). Antimicrobial agents from *Bletilla striata*. *Phytochemistry*, **22(4):** 1011-1015.

Tamehiro, N., Sato, Y., Suzuki, T., Hashimoto, T., Asakawa, Y., Yokoyama, S., Kawanishi, T., Ohno, Y., Inoue, K., Nagao, T., and Nishimaki-Mogami, T. (2005).

Riccardin C: a natural product that functions as a liver X receptor (LXR)alpha agonist and an LXRbeta antagonist. *FEBS Lett.*, **579(24):** 5299-5304.

Wu, Y.H., Gu, Y.H., Liu, P., and Zoltán, T. (2007). Effects of enhanced UV-B radiation on the growth of five bryophytes in Changbai Mountains. *Ying Yong Sheng Tai Xue Bao.*, **18(9):** 2139-2143.

Wu, X.Z., Cheng, A.X., Sun, L.M., and Lou, H.X. (2008). Effect of plagiochin E, an antifungal macrocyclic bis(bibenzyl), on cell wall chitin synthesis in *Candida albicans*. *Acta Pharmacologica Sinica*, **29(12)**: 1478-1485.

Xi, G.M., Sun, B., Jiang, H.H., Kong, F., Yuan, H.Q., and Lou, H.X. (2010). Bisbibenzyl derivatives sensitize vincristine-resistant KB/VCR cells to chemotherapeutic agents by retarding P-gp activity. *Bioorg Med Chem.*, **18(18):** 6725-6733.

Xiao, K., Zhang, H.J., Xuan, L.J., Zhang, J., Xu, Y.M., and Bai, D.L. (2008). Stilbenoids: chemistry and bioactivities. *In*: Studies in Natural Products Chemistry, Vol. 34, Ed. by Atta-ur-Rahman, Elsevier B.V., pp. 453-646.

Xie, C.F., and Lou, H.X. (2009). Secondary metabolites in bryophytes: an ecological aspect. *Chem Biodivers.*, **6(3):** 303-312.

Xie, C.F., Qu, J.B., Wu, X.Z., Liu, N., Ji, M., and Lou, H.X. (2010). Antifungal macrocyclic bis(bibenzyls) from the Chinese liverwort *Plagiochasma intermedlum* L. *Nat Prod Res.*, **24(6):** 515-520.

Xu, A.H., Hu, Z.M., Qu, J.B., Liu, S.M., Syed, A.K., Yuan, H.Q., and Lou, H.X. (2010). Cyclic bisbibenzyls induce growth arrest and apoptosis of human prostate cancer PC3 cells. *Acta Pharmacol Sin.*, **31(5):** 609-615.

Xue, X., Qu, X.J., Gao, Z.H., Sun, C.C., Liu, H.P., Zhao, C.R., Cheng, Y.N., and Lou, H.X. (2012a). Riccardin D, a novel macrocyclic bisbibenzyl, induces apoptosis of human leukemia cells by targeting DNA topoisomerase II. *Invest New Drugs*, **30(1):** 212-222.

Xue, X., Sun, D.F., Sun, C.C., Liu, H.P., Yue, B., Zhao, C.R., Lou, H.X., and Qu, X.J. (2012b). Inhibitory effect of riccardin D on growth of human non-small cell lung cancer: *In vitro* and *in vivo* studies. *Lung Cancer*, **76(3):** 300-308.

Zhang, L., Chang, W., Sun, B., Groh, M., Speicher, A., and Lou, H. (2011). Bisbibenzyls, a new type of antifungal agent, inhibit morphogenesis switch and biofilm formation through upregulation of DPP3 in *Candida albicans*. *PLoS One.*, **6(12):** e28953.

Zhang, X., Xu, J.K., Wang, J., Wang, N.L., Kurihara, H., Kitanaka, S., and Yao, X.S. (2007). Bioactive bibenzyl derivatives and fluorenones from *Dendrobium nobile*. *J Nat Prod.*, **70(1):** 24-28.

Zhang, Y.Z., Xu, G.B., and Zhang, T. (2008). Antifungal stilbenoids from *Stemona japonica*. *J Asian Nat Prod Res.*, **10(7-8):** 639-644.

Zheng, G.Q., Ho, D.K., Elder, P.J., Stephens, R.E., Cottrell, C.E., and Cassady, J.M. (1994). Ohioensins and pallidisetins: novel cytotoxic agents from the moss *Polytrichum pallidisetum*. *J Nat Prod.*, **57(1):** 32-41.

Bioactive Phytochemicals: Perspectives for Modern Medicine Vol. 2 (2013) *Pages* **137–153**
Editor: **V.K. Gupta**
Published by: **DAYA PUBLISHING HOUSE, NEW DELHI**

6

Heterophragma quadriloculare and *Heterophragma adenophyllum*: The Hope for Potential Bioactive Phytoconstituents from *Heterophragma* Genus

B.H. Satani[1*], V.S. Surana[1], R.H. Patel[2] and S.H. Mishra[3]

ABSTRACT

Heterophragma is a genus of family Bignoneaceae. Different species of the genus Heterophragma mentioned in literature are H. quadriloculare, H. adenophyllum, H. chelonoides, H. longipes, H. macrolobum, H. suaveolens, H. sulfureum and H. vestitum. There is confusion between reported species of Heterophragma genus for their synonyms. Hence, in this work we have tried to solve that ambiguity. We have also demonstrated world wide availability for authenticated species of Heterophragma genus using literatures. Two species Heterophragma quadriloculare and Heterophragma adenophyllum, are abundantly found in local forests. So we have selected these plants for our study. Morphological, phytochemical and biological information by various authors on both of these plants are compiled in this manuscript. At concluding point we have found that pharmacologically important compounds found in Heterophragma quadriloculare and Heterophragma adenophyllum are ursolic acid, oleanolic acid, α- and β-lapachone, lapachol, verbascoside, lupeol, quercitrin, phytosterole and tectol. We have correlated their traditional as well as

1 Department of Pharmacognosy, Maliba Pharmacy College, Uka Tarsadia University, Bardoli, Gujarat, India

2 Department of Pharmacognosy, Hariom Pharmacy College, Ambav, Kheda, Gujarat, India

3 Department of Pharmacy, Faculty of Technology and Engineering, The M.S. University of Baroda, Gujarat, India

* *Corresponding author*: E-mail: bhavik1231@gmail.com; bhavik.satani@utu.ac.in

ethnomedicinal claims with phytoconstituents reported from plants. Scientific study suggest that there are number of compound present in evaluated plants showing similar kind of activity that may give synergistic effect and that play a vital role while selection of plant for different purpose. Apart from correlating existing reported claims we have also found that plants studied can become important sources of novel drugs and lead compounds for antidiabetic, anti-hypertensive, anti-fungal, anti-viral and anti-cancer activities.

***Keywords*:** Bignoneaceae, Ethnomedicinal, *Heterophragma, Heterophragma adenophyllum, Heterophragma quadriloculare*, Lapachol, Phytoconstituents, Ursolic acid.

Introduction to Genus *Heterophragma*

Taxonomic hierarchy with name of author and year in which this genus has been placed in classification (Zhi-Yun and Thawatchai, 2010; The International Plant Names Index, 2007)

- ☆ **Domain:** Eukaryota–Whittaker and Margulis,1978–eukaryotes
  - ☆ **Kingdom:** Plantae–Haeckel, 1866–Plants
    - ☆ **Subkingdom:** Viridaeplantae–Cavalier-Smith, 1981
      - ☆ **Phylum:** Tracheophyta–Sinnott, 1935 Ex Cavalier-Smith, 1998–Vascular Plants
        - ☆ **Subphylum:** Euphyllophytina
          - ☆ **Infraphylum:** Radiatopses–Kenrick and Crane, 1997
            - ☆ **Class:** Magnoliopsida–Brongniart, 1843–Dicotyledons
              - ☆ **Subclass:** Lamiidae–Takhtajan Ex Reveal, 1992
                - ☆ **Superorder:** Lamianae–Takhtajan, 1967
                  - ☆ **Order:** Scrophulariales–Lindley, 1833
                    - ☆ **Family:** Bignoniaceae–Durande, 1782, Nom. Cons.–Trumpet-Creeper Family
                      - ☆ ***Genus:*** *Heterophragma*
                        - ☆ ***Specific epithet:*** *quadriloculare–K.Schum.*
                          - ☆ ***Botanical name:*** *–Heterophragma quadriloculare K.Schum.*

Description of Genus

Heterophragma is a genus of Bignoneace, containing few selected species from India. Plant from this genus is normally a large tree with opposite or ternate impair pinnate leaves and whites/purple flower in dense terminal downy panicle. The calyx is campalunated and three lobed; the corolla equally five parted, with the margin of divisions waved; there are for fertile stamens surrounded by a purple disk, and surmounted by simple style and two cleft stigma: the capsule is long and pointed;

and the seeds have broad wing, 2n = 40; contains flowering plants (John Lindley and Thomas Moore, 1866).

Species of the Genus *Heterophragma*

Different species of the genus *Heterophragma* mentioned in literature are *H. quadriloculare* K. Schum., *H. adenophyllum* Seem. ex Benth. and Hook.f., *H. chelonoides* Dalzell and A.Gibson, *H. longipes* Baker, *H. macrolobum* Backer ex K.Heyne, *H. suaveolens* Dalzell and A.Gibson, *H. Sulfureum* Kurz, *H. vestitum* Dop, *H. roxburghii* (Spreng.) DC, *Bignonia quadrilocularis* Roxb., *Spathodea roxburghii* Spreng., *Fernandoa adenophylla* Steenis, *Bignonia adenophylla* Wall., *Haplophragma adenophyllum* Wall. and *Spathodea adenophylla* DC. (Biodiversity Heritage Library, 2011; John Lindley and Thomas Moore, 1866).

Geographical Source of some Species of *Heterophragma* Genus

Biodiversity occurrence data (2011) published by *BeBIF Provider, Comisión nacional para el conocimiento y uso de la biodiversidad, Fairchild Tropical Botanic Garden, Missouri Botanical Garden, Museum national d'histoire naturelle et Reseau des Herbiers de France, National Museum of Natural History, SysTax, and The New York Botanical Garden* demonstrates world wide availability for different species of *Heterophragma* genus. This information is summarised in Table 6.1.

Clarification of Ambiguous Species of the *Heterophragma* Genus

Different species of the *Heterophragma* genus reported, but there is confusion between their synonyms. Reports are there that explaining synonyms of some of the species of this genus. As per that literatures there are total eight number of species reported from this genus and others are synonyms of different species. The information in Table 6.2 is showing name and synonyms of different species of this genus (USDA, ARS, National Genetic Resources Program, 2011; Zhi-Yun Zhang and Thawatchai Santisuk, 2010).

First two species *H. quadriloculare* and *H. adenophyllum* are abundantly found in local forests of India. Morphological, phytochemical and biological information by various authors on both of these plants are compiled in this manuscript.

Plant Description for *H. quadriloculare*

Its vernacular names provided by Andhra pradesh forest department are as follows (Andhra Pradesh forest department, 2011; Sandwith NY, 1967).

Ayurveda	: Waras
Hindi and Marathi	: Warras, Pullung, Paatang, Kusaga
Gujarati	: Varas
Telugu	: Bondugu, Barukoli-gottu, Kaligottu
Tamil	: Baro-kala-goru
Kannad	: Bechadi, Adwi-nuggi

Table 6.1: Reported availability of different species of *Heterophragma* genus.

Scientific Name	*Dataset*	*Institution Code*	*Collection Code*	*Catalogue Number*	*Basis of Record*	*Country*
Heterophragma adenophyllum	Missouri Botanical Garden	MO	Tropicos	1059913	Specimen	Thailand
Heterophragma adenophyllum	Fairchild Tropical Botanic Garden	FTG	Herbarium	128078	Specimen	United States
Heterophragma adenophyllum	Fairchild Tropical Botanic Garden	FTG	Herbarium	132809	Specimen	United States
Heterophragma adenophyllum	Fairchild Tropical Botanic Garden	FTG	Herbarium	132810	Specimen	United States
Heterophragma adenophyllum	NMNH Botany Collections	US	Botany	USw22930.2746560	Specimen	Myanmar
Heterophragma roxburghii	Herbario del Instituto de Ecología	IEX	XAL	14949	Unknown	India
Heterophragma roxburghii	NMNH Botany Collections	US	Botany	USw26564.2745491	Specimen	India
Heterophragma roxburghii	NMNH Botany Collections	US	Botany	USw8849.2745209	Specimen	India
Heterophragma roxburghii	NMNH Botany Collections	US	Botany	USw11132.2745461	Specimen	India
Heterophragma roxburghii	NMNH Botany Collections	US	Botany	USw5490.2760663	Specimen	India
Heterophragma sulfureum	Phanerogamie	MNHN	P	P00609746	Specimen	Cambodia
Heterophragma sulfureum	Phanerogamie	MNHN	P	P00609747	Specimen	Cambodia
Heterophragma sulfureum	Phanerogamie	MNHN	P	P00609748	Specimen	Cambodia
Heterophragma sulfureum	NMNH Botany Collections	US	Botany	USw21083.2779276	Specimen	Myanmar
Heterophragma sulfureum	NMNH Botany Collections	US	Botany	USw11253.2775832	Specimen	Myanmar
Heterophragma vestitum	Herbarium of The New York Botanical garden	NY	Herbarium	328856	Specimen	Cambodia

Table 6.2: Names and synonyms of various species of *Heterophragma* Genus.

Name	*Synonyms*
Heterophragma quadriloculare K. Schum.	*Heterophragma roxburghii* (Spreng.) DC. *Bignonia quadrilocularis* Roxb. *Spathodea roxburghii* Spreng.
Heterophragma adenophyllum Seem. Ex Benth. and Hook.f.	*Fernandoa adenophylla* Steenis *Bignonia adenophylla* Wall. *Haplophragma adenophyllum* Wall. *Spathodea adenophylla* DC.
Heterophragma chelonoides Dalzell and A.Gibson	Not Available
Heterophragma longipes Baker	Not Available
Heterophragma macrolobum Backer ex K.Heyne	Not Available
Heterophragma suaveolens Dalzell and A.Gibson	Not Available
Heterophragma sulfureum Kurz	Not Available
Heterophragma vestitum Dop	Not Available

Waras is a large deciduous tree, 5-15 m tall, with brown bark. Compound leaves are 1-2 ft long, crowded near the end of branches. Each leaf has 3-5 pairs of leaflets, and a terminal one. The elliptic leaflets are 5-12 cm long, unequal sided at the base. Flowers occur in panicles, which are densely velvet covered, at the end of branches. Flowers are 5-6 cm, white with a rosy tinge. The five petals are rounded with crinkled margins. Stamens are 4, with hairy filaments (Figure 6.1). Fruit is 20-30 cm long, and 3-5 cm broad, pointed (John Lindley and Thomas Moore, 1866).

In India, it is distributed throughout Gujarat, Madhya Pradesh, Maharashtra, Andhra Pradesh, Karnataka and Tamil Nadu (Khare, 2007). As per the information from Gazetteers of the Bombay precidency *Heterophragma roxburghii* found only or almost entirely on the hill-top and upper slopes but *Heterophragma roxburghii* is a synonym of *H. quadriloculare*. Therefore, we can conclude that this plant is in most the case found at alpine or sub-alpine region (Anonymous, 2006).

Phytochemical Reports for *H. quadriloculare*

Fat Obtained from the Root

Nigam *et al.* (1983) investigated the fat obtained from the root of the plant. The fat obtained from the petroleum ether extract of the root yielded palmitic acid 25.6 percent and stearic acid 2.7 percent. Unsaponifiable matter from the root was found to be a mixture of β-sitosterol and β-amyrin.

Lapachol from the Heartwood and its Conversion to Napthaquinone

Transformation of lapachol to a novel napthoquinone and related compound is reported by Jassbi *et al.* (2006). In boiling pyridine, lapachol is transformed into novel asymmetric naphthoquinone derivative, quadrilone, dehydro-α-lapachone, and adenophyllone. Lapachol used in the study was isolated from the heartwood of *Heterophragma quadriloculare* and found to be the most abundant naphthoquinone

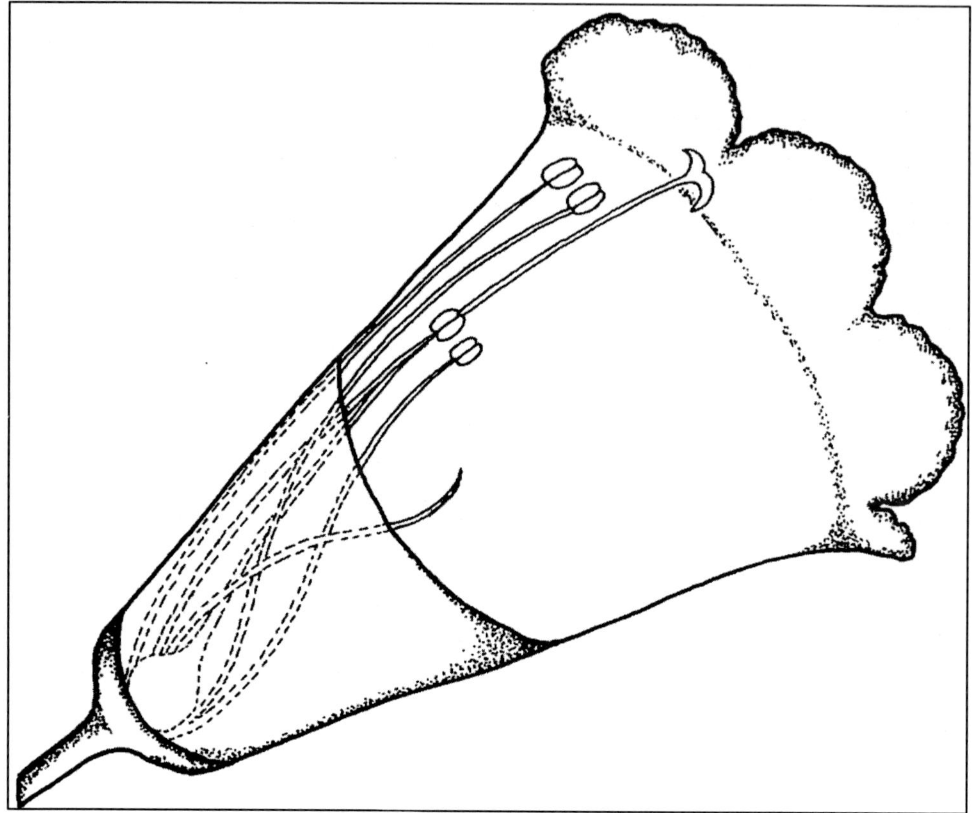

Figure 6.1: Flower of *Heterophragma quadriloculare* drawn to scale.

(Jassbi *et al.*, 2006).

Hentriacontane and Allantoin from Flowers

Flowers were extracted by Zirvi *et al.* (1972) with hot benzene for 48 hrs. The solvent was concentrated to give a waxy solid. The mother liquor was then chromatographed over active acidic alumina to give a hydrocarbon ***hentriacontane*** having molecular formula $C_{31}H_{64}$, melting point 62-63°C (from light petroleum), IR (KBr) υ 2933, 2877, 1471, 737, 720 cm^{-1}. ***Allantoin*** having molecular formula $C_4H_6N_4O_3$ was also isolated from ethanol extract of flowers. Yield of allantoin from ethanol extract of flowers was 0.01 per cent and it was recrystallized from warm ethanol. After isolation allantoin was identified by melting point, IR spectra, NMR spectra and positive furfuraldehyde- HCl colour test (Zirvi *et al.*, 1972).

Allantoin, Steroid, Terpenoids and Alkaloids from Fruits

There are literatures which show the presence of *Allantoin* (yield 0.05 per cent) in fruits of *Heterophragma quadriloculare*. Allantoin is isolated and identified as same as in the flowers. It is reported that medicinal value of the plant in skin diseases could be due to its high content of allantoin, which is a known dermatological agent (Zirvi *et al.*, 1972; Pullaiah, 2006b).

Mukherjee and Ray (1986) reported presence of steroid, terpenoids and alkaloids in fruits of the plant.

Phytochemical Reports on Seeds

Analysis of the fixed oil by Yadav (1981) from the seeds of *Heterophragma quadriloculare* represents presence of palmitic acid, stearic acid, linoleic acid, and oleic acid.

Yadav isolated new saponin from the seed of *H. quadroloculare* and that was methyl cholestenol glucopyranosyl rhamnoside. The structure of isolated saponins is as follow (Yadav, 1992).

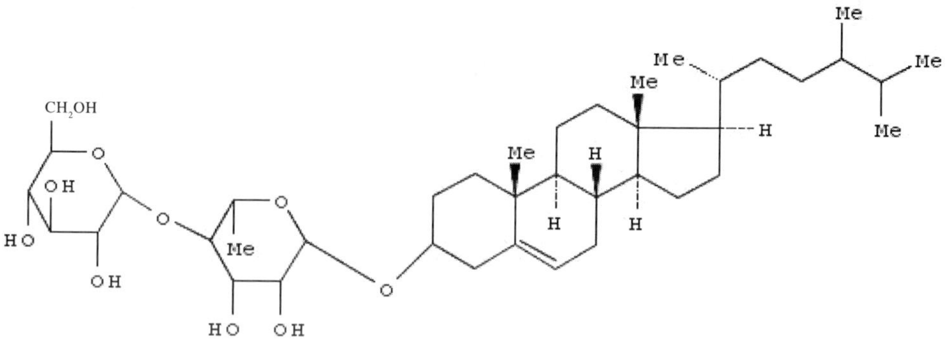

Methyl cholestenol glucopyranosyl rhamnoside

Chemical investigation of the seed performed by Saxena shows presence of lupeol, β-sitosterol, stigmasterol and subulin (Saxena, 1984).

Chemical evaluation of the seed for sugars and amino acids proved presence of nine amino acid and eight sugars. Amino acids found were arginine, cystine, aspartic acid, lysine, tyrosine, glycine, proline, leucine, isoleucine while sugars were lactose, glucose, fructose, L-rhamnose, galactose, xylose, ribose and arabinose (Saxena, 1984).

Phytochemical Investigation of Leaves

Literature shows leaves of *Heterophragma quadriloculare* contain sitosterol and ursolic acid (Zirvi *et al.*, 1972).

Phytochemical investigation of leaves by Mukherjee and Ray (1986) shows presence of steroid, terpenoids, alkaloids, flavanoids, tannins, phenolics. For that about 5 g of the dried crushed plant material has been extracted with 50 ml petroleum ether (60–80°C) in a 100 ml conical flask. The solvent has been then distilled off and the residue taken up in chloroform. The Liebermann-Burchard test has been performed. Bluish green or blue colour has taken to be positive for the presence of steroid and a pink violet colour, for terpenoid. Another 5 g of the dried crushed plant material has been soaked in 2 per cent aqueous HCl (25 ml) for 24 hrs and filtered. The filtrate has been divided into two parts for testing with Dragendorff's and Mayer's reagents.

Turbidity or precipitation with both the reagents has been taken as an evidence for the presence of alkaloid. About 5 g of the dried crushed plant material has been extracted with ethyl alcohol (50 ml) and Shinoda's test (Mg, conc. HCl) has been performed. Development of red colour indicated the presence of flavonoid compounds. About 0.5 g of alcoholic extract has been stirred with 10 ml distilled water and filtered. Addition of $FeCl_3$ reagent to the filtrate resulting in blue black, green or blue green coloration or precipitation has been taken as an evidence for the presence of tannin. About 0.5 g of 50 per cent alcoholic extract has been stirred with 10 ml distilled water and filtered. 10 per cent NaCl and 1 per cent gelatin solution reagent has been added to the filtrate. Resulting white precipitate indicated the presence of phenolics. For testing the presence of saponins about 0.5 g of the dried crushed plant material has been heated with water (10 ml) for 5 minutes at 100 °C in a test tube. It has been cooled and shaken vigorously. Formation of a froth (1-2 cm high) persisting for at least 30 minutes has taken as a positive test for saponins (Mukherjee and Ray, 1986).

Biological Investigations for *H. quadriloculare*

Anti-diabetic Activity

Aerial part extract of the plant in Indonesia is traditionally claimed as anti-diabetic in text reported by Amla Soumyanath (2006).

Hormonal Treatment

It has been found that root is to be used in mennorrhoea, premature ejaculation and night emission (Pullaiah, 2006a).

Activity as Antidote in Poisoning

It is traditionally claimed that root is useful as antidote (Pandey *et al.*, 2005). *Heterophragma quadriloculare* (Roxb.) *K. schum.* is abundant in Madhya Pradesh and frequently used to treat sanke-bite cases (Khare, 2007).

In Skin Disease

Review shows thick fluid-like tar extracted from wood is traditionally used for skin disease (Khare, 2007). Leaf juice is reported to be applied externally on toe sores and in chilblain (Khare, 2007).

Antimicrobial Activity

The essential oil obtained from flowers exhibits antimicrobial activity (Khare, 2007).

As a Food Material

Flowers of *Heterophragma quadriloculare* is reported to be used as food in some regions of Maharashtra (Punekar, 2002).

Unripe fruits of *Heterophragma quadriloculare* are also reported to be used as food in some regions of Maharashtra (Punekar, 2002).

Antifungal Activity

Antifungal activity of essential oil of the leaves of *Heterophragma quadroliculare* is reported. Essential oil from the leaves of *H. quadriloculare* was tested for antifungal properties against 5 skin pathogens (*Aspergillus fumigatus, A. niger, Microsporum gypseum, M. fulvum* and *Penicillium notatum*). Antifungal activity of the test samples was tested following the method of Maruzzella and Henry. Controls were run with 1000 PPM griseofulvin. All dishes were then incubated at 28° C for 3 days. Diameters of inhibition zones which developed around the filter paper discs were recorded. The test oil sample inhibited growth of all test fungi in almost all concentrations. The zone of inhibition was found to be larger with pure oil than griseofulvin (1000 PPM) against all fungi tested (Sahai and Shrivastava, 1984; Bayard group, Eureka Magazine).

Antiseptic Activity

Antiseptic activity of leaf of *Heterophragma quadriloculare* is reported by Suneeti K Jog in the list of plants used by plant healers of Amboli and Tillari area of Western ghat (Jog, 2009).

Plant Description for *H. adenophyllum*

Vernacular Name of the Plants

Local name	: Petthan
English	: Katsagon, Karenwood
Hindi	: Marodphali
Gujarati	: Padad
Bangoli	: Kau-a-turi (Chakma tribe) (John and Thomas, 1866)

Heterophragma adenophyllum seem, ex Benth. and Hook.f (Yingyong and Sunji, 1996) is middle sized deciduous tree, having opposite and pinnate leaves; 5-10 obovate, subsessile, pubescent beaneath leaflets. Flowers are pale brown with large panicles, 2.5 cm long campulunate tomentose calyx, 6 cm long yellowish brown outside tomatoes corolla. It has cylindrical loculicidally 2 valved capsules (Figure 6.2).

It is distributed in the hilly forest of the Assam, Tripura, Andaman. It is also found from Malaysia, Myanmar and Thailand.

Phytochemical Reports for *H. adenophyllum*

α-Lapachone and its Naphthoquinone Derivative

α-Lapachone was previously isolated from the wood of the Bignoniaceae tree *Haplophragma adenophyllum*. A new symmetric naphthoquinone dimer, dilapachone and a novel asymmetric naphthoquinone dimer, adenophyllone were isolated from the heartwood of *Heterophragma adenophyllum* (Kanchanapooma *et al.*, 2001).

Various authors have explained phytochemical investigation on different parts of the plant *Heterophragma adenophyllum*. All that literatures are compiled in Table 6.3. All the information which is given in literature regarding different parts of plant used, extract used and identified phytoconstituents is given in Table 6.3.

Figure 6.2: Morphology showing aerial part and fruits of *Heterophragma adenophyllum*.

Dilapachone

Adenophyllone

Biological Investigations for *H. adenophyllum*

Antifungal Activity

The dichloromethane extracts of *Fernandoa adenophylla* displayed antifungal activity against dermatophytes and wood rot fungi. The activity was shown probably due to lapachol and β-lapachone which were shown to be present in these plant extracts. Microscopical study has shown that these extracts inhibit the development of mycelium and conidia in the filamentous fungi (Marii *et al.*, 2004).

Table 6.3: Phytochemical reports on *Heterophragma adenophyllum.*

Plant Part	*Extract*	*Chemical Constituents*
Leaves	Light petroleum ether	Urosolic acid, Oleanolic acid, Beta amayrin, Triterpinoids, Quinones (Kanchanapooma *et al.*, 2001)
	Methanolic	Lignan glycoside (fernandoside) and Phenylpropanoid glycoside (2"-*O*-β-apiosylverbascoside) (Kanchanapooma *et al.*, 2001)
Root and Heartwood	Light petroleum ether	Benzoquinones, Lapachol, Lapachone, Dehyro alpha lapachone (Kanchanapooma *et al.*, 2001)
Bark	Light petroleum ether	Beta sitoserol, n-hentriacontarol (Jassbi *et al.*, 2006)
Root	Benzene extract	Dehydro alpha lapachone, Alpha lapachone, Lapachol, Beta sitosterol and Beta lapachone (Rizvi and Sultana, 1973)
Heartwood	Alcoholic extract	Dehydro alpha lapachone, Lapachol, Dehydro iso alpha lapachone, Dehydrotectol, Beta sitosterol and Tectol (Rizvi and Sultana, 1973)
Pod	N/A	Luaric acid, Myristic acid, Oleic acid, Linoleic acid, Palmitic acid, Stearic acid, Arachidic acid, Triterpinoid, Beta sitosterol (Willium, 2010)
Whole plant	N/A	Alkaloid, Iridoids, Anthraquinone, Shikimic acid, Verbascosides, Cornosides, Flavones, Quarcetrins, Urosolic acid, Saponins, Sapogenins, Oligosaccharides, Sucrose and other sugars (Ali *et al.*, 1998)

Antimicrobial Activity

The crude extracts of *Heterophragma adenophyllum* were found to show antibacterial activities towards the Gram-positive bacteria (*Bacillus subtilis, Staphyloccus aureus* and *Pseudomonas aeruginosa*), Gram-negative (*Escherichia coli*) and yeast (*Candida albicans*) (Rasadah and Houghton, 1998).

Leismanicidal Activity

Haplophragma adenophyllum (Wall.) is a rare species in the Bignoniaceae family. Three naphthoquinones (dehydro-α-lapachone, α-lapachone, and β-lapachone) have been isolated from heartwood and roots, and one (β -lapachone) has antitrypanosomal activity. In leishmanicidal activity against *L. major* methanolic extract shown the minimum lethal concentration 100 µg/ml (MLC) and minimum inhibitory concentration 50 µg/ml (MIC) (Marii *et al.*, 2004).

Activity of Lapachol and its Derivatives

Lapachol and its derivatives show various activities like: Anti-abscess, Anti-ulcer, Antileishmanial, Anticarcinomic, Antiedemic, Anti-inflammatory, Antimalarial, Antiseptic, Antitumor, Antiviral, Bactericidal, Fungicidal, Pesticide, Respiratory depressant (Anonymous, 2002).

Antiviral and Cardiovascular Activity

Hydro alcoholic extract of *Heterophragma adenophyllum* have been tested for a wide variety of biological activity. Anti viral and cardiovascular activity is confirmed by Bhakuni *et al.* (1971).

Ethnomedicinal Uses

In Thai traditional medicine, the leaves are used for external treatment of skin diseases. Fruits of *Fernandoa adenophylla*, used for cooking and Flowers were consumed as fresh vegetables (John and Thomas, 1866). Traditional uses of the *Heterophragma adenophyllum* are mentioned in Table 6.4. Roots of *Heterophragma adenophyllum* has reported to be used for the treatment of piles and constipation in Bangladesh (Rahmatullah *et al.*, 2010)

Table 6.4: Traditional uses of the *Heterophragma adenophyllum.*

Material Used	*Purpose*
Wood	In skin diseases
Leaf juice	Applied externally on toe sores and in chilblain.
Essential oil from flowers	As antimicrobial
Ethyl alcohol extract of aerial parts	As antiviral and hypotensive

Conclusions

Although there are literatures on availability record by different institution, still we are lacking for exact information on geographical source and abundance of both of studied plants. There are many researcher worked on both the plants and they collected these plant from different geographical sources. A notable number of bioactive compounds have been reported from both studied plants. This literature demonstrated a number of important activities, which are beneficial to human beings.

Among the various bio-active constituents reported from Bignoniaceae family plants, some of the more common but pharmacologically important compounds found in *Heterophragma quadriloculare* and *Heterophragma adenophyllum* are ursolic acid, oleanolic acid, α- and β-lapachone, lapachol, verbascoside, lupeol, quercitrin, phytosterole and tectol. Although a comprehensive evaluation of the activities of these compounds is beyond the scope of the present study, a brief review of some of the most recent literature on these compounds has revealed traditional and ethnomedicinal claims of studied plants.

Study of Lee J *et al.* (2010) suggests that low-dose ursolic acid may be used as a hypoglycemic agent and immune modulator in non-obese type 2 diabetic mice. The ursolic acid derivative enhanced insulin receptorphosphorylation in CHO/hIR cells and stimulate glucose uptake in L6 myotubes (Wei *et al.*, 2006). Ursolic acid can be used to treat obesity as demonstrated by its stimulation of lipolysis in primary-cultured rat adipocytes (Rahmatullah *et al.*, 2010). Anti-fungal activity has also been reported for ursolic acid (Shai *et al.*, 2008). Presence of ursolic acid made both of this plant to be useful as antidiabetic and anti-fungal agent. Use of studied plants as antidiabetic drug can further be justified by synergistic anti-hyperglycemic effect reported between *Syzygium cordatum*-derived oleanolic acid and insulin in streptozotocin-induced diabetic rats (Musabayane *et al.*, 2010). The compound reportedly increased urinary outputs and creatinine clearance of streptozotocin-induced diabetic rats (Rahmatullah *et al.*, 2010).

Inhibitory activities of lapachol and α- and β-lapachone derivatives have been reported against epimastigote and trypomastigote forms of *Trypanosoma cruzi*. Anti-neoplastic activity has been reported for monoarylhydrazones of α-lapachone (Rahmatullah *et al.*, 2010). The oxyrane derivative of α-lapachone has also been shown to be a potent growth inhibitor of *T. cruzi* epimastigote forms. This could be one of the reasons why both plants are more preferred for skin diseases. Allantoin has been shown to reduce oxidative stress and also found to be incorporated in human physiological mechanisms. Allantoin is reported to be beneficial in skin diseases so it can serve as a proof for the use of both plants in skin disease, toe sores and in chilblain (Scott *et al.*, 2001).

Verbascoside has been reported to demonstrate anti-inflammatory effects in THP-1 cells (human myelomonocytic leukemia). The compound, isolated from *Lepechinia speciosa* Benth. (Lamiaceae) reportedly showed inhibitory activity against herpes simplex virus, HSV-1 and HSV-2 *in vitro* (Rahmatullah *et al.*, 2010). Use of both plants as antiviral agent can be extended to presence of verbascoside and lapachol derivatives.

Naphthoquinones contributed to antifungal, antiseptic, antiviral, antimicrobial and leismonicidal activity of studied plants. Naphthoquinones are compounds present in various families of plants; their molecular structures endow them with redox properties which confer activity in various biological oxidative processes. In ethnomedicine, plants containing naphthoquinones have been employed for treatment of a number of diseases including cancer. Biological activities for naphthoquinones have been described such as being anti-inflammatory, bactericidal, fungicidal, virucidal, trypanocidal, anti-malarial, and snail fever, parasitic disease caused by blood flukes, trematode worms of the genus *Schistosoma*. Although naphthoquinones do appear to exhibit a wide spectrum of biological activities, the mechanism/s of action remains somewhat unclear (Habbal *et al.*, 2007).

Anti-arthritis, anti-inflammatory, antimalarial, antitumor, and hepatoprotective activities can be correlated to presence of lupeol (Margareth, Miranda, 2009).

Presence of sugars, amino acids and quarcitrin may be reason for both studied plants to be preferred as a food because quarcitrin has been suggested as a possible dietary nutritional supplement to prevent oxidation- related diseases (Rahmatullah *et al.*, 2010).

Bing *et al.* (2010) filled applications for use of ursolic acid saponin, oleanolic acid saponin in preparation of increasing leucocytes and/or platelet medicine. Tectol exhibited significant activity against human leukemia cell lines HL60 and CEM (Sonia *et al.*, 2001). So presence of ursolic acid, oleanolic acid and Tectol may become foundation for anti-cancer activity of plants studied in this chapter.

Scientific study suggest that there are number of compound present in evaluated plants showing similar kind of activity that may give synergistic effect and that play a vital role while selection of plant for different purpose especially for skin diseases, antifungal, anti-inflammatory, anti-bacterial.

Thus, plants can become important sources of novel drugs and lead compounds for antidiabetic, anti-hypertensive, anti-fungal, anti-viral and anti-cancer activities.

References

Ali, R.M., Houghton, P.J., and Hoo, T.S. (1998). Antifungal activity of some Bignoniaceae found in Malaysia. *Phytotherapy research,* **12(5):** 331-334.

Amla Soumyanath (2006). Traditional medicines for modern times: antidiabetic plants. CRC Taylor and Francis publication, pp. 19.

Andhra Pradesh forest department (2011). Retrieved on 17th October, 2010 from: http: //forest.ap.nic.in/Forest per cent 20Flora%20of%20Andhra%20Pradesh/ Family/Bignoniaceae.htm

Anonymous (2002). Proceedings of the 9th JIRCAS International Symposium-2002: Value-Addition to Agricultural Products, pp. 130–134.

Anonymous (2006). Gazetteers of the Bombay precidency, Thana place of interest, volume XIV, e-book edition, published by the executive editor and secretary, gazetteer department, govt. of Maharashtra, Bombay. Retrieved on 15th July, 2011 from: http: //www.maharashtra.gov.in/pdf/gazeetter_reprint/Thane-III/places_Matheran.html

Bayard group (2011). French Science Magazine- Eureka Magazine. Retrieved on 17th October, 2010 from: http: //eurekamag.com/research/004/764/antifungal-activity-essential-oil-leaves-heterophragma-quadriloculare.php

Bhakuni, D.S., Dhar, M.L., Dhar, M.M., Dhawan, B.N., Gupta, B., and Sharma R.C. (1971). Screening of Indian plants for biological activity: part III. *Indian Journal of Experimental Biology,* **9**: 91–102.

Bing Guang, Junjian Liu, Guangxin Dong, Xiangyang Peng, Xiaoxia Gong, Zhen Huang, Meirong Zhou, Wei Zhan, Jufang Yan, Yu Huang, Dongguang Qin, Jianxin J.l., and Bogang Li (2010). Use of ursolic acid saponin, oleanolic acid saponin in preparation of increasing leucocytes and/or platelet medicine. Patent application number: 20100197898. Retrived from: http: //www.faqs.org/ patents/app/20100197898#ixzz1tsNNE0o9 (04/05/2012).

Biodiversity Heritage Library (2011). Retrieved on 15th July, 2011 from: http: // zipcodezoo.com/Plants/H/Heterophragma per cent 5Fquadriloculare/

Biodiversity occurrence data (2011) published by BeBIF Provider. Retrieved from http: //data.gbif.org/datasets/provider/213

Habbal, O.A., Al-Jabri, A.A., and El-Hag, A.G. (2007). Antimicrobial properties of *Lawsonia inermis* (henna): a review. *Australian Journal of Medical Herbalism.* Retrieved May 03, 2012 from http: //www.thefreelibrary.com

Jassbi, A.R., Singh, P., Lamba, J., Jain, S., Baldwin, I.T., and, Tahara, S. (2006). Transformation of lapachol to a novel naphthoquinone and related compounds. *Zeitschrift für Naturforschung Section B-A Journal of Chemical Sciences*, **61(1):** 73-77.

Jog Suneeti, K. (2009). Sahyadris – flora and Ethnobotany. Report of William L. Brown Fellowship 2007-2008, Department of Biology, The University of Texas at Tyler.

John Lindley and Thomas Moore (1866). The Treasury of Botany: A Popular Dictionary of the Vegetable Kingdom. Volume 1. London: Longmans Green and Co., pp. 587.

Kanchanapooma, T., Kasaia, R., and Yamasaki, K. (2001). Lignan and phenylpropanoid glycosides from *Fernandoa adenophylla. Phytochemistry*, **57(8):** 1245-1248.

Khare, C.K. (2007). Indian medicinal plants. Springer publication, New Delhi, pp. 308.

Lee, J., Yee, S.T., Kim, J.J., Choi, M.S., Kwon, E.Y., Seo, K.I., and Lee, M.K. (2010). Ursolic acid ameliorates thymic atrophy and hyperglycemia in streptozotocin-nicotinamide-induced diabetic mice. *Chem Biol Interact.*, **188(3):** 635-42.

Rasadah, M. A., and Houghton, P. J. (1998). Antimicrobial activity of some species of Bignoniaceae. *ASEAN review of biodiversity and environmental conservation* (ARBEC). Article III: 1-3.

Margareth, B.C. Gallo and Miranda, J. Sarachine. (2009). Biological activities of lupeol. *Int J Bio Pharm Sci.*, **3(Special Issue1)**: 46-66.

Marii Takahashi, Hiroyuki Fuchino, Motoyoshi Satake, Yutaka Agatsuma and Setsuko Sekita (2004). *In vitro* screening of leishmanicidal activity in Myanmar timber extracts. *Biol. Pharm. Bull.*, **27(6):** 921-925.

Mukherjee Kaveri, and Ray, L.N. (1986). Phytochemical screening of some Indian medicinal plant species, Part II. *Int. J. Crude Drug Res.*, **24(4):** 187-205.

Musabayane, C.T., Tufts, M.A., and Mapanga, R.F. (2010). Synergistic antihyperglycemic effects between plant-derived oleanolic acid and insulin in streptozotocin-induced diabetic rats. *Renal Failure*, **32**: 832-839.

Nigam, S.S., Sahai Alok and Basis Seema (1983). Analysis of the fat from the root of *Heterophragma quadriloculare* (*Roxb.*) *Schum. BMEBR*: **4(3 and 4):** Sept. and Dec, published by CCRAS, Dept. of Ayush, Govt. of India: 162 – 164.

Pandey, C.N., Raval, B.R., Mali Seema and Salvi Harshad (2005). Medicinal plants of Gujarat, Gujarat Ecological Education and Research (GEER) Foundation, Gandhinagar, pp. 196.

Pullaiah, T (2006a). Biodiversity in India. Volume 4, Regency publications, New Delhi, pp. 1328.

Pullaiah, T (2006b). Encyclopedia of world medicinal plants. Volume 3, Regency Publication, New Delhi, pp. 1082.

Punekar Sachin, A. (2002). Some food plants of human langur *Semnopithecus entellus* (DUFRESNE) in the western ghats of Maharastra, India. *Zoos' print journal,* **17(6):** 797-801.

Rahmatullah, M., Samarrai, W., Jahan, R., Rahman, S., Sharmin, N., and Miajee, Z. (2010). An ethnomedicinal, pharmacological and phytochemical review of some Bignoniaceae family plants and a description of Bignoniaceae plants in folk medicinal uses in Bangladesh. *The Free Library*. Retrieved May 03, 2012 from http: //www.thefreelibrary.com/An ethnomedicinal, pharmacological and phytochemical review of some.-a0252944798

Rizvi, S. A. I., and Sultana, T. (1973). Chemical studies of a fixed oil, terpenoids and other compounds from the pods of *Heterophragma adenophyllum*. *Planta Med.,* **23(2):** 125-131.

Sahai, A., and Shrivastava, A. (1984). Indian-Perfume. *Essential Oil Association of India*, **28 (3/4):** 175-176.

Sandwith, N.Y. (1967). What Is *Heterophragma sulfureum* Kurz? In: Kew Bulletin 21(1). *Kew: Royal Botanic Gardens*, pp. 21-30 and 38.

Saxena, V.K. (1984). Chemical evaluation of the seeds of *Heterophragma quadriliculare*. *Acta cienc Indica Chem.,* **10(1)**: 35-36.

Scott, L.J., Dunn, C.J., and Goa, K.L. (2001). Calcipotriol Ointment: A review of its use in the management of psoriasis. *American Journal of Clinical Dermatology*, **2(2):** 95-120.

Shai, L.J., McGaw, L.J., Aderogba, M.A., Mdee, L.K., and Eloff, J.N. (2008). Four pentacyclic triterpenoids with antifungal and antibacterial activity from *Curtisia dentata* (Burm.f.) C.A. Sm. leaves. *Journal of Ethnopharmacology*, **119**: 238-244.

Sonia Maria, O., Costa Telma Leda, G., Lemos Otilia Deusdenial, L., Pessoa, Claudia Pessoa, Raquel, C. Montenegro, and Raimundo, Braz-Filho (2001). Chemical constituents from *Lippia sidoides* and cytotoxic activity. *J. Nat. Prod.,* **64 (6):** 792–795.

The International Plant Names Index, Jan 19, 2007. Retrieved from: http: // zipcodezoo.com/Plants/H/Heterophragma per cent 5Fquadriloculare/(on 14th June 2010).

USDA, ARS, National Genetic Resources Program. Germplasm Resources Information Network–(GRIN) [Online Database]. National Germplasm Resources Laboratory, Beltsville, Maryland. Retrieved from: http: //www.ars-grin.gov/ cgi-bin/npgs/html/exsplist.pl (on14th June 2010).

Wei Zhang, Di Hong, Yueyang Zhou, Yinan Zhang, Qiang Shen, Jing-ya Li, Li-hong Hu, and Jia Li. (2006). Ursolic acid and its derivative inhibit protein tyrosine phosphatase 1B, enhancing insulin receptor phosphorylation and stimulating glucose uptake. *Biochimica et Biophysica Acta*, **1760**: 1505-1512.

Willium Asiedu (2010). US patent. Patent no. US 7749544b2.

Yadav, R.N. (1981). Analysis of the fixed oil from the *Heterophragma quadriliculare. Indian drug pharm ind.,* **16(1):** 28.

Yadav, R.N. (1992). A new saponin from the seed of *H. quadriliculare. J Indian Chem Soc.,* **69(5):** 286-287.

Yingyong Paisooksantiyvatana and Sunji Kako (1996). Ethanobotany of the karn tribe in westarn Thailand. *Fortrop: tropical forestry in the 21st century,* pp 70.

Zhi-Yun Zhang and Thawatchai Santisuk (2010). Bignoniaceae, *in* Flora of China, Vol. 18 Published by Science Press (Beijing) and Missouri Botanical Garden Press, 213. Retrieved from: http: //zipcodezoo.com/Plants/H/Heterophragma per cent 5Fquadriloculare/ (on14th June 2010).

Zirvi Karimullah, A., Miana, G.A., and Jehangir, S. (1972). Constituents of *H. quadriliculare. phytochemistry,* **11(7):** 2349.

Bioactive Phytochemicals: Perspectives for Modern Medicine Vol. 2 (2014) *Pages* **155–163**
Editor: **V.K. Gupta**
Published by: **DAYA PUBLISHING HOUSE, NEW DELHI**

7

Modulatory Efficacy of Capsaicin on Polyamine Content, DNA, RNA Levels and DNA Damage During Experimental Lung Cancer

P. Anandakumar[1*], S. Kamaraj[2] and T. Devaki[3]

ABSTRACT

The protective effect of capsaicin on benzo(a)pyrene induced lung carcinogenesis in Swiss albino mice was studied. Benzo(a)pyrene was administered orally (50mg/kg body weight) to induce lung cancer in Swiss albino mice. Analysis of polyamines, DNA, RNA levels and DNA damage (by comet assay) was carried out. Benzo(a)pyrene induced lung cancer bearing mice showed elevations in polyamine content, DNA, RNA levels and induction of DNA damage. Capsaicin (10 mg/kg body weight) supplementation effectively counteracted all the above abnormalities, revealing its anti-cancer efficacy during experimental lung carcinogenesis.

Keywords: Benzo(a)pyrene, Capsaicin, Comet assay, DNA and RNA levels, Lung cancer, Polyamines.

1 Department of Biomedical Sciences, School of Health and Hospital, Asella Medical Campus, Adama Science and Technology University, Ethiopia

2 Department of Molecular Biology, Umeå University, Sweden

3 Department of Biochemistry, University of Madras, Chennai, India

* *Corresponding author*: E-mail: paksuccess@gmail.com

Introduction

Cancer chemoprevention is defined as the use of either natural or synthetic components to delay, inhibit or reverse the development of cancer in normal or pre-neoplastic conditions. There have been substantial efforts to search for naturally occurring substances that can curtail the multistage carcinogenesis. Different varieties of substances present in the medicinal herbs or dietary plants have been screened for their ability to prevent carcinogenicity. Natural products have been established to exert anti-cancer activities partially based on their ability to quench reactive oxygen species and protect critical cellular components like DNA, proteins and lipids from oxidative damage (Rafter, 2002). They may also interfere with intracellular signaling pathways, such as those, which regulate proliferation, induction of apoptosis, and response to oxidative stress (Loo, 2003).

Capsaicin (CAP) is a homovanillic acid derivative and the chief pungent principle found in hot red and chilli peppers derived from Capsicum fruit extracts. CAP has long been known as an irritant able to induce a special type of local inflammation called neurogenic inflammation by evoking release of substance P and calcitonin gene-related peptide by activating transient receptor potential vanniloid-1 (TRPV1) (Holzer, 1991). CAP pretreatment was also found to protect against the free radical induced pulmonary damage in rats exposed to gaseous chemical irritants such as sulfur dioxide and nitrogen dioxide (De and Ghosh, 1993). Several studies have proved the anti-carcinogenic activity of CAP against varieties of cancers (Modly *et al.*, 1986; Zhang *et al.*, 1993; Teel *et al.*, 1997). Studies from our laboratory have also proved the chemoprotective role of CAP in different dimensions (Anandakumar *et al.*, 2009a, b, c).

To gain further insight into the potential mechanism underlying the anticancer activity of capsaicin *in vivo*, we investigated the influence of CAP on polyamine content, DNA, RNA levels and DNA damage in benzo(apyrene (B(a)P) induced mice lung carcinogenesis.

Materials and Methods

Materials

Benzo(a)pyrene and capsaicin were purchased from M/s Sigma Chemicals, St. Louis, USA. All other chemicals were of analytical grade procured from M/s. SRL Chemicals Pvt. Ltd., Mumbai.

Animals

Healthy male Swiss albino mice weighing 20-25g (8-10 weeks old) obtained from Veterinary College, Chennai, were used throughout the experiment. This study was ethically approved by the Ministry of Social Justices and Empowerment, Government of India and by the Animal Ethics Committee guidelines of our Institution (IAEC No.01/024/08). The animals were housed under conditions of controlled temperature (26 ± 2°C) with 12-hr day/night cycle. They were fed standard rat/mice pellet diet (M/s. Hindustan Lever Ltd., Mumbai) under the trade name Amrut rat/mice feed and were given access to water *ad libitum*.

Experimental Design

Experimental animals were divided into four groups of six mice each as follows.

Group I: Control animals received olive oil throughout the course of the experiment.

Group II: Animals were treated with benzo(a)pyrene B(a)P (50 mg/kg body weight dissolved in olive oil) orally twice a week for four successive weeks.

Group III: Animals received CAP alone (10 mg/kg body weight dissolved in olive oil) intraperitoneally once in a week for 14 weeks to assess the cytotoxicity (if any) induced by CAP.

Group IV: Animals received B(a)P (as in Group II) along with CAP (10 mg/kg body weight dissolved in olive oil) intraperitoneally. CAP treatment was started one week prior to the first dose of B(a)P administration and continued for 14 weeks.

At the end of the experimental period (14th week), animals were sacrificed by cervical decapitation under ether anesthesia and lungs were excised immediately and washed with ice-cold saline. A 10 per cent homogenate of the washed tissue (lung) was prepared in 0.01 M phosphate buffer (pH 7.4). The homogenate was centrifuged at a speed of 12,000 × g for 30 min in a refrigerated high-speed centrifuge at 4°C. Blood was also collected and the serum was separated for other estimations. The following analysis was carried out in the supernatant.

Analysis of Polyamines, DNA and RNA

The levels of polyamines were assayed by the method of Endo (Endo, 1978). DNA was estimated as according to the method of Burton (Burton, 1956). RNA was estimated by the method of Rawal *et al.* (Rawal *et al.*, 1977).

Assessment of DNA Damage (Comet Assay)

DNA damage was assessed by using alkaline single cell gel electrophoresis (comet assay) (Singh *et al.*, 1988). 100 μl of 1 per cent NMPA in phosphate buffered saline was dropped on to frosted slides, immediately covered with cover slip and kept for 10 minutes in a refrigerator to solidify. Then, cover slips were removed and 100 μl of LMPA containing cells (100 μl of cell suspension in HBSS and 100 μl of LMPA) were added to the slides. The cover slips were replaced and the slides were kept in the refrigerator for another 10 minutes to solidify the LMPA. After this, the cover slips were removed and a top layer of 100 μl of LMPA was added and the slides were again cooled for 10 minutes. After removal of cover slips, the slides were immersed in cold lysing solution. The slides were kept in dark for at least 1 hr at 4°C. To prevent the occurrence of additional DNA damage, the following steps were performed under dim light. The slides were removed from the lysing solution and placed on a horizontal electrophoresis tank. The electrophoresis unit was filled with a freshly made electrophoresis buffer to a level of 0.25 cm above the slides. The cells were exposed to alkali for 20 minutes to allow DNA unwinding.

An electric current of 25 V and 300 mA was applied for 20 minutes. After electrophoresis, the slides were placed horizontally, and neutralized with Tris-HCl. Finally, 50 μl of ethidium bromide was added to each slide and covered with a cover slip and analysed using a fluorescence microscope (Nikon, Japan) with a calibrated scale in ocular. Images of 50 randomly selected cells were analyzed from each sample. For each cell, the length of the image (diameter of the nucleus plus migrated DNA) was measured.

Results

Analysis of Polyamines

The level of polyamines in the lung of control and experimental group of animals is shown in Table 7.1. The levels of putrescine, spermidine and spermine were found to be markedly ($P<0.05$) elevated in B(a)P induced lung cancer animals (Group II) when compared with control animals (Group I). On CAP supplementation these levels were brought back to near normal values in Group IV animals when compared to lung cancer animals (Group II).

Table 7.1: Levels of polyamines in the lung of control and experimental group of animals.

Parameters	*Group I (Control)*	*Group II (B(a)P)*	*Group III (CAP)*	*Group IV (B(a)P + CAP)*
Putrescine	19.81 ± 1.88	53.11 ± 4.91 [a]	19.80 ± 1.88	22.41 ± 2.15 [b]
Spermidine	848.2 ± 83.3	1521 ± 145.84 [a]	847.2 ± 83.19	1084.1 ± 107.20 [b]
Spermine	421.1 ± 38.83	732.2 ± 69.91 [a]	422.2 ± 39.83	548.2 ± 53.22 [b]

Results are expressed as mean ± S.D for six mice in each group.

Statistical significance at $P<0.05$, as compared with [a]Group I, [b]Group II.

Units: Polyamines – nmol/g of tissue

Assessment of DNA and RNA Levels

Table 7.2 shows the levels of DNA and RNA in control and experimental group of animals. Significantly ($P<0.05$) increased levels of DNA and RNA were noticed in lung cancer challenged Group II animals compared to control animals of Group I. The levels of DNA and RNA were markedly ($P<0.05$) brought back to near control values in CAP treated Group IV animals when compared to Group II animals.

Table 7.2: Levels of DNA and RNA in the lung of control and experimental group of animals.

Parameters	*Group I (Control)*	*Group II (B(a)P)*	*Group III (CAP)*	*Group IV (B(a)P + CAP)*
DNA	3.18 ± 0.28	5.27 ± 0.49 [a]	3.17 ± 0.28	3.76 ± 0.35 [b]
RNA	1.87 ± 0.19	3.13 ± 0.30 [a]	1.86 ± 0.19	2.10 ± 0.20 [b]

Results are expressed as mean ± S.D for six mice in each group.

Statistical significance at $P<0.05$, as compared with [a]Group I, [b]Group II.

Units: DNA and RNA – mg/g of tissue

Alkaline Single Cell Gel Electrophoresis (Comet assay)

Figure 7.1 shows the extent of DNA damage measured using comet assay in control and experimental group of animals. In lung cancer bearing animals (Group II), an increase in DNA tail length suggesting DNA damage was observed (Figure 7.1B) when compared to control animals (Figure 7.1A). A significant ($P<0.05$) reduction in the tail length was noted in CAP treated Group (Group IV) (Figure 7.1D) when compared with lung cancer bearing animals of Group II.

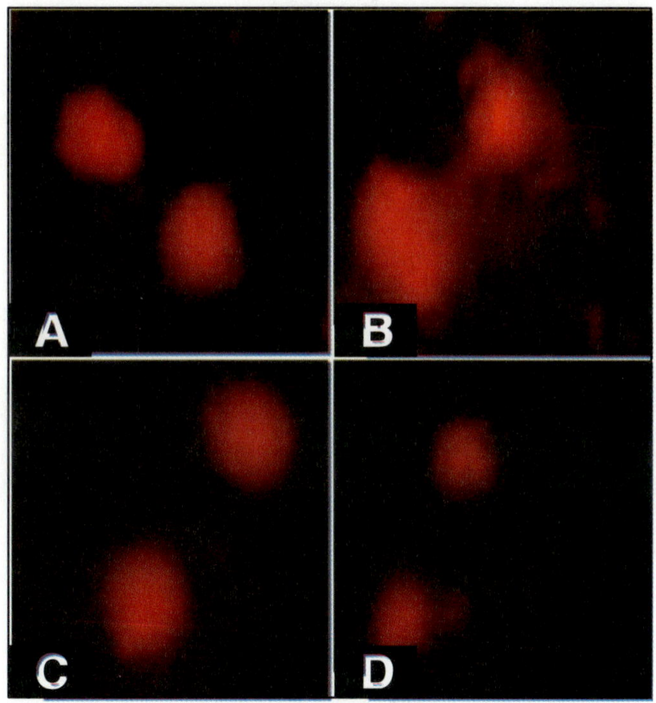

Figure 7.1: Alkaline single cell gel electrophoresis (Comet assay).
Plates A–D represent the lung sections of Groups I to IV of experimental animals respectively.

Discussion

Polyamines are essential for growth of cells and rapidly proliferating cells have higher levels of polyamines than do slowly growing or quiescent cells (Thirunavukkarasu *et al.*, 2005). Polyamines and their metabolites are considered to be involved in the process of cell multiplication and its regulation. In eukaryotic cells, polyamines play an essential part in proliferation, differentiation and neoplastic transformation (Selvendiran *et al.*, 2004; Farriol *et al.*, 2001). Earlier studies with different carcinoma tissues have demonstrated a marked increase in the levels of polyamines confirming their correlation with neoplastic growth and high rate of cell proliferation (Farriol *et al.*, 2001). The increased rate of synthesis of polyamines during cancerous condition is due to enhanced activity of the enzyme ornithine decarboxylase (ODC), which is the rate-limiting enzyme that is actively involved in the synthesis of polyamines (Pegg *et al.*, 1995). The fundamental concept of polyamine metabolism is that, tumors have a greater requirement of polyamine synthesis for continuous growth, whereas, normal tissues have a less requirement for polyamines because of their controlled growth behaviours (Pegg and Hu, 1995). Marton and Pegg (1995) cited the importance of polyamines in chemotherapy by demonstrating depleted levels of polyamines upon treatment with anticancer agents. Thus, regulating the abnormality of polyamine biosynthesis is a main strategy for cancer chemoprevention. In the present study, polyamines namely putrescine, spermidine and spermine levels were significantly increased in lung cancer bearing animals. Earlier reports demonstrated the potential of alkaloids and flavonoids to inhibit ODC in various cancer conditions (Thirunavukkarasu *et al.*, 2005; Selvendiran *et al.*, 2004). The above reports substantiate our present study, as CAP being an alkaloid inhibited ODC activity and restored the levels of polyamines to its near normal values suggesting its anti-proliferative nature against lung cancer.

Nucleic acids play an important role during neoplastic transformation. During neoplastic conditions, the abnormality in DNA and RNA increases with degree of malignancy (Palani *et al.*, 1999). DNA and RNA content of tumor are found to be an important indicator of prognosis, because they are well correlated with the size of the tumor in cancerous conditions (Gallagher, 1986). Since B(a)P has been shown to bind to cellular DNA, considerable evidences indicate that such B(a)P-DNA interactions are responsible for initiation of carcinogenic process and hence, the determination of DNA and RNA are more meaningful in tumorous conditions. Reports revealed abnormal amount of DNA in different cancers including lung cancer (Selvendiran *et al.*, 2004). Increase in DNA and RNA contents in lung cancer could be due to the increased expression of enzymes necessary for DNA synthesis in tumor cell with expression of many enzymes related to cell functions. CAP treatment effectively reduced DNA and RNA levels suggesting its protective efficacy.

The DNA damage recognition and repair machinery in mammalian cells is extensive and mounts a rapid and efficient response to the myriad forms of damage that occur, depending on the environmental or endogenous insult. Persistent exposure to exogenous carcinogens and other endogenous carcinogens could lead to DNA damage. Several studies have demonstrated that oxidative DNA damage is related to tumor initiation and progression of lung cancer (Wu *et al.*, 2005). A variety of studies have used DNA damage level as a predictive marker in cancer risk assessment (Perera *et al.*, 2002). Different mutagens can induce different types of DNA damage and initiate specific DNA damage response and repair pathways. The comet assay is a single-cell gel electrophoresis method used to measure DNA damage in individual cells. It is a sensitive and versatile method with the potential to be high throughput (Kassie, 2000). It is technically simple, relatively fast, and inexpensive, and DNA damage can be investigated in virtually all-mammalian cell types. Within the last decade, the comet assay has become increasingly popular in DNA damage detection and biomonitoring studies (Colleu-Durel *et al.*, 2004). Studies have also reported successful applications of the comet assay in risk assessment for different cancers including lung cancers (Smith *et al.*, 2003; Rajaee-Behbahani *et al.*, 2001). Benzo(a)pyrene diol epoxide (BPDE), a metabolite of procarcinogen B(a)P is well known to induce DNA damage by forming bulky DNA adducts (Selvendiran *et al.*, 2005). In the present study we have observed enhanced DNA damage in B(a)P induced lung cancer bearing animals. CAP treatment remarkably reduced DNA damage (single stranded DNA breaks) in lung cancer bearing animals revealing that CAP protects DNA against damage and thereby shows anti-mutagenic activity during lung cancer.

In conclusion, CAP exerted its beneficial effect by modulating the polyamine content, DNA, RNA levels and by preventing DNA damage. Hence CAP could be considered as a promising drug candidate for inhibiting lung carcinogenesis.

Acknowledgments

First author Anandakumar P wishes to thank Mr. Pandi, Mrs. Padmavathi, Mrs. Kalaivani and Ms. Prabhavathy for their moral support and encouragement throughout the period of study.

References

Anandakumar, P., Kamaraj, S., Jagan, S., Ramakrishnan, G., Naveenkumar, C., Asokkumar, S., and Devaki, T. (2009a). Capsaicin alleviates the imbalance in xenobiotic metabolizing enzymes and tumor markers during experimental lung tumorigenesis. *Molecular and Cellular Biochemistry,* **331(1-2):** 135-143.

Anandakumar, P., Kamaraj, S., Jagan, S., Ramakrishnan, G., and Devaki, T. (2009b). Effect of capsaicin on glucose metabolism studied in experimental lung carcinogenesis. *Natural Product Research,* **23(8):** 763-774.

Anandakumar, P., Kamaraj, S., Jagan, S., Ramakrishnan, G., and Devaki, T. (2009c). Chemopreventive task of capsaicin against benzo(a)pyrene induced lung cancer in Swiss albino mice. *Basic Clinical Pharmacology and Toxicology,* **104(5):** 360-365.

Burton, K. (1956). A study of the conditions and mechanism of the diphenylamine reaction for the colorimetric estimation of deoxyribonucleic acid. *Biochemical Journal,* **62:** 315-323.

Colleu-Durel, S., Guitton, N., Nourgalieva, K., Legue, F., Lévêque, J., Danic, B., and Chenal, C. (2004). Alkaline single-cell gel electrophoresis (comet assay): a simple technique to show genomic instability in sporadic breast cancer. *European Journal of Cancer,* **40:** 445–451.

De, A.K., and Ghosh, J.J. (1993). Capsaicin pretreatment protects free radical induced rat lung damage on exposure to gaseous chemical lung irritants. *Phytotherapy Research,***7:** 87-89.

Endo, Y. (1978). A simple and sensitive method of analysis of histamine, putrescine and polyamines without the use of an amino acid analyzer. *Annals of Biochemistry,* **89:** 235-246.

Farriol, M., Segovia-Silvestre, T., Castellanos, J.M., Venereo Y, and Orta, X. (2001). Role of putrescine in cell proliferation in a colon carcinoma cell line. *Nutrition,* **17:** 934-938.

Gallagher, R.E.(1986). Biochemistry of neoplasia. *In:* Comprehensive Text book of oncology. Ed. By Moosa, A.R., Robson, M.C., and Schimpff, S.C. Williams and Wilkins, Baltimore, USA, pp.36-45.

Holzer, P. (1991). Capsaicin: cellular targets, mechanisms of action, and selectivity for thin sensory neurons. *Pharmacology Review,* **43(2):** 143-201.

Kassie, F., Parzefall, W., and Knasmuller, S. (2000). Single cell gel electrophoresis assay: a new technique for human biomonitoring studies. *Mutation Research,* **463:** 13–31.

Loo, G. (2003). Redox-sensitive mechanisms of phytochemical mediated inhibition of cancer cell proliferation (review). *Journal of Nutritional Biochemistry,* **14:** 64-73.

Marton, L.J., and Pegg, A.E. (1995). Polyamines as targets for therapeutic intervention. *Annual Review of Pharmacology and Toxicology,* **35:** 55-91.

Modly, C.E., Das, M., Don, P.S., Marcelo, C.S., Mukhtar, H., and Bickers, D.R.(1986). Capsaicin as an *in vitro* inhibitor of benzo(a)pyrene metabolism and its DNA binding in human and murine keratinocytes. *Drug metabolism and disposal,* **14:** 413-416.

Palani, V., Senthilkumaran, R.K., and Govindasamy, S. (1999). Biochemical evaluation of antitumor effect of Muthu Marunthu (a herbal formulation) on experimental fibrosarcoma in rats. *Journal of Ethnopharmacology,* **65:** 257-265.

Pegg, A.E., and Hu, R.H. (1995). Effect of polyamine analogues and inhibition of polyamine oxidase on spermidine/spermine N1-acetyltransferase activity and cell proliferation. *Cancer Letters,* **95:** 247-252.

Pegg, A.E., Shantz, L.M., and Coleman, C.S. (1995). Ornithine decarboxylase as a target for chemoprevention. *Journal of Cellular Biochemistry, Supplement,* **22:** 132-138.

Perera, F.P., Mooney, L.A., Stampfer, M., Phillips, D.H., Bell, D.A., Rundle, A., Cho, S., Tsai, W.Y., Ma, J., Blackwood, A., and Tang, D. (2002). Associations between carcinogen-DNA damage, glutathione S-transferase genotypes, and risk of lung cancer in the prospective "Physicians" Health Cohort study. *Carcinogenesis,* **23:** 1641–1646.

Rafter, J. J. (2002).Scientific basis of biomarkers and benefits of functional foods for reduction of disease risk: cancer. *British Journal of Nutrition,* **88 (Suppl 2):** 219-224.

Rajaee-Behbahani, N., Schmezer, P., Risch, A., Rittgen, W., Kayser, K.W., Dienemann, H., Schulz, V., Drings, P., Thiel, S., and Bartsch, H.(2001). Altered DNA repair capacity and bleomycin sensitivity as risk markers for non-small cell lung cancer. *International Journal of Cancer,* **95:** 86–91.

Rawal, U.M., Patel, U.S., Rao, G.N., and Desai, R.R.(1977). Clinical and biochemical studies on cateractous human lens III. Quantitative study of protein, RNA and DNA. *Arogya Journal of Health Sciences,* **3:** 69-72.

Selvendiran, K., Banu, S.M., and Sakthisekaran, D. (2004). Protective effect of piperine on benzo(a)pyrene-induced lung carcinogenesis in Swiss albino mice. *Clinica Chimica Acta,* **350:** 73-78.

Selvendiran, K., Banu, S.M., and Sakthisekaran, D. (2005). Oral supplementation of piperine leads to altered phase II enzymes and reduced DNA damage and DNA-protein cross links in Benzo(a)pyrene induced experimental lung carcinogenesis. *Molecular and Cellular Biochemistry,* **268:** 141-147.

Singh, N.P., Mc Coy, M.T., Tice, R.R., and Schneider, E.L. (1988). A simple technique for quantitation of low levels of DNA damage in individual cells. *Experimental Cell Research,* **175:** 184-191.

Smith, T.R., Miller, M.S., Lohman, K.K., Case, L.D., and Hu, J.J. (2003). DNA damage and breast cancer risk. *Carcinogenesis,* **24:** 883–889.

Teel, R.W., Zhang, Z., Uynh, H., and Hamilton, S.M. (1997). Effects of capsaicin on the Metabolic activation of heterocyclic amines and on cytochrome P450 1A2 activity in hamster liver microsomes. *Proceedings of American Association of Cancer Research,* **38:** 363.

Thirunavukkarasu, C., Premkumar, K., Jagadeeswaran, R., and Sakthisekaran, D. (2005). The inhibitory effect of sodium selenite on N-nitrosodiethylamine-induced and phenobarbital promoted liver tumourigenesis in rats based on the modulation of polyamine levels. *Molecular and Cellular Biochemistry,* **280:** 165-172.

Wu, X., Roth, J.A., Zhao, H., Luo, S., Zheng, Y.L., Chiang, S., and Spitz, M.R. (2005). Cell cycle checkpoints, DNA damage/repair, and lung cancer risk. *Cancer Research,* **65:** 349–357.

Zhang, Z., Hamilton, S.M., Stewart, C., Strother, A., and Teel, R.W. (1993). Inhibition of liver microsomal cytochrome P-450 activity and metabolism of the tobacco-specific nitrosamine NNK by capsaicin and ellagic acid. *Anticancer Research,* **13:** 2341-2346.

Bioactive Phytochemicals: Perspectives for Modern Medicine Vol. 2 (2014) *Pages* **165–195**
Editor: **V.K. Gupta**
Published by: **DAYA PUBLISHING HOUSE, NEW DELHI**

8

An Insight in to Therapeutic Potentials of Imidazole Moiety with Special Reference to Alkaloids from Natural Sources

Papiya Bigoniya[1*] and Alok Shukla[1]

ABSTRACT

Alkaloids are natural products produced as secondary metabolites by plant and also by large variety of other organisms including bacteria, fungi and animals. Many alkaloids can be purified from crude extracts by acid-base extraction. Imidazoles are an important class of heterocycles which include many substances of both biological and chemical interest. The imidazole moiety is also contained in many histaminergic ligands for histamine H_1, H_2 and H_3 receptors. The important therapeutic properties of imidazole related drugs have encouraged the medicinal chemists to synthesize and test chemotherapeutic properties of several molecules with imidazole moiety. Alkaloids possessing imidazole nucleus are very rare. Imidazole alkaloids are found in some Rutaceae (leaves of pilocarpus and seeds of Casimiroa edulis), Cactaceae (Dolichotele) and in some Fabaceae plants. Synthetic as well as natural imidazoles have anticonvulsant, phosphodiesterase stimulation, central activity, chemotherapeutic, cardiovascular, anti-allergic and anti-inflammatory activity. The present chapter compiles the literature review of naturally occurring imidazole alkaloids viz. source, constituents, traditional use and reported pharmacological activities along with insight into the recent knowledge about research going on the synthetic compounds with imidazole moiety. This article also explores the possible therapeutic potential of natural imidazole alkaloids in the field of traditional medicine practice.

Keywords: Alkaloid, Histaminergic, Imidazole, Jaborandi, Lepidium, Maca, Phytopharmaceutical.

1 Radharaman College of Pharmacy, Fatehpur Dobra, Ratibad, Bhopal – 4620 02, M.P., India

* *Corresponding author*: E-mail: p_bigoniya2@hotmail.com

Introduction

Alkaloids are group of naturally occurring chemical compounds mostly containing basic nitrogen atoms. Some related compounds with neutral and weakly acidic properties are also included in alkaloid group. Alkaloids may also contain oxygen and sulfur along with carbon, hydrogen and nitrogen, and more rarely other elements such as chlorine, bromine and phosphorus. Alkaloids are natural products produced as secondary metabolites by plant and also by large variety of other organisms including bacteria, fungi and animals. Many alkaloids can be purified from crude extracts by acid-base extraction. The boundary between alkaloids and other nitrogen-containing natural compounds is not clear-cut. Alkaloids are usually very bitter and although may be poisonous, most of them are pharmacologically active like atropine, caffeine, coniine, morphine, nicotine, quinine, strychnine etc. Although alkaloids act on a diversity of metabolic systems in humans and other animals, they almost uniformly invoke a bitter taste. They often used as medications and in entheogenic rituals *viz.*

- ✰ Local anesthetic and stimulant cocaine
- ✰ Psychedelic psilocin
- ✰ Stimulant caffeine and nicotine
- ✰ Analgesic morphine
- ✰ Antibacterial berberine
- ✰ Anticancer compound vincristine
- ✰ Antihypertensive agent reserpine
- ✰ Cholinomimeric galatamine
- ✰ Spasmolytic atropine
- ✰ Vasodilator vincamine
- ✰ Anti-arrhythmic quinidine
- ✰ Anti-asthmatic ephedrine
- ✰ Antimalarial quinine etc.

Imidazole

Imidazoles are an important class of heterocycles which include many substances of both biological and chemical interest. They are part of a large number of highly significant biomolecules such as the essential amino acid histidine and related compounds, biotin and the imidazole alkaloids. The imidazoles are a class of antifungal azole derivatives and have a broad spectrum of activities both *in vitro* and *in vivo*, and also been shown to have antibacterial activity. The imidazole moiety is also contained in many histaminergic ligands for histamine H_1, H_2, and H_3 receptors. These are currently used as tools in pharmacological studies. The important therapeutic properties of imidazole related drugs have encouraged the medicinal chemists to synthesize and test chemotherapeutic properties of several molecules with imidazole moiety. Several P_{38} MAP Kinase inhibitors and 5-Lipoxygenase inhibitors containing the imidazole moiety have been synthesized recently (Luca, 2006; Brossi, 1984).

Imidazole alkaloid

Alkaloids containing one or more imidazole moieties as part of its structure are called imidazole alkaloids. Alkaloids possessing imidazole nucleus are very rare. Imidazole alkaloids are found in some Rutaceae (leaves of Pilocarpus and seeds of *Casimiroa edulis*), Cactaceae (Dolichotele) and in some Fabaceae plants. The amino acid L-histidine contains an imidazole ring and is thus the likely precursor of alkaloids containing this ring system. There are relatively few examples, however and definite evidence linking them to histidine is often lacking.

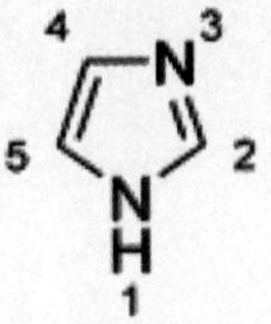

4 3 N 5 2 N H 1

H N CO_2H N NH_2

Imidazole moieties 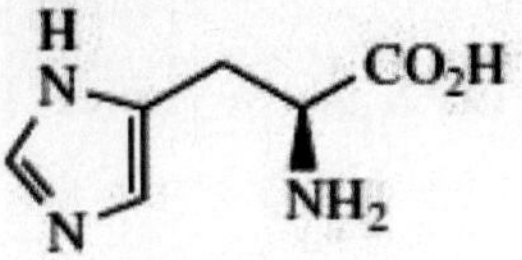 **L-histidine**

Histidine is a proven precursor of dolichotheline in *Dolichothele sphaerica* (Cactaceae), the remaining carbon atoms originating from leucine via isovaleric acid. The imidazole alkaloids found in Jaborandi leaves are also probably derived from histidine.

Plants Containing Imidazole Alkaloid

Pilocarpus Species (Rutaceae)

Pilocarpus or jaborandi consists of the dried leaflets of *Pilocarpus jaborandi, P. microphyllus*, or *P. pennatifolius*), small shrubs from Brazil and Paraguay. *P. microphyllus* is currently the main source. The alkaloid content (0.5–1.0 per cent) consists principally of the imidazole alkaloid pilocarpine, together with small amounts of pilosine and related structures. Isomers such as isopilocarpine and isopilosine are readily formed if base or heat is applied during extraction of the alkaloids. This is a result of enolization in the lactone ring followed by adoption of the more favourable trans configuration rather than the natural cis. However, the iso alkaloids lack biological activity. The alkaloid content of the leaf rapidly deteriorates on storage.

Pilocarpus microphyllus

Pilocarpine, an important imidazole alkaloid, is extracted from the leaves of *Pilocarpus microphyllus* (Rutaceae), known as jaborandi and used mainly for the

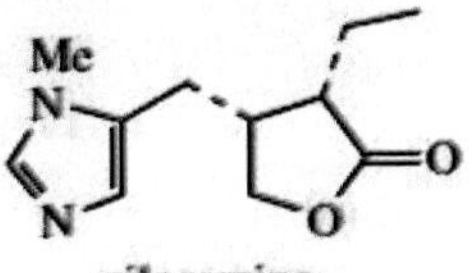

Me N N O O pilocarpine

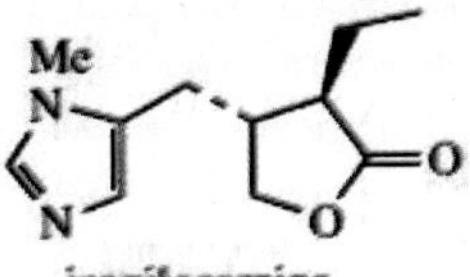

Me N N O O isopilocarpine

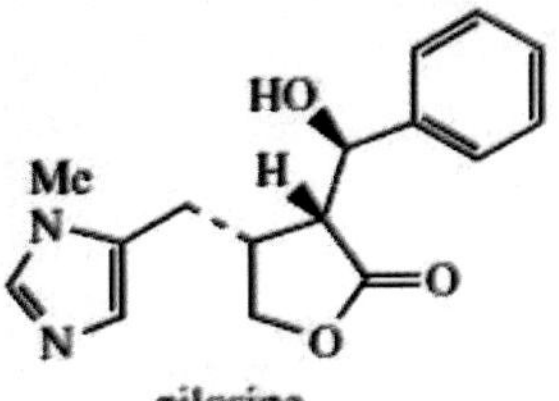

HO Me H N N O O pilosine

Pilocarpus microphyllus

treatment of glaucoma. Jaborandi leaves also contain other imidazole alkaloids like pilosine, isopilosine and epiisopilosine whose pharmacological and physiological properties are unknown.

Sawaya *et al.* (2008) reported a HPLC method coupled with ESI-MS(n) developed for qualitative and quantitative analysis. This method permits the chromatographic separation of the imidazole alkaloids found in extracts of jaborandi as well as the MS/MS analysis of the individual compounds. The method is suitable for the routine analysis of samples containing imidazole alkaloids as well as for the separation and identification of known and novel alkaloids from this family (Sawaya *et al.*, 2008).

Pilocarpus jaborandi

Jaborandi consists of the leaflets of *Pilocarpus jaborandi* Holmes a tropical American Rutaceous shrub indigenous to Brazil. The principal constituents are the three alkaloids, pilocarpine (not found in all species), isopilocarpine and pilocarpidine are extracted from dried leaflet of jaborandi.

Pilocarpidine

Pilocarpus trachyllophus

Roots of *Pilocarpus trachyllophus* yielded (–)-episesamin, hesperidin, tridecan-2-one, 1-hydroxytridecan-2-one, *N*-isobutyl-deca-2(E),4(E)-dienamide, pilocarpine and a new imidazolealkaloid, 13-nor-7(11)-dehydropilocarpine as reported by Andrade-Neto *et al.* (1996). Except for pilocarpine and tridecan-2-one, this is the first report of these substances in the genus *Pilocarpus*. Structural determinations were accomplished by spectroscopic analysis, particularly two-dimensional NMR, chemical interconversion and, where appropriate, comparison with literature data.

Pilocarpus jaborandi

Isopilosine

Epiisopilosine

Lepidium sativum (Cruciferae)

L. sativum commonly known as chandrasura is a small, herbaceous, glabrous annual, 15-45 cm high plant cultivated as salad supplement. *L. sativum* have been widely used to treat a number of ailments in traditional system of medicine throughout India. Cold infusions of seeds are used to relieve hiccough. The seeds are used in chronic enlargement of liver and spleen and also used as carminative adjunct to purgatives. Alkaloid content was found to be 0.29 per cent in the seed. The major

	R_1	R_2
Lepidine A	OCH_3	H
Lepidine B	OH	H
Lepidine C	H	OCH_3
Lepidine D	H	OH

	R_1	R_2
Semilepidine A	OH	H
Semilepidine B	OH	OCH_3

components of this alkaloid fraction are lepidine and semilepidine, a rare group of imidazole alkaloid. The active moiety in this alkaloid is 2- benzyl imidazole or imidazoline. (Maier *et al.*, 1998). Shukla *et al.* (2011) reported sedative, anxiolytic, myorelaxant and analgesic activity of *L. sativum* alkaloid with poly substituted phenylimidazoline basic nucleus.

Lepidium sativum

Lepidium meyenii (Brassicaceae)

For generations people of the rural community in Peru have used *Lepidium meyenii* Walpers (Maca), because of their belief that it improves fertility and sexual desire. Lentz *et al.* (2007) studied the effects of Maca on male sexual behavior and on chronic administration on anxiety. Maca treatment did not produce large changes in male sexual behavior. However, an increase in ejaculation latency and postejaculatory interval was observed after both acute and 7

Lepidium meyenii

days of treatment. After 21 days of treatment Maca had no effect on sexual behavior. Chronic administration of Maca did not increase locomotion or anxiety (Lentz *et al.*, 2007)

Two new imidazole alkaloids (lepidiline A and lepidiline B) have been isolated from the root extract of *L. meyenii* with the common name Maca and identified as 1,3-dibenzyl-4,5-dimethylimidazolium chloride and 1,3-dibenzyl-2,4,5-trimethylimidazolium chloride respectively. The structures of these two new compounds were determined by spectroscopic methods, as well as single-crystal X-ray diffraction (Cui *et al.*, 2003).

Butea monosperma (Fabaceae)

The seeds of *Butea monosperma* contain an alkaloid which has physostigmine like action as shown by specific atropine blockade, potentiation of acetylcholine responses and antagonism of curare action. The new alkaloids has been isolated and identified as

O, HN, N, Ac, H_3CO

Monospermin

Butea monosperma

monospermin (1-N-acetyl-2-oxo-4-methoxy-3H,5H-imidazole) (Singh and Panda, 2005).

Alchornea javanensis (Euphorbiaceae)

Bark and leaves of *Alchornea javanensis* contains alchorneine, isoalchorneine and alchorneidine a plant indigenous to rain forests of New Guinea (Hart *et al.*, 1970).

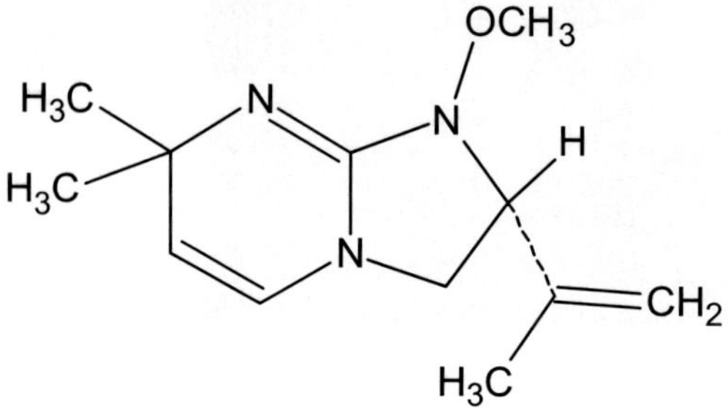

Alchorneine

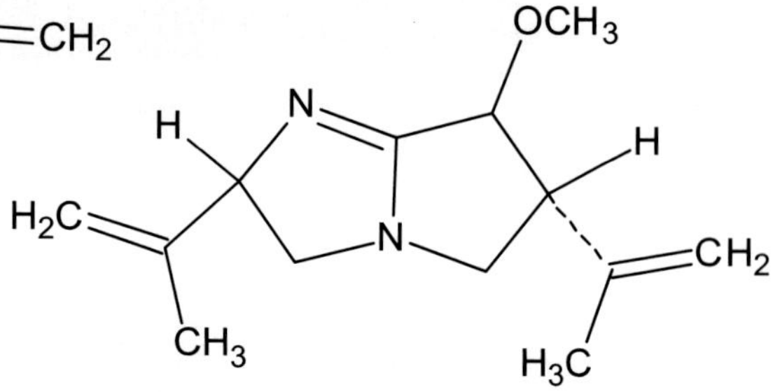

Isoalchorneine

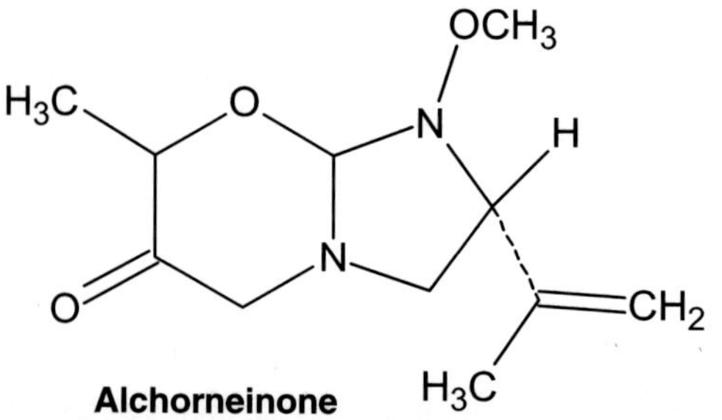

Alchorneinone

Casimiroa edulis (Rutaceae)

Casimiroedine and zapotidine is the principal alkaloid of the seeds of *Casimiroa edulis* widely distributed in Mexico and Central America (Amani *et al.*, 2007).

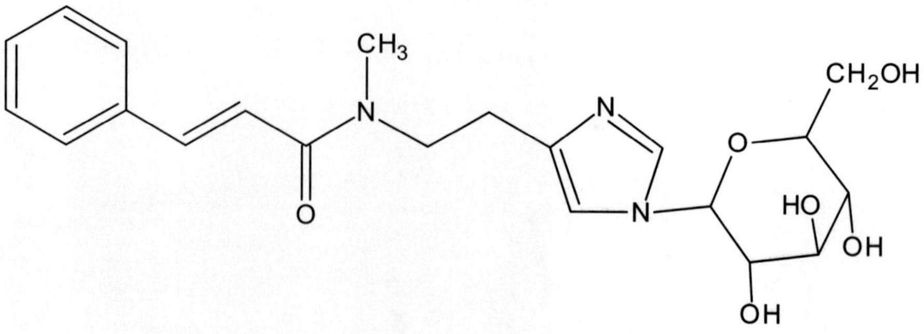

Casimicroedine

Casimiroa edulis

Cynometra hankei (Caesalpiniaceae)

Alkaloids anantine, isoanantine, cynometrine, isocynometrine and cynodine was isolated from stem bark and seed of *Cynometra hankei* (Peter *et al.*, 1981).

Anantine

Cynometrine

Isoanantine

Isocynometrine **Cynodine**

Cypholophus friesianus (Urticaceae)

Cypholophine and *O*-acetylcypholophine was isolated from *Cypholophus friesianus* (Hart *et al.*, 1971).

R= H Cypholophine
R= Ac *O*-acetylcypholophine

Dolichotheline sphaerica (Cactaceae)

Dolichotheline is a histamine derived alkaloid produced by the cactus *Dolichotheline sphaerica* native to Southern Texas and northern Mexico (Rosenberg and Paul, 1969).

Dolichotheline sphaerica

Dolichotheline

Glochidion philippicum (Euphorbiaceae)

Glochidine, glochidicine and *N*α-(4-oxodecanoyl) histamine has been isolated from leafs of *Glochidion philippicum* tree found in New Guinea (Johns and Lamberton, 1967).

Glochidion philippicum

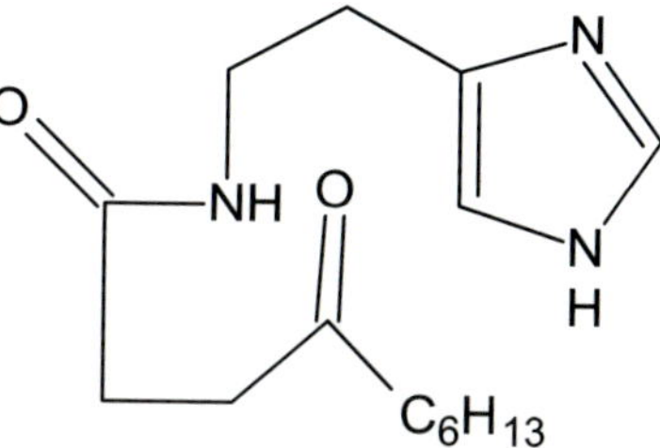

N^{α}- (4-Oxodecanoyl) histamine

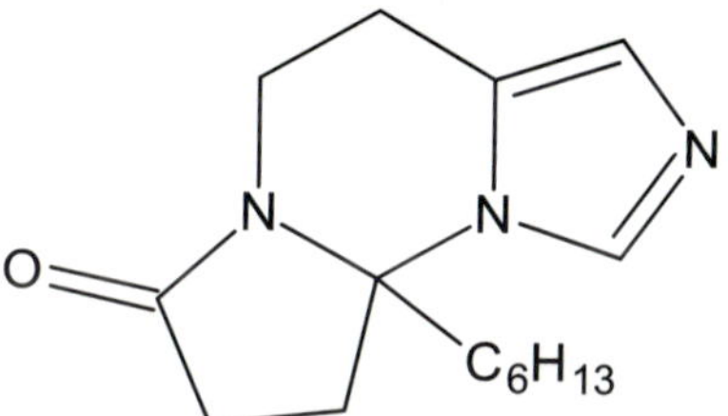

Glochidine

Glochidicine

Macrorungia longistrobus (Acanthaceae)

Macrorungia longistrobus shrub contains large number of imidazole alkaloids like macrorine, isomacrorine, macrorungine, normacrorine, longistrobine, isolongistrobine and dehydroisolongistrobine (Wuonola and Woodward, 1976).

Macrorungia longistrobus

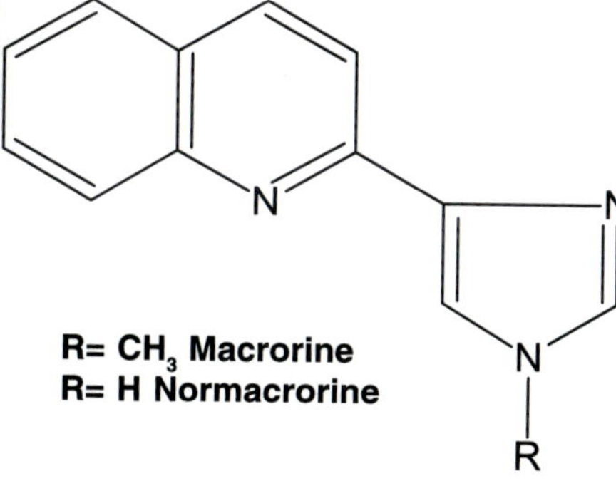

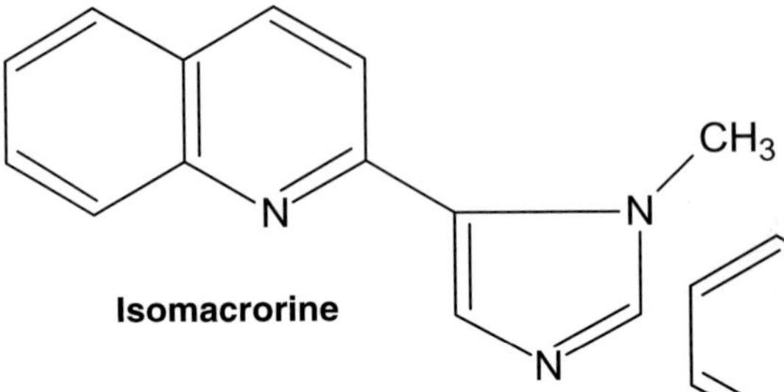

Isomacrorine

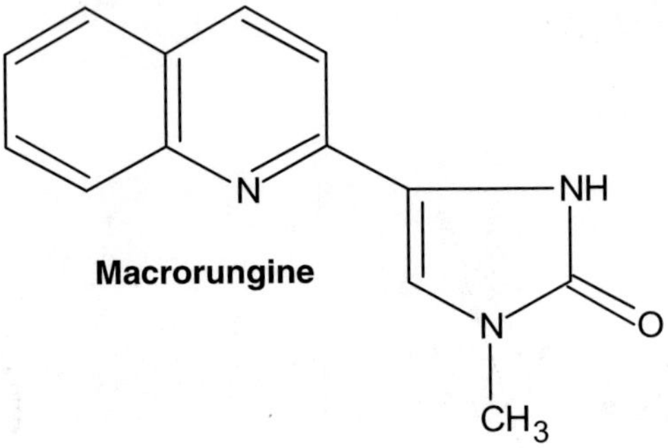

Macrorungine

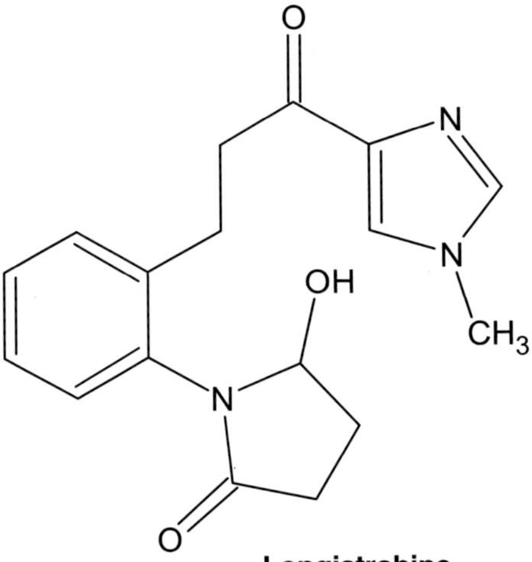

Longistrobine

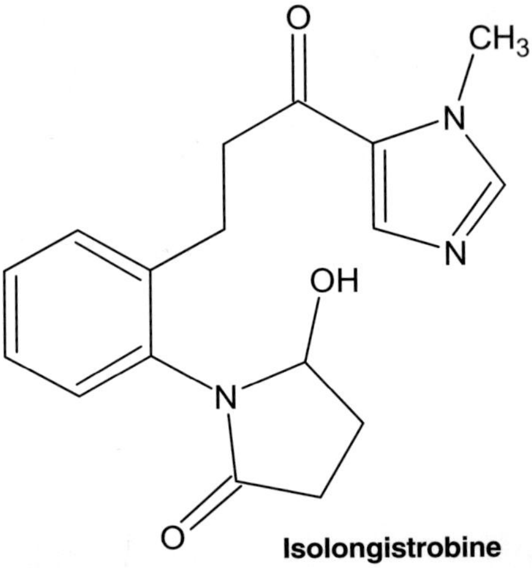

Isolongistrobine

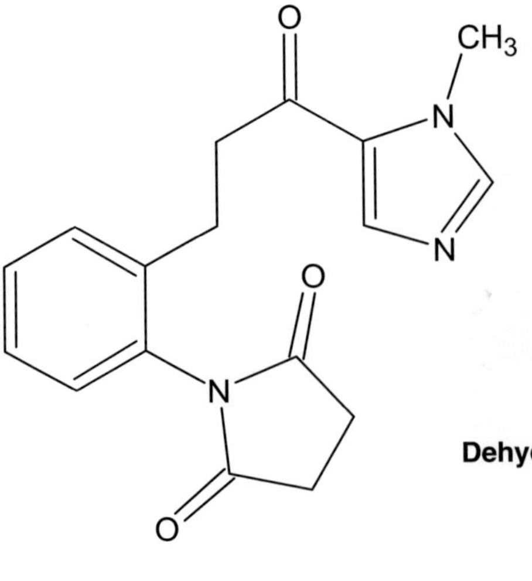

Dehydroisolongistrobine

Marine Source

Leucetta Species (Leucettidae)

New imidazole alkaloids naamidine F and clathridine C have been isolated from a sponge *Leucetta* sp., together with the known imidazole alkaloids naamidine A, pyronaamidine, clathridine and kealiiquinone by Carroll *et al.* (1993). The predatory nudibranch *Notodoris gardineri*, which was found feeding on the sponge, contained the known metabolite clathridine.

Kealiiquinone

Naamidine A

Pyronaamidine

Naamidine F

Agelas sventres **(Agelasidae)**

Brechle *et al.* (2003) isolated sventrin from *Agelas sventres* as the pyrrole-*N*-methylated analog of oroidin. Pyrrole-imidazole alkaloids are exclusively found in marine sponges of the orders Agelasida, Axinellida and Halichondrida.

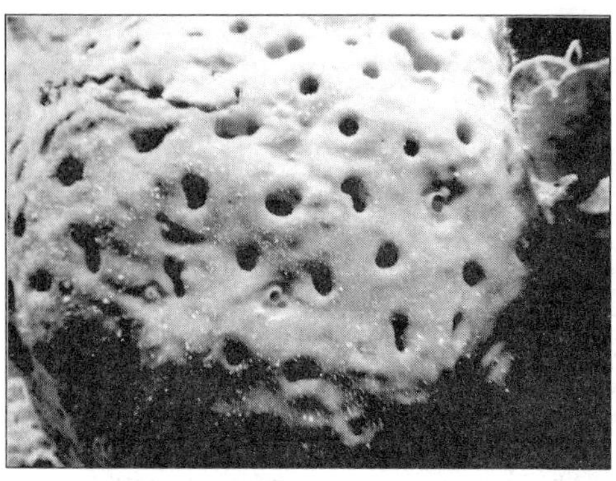

Sventrin (R = H)

Botryllus leachi **(Styelidae)**

Naturally occurring and biologically active alkaloids Botryllazine B and 2-(p-hydroxybenzoyl)-4-(p-hydroxyphenyl) imidazole was first isolated from the red ascidian *Botryllus leachi* (sea squirt) by Rosario *et al.* (1999).

2-(p-hydroxybenzoyl)-4-(p-hydroxyphenyl) imidazole

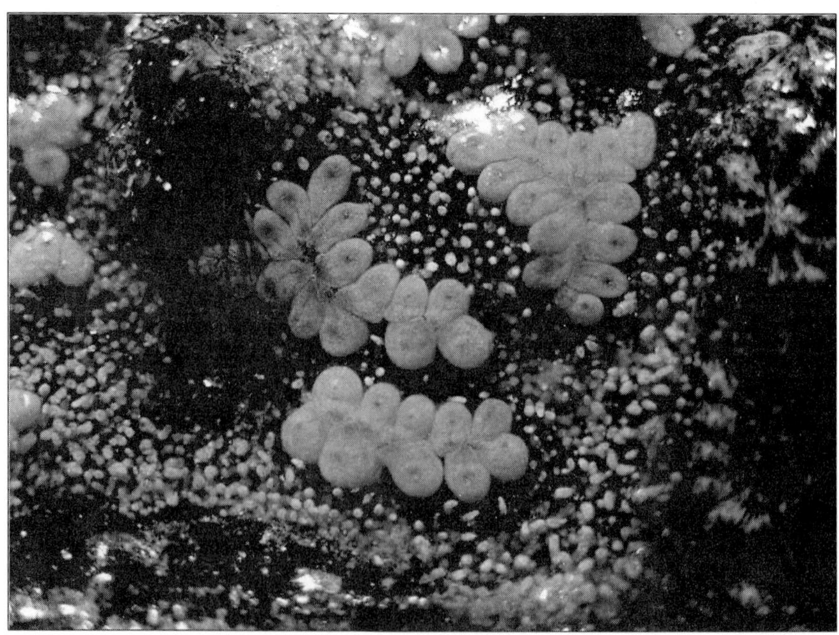

Parazoanthus axinellae **(Zoanthidae)**

Diimidazole alkaloid zoanthoxanthin was isolated from Mediterranean sea anemone *Parazoanthus axinellae*. Zoanthoxanthin contains aminohistidine unit can be derived from two histidine units (Strupi Suput *et al.*, 1996).

Parazoanthus axinellae

Zoanthoxanthin

Epizoanthus arenaceus **(Zoantharia)**

Epizoanthus arenaceus is a zoanthid closely related to *P. axinellae* contains alkaloid of same skeleton like zoanthoxanthin, they are epizoanthoxanthin A and B (Brossi, 1984).

R= H, R_1= CH_3 Epizoanthoxanthin A

R= CH_3, R_1= H Epizoanthoxanthin B

Epizoanthus arenaceus

Spinacine and spinaceamine have been isolated from skin of amphibian *Laptodactylus pentadactylus* (Brossi, 1984). Spinacine was also isolated from water soluble fraction of ethanolic root extract of *Panax ginseng* (Han *et al.*, 1987).

Spinacine **Spinaceamine**

Biologically Important Imidazole Alkaloid Pilocarpine

Jaborandi leaves are used for the extraction of pilocarpine. Pilocarpine is the lactone of pilocarpic acid having a glyoxaline nucleus. Pilocarpine is readily hydrolyzed in aqueous medium to pilocarpic acid and isomerized in to isopilocarpine.

An adaptable and rapid method is required for extraction due to lability of this compound. It also complicates the quality control and limits the storage time in solution resulting in loss of pharmacological activity.

Pilocarpine salts are valuable in ophthalmic practice and are used in eyedrops as miotics and for the treatment of glaucoma. The compound is used as its nitrate or as hydrochloride as eye drop Pilocarpine is a cholinergic agent (parasympathomimetic) and stimulates the muscarinic receptors in the eye, causing constriction of the pupil and enhancement of outflow of aqueous humour. Pilocarpine gives relief in both narrow and wide angle glaucoma. Pilocarpine causes the iris sphincter to contract and causes myosis which leads to the opening of the space between the iris and cornea in turn relieve narrow angle glaucoma. The contraction of the ciliary muscle on the other hand facilitates the flow of aqueous humor in case of wide angle glucoma. The ocular bioavailability of pilocarpine is low and it is rapidly eliminated, thus resulting in a rather short duration of action. The effects of pilocarpine are similar to physostigmine and the two agents are sometimes combined. The side effects include myosis, transient headache and increased tear secretion.

Pilocarpine is antagonistic to atropine. It has been found that pilocarpine gives relief for dryness of the mouth in patients undergoing radiotherapy for mouth and throat cancers. As muscarinic agonists, pilocarpine and its analogues are also being investigated for potential treatment of Alzheimer's disease. Pilocarpine is also used as sialagogue to relieve dryness of mouth in patients of radiotherapy with mouth cancer (Bruneton, 1999).

Promising Pharmacological Effects of Synthetic Imidazole Derivatives

Anticonvulsant Activity

The imidazole ring in nafimidone [1-(2-naphthyl)-2-(1-imidazolyl)ethanone] was substituted with various groups and anticonvulsant activity was investigated by Umal *et al.* (1988). Anticonvulsant activities of these compounds were determined by phase I tests of Antiepileptic Drug Development Programme.

Phosphodiesterase Stimulation

Imidazole has been reported to enhance the ability of phosphodiesterase to break down cyclic 3', 5'-AMP. Wilkenfeld and Levy (1969) have determined the effects of these agents on the inhibitory responses to isoproterenol and phenylephrine in the spontaneously contracting isolated rabbit ileum. The dose-response curve to isoproterenol was shifted to the right in the presence of imidazole, 10 μg/ml (1.5×10^{-4} M), and then back to control after removal of imidazole.

Central Activity

Caroon *et al.* (1982) prepared and evaluated several 2-[(1,4-benzodioxan-2-yl)alkyl]imidazoles for their blocking activity and relative selectivity on presynaptic (alpha 2) and postsynaptic (alpha 1) receptors in the isolated rat vas deferens. 1-Ethyl-2-[(1,4-benzodioxan 2-yl)methyl]imidazole was the most selective alpha

2-adrenoceptor antagonist of the series and was, for practical purposes, devoid of alpha 1-adrenoceptor antagonist activity. A unique combination of alpha 2-adrenoreceptor antagonist and serotonin-selective reuptake inhibitory activities has been identified in a series of 2-substituted 4,5-dihydro-1H-imidazole derivatives by Wentland *et al.* (1987). This combination of blocking activities has provided one of these derivatives, napamezole hydrochloride, with potential as an antidepressant.

Substitution in 2-position by alkyl, alkenyl, cycloalkenyl, and alkoxy groups in many cases gives compounds whose potencies and selectivity are significantly greater than those of the parent compound (Stillings *et al.*, 1985). A novel series of central nervous system (CNS) penetrant indane 2-imidazoles have been identified as potent, partial agonists of the α1A adrenergic receptor, having good selectivity over the α1B, α1D and α2 sub-types (Lee *et al.*, 2008).

Three moclobemide analogues were synthesized by replacing moclobem-mide phenyl ring with substituted imidazoles. N-[2-(4-morpholinyl)ethyl)]-1-benzyl-2-(methylthio)-1H-imidazole-5-carboxamide (7a), N-[2-(4-morpholinyl)ethyl)]-1-benzyl-2-(ethylthio)-1H-imidazole-5-carboxamide (7b) and N-[2-(4-morpholinyl)ethyl)]-1-benzyl-2-(benzylthio)-1H-imidazole-5-carboxamide (7c) were studied for the antidepresssant activity using forced swimming test in mice. Analogues 7a-c were found to be potent and minimum effective doses for moclobemide and analogues 7a-c were found to be 20, 2.5, 1.25 and 2.5 mg/kg i.p. respectively (Hadizadeha *et al.*, 2008). Imidazole inhibited (+/-) N-n-propylnorapomorphine (NPA) and BHT-920 induced penile erections, stretching and yawning in rats as well as apomorphine induced hypothermia in mice. Imidazole enhanced shock-elicited aggressiveness in rats and antagonized sleep induced by clonidine in chicks. Imidazole moreover displayed activity in behavioural tests used in specific screening for antidepressants, potentiating yohimbine toxicity in mice and antagonizing immobility time in the despair test, with a potency in some cases equal to imipramine. Imidazole per se, depressed motor activity in both mice and rats as reported by Ferrari (1985).

The synthesis and the 5-HT1A and 5-HT2 receptor affinity of 2-substituted 1-[3-(4-aryl-1-piperazinyl)propyl]-imidazoles has been described by Mokrosz and Duszynska, (1992). It has been shown that both the N-3 imidazole atom and the N-1 piperazine one should be considered as possible protonation centers under physiological conditions. Three different modes of interaction of the analyzed compounds with 5-HT1A and 5-HT2 receptor sites have been proposed.

Evidences support that imidazole 4-acetic acid (IMA) is a naturally occurring metabolite in brain an oxidation product of histamine. The biochemical pathways involved in its biosynthesis and breakdown are unknown. IMA has pronounced neuropharmacological properties consistent with an activation of $GABA_A$ receptors. IMA displays definite partial agonist characteristics as an enhancer of benzodiazepine binding to the $GABA_A$ receptor complex. In addition, it has an affinity for $GABA_C$ receptors to act as an antagonist and perhaps as a weak partial agonist. The third recognition site for IMA in brain is the imidazoline I_1 receptor. Parenteral administration to experimental animals leads to a sleep-like state which can often be accompanied by seizures. Central application of IMA is associated with a dose-related reduction in arterial pressure and sympathetic nervous discharge. IMA plays a role in regulating blood pressure (Tunnicliff, 1998).

Chemotherapeutic

More than 100 1-methyl-5-nitroimidazoles substituted in the 2-position via an oxymethyl group were synthesized and their structure-activity relationship toward various protozoa was investigated by Winkelmann and Raether, (1978). Among the derivatives substituted with an aromatic radical there are most of the compounds which are highly effective against trichomonads.

Four imidazole derivatives have undergone extensive open and comparative trials as topical agents in dermatomycoses and vaginal candidosis. They are chlormidazole (Chemie Grünenthal), clotrimazole (Bayer), miconazole (Janssen) and econazole (Janssen, CilagChemie) with added antibacterial activity. Many imidazoles have been marketed usually as antiprotozoal or anthelminthic agents and some of these have some anti-mycotic activity as well as other miscellaneous therapeutic properties. After prolonged topical application to animals and human subjects systemic absorption is negligible. All four agents which are available as cream, powder, lotion or vaginal tablets have many successful studies to their credit, often with clinical and mycological cure rates of over 80 per cent in a variety of dermatomycoses and in vaginal candidosis. (Holt, 1976).

Cardiovascular Effect

Imidazole increases the force of contraction of the isolated and intact heart as reported by Knope *et al.* (1973). The increased force of contraction is not always accompanied by an increased frequency of contraction. Propranolol in concentrations blocking the action of isoproterenol had no effect on the imidazole response. Similarly aminophylline in low concentrations failed to enhance the imidazole response. The data suggest that imidazole, unlike glucagon or epinephrine, may increase myocardial contractility by mechanisms other than those involved with cyclic 3′,5′-adenosine monophosphate but similar to those involved in the action of the cardiac glycosides.

Anti-allergic

Pinelli *et al.* (1989) reported that 2-substituted benzimidazole compounds exhibit anti H_2-histamine activity by decreasing gastric acid secretion, basal and stimulated with histamine or betazole. Iemura *et al.* (1986) tested a series of 2-(4-substituted-1-(homo)piperazinyl)benzimidazoles for H_1-antihistaminic activity *in vitro* and *in vivo*. Most of the compounds showed antihistaminic activity and some of the 1-[2-(substituted-oxy)ethyl] derivatives exhibited potent activity.

Anti-inflammatory

Pande *et al.* (1984) synthesized sixteen new 1,2-disubstituted-4-(indol-3'-yl) methylene imidazol-5-ones by the condensation of oxazolone with different aryl amines and evaluated their anti-inflammatory and antiproteolytic activities. These derivatives showed 2-38 per cent protection against carrageenin-induced paw oedema in albino rats at a dose of 100 mg/kg p.o. All compounds showed antiproteolytic activity.

The second-order rate constants, for quenching of molecular singlet oxygen O_2 by nonsteroidal anti-inflammatory imidazole drugs have been determined using time-resolved phosphorescence detection of singlet oxygen. A linear correlation was observed between log k delta (ranging from 7.90 to 8.50) and the anti- inflammatory activity of these compounds (ranging from ED_{50} = 15 to 300 mg/kg). The rate constant (k delta) can be considered as a quantitative expression of the electron-donating power of the imidazole drug as observed in the case of 2-substituted imidazoles (Bensasson *et al.*, 1992).

Pharmacology of Natural Imidazole Alkaloid Containing Plants

Jaborandi

Traditional Uses

In folk medicine systems of the tropics, jaborandi has been used as a natural remedy for epilepsy, convulsions, gonorrhea, fever, influenza, pneumonia, gastrointestinal inflammations, kidney disease, psoriasis, neurosis and as an agent to promote sweating. In Brazil jaborandi has been used by herbalists for bronchitis, asthma, pneumonia, diphtheria, colds and flu, laryngitis, renal insufficiency, hepatitis, diabetes, kidney diseases, oedema and fever. An infusion or cold maceration of the leaves induces sweating within 10 minutes–as much as 9 to 15 ounces of sweat can be excreted from a single dose. Externally it is used as a hair tonic which is believed to open pores and clean hair follicles, prevent hair loss and generally aid in the manageability of hair. Jaborandi leaves increases the secretion of saliva, sweat, gastric juices and tears, and stimulates the smooth muscle of the gastrointestinal tract, bronchi, bile duct and bladder.

Reported Pharmacological Effect

Pilocarpine possess CNS activity as a direct cholinergic which stimulates the parasympathetic system (bladder, tear ducts, sudoriferous and salivary glands). This alkaloid is the elected drug in the glaucoma treatment (Korolkovas, 1996) and the sialagogue property of pilocarpine has been exploited to treat the xerostomy (dry mouth) caused by the radio- or chemotherapy of throat cancer or those in the head area. Others alkaloids derived from *L*-tryptophan were also isolated from *Pilocarpus* species. Some examples are *N,N*-dimethyl-5-methoxy-triptamine and *N,N* dimethyl-triptamine (known as DMT) obtained from *P. organensis* leaves; both have a CNS activity causing hallucinations (Miller, 1993; Valdez *et al.*, 1993).

Maca

Traditional Uses

Maca is an herbaceous biennial plant or annual plant native to the high Andes of Peru and Bolivia. It is grown for its fleshy hypocotyl (actually a fused hypocotyl and taproot), which is used as a root vegetable and a medicinal herb. According to local folklore, Maca, (also called Peruvian Ginseng) was introduced to Spanish explorers by the native Incas sometime during the 16th century. Although their livestock survived the arduous journey, they were weak and infertile. Upon the advice of the Incas, the Spaniards fed Maca to their horses and other animals within a short period of time, the livestock regained its health and started to reproduce normally. Maca's history as a powerful strength and stamina enhancer and libido-fertility herb stretches back well over five hundred years. Maca's fertility- enhancing properties were reported as early as 1961, when researchers discovered that it increased fertility in rats. In addition to its use for fertility, Inca warriors would eat Maca before battles to promote strength and stamina. Maca is still a staple in the Peruvian diet. The root may be roasted like a potato or it may be used to make jam, pudding, juice or soup. As an aphrodisiac, Maca is typically formulated into a capsule. In Peruvian herbal medicine Maca is reported to be used as an immunostimulant, antianemic, antitubercular, in menstrual disorders, menopause symptoms, stomach cancer, sexual disorders and to enhance memory. Maca is a powerful adaptogen that can help balance our body's systems. Maca can stabilize blood pressure, boost immunity and harmonize the body's overall vitality.

Reported Pharmacological Effect

Maca contains alkaloids, steroids, glucosinolates, isothiocyanates and macamides (Dini *et al.*, 1994; Muhammad *et al.*, 2002). The Maca alkaloids, steroids, glucosinolates, isothiocyanates and macamides are probably responsible for its aphrodisiac, adaptogen, anabolic, immunostimulant and hormonal balancing properties (Gonzales *et al.*, 2003; Oshima *et al.*, 2003; Valentova and Ulrichova, 2003). Maca has reported antioxidant (Sandoval *et al.*, 2002), hepatoprotective (Valentova *et al.*, 2006) and immunostimulant effect (Lee *et al.*, 2004), and also has nutritional supplement activity and useful in joint diseases.

The literature reported pharmacological activities of Maca are CNS stimulant (Piacents *et al.*, 2002), anti-stress (Lopez *et al.*, 2004) and antidepressant (Rubio *et al.*, 2006) effects. In a study of the effect of yellow, red and black ecotypes of Maca on cognitive function and depression were assessed on ovariectomized mice that were fed each of three ecotypes for 21 days exhibited reduced water finding latency (a measure of cognitive function and learning) compared to controls ($p<0.05$), but those fed black Maca showed the most latency reduction (Rubio *et al.*, 2006). Maca has energizing properties as different doses of aqueous extracts of Maca (4, 10, 20 and 40 g/kg) have been shown to increase the swimming activity of mice. The extracts also aided recovery from muscle fatigue after strenuous physical activity, as measured by lactic acid and malonic acid production (Rubio *et al.*, 2006).

Sexual function effects of Maca are well reputated (Cicero *et al.*, 2001). Maca has estrogenic properties (Valentova *et al.*, 2006) and improve fertility in females. Administration of aqueous extract of lyophilized yellow Maca (1g/kg body weight) to adult female mice increases the litter size. This treatment also increased the uterine weight in ovariectomized animals. Maca reduces postmenopausal osteoporosis (Zhang *et al.*, 2006) and also prostrate cancer/enlargement (Gonzales *et al.*, 2005) in male rats. The mechanisms regulating endocrine are unclear at present, while some investigators suggest that Maca exerts hormonal balancing effect not through plant hormones or phytoestrogens, but through Maca alkaloids, which act on the hypothalamus-pituitary axis and the adrenals. Oshima *et al.* (2003) observed increased blood levels of testosterone and progesterone in mice fed Maca, which indicates a possible explanation for the steroid-like compounds improving osteoporosis in the ovariectomized rats.

High altitude testicular disturbance effects: Exposure to high altitude (4340m) results in a reduction in epididymal sperm count after seven days in adult male rats, and lower sperm values are maintained for up to 21 days. This altitude also reduced spermiation (stage VIII) to half and the onset of spermatogenesis (stages IX-XI) to a quarter on days 7 and 14. Treatment with an aqueous extract of Maca (666.6 mg per day) prevented these changes in spermiation and spermatogenesis. The Maca treatment was also able to prevent the altitude-induced reduction in sperm count. In the Maca treated group exposed to high altitude, epididymal sperm count was higher than in a non-treated group at sea level. It prevents erectile dysfunction activity (McKay 2004). Maca have steroid hormone (androgens) activity, and have spermatogenesis effects as evident by prevention of lead-induced infertility and also prevention of sperm damage by organophosphates (Bustos-Obregon *et al.*, 2005).

Chandrasura

Traditional Uses

Therapeutic importance of Chandrasura has been recognized in different system of traditional (Ayurveda, Unani, Sidha and Homeopathy) and modern medicine for the treatment of different diseases and ailments. The plant possesses the multitude of chemical compounds like flavonoids, coumarins, sulphur glycosides (glucosinolates), triterpines, sterols and imidazole alkaloids. This herb is cultivated as a salad plant throughout India, Europe and United States. It is an important medicinal plant since the Vedic era (Anonymous, 1962). The leaves are antiscorbutic, diuretic and stimulant. Seeds are bitter, thermogenic, depurative, galactagogue, emmenagogue, tonic, aphrodisiac, ophthalmic and diuretic. It is useful in leprosy, skin diseases, dysentery, diarrhoea, dyspepsia, eye diseases, leucorrhoea, scurvy, asthma, cough and cold, seminal weakness and also used as aperient, tonic, demulcent, carminative and ruvegacient (Khory, 1999). The polutice of seed is applied for hurt and sprains and for bronchial ashthama (Annonymous, 1962). The root is used in the treatment of secondary syphilis and tenesmus.

Reported Pharmacological Effect

Shinde *et al.* (2010) reported the protective effect of *L. sativum* aqueous extract against doxorubicin-induced nephrotoxicity in rats. The methanolic extract of *L. sativum* seeds showed antibacterial activity against the pathogens *Escherichia coli, Salmonella typhi, Pseudomonas aeruginosa, Staphylococcus aureus, Bacillus cereus* and *Micrococcus luteus* as reported by Gupta *et al.* (2010). Patel *et al.* (2009) investigated diuretic activity of aqueous and methanol extracts of *L. sativum* on rats. Mali *et al.* (2008) studied the bronchodilatory effect of *L. sativum* against allergen induced bronchospasm in guinea pigs. Afaf *et al.* (2008) investigated the hepatoprotective effect of *Lepidium sativum* methanolic seed extract against CCl_4 induced hepatic damage in rats. Radwn *et al.* (2007) reported the insecticidal activity of alcoholic seed extract (2 per cent), alcoholic herb extract (2 per cent), total glucosinolates (1 per cent), butanol fraction (2 per cent), glucotropaeoline (0.5 per cent), petroleum ether seed extract (2 per cent) and petroleum ether herb extract (2 per cent) carried on the white fly (*Bemisia tabaci*). Juma, (2007) revealed the effect of *L. sativum* seeds on fracture-induced healing in Rabbits. Archana *et al.* (2006) investigated the efficacy and safety of *L. sativum* in patients of bronchial asthma. Eddouks *et al.* (2005) reported aqueous extract exhibits a potent hypoglycaemic activity in rats without affecting basal plasma insulin concentrations. *L. sativum* seeds possess oral contraceptive activity as reported by Sharief *et al.* (2004). The chemoprotective effects of *L. sativum* juices towards benzo(a)pyrene B(a)P-induced DNA damage using the single cell gel electrophoresis (SCGE)/Hep G2 test system was reported by Kassie *et al.* (2003). *L. sativum* seeds showed abortifacient activity (Nath *et al.*, 1992). The seeds of *L. sativum* possess significant anti-inflammatory, antipyretic, analgesic and coagulant activities, and are free from serious side or toxic effects. Shukla *et al.* (2011) demonstrated sedative, anxiolytic, myorelaxant and analgesic activity of *L. sativum* alkaloid belongs to imidazole category with poly substituted phenylimidazoline basic nucleus.

Phytopharmacological Potential of Natural Imidazole Alkaloids

There is a great deal of interest and support available to the scientific community to carry out search for new and useful drugs from higher plants. Virtually every country of the world is active in this search and mostly in countries like People's Republic of China, Japan, India and the Federal Republic of Germany. Estimated the number of higher plants that have been described on the face of the Earth vary greatly–from about 250,000 to 150,000 out of these only a few have been studied as a source of new drugs. Most of the plants screened for bioactive compounds is targeted for one or two categories of diseases now the question arises, could any of these plants contain drugs effective for other disease states, such as arthritis, high blood pressure, acquired immune deficiency syndrome (AIDS), or heart trouble? Only a handful of plants that have been exhaustively studied for their potential value as a source of drugs, *i.e.*, tested for several effects instead of just only one.

Approximately 119 pure chemical substances extracted from higher plants are used in medicine throughout the world (Farnsworth *et al.*, 1985). For the most part, the discovery of the drugs originates from knowledge that their extracts are used to treat one or more diseases in humans. The potential plant extracts are then subjected to pharmacological and chemical exploration to determine the nature of the active components. This discovery of new drugs from plants is one of the few that has evolved from a random-selection broad pharmacological screening program. Therefore, it should be of interest to ascertain just how important plant drugs are throughout the world when used in the form of crude extracts. The 119 plant-derived drugs in use throughout the world today are obtained from less than 90 species of plants (Farnsworth *et al.*, 1985). How many more can be reasonably predicted to occur in the more than 250,000 species of plants on Earth?

Mega screening projects targeted at isolation and standardization of single bioactive marker compound yields moderately. The methods developed and reported by scientists for isolation and purification of active compounds is generally lengthy and time consuming (Cragg and Boyd, 1996). The yields always very largely depending on factors like laboratory conditions, researchers individual expertise, harvesting procedure, collection time, geographical conditions etc. These processes of isolation are often very costly with yield in mg's or μg's which is commercially not feasible for large scale production needed for clinical use. Presence of variable impurity profiles also makes storage and stability a difficult issue to resolve.

Most promising plants seem to be indigenous to developing countries and assurance that there will be a continued supply of the raw material needed to produce the useful drugs is difficult to assure. When the plant under consideration is rare and grows only in one isolated geographical region it is really difficult to maintain the source of raw material. As a remedy to this a cultivation programs should be initiated with production of drugs but not always possible or successful as the altitude, longitude, soil and other environmental conditions are needed to grow a plant essentially with required quantity of phytochemicals. In the early stages of development of vincristine and vinblastine, the plant source *C. roseus* it was collected from many different countries of the world and was also cultivated in eastern European countries and in the United States. But the ultimate solution is to develop a synthesis method in laboratory a ensure constant supply, consistent quality, reproducible quality and purity standards and most essentially preservation of biodiversity and the plant species from the risk of extinction.

Higher plants have been described as chemical factories that are capable of synthesizing unlimited numbers of highly complex and unusual chemical substances whose structures could escape the imagination of synthetic chemists forever. Considering that many of these unique gene sources may be lost forever through extinction and that plants have a great potential for producing new drugs of great benefit to mankind, some action should be taken to reverse the current apathy. The drugs can be chemically copied or synthesized by laboratories and no plant materials are used in the manufacture of the drug. A good example of this is the plant chemical quinine, which was discovered in a rainforest tree (*Cinchona ledgeriana*) over 100 years ago. For many years the quinine chemical was extracted from the bark of this tree and processed into pills to treat malaria. Then the scientists were able to synthesize or copy this plant alkaloid into a chemical drug without using the original tree bark for manufacturing the drug. Today, all quinine drugs sold are manufactured chemically without the use of any tree bark. However, another chemical in the tree called *quinidine* which was found to be useful for various heart conditions couldn't be completely copied in the laboratory and the tree bark is still harvested and used to extract this plant chemical from it. Quinidine extracted from the bark is still used today to produce quinidine-based drugs.

The three important and common natural imidazole alkaloid containing plants are Jaborandi, Maca and Chandrasura. Herbalists and natural health practitioners attribute the biological activities of the plant jaborandi as the main activities clinically validated for pilocarpine. CNS a stimulant, direct cholinergic/parasympathetic effect of pilocarpine has been reported after scientific exploration. The sialagogue property of pilocarpine to treat the xerostomy (dry mouth) caused by the radio or chemotherapy of throat cancer or those in the head area is also exploited scientifically. Maca (*L. meyenii*) many potential pharmacological activities are reported like adaptogenic, anabolic, immunostimulant and hormonal balancing properties, antioxidant, hepatoprotective, immunostimulant, sexual stimulant and antistress effect, but no studies have been done on isolated or enriched imidazole alkaloids. These wide promising effects of Maca shows its potential to a clinically effective drug which need phytopharmacological correlation studies. Chandrasura is reported to have nephroprotective, diuretic, antibacterial, antiasthmatic, hepatoprotective, wound-healing, potent hypoglycaemic, oral contraceptive and abortifacient activity. The only report of pharmacological activity of *L. sativum* alkaloid was by Shukla *et al.* (2011) demonstrating sedative, anxiolytic, myorelaxant and analgesic property.

These plants needs urgent attention of scientific community to work on pharmacological responses due to the presence of active principal imidazole alkaloids and clinical research to support them or to qualify them. Further as the structure of the imidazole alkaloids are not very complex or difficult to synthesize, they can be produced in the laboratory to establish the therapeutic potentials.

Conclusions

Imidazole drugs have broad applications in clinical medicine with antifungal, antibacterial, antiallergic, anti-inflammatory and central effects. Imidazoles are reported to be depressant of motor activity in mice and rats. Imidazole 4- acetic acid is a naturally occurring metabolite in brain, displays definite partial agonist characteristic as an enhancer of benzodiazepine binding to the $GABA_A$ receptor complex and imidazoline I_1 receptor. Alpha 2-adrenoreceptor antagonist and serotonin-selective reuptake inhibitory activities has been identified in a series of 2-substituted 4,5-dihydro-1H-imidazole derivatives and also have 5-HT1A and 5-HT2 receptor. Indane 2-imidazoles have been identified as potent, partial agonists of the α1A adrenergic receptor, having good selectivity over the α1B, α1D and α2 sub-types. Phenyl substituted imidazoles were found to be potent antidepresssant. Nitroimidazoles are reported to have as antiprotozoal, anthelminthic, antifungal and anti-mycotic activities. Analgesic and anti-inflammatory activities are reported in some 2-substituted and 1, 2-disubstituted imidazole derivatives. The natural imidazole alkaloid containing plants like *Pilocarpus jaborandi, P. microphyllus, Lepidium sativum* and *Lepidium meyenii* is reported to have potential central effects along with other like immunostimulant, aphrodisiac hepatoprotective, antidiabetic etc. Targeting these very well reported and validated effects of the extracts the plant can be explored scientifically to validate the active constituent responsible for the said activities. This can further lead to synthesis of the constituent for large scale exploration of pre-clinical and clinical validation and therapeutic use on plant scale production. Molecular manipulation of the structurally elucidated alkaloids by synthesis of analogues and derivatives may yield new compounds with potential advantages like improved activity and potency, reduced toxicity or side effects, and high selectivity.

References

Afaf, I., Nuha, H.S., and Mohammed, A.H. (2008). Hepatoprotective effect of *Lepidium sativum* against carbon tetrachloride induced damage in rats. *Research Journal of Animal and Veterinary Sciences,* **3:** 20-23.

Amani, S., Derek, A., Maitland, J., and Samar, M.M. (2007). New alkaloids from *Casimiroa edulis* fruits and their pharmacological activity. *Chemistry of Natural Compounds,* **43 (5):** 576-580.

Annonymous. (1962). The Wealth of India, (Raw Material), vol. 6. CSIR Publication; New Delhi, pp.71.

Archana, P.N., and Mehta, A.A. (2006). A study on clinical efficacy of *Lepidium sativum* seeds in treatment of bronchial asthma. *Iranian Journal of Pharmacology and Therapeutics,* **5 (1):** 55-59.

Bensasson, R.V., Frederiksen, J., Rougee, M., Lexa, D., and Harrit, N. (1992). Correlations between the rate constant of singlet oxygen quenching by imidazole derivatives and anti-inflammatory activity in rats. *Molecular Pharmacology,* **42(4):** 718-722.

Breckle, G., Polborn, K., and Lindel, T. (2003). Synthesis of the pyrrole-imidazole alkaloid sventrin from the marine sponge *Agelas sventres. Verlag der Zeitschrift für Naturforschung,* Tübingen, available online http: / /znaturforsch.com, 451-456, accessed on 10 May 2012.

Brossi A. The Alkaloids: Chemistry and Pharmacology. (1984). Vol. 22, Academic Press, New York, pp. 282-327.

Bruneton, J. (1999). Pharmacognosy Phytochemistry Medicinal Plants. 2nd edition, Intercept Ltd., New York, pp. 1045-1047.

Bustos-Obregon, E., Yucra, S., and Gonzales, G.F. (2005). *Lepidium meyenii* (Maca) reduces spermatogenic damage induced by a single dose of malathion in mice. *Asian Journal of Androogyl.,* **7(1):** 71-76.

Caroon, J.M., Clark, R.D., Kluge, A.F., Olah, R., Repke, D.B., Unger, S.H., Michel, A.D., and Whiting, R.L. (1982). Structure-activity relationships for 2-substituted imidazoles as alpha 2-adrenoceptor antagonists. *Journal of Medicinal Chemistry,* **25(6):** 666-70.

Carroll, A.F., Bowden, B.F., and Coll, J.C. (1993). New imidazole alkaloids from the sponge *Leucetta* sp. and the associated predatory Nudibranch *Notodoris gardineri. Australian Journal of Chemistry,* **b (8):** 1229 – 1234.

Cicero, A.F., Bandieri, E., and Arletti, R. (2001). *Lepidium meyenii* Walp. improves sexual behaviour in male rats independently from its action on spontaneous locomotor activity. *Journal of Ethnopharmacology,* **75(2-3):** 225-229.

Cragg, G. M., and Boyd, M. (1996). Drug discovery and development at the National Cancer Institute, in Medicinal Plant Resources of the Tropical Forest, Ed. By Balick, M.J., Elisabetsky, E., and Laird, S.A., Columbia University Press, New York, pp. 101.

Cui, B., Zheng, BL., He, K., and Zheng, Q.Y. (2003). Imidazole alkaloids from *Lepidium meyenii*. *Journal of Natural Products*, **66(8):** 1101-1103.

Dini, A., Migliuolo, G., Rastrelli, L., Saturnino, P., and Schettino, O. (1994). Chemical composition of *Lepidium meyenii*. *Food Chemistry*, **49:** 347–349.

Eddouks, M., Maghrani, M., Zeggwagh, N.A., and Michel, J.B. (2005). Study of hypoglycaemic activity of *Lepidium sativum* L. aqueous extract in normal and diabetic rat. *Journal of Ethnopharmacology*, **97:** 391-395.

Farnsworth, N.R., and Soejarto, D.D. (1985). Potential consequences of plant extinction in the United States on the current and future availability of prescription drugs. *Economic Botany*, **39(3):** 231-240.

Ferrari, F. (1985). Behavioural pharmacology of imidazole, a potential antidepressant agent. Archives internationales de pharmacodynamie et de thérapie. *Archives of International Pharmacodynamic Theapeutics*, **277(2):** 303-12.

Gonzales, G. F., Miranda, S., Nieto, J., Fernandez, G., Yucra, S., Rubio, J., Yi, P., and Gasco, M. (2005). Red maca (*Lepidium meyenii*) reduced prostate size in rats. *Reproductive Biology and Endocrinology*, **3(1):** 1-20.

Gonzales, G.F., Cordova, A., Vega, K., Chung, A., Villena, A., and Gonez, C. (2003). Effect of *Lepidium meyenii* (maca), a root with aphrodisiac and fertility-enhancing properties, on serum reproductive hormone levels in adult healthy men. *Journal of Endocrinology*, **176:** 163–168.

Gupta, P.C., Devki, P., Joshi, P., and Lohar, D.R. (2010). Evaluation of antibacterial activity of *Lepidium sativum* Linn. seeds against food-borne pathogens. *International Journal of Chemical and Analytical Science*, **1(4):** 74-75.

Hadizadeha, F., Hosseinzadeha, H., Motamed-Shariatyb, V.S., Seifib, M., and Kazemib, S. (2008). Synthesis and antidepressant activity of N-Substituted Imidazole-5-Carboxamides in forced swimming test model. *Iranian Journal of Pharmaceutical Research*, **7(1):** 29-33.

Han, Y.N., Han, B.H., Park, M.H., Ryu, S.Y., Woo, L.K. (1987). Alkaloids from Panax Ginseng. Natural Products Research institute, Seol, Korea. Available at: http: / /210.101.116.28/W_kiss3/0C800918_pv.pdf.

Hart, N.K., Johns, S.R., Lamberton, J.A., and Willing, R.I. (1970). Alkaloids of *Alchornea javanensis* (Euphorbiaceae): The isolation of Hexahydroimidazo [1, 2-a] pyrimidines and guanidines. *Australian Journal of Chemistry*, **23(8):** 1679 – 1693.

Hart, N.K., Johns, S.R., Lamberton, J.A., Loder, J.W., and Nearn, R.H. (1971). New imidazole alkaloids from *Cypholophus friesianus* (Urticaceae). The structures and synthesis of cypholophine and *O*-acetylcypholophine. *Australian Journal of Chemistry*, **24(4):** 857 – 864.

Holt, R.J. (1976). Topical pharmacology of imidazole antifungals. *Journal of Cutaneous Pathology*, **3(1)**: 45 – 59.

Iemura, R., Kawashima, T., Fukuda, T., Ito, K., and Tsukamoto, G. (1986). Synthesis of 2-(4-substituted-1-piperazinyl) benzimidazoles as H_1-antihistaminic agents. *Journal of Medicinal Chemistry,* **29(7):** 1178-1183.

Johns, S.R., and Lamberton, JA. (1967). New imidazole alkaloids from a Glochidion species (family Euphorbiaceae). *Australian Journal of Chemistry,* **20(3):** 555 – 560.

Juma, A.H. (2007). The effects of *Lepidium sativum* seeds on fracture induced healing in rabbits. *Medscape General Medicine,* **9(2):** 23 available online http: //www.ncbi.nlm.nih.gov/pmc/articles/PMC1994840/< accessed on 22 June 2010.

Kassie, F., Laky, B., Gminski, R., Mersch-Sundermann, V., Scharf, G., Lhoste, E., and Kansmuller, S. (2003). Effects of garden and water cress juices and their constituents, benzyl and phenethyl isothiocyanates, towards benzo(a)pyrene-induced DNA damage: a model study with the single cell gel electrophoresis/ Hep G2 assay. *Chemico- Biological Interactions,* **142(3):** 285-296.

Khory R. (1999). Materia medica of India and their therapeutics. Komal Prakashan; Delhi, India. pp. 63-64.

Knope, R., Moe, G.K., Saunders, J., and Tuttle, R. (1973). Myocardial effects of imidazole. *The Journal of Pharmacology and Experimental Therapeutics,* **185(1):** 29-34.

Korolkovas, A. (1995/1996). Dicionário Terapêutico Guanabara. Rio de Janeiro: Guanabara-Koogan, pp. 20.2- 20.3.

Lee, K.J., Dabrowski, K., and Rinchard, J. (2004). Supplementation of maca (*Lepidum meyenii*) tuber meal in diets improves growth rate and survival of rainbow trout *Oncorhynchus mykiss* (Walbaum) alevins and juveniles. *Aquactic Research,* **35:** 215-223.

Lentz, A., Gravitt, K., Carson, C.C., and Marson L. (2007). Acute and chronic dosing of *Lepidium meyenii* (Maca) on male rat sexual behavior. *Journal of Sexual Medicine,* **4 (2):** 332-40.

Lopez-Fando, A., Gomez-Serranillos, M.P., Iglesias, I., Lock, O., Upamayta, U.P., and Carretero, M.E. (2004). *Lepidium peruvianum* chacon restores homeostasis impaired by restraint stress. *Phytotherapy Research,* **18(6):** 471-474.

Luca, L.D. (2006). Naturally Occurring and synthetic imidazoles: Their chemistry and their biological activities. *Current Medicinal Chemistry,* **13(1):** 1-23.

Maier, U.H., Gundlach, H., and Zenk, M.H. (1998). Seven imidazole alkaloids from *Lepidium sativum. Phytochemistry,* **49(6):** 1791-1795.

Mali, G.R., Mahajan, G.S., and Mehta, A.A. (2008). Studies on bronchodilatory effect of *Lepidium sativum* against allergen induced bronchospasm in guinea pigs. *Pharmacognosy Magzine,* **4:** 189-192.

McKay, D. (2004). Nutrients and botanicals for erectile dysfunction: examining the evidence. *Alternative Medicine Review,* **9(1):** 4-16.

Miller, L.J. (1993). Oral pilocarpine for radiation-induced xerostomia. *Cancer Bulletin* (Houston), **45(6):** 549-550.

Mokrosz, J.L., and Duszynska, B. (1992). Structure-activity relationship studies of CNS agents. Part VII. The effect of the imidazole fragment in 2-substituted 1-[3-(4-aryl-1-piperazinyl) propyl] imidazoles on their interaction modes with 5-HT1A and 5-HT2 receptors. *Polish Journal of Pharmacology and Pharmacy*, **44(5):** 527-38.

Muhammad, I., Zhao, J.P., Dunbara, D.C., and Khan, I.A. (2002). Constituents of *Lepidium meyenii* 'maca'. *Phytochemistry*, **59:** 105–110.

Nath, D., Sethi, N., Singh, R.K., and Jain, A.K. (1992). Commonly used Indian abortifacient plants with special reference to their teratologic effects in rats. *Journal of Ethnopharmacology*, **36:** 147-154.

Oshima, M., Gu, Y., and Tsukada, S. (2003). Effects of *Lepidium meyenii* Walp. and *Jatropha macrantha* on blood levels of estradiol-17 beta, progesterone, testosterone and the rate of embryo implantation in mice. *Journal of Veterinary and Medical Sciences*, **65:** 1145–1146.

Pande, K., Tangri, K.K., Bhalla, T.N., Ahmad, S., Dhasmana, A., and Barthwal, J.P. (1984). Newer indolyl imidazolones as anti-inflammatory agents. *Pharmacology*, **29 (4)**: 193-197.

Patel, U., Kulkarni, M., Undale, V., and Bhosale, A. (2009). Evaluation of diuretic activity of aqueous and methanol extracts of *Lepidium sativum* garden cress (Cruciferae) in rats. *Tropical Journal of Pharmaceutical Research*, **8 (3):** 215-219.

Peter, G.W., and Dalton, F.F. (1981). Imidazole alkaloids from *Cynometra hankei. Phytochemistry*, **20(12):** 2765–2767.

Piacente, S., Carbone, V., Plaza, A., Zampelli, A., and Pizza, C. (2002) Investigation of the tuber constituents of maca (*Lepidium meyenii* Walp.). *Journal of Agriculture Food Chemistry*, **50(20):** 5621-5625.

Pinelli, A., Trivulzio, S., Malvezzi, L., Rossoni, G., and Beretta, L. (1989). 2-substituted derivatives of benzimidazole inhibiting gastric acid secretion in rats. *Arzneimittel-Forschung*, **39(4):** 467-9.

Radwan, H.M., El-Missiry, M.M., Al-Said, W.M., Ismail, A.S., Abdel, S.K.A., and Seif-El, N. (2007). Investigation of the glucosinolates of *Lepidium sativum* growing in Egypt and their biological activity. *Research Journal of Medical Science*, **2(2):** 127-132.

Rosario, D., Eva Z., María, J., Ortega, Santiago, N., and Javier, S. (1999). Novel alkaloids from the red ascidian *Botryllus leachi*. *Tetrahedron*, **55(46):** 13225–13232.

Rosenberg, H., and Paul, A.G. (1969). Dolichotheline, a novel imidazole alkaloid from *Dolichotheline sphaerica*. *Tetrahedron Letters*, **13:** 1039-1042.

Rubio, J., Caldas, M., Davila, S., Gasco, M., and Gonzales, G.F. (2006). Effect of three different cultivars of *Lepidium meyenii* (Maca) on learning and depression in ovariectomized mice. *BMC Complement and Alternate Medicine*, **6(1):** 6-23.

Sandoval, M., Okuhama, N.N., and Angeles, F.M. (2002). Antioxidant activity of the cruciferous vegetable maca (*Lepidium meyenii*). *Food Chemistry*, **79:** 207-213.

Sawaya, A.C., Abreu, I.N., Andreazza, N.L., Eberlin, M.N., and Mazzafera, P. (2008). HPLC-ESI-MS/MS of imidazole alkaloids in *Pilocarpus microphyllus*. *Molecules*, **30; 13(7):** 1518-29.

Sharief, M., and Zainab, H. (2004). Garden cress (*Lepidium sativum*) seeds as oral contraceptive plant in mice. *Saudi Medical Journal*, **25(7):** 965-966.

Shinde, N., Jagtap, A., Undale, V., Kakade, S., Kotwal, S. and Patil, R. (2010). Protective effect of *Lepidium sativum* against doxorubicin-induced nephrotoxicity in rats. *Research Journal of Pharmaceutical, Biological and Chemical Sciences*, **1(3):** 42.

Shukla, A., Singh, C.S., Bigoniya, P. (2011). Phytochemical and CNS activity of *Lepidium sativum* Linn. seed total alkaloid. *Der Pharmacia letter*, **3(2):** 226-237.

Singh, M.P., and Panda, H. (2005). Medicinal herbs with their formulations. Daya Publishing House, New Delhi, Vol 1, pp. 186-187.

Stillings, M.R., Chapleo, C.B., Butler, R.C., Davis, J.A., England, C.D., Myers, M., Myers, P.L., Tweddle, N., Welbourn, A.P., and Doxey, J.C. (1985). Alpha-adrenoreceptor reagents. 3. Synthesis of some 2-substituted 1, 4-benzodioxans as selective presynaptic alpha 2-adrenoreceptor antagonists. *Journal of Medicinal Chemistry*, **28(8):** 1054-62.

Strupi, S.J., Turk, T., Macek, P., and Suput, D. (1996). Pseudozoanthoxantin-like compound from *Parazoanthus axinellae* Adriaticus inhibits acetylcholinesterase. *Pflugers Archiv-European Journal of Physiology*, **431(6):** 315-6.

Tunnicliff, G. (1998). Pharmacology and function of imidazole 4-acetic acid in brain. *General Pharmacology*, **31(4):** 503-9.

Unsal, C., Sevim, D., Mevlut, E., and Rumeysa, S. (1988). The significance of the imidazole ring in anticonvulsant activity of (Arylalkyl) imidazoles. *Archiv der Pharmazie*, **321(12):** 841 – 846.

Valdez, I.H., Wolf, A., Atkinson, J.C., Macynsky, A.A., and Fox, P.C. (1993). Use of pilocarpine during head and neck radiation theraphy to reduce xerostomia and salivary dysfunction. *Cancer*, **71(5):** 1848-1851.

Valentova, K., and Ulrichova, J. (2003). *Smallanthus sonchifolius* and *Lepidium meyenii* prospective Andean crops for the prevention of chronic diseases. *Biomedical Papers*, **147:** 119–130.

Valentova, K., Buckiova, D., Kren, V., Peknicova, J., Ulrichova, J., and Simanek, V. (2006). The *in vitro* biological activity of *Lepidium meyenii* extracts. *Cell Biology and Toxicology*, **22(2):** 91-99.

Wentland, M.P., Bailey, D.M., Alexander, E.J., Castaldi, M.J., Ferrari, R.A., Haubrich, D.R., Luttinger, D.A., and Perrone, M.H. (1987). Synthesis and antidepressant properties of novel 2-substituted 4, 5-dihydro-1H-imidazole derivatives. *Journal of Medicinal Chemistry*, **30**(8): 1482-9.

Wilkenfeld, B.E., and Levy, B. (1969). The effects of theophylline, diazoxide and imidazole on isoproterenol induced inhibition of the rabbit ileum. *Journal of Pharmacology and Experimental Therapeutics*, **169(1):** 61-67.

Winkelmann, E., and Raether, W. (1978). Chemotherapeutically acitve nitro compounds. 4,5-Nitroimidazoles (Part III). *Arzneimittel-Forschung*, **28(5):** 739-749.

Wuonola, M.A., Woodward, R.B. (1976). Imidazole alkaloids of *Macrorungia longistrobus*: Revised structures and total syntheses. *Tetrahedron*, **32**(10): 1085–1095.

Zhang, Y., Longjiang, Yu., Mingzhang, A., and Wenwen, J. (2006). Effect of ethanol extract of *Lepidium meyenii* Walp. on osteoporosis in ovariectomized rat. *Journal of Ethnopharmacology*, **105:** 274–279.

Bioactive Phytochemicals: Perspectives for Modern Medicine Vol. 2 (2014) *Pages* **197–206**
Editor: **V.K. Gupta**
Published by: **DAYA PUBLISHING HOUSE, NEW DELHI**

9

Podophyllotoxin: A Second Generation Anticancer Drug

Dipto Bhattacharyya[1] and Sharmila Chattopadhyay[1*]

ABSTRACT

Podophyllotoxin belongs to the group of aryltetralin lignans and with two other classes the aryl naphthalene and dibenzocyclooctadiene falls under the group of cyclolignans. The 2, 7'cyclolignan podophyllotoxin is found in different plant families like Cupressaceae from Gymnosperms, and Berberidaceae, Polygalaceae, Lamiaceae, Linaceae from Angiosperms. Podophyllotoxin obtained from the resin of Podophyllum hexandrum and Podophyllum peltatum L. (Berberidaceae) was found to be too toxic and derivatized to produce the first clinically approved drug, etoposide. Due to its high cytotoxic effects, three of its derivatives are used in recent medical treatments which are comparatively less toxic. Etoposide phosphate inhibits topoisomerase II and causes DNA strands to break. It is administered orally or intravenously, singly or in combination with other drugs and is effective in chemotherapy of lung cancer, Ewing's sarcoma, testicular cancer, lymphoma, and is also used before bone marrow or blood stem cell transplantation. Etopophos produced by Bristol-Myers Squibb Co. is a water soluble ester of etoposide have shown metaphase arrest in chick fibroblasts and DNA-topoisomerase II dependent arrest in G2 portion of mammalian cell cycle. It is generally used for refractory testicular tumor and small cell lung cancer. Teniposide, the other drug has a substitution of a thenylidene group on the glucopyranoside ring. It is treated in patients having childhood acute lymphocytic leukemia.

P. hexandrum is a plant from the Himalayan region and has been on the centre point of focus for decades mainly due to the availability of antitumor and anticancer agent podophyllotoxin from its rhizomes. Ruthless over collection of this species from wild, in order

---

1 Drug Development/Diagnostics and Biotechnology Division, CSIR-Indian Institute of Chemical Biology, 4, Raja S.C.Mullick Road, Kolkata – 700 032, West Bengal, India

* *Corresponding author*: E-mail: sharmila@iicb.res.in

to extract podophyllotoxin has endangered its existence. There is extreme lack of information on cDNA of the enzymes involved in overall podophyllotoxin biosynthetic pathway especially in the later steps. Biotechnological approaches like transgenic approach, metabolic engineering etc. have also been considered to be attractive alternatives. Several attempts have also been directed to approach the unknown podophyllotoxin pathway gene/s identification through nontargetted global techniques combined with metabolite elicitation. Development of alternative and renewable sources of podophyllotoxin with superior pharmacological profile is an intensely pursued field.

Keywords: Cancer, Genomics, Proteomics, Podophyllotoxin, *Podophyllum hexandrum.*

Introduction

Cancer is the second leading cause of death throughout the world, second only to death by cardiovascular disease. There were an estimated 12.7 million cancer cases around the world in 2008, of these 6.7 million cases were in men and 6.0 million in women. This number is expected to increase to 21 million by 2030. In 2008 there were 12.7 million new cases of cancer diagnosed worldwide. Lung cancer was the most common cancer when men and women were analysed together and this represents 13 per cent of all cancers (excluding non-melanoma skin cancer). Nearly 1.4 million cases of breast cancer were diagnosed across the world in 2008, compared with about 500 000 cases in 1975. In India, around 555 000 people died of cancer in 2010, according to estimates published in *The Lancet on* March 28, 2012 (Dikshit *et al.*, 2012). In another report, India officially recorded over half a million deaths due to cancer in 2011- 5.35 lakhs as against 5.14 (2009) and 5.24 (2010). The Union health ministry says there are about 28 lakhs cases of cancer at any given point of time in India, with 10 lakhs new cases being reported annually. World Health Organization (WHO) says, the estimated cancer deaths in India are projected to increase to 7 lakhs by 2015.

Plants have been used as a dynamic source as food, fodder, shelter and health since time immemorial. Medicinal plants are an important source of diverse chemical compounds that have been used for the past several centuries in the treatment of cancer. About 25 per cent of drugs in the modern pharmacopoeia are derived from plants, including several anticancer drugs currently in clinical use such as vincristine, vinblastine, paclitaxel, podophyllotoxin, camptothecin and combretastatin.

In addition to these, several other plant-derived compounds are in clinical trials as potential cancer chemopreventive agents. Out of that list, few may be mentioned here like curcumin is under Phase II clinical trial for prevention of colorectal neoplasia (Carroll *et al.*, 2011). A study on Short-Term genistein intervention in patients with localized prostrate cancer is under Phase II clinical trials (Lazarevic *et al.*, 2011). Phase I randomized double-blind pilot study of micronized resveratrol in patients with hepatic metastases is under progress (Howells *et al.*, 2011).

Vinblastine and vincristine were isolated from *Catharanthus roseus* (L.) G. Don (Apocynaceae) and have been used clinically for over 40 years (van Der Heijden *et al.*, 2004). The vinca alkaloids and several of their semi-synthetic derivatives block mitosis with metaphase arrest by binding specifically to tubulin resulting in its depolymerization (Okouneva *et al.*, 2003). Vinflunine is a modification of vinblastine from *C. roseus*. Paclitaxel isolated from *Taxus brevifolia* Nutt. (Taxaceae) introduced in the early 1990s (Wall and Wani, 1996; Oberlies and Kroll, 2004). The taxanes, act by binding tubulin without allowing depolymerization or interfering with tubulin assembly (Schiff *et al.*, 1979; Horwitz, 2004). Camptothecin was isolated from *Camptotheca acuminata* Decne. (Nyssaceae) was found to act by selective inhibition of topoisomerase I involved in cleavage and reassembly of DNA (Cragg and Newman, 2004). Exatecan is an analog of camptothecin from *C. acuminata* Decne. may come up as a potential anticancer agent (Butler, 2004; Cragg and Newman, 2004).

Podophyllotoxin was isolated from the resin of *Podophyllum hexandrum* Royle and *Podophyllum peltatum* L. (Berberidaceae) but was found to be too toxic and derivatized to produce the first clinically approved drug, etoposide (Gordaliza *et al.*, 2004). The epipodophyllotoxins bind tubulin, and causes DNA strand breaks during the G2 phase of the cell cycle by irreversibly inhibiting DNA topoisomerase II (Gordaliza *et al.*, 2004).

Podophyllum hexandrum

Podophyllum hexandrum Royle syn. *P. emodi* Wall. ex Hook. f. and Thoms (Indian *Podophyllum*) is an erect, glabrous, succulent herb, 35-60cm high, and has creeping perennial rhizome bearing numerous roots and is found in the inner range of the Himalayas from Kashmir to Sikkim at altitudes of 3,000-4,200 m. Leaves are 2-3 in number, orbicular, reniform, palmate, peltate, with lobed segments. Flowers are white or pink, cup-shaped and solitary. Fruits are oblong or elliptic berry like and orange or red in colour. Specimens in eastern Himalayas differ from those in North Western Himalayas in nature of leaf insertion. The plant is generally associated with species of *Rhododendron*, *Salix*, *Juniperus*, and *Viburnum*. It is also found in open alpine meadows where its occurrence is less frequent (Wealth of India, 1969).

The resins from *P. peltatum* and *P. hexandrum* can be differentiated by treating their alcoholic solutions with aqueous potassium hydroxide, the resin from *P. peltatum* does not gelatinize but the resin from *P. hexandrum* produce a stiff gelly. A few drops of strong solution of copper acetate produce a bright green colouration with the alcoholic extract of *P. peltatum*, but a brown precipitate from *P. hexandrum* (Wealth of India, 1969). The resin content of American *Podophyllum* is 2-8 per cent, whereas the resin content of the Indian *Podophyllum* is 7-15 per cent. Various aryltetralin lignans and other compounds were reported from *P. hexandrum*.

Podophyllotoxin: The Active Principle

Podophyllotoxin belongs to the group of aryltetralin lignans and with two other classes the aryl naphthalene and dibenzocyclooctadiene falls under the group of cyclolignans. The 2, 7′cyclolignan podophyllotoxin is found in different plant families like Cupressaceae from Gymnosperms, and Berberidaceae, Polygalaceae, Lamiaceae, Linaceae from Angiosperms. The biological activity of these lignans and their derivatives is probably guided by the stereo chemistry of the 2β-H lignans with strong biological activity and their 2α-H isomers with relatively weaker biological activity

(Shuhui *et al.*, 2002). Their stereochemistry at C4-H position conveys their mechanism of action- the ones with 4β-H-2β-H inhibit microtubule polymerization whereas the isomers with 4α-H-2βH inhibit topoisomerase II enzyme and breakage of DNA strands. Due to its unacceptability for high cytotoxic effects, three of its derivatives are used in recent medical treatments which are comparatively less toxic. Etoposide inhibits topoisomerase II and causes DNA strands to break. D-ring of etoposide interacts with DNA in the ternary topoisomerase II cleavage complex (Pitts *et al.*, 2011). It is administered orally or intravenously, singly or in combination with other drugs and is effective in chemotherapy of lung cancer, Ewing's sarcoma, testicular cancer, lymphoma, and is also used before bone marrow or blood stem cell transplantation (Hande, 1998; Meresse *et al.*, 2004; Gordaliza *et al.*, 2004). Etopophos produced by Bristol-Myers Squibb Co. is a water soluble ester of etoposide is believed to have the same function as etoposide (Imbert, 1998; Shuhui *et al.*, 2002) and DNA-topoisomerase II dependent arrest in G2 portion of mammalian cell cycle. It is generally used for refractory testicular tumor and small cell lung cancer. Teniposide, the other drug has a substitution of a thenylidene group on the glucopyranoside ring. It is treated in patients having small cell lung cancer and childhood acute lymphocytic leukaemia (Gordaliza *et al.*, 2004). It has been found out in a study that teniposide could induce apoptosis of oral squamous cell carcinoma cells (Li *et al.*, 2006). Development of alternative and renewable sources of podophyllotoxin with superior pharmacological profile is an intensely pursued field (Canel *et al.*, 2000).

Podophyllotoxin Content

According to Foster (1993), 'conservation through cultivation' is the ultimate solution for medicinal plants like *Podophyllum*. Hence, extensive cultivation of *Podophyllum* may be the best approach to solve the growing demand of podophyllotoxin. Approaches by Pandey *et al.* (2007) appears to be the same as *P. hexandrum* have been grown on lower altitudes and podophyllotoxin content have been measured. Variation of podophyllotoxin concentration in rhizomes due to age and altitude of *P. hexandrum* (Nadeem *et al.*, 2007), have been looked in to but the net photosynthetic rate, stomatal conductance and carbon uptake also seems to increase with altitude (Vats and Kumar, 2006). Purohit *et al.* (1999), reported that below 3500 m, the podophyllotoxin content of the rhizomes of *P. hexandrum* were around 4 per cent by dry weight content and above 3500 m podophyllotoxin contents of around 10 per cent by dry weight was recorded. Striking finding from this report was the variation in podophyllotoxin contents with reference to the number of leaves. Also, the content of podophyllotoxin was found to be higher in plants during May-June than during September-October. Amount of podophyllotoxin also varied in different populations at the same altitude (Nadeem, 2001; Pandey, 2002). In a report, rhizome buds have been shown to possess significant amount of podophyllum lignans from *P. hexandrum* plants in Zanskar region of Himalayas (Kitchlu *et al.*, 2011). Different amounts of podophyllotoxin in differing geographical populations of *P. hexandrum* have been noted (Sultan *et al.*, 2009, 2010). Amount of podophyllotoxin in 2 year old cell lines varied from 0.15 per cent–0.45 per cent on dry weight basis whereas amount till fourth year cell lines decrease to 0.03-0.2 per cent (Majumder and Jha, 2009a). Thus age of the cell lines is crucial in maintaining a stable supply of podophyllotoxin.

Although a doubling of atmospheric carbon dioxide concentrations related to a significant increment in dry weight of the roots/rhizomes, however, the relation to the amount of podophyllotoxin have not been measured (Chaturvedi *et al.*, 2009).

Strategies to Stabilize Podophyllotoxin Supply

Ways to stabilize *Podophyllum* populations concern assessment of its genetic diversity. Tissue culture, propagation and maintenance of high altitude and low altitude *P. hexandrum* have been tried. The question of podophyllotoxin supply has been addressed in diverse ways by utilizing multifaceted scientific procedures. These include production of podophyllotoxin by plant cell and organ cultures and their optimization through manipulation of media, pH, growth hormones, elicitors and pathway precursors (Chakraborty *et al.*, 2009; Chattopadhyay *et al.*, 2001, 2002, 2003). Although these method is efficient, but failed to provide satisfactory upscale versions to supply podophyllotoxin. Strenuous efforts have been undertaken to decipher the biosynthetic pathway to podophyllotoxin, although mainly work was concentrated in other plants. Uptill now, through combined biochemical ways, the pathway from coniferyl alcohol to pinoresinol to secoisolariciresinol to matairesinol have been discovered. The proposed pathways signify more towards four or more unknown proteins to be characterized in order to reveal podophyllotoxin biosynthesis. Precursor feeding approaches combined with biochemical strategies seem not enough to unravel the whole pathway. Technologies of global significance and not targeted totally towards podophyllotoxin pathway like gene microarray, suppression subtractive hybridization, 2-D PAGE have been recorded. Barriers of an unsequenced genome still continue to unsolved questions on future aspects of how to solve this pathway.

Hypothesis on Podophyllotoxin Biosynthesis

Lignan was introduced by Hayworth (1936) is a dimer in which the two C_6-C_3 monomer units are linked by a β-β bond (Cole and Wiedhopf, 1978). The incorporation of DL-[β-^{14}C] phenylalanine into podophyllotoxin has been demonstrated. The subsequent stepwise degradation to 32, 42-methylenedioxy-3, 4, 5-trimethoxybenzophenone bearing a ^{14}C-carbonyl label establishes the *in vivo* synthesis of the carbon skeleton of this lignan. Feeding experiments in *P. hexandrum* have previously shown that the two phenyl propane precursors of podophyllotoxin have the same substitution pattern of ferulic acid (Jackson and Dewick, 1984a). Works on *P. hexandrum* further clarify the fact that desoxypodophyllotoxin is probably being converted to podophyllotoxin, and podophyllotoxin being converted to podophyllotoxone although the second reaction may be reversible (Jackson and Dewick, 1984b). There is extreme lack of information on cDNA of the enzymes involved in overall podophyllotoxin biosynthetic pathway especially in the later steps (Figure 9.1). Uptill now the complete CDS of CAD, dirigent protein oxidase and secoisolariciresinol dehydrogenase from *P. hexandrum* have been reported to the National Centre for Biotechnology Information (NCBI) (Acc. Nos. GU324975, DQ414685, **HQ268590).** In fact, lignan biosynthesis involves mechanisms for enantioselective dimerization. Dirigent protein oxidase affects the selective coupling of coniferyl alcohol radical giving (+)-pinoresinol (Davin *et al.*, 1997) and pinoresinol reductase converts pinoresinol to secoisolariciresinol via lariciresinol (Katayama *et*

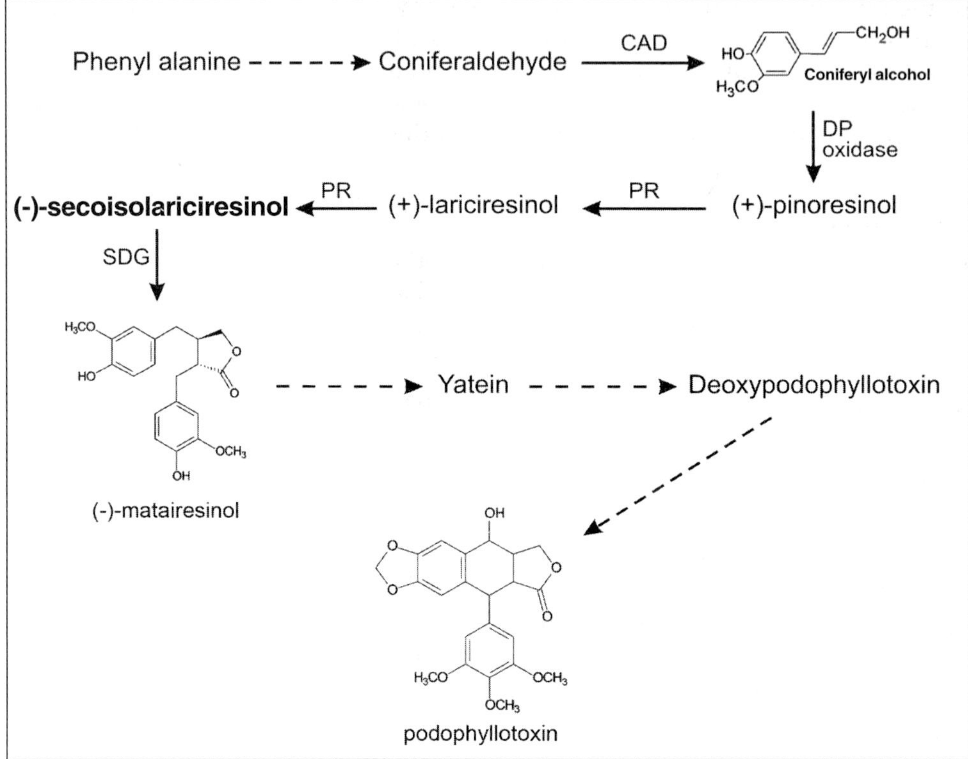

Figure 9.1: Schematic representation of podophyllotoxin biosynthesis pathway.

al., 1993). Further, (-)-secoisolariciresinol is converted to (-)–matairesinol by secoisolariciresinol dehydrogenase (Umezawa *et al.*, 1991). Matairesinol is a starting point for the biosynthesis of podophyllotoxin. One possible pathway is that matairesinol is converted to yatein and then to podophyllotoxin via deoxypodophyllotoxin (Dewick, 1989). Another direct pathway to podophyllotoxin from matairesinol via thujaplicatin has been proposed (Sakakibara *et al.*, 2003). The transformation from matairesinol to yatein involves hydroxylation, methylations and methylene dioxy bridge formation. Biotechnological approaches like transgenic approach, metabolic engineering etc. have also been considered to be attractive alternatives (Majumder and Jha, 2009b). Our attempts have been directed to approach the unknown podophyllotoxin pathway gene/s identification through nontargetted global techniques combined with metabolite elicitation. This has sketched out a pattern of like minded methyl jasmonate induced genes preferentially expressed at the specific time point of high podophyllotoxin accumulation. Genes and metabolites at same time points were not found to be actually correlated. A number of ESTs thought to be related to podophyllotoxin biosynthetic pathway were identified using the cutting-edge molecular tools like suppression subtractive hybridization and construction of a subtractive cDNA library. Out of which the full-length sequence of *CAD* was generated by RACE (Bhattacharyya, 2011). Furthermore, proteomic profiling followed by identification of candidate protein of podophyllotoxin biosynthetic pathway has been explored as well (Bhattacharyya *et al.*, 2012). Together, an enlightened molecular understanding at transcriptome, proteome and metabolome level of this frontline medicinal herb, *P. hexandrum*, with utmost therapeutic importance, is the need of the hour in order to attain the commercial scale production of podophyllotoxin, a second generation anticancer drug.

References

Bhattacharyya, D., Sinha, R., Ghanta, S., Chakraorty, A., Hazra, S., and Chattopadhyay, S. (2012). Proteins differentially expressed in elicited cell suspension culture of *Podophyllum hexandrum* with enhanced Podophyllotoxin content. *BMC Proteome Science*, **10**: 34.

Bhattacharyya, D. (2012). Evaluation of medicinal plants at genetic level. PhD thesis, Jadavpur University, Kolkata, India.

Butler, M.S. (2004). The role of natural product chemistry in drug discovery. *Journal of Natural Products*, **67:** 2141– 2153.

Canel, C., Moraes, R.M., Dayan, F.E., and Ferreira, D.(2000). Molecules of interest: Podophyllotoxin. *Phytochemistry*, **54:** 115-120.

Carroll, R.E., Benya, R.V., Turgeon, D.K., Vareed, S., Neuman, M., Rodriquez, L., *et al.* (2011). Phase IIa clinical trial of curcumin for the prevention of colorectal neoplasia. *Cancer Prevention Research*, **4**: 354-364.

Chaturvedi, A.K., Vashistha, R.K., Rawat, N., Prasad, P., and Nautiyal, M.C. (2009). Effect of CO_2 enrichment on photosynthetic behavior of *Podophyllum Hexandrum* Royle, an endangered medicinal herb. *Journal of American Science*, **5:** 113-118.

Chattopadhyay, S., Mehra, R.S., Srivastava, A.K., Bhojwani, S.S., and Bisaria, V.S. (2003). Effect of major nutrients on podophyllotoxin production in *Podophyllum hexandrum* suspension cultures. *Applied Microbiology and Biotechnology*, **60**: 541–546.

Chattopadhyay, S., Srivastava, A.K., Bhojwani, S.S., and Bisaria, V.S. (2001). Development of suspension culture of *Podophyllum hexandrum* for the production of podophyllotoxin. *Biotechnology Letters*, **23:** 2063-2066.

Chattopadhyay, S., Srivastava, A.K., Bhojwani, S.S., and Bisaria, V.S. (2002). Production of podophyllotoxin by plant cell cultures of *Podophyllum hexandrum* in bioreactor. *Journal of Bioscience and Bioengineering*, **93:** 215-220.

Chakraborty, A., Bhattacharya, D., Ghanta, S., and Chattopadhyay, S. (2010). An efficient protocol for *in vitro* regeneration of *Podophyllum hexandrum*, a critically endangered medicinal plant. *Indian Journal of Biotechnology*, **9**: 217–220.

Cole, J.R., and Wiedhopf, R.M. (1978). In: Rao C BS, ed. *Chemistry of Lignans* **39**. Andhra University Press, Visakhapatnam, India.

Cragg, G.M., and Newman, D.J. (2004). A tale of two tumor targets: topoisomerase I and tubulin. The Wall and Wani contribution to cancer chemotherapy. *Journal of Natural Products*, **67:** 232– 244.

Davin, L.B., Wang, H.B., Crowell, A.L., Bedgar, D.L., Martin, D.M., and Sarkanen. S. (1997). Stereoselective bimolecular phenoxy radical coupling by an auxiliary (dirigent) protein without an active center. *Science*, **275:** 362-367.

Dewick, P.M. (1989). Biosynthesis of lignans. In: Rahman A, ed. *Studies in natural products chemistry, Structure elucidation (part B)*, **V**, 459-503, Amsterdam, Elsevier.

Dikshit, R., Gupta, P. C., Ramasundarahettige, C., Gajalakshmi, V., Aleksandrowicz, L., Badwe, R., *et al.* (2012). Cancer mortality in India: a nationally representative survey. *The Lancet*, **379**: 1807-1816.

Foster, S. (1993). Medicinal plant conservation and genetic resources: Examples from the temperate northern hemisphere. *Acta Horticulturae*, **330:** 67-73.

Gordaliza, M., Garcia, P.A., del Corral, J.M., Castro, M.A., and Zurita, G. (2004). Podophyllotoxin: distribution, sources, applications and new cytotoxic derivatives. *Toxicon*, **44:** 441– 459.

Hande, K.R. (1998). Etoposide: four decades of development of a topoisomerase II inhibitor. *European Journal of Cancer*, **34:** 1514-1521.

Howells, L.M., Berry, D.P., Elliott, P.J., Jacobson, E.W., Hoffmann, E., Hegarty, B., *et al.* (2011). Phase I randomised double-blind pilot study of micronized resveratrol (SRT501) in patients with hepatic metastases–safety, pharmacokinetics and pharmacodynamics. *Cancer Prevention Research*, ***4:*** *1419–25.*

Horwitz, S.B. (2004). Personal recollections on the early development of taxol. *Journal of Natural Products*, **67**: 136– 138.

Imbert, T.F. (1998). Discovery of podophyllotoxins. *Biochimie*, **80:** 207-222.

Jackson, D.E., and Dewick, P.M. (1984b). Biosynthesis of podophyllum lignans-II. Interconversions of aryltetralin lignans in *Podophyllum hexandrum. Phytochemistry*, **23**: 1037-1042.

Jackson, D.E., and Dewick, P.M. (1984a). Biosynthesis of podophyllum lignans-I. Cinnamic Acid Precursors of Podophyllotoxin in *Podophyllum hexandrum. Phytochemistry*, **23**: 1029-1035.

Kitchlu, S., Ram, G., Koul, S., Koul, K., Gupta, K.K., and Ahuja, A. (2011). Podophyllum lignans array of *Podophyllum hexandrum* Royle populations from semi-desert alpine region of zanskar valley in Himalayas. *Industrial Crops and Products*, **33:** 584-587.

Katayama, T., Davin, L.B., Chu, A., and Lewis, N.G. (1993). Novel benzylic ether reductions in lignan biogenesis in *Forsythia intermedia. Phytochemistry*, **33**: 581–591.

Li, J., Chen, W., Zhang, P., and Li, N. (2006). Topoisomerase II trapping agent teniposide induces apoptosis and G2/M or S phase arrest of oral squamous cell carcinoma. *World Journal of Surgical Oncology*, **4:** 1-7.

Lazarevic, B., Boezelijn, G., Diep, L.M., *et al.* (2011). Efficacy and safety of short-term genistein intervention in patients with localized prostate cancer prior to radical prostatectomy: A randomized, placebo-controlled, double-blind phase 2 clinical trial. *Nutrition and Cancer*, **63:** 889-898.

Meresse, P., Dechaux, E., Monneret, C., and Bertounesque, E. (2004). Etoposide: discovery and medicinal chemistry. *Current Medicinal Chemistry,* **11**: 2443-2466.

Majumder, A., and Jha, S. (2009a). Characterization of podophyllotoxin yielding cell lines of *Podophyllum hexandrum. Caryologia,* **62:** -220-235.

Majumder, A., and Jha, S. (2009b). Biotechnological approaches for the production of potential anticancer leads podophyllotoxin and paclitaxel: An overview. *Journal of Biological Sciences,* **1**: 47-69.

Nadeem, M. (2001). Mass propagation and conservation strategies for two medicinally important plants of the Himalayan region. PhD thesis, Kumaun University, Nainital.

Nadeem, M., Palni, L.M.S., Kumar, A., and Nandi, S.K. (2007). Podophyllotoxin Content, above and belowground biomass in relation to altitude in *Podophyllum hexandrum* populations from kumaun region of the Indian Central Himalaya. *Planta Medica,* **73**: 388-391.

Okouneva, T., Hill, B.T., Wilson, L., and Jordan, M.A. (2003). The effects of vinflunine, vinorelbine, and vinblastine on centromere dynamics. *Molecular Cancer Therapeutics,* **2:** 427–436.

Oberlies, N.H., and Kroll, D.J. (2004). Camptothecin and taxol: historic achievements in natural products research. *Journal of Natural Products,* **67**: 129– 135.

Pitts, S.L., Jablonksy, M.J., Duca, M., Dauzonne, D., Monneret, C., Arimondo, P.B., *et al.* (2011). Contributions of the D-Ring to the activity of etoposide against human topoisomerase IIα: potential interactions with DNA in the ternary enzyme—drug—DNA complex. *Biochemistry,* **22:** 5058-5066.

Pandey, H., Nandi, S.K., Kumar, A., Palni, U.T., and Palni, L.S. (2007). Podophyllotoxin content in *Podophyllum hexandrum* Royle plants of known age of seed origin and grown at a lower altitude. *Acta Physiologiae Plantarum,* **29:** 121-126.

Purohit, M.C., Bahuguna, R., Maithani, U.C., Purohit, A.N., and Rawat, M.S.M. (1999). Variation in podophylloresin and podophyllotoxin contents in different populations of *Podophyllum hexandrum Current Science,* **77**: 1078-1079.

Pandey, H. (2002). Studies in relation to cotyledonary senescence and tissue culture of medicinally important alpine herbs. PhD thesis, Kumaun University, Nainital.

Schiff, P.B., Fant, J., and Horwitz, S.B. (1979). Promotion of microtubule assembly *in vitro* by taxol. *Nature,* **277:** 665– 667.

Shuhui, L., Xuan, T., Xingguo, C., and Zhide, H. 2002. Micellar electrokinetic capillary chromatographic separation of diastereoisomers of podophyllum lignans at the C4 position. *Chromatographia,* **56**: 687- 691.

Sultan, P., Shawl, A.S., Jan, A., Hamid, B., and Irshad, H. (2009). Germplasm conservation and quantitative estimation of podophyllotoxin and related glycosides of *Podophyllum hexandrum. Journal of Crop Science and Biotechnology,* **12**: 267-274.

Sultan, P., Shawl, A.S., Rehman, S., Ahmed, S.F., and Ramteke, P.W. (2010). Molecular characterization and marker based chemotaxonomic studies of *Podophyllum hexandrum* Royle. *Fitoterapia,* **81**: 243-247.

Sakakibara, N., Suzuki, S., Umezawa, T., and Shimada, M. (2003). Biosynthesis of yatein in *Anthriscus sylvestris. Organic and Biomolecular Chemistry,* **1**: 2474–2485.

Umezawa, T., Davin, L.B., and Lewis, N.G. (1991). Formation of lignans (")-secoisolariciresinol and (")-matairesinol with *Forsythia intermedia* cell-free extracts. *Journal of Biological Chemistry,* **266**: 10210–10217.

van der Heijden, R., Jacobs, D.I., Snoeijer, W., Hallard, D., and Verpoorte, R. (2004). The *Catharanthus* alkaloids: pharmacognosy and biotechnology. *Current Medicinal Chemistry,* **11:** 607–628.

Vats, S.K., and Kumar, S. (2006). Photosynthetic response of *Podophyllum hexandrum* Royle from different altitudes in Himalayan ranges. *Photosynthetica,* **44**: 136-139.

Wall, M.E., and Wani, M.C. (1996). Camptothecin and taxol: from discovery to clinic. *Journal of Ethnopharmacology,* **51:** 239–253.

Wealth of India. (1969). In: Ramanathan KR, ed. A dictionary of Indian Raw materials and industrial products, vol.6, Publication and Information directorate CSIR, New Delhi, India.

Bioactive Phytochemicals: Perspectives for Modern Medicine Vol. 2 (2014) *Pages* **207–230**
Editor: **V.K. Gupta**
Published by: **DAYA PUBLISHING HOUSE, NEW DELHI**

10

Phytopharmacological Review on *Calophyllum inophyllum* Linn.

Ch. Praveena[1] and C. Veeresham[1*]

ABSTRACT

Many plant derived, chemically disparate molecules have been evaluated for anti-HIV activity and among these various coumarins isolated from plants belonging to the genus Calophyllum have been found to act as potent non-nucleoside human immunodeficiency virus -1-reverse transcriptase inhibitors. Calophyllum inophyllum Linn. one of the most abundant plant of this genus is reported to contain anti-HIV active dipyranocoumarins Inophyllum B and Inophyllum P. Coumarins of this plant also exhibit anticoagulant, capillary permeability, anti-tumor, and piscicidal activities. Apart from pyranocoumarins the plant is reported to contain several other secondary metabolites like xanthones, chromanones, triterpenes and steroids. Xanthones showed anti-inflammmatory, antiplatelet aggregation, cytotoxic, and antiplasmodial activities. It possesses other bioactivities such as analgesic, healing property, antibacterial, molluscicidal, larvicidal, etc. In the present review the literature on phytochemical and pharmacological investigations on the C. inophyllum are summarized.

Keywords: Anti-HIV activity, *Calophyllum inophyllum*, Coumarins, Cytotoxic, Xanthones.

Introduction

Plant derived coumarins are important lead compounds for the development of anti-viral drugs. Dipyranocoumarins possessing phenyl or propyl group at C-4, which act as potent non-nucleoside human immunodeficiency virus -1-reverse transcriptase

1 Department of Medicinal Plant Biotechnology Laboratory, University College of Pharmaceutical Sciences, Kakatiya University, Warangal – 506 009, Andhra Pradesh, India

* *Corresponding author*: E-mail: ciddiveeresham@yahoo.co.in

inhibitors have been isolated from the genus *Calophyllum* family Guttiferae (Clusiaceae, Mangosteen family). Besides pyranocoumarins, the genus is a rich source of other secondary metabolites such as xanthones, chromanones, triterpenes and steroids. The genus consists of approximately 180-200 different species widespread in humid tropics. Although, a handful species have been identified in the new world, the genus is primarily found in the Indo-Pacific region, particularly Malaysia (Mc Kee *et al.*, 1998; Crane *et al.*, 2005).

Calophyllum inophyllum Linn. is the one of the most abundant species of this genus is reported to contain anti-HIV active dipyranocoumarins inophyllum B and inophyllum P. Apart from pyranocoumarins the plant is repoted to contain xanthones, chromanones, triterpenes and steroids and possesses variety of bioactivities such as, antiplatelet aggregation, cancer chemopreventive, anti-inflammatory, analgesic, piscicidal, antibacterial, molluscicidal etc. Different parts of the plant have been used in traditional and folk medicines for centuries and also grown as an ornamental plant. Several therapeutic properties have been attributed to all parts of the plant. The generic name *Calophyllum* of the plant is derived from Greek words 'kalos'–beautiful, 'phullon'- leaf and the specific epithet is derived from Greek words 'ino'- fiber, 'phullon'- leaf, alluding to the pronounced veins on the underside of the leaves. In other words this plant is named for its beautiful, fibrous leaves.

Synonyms

Preffered Syn. *Calophyllum bintagor* Roxb.

Unpreffered Syn. *Balsamaria inophyllum* (L.) Lour

Vernacular Names

English: Alexandrian laurel, Dilo oil tree, Oil nut tree; ***Fiji:*** Dilo; ***Hindi***: Nagakesar, Sultanachampa, Surpunka, Undi; ***Madagascar***: Vintanina; ***Sanskrit***: Nagachampa, Punnaga, Tunga; ***Sinhalese***: Domba, Dombagaha, Teldomba; ***Tulu***: Ponne; ***Tamil***: Pinnai, Punnagam, Surabunnagam; ***Telugu***: Ponnachettu(Tree), Ponnaginjalu(Seeds), Surapunnagamu; ***Kannada:*** Gadhavunate, Honne, Suragi.

Occurrence and Distribution

The plant is essentially a littoral species of the tropics. It is common on sandy beaches of the seashore but is sometimes found inland on sandy soils. It is native from east Africa to Australia and Malaysia and has been widely planted throughout the tropics, including many south and central Pacific islands, the Hawaiian Islands and the Caribbean islands (Orwa 2009; Friday and Okano, 2005; Kirtikar and Basu, 1987).

In India it is common on sandy coast and laterite soils just above high tide level in Orissa, Maharashtra, Karnataka, Tamil Nadu, Kerala, Andhra Pradesh, Lakshadweep and Andaman and Nicobar Islands (Parrotta, 2001).

Plant Description

It is a slow-growing, small or medium sized evergreen sub-maritime tree reaching up to 8-20 m in height. It has a broad, spreading crown, often with large, gnarled, horizontal branches with a crooked trunk. The bark is pale grey to dark brown, smooth, with longitudinal fissures. Bark yields an aromatic yellowish-green resin. The wood is reddish-white to reddish-brown. It has no buttress or other root modifications.

Leaves are opposite, dark green, shiny, thick, coriaceous, broadly elliptic-oblong rarely oblong, petiolate, 10-20 cm long and 7.5-10 cm wide. Apex is rounded to emarginated, with subrepand wavy margin and base is acute. Stipules are absent. Leaf veins are many, fine and very close, run parallel to each other and perpendicular to midrib, giving striate appearance to the blade. Midvein is raised abaxially, impressed adaxially. Petioles are robust, 0.9-1.5 cm long, stout and flat (Friday and Okano, 2005; Parrotta, 2001; Chattarjee and Pakrashi, 1992).

Inflorescence is axillary, racemose usually unbranched but occasionally with 3-flowered branches. Flowers are marble white, 2-2.5 cm in diameter, 8-14 mm long, sweetly scented. Pedicle is 1.5-4 cm long. Flowers are usually bisexual but sometimes functionally unisexual, with perianth of 8. Sepals are 4; outer 2: sub orbicular, much the smaller; inner 2: obovate. Petals are 4, oblong, obtruse and spreading (Orwa, 2009; Parrotta, 2001). Stamens are numerous and yellow in colour. Filaments united into 4-6 bundles. Anthers change from deep yellow through khaki to brown. Style is long, twisted; stigmas are large, mushroom shaped (Kirtikar and Basu, 1987).

Light green fruits grow in clusters. Fruit is globose, 2-5 cm in diameter and smooth. The skin, which turns yellow and then brown and wrinkled when the fruit is ripe, covers the thin pulp, the shell, a corky inner layer and a single seed kernel (Friday and Okano, 2005). Seed is globose, nearly as large as fruit, hard, with large cotyledons and radical pointing to the base of the fruit (Orwa, 2009).

The flowering and fruiting seasons vary. Flowering occurs twice a year, April–June and October-December (Friday and Okano, 2005). In India, flowers appear in July-August and sometimes again in November and fruits in September –October (Chattarjee and Pakrashi, 1992).

Phytochemistry

Extensive chemical investigation of this species has resulted in the isolation of a wide variety of natural products such as coumarins, xanthones, chromanones, triterpenes and steroids from different parts of the plant.

Coumarins

The known coumarins of the plant include pyranocoumarins [1-33], furo-pyranocoumarins [34-36], furocoumarins [37], and other coumarins [38-41]. Individual members of the groups vary with respect to the substituent at C*(4)* of the lactone ring of the coumarins, where Me, Pr, or Ph groups may be encountered.

Xanthones

The genus *Calophyllum* is considered as rich source of xanthone derivatives. Xanthones isolated from this plant include simple oxygenated such as monooxygenated [42-43], dioxygenated [44-48], trioxygenated [49-52], tetraoxygenated

Figure 10.1

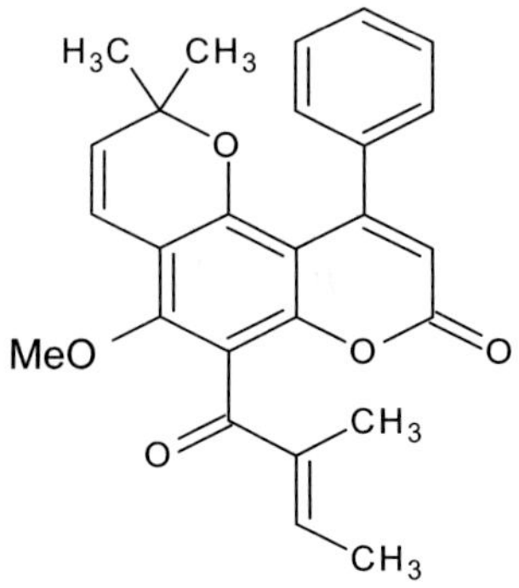

[2]

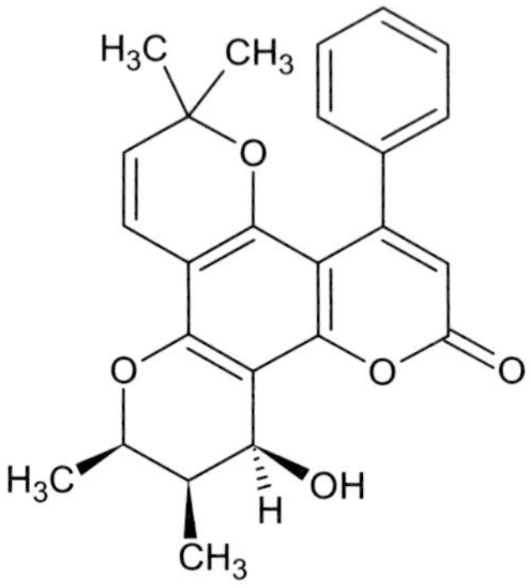

[7]

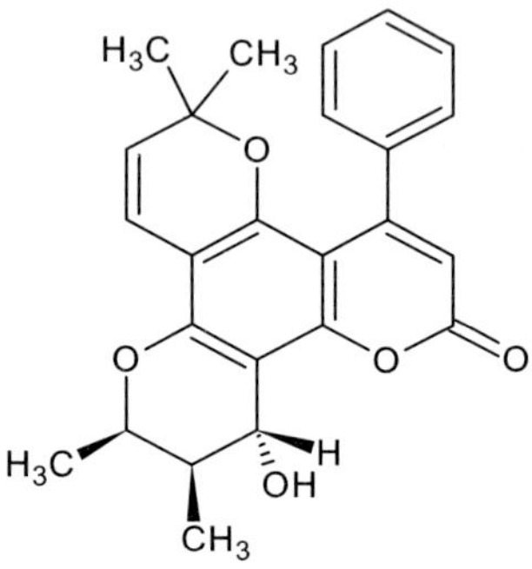

[8]

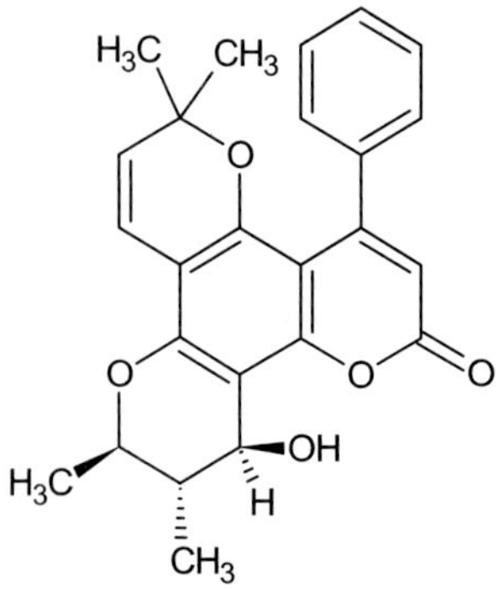

[9]

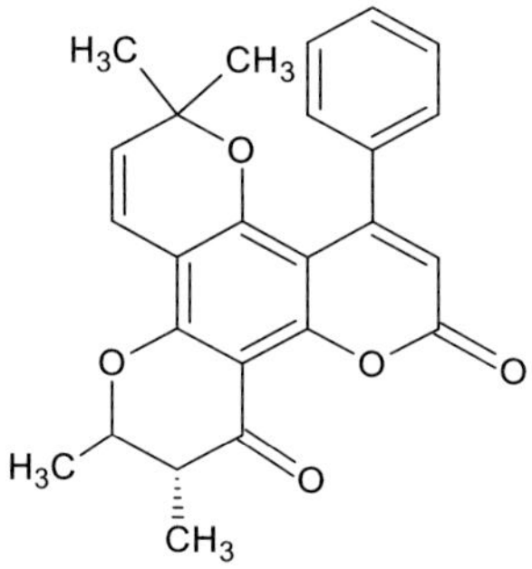

[10]

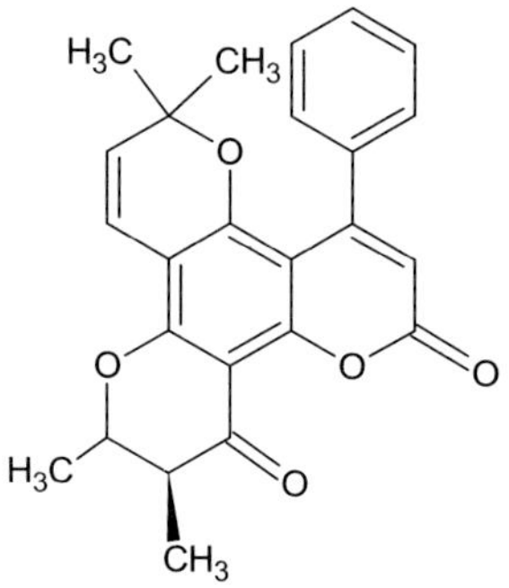

[11]

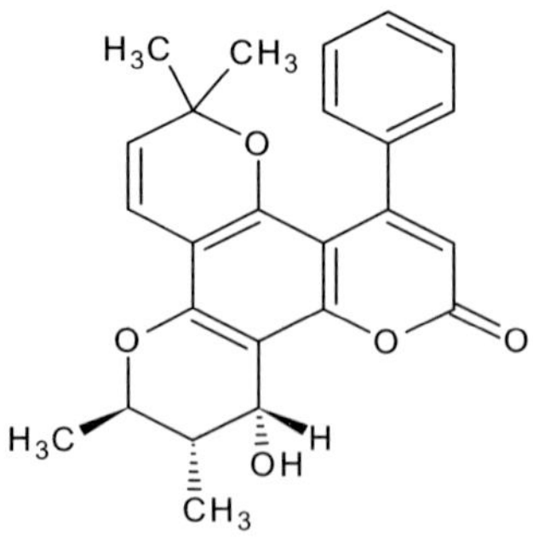

[12]

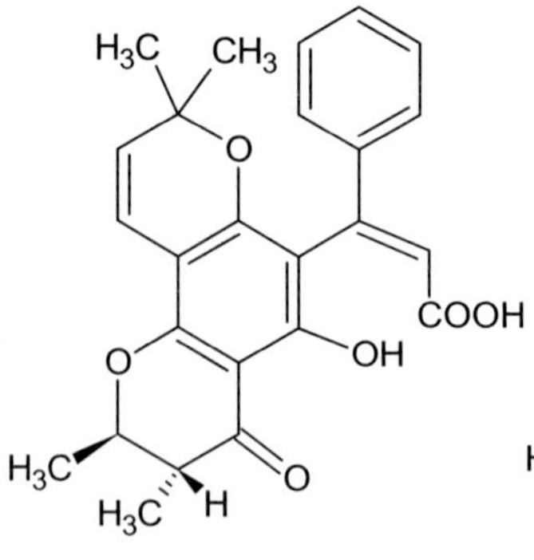

[14]

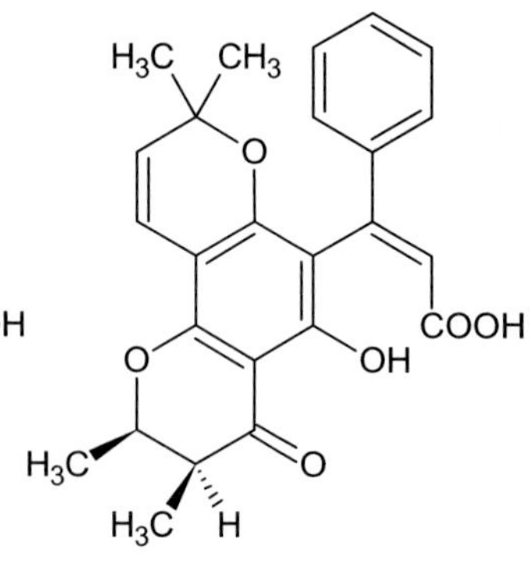

[15]

Contd...

Figure 10.1–*Contd...*

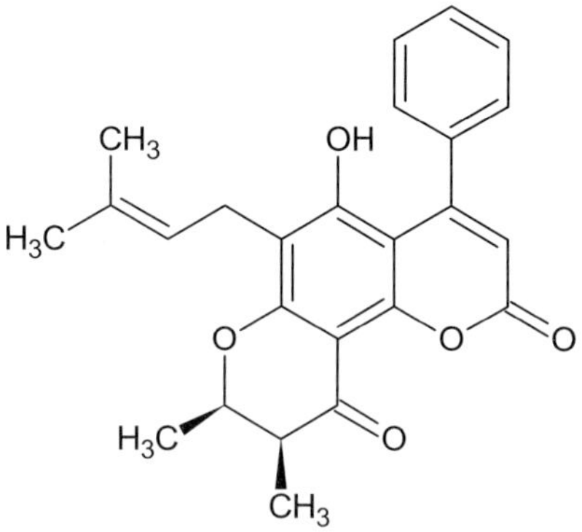

[19]

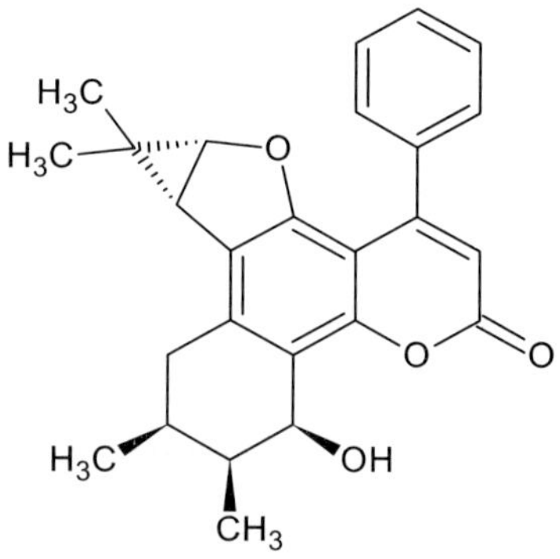

[34]

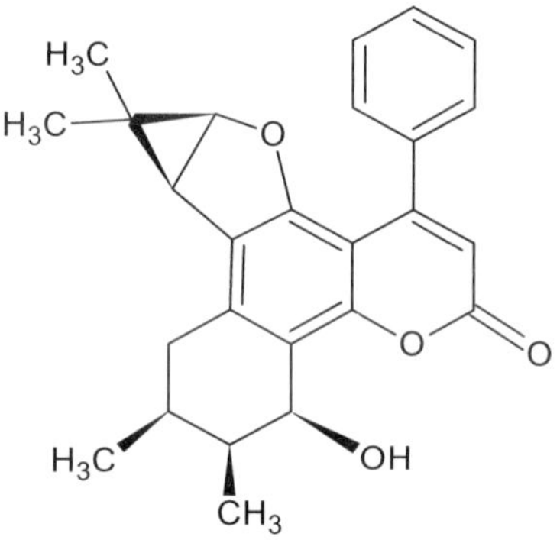

[35]

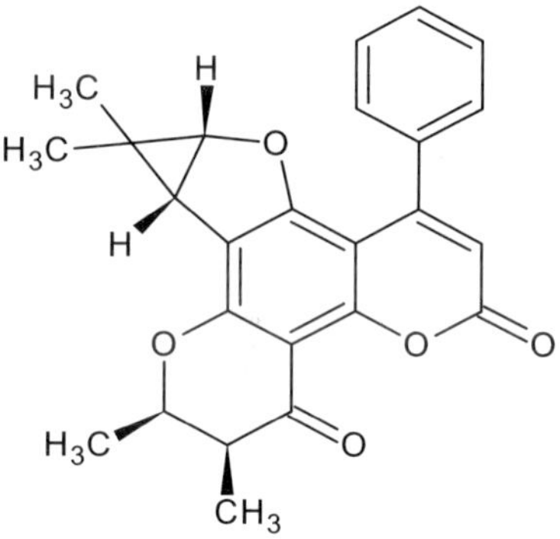

[36]

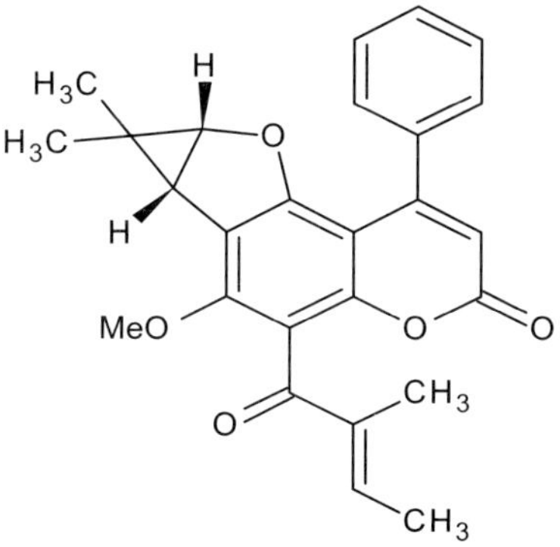

[37]

Contd...

Figure 10.1–*Contd...*

	R_1	R_2	R_3	R_4	R_5	R_6	R_7	R_8
[61]	OH	OCH_3	OH	H	OH	H	H	H
[62]	OH	H	OH	H	OH	OH	OCH_3	H
[64]	OH	H	H	H	OH	isoprenyl	H	H
[65]	OH	isoprenyl	OH	H	OH	H	H	H
[66]	OH	isoprenyl	OH	H	OH	OH	H	H

[57]

[58]

[63]

Contd...

Figure 10.1–*Contd...*

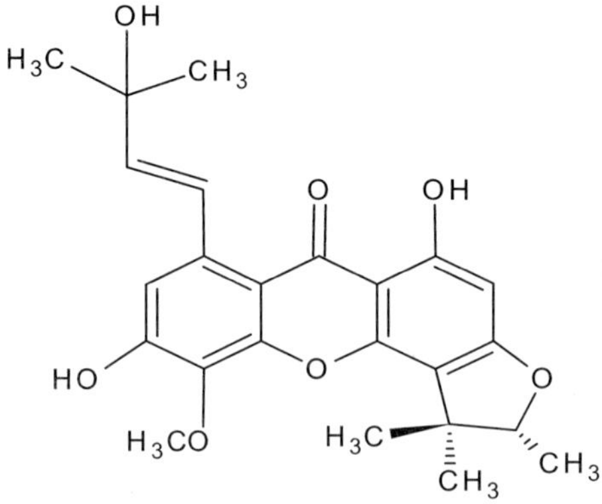

[67] 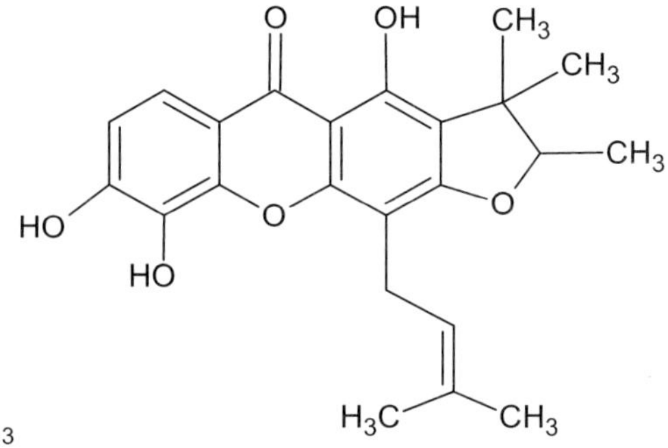 **[68]**

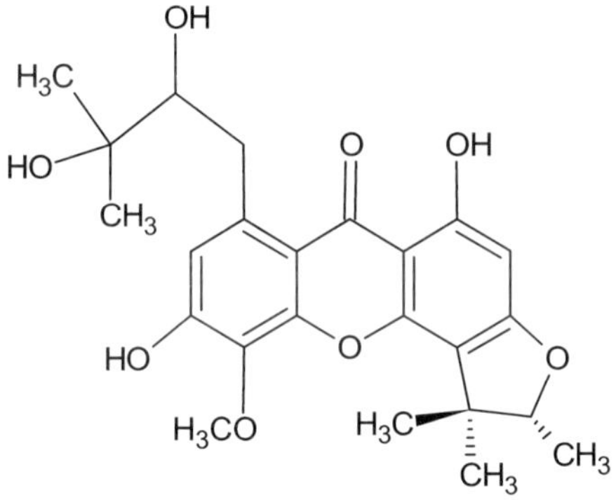

[69] 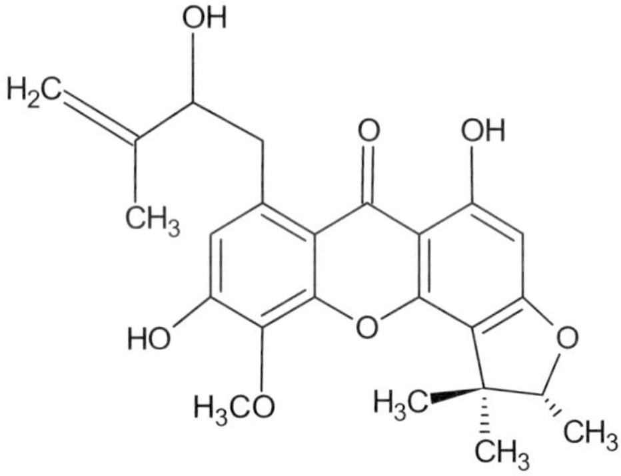 **[70]**

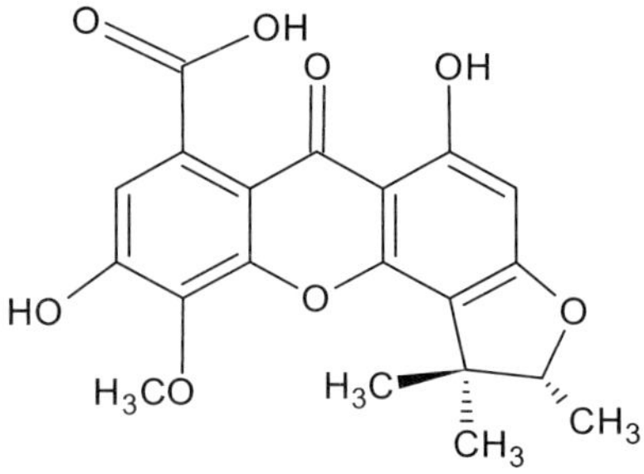

[71] 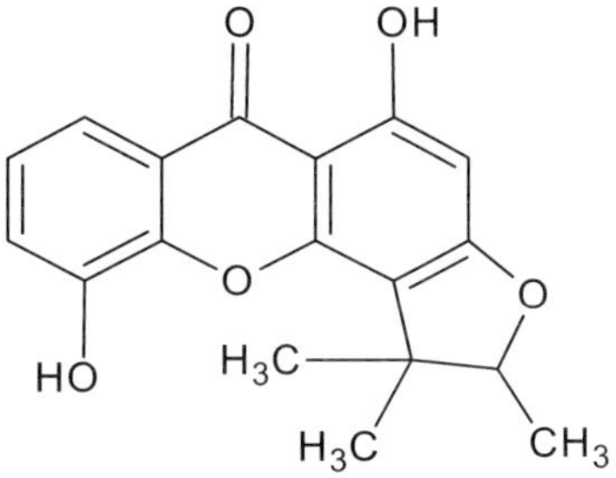 **[72]**

Contd...

Figure 10.1–*Contd...*

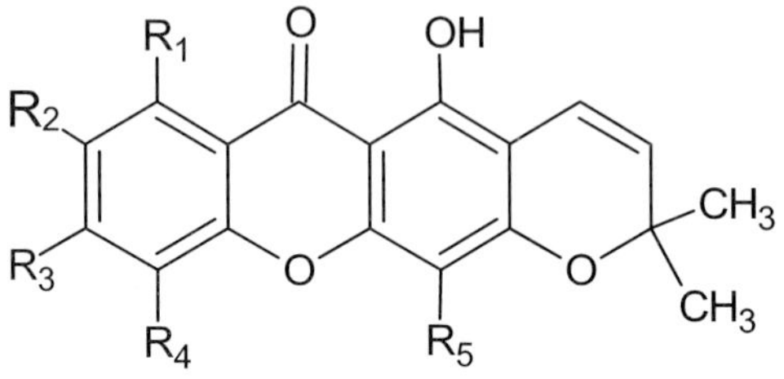

	R_1	R_2	R_3	R_4	R_5
[73]	H	H	H	OH	CH_2=CHC(Me)$_2$
[74]	H	H	H	OH	H
[75]	H	H	OH	OH	H
[77]	H	OH	OH	Isoprenyl	H

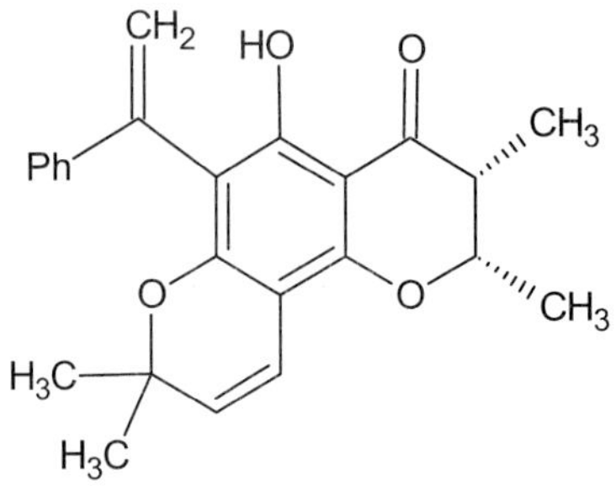

[89]

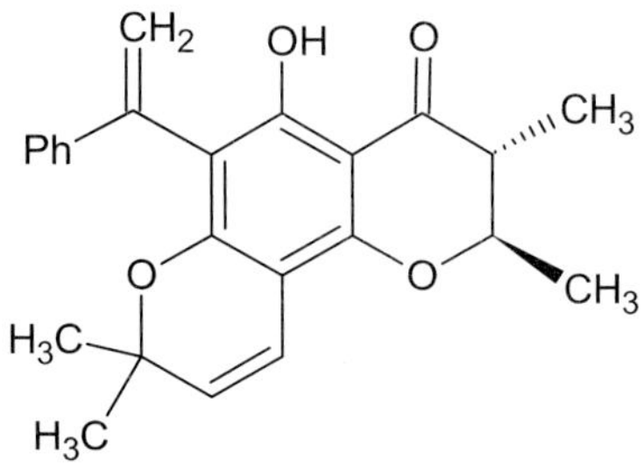

[90]

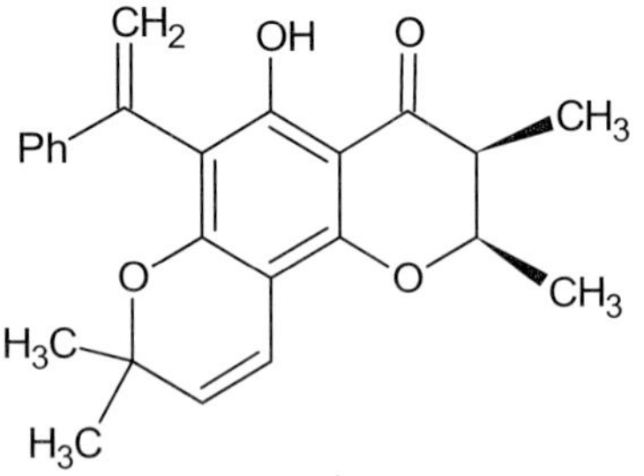

[91]

[53-61], pentaoxygenated [62-63], xanthones substituted with isoprenyl group [64-66], furanoxanthones [67-72] and pyranoxanthones [73-82].

Chromanones

Chromanone derivatives such as flavonoids [83-88], biflavanoid [89], tannins [90] and 1-benzopyran-4-one derivatives [91-93] have been reported.

Triterpenoids and Steroids

The most frequent triterpenes and steroids of the genus friedelin, canophyllol, sitosterol and also other compounds [94-110] are being reported.

Few other biologically active compounds from alcoholic extract of leaves were also isolated in antoher study. From the GC-MS analysis it was revealed that alcoholic extract mainly composed of oxygenated hydrocarbons, phenolic compounds and ketone group. A total of 17 compounds were identified and the major components were 1,2-Benzene dicarboxylic acid, diisooctyl ester, Androstan -1α–01-17- one, 2, 3 – isopropylidenedioxy-4β-methyl, 1-monolinoleoylglyceroltrimethylsilyl ether, squalene, oleic acid, n-Hexa decanoic acid, 3, 7, 11, 15- Tetramethyl -2 hexadecen- 1 – ol, 9,12 octadecadienoic acid methyl ester, (E.E), octadecanoic acid, Benzene(1-methyldodecyl) and phytol (Malarvizhi and Ramakrishnan, 2011).

Table 10.1

Sl.No.	*Compound Name*	*Plant Part and Reference*
	Coumarins	
1.	(±)-Inophyllolide	N (Polonsky, 1957)
2.	Calophyllolide	N (Polonsky, 1957), A.P.(Ito *et al.*, 1996), L (Patil *et al.*, 1993; Raharivelomanana *et al.*, 2005)
3.	Ponnalide	S (Adinarayana and Seshadri 1964)
4.	Brasiliensic acid	R.B. (Yimdjo *et al.*, 2004), St.B.R. (Stout *et al.*, 1968)
5.	Inophylloidic acid	R.B. (Yimdjo *et al.*, 2004), St.B.R. (Stout *et al.*, 1968)
6.	Isobrasiliensic acid	St.B.R.(Stout *et al.*, 1968)
7.	Inophyllum A	L (Patil *et al.*, 1993; Kawazu *et al.*, 1972),S (Joshi *et al.*, 2000)
8.	Inophyllum D	L (Patil *et al.*, 1993; Kawazu *et al.*, 1972),S (Joshi *et al.*, 2000)
9.	Inophyllum B	L (Patil *et al.*, 1993; Raharivelomanana *et al.*, 2005; Kawazu *et al.*, 1972), S (Joshi *et al.*, 2000; Spino *et al.*, 1998)
10.	Inophyllum C	L (Patil *et al.*, 1993; Raharivelomanana *et al.*, 2005; Kawazu *et al.*, 1972; Kawazu *et al.*, 1968), S (Joshi *et al.*, 2000; Spino *et al.*, 1998)
11.	Inophyllum E	L (Patil *et al.*, 1993; Kawazu *et al.*, 1972; Kawazu *et al.*, 1968), N (Yimdjo *et al.*, 2004), S (Spino *et al.*, 1998)
12.	Inophyllum P	L (Patil *et al.*, 1993; Raharivelomanana *et al.*, 2005), S (Joshi *et al.*, 2000; Spino *et al.*, 1998)
13.	Inophyllum F	L (Raharivelomanana *et al.*, 2005)
14.	Calophyllic acid	L (Patil *et al.*, 1993), S (Polonsky, 1957; Gautier *et al.*, 1972)
15.	Isocalophyllic acid	L (Raharivelomanana *et al.*, 2005), A.P.(Ito *et al.*, 1996)
16.	Pseudobrasilic acid	S (Gautier *et al.*, 1972)
17.	Calophynic acid	S (Gautier *et al.*, 1972), R.B. (Yimdjo *et al.*, 2004)
18.	Apetatolide	A.P. (Ito *et al.*, 1996)
19.	Calocoumarin A	A.P. (Ito *et al.*, 1996)
20.	Calaustralin	N (Yimdjo *et al.*, 2004; Bhushan *et al.*, 1975)

Contd...

Table 10.1–*Contd...*

Sl.No.	*Compound Name*	*Plant Part and Reference*
21.	12-methoxyinophyllum D	S (Joshi *et al.*, 2000)
23.	(3*E*)-isocalophyllic acid	L (Hang *et al.*, 2006)
24.	Tamanolide	S (Leu, *et al.*, 2009)
25.	Tamanolide D	S (Leu, *et al.*, 2009)
26.	Tamanolide P	S (Leu, *et al.*, 2009)
27.	Calanolide A	S (Leu, *et al.*, 2009)
28.	Calanolide B	S (Leu, *et al.*, 2009)
29.	Calanolide D	S (Leu, *et al.*, 2009)
30.	12-*O*-butylinophyllum D	L (Zou *et al.*, 2010)
31.	12-*O*-ethylinophyllum D	L (Zou *et al.*, 2010)
32.	Inophyllum H	L (Zou *et al.*, 2010)
33.	Inophyllum I	L (Zou *et al.*, 2010)
34.	Inophyllum G-1.	L (Patil *et al.*, 1993)
35.	Inophyllum G-2.	L (Patil *et al.*, 1993)
36.	Calocoumarin C	A.P. (Ito *et al.*, 1996)
37.	Calocoumarin B	A.P. (Ito *et al.*, 1996)
38.	Inocalophyllin A	S (Shen *et al.*, 2003)
39.	Inocalophyllin A methyl ester	S (Shen *et al.*, 2003)
40.	Inocalophyllin B	S (Shen *et al.*, 2003)
41.	Inocalophyllin B methyl ester	S (Shen *et al.*, 2003)
	Xanthones	
42.	2-hydroxyxanthone	St. and L (Li *et al.*, 2007), Tw (Xiao *et al.*, 2008)
43.	4-hydroxyxanthone	Tw. (Li *et al.*, 2007), R (Iinuma *et al.*, 1994a)
44.	1,7-dihydroxyxanthone/ Euxanthone	H.W. (Jeboury and Locksley1971), Tw. (Li *et al.*, 2007)
45.	1,5-dihydroxyxanthone	R.B. (Yimdjo *et al.*, 2004; Iinuma *et al.*, 1994a), Tw.(Li *et al.*, 2007)
46.	2-hydroxy-1-methoxyxanthone	R (Iinuma *et al.*, 1994b)
47.	1-hydroxy-2-methoxyxanthone	R (Iinuma *et al.*, 1994b)
48.	1,2-dimethoxyxanthone	R (Iinuma *et al.*, 1994b)
49.	1,5,6-trihydroxyxanthone/ (Mesuaxanthone B)	H.W. (Govindachari *et al.*, 1968)
50.	1,6-dihydroxy-5-methoxy-xanthone/buchanaxanthone	H.W. (Jackson *et al.*, 1969)
51.	6-hydroxy-1,5-dimethoxyxanthone	R.B. (Iinuma *et al.*, 1995)
52.	1,3,5-trihydroxyxanthone	R (Wu et al., 2003)

Contd...

Table 10.1–*Contd...*

Sl.No.	*Compound Name*	*Plant Part and Reference*
53.	1,3,8-trihydroxy-7-methoxyxanthone	R.B. (Iinuma *et al.,* 1995)
54.	1,3-dihydroxy-7,8-dimethoxyxanthone	R.B. (Iinuma *et al.,* 1995)
55.	1,3,5-trihydroxy-2-methoxyxanthone	R.B. (Iinuma *et al.,* 1995)
56.	1,7-dihydroxy-3,6-dimethoxyxanthone	T (Kumar *et al.,* 1976)
57.	Caloxanthone B	R.B. (Yimdjo *et al.,* 2004; Iinuma *et al.,* 1994a)
58.	Caloxanthone D	R.B. (Iinuma *et al.,* 1995)
59.	Macluraxanthone/3-Hydroxyblancoxanthone	R.B. (Yimdjo *et al.,* 2004; Iinuma *et al.,* 1994a)
60.	Caloxanthone M	R (Wu *et al.,* 2003)
61.	Tovopyrifolin C	R (Ee *et al.,* 2009)
62.	Caloxanthone E	R.H.W. (Iinuma *et al.,* 1995)
63.	Calophyllumine C	R (Wu *et al.,* 2003)
64.	Calophyllin B/6-(3,3-dimethylallyl)-1,5-dihydroxy-xanthone/Guanidin	H.W. (Govindachari *et al.,* 1968),T (Kumar *et al.,* 1976)
65.	2-(3,3-dimethylallyl)-1,3,5-trihydroxyxanthone	H.W. (Jeboury and Locksley, 1971)
66.	2-(3,3-dimethylallyl)-1,3,5,6-tetrahydroxyxanthone	H.W. (Jeboury and Locksley, 1971; Jackson *et al.,* 1969)
67.	Caloxanthone N	Tw. (Xiao *et al.,* 2008)
68.	Gerantoxanthone C	Tw. (Xiao *et al.,* 2008)
69.	Caloxanthone O	Tw. (Dai *et al.,* 2010)
70.	Caloxanthone P	Tw. (Dai *et al.,* 2010)
71.	Caloxanthone Q	Br. (Wei *et al.,* 2011), St. and L (Li *et al.,* 2011)
72.	2-deprenylrheediaxanthone B	Br. (Wei *et al.,* 2011)
73.	Caloxanthone C/Inoxanthone/Blancoxanthone	R.B. (Yimdjo *et al.,* 2004; Iinuma *et al.,* 1994b)
74.	6-deoxyjacareubin	T (Kumar *et al.,* 1976),H.W. (Jeboury and Locksley, 1971; Jackson *et al.,* 1969; Goh and Jantan, 1991),R (Iinuma *et al.,* 1994b),Tw. (Li *et al.,* 2007)
75.	Jacareubin	H.W. (Jackson *et al.,* 1969), T (Kumar *et al.,* 1976)
76.	Inophyllin A	R.B. (Ee *et al.,* 2006)
77.	Caloxanthone A	R.B. (Yimdjo *et al.,* 2004; Iinuma *et al.,* 1994a; Ee *et al.,* 2009)
78.	Calophinone	St.B. (Cheng *et al.,* 2004)
79.	Caloxanthone I	St.B. (Cheng *et al.,* 2004)

Contd...

Table 10.1–*Contd...*

Sl.No.	*Compound Name*	*Plant Part and Reference*
80.	Inophyllin B	R.B. (Ee *et al.*, 2004)
81.	Brasilixanthone B	R (Dai *et al.*, 2005), St.B.(Cheng *et al.*, 2004)
82.	Pyranojacareubin	R (Dai *et al.*, 2005), St.B.(Cheng *et al.*, 2004), A.P. (Li *et al.*, 2009)
	Chromanones	
83.	Leucocyanidin	S (Nair and Subramanian, 1964)
84.	Myricetin	Ad.Fl. (Subramanian and Nair, 1971)
85.	Myricetin-7-glucoside	Ad.Fl. (Subramanian and Nair, 1971)
86.	Quercetin	Ad.Fl. (Subramanian and Nair, 1971)
87.	Kaempferol-3-*O*-alpha-L-rhamnoside	Tw. (Li *et al.*, 2007)
88.	Quercetin-3-*O*-alpha-L-rhamnoside	Tw. (Li *et al.*, 2007)
89.	Amentoflavone	L (Dan *et al.*, 2008), Tw. (Li *et al.*, 2007)
90.	(-)-Epicatechin	R.B. (Iinuma *et al.*, 1994a)
91.	(2*S*, 3*R*)-2, 3-dihydro-5-hydroxy-2,3,8,8-tetramethyl-6-(1-phenylethenyl)-4H, 8H, benzo[1,2-b: 3,4-b'] dipyran-4-one	L (Khan *et al.*, 1996)
92.	Inophynone/(2*R*,3*R*)-2,3-dihydro-5-hydroxy-2,3,8, 8-tetramethyl-6-(1-phenyl-ethenyl)-4H,8H, benzo[1,2-b: 3,4-b']dipyran-4-one	L (Khan *et al.*, 1996; Ali *et al.*, 1999)
93.	Isoinophynone	L (Ali *et al.*, 1999)
	Triterpenoids and Steroids	
94.	Friedelin	L (Govindachari *et al.*, 1967), T(Kumar *et al.*, 1976), R.B. (Ee *et al.*, 2006)
95.	Canophyllol	L (Govindachari *et al.*, 1967)
96.	Canophyllal	L (Govindachari *et al.*, 1967)
97.	Canophyllic acid	L (Govindachari *et al.*, 1967; Ali *et al.*, 1999; Hang *et al.*, 2006)
98.	Friedelan-3β-ol	T (Kumar *et al.*, 1976)
99.	3,4-secofriedelane-3,28-dioic acid	L (Laure *et al.*, 2005)
100.	27-hydroxyacetate canophyllic acid	L (Laure *et al.*, 2005)
101.	3-oxo-27-hydroxyacetate friedelan-28-oic acid	L (Laure *et al.*, 2005)
102.	Campesterol	S.W. (Sampathkumar *et al.*, 1970)
103.	Friedelane-3-one	R.B. (Yimdjo *et al.*, 2004)

Contd...

Table 10.1–*Contd...*

Sl.No.	*Compound Name*	*Plant Part and Reference*
104.	Epifriedelanol	L (Hang *et al.*, 2006)
105.	3-oxo-28-friedelanoic aicd	L (Hang *et al.*, 2006)
106.	Taraxasterol	St and L (Dan *et al.*, 2008)
107.	Clovane-2,9-diol	St and L (Dan *et al.*, 2008)
108.	27-[(E)-p-coumaroyloxy] friedelin-28-carboxylic acid	L (Zou *et al.*, 2010)
109.	27-[(Z)-p-coumaroyloxy] friedelin-28-carboxylic acid	L (Zou *et al.*, 2010)
110.	3β-23-epoxy-friedelan-28-oic acid	L (Li *et al.*, 2010)

S: Seed; N: Nut; L: Leaves; R.B.: Root bark; St.: Stem; St.B.: Stem bark; St.B.R.: Stem bark resin; A.P.: Aerial parts; H.W.: Heart wood; Tw: Twigs; T: Timber; R.H.W.: Root heart wood; Ad. Fl.: Androecium of flower; Br.: Branches; S.W.: Sap Wood.

Composition of Seed Oil

Total kernel lipids of *C. inophyllum* were amounted to 60.1 per cent of the dry kernel. The total lipids consisted of three basic classes, neutral lipids 92 per cent, glycolipids 6.4 per cent and phospholipids 1.6 per cent. The oil contains a unique fatty acid calophyllic acid. The predominant fatty acids of total lipids were oleic acid (47.1 per cent) and linoleic acid (27.1 per cent) followed by palmitic acid (17.1 per cent) and stearic acid (8.3 per cent) (Lederer *et al.*, 1953; Hemavathy and Prabhakar, 1990).

Terpenoids of Volatile Oil

Steam distillation of the flower yielded pale yellow oil with pungent smell at 0.1 -1 per cent on fresh weight basis. Analysis of oil revealed the oil constituents into twenty-five components, of which 17 accounting for approximately 75 per cent of the oil had been identified. The most abundant component was found to be a naphthalene derivative, 1,2,3,4,4α,7-hexahydro-1,6-dimethyl-4-(1-methylethyl) naphthalene which accounts for 24.5 per cent of the oil. Other major components were α-cubebene, β-selinene, calarene, β-farnesene, δ-cadinene, β-bourbonene, zingiberene, copaene, murelene, sesquiphellandrene, octadecanal, hexadecane, nerolidol and farnesol (Samsudin *et al.*, 1996).

In essential oil obtained from twigs, twenty-six compounds were characterized and the major constituents were identified as δ-amorphene, β-caryophyllene, γ-cadinene, α-caryophyllene and α-farnesene (Mei *et al.*, 2006).

Pharmacological Activities

Anticoagulant Activity

Anticoagulant activity of calophyllolide was compared with dicumarol, ethylbiscoumacetate and nicoumalone in rabbits by single oral dose level of 50mg/

kg for hypoprothrombinaemic activity. Calophyllolide reduced coagulation valency 18 to 22 per cent after 48 hrs. Oral administration up to 1.75gm/kg was not fatal in rats. Its intraperitoneal LD_{50} in rats was 560mg/kg. Moderate increase in whole bleeding coagulation, bleeding and platelet clumping times was also observed by calophyllolide like dicumarol (Arora *et al.*, 1962).

Anti-inflammmatory and C.N.S. Deprressant Activity of Xanthones

Xanthones such as dehydrocycloguananidin, calophyllin B, 6-deoxyjacareubin, jacareubin, mesuaxanthone-B and euxanthone were screened for various pharmacological activities. They produced varying degrees of C.N.S. depression characterised by ptosis, sedation, decreased spontaneous motor activity, loss of muscle tone (in mice), potentiation of pentobarbitone sleeping time and ether anaesthesia in mice and rats. Also exhibited anti-inflammatory activity as tested by carrageenan induced rat-paw oedema, cotton pellet granuloma and granuloma pouch techniques, in normal and/as well as adrenalactomised rats. Jacareubin and 6-deoxyjacareubin exhibited anti-ulcer activity in rats (Gopalakrishnan *et al.*, 1980).

Calophyllolide showed potent anti-inflammatory activity in carrageenan-induced oedema in rats. It also showed anti-inflammatory and antiarthritic activity in formaldehyde –induced arthritis and adjuvant in rats. It was devoid of ulcerogenic activity up to twice the ED_{50} dose (Bhalla *et al.*, 1980).

Capillary Permeability

Pretreatment with calophyllolide (p.o.), a proved non-steroidal anti-inflammatory compound from the powdered kernel, showed significant protection against increased capillary permeability induced in mice by various chemical mediators involved in the inflammatory process *viz.* histamine, 5-hydroxytryptamine and bradykinin (Saxena *et al.*, 1982).

Inhibitory Activity on PAF Receptor Binding Ability

Xanthones such as 2-(3-methylbutyl-2-enyl)-1,3,5-trihydroxyxanthone, 6-deoxyjacareubin and 2-(3-methylbutyl-2-enyl)-1,3,5,6-tetrahydroxyxanthone exhibited strong inhibition of platelet activating factor (PAF) receptor binding in rabbit platelets (Jantan *et al.*, 2001).

Antiplatelet Aggregation Activity

Xanthones 2-(3-methylbut-2-enyl)-1,3,5-trihydroxyxanthone and 2-(3-methylbut-2-enyl)-1,3,5,6-tetrahydroxyxanthone showed strong inhibitory activity on platelet aggregation induced by Arachidonic acid in human whole blood in a dose-dependent manner. Macluraxanthone, pyranoamentoflavone and a neoflavonoid showed selective inhibitory activity on platelet aggregation induced by adenosine diphosphate (Jantan *et al.*, 2009).

Antiviral Activity

Antiviral activity of calophyllolide, inophyllum A, inophyllum B, inophyllum P, inophyllum D, inophyllum E, inophyllum C, inophyllum G-1, inophyllum G-2, calophyllic acid and isocalophyllic acid was evaluated against HIV-1 RT with scintillation proximity assay. Inophyllum B was consistently the most active compound followed by Inophyllum P. Inophyllum C and E were much less active or inactive. Calophyllic acids, isocalophyllic acid, calophyllolide, inophyllum G-1, inophyllum G-2 were inactive (Patil *et al.*, 1993)[l].

In addition, seed extract containing pyranocoumarins such as calophyllolide, inophyllum B, C, P and E was tested for anti-HIV activity and presumed that the activity was due to the presence of inophyllum P. Extract also exhibited cytotoxic activity (Spino *et al.*, 1998).

Anti-tumor Activity

Primary screening for anti-tumor-promoting activity of ten 4-phenylcoumarins isolated from aerial parts was studied by examining their inhibitory effects on Epstein–Barr virus early antigen (EBV-EA) activation induced by 12-*O*-tetradecnolylphorbol-13-acetate (TPA) in Raji cells. All the tested compounds showed an inhibitory effect on EBV activation, without showing any cytotoxicity. Calocoumarin-A, the pyran ring opened analogue of inophyllum-E exhibited the most potent inhibitory activity, suggesting that the prenyl side chain is a structure in increasing the anti-tumor effect. Furthermore, calocoumarin- A exhibited marked inhibitory effecton carcinogenesis in an *in vivo* two-stage test on skin papillomas of mice initiated by 7, 12-dimethylbenz[a]anthracene (DMBA) and prompted by TPA. The study suggested that some of the tested 4-phenylcoumarins might be valuable as potential cancer chemopreventive agents (Itoigawa *et al.*, 2001).

Cytotoxic Activity

Several xanthones and some derivatives identified as inoxanthone, caloxanthone A and B, macluraxanthone, 1, 5-dihydroxyxanthone, calophynic acid, brasiliensic acid, inophylloidic acid, friedelan-3-one, calaustralin, calophyllolide, and inophyllum C and E exhibited cytotoxic activity against KB-line cells and potential antimicrobial activity (Yimdjo *et al.*, 2004). Caloxanthone N and gerontoxanthone C exhibited cytotoxicity against chronic myelogenus leukemia cell line (K562) with IC_{50} values of 7.2 and 6.3 μg/ml, respectively (Xiao *et al.*, 2008) and caloxanthone O against human gastric cell lines (SGC-7901) with IC_{50} values of 22.4μg-mL^{-1} (Dai *et al.*, 2010).

The cell growth inhibitory effect of various concentrations of 3β, 23-epoxy-friedelan-28-oic acid, canophyllic acid, 3-oxo-friedelan-28-oic acid and oleanolic acid on human leukemia HL-60 cells were evaluated. All the compounds showed inhibitory activity but the compound 3-oxo-friedelan-28-oic acid was more effective than other three compounds (Li *et al.*, 2010).

Analgesic Activity

Analgesic and antibacterial activity of methanolic and chloroform extracts of dried stem barks were compared with standard drugs. The methanolic extract showed significant antibacterial property and analgesic activity than the chloroform extract (Mishra *et al.*, 2010).

Healing Properties

Nut oil appeared to reduce the size of the scars and hence improved their appearance. It was also found to have the healing property (Dweck and Meadows 2002; Chanson, 2006).

Antimicrobial Activity

An antibacterial principle isolated from root bark was highly active against many gram-positive bacteria (Potti and Kurup, 1970). Different concentrations of 80 per cent ethanolic extracts of powdered stem bark and leaf were showed pronounced antibacterial activity against gram positive bacteria than gram negative (Valsaraj *et al.*, 1997).

Methanol, hexane and dichloromethane extracts of leaf were screened for antibacterial activity, *in vitro* anti-tuberculosis activity against *Mycobacterium tuberculosis* H37Ra and antifungal activity against *Candida albican*. The methanolic extract of the concentration 10mg showed moderate inhibition against *Staphylococcus aureus* and *Salmonella typhi* and others have no activity. Hexane and dichloromethane extracts showed high activity in *in vitro* antituberculosis activity and all extracts showed no activity against *C. albican.* High potential of *C. inophyllum* leaf as antituberculosis against *M. tuberculosis* H37Ra and moderate antimicrobial potential against *S. aureus* and *S. typhi* may lead to use this medicinal plant as natural antimicrobial agents in the near future (Chaisawadi *et al.*, 2007).

Seed oil has antibacterial activity *in vitro* against gram-positive organisms (Bhat *et al.*, 1954). The antimicrobial phytochemical agents of oil include friedelin, canophyllol, canophyllic acid, and inophynone. Canophyllic acid demonstrated moderate antibacterial against *Proteus mirabilis.* Canophyllol demonstrated good activity against *Staphylococcus aureus*, *Cornebacterium diptheriae*, *Klebsiella pneumonia* and *Proteus mirabilis* as compared with the standard. Friedelin demonstrated good activity against *S. aureus.* In addition friedelin exhibited good antifungal activity against *Pseudallescheria boydii* and moderate activity against *Trichophyton schoenleinii.* The n-butanol fraction of oil showed significant antifungal activity against *Trichophyton semii* and *Trichophyton metagrophyte*. From the results of the study it was concluded that the antimicrobial agents in tamanu oil could be used effectively to treat a range of infections of the skin and eyes, and in cases of ring worm (Mahmud *et al.*, 1998).

Ethanol, butanol, aqueous and chloroform extracts of leaves and isolated pure compounds such as canophyllic acid, canophyllol, friedelin and inophynone were tested for antibacterial activity against gram positive and gram negative and antifungal activity against human, animal and plant pathogens. From the performed tests it was concluded that the chloroform extract can be used to inhibit the growth of tested human and animal pathogens. It was also observed that pure sample of friedelin was an excellent agent to inhibit the growth of tested human pathogens (Ali *et al.*, 1999).

The antimicrobial studies of the methanolic and *n*-hexane extracts of fruit peel prepared by the pressurized liquid extraction method demonstrated promising antibacterial activity against *Staphylococcus aureus* and *Mycobacterium smegmatis*. The preliminary phytochemical screening revealed the presence of phenolic compounds which was likely to be responsible for these activities (Ha *et al.*, 2009).

Different concentrations of ethanol and ethyl acetate extracts of various parts of the plant like leaf, fruit, seed, flower and root were evaluated for antimicrobial activity against clinically important bacterial and fungal species. Ethanol extracts were more effective and produced more zone of inhibition when compared to ethyl acetate extracts. The results of the study confirm the traditional usage of the plant and also suggest that some of the plant derived extracts can be used as antimicrobial agents (Saravanan *et al.*, 2011).

Role as Biological UV Filter in Ophthalmic Preparations

As most of the UV filters are cytotoxic for the eye, anti-UV property of *C. inophyllum* oil was investigated in order to propose it as a potential vehicle, free of toxicity, with a natural filter action in ophthalmic formulation. *C. inophyllum* oil exhibited antioxidant and cytoprotective properties and therefore might serve, for the first time, as a natural UV filter in ophthalmic preparations (Said *et al.*, 2007).

Ocular Burn

To study the effects of various rinsing and healing protocols on corneal wound repair and inflammation following alkali burn in rabbits *in vitro, in vivo* and *ex vivo* studies were performed. From the studies it was concluded that irrigation with marine solution followed by treatment with a mixture of *C. inophyllum* and *Aleurites moluccana* oils was found to be a promising treatment for ocular burns (Said *et al.*, 2009).

Insecticidal and Other Biological Properties

Piscicidal Activity

Piscicidal activity of five constituent's inophyllum A, B, C, D and E of the leaves was tested in comparision with pentachlorophenol using the killie-fish (*Oryzias loupes*). The ketones inophyllum C and E were more active than the alcohols inophyllum A, B and D. Among the alcohols inophyllum B which has all *trans* configuration was found to be the most active compound (Kawazu *et al.*, 1968; Kawazu *et al.*, 1972).

Molluscicidal Activity

Two types of crude seed extracts along with some natural neoflavanoids such as calophyllolide, calophyllic acid and pseudobrasiliensic acid were screened for the molluscicidal activity against snail *Biomphalaria glabrata.* The seed extracts and neoflavonoids such as calophyllic acid, pseudobrasiliensic acid showed molluscicidal activity at a concentration of 100 ppm and 10 ppm respectively (Ravelonjato *et al.*, 1992).

Antiplasmodial Activity

Antiplasmodial activity of compounds caloxanthone C, macluraxonthone, caloxanthones A and B, 1, 5-dihydroxanthone, calophynic acid, brasiliensic acid, inophylloidic acid and friedelan-3-one was investigated against *Plasmodium falsiparum* F32 (chloroquine sensitive) and FcM29 (chloroquine resistant).

Caloxanthone A, calophynic acid, brasiliensic acid and inophylloidic acid were found to be the most active (Azebaze *et al.*, 1980).

Larvicidal Activity

The larvicidal activity of seed and leaf extracts on the juveniles of *Culex quinquefasciatus*, *Anopheles stephensi* and *Aedes aegypti* was determined. Ethyl acetate soluble fractions of extracts showed very high larvicidal activity. The fractions also interfered with adult emergence. Experiments under laboratory and semi-field conditions showed that the activity of the extracts persisted for up to 10 days (Pushpalatha *et al.*, 1999). In another study, extracts and purified extracts of seeds exhibited larvicidal activity against the 2nd instar larvae of *Helicoverpa armigera* reared on synthetic diet (Pawar *et al.*, 2004).

Anti-repellant activity of 21 essential oils alone or in combination with *C. inophyllum* nut oil on female *Stomoxys calcitrans* (L.) was screened. The protection time was increased by binary mixtures of five essential oils (clove bud, clove leaf, thyme white, patchouli, and savory) and tamanu oil compared with that of either the constituted essential oil or tamanu oil alone. All the essential oils with the exception of savory essential oil, tamanu oil, and binary mixtures did not induce any adverse effects on the human volunteers. Thus, binary mixtures of essential oils and tamanu oil were described as potential repellants of *S. calcitrans* (Hieu *et al.*, 2010).

Toxicological Evaluation of Seed Oil

Toxicological effects of the seed oil of *C. inophyllum* on rat were evaluated by administering 5 per cent of the oil in normal rat feed and monitoring the effects of oil on physical appearance, feed intake, weight gain, plasma and tissue cholesterol and triglycerol levels. Mild, focal to severe and widespread lesions in the kidneys, hearts and livers, and significant raise in plasma cholesterol levels of the rats fed with oil were found when compared with rats fed with control feed. Thus it appeared not suitable for consumption (Ajayi *et al.*, 2008).

Conclusion

Plants play a vital role in maintaining human health and contribute towards improvement of human life. They are important source of novel drug compounds which are being used as therapeutic agents in different ailments. Modern civilization, uneducation of the people and several other factors led to the occurance of several diseases among which some are not curable *viz.*, AIDS. Despite the large number of synthetic medicines available for the treatment of disease, search for the effective therapies is still of great importance. Because of the adverse effects and limitations such as high cost, decreased sensitivity associated with the synthetic medicine, herbal medicines are frequently used by the individuals as an alternative therapy. Plant derived coumarins are important lead compounds for the development of anti-viral drugs.

Isolation of the anti-HIV active dipyranocoumarins from the genus *Calophyllum* (Guttiferae), which act as potent non-nucleoside human immunodeficiency virus-1-reverse transcriptase inhibitors is causing great interest in the scientific community about this genus and much attention has been paid to studies on the chemical components of this genus. *C. inophyllum* Linn., one of the most abundant plant of the genus is reported to contain anti-HIV active dipyranocoumarins, inophyllums B and P. The chemical literature reflects the existence of wide variety of secondary metabolites in this plant, such as coumarins, xanthones, chromanones and triterpenes. It also possess various bioactivities such as, antiplatelet aggregation, cancer chemopreventive, anti-inflammatory, analgesic, anticoagulant, piscicidal, antimicrobial, molluscicidal etc. In view of its phytochemical diversity and biological activities, the plant is drawing the attention of the researchers. And it is necessary to exploit its maximum potential in the field of medicinal and pharmaceutical sciences for novel and fruitful application.

References

Adinarayana, D., and Seshadri, T.R. (1964). Chemical components of the Indian seeds of *C.inophyllum*: The structure of a new 4-phenylcoumarin, ponnalide. Paper presented at the Symposium on Recent Advances in Plant Polyphenolics, National Institute of Sciences of India, New Delhi.

Ajayi, I.A., Oderinde, R.A., Taiwo, V.O., and Agbedana, E.O. (2008). Short-term toxicological evaluation of *Terminalia catappa, Pentaclethra macrophylla* and *C. inophyllum* seed oils in rats. *Food Chemistry*, **106**: 458-465.

Ali, M.S., Mahmud, S., Perveen, S., Ahmad, V., and Rizwani, G. (1999). Epimers from the leaves of *C. inophyllum*. *Phytochemistry*, **50**: 1385-1389.

Arora, R.B., Mathur, C.N., and Seth, S.D. (1962). Calophyllolide, a complex coumarin anticoagulant from *C. inophyllum* Linn. *Journal of Pharmacy and Pharmacoogy*, **14**: 534-535.

Azebaze, A.G., Yimdjo, M.C., Ouahouo, B.M.W., Meyer, M., Meddah, B., Valentin, A., and Nkengfack, A.E. (2007). Antiplasmodial constituents from two Cameroonian Guttiferae: *Mammea africana* and *C. inophyllum*. Absract submitted at Malaria 2007 Meeting, 21st-27th April, Dakar, Senegal.

Bhalla, T.N., Saxena, R.C., Nigam, S.K., Misra, G., and Bhargava, K.P. (1980). Calophyllolide- a new nonsteroidal anti-inflammatory agent. *Indian Journal of Medical Research*, **72**: 762-765.

Bhat, S.G., Kane, J.G., and Sreenivasan, A. (1954). The *in vitro* evaluation of the antibacterial activity of undi oil (*C. inophyllum* Linn.) *Journal of American Pharmaceutical Association*, **43**: 543-546.

Bhushan, B., Rangaswami, S., and Seshadri, T.R. (1975). Calaustralin, a new 4-phenylcoumarin from the seed oil of *C. inophyllum*. *Indian Journal of Chemistry*, **13**: 746-747.

Chaisawadi, A., Prasart Kittakup Kittakup, P., and Chaisawadi, S. (2007). Screening for antimicrobial activities of *C. inophyllum* Linn Leaf. In: 33rd Congress on Science and Technology of Thailand.

Chanson, N.M. (2006). Study of *Calophyllum* oil; clinical use demonstrates its healing properties. *Phytotherapie*, **4**: 167-171.

Chattarjee, A., and Pakrashi, S.C. (1992). The Treatise on Indian Medicinal Plants. Publications and Information Directorate, New Delhi, Vol.II, 151-153.

Cheng, H.C., Wang, L.T., Khalil, A.T., Chang, Y.T., Lin, Y.C., and Shen, Y.C. (2004). Pyranoxanthones from *C. inophyllum*. *Journal of the Chinese Chemical Society*, **51**: 431-435.

Crane, S., Aurore, G., Joseph, H., Mouloungui, Z., and Bourgeois, P. (2005). Composition of fattyacids triacylglycerols and unsaponifiable matter in *Calophyllum calaba* oil from Guadeloupe. *Phytochemistry*, **66:** 1825-1831.

Dai, H.F., Zeng, Y.B., Xiao, Q., Han, Z., Zhao, Y.X., and Mei, W.L. (2010). Caloxanthones O and P: Two new prenylated xanthones from *C. inophyllum*. *Molecules*, **15**: 606-612.

Dan, L.I.U., Zhan-lin, L.I., Yan-zhi, L.I., Hui-ming, H.U.A., and Ming-sheng, L.I.U. (2008). Chemical constituents of the stems and leaves of *C. inophyllum*. *Journal of Shenyang Pharmaceutical University*, **25**: 623-626.

Desai, P.D., Dutia, M.D., Ganguly, A.K., Govindachari, T.R., Joshi, B.S., Kamat, V.N, Prakash, D., Rane, D.F., Sathe, S.S., and Viswanathan, N. (1967). Chemical investigation of some Indian plants: Part III. *Indian Journal of Chemistry*, **5**: 523-524.

Dweck, A.C., and Meadows, T. (2002). Tamanu (*C. inophyllum*) – the African, Asian, Polynesian and Pacific panacea. *International Journal of Cosmetic Sciences*, **24**: 341-348.

Ee, G.C.L., Jong, V.Y.M., Sukari, M.A., Rahmani, M., and Kua, A.S.M. (2009). Xanthones from *C. inophyllum*. *Pertanika Jorurnal of Science and Technology*, **17**: 307– 312.

Ee, G.C.L., Kua, A.S.M., Cheow, Y.L., Lim, C.K., Jong, V., and Rahmani, M. (2004). A new pyranoxanthone Inophyllin B from *C. inophyllum*. *Natural Product Sciences*, **10**: 220-222.

Ee, G.C.L., Kua, A.S.M., Lim, C.K., Jong, V., and Lee, H.L. (2006). Inophyllin A, a new pyranoxanthone from *C. inophyllum*. *Natural Product Research*, **20**: 485-491.

Friday, J.B., and Okano, D. Species Profiles for Pacific Island Agroforestry. www.traditionaltree.org. 2006 April, 214-215.

Gautier, J., Kunesch, G., and Polonsky, J. (1972). Structure of calophynic acid, a novel constituent of *C. inophyllum*. *Tetrahedron Letters*, **27**: 2715-2718.

Goh, S.H., and Jantan, I. (1991). A xanthone from *C. inophyllum*. *Phytochemistry*; **30**: 366-367.

Gopalakrishnan, C., Shankaranarayanan, D., Nazimudeen, S.K., Viswanathan, S., and Kameswaran, L. (1980). Anti inflammatory and central nervous system depressant activities of xanthones from *C. inophyllum* and Mesua-ferrea. *Indian Journal of Pharmacology*, **12**: 181-192.

Govindachari, T.R., Pai, B.R., Muthukumaraswamy, N., Rao, U.R., and Rao, N.N. (1968). Chemical components of heartwood of *C. inophyllum*. I. Isolation of Mesuaxanthone B and a new xanthone Calophyllin B. *Indian Journal of Chemistry*, **6**: 57-59.

Govindachari, T.R., Vishwanathan, N., Pai, B.R., Ramadas Rao, U., and Srinivasan, M. (1967). Triterpenes of *C. inophyllum* Linn. *Tetrahedron Letters*, **23**: 1901-1910.

Ha, M.H., Nguyen, V.T., Nguyen, K.Q.C., Cheah, E.L.C., and Heng, P.W.S. (2009). Antimicrobial activity of *C. inophyllum* crude extracts obtained by pressurized liquid extraction. *Asian Journal of Traditional Medicine*, **4**: 141-146.

Hang, N.T.M., Chien, N.Q., and Hung, N.V. (2006). Triterpenes from the leaves of Vietnamese plant *C. inophyllum* L. *Journal of Chemistry*, **44**: 115-118.

Hemavathy, J., and Prabhakar, J.V. (1990). Lipid composition of *C. inophyllum* kernel. *Journal of American Oil Chemical Society*, **67**: 955-957.

Hieu, T.T., Kim, S-Il., Lee, S.G., and Ahn, Y.J. (2010). Repellency to *Stomoxys calcitrans* (Diptera: Muscidae) of plant essential oils alone or in combination with *C. inophyllum* nut oil. *Journal of Medical Entomology*, **47**: 575-580.

Iinuma, M., Tosa, H., Tanaka, T., and Yonemori, S. (1994a). Two xanthones from root bark of *C. inophyllum*. *Phytochemistry*, **35:** 527-532.

Iinuma, M., Tosa, H., Tanaka, T., and Yonemori, S. (1994b). Two new xanthones from roots of *C. inophyllum*. *Heterocycles*, **37**: 833-838.

Iinuma, M., Tosa, H., Tanaka, T., and Yonemori, S. (1995). Two xanthones from roots of *C. inophyllum*. *Phytochemistry*, **38**: 725-728.

Ito, C., Mishina, Y., Ohta, M., Furukawa, H., Tan, H., Tokuda, H., Nishino, H. (1996). Studies on anti-tumor promoter from tropical medicinal plants (8). Isolation and structure elucidation of new coumarins from *Calophyllum* species (Guttiferae). The 119th Annual Meetings of the Pharmaceutical Society of Japan, Tokushima, Abstract II, p.156.

Itoigawa, M., Ito, C., Tan, H.T.W., Kuchide, M., Tokuda, H., Nishino, H., and Furukawa, H. (2001). Cancer chemopreventive agents, 4-phenylcoumarins from *C. inophyllum*. *Cancer Letters*, **169**: 15-19.

Jackson, B., Locksley, H.D., and Scheinmann, F. (1969). The isolation of 6-Deoxyjacareubin, 2-(3, 3-Dimethylallyl)-1, 3, 5, 6-tetrahydroxy-Xanthone and Jacareubin from *C. inophyllum*. *Phytochemistry*, **8**: 927-929.

Jantan, I., Juriyati, J., and Warif, N.A. (2001). Inhibitory effects of xanthones on platelet activating factor on receptor binding *in vitro*. *Journal of Ethnopharmacology*, **75:** 287-290.

Jantan, I., Mohd Yasin, Y.H., Jalil, J., Murad, S., and Idris, M.S. (2009). Antiplatelet aggregation activity of compounds isolated from Guttiferae species in human whole blood. *Pharmaceutical Biology*, **47**: 1090-1095.

Jeboury, F.S.A., and Locksley, H.D. (1971). Xanthones in the heartwood of *C. inophyllum*; A geographical survey. *Phytochemistry*, **10**: 603-606.

Joshi, S.P., Deodhar, V.B., and Phalgune, U.D. (2000). A new coumarin from the seeds of *C. inophyllum* Linn. *Indian Journal of Chemistry*, 39: 560-561.

Kawazu, K., Ohigashi H, Takahashi, N., and Mitsui, T. (1972). Piscicidal constituents of *C. inophyllum*. *Bulletin of the Institute of Chemical Research*, **50**: 160-167.

Kawazu, K., Ohigashi, H., and Mitsui, T. (1968). The piscicidal constituents of *C. inophyllum* Linn. *Tetrahedron Letters*, **19**: 2383-2385.

Khan, N.U.D., Parveen, N., Singh, M.P., Singh, R., Achari, B., Dastidar, P.P.G., and Dutta, P.K. (1996). Two isomeric Benzodipyranone derivatives from *C. inophyllum*. *Phytochemistry*, **42**: 1181-1183.

Kirtikar, K.R., and Basu, B.D. (1987). Indian Medicinal Plants; 2nd Ed, vol. I: Dehradun, India International Book Distributors, 270-272.

Kumar, V., Ramachandran, S., and Sultanbawa, M.U.S. (1976). Xanthones and triterpenoids from timber of *C. inophyllum*. *Phytochemistry*, **15**: 2016-2017.

Laure, F., Herbette, G., Faure, R., Bianchini, J.P., Raharivelomanana, P., and Foglian, B. (2005). Structures of new secofriedelane and friedelane acids from *C. inophyllum* of French Polynesia. *Magnetic Resonance in Chemistry*, **43**: 65-68.

Lederer, E., Dietrich, P., and Polonsky, J. (1953). On the chemical constitution of calophylloide and calophyllic acid from the nuts of *C. inophyllum*. *Bulletin of the French Chemical Society*, **5**: 546-549.

Leu, T., Raharivelomanana, P., Soulet, S., Bianchini, J.P., Herbette, G., and Faure, R. (2009). New tricyclic and tetracyclicpyranocoumarins with an unprecedented C-4 substituent. Structure elucidation of tamanolide, tamanlide D, tamanolide P from *C. inophyllum* French Polynesia. *Magnetic Resonance in Chemistry*, **47**: 989-993.

Li, Y., Li, Z., Liu, M.S., Li, D.Y., Zhang, H., and Hua, H.M. (2009). Xanthones from leaves of *C. inophyllum*. *Acta Pharmacuetica Sinica*, **44**: 154-157.

Li, Y.Z., Li, L.Z., Hua, H.M., Li, Z.G., and Liu, M.S. (2007). Studies on flavonoids from stems and leaves of *C. inophyllum*. *Journal of Chinese Materia Medica*, **32:** 692-694.

Li, Y.Z., Li, Z.L., Yin, S.L., Shi, G., Liu, M.S., Jing, Y.K., and Hua, H.M. (2010). Triterpenoids from *C. inophyllum* and their growth inhibitory effects on human leukemia HL-60 cells. *Fitoterapia*, **81:** 586-589.

Li, Z.L., Liu, D., Li, D.Y., and Hua, H.M. (2011). A novel prenylated xanthone from stems and leaves of *C.inophyllum*. *Natural Product Research*, **25:** 905-908.

Mahmud, S., Rizwani, G.R., Ahmad, M., Ali, S., Perveen, S., and Ahmad, V.U. (1998). Antimicrobial studies on fractions and pure compounds of *C. inophyllum* Linn. *Pakistan Journal of Pharmacology*, **15:** 13-25.

Malarvizhi, P., and Ramakrishnan, N. (2011). GC-MS analysis of biologically active compounds in leaves of *C. inophyllum* L. *International Journal of ChemTech Research*, **3:** 806-809.

McKee, T.C., Covington, C.D., Fuller, R.W., Bokesch, H.R., Young, S., Cardellina II, J.H., *et al.* (1998). Pyranocoumarins from tropical species of the genus Calophyllum: A Chemotaxonomic study of extracts in the National Cancer Institute collection. *Journal of Natural Products*, **61:** 1252-1256.

Mei, W.I., Zeng, Y.B., Dai, H.F., and Zheng, X.Q. (2006).Chemical constituents of the volatile oil from *C. inophyllum* Linn. *Journal of Plant Resources and Environment*, **15:** 74-75.

Mishra UM, Murthy PN, Choudhury PK, Panigrahi G, Mohapatra S, and Pradhan D. (2010). Antibacterial and analgesic effects of the stem barks of *C. inophyllum*. *Interantional Journal ChemTech Research*, **2:** 973-979.

Nair, A.G.R., and Subramanian, S.S. (1964). (±) Leucocyanidin from the seedcoat of *C. inophyllum* Linn. *Current Sciences* (India), **11**: 336-337.

Orwa, C., Mutua, A., Kindt, R., Jamnadass, R., and Anthony, S. Agroforestry Database: a tree reference and selection guide version 4.0 (http: //www. worldagroforestry.org/sites/treedbs/treedatabases.asp)2009

Parrotta, J.A. (2001). Healing Plants of Peninsular India. UK, CABI Publishing, 214-215.

Patil, A.D., Freyer, A.J., Eggleston, D.S., Haltiwanger, R.C., Bean, M,F., Taylor, P.B., Caranfa, M.J., Breen, A.L., Bartus H.R., Johnson R.K., Hertzberg. R.P., and Westley, J.W. (1993). The inophyllums, novel inhibitors of HIV-1 reverse transcriptase isolated from the Malaysian tree, *C. inophyllum* Linn. *Journal of Medicinal Chemistry,* **36:** 4131-4138.

Pawar, P., Joseph, M., Tungikar, V., and Joshi, S. (2004). Bioactivity of non-edible oil seed extracts and purified extracts against *Helicoverpa armigera* (Hubner). *Indian Journal of Experimental Biology,* **42:** 91-95.

Polonsky, J. (1957). Structure chimique du calophyllolide, de l'inophyllolide et de l'acide calophyllique, constituants des noix de *C. inophyllum*. *Bulletin de la Societe Chimique de France*, 1079-1089.

Potti, G.R., and Kurup, P.A. (1970). Antibacterial principlae of the root bark of *C. inophyllum*: Isolation and Antibacterial activity. *Indian Journal of Experimental Biology,* **8**: 39-40.

Pushpalatha, E., and Muthukrishnan, J. (1999). Efficacy of two tropical plant extracts for the control of mosquitoes. *Journal of Applied Entomology,* **123**: 369–373.

Raharivelomanana, R., Laure, F., Charles, L., and Faure, R. (2005). LC/MS analysis of neoflavonoids: application for biodiversity investigation and bioactive molecule screening. In : 11th NAPRECA Symposium Book Proceedings, Antananarivo, Madagascar; 145-153.

Ravelonjato, B., Libot, F., Ramiandrasoa, F., Kunesch, N., Gayral, P., and Poisson, J. (1992). Molluscicidal constituents of *C. inophyllum* from Madagascar: activity of some natural and synthetic neoflavonoids and khellactones. *Planta Medica,* **58**: 51-55.

Said, T., Dutot, M., Labbe, A., Warnet, J.M., and Rat, P. (2009). Ocular Burn: Rinsing and healing with ionic marine solutions and vegetable oil. *Ophthalmologica,* **223**: 52-55.

Said, T., Dutot, M., Martin, C., Beaudex, J.L., Boucher, C., Enee, E., *et al.* (2007). Cytoprotective effect against UV-induced DNA damage and oxidative stress: role of new biological UV filter. *European Journal of Pharmaceutical Sciences,* **30**: 203-210.

Sampathkumar, P.S., Murthi, V.V.S., and Seshadri, T.R. (1970). Occurance of Erythroidal-3-acetate in sap wood of *C. inophyllum*. *Indian Journal of Chemistry*, **8**: 105.

Samsudin, M.W., Ibrahim, M.N., and Said, I.M. (1996). Composition of the steam volatile oil from *C. inophyllum*. ASEAN Review of Biodiversity and Environmental Conversion (ARBEC).

Saravanan, R., Dhachinamoorthi, D., Senthilkumar, K., and Thamizhvanan, K. (2011). Antimicrobial activity of various extracts from various parts of *C. inophyllum* L. *Journal of Applied Pharmaceutical Sciences*, **1**: 102-106

Saxena, R.C., Nath, R., Palit, G., Nigam, S.K., and Bhargava, K.P. (1982). Effect of Calophyllolide, a non-steroidal anti-inflammatory agent, on capillary permeability. *Planta Medica*, **44:** 246–248.

Shaiq Ali, M., Mahmud, S., Perveen, S., Rizwani, G., and Ahmad, V.U. (1999). Screening for the antimicrobial properties of the leaves of *C. inophyllum* Linn. (Guttiferae). *Journal of the Chemical Society of Pakistan*, **21**: 174- 178.

Shen, Y.C., Hung, M.C., Wang, L.T., and Chen, C.Y. (2003). Inocalophyllins A, B and their methyl esters from the seeds of *C. inophyllum*. *Chemical and Pharmaceutical Bulletin*, **51**: 802-806.

Spino, C., Marco, D., and Subramaniam, S. (1998). Anti-HIV coumarins from Calophyllum seed oil. *Bioorganic and Medicinal Chemistry Letters*, **8:** 3475-3478.

Stout, G.H., Krahn, M.M., and Breck, G.D. (1968). Calophyllum products. II. Brasilienic and Inophylloidic acids. *Tetrahedron Letters*, **9**: 3285-3290.

Subramanian, S.S., and Nair, A.G.R. (1971). Myrecitin-7-glucoside from the andraecium of the flowers of *C. inophyllum*. *Phytochemistry*, **10:** 1679-1680.

Valsaraj, R., Pushpangadan, P., Smitt, U.W., Adsersen, A., and Nyman, U. (1997). Antimicrobial screening of selected medicinal plants from India. *Journal of Ethnopharmacology*, **58**: 75–83.

Wei, D.J., Mei, W.L., Zhong, H.M., Zeng, Y.B., Wu, X.D., and Dai, H.F. (2011). A new prenylated xanthone from the branches of *C. inophyllum*. *Journal of Asian Natural Products Research*, **13**: 265-269.

Wu, Y., Zang, P.C., Chen, R.Y., Yu, D.Q., and Lian, X.T. (2003). Two new xanthones from *C. inophyllum*. *Acta Chimica Sinica*, **61**: 1047-1051.

Xiao, Qi., Zeng, Y.B., Mei, W.L., Zhao, Y.X., Deng, Y.Y., and Dai, H.F. (2008). Cytotoxic prenylated xanthones from *C. inophyllum*. *Journal of Asian Natural Products Research*, **10**: 993-997.

Yimdjo, M.C., Azebaze, A.G., Nkengfack, A.E., Meyer, A.M., Bodo, B., and Fomum, Z.T. (2004). Antimicrobial and cytotoxic agents from *C. inophyllum*. *Phytochemistry*, **65**: 2789-2795

Zou, J., Wu, J., Liu, S.Z., and Zhao, W.M. (2010). New Coumarins and Triterpenes from *C. inophyllum*. *Helvetica Chimica Acta*, **93:** 1812-1821.

Bioactive Phytochemicals: Perspectives for Modern Medicine Vol. 2 (2014) *Pages* **231–237**
Editor: **V.K. Gupta**
Published by: **DAYA PUBLISHING HOUSE, NEW DELHI**

11

L-DOPA: An Anti Parkinsonian Agent from *Mucuna pruriens*

Archana P. Raina[1]* and M. Dutta[1]

ABSTRACT

Mucuna pruriens Linn. (Fabaceae) is commonly known as velvet bean, cowhage, kewanch and atmagupta. The plant is widely distributed throughout the India and in other parts of the tropics including Central and South America. It is an important medicinal plant used for treatment of Parkinson's disease and many others in ancient Indian Medical System. In Ayurvedic system of medicine M. pruriens was used for the management of male fertility, nervous disorders and also as an aphrodisiac. It is a constituent of more than 200 indigenous drug formulations. In India, 14 species of Mucuna are found in the foothills of the Himalayas, the plains of West Bengal, Madhya Pradesh, Karnataka, Kerala, Andhra Pradesh, Uttar Pradesh and the Andaman and Nicobar islands. The plant is reported to have L-DOPA (L-3,4-dihydroxy phenylalanine), as a major constituent in seeds, which is the intermediate precursor of the neurotransmitter dopamine. L-DOPA extracted from seeds of Mucuna is more effective drug than the synthetic counterpart. The present study was undertaken with the objective of phytochemical evaluation of M. pruriens germplasm collected from different geographical regions of India for bioactive compound L-DOPA. A sensitive and rapid densitometric high performance thin-layer chromatography (HPTLC) method was developed for quantification of L-DOPA content present in the seeds. The method involves separation of L-DOPA on precoated silica gel G 60 F 254 HPTLC plates using n-butanol-acetic-acid-water (4: 1: 1, v/v) as the mobile phase and it was quantified at 280 nm. Results indicated wide variability among M. pruriens germplasm for seed characters like seed colour, size and shape. The L-DOPA content in seeds of M. pruriens germplasm ranged between 3.29 per cent to 5.44 per cent (on dry weight basis). Promising accessions having high L-DOPA content were identified. Due to the economic importance of the drug, these promising M. pruriens collections

1 Germplasm Evaluation Division, National Bureau of Plant Genetic Resources, New Delhi, India

* *Corresponding author*: E-mail: aprraina@yahoo.co.in

can be exploited for large scale cultivation for their pharmacological important constituents mainly L-DOPA content.

Keywords: Germplasm, High performance thin layer chromatography, L-DOPA, Variability, *Mucuna pruriens*, Parkinsonian.

Introduction

Mucuna pruriens Linn. (Fabaceae) is a leguminous, twining annual, having wide distribution in tropical and subtropical areas of the world. Plant is endemic in India and in other parts of the tropics including Central and South America. It is commonly known as Kewanch or Kaunch, Atmagupta and Velvet bean (Anonymous, 1985). About fourteen species of *Mucuna* are found in the forests and plains of India. It is a popular medicinal plant of India and is constituent of more than 200 indigenous drug formulations. A screening survey from more than 1000 species has shown that the seeds of genera *Mucuna* are possible commercial source of L-DOPA (Daxenbichler *et al.*, 1971; 1972). All parts of *M. pruriens* possess valuable medicinal properties and there is a heavy demand of *Mucuna* in Indian and International drug markets (Lorenzetti *et al.*, 1998). In the Ayurveda system seed powder of *Mucuna* is used for treating Parkinson's disease and is subjected to special processing. This plant is pharmacologically important for various activities like antidiabetic, aphrodisiac, antineoplastic, antiepileptic, antimicrobial activities etc (Sathiyanarayanan and Arulmozhi, 2007). *M. pruriens* seeds contain high amount of L-DOPA (L-3, 4 dihydroxy phenylalanine), which is a neurotransmitter precursor and effective remedy for the relief in Parkinson's disease. In addition seeds contain tryptamine, 5-hydroxytryptamine (5-HT), mucunine, mucunadine, prurienine and prurieninine. Besides its medicinal uses, the plant being a legume is also important in soil amelioration, conservation and fertility improvement. It is grown as a green manure and cover crop and is also grown for its pods and young leaves, which are used as vegetable and fodder.

Plant Distribution

M. pruriens is indigenous to tropical regions, especially Africa, India, and West Indies. In India, 14 species of *Mucuna* are found in the foothills of the Himalayas, the plains of West Bengal, Madhya Pradesh, Karnataka, Kerala, Andhra Pradesh, Uttar Pradesh and the Andaman and Nicobar islands (Chopra *et al.*, 1958). It is found in bushes, hedges, and dry-deciduous, low forests throughout the plains of India.

Plant Description

M. pruriens is an annual herbaceous twining, climbing legume that grows 3-18 m in height with long, thin branches and opposite, trifoliolate, lanceolate leaves 15 to 30 cm length. Leaflets broadly ovate, elliptic or rhomboid ovate, unequal at base. Its flowers are white to dark purple and hang in long clusters or pendulous racemes, grow in racemes in 2 or 3. The fruit of the plant is pod, which is thick and leathery. It is covered with reddish-orange coloured long stiff hairs that are readily dislodged and responsible for itching in workers involved in collection of the plant. The species name "*pruriens*" (from Latin, "itching sensation") refers to the results from contact with the seedpod hairs. Pods curved, 5-10cm long, longitudinally ribbed, turgid, densely clothed with persistent pale brown or gray, irritant bristles. Seeds are black/ white, off white/brown, 4-6 in pod, ovoid (6-12 mm long) with funicular hilum.

Parts Used

Seeds, leaves and roots

Traditional Uses

M. pruriens finds traditional use in number of diseases and its various parts are used for various purposes. Root is bitter, thermogenic, emollient, stimulant, purgative, and aphrodisiac, emmenagogue, anthelmintic, febrifuge and tonic. Leaves are aphrodisiac, anthelmintic, tonic, and are useful in ulcers, inflammation and general debility. Seeds have a long history of use in Indian Ayurveda medicine, where it is used for worms, dysentery, diarrhoea, snakebite, sexual debility, cough, tuberculosis, impotence, rheumatic disorders, diabetes, and cancer. In India, it is considered an aphrodisiac, emmenagogue, uterine stimulant nerve tonic, diuretic, and blood purifier.

Chemical Constituents

M. pruriens seed powder contain high amount of L-DOPA (Figure 11.1) ranging from 2 to 7 per cent. In addition, seeds contain mucunadine, mucunine, prurienidine, prurienine, beta–sitosterol, gluatathione, lecithin, fatty acids, gallic acid, minerals and amino acids (Rastogi and Mehrotra, 1991). Seed oil contains stearic, palmitic, myristic, arachidic, oleic and linoleic acid. L-DOPA is used for treatment of human disease like Parkinson's disease and as an anti-hypertensive. L-DOPA stimulates human growth hormone released by the pituitary gland.

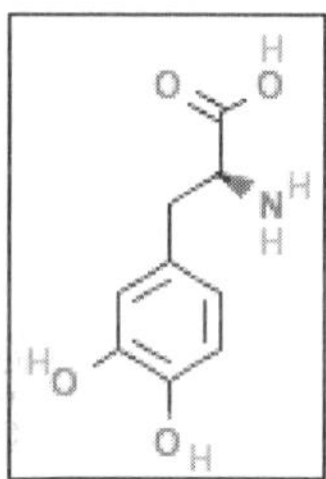

Figure 11.1: Chemical structure of 3,4-dihydroxyphenyl-L-alanine (L-DOPA).

Materials and Methods

Chemicals

Standard L-DOPA was obtained from Sigma-Aldrich. All other chemicals used in the experiment were of analytical grade. A stock solution of L-DOPA (1mg/ml) was prepared by dissolving an accurately weighed 10 mg of L-DOPA in 10 ml of 0.1N HCl in a volumetric flask. Standard working solution was prepared by diluting stock solution with 0.1 N HCl with concentration of 0.1µg/µl. HPTLC plates silica gel $60F_{254}$ (20 cm × 20 cm) purchased from E. Merck were used in the studies.

Plant Material

Seeds of thirty-eight accessions of *M. pruriens* germplasm augmented from different geographical parts of India along with three exotic accessions were used for the present study to see phytochemical variation in L-DOPA content among germplasm.

Preparation of Seed Extract

Seeds of *M. pruriens* were dried and powdered after removing seed coats. 1g of seed powder was refluxed with 50 ml 0.1N HCl on boiling water bath for 30 min. The extract obtained was filtered using whatman filter paper and the residue was again refluxed with fresh 50 ml of 0.1N HCl for another 30 min. The extract was filtered and both the filtrates were pooled to make total volume of the extract 100 ml with 0.1N HCl for estimation of L-DOPA by HPTLC.

HPTLC Analysis

A Camag HPTLC system equipped with linomat V automatic TLC sample spotter, glass twin trough chamber (20×10 cm), scanner 3 and integrated Wincats 4.03 Software was used for the analysis. The samples were spotted in the form of 8 mm bands on HPTLC plates (20×10 cm) using Linomat V under a flow of N_2 gas, 10 mm from the bottom. The linear ascending development was carried out in a twin trough chamber which was presaturated with mobile phase of n-butanol: acetic acid glacial: water (4: 1: 1, v/v) for 30 min. The plates were developed upto 8 cm and then air dried. Quantitative evaluation of the plate was performed in absorption-reflection mode at 280 nm, using a Camag TLC scanner 3 integrated with WINCATS software. Stock solution of L-DOPA (0.1µg/µl) was used in the different concentrations (100-1000ng/spot) for preparing calibration graph of peak area versus concentration. The concentration of L-DOPA in unknown sample was determined using this calibration equation.

Results and Discussion

Different combinations of the mobile phase were tried, a good resolution of L-DOPA was achieved by n-butanol-acetic-acid-water (4: 1: 1, v/v) mobile phase with chamber saturation for 30 min. A peak corresponding to pure L-DOPA was seen in sample at R_f value of 0.35 ± 0.02 at 280nm. It was confirmed by comparing spectra of the spot with that of the standard L-DOPA. The method was found highly specific as it well resolved L-DOPA in the presence of other compounds in the extract. A linear calibration equation was obtained for L-DOPA in the concentration range of 100-1000 ng/spot with the correlation coefficient of 0.9980. Results of regression analysis on calibration curve and other method validation parameters are presented in Table 11.1.

Table 11.1: Results of method validation parameters for quantification of L-DOPA by HPTLC.

Parameter	*Results*
R_f	0.35 ± 0.02
Linearity range (ng/spot)	100-1000
Regression equation	Y= 404.073 + 2.762* X
Slope	2.762
Intercept	404.073
Limit of detection (LOD)	3.41 ng
Limit of quantification (LOQ)	10.35 ng
Linearity (correlation coefficient)	0.9980
Repeatability (CV, per cent) (n=6)	1.121 per cent

Figure 11.2: Variability in seeds of *Mucuna pruriens* germplasm collected from different agro-climatic regions of India.

A preliminary screening of *M. pruriens* germplasm showed large variation in seed characters. The seed colour varied from off white to brown, brown black, black (Figure 11.2). The length of the seeds ranged from 0.80 to 1.70 cm, thickness 0.4 to 1.1 cm and width from 0.5 to 1.3 cm, 100 seed weight from 20.8 to 170.4 g was recorded. The L-DOPA content in seeds of *M. pruriens* germplasm was found to vary significantly between 3.29 per cent to 5.44 per cent in dry seeds with mean L-DOPA content of 4.53 per cent and standard deviation of 0.57 (Figure 11.3). Promising accessions identified for high L-DOPA content were IC551549 (5.44 per cent), IC551514 (5.41 per cent), IC331536 (5.40 per cent), IC552857 (5.36 per cent), IC385844 (5.32 per cent) and IC439197 (5.32 per cent). Present studies indicated L-DOPA content in Indian germplasm ranged from moderate to high levels similar to other reports (Bell and

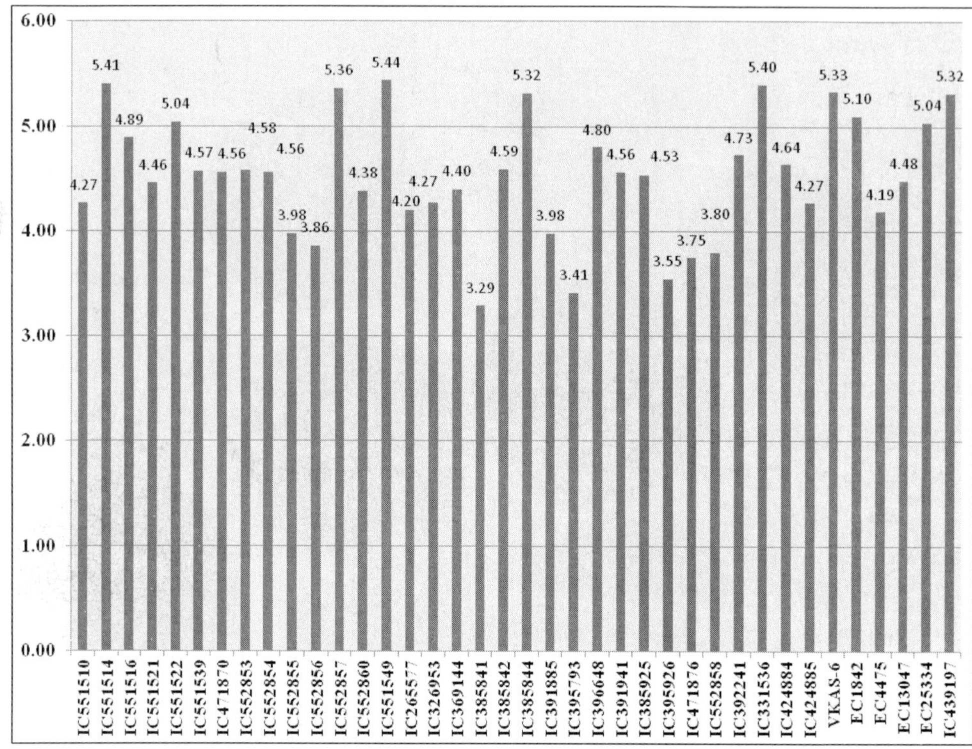

Figure 11.3: Variation in L-DOPA content (per cent) in seeds of *Mucuna pruriens* germplasm.

Jansen 1971; Prakash and Tiwari, 1999; Singh *et al.*, 2008). All these high L-DOPA yielding *M. pruriens* accessions can be exploited for extraction of L-DOPA contents.

Conclusion

There is a high demand for the L-DOPA in drug industry which has stipulated considerable interest in the search for a rich source of this drug. Therefore, promising collections of *M. pruriens* identified for high L-DOPA content in the present studies can be exploited for their pharmacological properties. The HPTLC method developed here for the quantification of L-DOPA in *M. pruriens* seeds allows a large number of samples to be measured simultaneously with a very good accuracy, sensitivity, and precision. It is simple, rapid, cost-effective and easily adaptable for screening and quantitative determination of L-DOPA in large number of *M. pruriens* germplasm. This method can also be used for the quantitative determination of L-DOPA in herbal extract, its formulations and can serve effective tool for quality control.

References

Anonymous (1985). *The Wealth of India*. Raw Materials. Vol 6, CSIR, New Delhi, p. 442

Bell, E.A., and Jansen, D. S. (1971). Medical and ecological considerations of L-DOPA and 5-HTP in seeds. *Nature*, **229**: 136-137.

Chopra, R.N., Chopra, I.C., Handa, K.L., and Kapoor, L.D. (1958). Indigenous Drugs of India. 2nd ed. Calcutta, Academic Publishers.

Daxenbichler, M.E., Vanetten, C. H., and Hallinan, F.A. (1971). Seeds as sources of L-DOPA. *Journal of Medicinal Chemistry,* **14**: 463-466.

Daxenbichler, M.E., Vanetten, C.H., Earle, R.E., and Tallent W.H. (1972). L-DOPA recovery from *Mucuna* seed. *Journal of Agricultural Food Chemistry,* **20**: 1046-1048.

Farooqi, A.A., Khan, M. M., and Sundhara, M.A. (1999). *Production Technology of Medicinal and Aromatic Crops,* Natural Remedies Pvt. Ltd., Bangalore, India, pp 26-28.

Lorenzetti, F., Maclsaac, S., Arnason, J.T., Awang, D.V.C., and Buckles, D. (1998). The phytochemistry, toxicology and food potentials of velvet bean (*Mucuna adans. spp fabaceae*). In: Buckles D, Eteka A, Osiname O, Galiba M and Galiba G (Eds) *Cover Crops in West Africa: Contributing to Sustainable Agriculture.* IDRC, Ottawa, Canada pp. 67-84

Rastogi, R.P., and Mehrotra, B. N. (1994). Compendium of Indian Medicinal Plants. Vol 5. CDRI, Lucknow, p.554.

Sathiyanarayanan, L., and Arulmozhi, S. (2007). *Mucuna pruriens* Linn.–A comprehensive review. *Pharmacognosy Review,* **1**: 157-162.

Mennickent, S., Nail, M., Vega, M., and Dediego, M. (2007).Quantitative determination of L-DOPA in tablets by high performance thin layer chromatography. *Journal of Separation Science,* **30**: 1893-1898.

Modi, K. P., and Patel, N.M. (2008). Estimation of L-DOPA from *Mucuna pruriens* Linn. and formulations containing *Mucuna pruriens* by HPTLC method. *Chemical Pharmaceutical Bulletin,* **56**: 357-359.

Prakash, D., and Tewari, S. K. (1999). Variation of L-DOPA contents in *Mucuna* species. *Journal Medicinal Aromatic Plant Sciences,* **21**: 343-346.

Singh, A.K., Malik, S.S., and Tomar Y.S. (2008). Studies on medicinal compound L-DOPA in *Mucuna pruriens* Bak. *Indian Journal Plant Genetic Resources,* **21**: 217-220.

Bioactive Phytochemicals: Perspectives for Modern Medicine Vol. 2 (2014) *Pages* **239–293**
Editor: **V.K. Gupta**
Published by: **DAYA PUBLISHING HOUSE, NEW DELHI**

12

Biological Effectiveness and Recent Advancing of Natural Products on the Discovery of Anticancer Agents

Maria Aparecida M. Maciel[1,2*], Fabiano E. S. Gomes[2], Breno Almeida Soares[2], Noema Fraga Grynberg[3], Aurea Echevarria[3], Ilce Mara S. Cólus[4], Carlos R. Kaiser[5], Waldenice A. Morais[6] and Nereide S. Santos Magalhães[7]

ABSTRACT

This topical article presents a brief historic of bioactive natural products against cancer. In that, some aspects of current research in pharmacological and biochemical profiling of lead compounds obtained from traditional medicinal plants are herein presented. Emphasize some important natural anticancer agents such as vinca alkaloids, epipodophyllotoxis, taxanes,

1 Universidade Potiguar, Área da Saúde, Campus Salgado Filho, 59075-000, Natal, RN

2 Instituto de Química, Universidade Federal do Rio Grande do Norte, Campus Universitário, 59078-970, Natal, RN, Brazil

3 Departamento de Química, Instituto de Ciências Exatas, Universidade Federal Rural do Rio de Janeiro, 23851-970, Seropédica, RJ, Brazil

4 Departamento de Biologia Geral, Centro de Ciências Biológicas, Universidade Estadual de Londrina, 86051- 970, Londrina, PR, Brazil

5 Instituto de Química, Universidade Federal do Rio de Janeiro, Centro de Tecnologia, Cidade Universitária, 21945- 970, Rio de Janeiro, RJ, Brazil

6 Departamento de Farmácia, Prédio da Faculdade de Farmácia, Universidade Federal do Rio Grande do Norte, Campus Universitário, 59012-1570, Natal, RN, Brazil

7 Departamento de Bioquímica, Laboratório de Imunopatologia, Keizo-Asami (LIKA), Universidade Federal de Pernambuco, Cidade Universitária, 50670-901, Recife, PE, Brazil

* *Corresponding author*: E-mail: mammaciel@hotmail.com

campothecins among other, as well as present a general chemotherapeutic approach, including mutagenic investigation. Moreover, the bioavailability of these chemotherapic agents into drug delivery systems are herein discussed aiming to present both evolutionary perspectives and how far some important commercialized drugs become. Additionally, the very efficient Brazilian medicinal plant Croton cajucara Benth, is cited carrying the responsibility for answering in sequential fashion critical questions on both chemical aspects and anticancer properties. In this context, describes results on the chemotherapic effectiveness of the 19-nor-clerodane trans-dehydrocrotonin (DCTN) isolated from this Croton and also presents insights into the use of DCTN and its semi-synthetic derivatives as a therapeutic agent as well as the potential advantages of DCTN incorporation into drug delivery systems.

Keywords: Medicinal plants, Natural products, Antitumoral and Antimutagenic agents, Drug delivery systems, Trans-dehydrocrotonin.

Introduction

Currently, the search for chemopreventive agents acting against human cancer development has become an important high-priority subject and has been improved by development of bioassay methods and bioavailability into drug delivery systems. In order to improve the quality of medicinal prolonged treatments (safely administration over long periods) the continuing progress represent a rich tool for antitumor (or antineoplastic), cytotoxic and anticancer new drug discovery, in which nanoformulation systems has been largely utilized. Regarding cancer terminology an old description suggested by Suffness and Dours (1982) become appropriated, anticancer refers to compound that is active against human cancer, cytotoxicity for compound (or plant extract or bioactive plant fraction) that shows activity against tumor cell lines, and antitumor (or antineoplastic) when the effectiveness of the tested materials is correlated to *in vivo* experiments. Taking in account those above comments, in this present topical article, the authors emphasize the progress on the pharmacological effects anticancer chemotherapies. For that, some of the most important anticancer agents will be distinctly pointed, including its significant role in the bioavailability and development into drug delivery process aiming the safety human drug administration. Additionally, highlights pharmacological procedures on chemotherapy research and gives information on the scientific results of the some medicinal plants around the world, which includes a native plant from the Amazon region of Brazil named *Croton cajucara* Benth (Euphorbiaceae). Since its major natural compound *trans*-dehydrocrotonin (DCTN) [a 19-*nor*-clerodane diterpene-type isolated from the steam bark of *Croton cajucara*] presented large pharmacological importance including cytotoxic, antitumor and antimutagenic effects, will be included herein. Modified related DCTN clerodanes compounds, as well as the nanoformulation systems progress performed with DCTN are also considered.

Partial Survey on the Bioactive Natural Product

The information on the use of medicinal plants and their therapeutic values has been accumulating over the centuries, and several vegetal species were incorporated to medicine barely by chance. For a long time, medicinal plants established the grounding in therapy. Pure substances, isolated or synthesized, had just been incorporated to the medicinal achievements in the middle of the last century, when they were named pharmaceuticals. In the 16th century, the Swiss doctor, Paracelsus was the first to propose the extraction of bioactive natural products from vegetal sources, originating the idea of the bioactive principle. However, the identification process of these substances only started in the 18th century, and consequently, the transition from medicinal plants to modern pharmaceutical products (Samuelsson, 1999; Maciel *et al.*, 2007a; Mehta *et al.*, 2010). Just quoting some examples, one of the first substances which were isolated from a plant was the morphine (analgesic), followed by aconitine and emetine (amebacide/emetic); atropine (anticholinergic); digitoxin (cardio tonic); vincristine (antitumor); reserpine (antihypertensive); quinine and artemisinine (antimalarial) (Farnsworth and Morris, 1976). Nowadays, many of these substances are synthesized and traded in a large scale, yet generally speaking, natural medicines had been gradually substituted for analogues or semi-synthetic derivatives along the years (Walton *et al.*, 2003). A well known example is the acetylsalicylic acid (aspirin), a substitute for the salicylic acid (Figure 12.1), which was isolated from the "White-Willow" (*Salix alba*); a commonly used specie in the malaria treatment during the 19th century. Since 1997, aspirin compound turned 100 years and become the most well known semi-synthetic compound and bought medicines in the world. More than 30 billion aspirin tablets a year are consumed in the U.S.A. The *Salix alba* pharmacological efficiency is very well documented; starting with Hypocrites (the Greek doctor and philosopher regarded as the father of modern medicine), in the 5th century BC. This philosopher used to prescribe preparations made from vegetal species, among them the "White-Willow" (*Salix alba*) leaves and barks were used to treat higher temperatures (fever) and to relieve birth pain. Another remarkable doctor from the ancient times was Dioscorides, who lived in Greece in the 1st century AD, prescribed plasters made from white willow barks and leaves for rheumatic pain treatment. Additionally, he was the author of "De Materia Medica", in which more than 600 medicinal plants were described (Agosta, 1997; Samuelsson, 1999; Shoeb, 2006; Maciel *et al.*, 2007a).

Taking in account that Paracelsus pointed out in the 16th century that "All substances are poisons; there is none which is not poison. The right dose differentiates a poison from a remedy". Thus, it should be kept in mind that the dose of any natural product determines the mode of action and the doses makes a poison. As an example, low-doses of oleanolic acid showed hepatoprotective effect [60-90 mg/day, for 30 days/in 70 cases of clinical trial or long-term use (more than 3 months)/in 188 cases of chronic hepatitis] (Xu and Wan, 1980; Xu, 1985). Meanwhile, the high-dose could produce cholestasis and hepatotoxicity (Liu, 2005). Because of this notably hepatoprotection oleanolic acid was marketed in China for human liver disorders (Liu, 1995).

Regarding to terpenoid-type natural product, during the last decades over 900 research articles have been published on ursolic acid (UA) as well as its isomer oleanolic acid (OA), reflecting tremendous interest and progress in their pharmacological importance, including in those studies UA and OA chemical modifications to make these natural products more effectives and water soluble

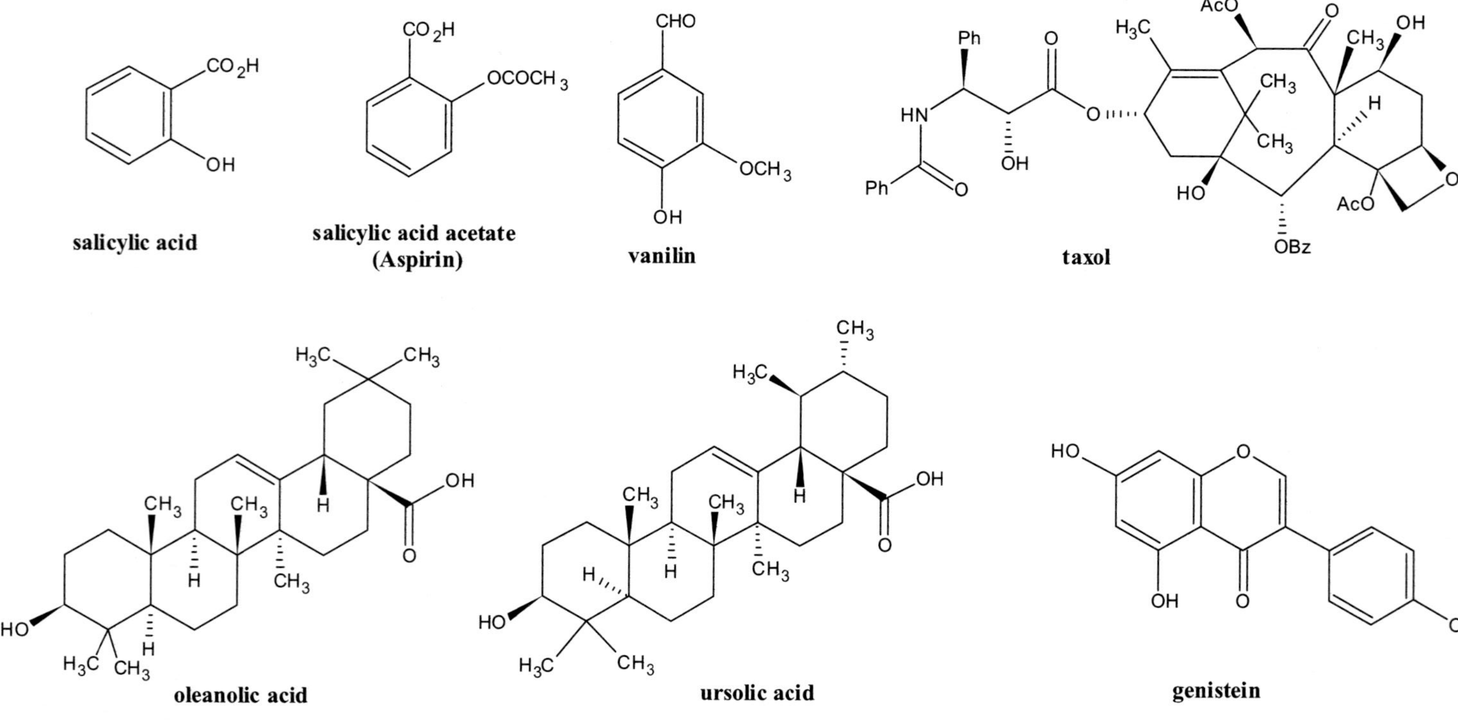

Figure 12.1: Chemical structures of some bioactive natural products.

derivatives (Maciel *et al.*, 2010a). Ursolic acid, a natural pentacyclic triterpenoid carboxylic acid, is almost the major component of some traditional medicine herbs and it is well known to possess a wide range of biological functions, as well as oleanolic acid. For both UA and OA pharmacological research include pre-clinical, toxicity evaluations and human clinical use on the treatment of various diseases. Among many important pharmacological effects anticancer chemotherapies, hypolipidemic, hypoglycemic, and hepatoprotective effects; antiatherosclerotic and antidiabetogenic properties; cardiovascular protective effect; gastrointestinal transit modulating activity; antiulcer, antiinflammatory, antibacterial, and antifungal activities; additionally to immunomodulatory, and diuretic properties were all confirmed (Huang *et al.*, 1994; Liu *et al.*, 1994; 1998; Liu, 1995; 2005; Sattar *et al.*, 1995; Hsu *et al.*, 1997; Jeong *et al.*, 1999; Latha and Panikkar, 1998; 1999; 2001; Li *et al.*, 1999; 2002; Mizushina *et al.*, 2000; Tang *et al.*, 2000; Yoshikawa and Matsuda, 2000; Marquina *et al.*, 2001; Mix *et al.*, 2001; Álvares *et al.*, 2002; Somova *et al.*, 2003; Murakami *et al.*, 2004; Vechia *et al.*, 2009; Ikeda *et al.*, 2008; Maciel *et al.*, 2010a; Sultana, 2011). Additionally, the potent inhibitory effect of ursolic acid against HIV-1 protease (with IC_{50} values of 8.0 μM) was reported by Min *et al.* (1999) and confirmed by Ma *et al.* (2000) and Cassels and Asencio (2011). Kashiwada *et al.* (2000) confirmed anti-HIV effect for its derivatives. In contrast to those beneficial properties, it was considered that UA could induce pro-inflammatory effect, on normal cells or tissues. Meanwhile, comprehensive studies of the pro-inflammatory action mechanisms of this terpenoid are poorly investigated (You *et al.*, 2001; Ikeda *et al.*, 2005, 2006, 2007).

Both ursolic acid and oleanolic acid exist widely in plant kingdom, medicinal herbs, and also are integral part of the human diet. These two triterpenes may occur in the form of free acids (Figure 12.1) or aglycones of saponins (Li *et al.*, 1999; Marquina *et al.*, 2001; Janicsák *et al.*, 2006; Kowalski, 2007; Razboršek *et al.*, 2008), which represent a source of materials for the pharmaceutical industry. The fruiting spikes of *Prunella vulgaris* L. (Labiatae), known as "Hsia ku Tsao" in Chinese folklore, are used as herbal remedies for human treatment of tuberculosis, jaundice, infectious hepatitis, bacillary dysentery, pleuritis with effusion, and cancer (Zurcher *et al.*, 1954; Lee *et al.*, 1988). The triterpenoids UA and OA, among other constituents, were isolated from the whole herb. From this other herbal remedies "Khin Piah Leng" obtained from *Psychotria serpens* (Rubiaceae) and/or *Hyptis capitata* (Labiatae), a specific bioassay-directed fractionation led to the isolation and characterization of ursolic acid (UA) as one of the bioactive principle of this herbal remedies (Zurcher *et al.*, 1954; Lee *et al.*, 1988).

Continuing discovery has also been facilitated by the recent development of new bioassay methods and become an important high-priority subject. Indeed, natural products should be safely used, including food components taken in daily life. In this context, in addition to ursolic acid and oleanolic acid, the natural compound vanilla is other significant example. In fact, about 12.000 ton per year of vanilla are consumed as food additive all over the world. Just 1 per cent corresponds to the consumption of the natural product (*Vanilla planifolia* Andrews). The synthesized product costs 15 dollars/kg while the natural vanilla price ranging from 1.200 to 4.000 dollars/kg (Farnsworth and Morris, 1976). Citing other examples of widely known integrant of human diet acting as chemopreventive agents there are retinol (vitamin A), α- and β-carotenes, omega-3 fatty acids, polyphenol-type compounds such as epigallocatechin-3-gallate, epigallin, epicatechin and tocopherols present in the green, black and black raspberry tea (Malone, 1991; Murakoshi *et al.*, 1992; Yamane *et al.*, 1995; Fukutake *et al.*, 1996; Metha *et al.*, 2010; Zhang *et al.*, 2011). In other hand, within the sphere of cancer chemotherapy, most of the available drugs, injure rapidly dividing normal cells and, therefore, have substantial side effects. Nowadays, some approved drugs are being abandoned for their unexpectedly low-response rates or undesirable side effects when administered to patients. The natural product retinol, as an example, was not as effective as *trans*-retinoic acid which was associated to possess teratogenic properties and toxicity (Metha *et al.*, 2010; Napoli, 2011). In a large context natural substances with complex chemical structures, which make their synthetic production impractical, usually need the isolation from the natural source for a later medicinal use, mostly of them with low isolation yield made the clinical use difficult. This is the case of taxol (Figure 12.1), a compound with significant anticancer activities, initially isolated from the *Taxus brevifolia* Nutt and later from several *Taxus* species. In 1977 it was chosen as a strong candidate for the clinical use, resulting from its high antitumor activity against the B16 melanoma. However, its low isolation yield (varying from 0.00003 to 0.069 per cent), difficult taxol clinical prescription, since 27.300kg of barks were necessary to produce 1.900g of taxol, involving 6.000 units of trees in the gathered process (Kingston, 1991; Nadeem *et al.*, 2002). Despite of its limitation, in December, 1992 the natural product taxol was approved by FDA for ovarian cancer treatment; and in 1994 for breast cancer treatment (Holmes *et al.*, 1991; Potier, 1992; Borman, 1994; Nadeem *et al.*, 2002). At that time, it was estimated that 25kg of the drug would be necessary to be distributed in clinical centers in the U.S.A (Holmes *et al.*, 1991; Potier, 1992; Borman, 1994). Owing to the clinical importance of taxol, researchers from all over the world contributed to the synthetic studies of this molecule. This scientific effort has been evidentially noticed during the last 20 years, when about 40 research groups have been working on the partial or total synthesis of taxol, as well as other natural products which belong to the taxan class (Hongjie *et al.*, 1994; Ma *et al.*, 1994; Voronkov *et al.*, 2003; Horiguchi *et al.*, 2003a,b; Itokawa *et al.*, 2008; Fu *et al.*, 2009; Ma and Wang, 2009; Nobili *et al.*, 2009; Ndungu *et al.*, 2010; Paterson *et al.*, 2010, Nikolic *et al.*, 2011, Mendoza *et al.*, 2012). Several clinical studies showed that taxol is an effective chemotherapy agent for various types of cancer (ovarian, breast, lung, brain, throat, prostate, among others) (Kris *et al.*, 1986; Chang *et al.*, 1993; Murphy *et al.*, 1993; Suffness, 1993; Borman, 1994; He *et al.*, 2003; Bajaj, 2010; Kosmidis, 2011). After researches in many kinds of cancer, the combination of taxol with other chemotherapy agents, as well as in various formulas, brought about new options of treatment. Nowadays, the trade of paclitaxel^{Ò} [the first commercial product, a nutritional supplement of taxol/Bristol-Myers-Squibb (isolated from the barks of *Taxus brevifolia*)] and its docetaxel^{Ò} derivatives (taxotere/Rhone-Poulenc Rorer) are still depending on natural sources (Maciel *et al.*, 2007a).

Medicinal advances of semi-synthetic derivatives, obtained from natural products which are clinically used in the treatment of many cancer types, are evidenced by the taxol derivatives, and also by the following chemotherapics: etoposide and teniposide (*Podophyllum peltatum*); irinotecan, topotecan and camptothecin (*Camptotheca acuminate*); vinblastine and vincristine (*Catharanthus roseus*); and homoharringtonine (*Cephalotaxus harringtonia*) (Figure 12.2) (Visani *et al.*, 1997; Keller *et al.*, 2001; Hitt, 2002; Castro *et al.*, 2003; Cragg and Newman, 2005; Kudoh *et al.*, 2005; Miller *et al.*, 2005; Quoix *et al.*, 2005; Reddy *et al.*, 2005; Ziotopoulos *et al.*, 2005; Shoeb, 2006; Itokawa *et al.*, 2008; Ma and Wang, 2009; Nobili *et al.*, 2009; Ngo *et al.*, 2010; Zhang *et al.*, 2011; Gao *et al.*, 2012; Mendoza *et al.*, 2012). The relevant use of these chemotherapics is exemplified by a research published in the Journal of the American Medical Association (JAMA). It was quoted that 42 per cent of Americans had used at least one alternative therapy in 1997, while in 1990 they were only 34 per cent (Solomon *et al.*, 2002). In other hand, among HIV-infected patients, the current percentage has reached the relevant index of 78 per cent (WHO, 2002). During the period of 1987 to 1996, the U.S. National Cancer Institute (NCI) provided the appropriate infrastructure development for the execution of studies searching for drug discovery against HIV (WHO, 2002). Many natural products obtained from plants, animals, micro-organisms or from substances produced through enzymes or by semi-synthesis, were able to inhibit this virus (Yang, 2001; Balunas and Kinghorn, 2005; Cassels and Asencio, 2011). In other hand, cancer is the predominant in both developed and developing countries, in which was evidenced 10.9 million new cases, 24.6 million persons living with cancer and 6.7 million deaths around the word (Parkin *et al.*, 2005). Indeed, NCI collected about 35.000 plant samples from 20 countries and has screened around 114.000 extracts for anticancer activity (Shoeb, 2006).

The search for metabolites with anticancer activity in plants, led to the discovery of potential antiestrogenic substances, which have been used in the breast cancer treatment. Among all the bioactive natural products that show estrogenic activity, the steroid alkaloid, coumestan and isoflavone stand out (Laekeman, 1986; Saito *et al.*, 1993; Harbone, 1994; Pupo *et al.*, 2002; Chaves *et al.*, 2004; Lorand *et al.*, 2010). Setting an example for flavonoid class, a long time ago, a curious event happened to a flock of lambs from Western Australia, the flock had been fed a clover named *Trifolium subterraneum*. The animals, independently if male or female, started to have pregnant disturbance, developing mammary glands and milk secretion. This estrogenic activity was attributed to the presence of an isoflavone, the genistein (Figure 12.1), which was part of the clover chemical composition (Bradbury and White, 1951). At the present time, it is known that soybeans contain a relatively high percentage of this substance (about 3mg/g), and it also shows a significant pharmacological action (prevents cancer and performs immune protection and cardio protection, antimutagenic, antioxidant and antiosteoporosis action) (Maciel *et al.*, 2007a). The broad bioactivity of genistein has been attributed to its estrogenic action (Barnes, 1995; Fukutake *et al.*, 1996; Miyazawa *et al.*, 1999; Setchell and Cassidy, 1999; Conklin, 2000; Nakagawa *et al.*, 2000; Polkowski and Mazurek, 2000; Arliss and Biermann, 2002; Dixon and Ferreira, 2002; Klein and King, 2007; Boeheme *et al.*, 2009). As this isoflavone, other flavones have been showing benefits to human health, such as bactericide, antifungal, anticarcinogenic and antimutagenic activities, besides the immune protection, antiallergic and antiatherosclerosis effects (Brandi, 1997; Setchell and Cassidy, 1999; Polkowski and Mazurek, 2000; Yao *et al.*, 2011).

Figure 12.2: Chemical structures of some anticancer agents.

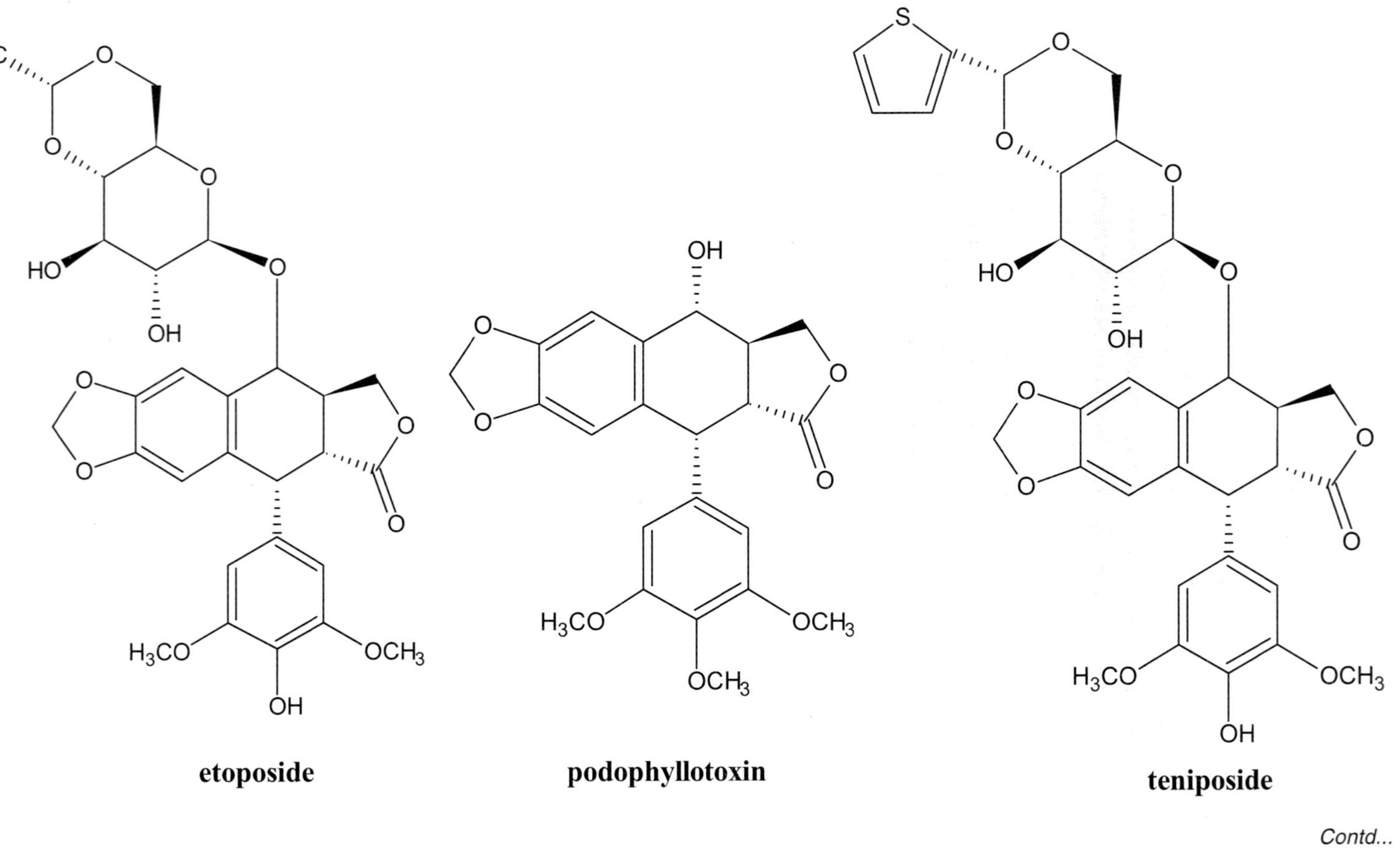

Contd...

Figure 12.2–*Contd...*

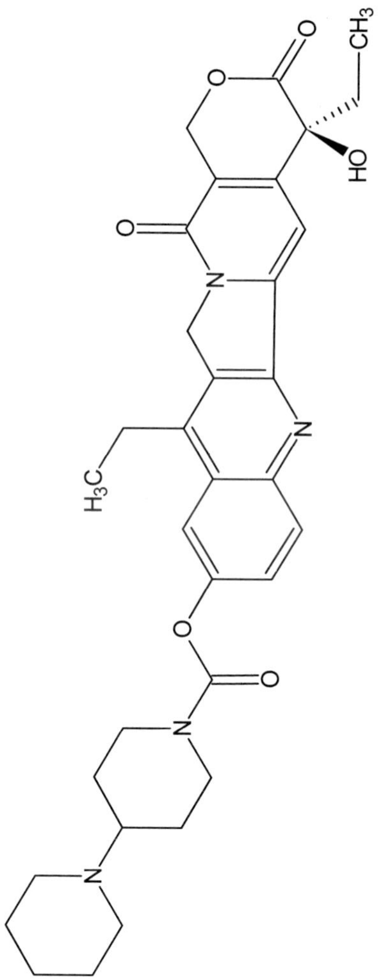

irinotecan

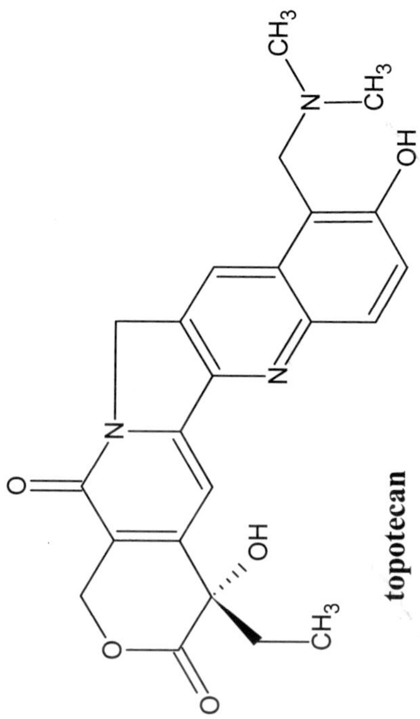

topotecan

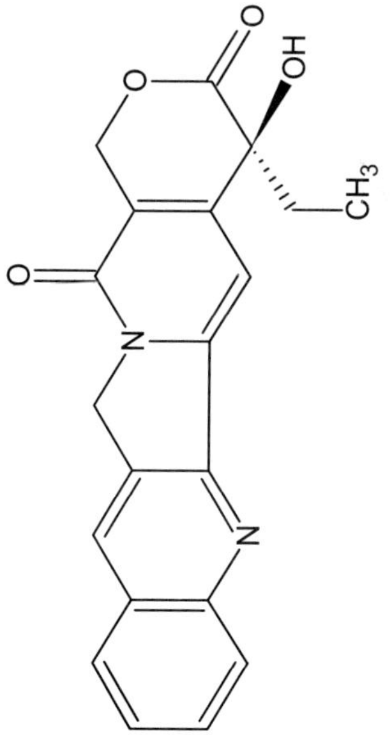

camptothecin

Contd...

Figure 12.2–*Contd...*

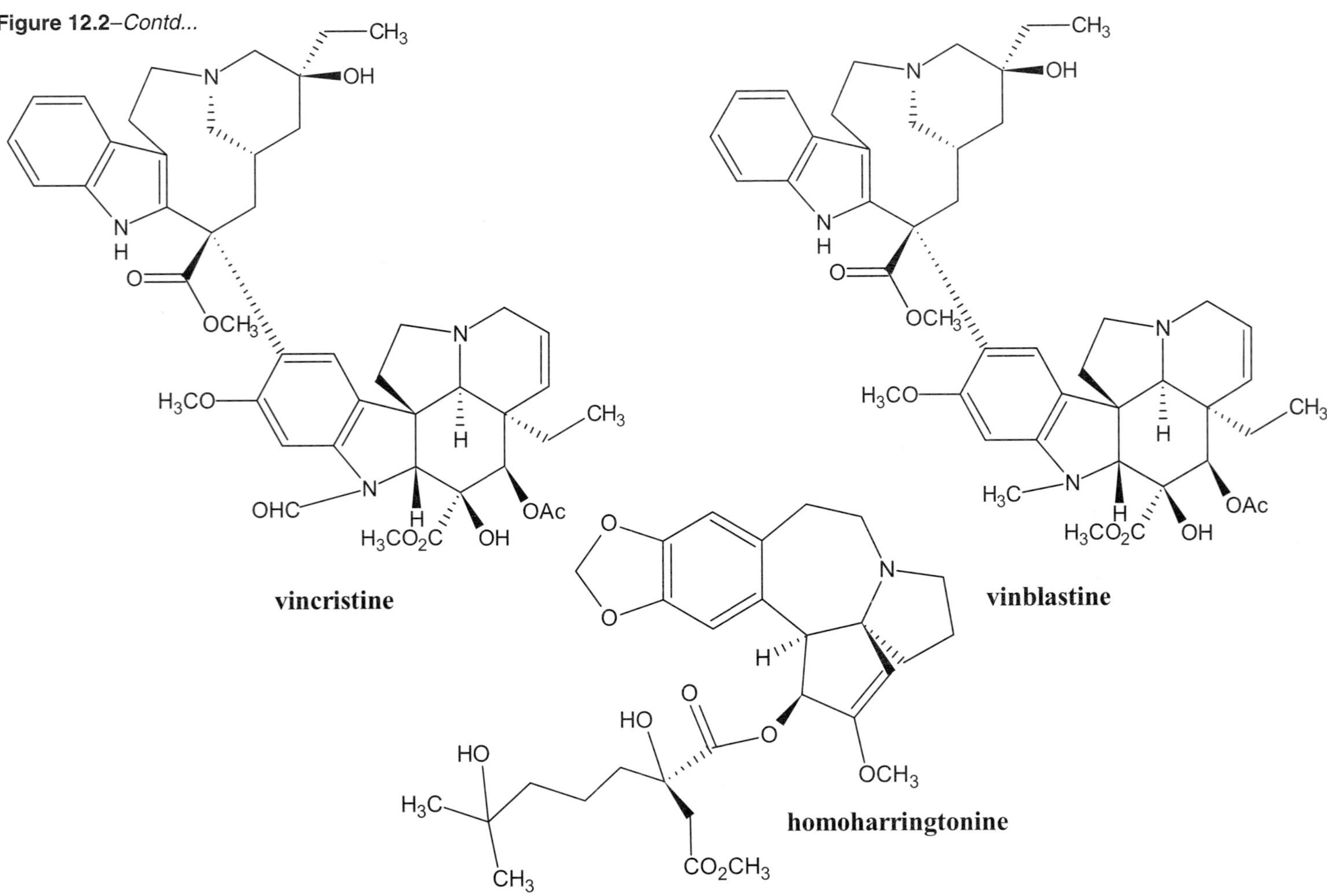

Whenever the subject is medicine related to natural product for cancer treatment, there are many substances with great chance to become market leaders, since they may produce semi-synthetic derivatives and/or synthetic products (from total or partial synthesis), it is the case of colchicines (*Colchium autumnale*), etoposide, vinblastine and vincristine (Figure 12.2) (Kinghorn and Seo, 1996; Gidding *et al.*, 1999; Itokawa *et al.*, 2008; Ma and Wang, 2009; Nobili *et al.*, 2009; Ngo *et al.*, 2010). The pharmacological industry has demonstrated its extreme interest in the use of efficient natural chemotherapics, but after facing the complexity of these metabolites molecular structures, almost impossible to be synthesized under profitable costs, semi-synthetic analogues have been chosen. These derivatives have been currently inserted with great success, substituting the originally natural ones, since they can be less toxic, more active or, simply, easier to be acquired. Vumon, for example, has been recently introduced in leukaemia treatment (Kinghorn and Seo, 1996). It is an analogue of podophyllotoxin, which is related to the discovery of etoposide and teniposide (Figure 12.2), as well as etopophos (Imbert, 1998; Castro *et al.*, 2003; Chunqi *et al.*, 2010). The chemical modifications that led to etoposide and its derivatives, led also to a change in the mechanism of action. In fact different from podophyllotoxin which is an antimicrotubule agent, etoposide and derivatives were antitopoisomerase II (Castro *et al.*, 2003). Vinorelbine (or navelbine), a derivative of vinblastine and vincristine bis-indolic alkaloids, has been used as an antitumor agent for about 40 years; it was developed for the specific treatment of lung tumors and for the treatment of canine mast cell tumors (MCT) (Kinghorn and Seo, 1996; Grant *et al.*, 2008). However, among the best succeeded pharmacological products for cancer treatment, taxol (paclitaxel) (Figure 12.1) stands out by its efficiency to cure various types of cancer, together with camptothecin (Figure 12.2), one of the firstly, secondary metabolites from plants, inserted into the American market in 30 years (Kinghorn and Seo, 1996). The process of development for these chemotherapics and their derivatives clinical use dates back to the beginning of 1960's, until their consolidation in the 1990's (Cragg and Newman, 2005). Another example of an efficient semisynthetic compound is the epothilones derivatives, which also prevent cancer cells from dividing by interfering with tubulin. In early trials epothilones have better efficacy and milder adverse effects than taxanes (Altmann *et al.*, 2011).

In spite of the huge medical *viability* confirmed to bioactive substances obtained from vegetal species, in 1991 it was estimated that 12 years of research were necessary until FDA could approve their commercialization. It represents a 231 million dollars research cost (Vagelos, 1991). According to an early estimation, for each 10.000 synthesized and evaluated substances in *in vitro* pharmacological assays, only 20 will be tested in *in vivo* experiments and, half of them, in human assays. From this number just one might be approved by FDA (Vagelos, 1991).

General Procedure for the Search of New Chemotherapy Agents

Medicinal plant extracts are source for several chemical constituents; they can show synergic effects when tested, because of the different bioactive principles from their chemical composition. The isolation of components present on the extracts plants, the discovery of their activities, as well as their mechanisms of action, are still one of the greatest challenges for phytochemistry, pharmacology, biochemistry, and medicinal chemistry researchers. One of the first steps to study the biological activity of the plants involves the bioassay selection, aiming at the detection of specific biological effects, related to the therapeutic indication of each species. The presence of different class substances or the same one, that presents small structural variations, can usually contribute to the same biological activity. In this way, an extract can show certain pharmacological action, and its isolated components can be less active (higher doses for a significant biological response) or less inactive. Biological assays must be simple, sensitive and reproductive; they can involve inferior organisms (micro-organisms and micro-crustaceans, for example), biochemical assays aiming molecular targets (enzymes and receptors) and human or animal cell culture. Any research aiming bioactive drug development needs appropriate models for the identification of molecular targets, which are fundamental for the cellular growth, *in vivo* or *in vitro*. Among the mainly intra-cellular targets, there is the DNA, RNA, microtubules and enzymes.

Cancer therapy is usually associated to adverse effects and, historically, chemotherapy agents have been selected based on their toxicity, which does not consider their sophisticated intervention in the specific biology of the tumor. New chemotherapy agents are evaluated based on their toxic effects in cancer cell lines and the fast-growth tumors in rodents. This strategy leads to agents which present toxicity to normal cells and tissues, and it shows tumor cell characteristics, such as high cellular turn-over. Therefore, there is a promising future in this research area, since there is a great interest in the development of therapies that protect the normal tissue against the chemotherapy agent effects, avoiding an anticancer action interface. Nowadays, an important number of antimutagenic drugs, from synthetic or natural origin, are in various stages of clinical selection as chemoprotective agents. In these studies, tests of short duration in eukaryotic and prokaryotic organisms have been shown to be essential, since some results led to clinical selections, in which the human use is carefully being carried out (Mitscher *et al.*, 1996; Giri *et al.*, 1999; Hammons *et al.*, 1999; Lohmam, 2002; Reinchenbach and Hofle, 2008; Mhaidat *et al.*, 2009; Caneguim *et al.*, 2011).

Mutagenic and Antimutagenic Activities

DNA is the precursor macro-molecule of all life characteristics. It is the informational component of all living beings, except for some virus which have the ribonucleic acid (RNA) as genetic material. In this way, from bacteria to a human being, the DNA has the same characteristics; the mechanisms used to stock information and gene expression are very similar in all organisms. During the evolution of every life form, the DNA had passed for mutations that led it to the optimization of the reproductive capability in our ancestral environment. However, there are some mutations with harmful effects, since they can involve relatively permanent alterations in the DNA linear sequence (Shamon and Pezzuto, 1994). And when the gene composition is changed, it can result in an inactive gene product, or maybe even in a cellular death (Brusick, 1987). Mechanisms that induce mutagenic processes vary from indirect (transitions, transversions and frameshift mutations, among other) to those which lead to an error production, during the DNA replication or repairing processes. Ionic radiations and chemical substances with genotoxic action are examples of mutagenic agents. These substances have chemical and physical properties in common, which allow interactions with the nucleic acids (Shugart, 1995). Because of their high level of reactivity, they can lead to hereditary defect through germinative cell mutations (Mitchelmore and Chipman, 1998). The most usual consequence, when the mutation occurs in the material, is the formation of benign or malign tumors, or even the development of other diseases, such as arterioscleroses; cellular death, that can induce to the precocious aging; teratogenic effects or bad formations, besides abortions (Bridges, 1987, 1990, 1991; Penn, 1990; Hansen, 1990; Wakabayashi, 1990). The process of carcinogenesis in which normal cells become malignant, is quite complex. In a simplified manner, the process is accompanied by changes in structure and function of genic information coded in the DNA. Moreover, multiple signaling pathways become deregulated during the development of cancer (Mehta *et al.*, 2010). Somatic mutations are the primary cause of cancer development in patients of all ages and in all tissue types; this process explains most of the variation in the age of cancer incidence. DNA mutation are minimized by repair systems that can recognize and correct the damage. When a repair system corrects DNA damage, there is no consequence. Mutations can result, however, when the repair system fails and can lead to genomic instability or apoptosis, which is widely believed to be major anti-proliferative mechanism of DNA-damaging anticancer drugs (Ma and Wang, 2009; Nobili *et al.*, 2009; Colin and Dubrova, 2012).

Genotoxicity evaluation involves different interaction types of the mutagenic agent and the DNA, and their effects, which can be separately examined. In this way, the connective capability between genotoxic agent and the genetic material can be evaluated through the observation of adult DNA. The potency in the DNA break can be estimated by alkaline eluted or by the Comet assay; and the chromosome and/or genomic mutations, which are generated by the genotoxic agent, can be quantified by micronucleus tests or chromosome aberrations (Van-Goethem *et al.*, 1997).

Mutations are detectable using micro-organisms, as well as, mammalian or plant cells. The examination of mutagenicity/genotoxicity, through short duration tests in mammalian cells (*in vivo* or *in vitro*), enables the potential estimate for genetic damage from genotoxic chemical compounds, and the evaluation of their risks to human health given to the short phylogenetic distance between the organisms. However, it is important to emphasize that animal species differ from each other because of their functional, biochemical and morphological characteristics, and consequently, the reaction to drug actions is also variable, quantitatively or qualitatively. Every variation must be considered whenever the obtained results are extrapolated from experimental animals to human beings (Jacob, 1968). It is already known that a great variety of mechanisms can lead to mutagenesis; on the other hand, there are biological mechanisms, facilitated by anti-mutagenic substances, which work to impede the action of genotoxic agents (Van-Goethem *et al.*, 1997). Recent researches show many natural products presenting properties which, directly or indirectly, reduce the mutagenic action of toxic substances (Slupphaug *et al.*, 2003; Reddy *et al.*, 2003; Verhagen *et al.*, 2003). Several kinds of experiments to evaluate anti-mutagenic activities have been carried out for the last years, producing lots of revision works. Various examples can be highlighted, among them, there is the antimutagenic effect of *chilli*, an important Indian spice, observed in the Ames test; as well as the effect of *Curcuma longa* (tumeric), a spice used in Asia, tested in rodents; and also both black and green teas, tested in the *Salmonella* assay, in which the green tea showed a more efficient protective activity (Lahiri *et al.*, 1993; Mukundan *et al.*, 1993; Tanaka *et al.*, 1994; Hernaez and Dashwood, 1998).

Principal Action Mechanisms of Natural Chemotherapy agents

The study of the principal action mechanisms of natural substances arouses great interest in the search for new chemotherapy agents. The advancements in this study show that the main molecular targets are these mostly investigated ones microtubules, DNA via apoptosis and free-radicals, membrane through free radicals and signalling molecules. Some revision articles show global perspectives to the prevention and treatment of several cancer types (Cortesi *et al.*, 1997; Reddy *et al.*, 2003; Slupphaug *et al.*, 2003; Verhagen *et al.*, 2003; Vitaglione and Fogliano, 2004; Commane *et al.*, 2005; Flora and Ferguson, 2005; Steele and Kelloff, 2005; Ramsay and Leung, 2009; Schwartz, 2009; Demain and Vaishnav, 2011; Qurishi *et al.*, 2011).

Microtubules

Microtubules are components of the vital mitotic fuse for many cellular functions, among them are included cellular form maintenance, intracellular transport, secretion processes and mitosis. Microtubules are formed by structural proteins, the tubulines α and β, which exist in a dynamic equilibrium, involving the polymerization and depolymerization in answer to physiological messages (De Vita, 1997) (Scheme 12.1). Microtubules are the key components of the mitotic spindle of eukaryotic cells and they play an important role in cell division. The discovery of natural and synthetic substances capable of interfering with the polymerization or depolymerization of microtubules has gained much attention because microtubules are an attractive pharmacological target for the design of anticancer drugs. Indeed, tubulin binding molecules interfere with microtubules assembly and functions, thus resulting in mitotic arrest of eukaryotic cells. Actually a large number of tubulin agents are in clinical development and represent one of the most important classes of anticancer agents with major drugs on the market and many promising compounds in clinical trials (Schwartz, 2009; Ngo *et al.*, 2010; Passarella *et al.*, 2010). As an example, vinca alkaloids (vincristine and vinblastine, Figure 12.2), act through a specific active site of tubuline, inhibiting the polymerization in metaphase microtubules (Ngo *et al.*, 2010). Other antimicrotubule agents, such as colchicines, podophyllotoxin and taxol (Figures 12.1 and 12.2, respectively) act in different sites (Nobili *et al.*, 2009; Passarella *et al.*, 2010; Paterson *et al.*, 2010). Taxans, for example, bind to the microtubules in the tubuline subunit β, become stabilized against the depolymerization, showing an antimitotic effect (Nobili *et al.*, 2009).

Apoptosis

Apoptosis is a normal aspect of human physiology ensuring tissue homeostasis and represent the cell mechanism of self-destruction in a genetically programmed

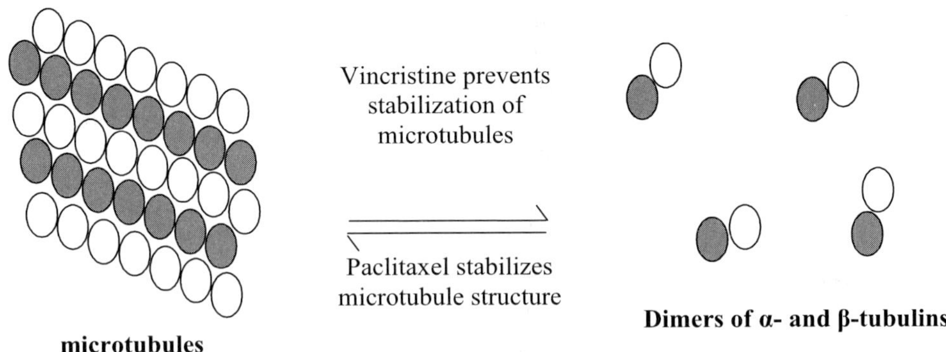

Scheme 12.1: Simplified equilibrium process between dimeric tubulin and microtubules.

sequence of biochemical events resulting in cellular death. So it is a vital process for embryologic development and the maintenance of homeostasis in multicellular organism. In other hand, it is also involved in a wide range of pathological processes, including cancer, which request both energy and protein synthesis for its occurrence. Evasion of endogenous cell death process, including apoptosis, represents one of the characteristics of cancer. In fact, impaired apoptosis is associated with hyperproliferation and tumorigenesis. Apoptosis can be induced by cytotoxic agents, heat and irradiation, when in low doses, and high doses could cause cellular death by necrosis (disordered disintegration of the damage cells), in which there is a rupture of the cellular membrane, usually resulting in an inflammatory response. Defects in the physiological mechanisms of apoptosis contribute to the pathological cell expansion and to the development and progression of cancer. (Jabs, 1999; Rang *et al.*, 2001; Burz *et al.*, 2009; Chen *et al.*, 2010; Zivny *et al.*, 2010; Pinon *et al.*, 2011; Manikandan *et al.*, 2012). Apoptosis is described by morphological and biochemical criteria. The morphological criteria include the chromatin aggregation, cytoplasm and nucleus condensation, and the formation of apoptotic bodies. Chemical modifications, as the chromatin cleavage, result in DNA fragments, in multiples of about 180 base pairs, which are detectable by agarose gel electrophoresis, resulting in DNA pattern (Rang *et al.*, 2001). The apoptosis in animal and plants occurs in 3 phases: i) induction phase – the main signalling ways, that trigger off the apoptosis, are taken from growth factors, TNF (tumoral necrosis factor) bond with receptors, activation of the p53 tumor suppressing gene, oxidant stress agent, ceramides, virus, inhibitors of protein synthesis, hypoxia, nitric oxide; ii) effector phase – involves family members of *Bcl-2* mitochondrial proteins which can have pro and anti-apoptosis functions; iii) execution phase – it is the apoptosis final way in which signalling and regulatory mechanisms converge on a proteases cascade, called caspases. Capases exist as zymogens. After the activation of a triggering caspase, the death program is sequentially executed; the protein hydrolysis of the cytoskeleton and nuclear matrix leads to the cell destruction (Scheme 12.2). Although many of the proteins vital for apoptosis have been identified, the molecular pathways of cellular death still remain to be elucidated (Jabs, 1999; Rang *et al.*, 2001; Burz *et al.*, 2009).

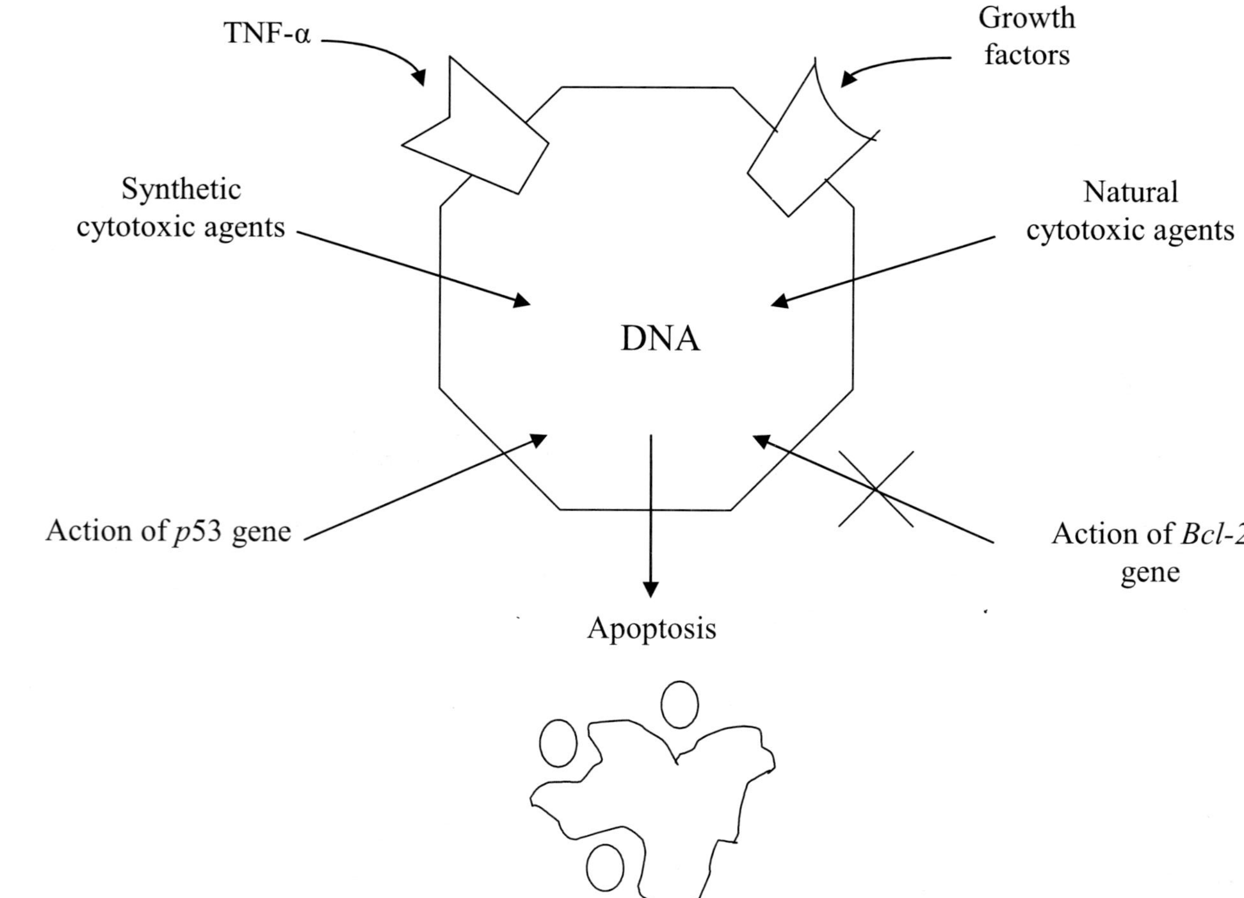

Scheme 12.2: Simplified sequence and signalling apoptosis process.

Kaurene diterpene-type are found in numerous plants, which are popularly used in the treatment of inflammatory processes and cancer (Lee *et al.*, 2002). The antiinflammatory action of these terpenoid class involves the inhibition of the nuclear transcription factor-κB (NF-κB). This factor is an apoptosis mediator, and it also mediates an immune response. NF-κB promotes the gene expression by the macrophages involved in the host defense, such as NOS-2 (nitric oxide synthase-2), cyclooxygenase 2, adhesion molecules and several matrix metalloproteinases. Moreover, those diterpenes are involved in the gibberellin synthesis and they have also cytotoxic activity against different human cell lines (Castrillo *et al.*, 2001). Kaurene and clerodane-type diterpenes might suffer Michael addition reaction, whose mechanism is important in the inhibition of the NF-κB activity (Castrillo *et al.*, 2001). The cell treatment of the MCF-7 human breast cancer with kaurenes kamebakaurin, for example, increased the elevation of susceptibility to apoptosis, induced by TNF-α via caspase-8 activity (Lee *et al.*, 2002). On one hand, Labdane-type diterpenes also induce apoptosis in human leukaemia cells (MOLT3) by the reduction of the proto-oncogene expression, but on the other the *bcl-2* gene is not altered (Dimas *et al.*, 2001). Besides being an antimitotic agent, taxol also causes apoptosis in tumor cells, it possibly occurs through the activation of the NF-κB cascade; yet some tumors, such as MCF-7 breast cancer cells and the prostate tumor, are resistant to apoptosis, since they do not have this activation. However, they are sensitive to the action of taxol on the microtubules (Miller *et al.*, 1999).

Free Radicals

Metabolic processes in plants and human body inevitably produce highly reactive free radicals, especially oxygen derivatives, which oxidize biomolecules resulting in cellular death and tissue damage (Cheeseman and Slater, 1993). Oxidative damages have a significant role in the etiology of some diseases, such as atherosclerosis, cancer, arthritis and cirrhosis. The organism defense, that protects it from the cellular damage caused by free radicals, is constituted by antioxidant agents which can have an enzymatic character or not. Enzymes like superoxide dismutase, catalase, glutathione peroxidase and endogenous antioxidant substances, such as the glutathione, ascorbic acid, carotenoids and α-tocoferol, as well as exogenous antioxidants such as flavonoid class, all of them possibly to found in plants. Green tea can be cited as an example, since it has the capacity to capture free radicals; this is attributed to the flavon-3-ol and gallate-type compounds, which inhibit the cell oxidant stress (Middleton *et al.*, 2000).

In order to exemplify the phenolic-type diterpenes antioxidant agents, carnosol, rosmanol and epirosmanol can be cited, they are extracted from the *Rosmarinus officinalis*, a Mediterranean plant (Zeng *et al.*, 2001). This species extract has been used to prevent cardiovascular diseases as well as in the cosmetic industry. These derivatives contain phenolic hydroxyls in *ortho* positions that easily donate hydrogen radical, thus facilitating the reaction with other endogenous or exogenous free radicals, forming stable products. Diterpenes from *Podocarpus nagi* like totarol and its analogues, inhibit the lipidic peroxidation, since present isopropyl group in an *ortho* position in the phenolic ring, acting as antioxidants, producing free radicals (Haraguchi *et al.*, 1997; Chenqi *et al.*, 2012).

Signalling Molecules

There are complex protein systems in animal cells which allow the interaction with other cells through a signal sending, that might, or not, permeate in the membrane. The signalling molecules belong to two classes: i) very large or hydrophilic molecules, which interact with protein receptors from the cellular surface involving protein-kinases, phosphatase, G-protein coupled proteins and several cytokines; ii) small and hydrophobic molecules that diffuse through the cellular membrane. This class involves steroid hormones, thyroid hormones, vitamin D, and retinoic acid, among other that interact with receptors that regulate the gene transcription, which are cytosolic or soluble intranuclear proteins (Rang *et al.*, 2001). It is a protein mainly produced by macrophages and monocytes, among other cells; after the connection to the membrane receptor, it promotes a variety of biological activities in organs and tissues, for example: in endothelial cells, it induces the adhesion (ICAM-1), iNOS, and stimulates the T cells (tumor cells) to produce IL-2 (interleukin) and IFN-γ (interferon). TNF (tumor necrosis factor, also known as TNF α) is a pro-inflammatory cytokine that has play important roles in diferrent biological processes, including the induction of other cytokines (Li and Lin, 2008). TNF cytotoxic or cytostatic activity that results in cellular death through apoptosis, while in high doses, it can originate various diseases, such as the cachexia associated with cancer and also AIDS, death from septic shock and also immunological illnesses (De Vita, 1997). The necrosis factor, one of the signalling molecules, was discovered through its antitumor activity and it is also related to the immunological and inflammatory response. Receptor interactions can become effective for treating diseases in which T cell have an important role, including inflammatory conditions, autoimmunity and cancer (Croft, 2009). One of the most important cownstream signaling targets activated by TNF is the NF-κB transcription factor, which has been identified to be involved in inflammatory, anti-apoptotic, and immune responses (Croft, 2009; Li and Lin, 2008). According to Schetter *et al.* (2010), chronic inflammation and infection are major causes of cancer. It has been proved that 25 per cent of all cancers are due to chronic infection or other type of chronic inflammation, which can become oncogenic by several different mechanism-types. It includes induction of genomic instability, increasing angiogenesis, altering the genomic epigenetic state and increasing cell proliferation. Additionally, inflammation can alter the expression of oncogenes and tumor suppressor genes (including both protein coding genes and non-coding microRNA genes) to promote neoplastic transformation. According to Croft (2009) a recognized triumph in immunotherapy has been the development of neutralizing antibodies and Fc fusion proteins that inhibit the binding of tumor necrosis factor to none or bothof its receptors – TNF receptor 1 (TNFR1; also known as TNFRSF1A) and TNFR2 (also known as TNFRSF1B). Multiple experimental approaches have shown that TNFR1 initiates the majority of biological functions of TNF. As a primary function of the TNF superfamily molecules is to regulate cell survival, inhibition of these interactions prevents the activation of signaling pathway downstream of the TNFRs, thereby minimizing the pro-inflammatory programme they initiate in immune and inflammatory diseases (Li and Lin, 2008; Schetter *et al.*, 2010; Teiten *et al.*, 2012).

Croton cajucara Benth: Focusing on Medicinal Potential Advantages of its Chemotherapic Agents and Evolutionary Perspectives

Croton cajucara Benth (Euphorbiaceae) popularly known as "sacaca", is broadly being used in the Amazon region, and other regions of Brazil, to control diabetes, high levels of cholesterol and gastrointestinal disturbances (gastritis, ulcer). It also treats malaria and fights against liver, kidney and gall bladder inflammations (Maciel *et al.*, 2000; 2002a; 2006a; 2007a). *Croton cajucara* phytochemical studies led to the isolation of the 19-*nor*-clerodane diterpene *trans*-dehydrocrotonin (DCTN, Figure 12.3) additionally to other clerodanes such as *trans*-crotonin (CTN), *cis*-cajucarin B (c-CJC-B), *trans*-cajucarin B (t-CJC-B), cajucarin A (CJC-A), cajucarinolide (CJCR) and isocajucarinolide (ICJCR) (Figure 12.3) (Maciel *et al.*, 1998; 2003; 2010b). Maciel *et al.*, largely investigated this *Croton* and covered studies targeting ethnobotanical, phytochemical, pharmacological and synthetic approaches (2000; 2002a; 2002b; 2006a; 2006b; 2010b).

Phytochemical studies carried out with this specie had involved plants ageing from 1 ½ to 6 years old (native and cultivated plants); it was proved that the stem barks from *Croton cajucara* are source of clerodane-type diterpenes. Among them, the DCTN presented major percentages in trees ageing up to 3 years old (Maciel *et al.*, 2000). Pharmacological studies performed by our research group, as well as, by other Brazilian researchers, have been contributing to biochemical studies of this specie since 1998, they represent a scientific effort that might lead to the *Croton cajucara* phytotherapy validation; it is very well represented by its major bioactive component DCTN, which showed striking correlation with the folk traditional therapeutic use of this *Croton* (Maciel *et al.*, 2002a; 2002b; 2006a; 2006b). This compound provided to possess antiulcerogenic (Souza Brito *et al.*, 1998; Hiruma-Lima *et al.*, 1999; Maciel *et al.*, 2000), hypoglycemic (Farias *et al.*, 1997), antiinflammatory and antinociceptive (Carvalho *et al.*, 1996; Perazzo *et al.*, 2007; Khan *et al.*, 2009) activities. In addition, DCTN showed antiestrogenic, antiatherogenic and hypolipidaemic effectiveness (Luna Costa *et al.*, 1999; Silva *et al.*, 2001a; 2001b; 2001c). The cardiovascular action of this natural product was also evidenced. DCTN administered in low doses (10 and 15 mg/kg iv.) showed reduction of the blood pressure and cardio frequency. In addition to the induction of a negative chronotropic effect, it was observed a vasorelaxing effect (*in vitro* experiments) in the aorta artery segment isolated from rats, and in the right atrium also isolated from them. The hypotensive and bradicardia effects of the DCTN were correlated to the nitric oxide liberation, as well as the effects caused in the smooth vascular musculature and in the cardio rhythm activity (Silva *et al.*, 2005).

This significant amount of chemopharmacological researches demonstrated that *Croton cajucara* (and specifically its major compound DCTN) is an effective medicinal agent to various kinds of diseases, among them, cancer is highlighted. Since 1999, the antitumor, antigenotoxicity and antimutagenic actions of the DCTN have been reported and will shortly be focused herein. Santos *et al.* (2006a; 2008) observed the absence of mutagenicity of the stem bark methanolic extract (ME) of *Croton cajucara*, in

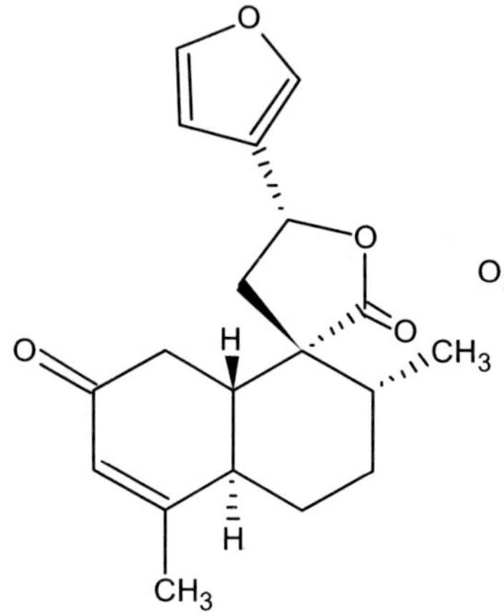

DCTN

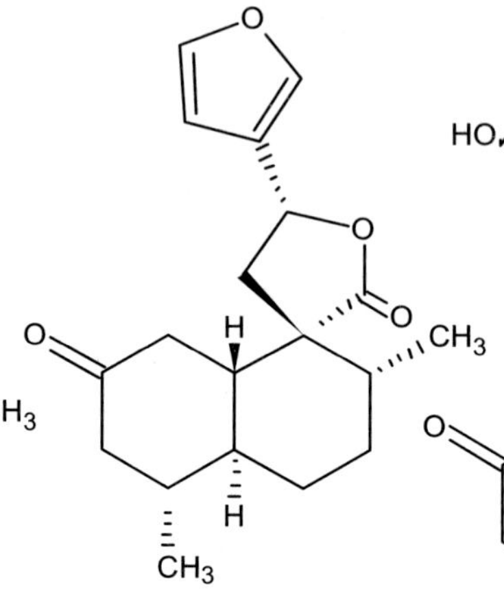

CTN

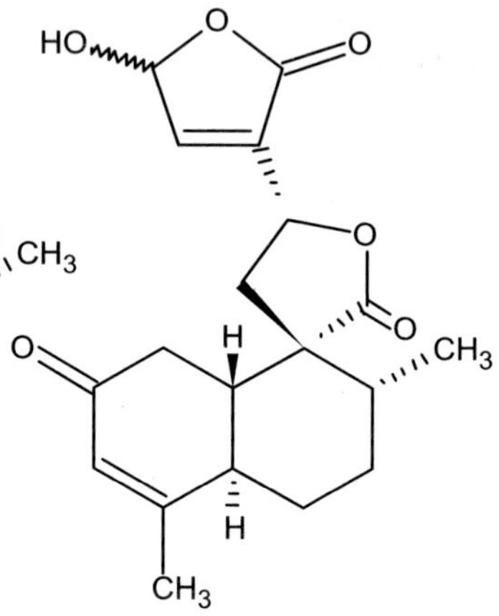

CJCR

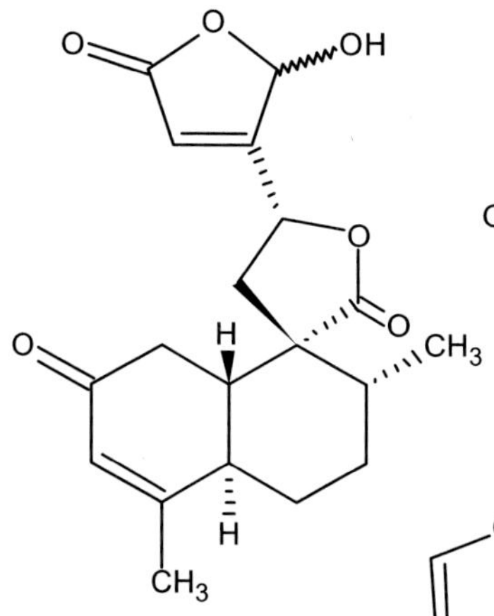

ICJCR

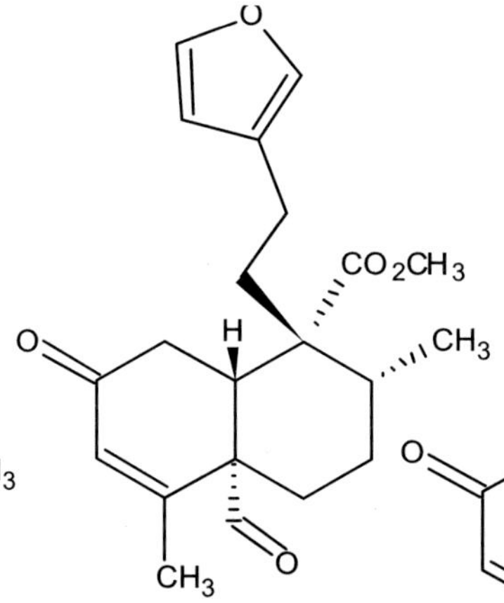

CJC-A

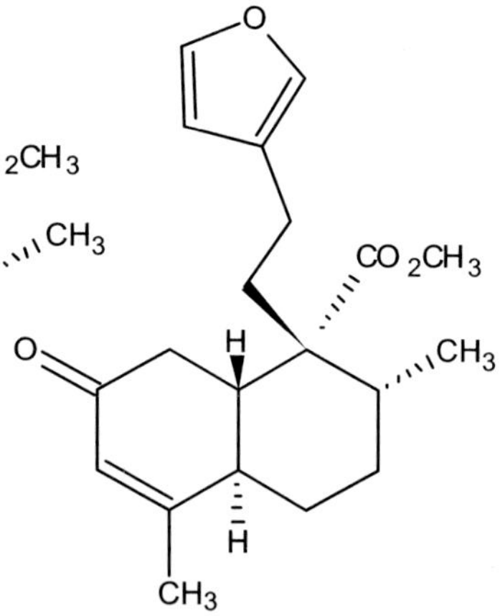

t-CJC-B

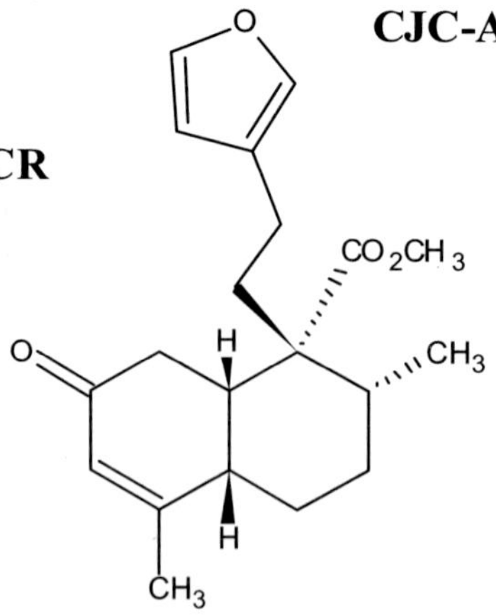

c-CJC-B

Figure 12.3: Chemical structures of clerodane-type diterpenes from *Croton cajucara* Benth.

bone marrow cells of Swiss albino mice, in which the animals under an acute treatment, via gavage did not show micronucleus frequency increase. Moreover, in a previously investigation Agner *et al.* (1999, 2001) tested DCTN in Swiss mice (under acute treatments, via gavage and intraperitoneal route). The performed assays showed that DCTN does not present any mutagenic effect, since the mice did not show MN frequency increase. Encouraged by these results, and considering the wide use of *Croton cajucara* in the traditional medicine of the Amazon region of Brazil, Agner *et al.* (2001) and Santos *et al.* (2006a) evaluated, respectively, the chemoprotective action of the ME extract and DCTN, correlating them to cyclophosfamide, a cytotoxic chemotherapy agent with known DNA alkylation action. The animals underwent a subchronic treatment with the methanol extract (ME = 312.5; 625.0 and 1,250.0 mg/kg ea.), for 4 consecutive weeks. The ME (diluted in DMSO and water) was administered to the animals via gavage, an hour before the administration of cyclophosfamide (150 mg/kg; via intraperitoneal). The results revealed that the animals [treated in the lowest administered concentration of ME (312.5 mg/kg p.c.)] showed in the last two weeks of treatment, significant reduction of the MN cell frequency, which resulted from the cyclophosfamide treatment. Meanwhile, higher concentrations of the method extract so called ME (625 and 1.250 mg/kg p.c.) revealed a protective effect during the entire period of treatment, with a significant reduction of the MN frequency produced by the cyclophosfamide action. The cyclophosfamide effect was evaluated in bone marrow cells of Swiss mice, using chromosome (CA) and micronucleus aberration tests. The animals underwent intraperitoneal treatment and gavage injection, with a concomitant use of 2 drugs (DCTN and cyclophosfamide); they were sacrificed 24 hours after the treatment. The 25 per cent dose of the LD_{50} had just showed antimutagenic effect in relation to cyclophosfamide for animals treated via gavage. The DCTN, in the doses of 50 and 75 per cent of the LD_{50}, reduced the genotoxicity performed by cyclophosfamide (in both, intraperitoneal and gavage administration), in the two types of cytogenetic tests evaluated (MN and CA). These results confirmed both DCTN and ME as antimutagenic agents, in blood and bone marrow cells (Agner *et al.*, 2001; Santos *et al.*, 2006a). Additionally, the possible clastogenic, apoptotic and cytotoxic effects of protective effect of DCTN as well as its protective effect against three mutagenic agents using the micronucleus test (MN) and apoptosis index in CHO-K1 cells *in vitro* were assayed, been confirmed the protective effect of DCTN (Poersch *et al.*, 2007).

Concerning to *Croton cajucara* bioactive fractions, Farias *et al.* (2005) confirmed that several of bioactive chromatographic fractions from the stem bark of this *Croton* inhibit cell proliferation and induce differentiation-inducing properties in a human promyelocytic leukemia HL-60 cells.

The DCTN antitumor action was proved in tests that evaluated the Sarcoma 180 tumor (S180) and Ehrlich carcinoma in ascitic forms, which had been kept in mice (DBA/2 males) by weekly transplants (via intraperitoneal i.p.). In these experiments, the mice were injected on the zero-day with 5×10^5 ascitic cells of the mentioned tumors (Grynberg *et al.*, 1999). The DCTN action over the inoculated tumor development showed significant inhibition in the dose of 84 mg/kg (equivalent to the total dose during the entire treatment) with per cent T/C value of 137 (per cent T/

C values ≥ 125 indicates significant antitumor activity) for S180; and 120 mg/kg with per cent T/C = 128 for Ehrlich carcinoma. These results can be compared to the ones observed to 5-fluorouracil, a clinically used chemotherapy agent, it presented a per cent T/C value = 140, in the dose of 38 mg/kg (total dose). Considering the molecular weight of the assayed drugs, 292.1 μmol/kg of the patterned substance and 267.2 μmol/kg of DCTN were inoculated in order to have similar effects over S180. Therefore, the DCTN was shown in this study as a promising antitumour agent (Grynberg *et al.*, 1999). Additionally, Melo *et al.* (2004), have optimized experiments for the Ehrlich carcinoma and found significant inhibition of the DCTN in a lower dose (20 mg/kg). According to their observed results DCTN in *in vivo* treatment on total natural killer (NK) cell activity of tumor-bearing mice, revealed that DCTN improved NK cytotoxicity against YAC-1, a Moloney virus-induced mouse T-cell lymphoma of A/SN origin. DCTN exhibits both antitumor efficacy and immunomodulatory actions (Melo *et al.*, 2004). Moreover, Grynberg *et al.* (1999) evidenced that DCTN administered in higher dose (120 mg/kg) in the treatment of Ehrlich carcinoma resulted in a significant increase of the TNFα level.

Point out some critical aspects of current cancer investigations performed with DCTN, it is evident that investigations on DCTN mechanisms of action including for example, interaction with microtubules, inhibition of topoisomerases I and II, alkylation of DNA, and interference with tumor signal transduction still not concluded at all. Certainly, new screening methods for both antitumor and antimutagenic potential mechanisms aspects involved in the DCTN action is strongly necessary. In other hand, novel technologies, such as nanoformulations, along with a better understanding of DCTN molecular mechanisms of its biological activities represent now the current advancement of DCTN studies. Aiming to point out the evolutionary perspectives to this target natural product DCTN, the actual status of its recent bioformulations and some bioactive DCTN-derivatives, as well as a general comment on *Croton cajucara* Benth toxicity, will be cited herein.

Croton cajucara Benth Toxicity

The subacute toxicity of DCTN was researched in adult Wistar rats. The animals (40 males and 40 females) were randomly distributed among 4 groups (1 as a control and 3 treated). DCTN was orally administered diluted in Tween 80 (12 per cent) in the doses of 25, 50 and 100 mg/kg for 35 days. After this period, a significant increase in the liver weight was noticed for every animal which had been under higher doses. The level reduction of the alkaline phosphatase and cholesterol, as well as an increase of the γ-glutamil transpeptidase (at 100 mg/kg dose) was observed in female rats. DCTN has caused histopathological changes in the liver, which included dim tumefaction, microvascular degeneration and nuclear alterations (Rodríguez *et al.*, 2004).

The DCTN acute toxicity was observed for 72 hours after the oral administration. The DCTN LD_{50} was 555.0 mg/kg (o.v.) in mice, 10 times higher than the ED_{50} (Carvalho *et al.*, 1996). This study did not detect any symptoms that could justify the toxic action to the level of the central nervous system, such as stereotypia, ataxia or convulsion. Additionally, the DCTN acute toxicity was also evaluated in male Swiss mice. In which, after 12h of fast, the animals were treated with DCTN, orally (125, 250, 500, 750 and 1000 mg/kg) and via intraperitoneal injection (31, 25, 62.5, 125, 250 and 500 mg/kg). The number of survivals was monitored twice a day for 14 days. The obtained LD_{50} values, from oral and intraperitoneal administration, were 876 mg/kg (12h) and 47.2 mg/kg, respectively. The achieved results showed the low toxicity of the DCTN (Souza-Brito *et al.*, 1998) as well as in other *in vitro* experiments using Ehrlich carcinoma, K562 and HL60 leukaemia cells (Freire *et al.*, 2003a; 2003b; Anazetti *et al.*, 2003; 2004).

Although the studies of the DCTN acute and subchronic toxicity in animals have shown low levels of toxicity, the wrong use of *Croton cajucara* Benth leaves and barks in long treatments, of alternative medicine, might cause toxic hepatitis. This plant use has already shown it can cause irreversible hepatotoxic effects, at least, in two Brazilian cities (Belém and Rio de Janeiro) (Maciel *et al.*, 2002a; Veiga Jr. *et al.*, 2005; Soares, 2004). Acute, chronic and massive hepatitis cases were notified in the Amazon region and in other regions of the country, in patients who used this *Croton* to control weight and reduce cholesterol levels. There is a significant history of liver disease victims in the hospital of the Universidade Federal do Pará [located at Belém city, state of Pará (Brazil)], people who were ingesting *Croton cajucara* Benth leaves on long diets to lose weight. Although the population in Belém is well warned of these leaves toxicity, it is still consuming this *Croton* leaves on diets to lose weight, the users said to be encouraged by the percentage of people who guarantee to have healthily lost weight (Maciel *et al.*, 2002a).

Recent studies carried out in patients of the Amazon region, who had used *Croton cajucara* for 36 months, registered 25 cases of hepatotoxicity attributed to it; 21 people had acute hepatitis, 3 chronic and 1 massive (Soares, 2004).

In order to evaluate a probable toxic hepatitis as a side effect of *Croton cajucara* Benth stem bark infusions (corresponding to a tea preparation, AE) this plant was administrated to Swiss albino mice with initial weight between 28 – 35 g (Maciel *et al.*, 2006b). The treated animals were divided in three groups of ten, one of them being the control group, the remained groups were treated with 12.5 g/L and 25 g/L of AE, in the wells exclusively. Their weight variation was average once a week during 42 days of the whole treatment, in which the average dosages were strictly correlated to hepatic investigations, based in the biologic action of GPT, an enzyme that has a predominant hepatic origin and possesses remarked alterations in acute hepatitis case. The obtained results proved that the control group animals as well the animals treated with 12.5 g/L AE got weight regularly, while the animals treated with 25 g/L AE showed initially remarked weight loss been regular in the rest of the period of treatment. Although the results were found in intervals considered normal for male mice and the dosage-response relation observed, suggest other investigations such as an anatomopathological evaluation of mice liver, in order to confirm the absence of hepatic toxicity of AE (Maciel *et al.*, 2006b).

This pharmacological result jointly with the significant contents of DCTN in AE, proved that drinking *Croton cajucara* tea instead to take capsules is an unambiguously way to be protected against DCTN over doses side effects.

Within the program of chemical transformation of the target molecule DCTN, which means preparing DCTN derivative compounds for screening as pharmacological agents, the semi-synthetic *trans*-crotonin CTN, cajucarinolide (CJCR) and isocajucarinolide (ICJCR) (Figure 12.3) were already obtained (Maciel *et al.*, 2006a; 2007b) from the target molecule DCTN. Semi-synthetic CTN was obtained in great amount, by hydrogenation of DCTN, allowing pharmacological studies, among them antiinflammatory, antinociceptive, antiulcerogenic, cytotoxicity action were evaluated (Maciel *et al.*, 2000; 2007b; Almeida *et al.*, 2003; Perazzo *et al.*, 2007). Concerning to cytotoxic effect, it was evidenced that semi-synthetic CTN was less toxic than the target molecule DCTN (Almeida *et al.*, 2003). Related to cajucarinolides-derivatives (CJCR and ICJCR-derivatives), they were obtained by regiospecific oxidation of substituted furans, which gives the corresponding hydroxybutenolides (Maciel *et al.*, 2007b). Using this method, which was previously describe by Kernan and Faulkner (1988), these cajucarinolides-derivatives were synthesized in good yields by synthetic transformation of the DCTN furan moiety, with singlet oxygen, generated from molecular oxygen by irradiating a polymer-bound rose bengal catalyst in dichloromethane solution. The semi-synthetic obtained hydroxybutenolides as well as CTN had their purity confirmed by NMR 1D and 2D analysis (Maciel *et al.*, 2007b).

In order to identifying the biological antiproliferative action of all-clerodane presented in the stem bark of *Croton cajucara* Benth (Figure 12.3), the effects of the natural compounds DCTN, CTN, *cis*-cajucarin B (c-CJC-B) and *trans*-cajucarin B (t-CJC-B) against the growth of ascitic Ehrlich carcinoma cells and human K562 leukemia, were performed additionally to the semi-synthetic CTN, CJCR and ICJCR. The experiments involved MTT [3-(4,5-dimethylthiazole-2-yl)-2,5-biphenyl tetrazolium bromide] assay, quercetin as positive controls for Ehrlich carcinoma and vincristine for human K562 leukemia. Comparatively, the higher antiproliferative action was correlated to the semi-synthetic hydroxybutenolides (CJCR and ICJCR) (Maciel *et al.*, 2007b).

Other derivatives compounds such as DCTN-OL, DCTN-tri-OL and DCTN-20 Amide (Figure 12.4) obtained from natural DCTN were cited in literature as bioactive derivatives, which were evaluated in cytotoxicity on V79 Chinese hamster cells, promyelocytic leukemia cell line (HL60), human peripheral blood mononuclear cell experiments and also *Escherichia coli* (Melo *et al.*, 2001, 2004; Anazetti *et al.*, 2003, 2004). Additionally, the gastroprotective effect these synthetic derivatives were investigated (Melo *et al.*, 2003). The obtained results of these pharmacologic investigations could not be focused in one pure derivative as the authors described, at least for two of them DCTN-OL and DCTN-tri-OL. As a general comment, since it is widely known that $LiAlH_4$ is a strong, unselective reducing agent for polar double bonds [most easily thought of as a source of hydride ions (H^-)]. It will reduce aldehydes, ketones, esters, carboxylic acid chlorides, carboxylic acids and even carboxylate salts to alcohols.

$NaBH_4$ is less reactive than $LiAlH_4$ but reaction with DCTN demonstrated that they were quite similar. In fact, in our work we found that the reductive ability of $NaBH_4$ in alcohol solution (EtOH, ranging from 20 to 40 min., at room temperature) gave a complex mixture of DCTN-derivatives. Reduction of the reaction time (DCTN with $NaBH_4$; ethanol, r.t., 15 min.) produced a mixture containing DCTN-OL (OH-C2 at β-position) and CTN-OL (epimeric mixture). It is important to cite that at lower time reaction (ranging from 5 to 10 min.) it is possible to have DCTN-OL (epimeric at C2) in small amounts, been the target molecule the contaminant on reaction product (the major amount of the non reacted DCTN was evidenced by GC-MS analyzes as well as

Figure 12.4: Chemical structures of DCTN-derivatives.

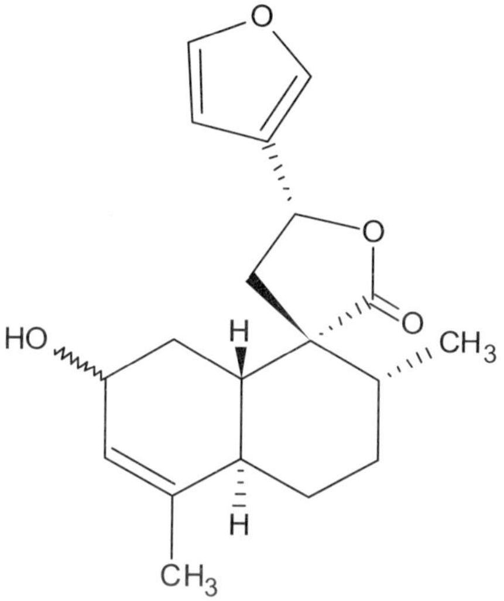

DCTN-OL

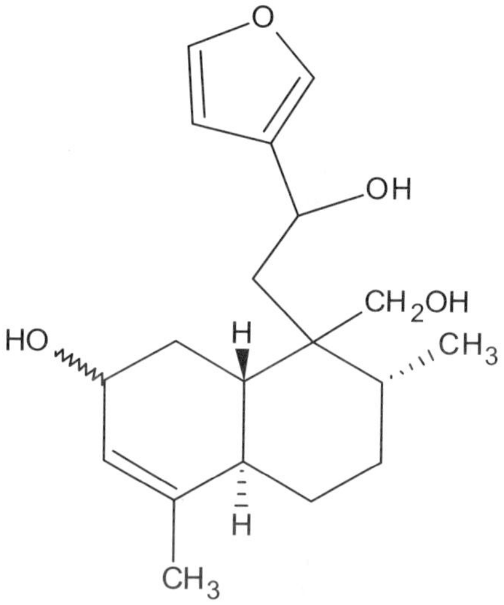

DCTN-tri-OL

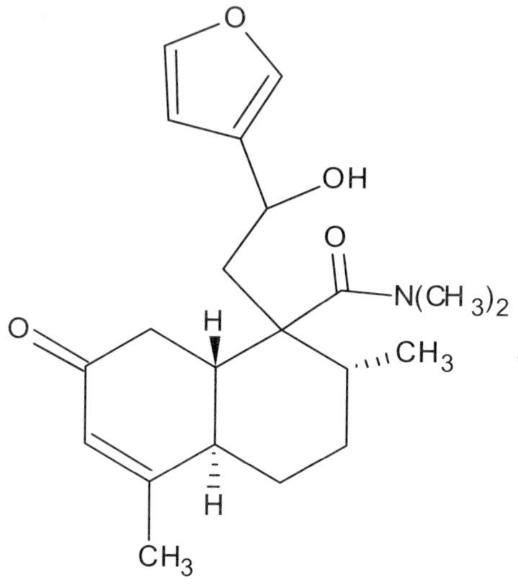

DCTN-20 Amide

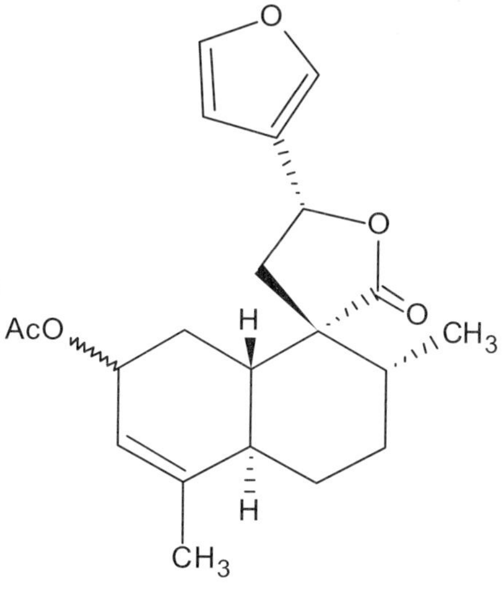

DCTN-EtOAc

Contd...

Figure 12.4–*Contd...*

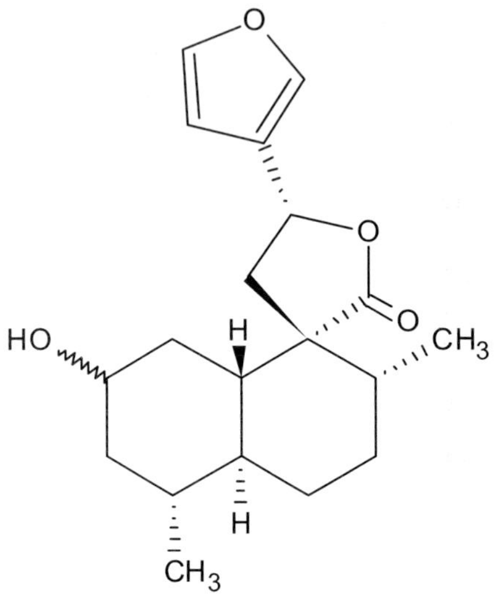

CTN-OL

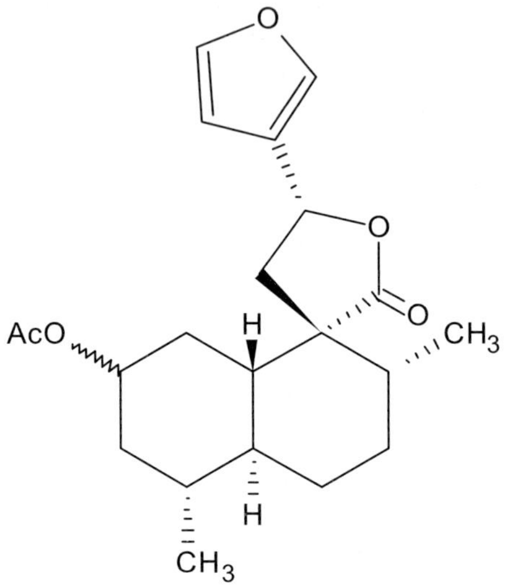

CTN-EtOAc

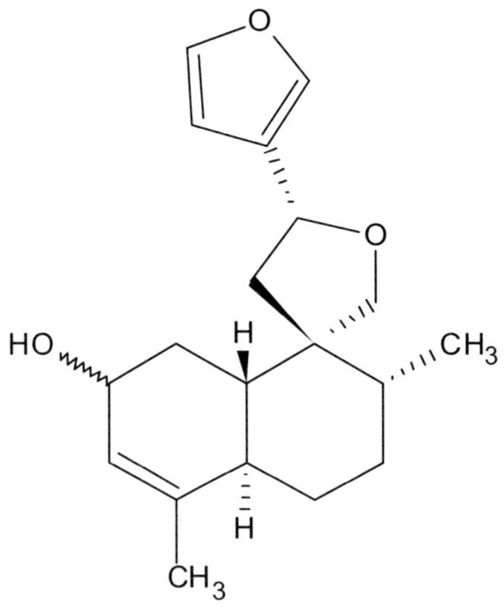

DCTN-2-OL-12,20 Oxolane

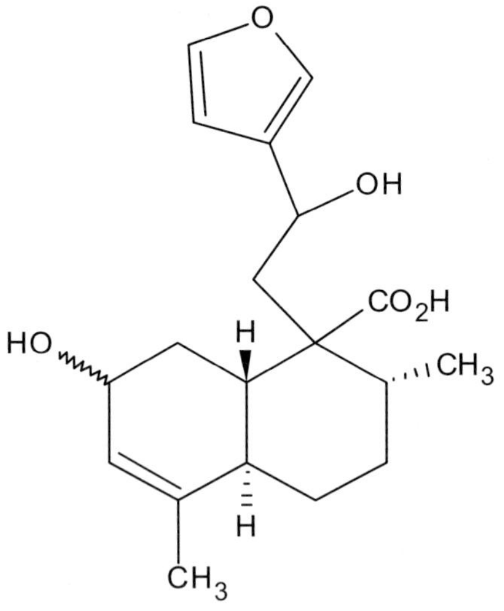

DCTN-2-OL-20 Acid

NMR data). Analysis from NMR data of the obtained products with time reaction above 30 minutes revealed that the major compounds produced are β-DCTN-OL and β-CTN-OL and in minor proportion: the stereoisomeric derivatives α-CTN-OL, DCTN-tri-OL (2,12,20-OL), DCTN-2-OL-12,20-oxolane, and DCTN-2-OL-20 acid (Figure 12.4). Considering that each of obtained derived compound been as stereoisomeric pairs (in which the hydroxyl group at C2 is epimeric), $NaBH_4$ produced at least a whole of

10 compounds [β- and α-DCTN-OL; β- and α-CTN-OL; β- and α-DCTN-tri-OL; β- and α-DCTN-OL-12,20-Oxolane; β- and α-DCTN-OL-20 Acid]. After a very hard chromatography work performed aiming purification of these derivatives, the chemical structures of each stereoisomeric pair were confirmed by spectroscopy data focusing in the NMR 1D and 2D experiments (Maciel *et al.*, 2010c). Following the data analyzes we conclude that $NaBH_4$ in alcohol solution is a strong reducing agent with does not show selectivity towards reduction of the target molecule DCTN. This complex mixture of reduced-DCTN (using $NaBH_4$) was evaluated in cytotoxic experiments against tumor cells. The found results encourage us to prepare new compounds exploring other reactions involving DCTN, in this regard various modifications have been made to produce a reactive but more stable reducing agent (Maciel *et al.*, 2010c).

Nanotechnology: A Powerful Therapeutic Strategy for the Clinical Use of DCTN and Further Anticancer Agents

The pharmaceutical nanotechnology is one of the cutting edges of the Pharmaceutical Sciences, which is concerned with multidisciplinary aspects in the design of modern nanocarrier systems for drugs. Nanocarriers have been shown to be able to improve the bioavailability of bioactive compounds, trial products or drugs through a controlled release with low toxicity by means of a decreasing in their therapeutic dose (Ravi-Kumar, 2000). By the ending of 1960's, a huge progress took place in the pharmaceutical technology, especially due to the research and the industrial development of drug delivery systems in micro and nanometric scales, namely: liposomes, nanoparticles and microparticles (Table 12.1), as well as implants and patches. Nevertheless, many challenges have been faced related to the manufacturing, the efficiency of the processes and mostly the high cost of the vectorized and controlled drug delivery systems. The rational drug design consider parameters of optimization such as biocompatibility, stability, size, sterility, control of the release rate, and site-specific approaches that can produce well-succeed systems (Ratner, 2002; Talekar *et al.*, 2011).

Table 12.1: Nanocarrier system-types for drug controlled release.

Controlled Release Systems		*Description*
Liposomes		Microscopic aqueous vesicles composed by one or more phospholipidic bilayers (Al-Angary *et al.*, 1995).
Nanoparticles	Nanocapsules	Reservoir polymeric systems containing a cavity enveloped by a thin polymeric wall, size less that 1 μm (Soppimath *et al.*, 2001, Kim and Kim, 2003).
	Nanospheres	Matricial polymeric systems in which the drug is physically and uniformly dispersed in the polymeric matrix, size less than 1 μm (Al-Angary *et al.*, 1995).
Microparticles	Microcapsules	Reservoir polymeric systems containing a cavity enveloped by a thin polymeric wall, size between 1 and 1000 μm (Singh *et al.*, 2001).
	Microspheres	Matricial polymeric systems in which the drug is physically and uniformly dispersed in the polymeric matrix, size ranging from 1 to 1000 μm (Berkland *et al.*, 2001, 2002).

In the most recent years, there has been a massive increase in the concerning to pharmaceutical industry interest in the field of nanotechnology, partially motivated by the commercial success achieved by a significant number of products bioavailable by nanocarriers systems, and also due to several advantages such as the assortment of manufacture processes, the multiplicity of polymeric materials, the accomplishment attained for countless drugs, and the versatility in the production of such systems for different routes of administration. The use of silicon rubber or polyethylene in drug delivery systems is very common. However, the lack of biodegradability of these materials impairs their biological use. In the 1970 decade, the first biodegradable polymers were applied, which biocompatibility can be reached by the use of natural polymers, synthetic polymers, or polymers made by the use of natural monomers. (Freiberg and Zhu, 2004).

The most common natural polymers used in drug delivery systems are bovine or human albumin, gelatin, collagen, alginate, and chitosan (Benoit *et al.,* 2000). However, these natural polymers may present drug incompatibility, lack of purity, or need of reticulation, that can impair their use. In this context, biodegradable synthetic polymers have been widely applied for drug delivery. These polymers may be enzymatically degradable or not, but must present toxicological safe and not produce toxic metabolites. (Hans and Lowman, 2002; Jain, 2000). Among the most used types of synthetic polymers are included: polyamides, polyamine acids, polyesters, polyorthoesters, polyurethanes, and polyacrylamides (Jain, 2000). Polyesters like poly(lactic acid) (PLA), poly(glycolic acid) (PGA) and their copolymer poly(D,L-lactic-*co*-glycolide) (PLGA) have attracted great attention mainly due to their good biocompatibility, biodegradability, and lack of toxicity. All these polymers cited above have been approved by Food and Drug Administration (FDA) for use in drug delivery systems (Feng *et al.,* 2004 Jain, 2000; Birnbaum and Brannon-Peppas, 2003; Santos-Magalhães *et al.,* 2000; Ravi-Kumar, 2000).

PLGA microparticles showed encapsulation efficiency (82.5–90.0 per cent) for paclitaxel and for lidocaine (hydrophilic drugs) 85–88 per cent of efficiency (Ain *et al.,* 2002; Elkharraz *et al.,* 2006). Polymeric micro- and nanoparticles can form di-, tri-, multiblock copolymers, when conjugated with polymers like PEG (polyethyleneglycol) that possess the ability to extend its resident time into blood circulation, allowing precise time and local delivery. PEG is a biodegradable, biocompatible, nontoxic, flexible and hydrophilic compound approved by FDA for use in humans (Brannon-Peppas, 1995; Kumar *et al.,* 2001; Brannon-Peppas and Blanchette, 2004). These long term nanoparticles, also called stealth nanoparticles, have been extensively used for treatment of tumors (Fang *et al.,* 2006). PEG seems to act enhancing hydrophilic character and reducing the absorption of opsonins (proteins for cell reconnaissance) by becoming the polymeric conjugated less accessible for cell fagocytosis (Moffatt and Cristiano, 2006; Fang *et al.,* 2006). This association seems also enhancing gastrointestinal stability, as showed by studies using PEG-PLA and PEG-PLGA micro- and nanoparticles (Gref *et al.,* 2001).

Micelle like nanoparticles are smaller than typical blood cells, such as erythrocytes or lymphocytes (ca. 7-10 μm), hence may be injected into the bloodstream. After intravenous injection, such particles may be able to achieve site- or tissue-specific delivery by exploiting physiological clearance mechanisms in the body (Venkatraman *et al.*, 2005). Paclitaxel was loaded into the micellar nanospheres of PLA-PEG-PLA triblock copolymer as chemotherapeutic carrier, in that the drug release profile showed good control over the release of paclitaxel (Venkatraman *et al.*, 2005). In order to offer efficient controlled release, several studies have been developed with paclitaxel as well as other anticancer agents [loaded into formulations containing or not PGA, PLA or PLGA (including on it microemulsions nanoformulations-type)], aiming several advantages such as dose reduction, enhanced efficacy and reduce side effects (Cortesi *et al.*, 1997; Jeong *et al.*, 1999; Singer *et al.*, 2001; Güthlein *et al.*, 2002; Fonseca *et al.*, 2002; Liu *et al.*, 2002; Junping *et al.*, 2003; He *et al.*, 2003; Kang *et al.*, 2004; Le Garrec *et al.*, 2004; Ruel-Gariépy *et al.*, 2004; Reddy *et al.*, 2005; Drummond *et al.*, 2006; Oostendorp *et al.*, 2010; Qin *et al.*, 2010; Kanai *et al.*, 2012).

Due to some problems related to PEG, other molecules have been studied to be incorporated at the surface of long term drug delivery systems. Among these molecules we can cite anphiphilic polyethers used in the preparation of cationic lipoplexes for DNA incorporation. These systems can enhance up 5 times the resident time in blood circulation when compared with PEG-lipoplexes (Garinot *et al.*, 2007). Particulate polymeric systems, like micro- and nanoparticles, present interesting properties like good stability and encapsulation efficiency, and because this, they have been extensively studied for oral, transdermal, intramuscular, subcutaneous, and injectable administration of drugs, in which the development was obtained from poly(alkylcyanoacrylate) (Hans and Lowman, 2002).

Microparticles are spherical particles with average size 1-1000 μm, and may be classified in microcapsules and microspheres. Microcapsules are reservoir systems that present a polymeric layer that have the function to protect and isolate the drug into the inner core. Microspheres are monolithic systems in which the drug is dispersed into a polymeric matrix (Mandal *et al.*, 2001; Berkland *et al.*, 2001; Singh *et al.*, 2001).

The prospect is adopting this innovative therapeutic reality, since the incorporation of well-established drugs in nanocarrier-based systems could provide a 20 per cent cut back in cost, besides shrinking by half the time required to launch a new medicine from the discovery of new bioactive molecules. These systems have many advantages, including the improvement in an effective and rational way of the therapeutic efficacy as well as low toxicity and side-effects by reducing the high levels of drug plasmatic concentration, thereby maintaining a more constant level of the drug in the blood; the controlled and prolonged drug release, reducing the rate of recurrence of doses; the optimization of the biological activity yielding lower doses owing to their site-specificity and the protection of the therapeutic substance from deleterious effects of the fluids in the organism (Berkland *et al.*, 2002; Tao and Desai, 2003; Feng *et al.*, 2004; Singh and Lillard, 2009).

Particulate systems can be modeled aiming a specific releasing profile (from hours to years) by changing some chemical properties of the polymers used, such as: crystal lattices, monomer ratio, and polymerization conditions (Brannon- Peppas, 1995). These systems have been employed with success for a great variety of drugs, including: antibiotics, analgesics, antifungal, anticancer, antiinflammatory, antirheumatic, antimalarial, antivirus agents, and also enzymes, hormones, and peptides (Ravivarapu *et al.*, 2000; Santoyo *et al.*, 2002; Freitas *et al.*, 2005).

Microemulsions and nanoemulsions have also been extensively studied as drug carrier systems. The main advantage of these systems is its enhanced solubilization capacity for both hydrophilic and lipophilic drugs (Lawrence and Rees, 2000; Cortesi *et al.*, 1997). Microemulsions are thermodynamically stable and isotopic colloidal systems constituted by surfactants, an oil phase and an aqueous phase, and may also contain a cosurfactant. Several microemulsion-based drug delivery systems have been studied for enhancing drug bioavailability (Kang *et al.*, 2004; Junping *et al.*, 2003), improvement of skin permeation of transdermal drugs (Liu *et al.*, 2011), reduction of adverse effects, among others.

Microemulsions found application for delivery of a various class of drugs by several routes of administration. For example, in oral administration self-microemulsifying drug delivery systems (SMEDDS) are a promising carrier system for delivery of antibiotics and anticancer agents (Goyal *et al.*, 2012; Cui *et al.*, 2009; Yang *et al.*, 2004; Kang *et al.*, 2004), in special after the commercial success of some marketed products, like Sandimmune Neoral®, Fortovase®, and Norvir®. For transdermal delivery, microemulsions and microemulsion-based hydro- and organogels have been extensively studied for antiinflammatory and anesthetics drugs (Neubert, 2011; Tavano *et al.*, 2011; Liu *et al.*, 2011). Parenteral route is another route in which microemulsions have found application, mainly for the delivery of highly lipophilic drugs (Date and Nagarsenker, 2008; Brime *et al.*, 2002). In addition, microemulsion-based systems are also studied for drug delivery by other routes, like ocular (Hasse and Keipert, 1997), nasal (Zhang *et al.*, 2004), and vaginal (D'Cruz and Uckun, 2001).

The recent advances in drug delivery technology have significantly contributed to the development of new tools for human health. The aim of drug delivery systems is optimize drug administration by offer some well documented advantages in pre-clinical (Lawrence and Rees, 2000) and clinical experiments (Brannon-Peppas and Blanchette, 2004), such as: solubilization of lipophilic drugs, drug protection against degradation, low fluctuations in drug plasmatic concentration, reduction of number of ministered doses and reduction of costs (Brannon-Peppas, 1995). It is estimated that the use of drug delivery systems is able to reduce in 20 per cent the costs and halves the time expended to the development of new bioactive compounds or lead compounds (Feng *et al.*, 2004). In fact, several studies corroborated the significant improvement in the efficiency and the cutback of toxicity of several drugs or bioactive molecules, both from natural or synthetic sources, when administered as liposomes or nanoparticulate systems. Latest reports provide ultimate evidence of the cytotoxicity and antitumor activity of the taxol and the usnic acid from *Cladonia substellata* in microspheres and nanocapsules, the doxorubicin in liposomes and nanoparticles, the lectin from *Cratylia mollis* encapsulated into liposomes; the xantones; the 5-floururacil into liposomes, and also the antiparasitic activity of praziquatel into liposomes; the antibiotic activity of vancomicin into microparticles; antiestrogen encapsulated into liposomes and nanoparticles for the treatment of the breast cancer (Barraud *et al.*, 2001; Andrade *et al.*, 2004; Gupte and Ciftci, 2004; Ribeiro-Costa *et al.*, 2004; Bozkir and Saka, 2005; Le Ray *et al.*, 2005; Mourão *et al.*, 2005; Teixeira *et al.*, 2005; 2006; Santos *et al.*, 2006b).

Another example of recent advances in nanotechnology are the carbon nanotubes (CNTs). They are among the most interesting nanovectors currently under investigation. Functionalized CNTs have shown great promise as novel delivery systems based on their ability to cross biological barriers. In fact, even though the specic mechanism of internalization (endocytosis or needle like penetration) is still not fully elucidated, it is generally recognized that CNTs are able to enter cells, independent of cell type and functional groups at their surface (Kostarelos *et al.*, 2007). Fabbro and co-workers (2012) had described this property against cancer.

Anchored in the promising findings reached for the antitumor activity of *trans*-dehydrocrotonin (DCTN) and the likelihood of its severe hepatotoxicity supported by toxicological reports in humans (Maciel *et al.*, 2002a, 2006a; Soares, 2004), this natural product become, as a consequence, a drug prototype to be taken on the development of nanosystems for controlled release (liposomes, nanoparticles and microparticles). In this way, new DCTN formulations incorporated into microparticles of copolymer of lactic and glycolic acid (PLGA) have been developed aiming ameliorate the therapeutic efficiency and minimizing possible hepatotoxicity effects produced by this drug when administered in its free form. The hypoglycemic activity of both free and encapsulated DCTN is currently been evaluated in normal and hyperglycemic animals and the promising results point toward an improved bioavailability of this natural product, in which studies carried out with PLGA microparticles showed encapsulation efficiency for lipophilic DCTN of 85.5 per cent (Morais *et al.*, 2008; Nascimento-Filho *et al.*, 2010). Additionally, a complex of β-cyclodextrin with DCTN was developed to improve the delivery of this compound. It was assayed for both freshly isolated rat hepatocytes and in V79 cell. In all assays, the lower evidenced cytotoxicity of DCTN-tested complex suggests that such complex may be useful for the administration of DCTN *in vivo* (Corrêa *et al.*, 2005).

Since the stability of *t*-DCTN loaded in different biological formulations had been investigated aiming its anticancer pharmacologic improvement (Canegim *et al.*, 2011; Lapenda *et al.*, 2012, 2013) new biochemical and nanocarrier systems had been assessed, including experimental and NMR theoretical methodology applied to the geometric analysis of this bioactive clerodane (Soares *et al.*, 2013). The aiming of this work was the development of a new theoretical quantum chemistry methods applied to *t*-DCTN analysis on its magnetic and geometrical properties, attracting significant importance on physico-chemical characteristics of biocompounds that could be assessed in advancement of drug use in therapy. In fact, the *in vitro* kinetic profiles of *t*-DCTN loaded (liposomes and its inclusion complexes with hydropropyl-β-cyclodextrin) improve its antitumor activity (Lapenda *et al.*, 2013).

Concerning to pharmaceutical industry, drug delivery systems have attracted attention of mainly due to those cited above advantages as well other ones (Lu *et al.*, 2008; Qin *et al.*, 2010). Nowadays several drug delivery formulations have been marketed, such as ALPRESSä LP (prazosin); CARDURAâ XL (doxazosin mesylate); CONCERTAâ (methylphenidate); COVERA-HSâ (verapamil); DITROPANâ XL (oxybutynin); DYNACIRCâ CR (isradipine); EFIDACâ 24 (chlorpheniramine); GLUCOTROLâ XL (glipizide); SUDAFEDâ 24h (pseudoephedrine); PROCARDIAâ XL (nifedipine); CATAPRESS-TTSâ (clonidine); DURAGESICâ (fentanyl); ESTRADERMâ (estradiol); NICODERMâ CQ, CLEAR NICODERMâ CQ (nicotine); TESTODERMâ TTS, TESTODERMâ (testosterone); TRANSDERM-NITROâ (nitroglycerin); TRANSDERMâ SCOP (scopolamine); DOXILâ (doxorubicin) (http//www.alza.com).

As final comments, among the nonbiological anticancer agents, only 11 per cent of them are natural products isolated from natural sources, including plants, marine organisms and micro-organisms. From these sources effective anticancer agents represents 60 per cent. However, 77.8 per cent are either natural product per se, or are based thereon, while only 22 per cent of all anticancer are totally synthetic drugs, found by random screening or by modification of an existing agent (Cragg *et al.*, 2005; Newman and Cragg, 2007). An important aspect of phytotherapeuticals and its analogues is the problem of toxic responses, ranging from fatigue, aches, gastrointestinal disturbance to drastic toxic side effects such as toxic hepatitis, genotoxicity and mutagenicity. It has been recognized that plant-derived chemotherapeutic-agents commonly classified into major classes such as taxanes (paclitaxel and docetaxel), vinca alkaloids (vinblastin, vincristine and vindesine), epipodophyllotoxins (etoposide and teniposides) and campothecins (campothecin and irinotecan) had some accepted toxicity. Joined to that, genetically modified organisms (GMO) has been largely emphasized focusing their development, been an additional work to be considered in the field of chemotherapeutic agents. From this point of view, some natural products act as prevention of chemical carcinogenesis, been important chemopreventive for human dietary, food components taken in daily life are good candidates for this purpose and become an increasingly attractive area of study.

References

Agner, A.R., Maciel, M.A.M., Pinto, A.C., Pamplona, S.G.S.R., and Cólus, I.M.S.(1999). Investigation of genotoxicity activity of *trans*-dehydrocrotonin, a clerodane diterpene from *Croton cajucara. Teratogenesis Carcinogenesis and Mutagenesis*, **19(6)**: .377-384.

Agner, A.R., Maciel, M.A.M., Pinto, A.C., and Cólus, I.M.S. (2001). Antigenotoxicity of *trans*—dehydrocrotonin, a clerodane diterpene from *Croton cajucara. Planta Medica*, **67(9)**: 815-819.

Agosta, W.C. (1997). Medicines and drugs from plants. *Journal of Chemical Education*, **74(7):** 857-860.

Ain, Q., sharma, S., Garg, S.K., and Khuller, G.K. (2002). Role of poly[DL-lactide-co-glycolide] in development of a sustained oral delivery system for antitubercular drug(s). *International Journal of Pharmaceutics*, **239(1-2):** 37-46.

Al-Angary, A.A., Bayomi, M.A., Khidr, S. H., Al-Meshal, M.A., and Al-Dardiri, M. (1995). Characterization, stability and *in vivo* targeting of liposomal formulations containing cyclosporine. *International Journal of Pharmaceutics*, **114(2): 221-225.**

Almeida, A.B.A., Melo, P.S., Hiruma-Lima, C.A., Gracioso, J.S., Carli, L., Nunes, D.S., Haun, M., and Souza Brito, A.R.M. (2003). Antiulcerogenic effect and cytotoxic activity of semi-synthetic crotonin obtained from *Croton cajucara* Benth. *European Journal of Pharmacology*, **472(3):** 205-212.

Álvarez, M.E., María, A.O.M., and Saad, J.R. (2002). Diuretic activity of *Fabiana patagonica* in rats. *Phytotherapy Research*, **16(1):** 71-73.

Altmann, K.H., Gaugaz, F.Z., and Schiess, R. (2011). Diversity through semisynthesis: the chemistry and biological activity of semisynthetic epothilonederivatives. *Molecular Diversity*, **15(2):** 383-399.

Anazetti, M.C., Melo, P.S., Durán, N., and Haun M. (2003). Comparative cytotoxicity of dimethylamide-crotonin in the promyelocytic leukemia cell line (HL60) and human peripheral blood mononuclear cells. *Toxicology*, **188(2-3):** 261-274.

Anazetti, M.C., Melo, P.S., Durán, N., and Haun M. (2004). Dehydrocrotonin and its derivative, dimethylamide-crotonin induce apoptosis with lipid peroxidation and activation of caspases-2, -6, and -9 in human leukemic cells HL60. *Toxicology*, **203(1-3):** 123-137.

Andrade, C.A., Correia, M.T., Coelho, L.C., Nascimento, S.C., and Santos-Magalhães, N.S. (2004). Antintumor activity of *Cratylia mollis* lectin encapsulated into liposomes. *International Journal of Pharmaceutics*, **278(2)**: 435-445.

Andrade, C.A., Correia, M.T., Coelho, L.C., Nascimento, S.C., and Santos-Magalhães, N.S. (2005). Usnic acid-loaded nanocapsules: An evaluation of cytotoxicity. *Journal of Drug Delivery Science and Technology*, **15(5): 355-361.**

Arliss, R.M., and Biermann, C.A. (2002). Do soy isoflavones lower cholesterol, inhibit atherosclerosis, and play a role in cancer prevention? *Holistic Nursing Practice*, **17(1):** 40-48.

Bajaj, N.R., and Garratt, K.N. (2010). Clinical studies with paclitaxel–eluting stent systems. *Current Pharmaceutical Design*, **16(36):** 4025-4036.

Balunas, M.J., and Kinghorn, A.D. (2005). Drug discovery from medicinal plants. *Life Sciences*, **78(5):** 431-441.

Barnes, S. (1995). Effect of genistein on *in vitro* and *in vivo* models of cancer. *Journal of Nutrition*, **125(supl.3)**: 777S-783S.

Barraud, L., Merle, P., Soma, E., Lefrançois, L., Guerret, S., chevallier, M., dubernet, C., Couvreur, P., Trepo, C., and Vitvitski, L. (2001). Increase of doxorubicin sensitivity by doxorubicin-loading into nanoparticles for hepatocellular carcinoma cells *in vitro* and *in vivo*. *Journal of Hepatology*, **42(5):** 736-743.

Benoit, J.-P., Faisant, N., Venier-Julienne, M.-C., and Menei, P. (2000). Development of microspheres for neurobiological disorders: From basics to clinical applications. *Journal of Controlled Release*, **65(1-2):** 285-296.

Berkland, C., King, M., Cox, A., Kim, K., and Pack, D.W. (2002). Precise control of PLG microsphere size provides enhanced control of drug release rate. *Journal of Controlled Release*, **82(1):** 137-147.

Berkland, C., Kim, K.K., and Pack, D.W. (2001). Fabrication of PLG microspheres with precisely controlled and monodisperse size distributions. *Journal of Controlled Release*, **73:** 59-74.

Birnbaum, D.T., and Brannon-Peppas, L. (2003). Molecular weight distribution changes during degradation and release of PLGA nanoparticles containing epirubicin HCl. *Journal of Biomaterials Science, Polymer Edition*, **14(1):** 87-102.

Boeheme, K., Simon, S., and Mueller, S.O. (2009). Gene expression profiling in ishikawa cells: A fingerprint for estrogen active compounds. *Toxicology and Applied Pharmacology*, 236(1): 85-96.

Borman, S. (1994). Total synthesis of anticancer agent taxol achieved by two different routes. *Chemical and Engineering News*, **72**: 32-34.

Bozkir, A., and Saka, O.M. (2005). Formulation and investigation of 5-FU nanoparticles with factorial design-based studies. *Farmaco*, **60(10):** 840–846.

Bradbury, R. B., and White, D. E. (1951). Chemistry of subterranean clover. I. Isolation of formononetin and genistein. *Journal of Chemical Society*, 3447-3449.

Brandi, M.L. (1997). Natural and synthetic isoflavones in the prevention and treatment of chronic diseases. *Calcified Tissue International*, **61(suppl.1)**: S5-S8.

Brannon-Peppas, L. (1995). Recent advances on the use of biodegradable microparticles and nanoparticles in controlled drug delivery. *International Journal of Pharmaceutics*, **116(1):** 1-9.

Brannon-Peppas, L., and Blanchette, J.O. (2004). Nanoparticle and targeted systems for cancer therapy. *Advanced Drug Delivery Reviews*, **56(11):** 1649-1659.

Bridges, B.A. (1987). International Commission for Protection against Environmental Mutagens and Carcinogens. ICPEMC Topic No. 1. Are somatic mutations involved in atherosclerosis? *Mutation Research*, **182(6):** 301-302.

Bridges, B.A., Bowyer, D.E., Hansen, E.S., Penn, A., and Wakabayashi, K. (1990). International Commission for Protection Against Environmental Mutagens and Carcinogens. The possible involvement of somatic mutations in the development of atherosclerotic plaques. Report of ICPEMC Subcommittee 7/1. Conclusions and recommendations. *Mutation Research*, **239(3):** 143-148.

Bridges, B.A. (1991). International Commission for Protection Against Environmental Mutagens and Carcinogens. Report of ICPEMC Subcommittee 7/1. The possible involvement of somatic mutations in the development of atherosclerotic plaques. *Atherosclerosis*, **89(2-3):** 255-261.

Brusick, D. (1987). *Principles of Genetic Toxicology*. Nova Iorque: Plenum Press, 2.ed., p.13.

Burz, C., Berindan-Neagoe, I., Balacescu, O., and Irimie, A. (2009). Apoptosis in cancer: key molecular signaling pathways and therapy targets. *Acta Oncologica*, **48(6):** 811-821.

Canegim, B.H., Serpeloni, J.M., Maciel, M.A.M., Colus, I.M.S., and Mesquita, S.F.P. (2011). Reduction of DNA-damage by *Croton cajucara* methanolic extract but not the testicular alterations induced by doxorubicin. *Journal of Medicinal Plants Research*, **5(14)**: 3277-3285.

Carvalho, J.C.T., Silva, M.F.C., Maciel, M.A.M., Pinto, A.C., Nunes, D.S., Lima, R.M., Bastos, J.K., and Sarti, S.J. (1996). Investigation of anti-inflammatory and antinociceptive activities of *trans*-dehydrocrotonin, a 19-*nor*-clerodane diterpene from *Croton cajucara*. Part 1. *Planta Medica*, **62(5)**: 402-404.

Cassels, B.K., and Asencio, M. (2011). Anti-HIV activity of natural triterpenoids and semisynthetic derivatives 2004-2009. *Phytochemistry Reviews*, **10(4):** 545-564.

Castrillo, A., De las Heras, B., Hortelano, S., Rodríguez, B., Villar, A., and Boscá, L. (2001). Inhibition of the nuclear factor κB (NF-κB) pathway by tetracyclic kaurene diterpenes in macrophages. Specific effects on NF-κB-inducing kinase activity and on the coordinate activation of ERK and p38 MAPK. *Journal of Biological Chemistry*, **276(19):** 15854-15860.

Castro, A., Del Corral, J.M.M., Gordaliza, M., Grande, C., Gómez-zurita, A., García-grávalos, D., and Feliciano, A.S. (2003). Synthesis and cytotoxicity of podophyllotoxin analogues modified in the A ring. *European Journal of Medicinal Chemistry*, 38(1): 65-74.

Chang, A.Y., Kim, K., Glick, J., Anderson, T., Karp, D., and Johnson, D. (1993). Phase II study of taxol, merbarone, and piroxantrone in stage IV non-small-cell lung cancer: The Eastern Cooperative Oncology Group Results. *Journal of National Cancer Institute*, **85(5):** 388-394.

Chaves, M.H., Roque, N.F., and Ayres, M.C.C. (2004). Steroids and flavonoids of *Porcelia macrocarpa*. *Journal of the Brazilian Chemical Society*, **15(4):** 608-613.

Cheeseman, K.H., and Slater, T.F. (1993). An introduction to free radical biochemistry. *British Medical Bulletin*, **49(3):** 481-493.

Chen, H., Bi, W., Cao, B., Yang, Z., Chen, S., Shang, H., Yu, P., and Yang, J. A. (2010). Novel podophyllotoxin derivative (YB-1EPN) induces apoptosis and down-regulates express of P-glycoprotein in multidrug resistance cell line KBV2000. *European Journal of Pharmacology*, **627**: 69-74.

Commane, D., Hughes, R., Shortt, C., and rowland, I. (2005). The potential mechanisms involved in the anti-carcinogenic action of probiotics. *Mutation Research*, **591(1-2):** 276-289.

Conklin, K.A. (2000). Dietary antioxidants during cancer chemotherapy: impact on chemotherapeutic effectiveness and development of side effects. *Nutrition and Cancer*, **37(1):** 1-18.

Corrêa, D.H.A., Melo, P.S., Carvallho, C.A.A., Azevedo, M.B.M., Durán, N., and Haun, M. (2005). Dehydrocrotonin and its beta-cyclodextrin complex: Cytotoxicity in V79 fibroblasts and rat cultured hepatocytes. *European Journal of Pharmacology*, **510(1-2):** 17-24.

Cortesi, R., Esposito, E., Maietti, A., Menegatti, E., and Nastruzzi, C. (1997).Formulation study for the antitumor drug camptothecin: liposomes, micellar solutions and a microemulsion. *International Journal of Pharmaceutics*, **159(1):** 95-103.

Cragg, G. M., and Newman, D.J. (2005). Plants as a source of anti-cancer agents. *Journal of Ethnopharmacology*, **100(1-2):** 72-79.

Croft, M. (2009). The role of TNF superfamily members in T-cell function and diseases. *Nature Reviews Immunology*, **9(4):** 271-285.

Cui, S. X., Nie, S. F., Li, L., Wang, C. G., Pan, W. S., and Sun, J. P. (2009). Preparation and evaluation of self-microemulsifying drug delivery system containing vinpocetine. *Drug Development an Industrial Pharmacy*, **35(5):** 603-611.

Date, A.A., and Nagarsenker, M.S. (2008). Parenteral microemulsions: An overview. *International Journal of Pharmaceutics*, **355(1-2):** 19-30.

D'cruz, O.J., and Uckun, F.M. (2001). Intravaginal toxicity studies of a gel-microemulsion formulation of spermicidal vanadocenes in rabbits. *Toxicology and Applied Pharmacology*, **170(2):** 104-112.

De Vita, J.E., Hellman, S., and Rosenberg, S.A. (1997). *Cancer, Principles and Practice of Oncology*. Filadelfia: Lippincott-Raven Publishers, 5.ed., v.1.

Dimas, K., Demetzos, C.; Vaos, V., Ioannidis, P., and Trangas, T. (2001). Labdane type diterpenes down-regulate the expression of *c-myc* protein, but not of *bcl-2*, in human leukemia T-cells undergoing apoptosis. *Leukemia Research*, **25(6):** 449-454.

Dixon, R.A., and Ferreira, D. (2002). Genistein. *Phytochemistry*, **60(3):** 205-211.

Drummond, T.C., Noble, C.O., Guo, Z., Hong, K., Park, J.W., and Kirpotin, D.B. (2006). Development of a highly active nanoliposomal irinotecan using a novel intraliposomal stabilization strategy. *Cancer Research*, **66(6):** 3271-3277.

Elkharraz, K., Faisant, N., Guse, C., Siepmann, F., Arica-Yegin, B., Oger, J.M., Gust, R. Goepferich, A., Benoit, J.P., and Siepmann, J. (2006). Paclitaxel-loaded microparticles and implants for the treatment of brain cancer: Preparation and physicochemical characterization. *International Journal of Pharmaceutics*, **314(2):** 127-136.

Fabbro, C., Ali-Boucetta, H., Da Ros, T., Kostarelos, K., Bianco, A., and Prato, M. (2012). Targeting carbon nanotubes against cancer. *Chemical Communications*, DOI: 10.1039/c2cc17995d.

Fang, C., Shi, B., Pei,Y. Y., Hong, M.H., Wu, J., and Chen, H. Z. (2006). *In vivo* tumor targeting of tumor necrosis factor-α-loaded stealth nanoparticles: Effect of MePEG molecular weight and particle size. *European Journal of Pharmaceutical Sciences*, **27(1):** 27-36.

Farias, R.A.F., Rao, V.S.N., Viana, G.S.B., Silveira, E.R., Maciel, M.A.M., and Pinto, A.C. (1997). Hypoglycemic effect of *trans*-dehydrocrotonin, a *nor*-clerodane diterpene from *Croton cajucara*. *Planta Medica*, **63(6):** 558-560.

Farias, R.A.F., González, R.P., Leyva, A., Maia, L.S., Maciel, M.A.M., Pinto, A.C., Veiga Jr., V.F., Pessoa, C., and Moraes, M.O. (2005). Chromatographic fractions from *Croton cajucara* inhibit cell proliferation and induce differentiation in a human leukemia cell line. *Journal of Cancer Integrative Medicine*, **3(2):** 75-80.

Farnsworth, N.R., and Morris, R.W. (1976). Higher plants – the sleeping giant of drug development. *American Journal of Pharmacy*, **148(2):** 46-52.

Feng, S.S., Ruan, G., and Li, Q.T. (2004). Fabrication and characterization of a novel drug delivery device liposomes-in-microspheres (LIM). *Biomaterials*, **25(21):** 5181-5189.

Flora, S., and Ferguson, L.R. (2005). Overview of mechanisms of cancer chemopreventive agents. *Mutation Research*, **591(1-2):** 8-15.

Fonseca, C., Simões, S., and Gaspar, R. (2002). Paclitaxel-loaded PLGA nanoparticles: Preparation, physicochemical characterization and *in vitro* anti-tumoral activity. *Journal of Controlled Release*, **83(2):** 273-286.

Freiberg, S., and Zhu, X.X. (2004). Polymer microspheres for controlled drug release. *International Journal of Pharmaceutics*, **282(1/2):** 1-18.

Freire, A.C.G., Assis, C.F., Frick, A.O., Melo, P.S., Haun, M., Aoyama, H., Durán, N., Sauer, M.M., Kallás, *E.G.*, and Ferreira, C.V. (2003a). Influence of protein phosphatase inhibitors on HL60 cells death induction by dehydrocrotonin. *Leukemia Research*, **27(9):** 823-829.

Freire A.C.G., Silva M.P., Aoyama H., Haun M., Duran N., and Ferreira C.V. (2003b). Cytotoxic effect of the diterpene lactone dehydrocrotonin from *Croton cajucara* on human promyelocytic leukemia cells. *Planta Medica*, **69(1):** 67-69.

Freitas, S., Merkle, H.P., and Gander, B. (2005). Microencapsulation by solvent extraction/evaporation: reviewing the state of the art of microsphere preparation process technology. *Journal of Controlled Release*, **102(2):** 313-332.

Fu, Y., Li, S., Zu, Y., Yang, G., Yang, Z., Luo, M., Jiang, S., Wink, M., and Efferth, T. (2009). Medicinal chemistry of paclitaxel and its derivatives. *Current Medicinal Chemistry*, **16(30):** 3966-3985.

Fukutake, M., Takahashi, M., Ishida, K., Kawamura, H., Sugimura, T., and Wakabayashi, K. (1996). Quantification of genistein and genistin in soybeans and soybeans products. *Food and Chemical Toxicology*, **34(5):** 457-461.

Gao, F., Yang, Z.K., Chen, Q.H., Chen, X.G., and Wang, F.P. (2012). A novel D-ring modified taxoid: synthesis and biological evaluation of a gamma-lactone analogue of docetaxel. *Organic and Biomolecular Chemistry*, **10(2):** 361-366.

Garinot, M., Mignet, N., Largeau, C., Seguin, J., Scherman, D., and Bessodes, M. (2007). Amphiphilic polyether branched molecules to increase the circulation time of cationic particles. *Bioorganic and Medicinal Chemistry*, **15(9):** 3176-3186.

Gidding, C.E.M., Kellie, S.J., Kamps, W.A., and De Graaf, S.S.N. (1999). Vincristine revisited. *Critical Reviews in Oncology/Hematology*, **29**: 267-287.

Giri, A.K., Mukhopadhyay, A., Sun, J., Hsie, A.W., and Ray, S. (1999). Antimutagenic effects of centchroman-a contraceptive and a candidate drug for breast cancer in multiple mutational assays. *Mutagenesis*, **14(6):** 613-619.

Glen, C.D., and Dubrova, Y.E. (2012). Exposure to anticancer drugs can result in transgenerational genomic instability in mice. *Proceeding of the National Academy of Science of the United States of America*, **10(8)**: 2984-2988.

Grant, I.A., Rodriguez, C.O., Kent, M.S., Sfiolgoi, G., Gordon, I., Davis, G., Lord, L., and London, C.A. (2008). A phase II clinical trial of vinorelbine in dogs with cutaneous mast cell tumors. *Journal of Veterinary Internal Medicine*, **22(2):** 388-393.

Gref, R., Quellec, P., Sanchez, A., Calvoc, P., Dellacherie, E., and Alonso, M.J. (2001). Development and characterization of CyA-loaded poly(lactic acid)-poly(ethylene glycol)PEG micro- and nanoparticles. Comparison with conventional PLA particulate carriers. *European Journal of Pharmaceutics and Biopharmaceutics*, **51(2):** 111-118.

Grynberg, N.F., Echevarria, A., Lima, J.E., Pamplona, S.G.S.R., Pinto, A.C., and Maciel, M.A.M. (1999). Anti-tumour activity of two 19-*nor*-clerodane diterpenes, *trans*-dehydrocrotonin and *trans*-crotonin, from *Croton cajucara*. *Planta Medica*, **65(8):** 687-689.

Gupte, A., and Ciftci, K. (2004). Formulation and characterization of Paclitaxel, 5-FU and Paclitaxel + 5-FU microspheres. *International Journal of Pharmaceutics*, **276(1-2):** 93-106.

Güthlein, F., Burger, A.M., Brandl, M., Fiebig, H.H., Schubert, R., Unger, C., and Massing, U. (2002). Pharmacokinetics and antitumor activity of vincristine entrapped in vesicular phospholipid gels. *Anti-cancer Drugs*, **13(8):** 797-805.

Hansen, E.S. (1990). International Commission for Protection Against Environmental Mutagens and Carcinogens. ICPEMC Working Paper 7/1/2. Shared risk factors for cancer and atherosclerosis – a review of the epidemiological evidence. *Mutation Research*, **239(3):** 163-179.

Hammons, G.J., Fletcher, J.V., Stepps, K.R., Smith, E.A., Balentine, D.A., Harbowy, M.E., and Kadlubar, F.F. (1999). Effects of chemoprotective agents on the metabolic activation of the carcinogenic arylamines PhIP and 4-aminobiphenyl in human and rat liver microsomes. *Nutrition and Cancer*, **33(1):** 46-52.

Hans, M.L., and Lowman, A.M. (2002). Biodegradable nanoparticles for drug delivery and targeting. *Current Opinion in Solid State and Materials Science*, **6(4):** 319-327.

Harbone, J.B. (1994). *The Flavonoids, Advances in Research Since 1986*. London: Chapman and Hall.

Haraguchi, H., Ishikawa, H., and Kubo, I. (1997). Antioxidative action of diterpenoids from *Podocarpus nagi*. *Planta Medica*, **63(3):** 213-215.

Hasse, A., and Keipert, S. (1997). Development and characterization of microemulsions for ocular application. *European Journal of Pharmaceutics and Biopharmaceutics*, **43(2):** 179-183.

He, L., Wang, G.I., and Zhang, Q. (2003). An alternative paclitaxel microemulsion formulation: hypersensitivity evaluation and pharmacokinetic profile. *International Journal of Pharmaceutics*, **250(1):** 45-50.

Hernaez, J.F., Xu, M., and Dashwood, R.H. (1998). Antimutagenic activity of tea towards 2-hydroxyamino-3-methylimidazo[4,5-f]quinoline: effect of tea concentration and brew time on electrophile scavenging. *Mutation Research*, **402(1-2):** 299-306.

Hiruma-Lima, C.A., Spadari-Bratfisch, R.C., Grassi-Kassisse, D.M., and Souza-Brito, A.R.M. (1999). Antiulcerogenic mechanisms of dehydrocrotonin, a diterpene lactone obtained from *Croton cajucara*. *Planta Medica*, **65(4):** 325-330.

Hitt, E. (2002). Homoharringtonine effective in CML patients. *Lancet Oncology*, **3(5):** 259.

Holmes, F.A., Walters, R.S., Theriault, R.L., Forman, A.D., Newton, L.K., Raber, M.N., Buzdar, A.U., Frye, D.K., and Hortobagyi, G.N. (1991). Phase II trial of taxol, an active drug in the treatment of metastatic breast cancer. *Journal of the National Cancer Institute*, **83(24):** 1797-1805.

Hongjie, Z., Takeda, Y., Matsumoto, T., Minami, Y., Yoshida, K., Wei, X., Qing, M., and Handong, S. (1994). Taxol related diterpenes from the roots of *Taxus yunnanensis*. *Heterocycles*, **38(5):** 975-980.

Horiguchi, T., Oritani, T., and Kiyota, H. (2003a). Synthesis and antitumor activity of 2-(*m*-substituted-benzoyl)baccatin III analogs from taxinine. *Tetrahedron*, **59(9):** 1529-1538.

Horiguchi, T., Rithner, C.D., Croteau, R., and Williams, R.M. (2003b). Studies on taxol biosynthesis. Preparation of 5α-acetoxytaxa-4(20),11-dien-2α,10β-diol derivatives by deoxygenation of a taxadiene tetra-acetate obtained from Japanese yew. *Tetrahedron*, **59(2):** 267-273.

Hsu, H.Y., Yang, J.J., and Lin, C.C. (1997). Effects of oleanolic acid and ursolic acid on inhibiting tumor growth and enhancing the recovery of hematopoietic system postirradiation in mice. *Cancer Letters*, **111(1-2):** 7-13.

Hu, C.Q., Xu, D.Q., Du, W.T., Qian, S.J., Wang, L., Lou, J.S., He, Q.J., Yang, B., and Hu, Y.Z. (2010). Novel 4 beta-anilino-podophyllotoxin derivatives: design synthesis and biological evaluation as potent DNA-topoisomerase II poisons and anti-MDR agents. *Molecular Biosystems*, **6(2)**: 410-420.

Hu, C.Q, Nikolic, D., Eggler, A.L., Mesecar, A.D., and Van Breemen, R.B. (2012). Screening for natural chemoprevention agents that modify human Keapl. *Analytical Biochemistry*, **421(1):** 108-114.

Huang, M.T., Ho, C.T., Wang, Z.Y., Ferraro, T., Lou, Y.R., Stauber, K., Ma, W., Georgiadis, C., Laskin, J.D., and Conney, A.H. (1994). Inhibition of skin tumorigenesis by rosemary and its constituents carnosol and ursolic acid. *Cancer Research*, **54(3):** 701-708.

Ikeda, Y., Murakami, A., and Ohigashi, H. (2005). Ursolic acid promotes the release of macrophage migration inhibitory factor via ERK2 activation in resting mouse macrophages. *Biochemical Pharmacology*, **70(10):** 1497-1505.

Ikeda, Y., Murakami, A., Nishizawa, T., and Ohigashi, H. (2006). Ursolic acid enhances cyclooxygenases and tumor necrosis factor-α expression in mouse skin. *Bioscience, Biotechnology and Biochemistry*, **70(4):** 1033-1037.

Ikeda, Y., Murakami, A., Fujimura, Y., Tachibana, H., Yamada, K., Masuda, D., Hirano, K.I., Yamashita, S., and Ohigashi, H. (2007). Aggregated ursolic acid, a natural triterpenoid, induces IL-1β release from murine peritoneal macrophages: Role of CD36. *Journal of Immunology*, **178(8):** 4854-4864.

Ikeda, Y., Murakami, A., and Ohigashi, H. (2008). Ursolic acid: An anti- and pro-inflammatory triterpenoid. *Molecular Nutrition and Food Research*, **52(1):** 26-42.

Imbert, T.F. (1998). Discovery of podophillotoxins. *Biochimie*, **80:** 207-222.

Itokawa, H., Morris-Natschk, S.L., Akiyama, T., and Lee, K. (2008). Plant-derived natural product research aimed at new drug discovery. *Journal of Natural Medicine*, **62:** 263-280.

Jabs, T. (1999). Reactive oxygen intermediates as mediators of programmed cell death in plants and animals. *Biochemical Pharmacology*, **57(3):** 231-245.

Jacob, J. (1968). *The influence of variations in species and strains of experimental animals on drug responses*. In: Tedeschi, D.H.; Tedeschi, R.E. (eds.). *Importance of fundamental principles in drug evaluation*. Nova Iorque: Raven Press.

Jain, R.A. (2000). The manufacturing techniques of various drug loaded biodegradable poly(lactide-co-glycolide) (PLGA) devices. *Biomaterials*, **21(32):** 2475-2490.

Janicsák, G., Veres, K., Kakasy, A.Z., and Máthé, I. (2006). Study of the oleanolic and ursolic acid contents of some species of the Lamiaceae. *Biochemical Systematics and Ecology*, **34(5):** 392-396.

Jeong, H.G. (1999). Inhibition of cytochrome P450 2E1 expression by oleanolic acid: Hepatoprotective effects against carbon tetrachloride-induced hepatic injury. *Toxicology Letters*, **105(3):** 215-222.

Junping, W., Takayama, K., Nagai, T., and Maitani, Y. (2003). Pharmacokinetics and antitumor effects of vincristine carried by microemulsions composed of PEG-lipid, oleic acid, vitamin E and cholesterol. *International Journal of Pharmaceutics*, **251(1-2):** 13-21.

Kanai, R., Rabkin, S.D., Yip, S., Sgubin, D., Zaupa, C.M., Hirose, Y., Louis, D.N., and Wakimoto, H. (2012). Oncolytic virus-mediated manipulation of DNA damage responses: synergy with chemotherapy in killing glioblastoma stem cells. *Journal of the National Cancer Institute*, **104(1):** 42-55.

Kang, B.K., Chon, S.K., Kim, S.H., Jeong, S.Y., Kim, M.S., Cho, S.H., Lee, H.B., and Khang, G. (2004). Controlled release of paclitaxel from microemulsion containing PLGA and evaluation of anti-tumor activity *in vitro* and *in vivo*. *International Journal of Pharmaceutics*, **286(1-2):** 147-156.

Kashiwada, Y., Nagao, T., Hashimoto, A., Ikeshiro, Y., Okabe, H., Cosentino, L.M., and Lee, K.H. (2000). Anti-HIV activity of 3-*O*-acyl ursolic acid derivatives. *Journal of Natural Products*, **63(12):** 1619-1622.

Keller, L., Dumas, F., and D´Angelo, J. (2001). Enantioselective synthesis of the ester side chain of homoharringtonine. *Tereahedron Letters*, **42(10):** 1911-1913.

Kernan, R.M., and Faulkner, D.J. (1988). Regioselective oxidation of 3-alkylfurans to 3-alkyl-4-hydroxybutenolides, *Journal of Organic Chemistry*, **53**: 2773-2776.

Khan, M.T.H., Ather, A., Pinto, A.C., and Maciel, M.A.M. (2009). Potential benefits of the 19-*nor*-clerodane *trans*-dehydrocrotonin on the central nervous system. *Brazilian Journal of Pharmacognosy*, **19(1):** 7-13.

Kim, I.S., and Kim, S.H. (2003). Development of polymeric nanoparticulate drug delivery systems: Evaluation of nanoparticles based on biotinylated poly(ethylene glycol) with sugar moiety. *International Journal of Pharmaceutics*, **257(1-2):** 195-203.

Kingston, D.G.I. (1991). The chemistry of taxol. *Pharmacology and Therapeutics*, **52(1):** 1-34.

Kinghorn, A.D., and Seo, E.K. (1966). Cultivating the pharmacopoeia. *Chemthec*, **26(7):** 46-54.

Klein, C.B., and King, A.A. (2007). Genistein genetoxicity: Critical considerations of *in vitro* exposure dose. *Toxicology and Applied Pharmacology*, **224(1):** 1-11.

Kosmidis, P.A., Fountzilas, G., Eleftheraki, A.G., Kalofonos, H.P., Pentheroudakis, G., Skarlos, D., Dimopoulos, M.A.; Bafaloukos, D., Pectasides, D.; Samantas, E., Boukovinas, J., Lambaki, S., Katirtzoglou, N., Bakogiannis, C., and Syrigos, K.N. (2011). Paclitaxel and gemcitabine versus paclitaxel and vinorelbine in patients with advanced non-small-cell lung cancer. A phase III study of the Hellenic Cooperative Oncology Group (HeCOG). *Annals of Oncology*, **22(4):** 827-834.

Kostarelos, K., Lacerda, L., Pastorin, G., Wu, W., Wieckowski, S., Godefroy, S., Pantarotto, D.; Briand, J.P., Muller, S., Prato, M., and Bianco, A. (2007). Cellular uptake of functionalized carbon nanotubes is independent of functional group and cell type. *Nature Nanotechnology*, **2(2):** 108-113.

Kowalski, R. (2007). Studies of selected plant raw materials as alternative sources of triterpenes of oleanolic and ursolic acid types. *Journal of Agricultural and Food Chemistry*, **55(3):** 656-662.

Kris, M.G., O´Connell, J.P., Gralla, R.J., Wertheim, M.S., Parente, R.M., Schiff, P.B., and Young, C.W. (1986). Phase I trial of taxol given as a 3-hour infusion every 21 days. *Cancer Treatment Reports*, **70(5):** 605-607.

Kudoh, S., Nakamura, S., Nakano, T., Komuta, K., Isobe, T., Katakami, N., Fukuda, Y., Takada, Y., Takada, M., Fukuoka, M., and Ariyoshi, Y. (2005). Irinotecan and etoposide for previously untreated extensive-disease small cell lung cancer: a phase II trial of West Japan Thoracic Oncology Group. *Lung Cancer*, **49(2):** 263-269.

Kumar, N., Ravikumar, M.N.V., and Domb, A.J. (2001). Biodegradable block copolymers. *Advanced Drug Delivery Reviews*, **53(1):** 23-44.

Laekeman, G.M., Claeys, M., Rwangabo, P.C., Herman, A.G., and Vlietinck, A.J. (1986). Cardiovascular effects of 3-methylquercetin. *Planta Medica*, **6**: 433-437.

Lahiri, M., Maru, G.B., and Bhide, S.V. (1993). Effect of plant phenols, β-carotene and α-tocopherol on benzo[a]pyrene-induced DNA damage in the mouse forestomach mucosa (target organ) and bone marrow polychromatic erythrocytes (non-target organ). *Mutation Research*, **303(2):** 97-100.

Lapenda, T.L.S., Morais, W.A., Lira, M.C.B., Maciel, A.A.M., and Santos-Magalhaes, N.S. (2012). Valiation of a UV spectrophotometric method for determining *trans*-dehydrocrotonin in inclusion complex with hydroxypropyl-β-cyclodextrin. *Latin American Journal of Pharmacy*, **31(1):** 97-103.

Lapenda, T.L.S., Morais, W.A., Almeida, F.J.F., Ferraz, M.S., Lira, M.C.B., Santos, N.P.S., Maciel, M.A.M. and Santos-Magalhaes, N.S. (2013). Encapsulation of *trans*-dehydrocrotonin in liposomes: Na enhancement of the antitumor activity. *Journal of Biomedical Nanotechnology*, **31(1):** 499-510.

Latha, P.G., and Panikkar, K.R. (1998). Cytotoxic and antitumor principles from *Ixora coccinea* flowers. *Cancer Letters*, **130(1-2):** 197-202.

Latha, P.G., and Panikkar, K.R. (1999). Modulatory effects of *Ixora coccinea* flower on cyclophosphamide-induced toxicity in mice. *Phytotherapy Research*, **13(6):** 517-520.

Latha, P.G., and Panikkar, K.R. (2001). Chemoprotective effect of *Ixora coccinea* L. flowers on cisplatin induced toxicity in mice. *Phytotherapy Research*, **15(4):** 364-366.

Lawrence, M.J., and Rees, G.D. (2000). Microemulsion-based media as novel drug delivery systems. *Advanced Drug Delivery Reviews*, **45(1):** 89-121.

Le Garrec, D., Gori, S., Luo, L., Lessard, D., Smith, D.C., Yessine, M.A., Ranger, M., and Leroux, J.C. (2004). Poly(*N*-vinylpyrrolidone)-block-poly(D,L-lactide) as a new polymeric solubilizer for hydrophobic anticancer drugs: *In vitro* and *in vivo* evaluation. *Journal of Controlled Release*, **99(1):** 83-101.

Lee, J.H., Koo, T.H., Hwang, B.Y., and Lee, J.J. (2002). Kaurane diterpene, kamebakaurin, inhibits NF-κB by directly targeting the DNA-binding activity of p50 and blocks the expression of antiapoptotic NF-κB target genes. *Journal of Biological Chemistry*, **277(21):** 18411-18420.

Lee, K.H., Lin, Y.M., Wu, T.S., Zhang, D.C., Yamagishi, T., Hayashi, T., Hall, I.H., Chang, J.J., Wu, R.Y., and Yang, T.H. (1988). The cytotoxic principles of *Prunella vulgaris*, *Psychotria serpens*, and *Hyptis capitata*: Ursolic acid and related derivatives. *Planta Medica*, **54(4):** 308-311.

Le-Ray, A.M., Gautier, H., Laty, M.K., Daculsi, G., Merle, C., Jacqueline, C., Hamel, A., and Caillon, J. (2005). *In vitro* and *in vivo* bactericidal activities of vancomycin dispersed in porous biodegradable poly(α-caprolactone) microparticles. *Antimicrobial Agents and Chemotherapy*, **49(7):** 3025-3027.

Li, Y., Matsuda, H., and Yoshikawa, M. (1999). Effects of oleanolic acid glycosides on gastrointestinal transit and ileus in mice. *Bioorganic and Medicinal Chemistry*, **7(6):** 201-1205.

Li, J., Guo, W.J., and Yang, Q.Y. (2002). Effects of ursolic acid and oleanolic acid on human colon carcinoma cell line HCT15. *World Journal of Gastroenterology*, **8(3):** 493-495.

Li, H., and Lin, X. (2008). Positive and negative signaling components involved in TNF a-induced NF-κB activation. *Cytokine*, **41:** 1-8.

Liu, J., Liu, Y., and Klaassen, C.D. (1994). The effect of Chinese hepatoprotective medicines on experimental liver injury in mice. *Journal of Ethnopharmacology*, **42(3):** 183-191.

Liu, J. (1995). Pharmacology of oleanolic acid and ursolic acid. *Journal of Ethnopharmacology*, **49(2):** 57-68.

Liu, Y., Hartley, D.P. and Liu, J. (1998). Protection against carbon tetrachloride hepatotoxicity by oleanolic acid is not mediated through metallothionein. *Toxicology Letters*, **95(2):** 7-85.

Liu, J.J., Hong, R.L., Cheng, W.F., Hong, K., Chang, F.H., and Tseng, Y.L. (2002). Simple and efficient liposomal encapsulation of topotecan by ammonium sulfate gradient: Stability, pharmacokinetic and therapeutic evaluation. *Anti-Cancer Drugs*, **13(7):** 709-717.

Liu, J. (2005). Oleanolic acid and ursolic acid: Research perspectives. *Journal of Ethnopharmacology*, **100(1-2):** 92-94.

Liu, C.H., Chang, F.Y., and Hung, D.K. (2011). Terpene microemulsions for transdermal curcumin delivery: Effects of terpenes and cosurfactants. *Colloids and Surfaces B: Biointerfaces*, **82(1):** 63-70.

Lohman, P.H.M. (2002). International Commission for the Protection of the Environment against Mutagens and Carcinogens: A historical perspective. *Mutatation Research*, **511(1):** 63-71.

Lorand, T , Vigh, E., and Garai, J. (2010). Hormonal action of plant derived and anthropogenic non-steroidal estrogenic compounds: phytoestrogens and xenoestrogens. *Current Medicinal Chemistry*, **17(30):** 3542-3574.

Lu, J., Choi, E., Tamanoi, F., and Zink, J.I. (2008). Light-activated nanoimpeller-controlled drug release in cancer cells. *Small*, **4(4):** 421-426.

Luna-Costa, A.M., Silva, J.C.R., Campos, A.R., Rao, V.S.N., Maciel, M.A.M., and Pinto, A.C. (1999). Antiestrogenic effect of *trans*-dehydrocrotonin, a *nor*-clerodane diterpene from *Croton cajucara*. *Phytotherapy Research*, **13(8):** 689-691.

Ma, C.M., Nakamura, N., Hattori, N., Kakuda, H., Qiao, J.C., and Yu, H.L. (2000). Inhibitory effects on HIV-1 protease of constituents from the wood of *Xanthoceras sorbifolia*. *Journal of Natural Products*, **63(2):** 238-242.

Ma, X., and Wang, Z. (2009). Anticancer drug discovery in the future: an evolutionary perspective. *Drug Discovery Today*, **14(23/24):** 1136-1142.

Ma, W., Park, G.L., Gomez, G.A., Nieder, M.H., Adams, T.L., Aynsley, J.S., Sahai, O.P., Smith, R.J., Stahlhut, R.W., Hylands, P.J., Bitsch, F., and Shackleton, C. (1994). New bioactive taxoids from cell cultures of *Taxus baccata*. *Journal of Natural Products*, **57(1):** 116-122.

Maciel, M.A.M., Pinto, A.C., Brabo, S.N., and Silva, M.N. (1998). Terpenoids from *Croton cajucara*. *Phytochemistry*, **49(3):** 823-828.

Maciel, M.A.M., Pinto, A.C., Arruda, A.C., Pamplona, S.G.S.R., Vanderline, F.A., Lapa, A.J., Echevarria, A., Grynberg, N.F., Côlus, I.M.S., Farias, R.A.F., Luna Costa, A.M., and Rao, V.S.N. (2000). Ethnopharmacology, phytochemistry and pharmacology: a successful combination in the study of *Croton cajucara*. *Journal of Ethnopharmacology*, **70(1):** 41-55.

Maciel, M.A.M., Pinto, A.C., Veiga Jr., V.F., Echevarria, A., and Grynberg, N.F. (2002a). Plant as medicinais: a necessidade de estudos multidisciplinares. *Quimica Nova*, **25(3):** 429-438.

Maciel, M.A.M., Pinto, A.C., Veiga Jr., V.F., Martins, J.R., Grynberg, N.F., Echevarria, A., Lapa, A.J., and Vanderlinde, F.A. (2002b). *Croton cajucara* as an alternative to traditional medicine in a modern health system, Phytochem. Pharmacol. II Ser. *Recent Prog. Med. Plants*, **8**: 502-517.

Maciel, M.A.M., Pinto, A.C., and Kaiser, C.R. (2003). NMR and structure review of some natural fluroclerodanes. *Magnetic Resonance in Chemistry*, **41(4):** 278-282.

Maciel, M.A.M., Dantas, T.N.C., Câmara, J.K.P., Pinto, A.C., Veiga Jr., V.F., Kaiser, C.R., Pereira, N.A., Carneiro, C.M.T.S., Vanderlinde, F.A., Lapa A.J., Agner, A.R., Cóllus, I.M.S., Echevarria-Lima, J., Grynberg, N.F., Esteves-Souza, A., Pissinate, K., and Echevarria, A. (2006a). Pharmacological and biochemical profiling of lead compounds from traditional remedies: the case of *Croton cajucara*. In: *Advances in Phytomedicine* (*Lead molecules from natural products, Discovery and New Trends*), Eds. Khan, M.T.H.; Ather, A., Vol. 2, chapter 14, p.229-257.

Maciel, M.A.M., Cortez, J.K.P.C., and Gomes, F.E.S. (2006b). *Croton* genus and relevant aspects of clerodane diterpenes. *Fitos*, **2(3):** 54-73.

Maciel, M.A.M., Gomes, F.E.S., Pinto, A.C., Cólus, I.M.S., Magalhães, N.S.S., Grynberg,N.F., and Echevarria, A. (2007a). Aspects of natural products on the discovery of new antitumoral and antimutagenic agents. *Fitos*, **3(1):** 38-59.

Maciel, M.A.M., Martins, J.R., Pinto, A.C., Kaiser, C.R., Esteves-Souza, A., and Echevarria, A. (2007b). Natural and semi-synthetic clerodanes of *Croton cajucara* and their cytotoxic effects against Ehrlich carcinoma and human K562 leukemia cells. *Journal of Brazilian Chemical Society*, **18(2):** 391-396.

Maciel, M.A.M., Echevarria, A., Monteath, S.A.F.A., Veiga Jr., V.F., Kaiser, C.R., Gomes, F.E.S., Silveira, J.W.S., Sousa, R.H.C., and Vanderlinde, F.A. (2010a). Ethnobotany, Chemistry and Pharmacology Studies of the medicinal specimen *Ixora coccinea* Linn. In: *Medicinal Plants: Phytochemistry, Pharmacology and Therapeutics*, Eds.

Gupta, V.K.; Singh, G.D.; Singh, S.; Kaul, A. Daya Publishing House: New Delhi, v.1, Chapter 2, p.32-50.

Maciel, M.A.M., Veiga Jr., V.F., Pinto, A.C., Gomes, F.E.S., and Dantas, T.N.C. (2110b). Chromatographic techniques applied to the natural products chemistry. In: *Compendium of Bioactive Natural Products* (*Quality Control and Standardization*), Eds. Gupta, V.K.; Taneja, S.C.; Gupta, B.D., Stadium Press: USA, v.8, Chapter 1, p.1-40.

Maciel, M.A.M. *et al.* (2010c), unpublished results. Meanwhile, those cited results were divulged at Brazilian scientific meetings [Esteves-Souza, A.; Echevarria, A.; Maciel, M.A.M.; Pinto, A.C. The Synthesis and Cytotoxicity of *trans*-Dehydrocrotonin Derivatives. In: 11th Brazilian Meeting on organic Synthesis, 2005. (In that, DCTN-derivatives were obtained and were assayed for cytotoxicity against tumor cells (Ehrlich carcinoma and K562). The obtained results showed that DCTN-derivatives were more activity than natural target DCTN)]; [Cortez, J.K.P.C.; MACIEL, M.A.M.; Dantas, T.N.C.; Pinto, A.C. Kaiser, C.R. Redução do clerodano DCTN e estudo de RMN-2D de seus derivados (Synthetic reductions of the clerodane DCTN and the 2D-NMR study of it derivatives). In: *29ª Reunião Anual da Sociedade Brasileira de Química*, 2006 (In that, the structural elucidation of DCTN-derivatives were achieved by spectroscopic measurements including 2D-NMR experiments). In this study reduction of DCTN with $NaBH_4$ produced a mixture containing β-DCTN-OL (OH-C2 at β-position) and CTN-OL (epimeric mixture). Analysis from NMR data of the obtained products with time reaction of 30 minutes revealed that the major compounds produced are β-DCTN-OL and β-CTN-OL and in minor proportion their derivatives α-CTN-OL, DCTN-tri-OL (2,12,20-OL), DCTN-2-OL-12,20-oxolane, and DCTN-2-OL-20-acid. ^{1}H NMR [$CDCl_3$, 300 MHz, d (ppm), J (Hz)] spectrum allowed attribution of the characteristic absorptions such as: **β-DCTN-OL** 1,32-1,25 (H1α,*m*); 1,96(H1β,*s*); 3,73 (H2α, *s*); 5,52 (H3,*s*,); 2,63 [H5α,*t*,J=10,6]; 1,06 [H6β,*ddd*,J=12,7;3,4)]; 2,09 [H6α,*m*,J=12,7;3,2];1,80-1,72(H7α);1,67-1,52 (H7β); 1,80-1,72(H8β); 1,54 (H10β,*m*,J=2,7,10,7); 2,47 (H11A,*dd*, J=13,8;8,8); 2,35 (H11B, *dd*, J=13,7;8,4); 5,42 (H12,*t*, J=8,6); 6,42 (H14,*d*, J=0,8); 7,45 (H15,*m*,J=1,6]; 7,45 (H16,*m*,J=1,6); 1,11 [Me-17, *d*, J=6,4); 1,71 [Me-18,*d*, J=0,8); β-**CTN-OL**: 0,90-0,80 (1α); 2,01 (H1β,*d*, J=11,6); 3,33 (H2α,*s*]; 0,90-0,80 (H3β); 1,77 (H3α),*d*,J=11,7); 1,04 (H4β,*m*); 1,65-1,53(H5α); 0,90-0,71(H6β); 1,90(H6α,*dd*, J=12,8; 3,2); 1,59-1,42(H7a); 1,59-1,42(H7e);1,37-1,25 (H8); 1,09 (H10β,*dd*,J=10,3); 2,41 (H11A,*dd*,J=13,9;8,6); 2,31 (H11B,*dd*,J=13,8; 8,8); 5,44 (H12,*t*,J=8,6); 6,53 (H14, *dd*, J=1,8;0,8); 7,75 (H15,*m*); 7,75 (H16,*m*); 1,00 [Me-17,*d*,J=6,3); 0,87 [Me-18,*d*,J=6,2); 4,53 (OH,*s*); **α-CTN-OL:** 177,07 (H20); 144,08 (H15); 140,21 (H16); 125,88 (H13); 44,60 (H3); 108,29 (H14); 72,16 (H12); 72,16 (H2); 51,75 (H9); 42,18 (H8); 44,60 (H10); 41,33 (H11); 41,76 (H5); 30,63 (H7); 39,04 (H1); 28,49 (H6); 21,07 (H18); 17,66 (H17); **β-DCTN-2-OL-12,20-oxolane**: 7,44 (H15); 7,44 (H16); 6,40 (H14); 5,58 (H3); 5,36 (H12); 3,77 (H2); 3,78-3,51 (H20); 1,69-1,63 (H18); 1,25 (H17); **β-DCTN-2-OL-20-acid**: 7,78 (H15); 7,73 (H16); 6,53 (H14); 5,43 (H3); 3,60 (H2); 2,49 (H18,s);1,65 (H17,s); 8,48 (OH,s); **β-DCTN-tri-OL**: 7,42 (H15); 7,26 (H16); 6,32 (H14); 5,53 (H3); 5,45-5,30 (H12); 4,18 (H2); 3,78-3,51 (H20); 1,79-1,63 (H18); 1,14-1,09 (H17). Other ^{1}H

NMR as well as ^{13}C NMR absorptions were evidenced using 2D NMR (HOMOCOSY, HMQC e HMBC) data].

Malone, W.F. (1991). Studies evaluating antioxidants and β-carotene as chemopreventives. *American Journal of Clinical Nutrition*, **53**: 305S-313S.

Mandal, T.K., Bostanian, L.A., Graves, R.A., Chapman, S.R., and Idodo, T.U. (2001). Porous biodegradable microparticles for delivery of pentamidine. *European Journal of Pharmaceutics and Biopharmaceutics*, **52(1):** 91-96.

Manikandan, R., Beulaja, M., Arulvasu, C., Sellamuthu, S., Dinesh, D., Prabhu, D., Babu, G., Vaseeharan, B., and Prabhu, N.M. (2012). Synergistic anticancer activity of curcumin and catechin: An *in vitro* study using human cancer cell lines. *Microscopy Research and Technique*, **75(2):** 112-116.

Marquina, S., Maldonado, N., Garduño-Ramírez, M.L., Aranda, E., Villarreal, M.L., Navarro, V., Bye, R., Delgado, G., and Alvarez, L. (2001). Bioactive oleanolic acid saponins and other constituents from the roots of *Viguiera decurrens*. *Phytochemistry*, **56(1):** 93-97.

Melo, P.S., Durán, N., and Haun, M. (2001). Cytotoxicity of derivatives from dehydrocrotonin on V79 cells and *Escherichia coli*. *Toxicology*, **159(3):** 135-141.

Melo, P.S., Duran N., and Haun, M. (2002). Derivatives of dehydrocrotonin, a diterpene lactone isolated from *Croton cajucara*: cytotoxicity in rat cultured hepatocytes and in V79 cells. *Human and Experimental Toxicology*, **21(5):** 281-288.

Melo, P.S., Durán, N., Hiruma-Lima, C.A., Souza Brito, A.R.M., and Haun, M. (2003). Comparison of the gastroprotective effect of a diterpene lactone isolated from *Croton cajucara* with its synthetic derivatives. *Journal of Ethnopharmacology*, **87(2-3):** 169-174.

Melo, P.S., Justo, G.Z., Durán, N., and Haun, M. (2004). Natural killer cell activity and anti-tumor effects of dehydrocrotonin and its synthetic derivatives. *European Journal of Pharmacology*, **487(1-3):** 47-54.

Mendoza, A., Ishihara, Y., and Baran, P.S. (2012). Scalable enantioselective total synthesis of taxanes. *Nature Chemistry*, **4(1):** 21-25.

Metha, R. G., Murillo, G., Naithani, R., and Peng, X. (2010). Cancer chemoprevention by natural products: how far have we come? *Pharmacology Research*. (*In press*).

Mhaidat, N.M., Alali, F. Q., Matalqah, S.M., Matalka, I.I., Jaradat, S.A., Al-Sawalha, N.A., and Thorne, R.F. (2009). Inibition of MEK sensitizes paclitaxel-induced apoptosis of human colorectal cancer cells by down regulations of GRP78. *Anti-cancer Drugs*, **20(7):** 601-606.

Middleton Jr., E., Kandaswami, C., and Theoharides, T.C. (2000). The effects of plant flavonoids on mammalian cells: Implications for inflammation, heart disease, and cancer. *Pharmacological Reviews*, **52(4):** 673-751.

Miller, D.S., Blessing, J.A., Schilder, J., Munkarah, A., and Lee, Y.C. (2005). Phase II evaluation of topotecan in carcinosarcoma of the uterus: A gynecologic oncology group study. *Gynecologic Oncology*, **98(2):** 217-221.

Miller III, M.C., Johnson, K.R., Willingham, M.C., and Fan, W. (1999). Apoptotic cell death induced by baccatin III, a precursor of paclitaxel, may occur without G2/M arrest. *Cancer Chemotherapy and Pharmacology*, **44(6):** 444-452.

Min, B.S., Jung, H.J., Lee, J.S., Kim, Y.H., Bok, S.H., Ma, C.M., Nakamura, N., Hattori, M., and Bae, K. (1999). Inhibitory effect of triterpenes from *Crataegus pinatifida* on HIV-I protease. *Planta Medica*, **65(4):** 374-375.

Mitscher, L.A., Telikepalli, H., Mcghee, E., and Shankel, D.M. (1996). Natural antimutagenic agents. *Mutation Research*, **350(1):** 143-152.

Mitchelmore, C.L., and Chipman, J.K. (1998). DNA strand breakage in aquatic organisms and the potential value of the comet assay in environmental monitoring. *Mutation Research*, **399(2):** 135-147.

Mix, K.S., Mengshol, J.A., Benbow, U., Vincenti, M.P., Spom, M.B., and Brinckerhoff, C.E. (2001). A synthetic triterpenoid selectively inhibits the induction of matrix metalloproteinases 1 and 13 by inflammatory cytokines. *Arthritis Reumatology*, **44(5):** 1096-1104.

Miyazawa, M., Sakano, K., Nakamura, S., and Kosaka, H. (1999). Antimutagenic activity of isoflavones from soybean seeds (*Glycine max* Merrill). *Journal of Agricultural and Food Chemistry*, **47(4):** 1346-1349.

Mizushina, Y., Iida, A., Ohta, K., Sugawara, F., and Sakaguchi, K. (2000). Novel triterpenoids inhibit both DNA polymerase and DNA topoisomerase. *Biochemical Journal*, **350(3):** 757-763.

Moffatt, S., and Cristiano, R.J. (2006). Uptake characteristics of NGR-coupled stealth PEI/pDNA nanoparticles loaded with PLGA-PEG-PLGA tri-block copolymer of targeted delivery to human monocyte-derived dendritic cells. *International Journal of Pharmaceutics*, **321(1-2):** 143-154.

Morais, W.A., Costa, M.P., Paixão, A.D., Maciel, M.A.M., and Santos-Magalhães, N.S. (2008). Encapsulation and release characteristics of DCTN/PLGA microspheres. *Journal of Microencapsulation*, **26:** 187-191.

Mourão, S.C., Costa, P.I., Salgado, H.R., and Gremião, M.P. (2005). Improvement of antischistosomal activity of praziquantel by incorporation in to phosphatidylcholine-containing liposomes. *International Journal of Pharmaceutics*, **295(1) :** 157-162.

Mukundan, M.A., Chacko, M.C., Annapurna, V.V., and Krishnaswamy, K. (1993). Effect of turmeric and curcumin on BP-DNA adducts. *Carcinogenesis*, **14(3):** 493-496.

Murakami, S., Takashima, H., Sato-Watanabe, M., Chonan, S., Yamamoto, K., Saitoh, M., Saito, S., Yoshimura, H., Sugawara, K., Yang, J., Gao, N., and Zhang, X. (2004). Ursolic acid, an antagonist for transforming growth factor (TGF)-β1. *FEBS Letters*, **566(1-3):** 55-59.

Murakoshi, M., Nishino, H., Satomi, Y., Takayasu, J., Hasegawa, T., Tokuda, H., Iwashima, A., Okuzumi, J., Okabe, H., Kitano, H., and Iwasaki, R. (1992). Potent preventive action of α-carotene against carcinogenesis: spontaneous liver carcinogenesis and promotion stage of lung and skin carcinogenesis in mice are suppressed more effectively by α-carotene than β-carotene. *Cancer Research*, **52**: 6583-6587.

Murphy, W.K., Fossella, F.V., Winn, R.J., Shin, D.M., Hynes, H.E., Gross, H.M., Davilla, E., Leimert, J., Dhingra, H., Raber, M.N., Krakoff, I.H., and Hong, W.K. (1993). Phase II study of taxol in patients with untreated advanced non-small-cell lung cancer, *Journal of National Cancer Institute*, **85(5):** 384-388.

Nadeem, M., Rikhari, H.C., Kumar, A., Palni, L.M.S., and Nandi, S.K. (2002). Taxol content in the bark of Himalayan yew in relation to tree age and sex. *Phytochemistry*, **60(6):** 627-631.

Nakagawa, H., Yamamoto, D., Kiyozuka, Y., Tsuta, K., Uemura, Y., Hioki, K., Tsutsui, Y., and Tsubura, A. (2000). Effects of genistein and synergistic action in combination with eicosapentaenoic acid on the growth of breast cancer cell lines. *Journal of Cancer Research and Clinical Oncology*, **126(8):** 448-454.

Napoli, J.L. (2011). Effects of ethanol on physiological retinoic acid levels. *IUBMB LIFE*, **63(9):** 701-706.

Nascimento-Filho, J.M., Melo, C.P., Santos-Magalhães, N.S., Rosilio, V., Maciel, M.A.M., and Andrade, C.A.S. (2010). Thermodynamic investigation of mixed monolayers of *trans*-dehydrocrotonin and phospholipids. *Colloids and Surfaces*, **358**: 42-49.

Ndungu, J.M., Lu, Y.J., Zhu, S., Wang, X., Chen, G., Shin, D.M., Snyder, J.P., Shoji, M., and Sun, A. (2010). Targeted delivery of paclitaxel to tumor cells: synthesis and *in vitro* evaluation. *Journal of Medicinal Chemistry*, **53**: 3127-3132.

Neubert, R.H.H. (2011). Potentials of new nanocarriers for dermal and transdermal drug delivery. *European Journal of Pharmaceutics and Biopharmaceutics*, **77(1):** 1-2.

Newman, D.J., and Cragg, G.M. (2007). Natural products as source of new drugs over the last 25 years. *Journal of Natural Products*, **70(3):** 461-477.

Ngo, Q.A.; Roussi, F.; Thoret, S., and Guéritte, F. (2010). Elaboration of simplied vinca alkaloids and phomopsin hybrids. *Chemical Biology Drug Design*, **75:** 284-294.

Nikolic, L.B., Savic, I.M, Stankovic, M.Z., and Marinkovic, V.D. (2011).Paclitaxel as an anticancer agent: isolation, activity, synthesis and stability. *Central European Journal of Medicine*, **6(5):** 527-536.

Nobili, S., Lippi, D., Witord, E., Donnini, M., Bausi, L., Mini, E., and Capaccioli, S. (2009). Natural compounds for cancer treatment and prevention. *Phamacological Research*, **59**: 365-378.

Oostendorp, R.L., Buckle, T., Lambert, G., Garrigue, J.S., Beijnen, J.H., Schellens, J.H.M., and Tellingen, O. (2010). Paclitaxel in self-micro emulsifying formulations: oral bioavailability study in mice. *Invest. New Drug*, (in press).

Parkin, D.M.; Bray, F.; Ferlay, J., and Pisani, P. (2005). Global cancer statistics, 2002. *CA: A Cancer Journal Clinicians*, **55**: 74-108.

Paterson, I., Naylor, G.J., Fujita, T., Guzman, E., and Wright, A.E. (2010). Total synthesis of a library of designed hybrids of the microtubule-stabilising anticancer agents taxol, discodermolide and dictyostatin. *Chemical Communications*, **46:** 261-263.

Passarela, D., Peretto, B., Blasco Y Yepes, R., Cappelletti, G., Catelli, D., Rochi, C., Snaith, J., Fontana, G., Danieli, B., and Borlak, J. (2010). Synthesis and biological evaluation of novel thiocolchicine–podophyllotoxin conjugates. *European Journal of Medicinal Chemistry*, **45:** 219-226.

Penn, A. (1990). ICPEMC working paper 7/1/1: mutational events in the etiology of arteriosclerotic plaques. *Mutation Research*, **239(3):** 149-162.

Perazzo, F.F., Carvalho, J.C.T., Rodrigues, M., Morais, E.K.L., and Maciel, M.A.M. (2007). Comparative anti-inflammatory and antinociceptive effects of terpenoids and an aqueous extract obtained from *Croton cajucara* Benth. *Brazilian Journal of Pharmacognosy*, **17(4):** 521-528.

Pinon, A., Limami, Y., Micallef, L., Cook-Moreau, J., Liagre, B., Delage, C., Duval, R.E., and Simmon, A. (2011). A novel form of melanoma apoptosis resistance: Melanogenesis up-regulation in apoptotic B16-F0 cells delays ursolic acid-triggered cell death. *Experimental Cell Research*, **317(12):** 1669-1676.

Poersch, A., Santos, F.V., Maciel, M.A.M., Câmara, J.K.P., Dantas, T.N.C., and Cólus, I.M.S. (2007). Protective effect of DCTN (*trans*-dehydrocrotonin) against induction of micronuclei and apoptosis by different mutagenic agents *in vitro*. *Mutation Research*, **629:** 14-23.

Polkowski, K., and Mazurek, A.P. (2000). Biological properties of genistein: A review of *in vitro* and *in vivo* data. *Acta Poloniae Pharmaceutica*, **57(2):** 135-155.

Potier, P. (1992). Search and discovery of new antitumor compounds. *Chemical Society Reviews*, **21(2):** 113-119.

Pupo, M.T., Adorno, M.A.T., Vieira, P.C., Fernandes, J.B., Silva, M.F.G.F., and Pirani, J.R. (2002). Terpenoids and steroids from *Trichilia* species. *Journal of the Brazilian Chemical Society*, **13(3):** 382-388.

Qin, L., Xue, M., Wang, W., Zhu, R., Wang, S., Sun, J., Zhang, R., and Sun, X. (2010). The *in vitro* and *in vivo* anti-tumor effect of layered double hydroxides nanoparticles as delivery for podophyllotoxin. *International Journal of Pharmaceutics*, **388:** .223-230.

Qurishi, Y., Hamid, A., Majeed, R., Hussain, A., Qazi, A.K., Ahmed, M., Zargar, M.A., Singh, S.K., and Saxena, A.K. (2011). Interaction of natural products with cell survival and signaling pathways in the biochemical elucidation of drug targets in cancer. *Future Oncology*, **7(8):** 1007-1021.

Quoix, E., Breton, J., Gervais, R., Wilson, J., Schramel, F., Cardenal, F., Ross, G., Preston, A., Lymboura, M., and Mattson, K. (2005). A randomised phase II study of the efficacy and safety of intravenous topotecan in combination with either cisplatin or etoposide in patients with untreated extensive disease small-cell lung cancer. *Lung Cancer*, **49(2):** 253-261.

Ramsay, A.K., and Leung, H.Y. (2009). Signalling pathways in prostate carcinogenesis: potentials for molecular-targeted therapy. *Clinical Science*, **117:** 209-228.

Rang, H.P., Dale, M.M., and Ritter, J.M. (2001). *Farmacologia*. Rio de Janeiro: Guanabara Koogan, 4.ed.

Ratner, B.D. (2002). Reducing capsular thickness and enhancing angiogenesis around implant drug release systems. *Journal of Controlled Release*, **78(1-3):** 211-218.

Ravi-Kumar, M.N.V. (2000). Nano and microparticles as controlled drug delivery devices. *Journal of Pharmacy and Pharmaceutical Sciences*. **3(2):** 234-258.

Ravivarapu, H.B., Burton, K., and Deluca, P.P. (2000). Polymer and microsphere blending to alter the release of a peptide from PLGA microspheres. *European Journal of Pharmaceutics and Biopharmaceutics*, **50(2):** 263-270.

Razboršek, M.I, Vonèina, D.B., Doleèek, V., and Vonèina, E. (2008). Determination of oleanolic, betulinic and ursolic acid in Lamiaceae and mass spectral fragamentation of their trimethylsilylated derivatives. *Chromatographia*, **67(5-6):** 433-440.

Reddy, L., Odhav, B., and Bhoola, K.D. (2003). Natural products for cancer prevention: A global perspective. *Pharmacology and Therapeutics*, **99(1):** 1-13.

Reddy, L.H., Sharma, R.K., Chuttani, K., Mishra, A.K., and Murthy, R.S.R. (2005). Influence of administration route on tumor uptake and biodistribution of etoposide loaded solid lipid nanoparticles in Dalton's lymphoma tumor bearing mice. *Journal of Controlled Release*, **105(3):** 185-198.

Reinchenbach, H., and Hofle, G. (2008). Discovery and development of the epothilones: a novel class of antineoplastic drugs. *Drugs in R&D*, **9(1):** 1-10.

Ribeiro-Costa, R.M., Alves, A.J., Santos, N.P., Nascimento, S.C., Gonçalves, E.C.P., Silva, N.H., Honda, N.K., and Santos-Magalhães, N.S. (2004). *In vitro* and *in vivo* properties of usnic acid encapsulated into PLGA-microspheres. *Journal of Microencapsulation*, **21(4):** 371-384.

Rodríguez, J.A., Hiruma-Lima, C.A., and Souza Brito, A.R.M. (2004). Antiulcer activity and subacute toxicity of *trans*-dehydrocrotonin from *Croton cajucara*. *Human and Experimental Toxicology*, **23(9):** 455-461.

Ruel-Gariépy, E., Shive, M., Bichara, A., Berrada, M., Le Garrec, D., Chenite, A., and Leroux, J.-C. (2004). A thermosensitive chitosan-based hydrogel for the local delivery of paclitaxel. *European Journal of Pharmaceutics and Biopharmaceutics*, **57(1):** 53-63.

Saito, K., Koike, Y., Suzuki, H., and Murakoshi, I. (1993). Biogenetic implication of lupine alkaloid biosynthesis in bitter and sweet forms of *Lupinus luteus* and *L. albus*. *Phytochemistry*, **34(4):** 1041-1044.

Samuelsson, G. (1999). Drugs of natural origin: A textbook of pharmacognosy, 4th ed., Stockolm, Swedish Pharmaceutical Press.

Santos, F.V., Mesquita, S.F.P., Faria, M.J.S.S., Poersh, A., Maciel, M.A.M., Pinto, A.C., Morimoto, H.K., and Cólus, I.M.S. (2006). Absence of mutagenicity in somatic and germ cells of mice submitted to subchronic treatment with an extract of *Croton cajucara* Benth (Euphorbiaceae). *Genetics and Molecular Biology*, **29(1):** 59-165.

Santos, H.M.L.R., Queiroz, F. B., Lira, M.C.B., Nascimento, S.C., Maior, R.M.S., and Santos-Magalhães, N.S. (2006b). Cytotoxicity of doxorubicin-loaded Con A-liposomes. *Drug Development Research*, **67(5):** 430-437.

Santos, F.V., Santos, V.J.D.S.V., Farias, M.J., Mesquita, S.F.P., Maciel, M.A.M., Angelo, C.P., and Cóllus, I.M.S. (2008). Mutagenicity and antimutagenicity of *Croton cajucara*. *Biologia*, **63(3):** 327-331.

Santos-Magalhães, N.S., Pontes, A., Pereira, V.M.W., and Caetano, M.N.P. (2000). Colloidal carriers for benzathine penicillin G: Nanoemulsions and nanocapsules. *International Journal of Pharmaceutics*, **208(1-2):** 71-80.

Santoyo, S., De Jalón, *E.G.*, Ygartua, P., Renedo, M.J., and Blanco-Príeto, M.J. (2002). Optimization of topical cidofovir penetration using microparticles. *International Journal of Pharmaceutics*, **242(1-2):** 107-113.

Sattar, A.A., Bankova, V., Kujumgiev, A., Galabov, A., Ignatova, A., Todorova, C., and Popov, S.S. (1995). Chemical composition and biological activity of leaf exudates from some Lamiaceae plants. *Pharmazie*, **50(1):** 62-65.

Setchell, K.D.R., and Cassidy, A. (1999). Dietary isoflavones: Biological effects and relevance to human health. *Journal of Nutrution*, **129(3):** 758S-767S.

Schetter, A.J., Heegaard, N.H.H., and Harris, C.C. (2010). Inammation and cancer: interweaving microRNA, free radical, cytokine and p53 pathways. *Carcinogenesis*, **31(1):** 37-49.

Schwartz, E.L. (2009). Antivascular Actions of microtubule-binding drugs. *Clinical Cancer Research*, **15(18):** 2594-2601.

Shamon, L.A., and Pezzuto, J.M. (1994). *Plant antimutagens: Review and strategy for the identification of therapeutically useful agents*. In: *Economic and Medicinal Plant Research*. Nova Iorque: Academic Press, p.235.

Shoeb, M. (2006). Anticancer agents from medicinal plants. *Blangadesh Journal of Pharmacology*. **1:** 35-41.

Shugart, L.R. (1995). *Environmental genotoxicology*. In: RAND, G.M. *Fundamentals of aquatic toxicology: Effects, environmental fate and risk assessment*. Washington: Taylor and Francis, 2.ed., p.405.

Silva, R.M., Santos, F.A., Rao, V.S.N., Maciel, M.A.M., and Pinto, A.C. (2001a). Blood glucose- and triglyceride-lowering effect of *trans*-dehydrocrotonin, a diterpene from *Croton cajucara* Benth., in rats. *Diabetes, Obesity and Metabolism*, **3(6):** 52-456.

Silva, R.M., Santos, F.A., Maciel, M.A.M., Pinto, A.C., and Rao, V.S.N. (2001b). Effect of *trans*-dehydrocrotonin, a 19-*nor*-clerodane diterpene from *Croton cajucara* on experimental hypertriglyceridaemia and hypercholesterolaemia induced by triton WR 1339 (tyloxapol) in mice. *Planta Medica*, **67(8):** 763-765.

Silva, R.M., Santos, F.A., Rao, V.S.N., Maciel, M.A.M., and Pinto, A.C. (2001c). The lipid-lowering effect of *trans*-dehydrocrotonin, a clerodane diterpene from *Croton cajucara* Benth. in mice fed on high-fat diet. *Journal of Pharmacy and Pharmacology*, **53(4):** 535-539.

Silva, R.M., Oliveira, F.M., Cunha, K.M.A., Maia, J.L., Maciel, M.A.M., Pinto, A.C., Nascimento, N.R.F., Santos, F.A., and Rao, V.S.N. (2005). Cardiovascular effects of *trans*-dehydrocrotonin, a diterpene from *Croton cajucara* in rats. *Vascular Pharmacology*, **43(1):** 11-18.

Singer, J.W., Bhatt, R., Tulinski, J., Buhler, K.R., Heasley, E., Klein, P., and De Vries, P. (2001). Water-soluble poly-(L-glutamic acid)-Gly-camptothecin conjugates enhance camptothecin stability and efficacy *in vivo*. *Journal of Controlled Release*, **74(1-3):** 243-247.

Singh, M., Shirley, B., Bajwa, K., Samara, E., Hora, M., and O'hagan, D. (2001). Controlled release of recombinant insulin-like growth factor from a novel formulation of polylactide-co-glycolide microparticles. *Journal of Controlled Release*, **70(1/2)**: 21-28.

Singh, R., and Lillard Jr, J.W. (2009). Nanoparticle-based targeted drug delivery. *Experimental and Molecular Pathology*, **86(3):** 215-223.

Slupphaug, G., Kavli, B., and Krokan, H.E. (2003). The interacting pathways for prevention and repair of oxidative DNA damage. *Mutation Research*, **531(1-2):** 231-251.

Soares, B.A., Firme, C.L., Maciel, M.A.M., Kaiser, C., Schilling, E., and Bortoluzzi, A.. (2013). *Journal of Organic Chemistry* (manuscript submitted for publication).

Soares, M.P. (2004). Would sacaca, *Croton cajucara* Benth. (Euphorbiaceae) be an hepatotoxic plant like Germander, *Teucrium chamaedrys* L. (Labiatae)? *Revista da Sociedade Brasileira de Medicina Tropical*, **37(2):** 96-97.

Somova, L.O., Nadar, A., Rammanan, P., and Shode, F.O. (2003). Cardiovascular, anti-hyperlipidemic and antioxidant effects of oleanolic and ursolic acids in experimental hypertension. *Phytomedicine*, **10(2-3):** 115-121.

Solomon, P.R., Adams, F., Silver, A., Zimmer, J., and Deveaux, R. (2002). Ginkgo for memory enhancement: a randomized controlled trial. *Journal of the American Medical Association*, **288(7):** 835-840.

Soppimath, K.S., Aminabhavi, T.M., Kulkarni, A. R., and Rudzinski, W.E. (2001). Biodegradable polymeric nanoparticles as drug delivery devices. *Journal of Controlled Release*, **70(1/2)**: 1-20.

Souza-Brito, A.R.M., Rodríguez, J.A., Hiruma-Lima, C.A., Haun, M., and Nunes, D.C. (1998). Antiulcerogenic activity of *trans*-dehydrocrotonin from *Croton cajucara*. *Planta Medica*, **64(2):** 126-129.

Steele, V.E., and Kelloff, G.J. (2005). Development of cancer chemopreventive drugs based on mechanistic approaches. *Mutation Research*, **591(1-2):** 16-23.

Suffness, M., and Douros, J. (1982). Current status of the NCI plant and animal product program. *Journal of Natural Products*, **45(1):** 1-14.

Suffness, M. (1993). Taxol: from discovery to therapeutic use. *Annual Reports in Medicinal Chemistry*, **28**: 305-314.

Sultana, N. (2011). Clinically useful anticancer, antitumor, and antiwrinkle agent, ursolic acid and related derivatives as medicinally important natural product. *Journal of Enzime Inhibition and Medicinal Chemistry*, **26(5):** 616-642.

Tanaka, T., Kawamori, T., Ohnishi, M., Okamoto, K., Mori, H., and Hara, A. (1994). Chemoprevention of 4-nitroquinoline 1-oxide-induced oral carcinogenesis by dietary protocatechuic acid during initiation and postinitiation phases. *Cancer Research*, **54(9):** 2359-2365.

Tang, H.Q., Hu, J., Yang, L., and Tan, R.X. (2000). Terpenoids and flavonoids from *Artemisia* species. *Planta Medica*, **66(4):** 391-393.

Tao, S.L., and Desai, T.A. (2003). Microfabricated drug delivery systems: From particles to pores. *Advanced Drug Delivery Reviews*, **55(3)**: 315-328.

Talekar, M., Kendall, J., Denny, W., and Garg, S. (2011). Targeting of nanoparticles in cancer: drug delivery and diagnostics. *Anti-Cancer Drugs*, **22(10):** 949-962.

Tavano, L., Alfano, P., Muzzalupo, R., and Cindio, B. (2011). Niosomes vs microemulsions: New carriers for topical delivery of capsaicin. *Colloids and Surfaces B: Biointerfaces*, **87(2)**: 333-339.

Teiten, M.H., Gaascht, F., Dicato, M., and Diederich, M. (2012). Targeting the wingless signaling pathway with natural compounds as chemopreventive or chemotherapeutic agents. *Current Pharmaceutical Biotechnology*, **13(1):** 245-254.

Teixeira, M., Alonso, M.J., Pinto, M.M.M., and Barbosa, C.M. (2005). Development and characterization of PLGA nanospheres and nanocapsules containing xanthone and 3-methoxyxanthone. *European Journal of Pharmaceutics and Biopharmaceutics*, **59(3):** 491-500.

Teixeira, M., Alonso, M.J., Pinto, M.M.M., and Barbosa, C.M. (2006). Nanoencapsulation of usnic acid: An attempt to improve antitumour activity and reduce hepatotoxicity. *European Journal of Pharmaceutics and Biopharmaceutics*, **64(2):** 154-160.

Van-Goethem, F., Lison, D., and Kirsch-Volders, M. (1997). Comparative evaluation of the *in vitro* micronucleus test and the alkaline single cell gel electrophoresis assay for the detection of DNA damaging agents: genotoxic effects of cobalt powder, tungsten carbide and cobalt-tungsten carbide. *Mutation Research*, **392(1-2):** 31-43.

Vagelos, P.R. (1991). Are prescription drug prices high? *Science*, **252(5010):** 1080-1084.

Vechia, L.D., Gnoato, S.C.B., and Gosmann, G. (2009). Derivados oleananos e ursanos e sua importância na descoberta de novos fármacos com atividade antitumoral, anti-inflamatória e antioxidante. *Química Nova*, **32(5):** 1245-1252.

Veiga Jr., V.F., Pinto, A.C., and Maciel, M.A.M. (2005). Plantas Medicinais: Cura segura? *Química Nova*, **28(3):** 519-528.

Venkatraman, S.S., Jie, P., Min, F., Freddy, B.Y.C., and Leong-Huat, G. (2005). Micelle-like nanoparticles of PLA-PEG-PLA triblock copolymer as chemotherapeutic carrier. *International Journal of Pharmaceutics*, **298(1):** 219-232.

Verhagen, H., Aruoma, O.I., Van Delft, J.H.M., Dragsted, L.O., Ferguson, L.R., Knasmüller, S., Pool-Zobel, B.L., Poulsen, H.E., Williamson, G., and Yannai, S. (2003). The 10 basic requirements for a scientific paper reporting antioxidant, antimutagenic or anticarcinogenic potential of test substances in *in vitro* experiments and animal studies *in vivo*. *Food and Chemical Toxicology*, **41(5):** 603-610.

Visani, G., Russo, D., Ottaviani, E., Tosi, P., Damiani, D., Michelutti, A., Manfroi, S., Baccarani, M., and Tura, S. (1997). Effects of homoharringtonine alone and in combination with alpha interferon and cytosine arabinoside on "*in vitro*" growth and induction of apoptosis in chronic myeloid leukemia and normal hematopoietic progenitors. *Leukemia*, **11(5):** 624-628.

Vitaglione, P., and Fogliano, V. (2004).Use of antioxidants to minimize the human health risk associated to mutagenic/carcinogenic heterocyclic amines in food. *Journal of Chromatography B*, **802(1):** 189-199.

Voronkov, M.V., Gontcharov, A.V., and Wang, Z.M. (2003). Improved large-scale synthesis of phenylisoserine and the taxol C-13 side chain. *Tetrahedron Letters*, **44(2):** 407-409.

Wakabayashi, K. (1990). International Commission for Protection Against Environmental Mutagens and Carcinogens. ICPEMC Working Paper 7/1/3. Animal studies suggesting involvement of mutagen/carcinogen exposure in atherosclerosis. *Mutation Research*, **239(3):** 181-187.

Walton, N.J., Mayer, M.J., and Narbad, A. (2003). Vanillin. *Phytochemistry*, **63(5):** 505-515.

WHO (2002). Traditional Medicines Strategy 2002-2005, Genebra.

Xu, L.Z., and Wan, Z.X. (1980). The effect of oleanolic acid on acute hepatitis (70 cases). *Human Medicine*, **7:** 50-52.

Xu, S.L. (1985). Effects of oleanolic acid on chronic hepatitis: 188 case reports. *Symposium on oleanolic acid*, p.23-25.

Yao, N., Chen, C., Wu, C., Motonishi, K., Kung, H., and Lam, K.S. (2011).Novel flavonoids with antiproliferative activities against breast cancer cells. *Journal of Medicinal Chemistry*, **54(13)**: 4339-4349.

Yamane, T., Takahashi, T., Kuwata, K., Oya, K., Inagake, M., Kitao, Y., Suganuma, M., and Fujiki, H. (1995). Inhibition of N-methyl-N-nitro-N-nitrosoguanidine-induced carcinogenesis by (-)-epigallocatechin gallate in the rat glandular stomach. *Cancer Research*, **55:** 2081-2084.

Yang, S.S., Cragg, G.M., Newman, D.J., and Bader, J.P. (2001). Natural product-based anti-HIV drug discovery and development Facilitated by the NCI developmental therapeutics program. *Journal of Natural Products*, **64(2):** 265-277.

Yang, S., Gursoy, R.N., Lambert, G., and Benita, S. (2004). Enhanced oral absorption of paclitaxel in a novel self-microemulsifying drug delivery system with or without concomitant use of P-glycoprotein inhibitors. *Pharmaceutical Research*, **21(2):** 261-270.

Yoshikawa, M., and Matsuda, H. (2000). Antidiabetogenic activity of oleanolic acid glycosides from medicinal foodstuffs. *BioFactors*, **13(1-4):** 231-237.

You, H.J., Choi, C.Y., Kim, J.Y., Park, S.J., Hahm, K.-S., and Jeong, H.G. (2001). Ursolic acid enhances nitric oxide and tumor necrosis factor-α production via nuclear factor=κB activation in the resting macrophages. *FEBS Letters*, **509(2):** 156-160.

Zeng, H.H., Tu, P.F., Zhou, K., Wang, H., Wang, B.H., and Lu, J.F. (2001). Antioxidant properties of phenolic diterpenes from *Rosmarinus officinalis*. *Acta Pharmacologica Sinica*, **22(12):** 1094-1098.

Zhang, Q., Jiang, X., Jiang, W., Lu, W., Su, L., and Shi, Z. (2004).Preparation of nimodipine-loaded microemulsion for intranasal delivery and evaluation on the targeting efficiency to the brain. *International Journal of Pharmaceutics*, **275(1-2):** 85-96.

Zhang, X., Chen, K., Chen, J., Liu, Y.X., Qu, P., and Li, T.Y. (2011). Effect of marginal vitamin A deficiency during pregnancy on retinoic acid receptors and N-methyl-D-aspartate receptor expression in the offspring of rats. *Journal of Nutrition Biochemistry*, **22(12):** 1112-1120.

Zhang, Y.Q., Zhang, H.J., Zhang, J., Wang, J., Yao, J.Z., Zhu, L.J., Zhuang, C.L., Wang, S.Z., Dong, G.Q., Sheng, C.Q., Miao, Z.Y., and Zhang, W.N. (2012). Antitumor activities and structure-activity relationship of phosphotriester derivatives of homocamptothecin based on a semisynthetic route. *Australian Journal of Chemistry*, **10(2):** 361-366.

Ziotopoulos, P., Androulakis, N., Mylonaki, E., Chandrinos, V., Zachariadis, E., Boukovinas, I., Agelidou, A., Kentepozidis, N., Ignatiadis, M., Vossos, A., and Georgoulias, V. (2005). Front-line treatment of advanced non-small cell lung cancer with irinotecan and docetaxel: a multicentre phase II study. *Lung Cancer*, **50(1):** 115-122.

Zivny, J., Klener, J.R., P., Pytlik, R., and Andera, L. (2010). The role of apoptosis in cancer development and treatment: focusing on the development and treatment of hematologic malignancies. *Current Pharmaceutical Design*, **16(1)**: 11-33,.

Zurcher, A., Jerger, O., and Ruzicka, L. (1954). Zur kenntnis der triterpene. 180. Uberfuhrung der chinovasaure in phyllanthol und uvaol: Urber die konstitution der ursolsaure und des uvaols. *Helvetica Chimica Acta*, **37:** 2145-2152.

Bioactive Phytochemicals: Perspectives for Modern Medicine Vol. 2 (2014) *Pages* **295–306**
Editor: **V.K. Gupta**
Published by: **DAYA PUBLISHING HOUSE, NEW DELHI**

13

The Current Status on Imperative Medicinal Plant: *Catharanthus roseus* (Linn.) G. Don

Sandeep Dhankhar[1] and Anil Kumar Chhillar[1*]

ABSTRACT

Medicinal plants are building block and pack up of human society to fight diseases, from the first light of civilization. In the search for environmentally safe medicines, much research has been conducted on the use of plants for the protection of human race from various diseases. The base of this study is to make the society aware of medicinal values of Catharanthus roseus (Linn.) G. Don which originates from Madagascar, but has now been spread throughout the tropics and subtropics. The Catharanthus roseus provides many useful compounds such as Vincristine and Vinblastine that are used in the treatment of cancer. The other alkaloids such as two semi-synthetic derivatives, vindesine and vinorelbine, are also used as anticancer compounds. However, in addition to possible beneficial health effects, such as blood sugar lowering properties, anti-cancer, antioxidant, antimicrobial and hypotensive effects, also toxic effects are described. The present article aims to compile the morphology, distribution, phytochemistry of Catharanthus roseus with compounds metabolism and pharmacokinetics of vinca alkaloids, as well as, their future prospects for the advance scientific investigation for the development of effective therapeutic compounds.

Keywords: Anti-cancer, Antimicrobial, Antioxidant, *Catharanthus roseus*, Vinblastine, Vincristine.

1 Centre for Biotechnology, Maharshi Dayanand University, Rohtak – 124 001, Haryana, India

* *Corresponding author*: E-mail: anil.chhillar@gmail.com

Introduction

Drugs of plant origin and medicinal herbs have occupied a fundamental place in therapeutics for centuries. Therefore, in recent time there has been a visible shift towards herbal cures because of the marked cumulative and irreversible reactions of current drugs. Herbal medicine sometimes referred to as Herbalism or Botanical Medicine, is the use of herbs for their therapeutic or medicinal value. An herb is a plant or plant part valued for its medicinal, aromatic or savory qualities. Herbal drugs represent a major division in all traditional systems of medicines. Herbal medicine is an achievement of admired therapeutic diversity. Therefore, the common use of natural remedies in India led to the development of the Ayurvedic system of medicine, and look for new plant-derived chemicals. There are some Ayurvedic herbs that are very useful for reducing cholesterol, diabetes and cancer etc.

One of the many plants which are being evaluated for their therapeutic efficacies is *Catharanthus roseus* (Linn.) G. Don which is a renowned medicinal plant, belonging to the family Apocynaceae; and is an herbaceous sub-shrub also known as Madagascar periwinkle, Sadabahar, *Vinca rosea* or *Lachnera rosea* worldwide. *C. roseus* has gained interest from the pharmaceutical industry; the alkaloids Vincristine and Vinblastine from its sap have been shown to be an effective treatment for leukaemia and lymphoma. The phytochemical alkaloids of *C. roseus* varies considerably in various parts; the maximum being in the root bark which ranges from 0.15 to 1.34 per cent and even up to 1.79 per cent in some strains (Singh and Jagdev, 1996). *C. roseus* is widely studied as a model of medicinal plants for its highly economic and pharmaceutical value which comes from its diversity of useful terpenoid indole alkaloids of therapeutic interest, including monomers such as ajmalicine and serpentine, used in the treatment of antiarrythmic heart disorder and anxiety, and heterodimers such as vinblastine and vincristine, used especially for their anti-tumor activities (Van der *et al.*, 2004; Ziegler *et al.*, 2008; Guirimand *et al.*, 2010). The biosynthesis of vinblastine has been shown to be highly complex, involving more than twenty enzymatic steps, and a great deal is already known about the pathway, the enzymes and genes involved, and about their regulation (Meijer *et al.*, 1993; Gantet and Memelink, 2002). The discovery of vincristine is credited with raising the survival rate of childhood leukemia from under 10 per cent to over 90 per cent. Thousands of children's lives have therefore been saved by an extract of this humble garden plant. The plant has been used for centuries to treat diabetes, high blood pressure, asthma, constipation and menstrual problems. More recently, extracts from Madagascar periwinkle have been shown to be effective in the treatment of various kinds of leukemia, skin cancer, lymph cancer, breast cancer and Hodgkin's disease. However, it is poisonous if ingested or smoked. It has caused poisoning in grazing animals. Even under a doctor's supervision for cancer treatment, products from Madagascar periwinkle produce undesirable side effects. Therefore the present article deals with the botany and the composition, the structure of main constituents, biological activity of this plant with its insignificant toxicity.

Morphology and Distribution

C. roseus is a semi woody evergreen perennial, moderate growing plant to 1 or 2 feet high. It has glossy, dark green, oval leaves (1-2 inches long) with graceful pink or white salver form flowers (Jayanthi *et al.*, 2010). The blooms of the natural wild plants are a pale pink with a purple "eye" in their centers. It is native to southeastern and eastern Madagascar, the plant is easily cultivated, and European colonists exported it widely as an ornamental. It is now grown almost worldwide, and is found naturalized in most tropical and subtropical regions following escapes from cultivation. Propagation is usually grown from seeds. It can also be grown from stem cuttings. The plant prefers high humidity. The plant is very sensitive to over watering, excessive fertilizer, cool growing conditions, and low light intensity. It occasionally suffers from leaf blights.

Phytochemistry

There has been a resurgence of interest in natural products with a focus on plants as an alternative source of therapeutic compounds. Bioactive constituents from plant sources include plant secondary metabolites like alkaloids, polyphenolics (flavonoids, phenolic acids, anthraquinones) and terpenoids. *C. roseus* (L.) G. Don is one of the most extensively investigated medicinal plants and is known mainly for its pharmacologically important alkaloids. At present, the *Catharanthus* alkaloids comprise a group of about 130 terpenoid indole alkaloids (TIAs) (Toki *et al.*, 2008). Regulation of alkaloid biosynthesis (Figure 13.1) is accomplished spatially and temporally, by restricting product biosynthesis and accumulation to particular cells and to defined stages of plant development (De Luca *et al.*, 1987). The presence of different alkaloids and expressed genes within the above-ground and underground tissues of the plant confirms the expression of distinct tissue-specific pathways responsible for the precise tissue- and organ-specific compartmentation of TIA biosynthesis. The roots of the plant accumulate ajmalicine and serpentine (Rodriguez *et al.*, 2003), which are important components of medicines for controlling high blood pressure and other types of cardio-vascular maladies. However, the plant is particularly known for its economically important leaf-specific dimeric indole alkaloids, vinblastine and vincristine, which are potent antineoplastic agents and indispensable constituents of most cancer chemotherapies (Mersey and Cutler, 1986). At least 35 enzymatic conversions are required to produce TIAs from the primary metabolites tryptophan and geraniol. All of the known TIA biosynthesis genes are induced by the defense hormone jasmonic acid, and many are regulated by the jasmonate-responsive AP2/ERF-domain transcription factors ORCA2 and ORCA3. The expression of the *ORCA* genes in turn is controlled by the jasmonate-responsive basic helix-loop-helix transcription factor CrMYC2.

The monoterpenoid indole alkaloids (MIAs) of *C. roseus* also continue to be the most important source of natural drugs in chemotherapy treatments for human cancers (Facchini and De Luca, 2008). The vinblastine and vincristine are derived from the coupling of catharanthine and vindoline to yield powerful dimeric indole alkaloids that prevent cell division. However, the precise mechanisms (Figure 13.2) for their assembly within plants remain difficult to understand. Although the entire production of catharanthine and vindoline occurs in young developing leaves, catharanthine accumulates in leaf wax exudates of leaves, whereas vindoline is found within leaf cells (Mahroug *et al.*, 2007).The spatial separation of these two MIAs provides a

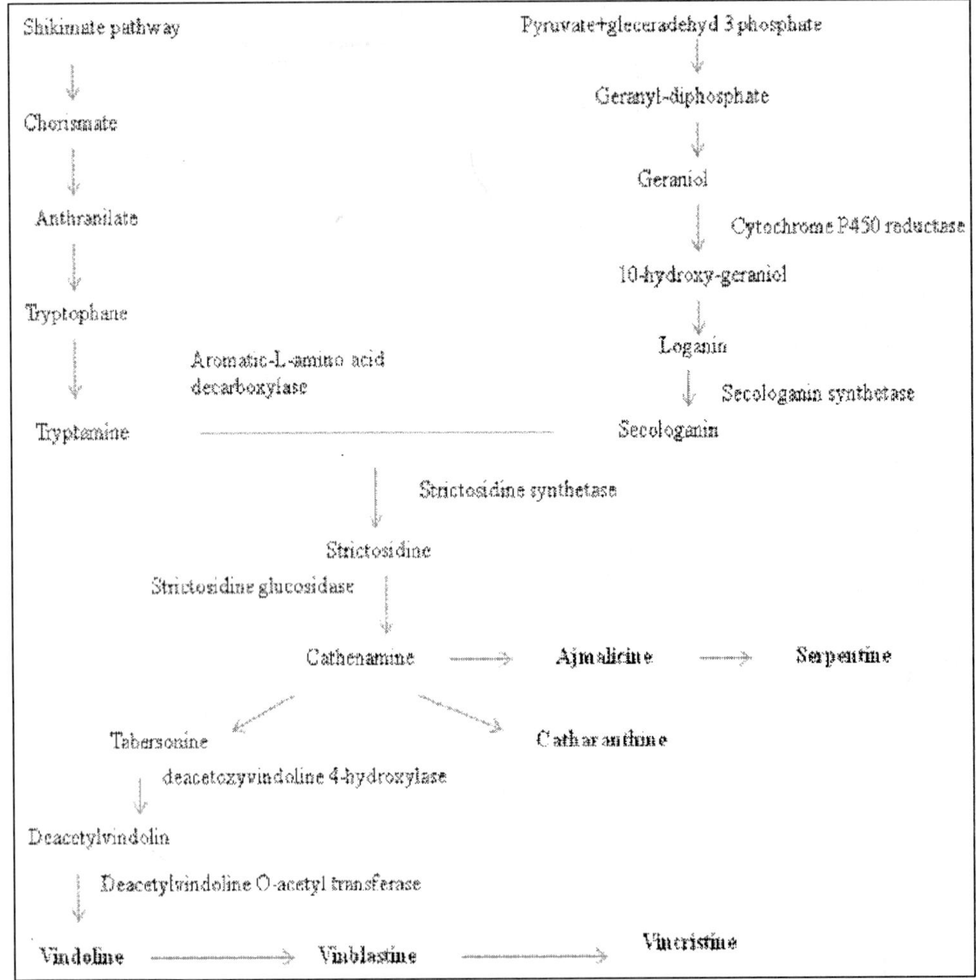

Figure 13.1: Biosynthetic pathway for some important alkaloids (TIA).

biological explanation for the low levels of dimeric anticancer drugs found in the plant that result in their high cost of commercial production (Heijden *et al.*, 2004).

Description of the Compounds

The vinca alkaloids are a subset of drugs, called spindle inhibitors, which are derived from the periwinkle plant, *Catharanthus roseus*. These alkaloids have a common chemical structure (Figure 13.3), consisting of an asymmetric dimer composed of a vindoline ring connected to a catharanthine ring through carbon-carbon bonds (Noble, 1990). There are four major Vinca alkaloids in clinical use: (a) vincristine, (b) vinblastine, (c) vindesine and (d) vinorelbine.

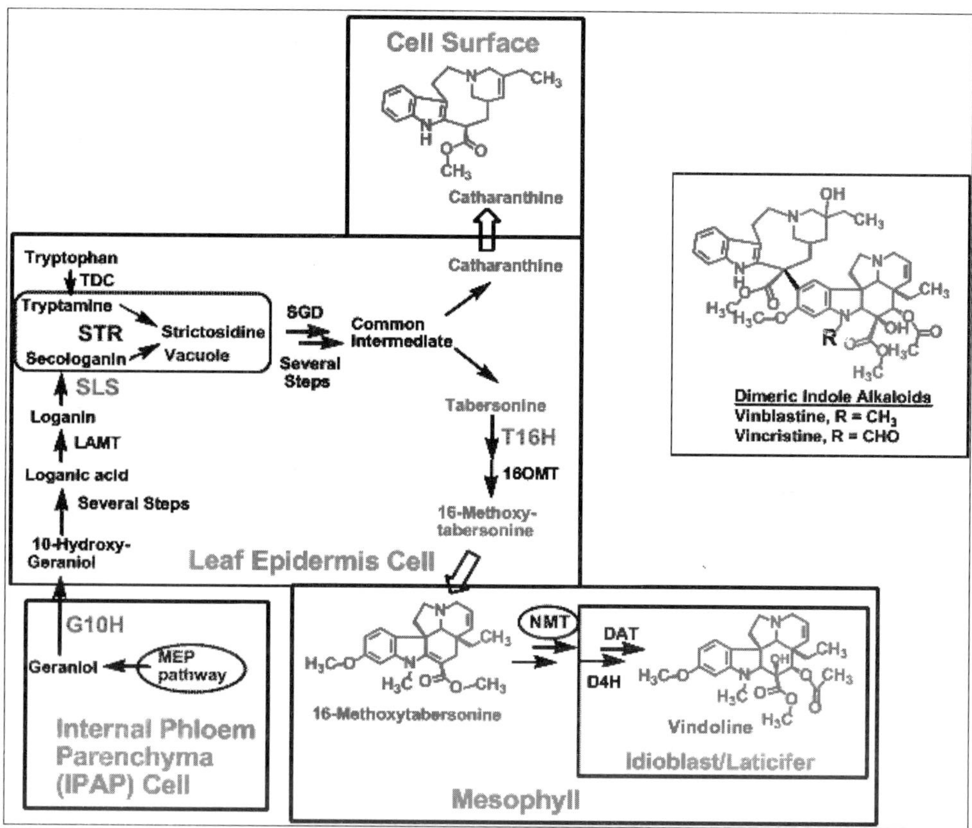

Figure 13.2: Mechanism of biosynthesis for monomeric Indole alkaloids (MIA) in leaves of *Catharanthus roseus.*

Molecular Mechanism

Vinca alkaloids inhibit cell proliferation by binding to microtubules (Figure 13.4.), which lead to a mitotic block and apoptosis (Yang *et al.*, 2010). Vincristine and related compounds induce destabilization of microtubules by binding to tubulin and blocking the polymerization. Exposure of cells to vinca alkaloids results in an induction of tumor protein p53 and cyclin-dependent kinase inhibitor 1A (p21) and in rapid alterations of protein kinase activities. These protein kinases are directly or indirectly responsible for BCL-2 phosphorylation, which results in the inactivation of BCL-2 function. The induction of BCL-2 phosphorylation is followed by the loss of its ability to form heterodimers with BAX. The loss of BCL -2 function together with the elevations of p53 and p21 lead to apoptosis. They also play a competitive role in the transport of amino acids in the intracellular compartment, in the inhibition of purine, of DNA and RNA, inhibition of protein synthesis and in the release of histamine from mast cells. They can also crystallize in the cell causing direct mechanical damage. They do not selectively inhibit cancer cell division. They can also halt the division of some healthy cells, leading to some of the common side effects.

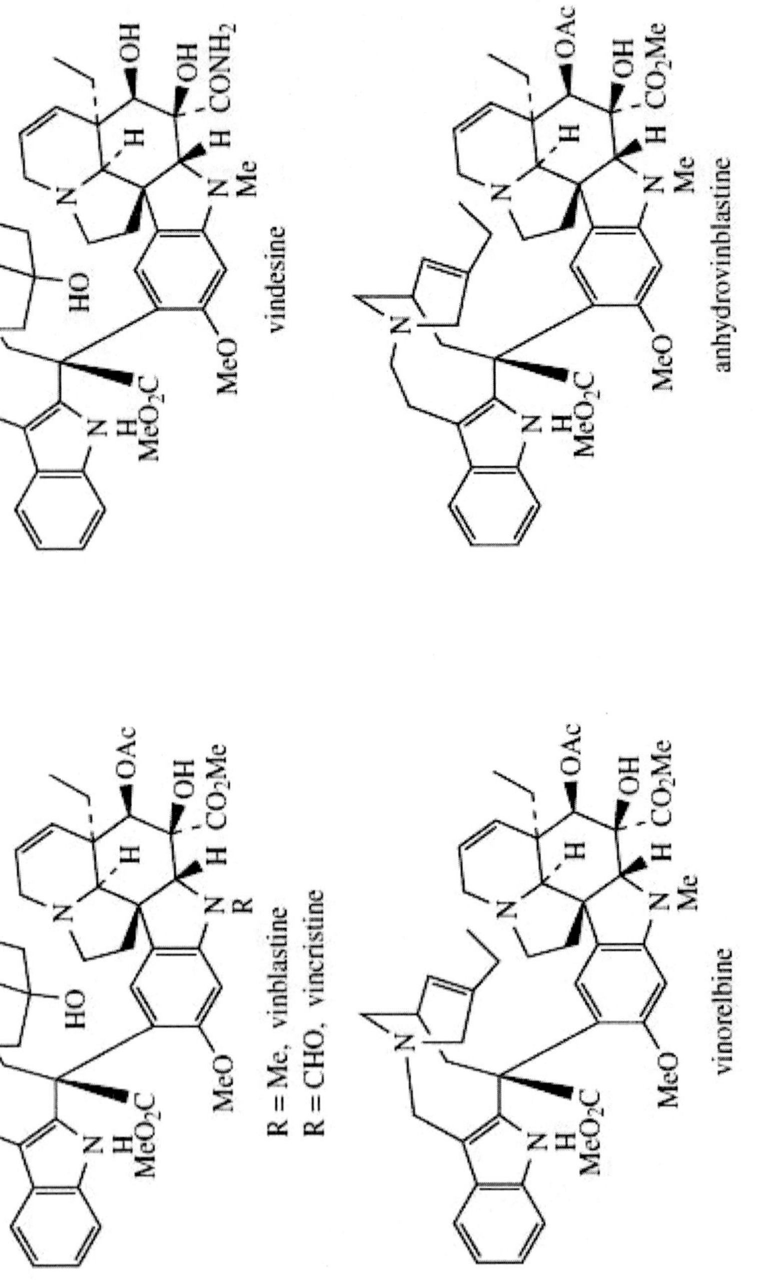

R = Me, vinblastine
R = CHO, vincristine

vindesine

vinorelbine

anhydrovinblastine

Figure 13.3: Structure of various imperative bioactive phytochemicals isolated from *C. roseus*.

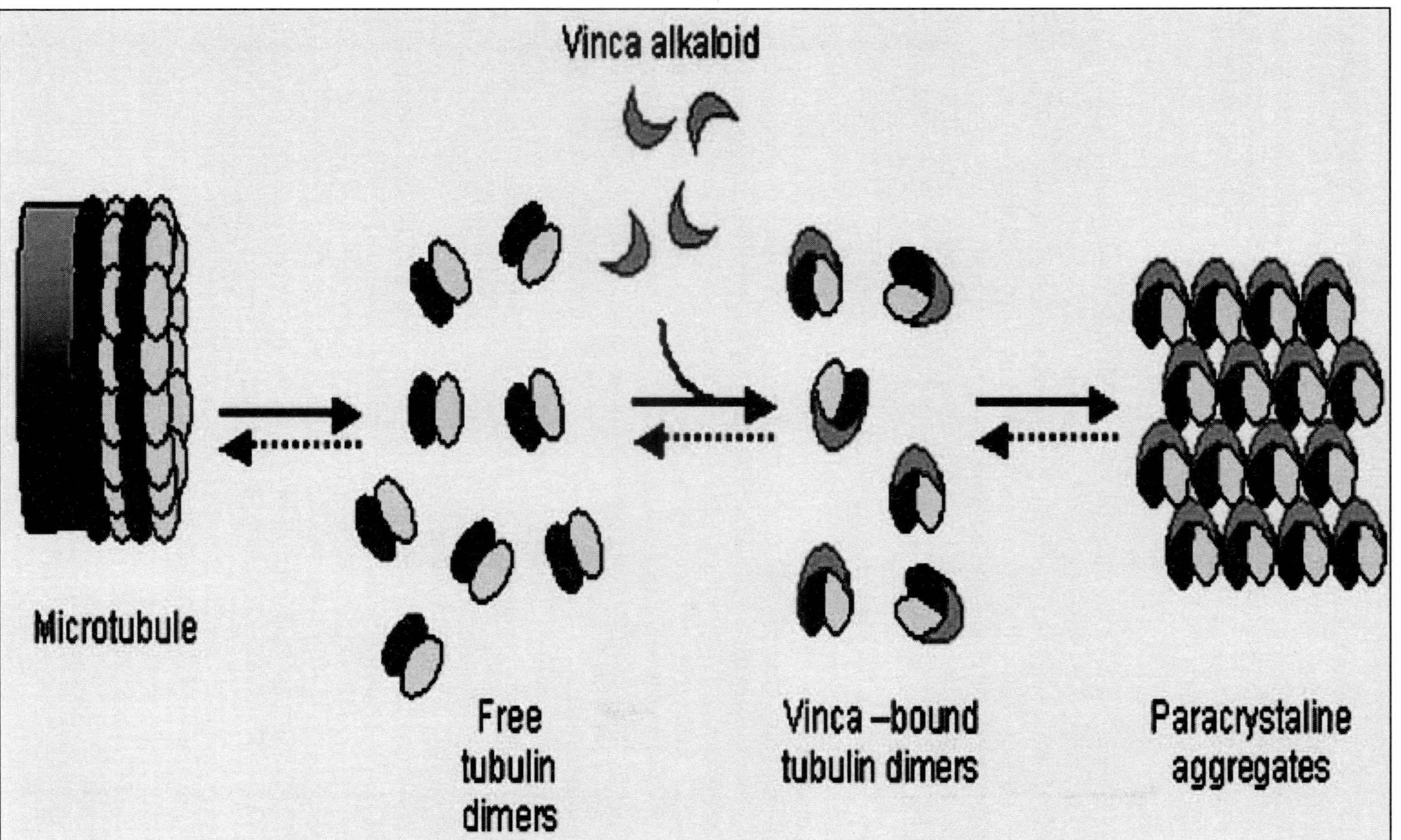

Figure 13.4: Mechanism of binding Vinca alkaloids with tubulin-dimers to inhibit cell proliferation.

Metabolism

All vinca alkaloids are administered intravenously and require special attention because they are vessels-sclerotics. These alkaloids are characterized by a large total distribution volume, rapid total plasma clearance and a long terminal half-life. After injection, they are eventually metabolized by the liver. It is therefore essential to monitor liver function and, if necessary, reduce the dose up to 50 per cent. In humans the main elimination route of vinca alkaloids is via faecal excretion.

Toxicity

The vinca alkaloids have the common ability to induce thrombocytopenia, due to the abundance of tubulin which is present in thrombocytes. Moreover, vinca alkaloids commonly cause reversible peripheral neurological side effects, which include tingling of the fingers or toes, abdominal pain, constipation, and muscle weakness. Other side effects include bone marrow suppression and reversible hair loss. There may also be pain and irritation at the injection site.

(A) Vincristine

Vincristine also known as Leurocristine, it is used to treat many different types of cancers. It has the best affinity for tubulin compared to the other vinca alkaloids. It is the vinca alkaloid with longer terminal half-life and lower clearance. It is frequently used in combination with other drugs. Malignancies in which vincristine is used include: acute leukemia; rhabdomyosarcoma; neuroblastoma; Hodgkin's disease and other lymphomas; lymphorecticular neoplasms; childhood leukemias. Being carcinogenic, mutagenic and teratogenic, it causes mainly peripheral neurotoxicity, gastrointestinal toxicity (more rarely), but mild myelosuppression. Myelosuppression may be severe in races with mutations in the gene coding for the *p*-glycoprotein.

Pharmacokinetic

High protein binding (75 per cent); (b) hepatic metabolism; (c) biliar (90 per cent) and renal (10 per cent) excretion; (d) half-life: from 19 to 155 hours (Jeannie *et al.*, 1998).

Side Effects

Unlike some of the other vinca alkaloids, vincristine does not cause severe bone marrow suppression (decreased blood cell counts). Common side effects include: hair loss; pain/redness at location of injection; nausea/stomach pain/vomiting; lowered blood cell count; numbness; headache; constipation; nervous system problems such as neuropathy or sensory impairment; blurred or double vision; back pain and overall weakness (Qweider *et al.*, 2007).

Moreover, this drug should not be taken by a patient who is pregnant, planning a pregnancy, or breast-feeding as it may cause birth defects. Patients should not receive any immunizations (vaccinations) while taking this medication. Vincristine may cause immunosuppression (weakened immunity) and vaccinations could lead to an illness.

(B) Vinblastine

Initially, vinblastine was also known as vincaleukoblastine, it differs from vincristine only for one CH_3 group at the vindoline ring and has similar pharmacokinetic characteristics. It can also inhibit angiogenesis. The main toxicity is non-cumulative myelosuppression, and then there is gastrointestinal toxicity, while neurotoxicity is rare. Cancers for which vinblastine is used include breast cancer, testicular cancer, lymphomas (Yang *et al.*, 2010).

Pharmacokinetic

(a) Hepatic metabolism (CYP3A4-mediated); (b) biliar and renal excretion; (c) half-life: 24.8 hours. Vinblastine is metabolized to desacetilvinblastine and this compound is more biologically active than the parent drug.

Side Effects

Common side effects include: bone or muscle pain; nausea and vomiting; alopecia; sores in mouth and on lips; swelling of feet or lower legs. Less common side effects include: dizziness, double vision, drooping eyelids, headache, jaw pain, mental depression, numbness or tingling in fingers and toes, pain finger and toe pain, testicle pain, waking difficulties, weakness, blood in urine or stools, and unusual bleeding or bruising (Goppi *et al.*, 2003).

(C) Vindesine

It is a semi-synthetic derivative of vinblastine. Cancers for which vindesine is used include: acute lymphocytic leukemia; lung carcinomas; breast cancer; chronic myelogenous leukemia and colorectal cancer (Rhomberg *et al.*, 2008).

Pharmacokinetic

(a) High protein binding (65-75 per cent); (b) hepatic metabolism (CYP3A4-mediated); (c) biliar and renal excretion; (d) half-life: 24 hours.

Side Effects

Common side effects include: hair loss; rash; nausea and vomiting; constipation; stomach cramps; jaw pain; vein inflammation (phlebitis); decreased blood cell counts (also called bone marrow suppression or, myelosuppression) and decrease in blood platelet number (thrombocytopenia) (Gaukroger *et al.*, 1985).

(D) Vinorelbine

It is also a semi-synthetic derivative of vinblastine. It is unique among the alkaloids because it has a very liposoluble structure, which ensures the complete depolymerization of microtubules. It can also be administered orally. The major metabolite is the desacetylvinorelbine. Vinorelbine is metabolized to its active metabolite, 4-*O*-deacetyl vinorelbine and two other minor metabolites, 20′-hydroxyvinorebine and vinorelbine 6′-oxide. *In vitro* data in human hepatic microsomes indicated that CYP3A4 is the main enzyme involved in the hepatic metabolism of vinorelbine. Cancers for which vinorelbine is used include: breast cancers; testicular cancers; epithelial ovarian cancers and non-small cell lung cancers (Casanova *et al.*, 2002).

Pharmacokinetic

(a) Bioavailability: 43 +/- 14 (oral); (b) protein binding: 79-91; (c) hepatic metabolism (CYP3A4-mediated); (d) fecal (46 per cent) and renal (18 per cent) excretion; (e) half-life: 27.7–46.3 hours (Marty *et al.*, 2001).

Side Effects

The side effects and toxicity of vinorelbine are generally milder than other vinca alkaloids. However, various side effects do frequently occur with treatment. Common side effects include: myelosuppression, which results in a reduced number of platelets, red blood cells and white blood cells; shortness of breath; constipation or loose stools; alopecia; numbness in the hands or feet. Birth defects are possible so patients who are pregnant or considering becoming pregnant should not take this drug. Breast-feeding should also be avoided while using vinorelbine. It is also critical that no immunizations are administered while taking vinorelbine. As a result of a lowered number of white blood cells, the body's natural defense mechanisms are weakened and often unable to fight infection. Likewise, it is important to avoid interaction with individuals recently given an oral polio vaccination.

Conclusion

The clinical interest of vinca alkaloids was clearly identified as early as 1965 and so this class of compounds has been used as anticancer agents for more than 30 years. Today, two natural compounds, vinblastine and vincristine and two semi-synthetic derivatives, vindesine and vinorelbine, have been registered and thus Vinca alkaloids can be considered to represent a chemical class of definite utility in cancer chemotherapy. Only two of these compounds, vinblastine and vincristine, have received extensive clinical evaluation. In spite of their close similarity, chemically, a somewhat different group of human neoplasms responds to these compounds, and there has been a singular lack of cross-resistance between these two drugs and any other oncolytic drug now in wide use. Vinblastine has proved effective in chorioepithelioma, Hodgkin's disease, and other lymphomas, and a number of beneficial results have been obtained in carcinoma of the breast and bronchus. In addition, vincristine has been striking in its ability to induce complete hematological remission of the acute leukemias of childhood, both lymphocytic and myelogenous in type. Moreover, recent studies have demonstrated that certain newly identified properties, such as antiangiogenic activities, could enlarge the therapeutic usage of natural and semi-synthetic vinca alkaloids. Thus, vinca alkaloids remain a drug family with a continuing interest for future anticancer therapy.

Future Prospects

There remains a great interest in the search for natural products from terrestrial plants as potential drug chemical leads for the treatment of a wide range of disease conditions. There is an especially strong interest in developing new natural product anticancer agents from natural sources, because of the many past successes in this endeavor. However, screening efforts of many natural products laboratories have concentrated more recently on attempts to find cures for the most common forms of cancer (*e.g.*, breast, colorectal, lung and prostate) rather than rarer acute and chronic forms of leukemia. Clearly, there is copious scope for new discoveries to be made of natural products from plants and other organisms with promising activities against hematological malignancies. In addition to random screening in this regard, it is likely that the targeted investigation of plants used in systems of traditional medicine may be particularly beneficial. The value of collaboration between laboratories specializing in natural products isolation and hematology in this type of oncology drug discovery must be emphasized.

Acknowledgements

This work has been supported by Council of Scientific and Industrial Research (CSIR) and University Grants Commission (UGC).

References

Casanova, M., Ferrari, A., Spreafico, F., Terenziani, M., Massimino, M., Luksch, R., Cefalo, G., and Polastri, D. (2002). Vinorelbine in previously treated advanced childhood sarcomas: Evidence of activity in rhabdomyosarcoma. *Cancer*, **94**: 3263–3268.

De Luca, V., and Cutler, A.J. (1987). Sub cellular localization of enzymes involved in indole alkaloid biosynthesis in *Catharanthus roseus*. *Plant Physioliology*, **85**: 1099-1102.

Facchini, P.J., and De Luca, V. (2008). Opium poppy and Medagascar periwinkle: model non-model system to investigate alkaloids biosynthesis in plants. *Plant Journal*, **54:** 763-784.

Gantet, P., and Memelink, J. (2002). Transcription factors: Tools to engineer the production of pharmacologically active plant metabolites. *Trends in Pharmacology Science*, **23**: 563–569.

Gaukroger, L.W. and Mackie, R. (1985). Cytotoxicity of etretinate and vindesine. *British Journal of Cancer*, **52**: 369–375.

Goppi, P.G., Broglia, C., Merli, F., Dell'Olio, M., Stelitano, C., Iannitto, E., Federico, M., Bertè, R., Luisi, D., Molica, S., Cavalli, C., Dezza, L., and Ascari, E. (2003). Vinblastine, bleomycin, and methotrexate chemotherapy plus irradiation for patients with early-stage, favorable hodgkin lymphoma. *Cancer*, **98**: 2393–2401.

Guirimand, G., Courdavault, V., St-Pierre, B., and Burlat, V. (2010). Biosynthesis and regulation of alkaloids. *In*: Plant developmental biology–biotechnological perspectives, vol 2, Ed. by Pua, E.C. and Davey, M. Springer, Berlin, pp. 139–160.

Heijden R.V.D., Jacobs, D.I., Snoeijer, W., Hallard, D., and Verpoorte, R. (2004). The *Catharanthus* alkaloids: Pharmacognosy and biotechnology. *Current Medicinal Chemistry*, **11:** 607-628.

Jayanthi, M., Sowbala, N., Rajalakshmi, G., and Kanagavalli, S.V. (2010). Study of antihyperglycemic effect of *Catharanthus roseus* in alloxan induced diabetic rats. *International Journal of Pharmacy and Pharmaceutical Science*, **2:** 114-116.

Jeannie, D., and Chan, P.D. (1998). Pharmacokinetic drug interactions of Vinca alkaloids: Summary of case reports. *Pharmacotherapy: The Journal of Human Pharmacology and Drug Therapy*, **18**: 1304–1307.

Mahroug, S., Burlat, V., and St-Pierre, B. (2007). Cellular and sub-cellular organization of the monoterpenoid indole alkaloid pathway in *Catharanthus roseus*. *Phytochemistry Reviews*, **6:** 363-381.

Marty, M., Fumoleau, P., Adenis, A., Rousseau, Y., Merrouche, Y., Robinet, G., Senac, I., and Puozzo, C. (2001). Oral vinorelbine pharmacokinetics and absolute bioavailability study in patients with solid tumors. *Annals of Oncology*, **12**: 1643–1649.

Meijer, A.H., Verpoorte, R., and Hoge, J.H.C. (1993). Regulation of enzymes and genes involved in terpenoid indole alkaloid biosynthesis in *Catharanthus roseus*. *Journal of Plant Research*, **3:** 145–165.

Mersey, B.G., and Cutler, A.J. (1986). Differential distribution of specific indole alkaloids in leaves of *Catharanthus roseus*. *Canadian Journal of Botany*, **64**: 1039–1045.

Noble, R.L. (1990). The discovery of the vinca alkaloids-chemotherapeutic agents against cancer. *Biochemistry and Cell Biology*, **68:** 1344-1351.

Qweider, M., Gilsbach, J.M., and Rohde, V. (2007). Inadvertent intrathecal vincristine administration: A neurosurgical emergency. Case Report. *Journal of Neurosurgery: Spine*, **6**: 280–283.

Rhomberg, W., Eiter, H., Schmid, F., and Saely, C.H. (2008). Combined vindesine and razoxane shows antimetastatic activity in advanced soft tissue sarcomas. *Clinical and experimental metastasis*, **25:** 75-80.

Rodriguez, S., Compagnon, V., Crouch, N.P., St-Pierre, B., and De Luca, V. (2003). Jasmonate-induced epoxidation of tabersonine by a cytochrome P-450 in hairy root cultures of *Catharanthus roseus*. *Phytochemistry*, **64:** 401-409.

Singh, V.P., and Jagdev, R.S.D. (1996). Ajmalicine (raubacine); A medicinally important alkaloid from *Catharanthus roseus* (*Vinca rosea*). *In*: supplement to cultivation and utilization of medicinal plants. Ed. by Handa, S.S. and Kaul, M.K. RRL, Jammu, pp. 199-206.

Toki, K., Saito, N., Irie, Y., Tatsuzawa, F., Shigihara, A., and Honda, T. (2008). 7-O-Methylated anthocyanidin glycosides from *Catharanthus roseus*. *Phytochemistry*, **69**: 1215–1219.

Van der, H.R., Jacobs, D., Snoeijer, W., Hallard, D., and Verpoorte, R. (2004). The *Catharanthus* alkaloids: pharmacognosy and biotechnology. *Current Medicinal Chemistry*, **11**: 607–628.

Yang, H., Ganguly, A., and Cabral, F. (2010). inhibition of cell migration and cell division correlate with distinct effects of microtubule inhibiting Drugs. *Journal of Biological Chemistry*, **285**: 32242–32250.

Ziegler, J., and Facchini, P.J. (2008). Alkaloid biosynthesis: metabolism and trafficking. *Annual Review Plant Biology*, **59:** 735–769.

Bioactive Phytochemicals: Perspectives for Modern Medicine Vol. 2 (2014) *Pages* **307–325**
Editor: **V.K. Gupta**
Published by: **DAYA PUBLISHING HOUSE, NEW DELHI**

14

Endophytic Fungi: Novel and Potential Source of Bioactive Natural Products

Seema Dhankhar[1] and J.P. Yadav[1*]

ABSTRACT

Endophytic fungi are a diverse group of ascomycetous (primarily) fungi functionally defined by their occurrence within asymptomatic photosynthetic tissues of plants. It occurs in all major lineages of land plants ranging from the arctic to the tropics. The search for bioactive natural products of endophytic fungi, isolated from higher plants, are attracting considerable attention from researchers worldwide, as indicated by the increase of work and publications on therapeutic potential during recent years. The diverse range of biosynthetic pathways in plants and endophytic fungi has provided an array of lead structures that have been used in drug development. Endophytes contain different bioactive compounds for commercial and vital therapeutic emergencies, which mainly include alkaloids, benzopyranones, chinones, cytochalasines, depsipeptides, enniatines, flavonoids, furandiones, isocumarines, peptides, polyketones, phenols, quinols, terpenoids, tetralones and xanthones. The above bio-therapeutic compounds are selected based on their anticancerous, anti-oxidant, anti-viral and antimicrobial activities. The present review chapter focuses on the pharmaceutical potentials of endophytic fungi, in novel natural products screening series.

Keywords: Bioactivities, Drug discovery, Fungal endophytes, Natural products plants.

Introduction

The phytosphere is an attractive habitat for microorganisms because of the availability of abundant nutrient and the environmental stability. Among these diverse

1 Department of Genetics, Maharshi Dayanand University, Rohtak – 124 001, Haryana, India

* *Corresponding author*: E-mail: yadav1964@rediffmail.com

microorganisms are endophytes and epiphytes. Endophytes defined by Petrini (1991) include all organisms inhabiting plant organs that at some time in their life can colonize internal plant tissues without causing apparent harm to the host. Several decades of research and numerous articles on endophytic fungi in plants have resulted in a plethora of knowledge of the endophytes and their natural products.

Endophytes are also known to affect the interactions of plants with their environment, and to alter the course of their interactions with plant pathogens. Endophytes are a poorly investigated group of microorganisms that represent an abundant and dependable source of bioactive and chemically novel compounds with potential for exploitation in a wide variety of medical, agricultural and industrial arenas. The mechanisms through which endophytes exist and respond to their surroundings must be better understood in order to be more predictive about which higher plants to seek, study, and spend time isolating micro floral components. This may facilitate the product discovery processes. Although work on the utilization of this vast resource of poorly understood microorganisms has been initiated, it has already become obvious that an enormous potential for organism, product, and utilitarian discovery in this field holds exciting promise. This is witnessed by the discovery of a wide range of products, and microorganisms. There are other characteristics of endophytic fungi that make them desirable for manipulation in an industrial screening program. Most screening has focused on soil dwelling fungi; little attention has been directed toward endophytes. As a consequence, they have not been subjected to intensive screening programs, which suggest the vast majority largely remain undiscovered. In spite of this, fungal endophytes are relatively untapped source of biodiversity.

The worldwide people are being threatened by more and more advent diseases such as cancers, AIDS, diabetes etc., accompanied with an increasing extent of environmental degradation, loss of biodiversity. Therefore it is a must for isolating new and beneficial compounds to provide assistance and relief in all aspects of human condition. Because of the great biodiversity of the fungal endophytes inhabiting sometimes unusual niches provides an opportunity to find new natural products for novel drug discovery. In addition to represent a source of organisms for disease control and plant improvement, the study of endophytes may have an important influence in the conceptual framework where plant-pathogen interactions are interpreted and investigated. The study of fungal endophyte diversity is hampered by little sampling, isolating and lack of characterization. Sampling and characterizing fungal endophyte diversity is an emerging challenge and promises to lead to the discovery of new species, novel compounds and a better understanding of their role in ecosystems (Arnold and Lutzoni, 2007; Saikkonen, 2007; Sieber, 2007; Rodriguez *et al.*, 2009).

Intelligent and efficient industrial screening of microorganism requires that a high diversity of organisms be maintained, while simultaneously minimizing redundancy among the taxa screened. This can only be achieved through an understanding of the floristic composition and pattern of colonization of the microorganisms within the particular ecological niche being sampled, *i.e.*, what species are likely to inhabit the particular endophytic host and what is their relative abundance? How many species are likely to be found by sampling a single plant or individual trees? Does species richness and floristic composition vary among samples throughout a landscape? Does the isolation media or sampling of a particular plant organ influence the species recovered? (Tejesvi *et al.*, 2007)

Endophytes have proven to be rich sources of novel natural compounds with a wide-spectrum of biological activities and a high level of structural diversity. Endophytic fungi are one of the largely unexploited resources that likely have several, economically important applications in the future. Both pharmaceutical industry and plant breeders are interested in endophytes for production of enzymes, medicines and biological control agents. In addition to exploring endophytes for pharmacologically active natural products, which might prove to be suitable for specific medicinal or agrochemical applications, a field of major interest is examining the ecological function of secondary metabolites from endophytes. In many cases endophytic fungi might be involved in the biosynthesis of plant products, but they might by themselves be the producers of a wide array of new biologically active and extremely unusual secondary metabolites. This review chapter provides an insight into the diversity of fungal endophytes particularly focuses on the pharmaceutical potential of these mycobiota for human use.

Approaches for Selection of Host Plant

Herbal medicine is one of the oldest forms of health care known. It is believed that every plant on earth is known to harbour at least one endophytic microbe. While selecting the medicinal plants for the study, their unique mycoflora and the bioactive produced by them should be considered. Keep the following important points in mind:

1. Plant species already known to produce one or more novel compounds has been taken into account or that is of ethno botanical or importance in traditional medicine.
2. Plants located in ecological settings that suggest microorganisms playing a role in protection against other microorganisms.
3. Plants growing in extreme conditions, *e.g.*, extreme temperatures, aquatic environments, or high exposure to radiation or salt concentrations (Strobel and Long, 1998).

Isolation of Endophytic Fungi

After a plant has been selected for study, it was identified, and its location was plotted using a global positioning device. Small leaf, stem and root pieces were cut from the plant and placed in sealed plastic bags after excess moisture was removed. The materials were stored at 4°C until isolation procedures begin. In the laboratory, plant materials were thoroughly surface sterilized with 70 per cent ethanol and sodium hypochlorite bleach for few minutes. The ethanol acts as a surfactant and the NaOCl is the actual sterilizing agent. The dilutions and immersion times in NaOCl vary with the type of tissue and host. Sometimes they were flamed, and ultimately they were air dried under a laminar-flow hood. This was done in order to eliminate surface -contaminating microbes. Then, with a sterile knife blade, outer tissues were removed from the samples and the inner tissues were carefully excised by scalpel and placed on water agar plates. After several days of incubation, hyphal tips of the fungi were removed and transferred to potato dextrose agar culture medium.

The endophytes were encouraged to sporulate on specific plant materials and are eventually identified via standard morphological and molecular biological techniques and methods.

Approaches for the Selection of Fungal Endophytes

Once a biotechnological target has been identified, two questions follow: first, what might be the best organism or group of organisms to investigate? Second, what screening procedures should be used in order to detect the desired activity?

The following approaches can be used for organism selection: (i) play the percentage game, *e.g.*, endophytic fungi and actinomycetes for biopharmaceutins; (ii) make reference to taxa-chemistry and taxon-property databases of known metabolites; (iii) focus on novel and neglected taxa; (iv) highlight isolates from unusual or little-explored ecosystems; (v) match the target with members of previously unscreened but known taxa (Tejesvi *et al.*, 2007).

The flow chart for isolation and characterisation of bioactive compounds from endophytic fungi has been illustrated in Figure 14.1.

Bioactive Secondary Metabolites from Endophytic Fungi

Secondary metabolites, defined as low molecular-weight compounds not required for growth in pure culture, are produced as an adaptation for specific functions in nature. They play vital role *in vivo*, *e.g.*, numerous metabolic interactions between fungi and their plant hosts, such as signaling, defense and regulation of the symbiosis (Schulz *et al.*, 2005).

Endophytic fungi are source for screening of natural products and in optimizing the search for new bioactive secondary metabolites; it is pertinent to consider that:

1. The secondary metabolites a fungus synthesizes may correspond with its respective ecological niche and
2. Continual metabolic interactions between fungus and plant may enhance the synthesis of secondary metabolites.

A large number of bioactive metabolites of endophytic fungi have been extracted and characterized over years, and those isolated belong to diverse structural groups (Figure 14.2.) *i.e.*, alkaloids, steroids, terpenoids, peptides, polyketones, flavonoids, quinols, phenols, xanthones, chinones, isocumarines, benzopyranones, tetralones, cytochalasines, perylene derivatives, furandiones, depsipeptides and enniatines. Some of them represent novel structural groups, *e.g.*, the palmarumycins and a new benzopyroanone (Tan and Zou, 2001). The secondary metabolites produced by endophytes associated with medicinal plants can be exploited for curing diseases (Tejesvi *et al.*, 2007). Endophytes are chemical synthesizer inside plants (Owen and Hundley, 2004), in other words, they play a role as a selection system for microbes to

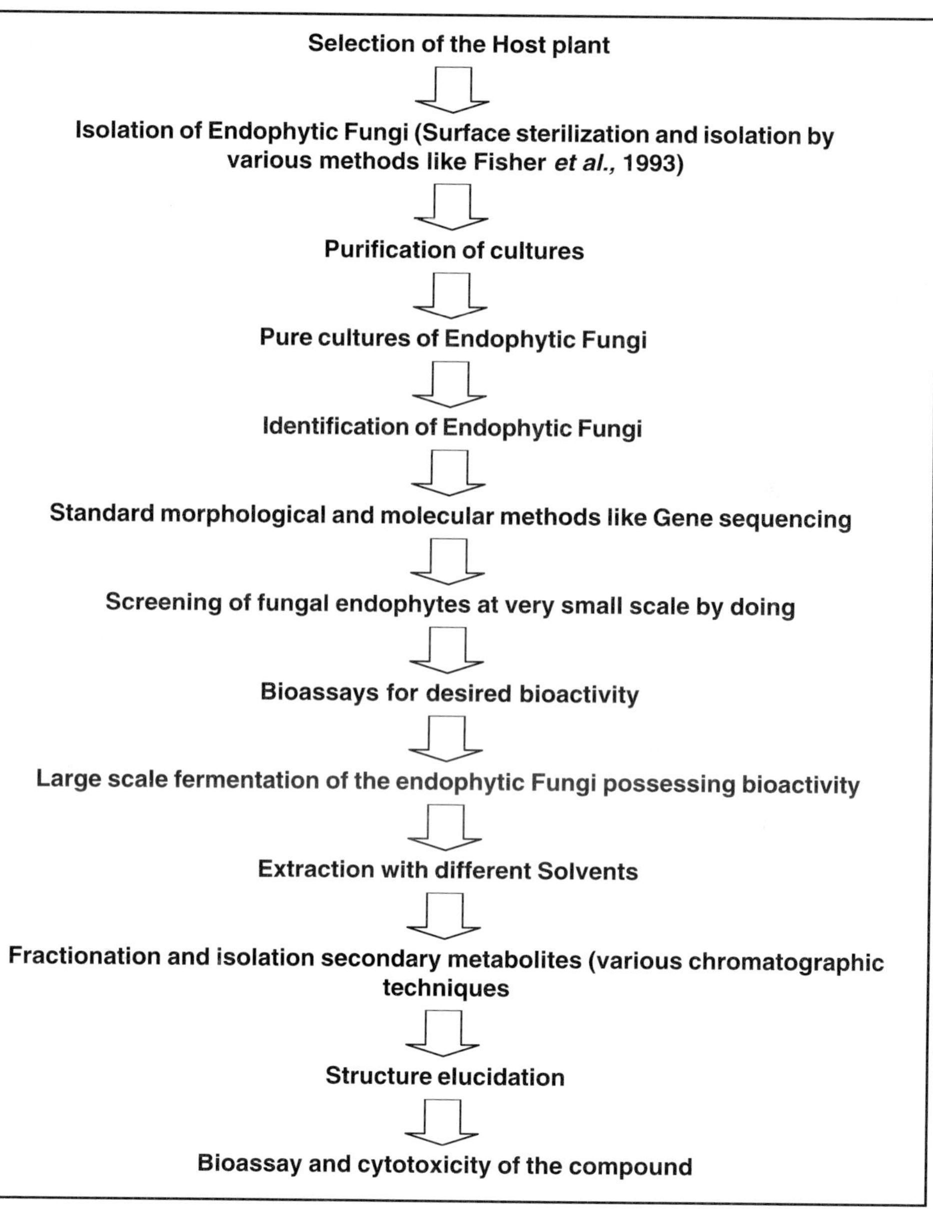

Figure 14.1: Method for isolation of endophytic fungi from ethnopharmaceutically important plant for therapeutic lead compound(s).

produce bioactive substances with low toxicity toward higher organisms (Strobel, 2003). These secondary metabolites of endophytes are synthesized via various metabolic pathways *e.g.*, polyketide, isoprenoid, amino acid derivation. Table 14.1 describes some of the secondary metabolites isolated from endophytic fungi possessing different bioactivities.

Figure 14.2: Structures of some important secondary metabolites isolated from fungal endophytes.

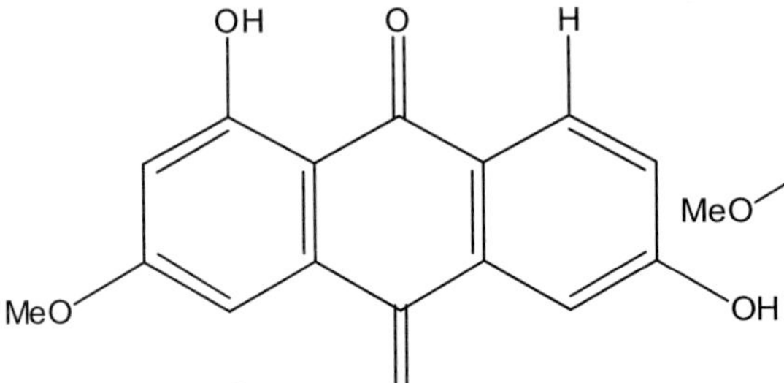

Lunatin

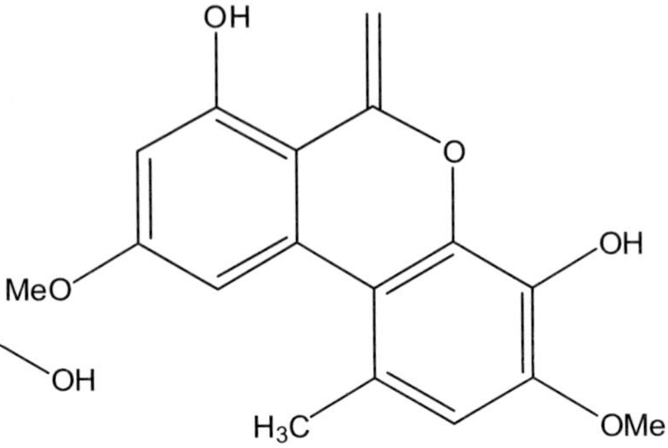

Graphislactone

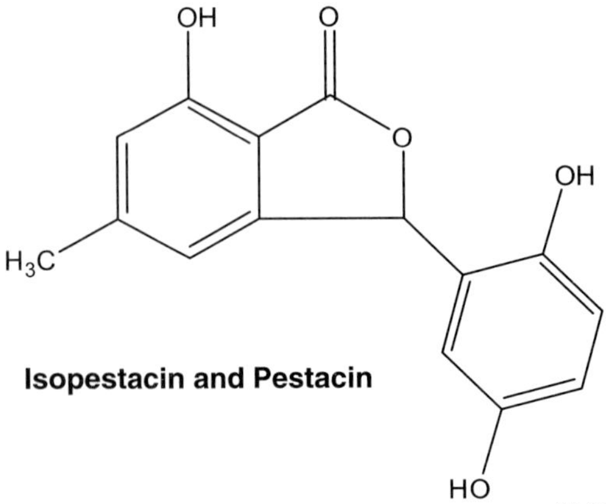

Isopestacin and Pestacin

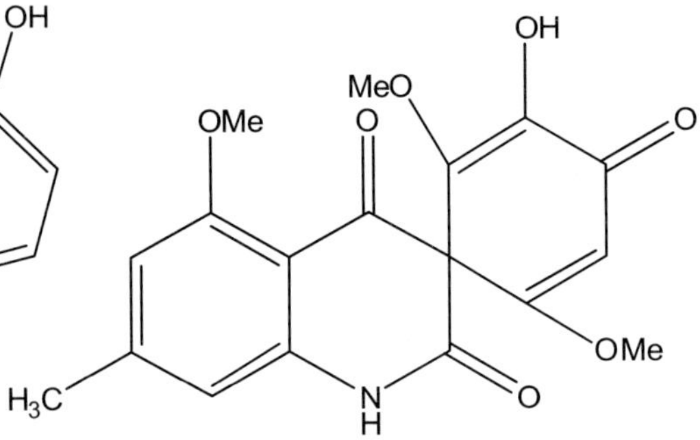

Asperfumoid

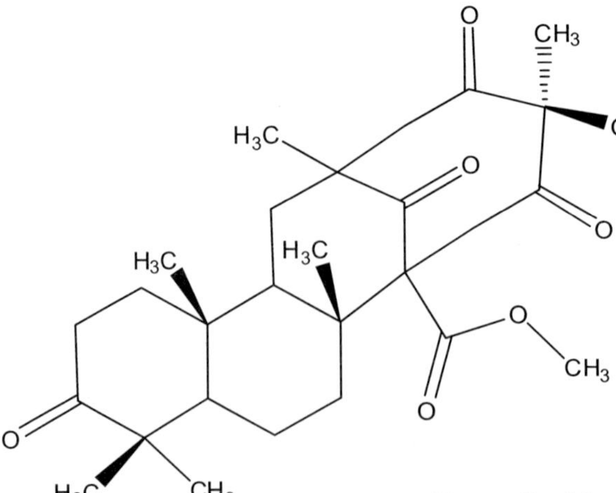

Preaustinoid

Contd...

Figure 14.2–*Contd...*

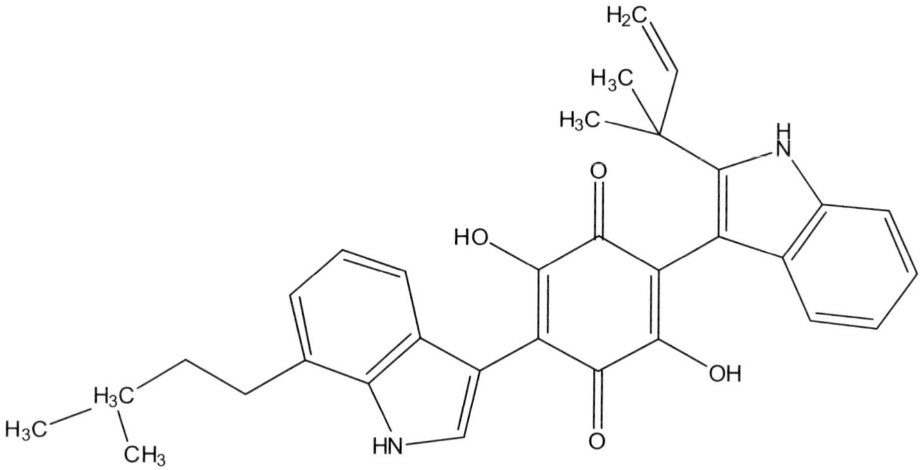

Dimethyl-asterriquinone B1

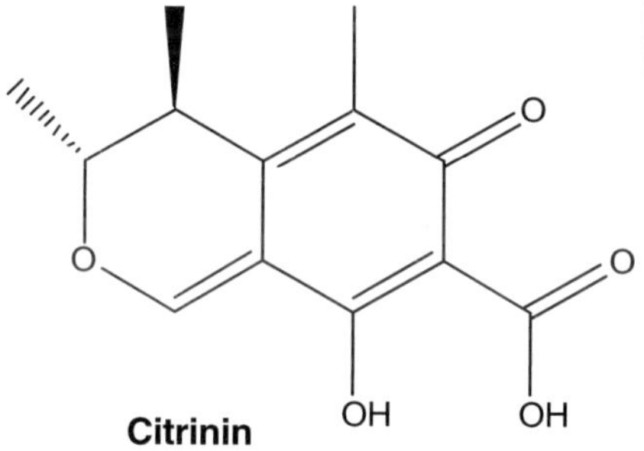

Citrinin

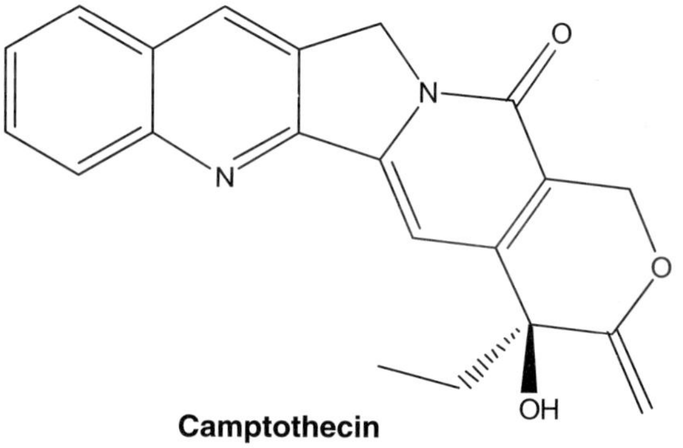

Camptothecin

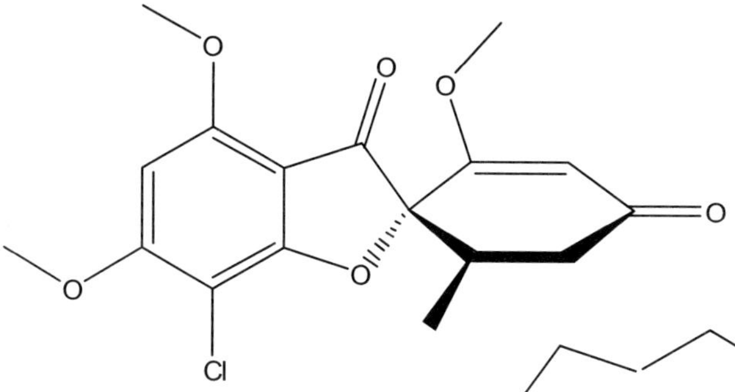

Griseofulvin

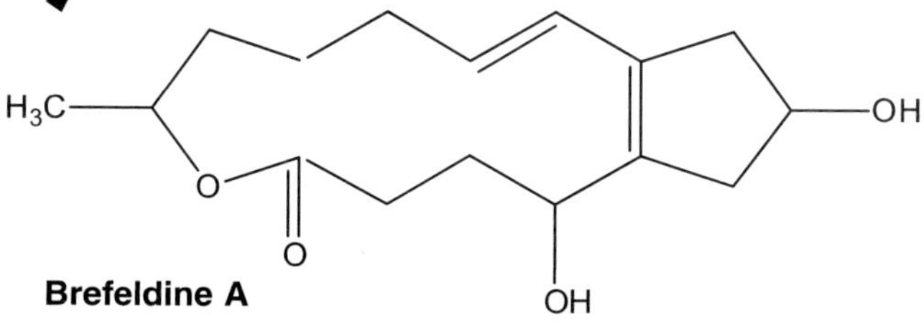

Brefeldine A

Contd...

Figure 14.2–*Contd...*

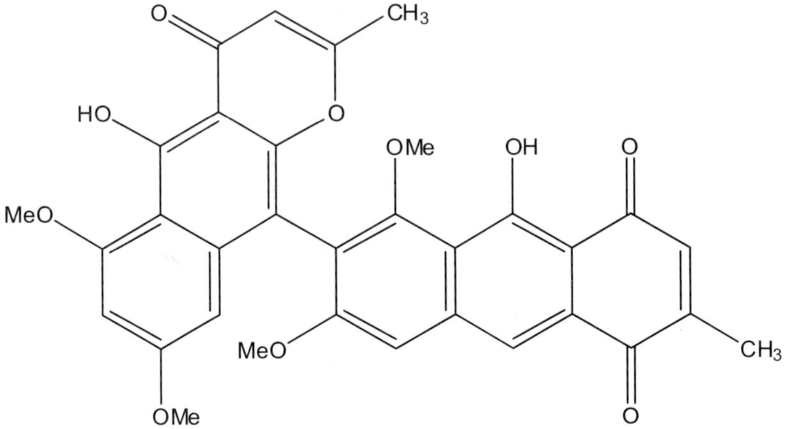

Rubrofusarin B

Several advantages of fermentation of fungal endophytes producing bioactive substances are: (i) Industrial production of bioactive substances (pharmaceutical drugs) require reproducible, dependable productivity. If the microbe is the source organism, it can be grown in tank fermentors as needed, producing a virtually inexhaustible supply; (ii) Microorganisms typically respond favourably to routine culture techniques and tissue culture or growing plants requires either specialized techniques or months of growth before harvesting is feasible; (iii) Product escalation is relatively easy in microorganisms. Directed changes in culture conditions can be explored indefinitely as a method of optimizing various biosynthetic pathways, which may lead to even more effective derivatives of novel compounds.

Apparently, number of significant work has accumulated since the first discovery of endophytic fungus in darnel in 1904. In fact, since the publication of the report by Stierle and his co-workers, there has been an increase in the number of US patents filed on endophytic fungi producing important metabolites with diverse biological activities (Priti *et al.*, 2009). An array of natural products has been characterized from endophytic fungi, which includes anti-cancerous, antioxidants, anti-viral, anti-insecticidal, immunosuppressant, anti-microbial, anti-malarial and anti-mycobacterial.

Antimicrobial Compounds

The serendipitous discovery of penicillin from the filamentous fungi *Penicillium notatum* by Flemming in 1929, and the observation of the broad spectrum therapeutic use of these agents in the 1940's, ushered in a new era in medicine, the so called 'Golden age of Antibiotics'. Metabolites bearing antibiotic activity can be defined as low molecular weight organic natural substances made by microorganisms that are active at low concentrations against other microorganisms (Guo *et al.*, 2008).

Three metabolites named phomopsin A (1), B (2) and C (3), together with two known compounds cytosporone B (4) and C (5), were isolated from the mangrove endophytic fungus, *Phomopsis* sp. ZSU-H76 obtained from the South China Sea. Their structures were elucidated by spectroscopic methods, mainly by 1D and 2D NMR spectroscopic techniques. The medium-sized cyclic phenol ether based on 1 or 2 is rare in natural products. In bioassays, compounds 1, 2, and 3 had no significant antibiotic activities, but compounds 4 and 5 inhibited two fungi *Candida albicans* and *Fusarium oxysporum* (Huang *et al.*, 2008). Endophytic *Fusarium* sp. isolated from plant *Selaginella pollescens* collected from Guanacaste conservation area of Costa Rica was screened for antifungal activity which showed potent activity against *Candida albicans* in agar diffusion assay (Strobel and Daisy, 2003). Jesterone an antifungal compound which was isolated from fungus *Pestalotiopsis jesteri.* Jesterone is only compound from the endophytes in which total synthesis was reported (Li and Strobel, 2001). The endophytic fungus *Chloridium* sp. produces Javanicin (Verma *et al.*, 2007). This highly functionalized napthaquinone exhibits strong antibacterial activity against *Pseudomonas* sp.Three metabolites: Phomoenamide (Aly *et al.*, 2008), Phomonitroester and Deacetylphomoxanthone B have been reported from endophytic fungus *Phomopsis* sp PSU D (Sappapan *et al.*, 2008).

Hoffman *et al.* (2008) isolated a group of phenolic acids from culture broth of a *Phoma* sp., discovered as an endophyte on a Guinea plant (*Saurauia scaberrinae*). The compounds exhibit antibacterial activities against gram-positive bacteria (*S. aureus, Sclerotinia sclerotiorum* and *Rhizoctonium solani*).

Dhankhar *et al.* (2013) reported antimicrobial potential of the acetonic extract of *Aspergillus* sp. JPY1 isolated from *Salvadora oleoides* against many pathogenic microbes (*Aspergillus niger, A. flavus, A. fumigatus, Bacillus subtilis, Escherichia coli, Vibrio cholerae, Staphylococcus aureus* and *Klebsiella pneumonia*) with very low cytotoxicity. GCMS analysis of the acetonic extract of this fungi reveals the presence of Phenol, 2, 6-bis(1,1-dimethylethyl)-4 acid and 1,2-Benzenedicarboxylic acid, bis(2-methylpropyl) ester as major compounds; these may be responsible for its antimicrobial activity.

Many antibiotic compounds isolated from the endophytic fungi have been given in Table 14.1.

Antiviral compounds

Many antiviral agents are reported from endophytic fungi. Two novel compounds cytonic acid A and B have been isolated from the endophytic fungus *Cytonaema* sp. These compounds are inhibitor of human cytomegalovirus (hCMV) protease (Guo *et al.*, 2000).

The microbial flora characteristic of oak trees resulted in the isolation of a potentially valuable fungal specimen from the leaves of *Quercus coccifera*. These fungal endophyte synthesize a compound hinnuliquinone, a potent inhibitor of the HIV-1 protease (Singh *et al.*, 2004).

Four new cyclohexadepsi peptides, pullularins A–D, were isolated from the endophytic fungus *Pullularia* sp. BCC 8613. Pullularin A exhibited activities against the malaria parasite *Plasmodium falciparum* K1 and Herpes Simplex Virus type 1

Table 14.1: Overview of some important endophytic secondary metabolites with antimicrobial, anticancerous, antioxidant and antidiabetic activities.

Fungal Endophyte	*Host Plant*	*Compound Isolated*	*Activity*	*Reference*
Unidentified ascomycete	*Arbutus unedo*	Anofinic acid	Antibacterial and antifungal against *Eschrichia coli, Bacillus megaterium, Microbotryum violaceum*	Qin *et al.* (2011)
Nigrospora oryzae	*Nyctanthes arbor-tristis*	Bioactive extract	Antibacterial activity against *Shigella* sp. and *Pseudomonas aeruginosa*	Gond *et al.* (2011)
A. niger, A. alternate	*Tabebuia argentea*	Naphthaquinone	Antimicrobial activity against an array of Pathogenic fungi and bacteria	Sadananda *et al.* (2011)
Penicillium brasililanum	*Melia azedarach*	Phenylpropanoid amides	Anticancer, antioxidant, antimicrobial, anti inflammatory and immunosuppressive agent	Fill *et al.* (2010)
Nodulisporium sp.	*Erica arborea*	Nodulisporins D and F	Antibacterial and Antifungal	Dai *et al.* (2009)
Aspergillus fumigatus	*Juniperus communis*	Podophyllotoxin	Potent anticancerous activity	Kusari *et al.* (2009)
Phomopsis sp.	*Laurus azorica*	Cycloepoxytriol B and Cycloepoxylactone	Antibacterial and antifungal activity against *Bacillus megaterium* and *Microbotryum violaceum*	Hussain *et al.* (2009)
Alternaria alternate	*Coffea arabica*	Compound isolated	Cytotoxic activity towards HeLa cells	Fernandes *et al.* (2009)
Verticillium sp.	*Rehmannia glutinosa*	2,6-Dihydroxy-2-methyl-7-(prop-1 E-enyl)-1-benzofuran-3 (2H)-one	Active against *Verticillium* sp.	You *et al.* (2009)
Xylaria sp.	*Ginko biloba*	7-amino-4-methyl-coumarin	Antibacterial and antifungal activity against *Staphylococcus aureus, E. coli, Salmonella typhi, S. typhimurium, S. enteritidis, Aspergillus hydrophila, Yersinia* sp., *Verticillium albicans, A. niger*	Liu *et al.* (2008)
Ampelomyces sp.	*Urospermum picroides*	6-O-methylalaternin and altersolanol	Antimicrobial activity against gram positive pathogens, *Staphylococus aureus, S. epidermidis, Enterococcus faeccalis*	Aly *et al.* (2008)

Contd...

Table 14.1–*Contd...*

Fungal Endophyte	*Host Plant*	*Compound Isolated*	*Activity*	*Reference*
Phomopsis sp.	*Mangrove*	Cytosporone B and C	Antifungal activity against *Candida albicans* and *Fusarium oxysporum*	Huang *et al.* (2008)
Cladosporium sp.	*Quercus viriabilis*	Brefeldin A	antibiotic	Wang *et al.* (2007)
Muscodor albus	Tropical tree	Tetrohydofuran, 2-methyl furan, 2-butanone, aciphyllene	Active against *Stachybotrys chartarum*	Atmosukarto *et al.* (2005)
Phomopsis cassia	*Cassia spectabilis*	Ethyl 2,4-dihydroxy-5, 6-dimethylbenzoate and phomopsilactone	Antifungal activity against two phytopathogenic fungi *Cladosporium cladosporioides* and *C. sphaerospermum*	Silva *et al.* (2005)
Xylaria sp. *F0010*	*Abies halophylla*	Griseofulvin	Antibiotic agent used against the treatment of human mycotic diseases	Park *et al.* (2005)
Penicillium janthinellum	*Meliaazedarach*	Citrinin	100 per cent antibacterial activity against *Leishmania* sp.	Marinho *et al.* (2005)
Cephalosporium sp. IFB-E001	*Trachelospermum*	Graphislactone A	Free radical scavenging and antioxidant activities	Song *et al.* (2005)
Acremonium zeae	*Maize*	Pyrrocidines A, B	Active against *Aspergillus flavus, Fusarium verticillioides*	Wicklow *et al.* (2005)
Aspergillus fumigatus CY018	*Cyndon dactylon*	Asperfumoid, Fumigaclavin C, Fumitremorgin C, Physcion, Helvoic acid	Inhibit *C. albicans*	Liu *et al.* (2004)
Aspergillus niger IBF-E003	*Cynodon dactylon*	Rubrofusarin B, Aurasperone A	Strong co inhibitors on xantin oxidase, colon cancer cell and some microbial pathogens	Song *et al.* (2004)
Phoma medicaginis	*Medicago sativa, Medicago lupulina*	Brefelidine A	Antibiotic activity, initiation of apoptosis in cancer cells	Weber *et al.* (2004)
Phomopsis sp.	*Erythrina cristagalli*	Mevinic acid	Anti-inflammatory activity	Weber *et al.* (2004)

Contd...

Table 14.1–*Contd...*

Fungal Endophyte	*Host Plant*	*Compound Isolated*	*Activity*	*Reference*
Pestalotiopsis microspora	*Terminalia morobenesis*	Pestacin and isopestacin	Antioxidant, antimycotic and antifungal activity	Harper *et al.* (2003); Strobel *et al.* (2002)
Curvularia lunata	*Niphates olemda*	Cytoskrins	Antibacterial activity, potential anticancer agent	Brady *et al.* (2000); Jadulco *et al.* (2002)
Curvularia lunata	*Niphates olemda*	Lunatin	Active against *Bacillus subtilis, Escherichia coli, Staphylococcus aureus, Cladosporium herbarum*	Jadulco *et al.* (2002)
Pestalotiopsis microspore	*Torreya taxifolia*	Torreyanic acid	Antifungal agent	Liu *et al.* (2001)
Rhinocladiella sp. *Monochaetia* sp.	*Tripterygium wilfordii*	22-oxa-[12]-Cytochalasin	Antitumour	Wagenaar *et al.* (2000)
Cytonaema sp.	*Quercus* sp.	Cytonic acid A and B	Human cytomegalovirus protease inhibitor (HCMV)	Guo *et al.* (2000)
Pseudomassaria sp.	From African rainforest near Kinshasha in the Democratic Republic of the Congo	Nonpeptidal fungal metabolite	Insulin mimetic compound (L-783,281)	Zhang *et al.* (1999)

(Isaka *et al.*, 2007). An endophytic fungus, *Pestalotiopsis theae* isolated from an unidentified tree of Jianfeng Mountain, China, was capable of producing Pestalotheol C with anti-HIV properties (Li *et al.*, 2008).

Anticancer Compounds

Cancer is one of the major causes of death in the world. This disease is characterized by unregulated proliferation of cells. Large number of plants, microorganisms and marine sources of natural products, have been isolated and identified as pure anti cancer compounds. The taxol has generated more attention and interest than any other new drug since its discovery; it is mainly due to its unique mode of action (Gangadevi and Muthumary, 2008) and excellent clinical activity against breast and ovarian cancer. Several reports about anticancer properties (Liu *et al.*, 2007; Kakolyris *et al.*, 2006) of taxol, as well as other sources of its production have been investigated in the last decade. The isolation of taxol–producing endophyte *Taxomyces andreanae* has provided an alternative approach to obtain a cheaper and more available product via microorganism fermentation (Stierle *et al.*, 1993). After that, taxol has also been found in a number of different genera either associated or not to Yews, such as *Taxodium distichum* (Li *et al.*, 1996), *Wollemia nobilis* (Strobel *et al.*, 1997), *Bartalinia robillardoides* (Gangadevi and Muthumary, 2008) and *Botryodiplodia theobromae* (Pandi *et al.*, 2010).

The compound isolated from the endophytes have attracted much interest and attention; these are phenylpropanoides, camptothecin and its analogues, ergoflavins, podophyllotoxin and its analogue, cytochalasins, dimeric quinine. Some other compounds with anticancer properties isolated from endophytic microbes were reported such as cytoskyrins. Phomoxanthones A and B, Photinides A-F, Rubrofusarin B and (+)-epiepoxydon (Brady *et al.*, 2000; Song *et al.*, 2004; Klemke *et al.*, 2004).

Antioxidant Compounds

The discoveries of two antimicrobial and antioxidant compounds as pestacin (1,3-dihydrovisobenzofuran) and isopestacin (isobenzofuranone) from an endophytic fungus *P. microspora* residing in combretaceaous plant, *Terminalia morobensis*, growing in the Sepik river drainage of Papua New Guinea, led to the search for endophytic fungi with an antioxidant potential (Harper *et al.*, 2003; Strobel *et al.*, 2002). Graphislactone A (phenolic metabolite) isolated from the endophytic fungus *Cephalosporium* sp. IFB-E001 residing in *Trachelospermum jasminoides*, confirmed to have free radical-scavenging and antioxidant activities *in vitro* (Song *et al.*, 2005). The methanol extract of an endophyte *Xylaria* sp. isolated from the medicinal plant *Ginkgo biloba* exhibited strong antioxidant capacity due to the presence of "phenolics" and "flavonoids" compounds among 41 identified compounds (Liu *et al.*, 2007). Polysaccharides from plants and microorganisms have been extensively studied and considered as potent natural antioxidants (Sun *et al.*, 2004; Kardošová *et al.*, 2006; Yu *et al.*, 2007; Luo *et al.*, 2008; Chen *et al.*, 2009). Recently, endophytic fungus *Phyllosticta* sp. isolated from *Guazuma tomentosa* exhibited strong antioxidant activity (Srinivasan *et al.*, 2010).

Crude extracts of four fungal endophytes: *Aspergillus* sp. JPY2, *Aspergillus* sp.JPY1, *Penicillium chrysogenum, Phoma* sp. possesses antioxidant potential. These endophytic fungi were isolated from the *S. oleoides*. (Dhankhar *et al.*, 2012).

Antidiabetic Compounds

Management of diabetes with an agent devoid of any side effects is still a challenge in the medical science. This concern has led to an increase in the demand for natural products with antihyperglycemic activity. In the recent past many hypoglycemic agents are introduced, still the diabetes and the related complications continue to be a major medical problem not only in developed countries but also in developing countries. A nonpeptidal fungal metabolite (L-783,281) was isolated from an endophytic fungus (*Pseudomassaria* sp.) collected from an African rainforest near Kinshasha in the Democratic Republic of the Congo (Zhang *et al.*, 1999) which acts as insulin mimetic and unlike insulin, is not destroyed in the digestive tract.

Our study on an endophytic fungus (unidentified) isolated from the *Salvadora oleoides*, synthesizes potent antidiabetic metabolites with extremely low toxicity to human cells. Viewing the potential of this fungal isolate 2, 6-di-tert-butyl-p-cresol (major compound as determined by the GCMS of the aqueous extract of unidentified fungus) may be exploited for, commercial production of antidiabetic compound (unpublished data).

Conclusions

Endophytic fungi represent an abundant source of diversity. These fungal endophytes proven themselves as a dependable source of bioactive and chemically novel compounds with potential for exploitation in a wide variety of medical, agricultural, and industrial arenas. Novel endophytes usually have associated with them novel secondary natural products and/or processes. It can be concluded that the vast majority of plants have not been studied for their endophytes. Thus, enormous opportunities exist for the recovery of novel fungal forms, taxa, and biotypes.

Moreover, we stand at the cross-roads of time when the world's biodiversity is declining at an alarming rate. Many endemic, endangered and medicinally valuable plants are on the verge of extinction. Along with these plants, the endophytes harbored in them are also threatened. If, phytochemials are prospected and produced from microorganisms the unwanted destruction of medicinal plants can be prevented protecting the green cover. The products can also be produced at large scale at economic pricings.

But question remains as to how endophytes resides in a plant forms a relationship with it? Also how are secondary metabolite product synthesized? What genetic control exists to control product formation and synthesis of secondary metabolites? Much more work is essential to understand endophytes physiology, biochemical pathways, defensive role, secondary metabolite production, motivation and encouragement of researcher from life sciences to contribute research related to endophytes.

Acknowledgements

The authors are thankful to Department of Science and Technology, India for providing INSPIRE-DST-JRF to Mrs. Seema Dhankhar. The authors are also very thankful to University Grant Commission, Govt of India, for financial support in the form of Major Research Project on study of fungal endophytes isolated from *Salvadora oleoides*.

References

Aly, A.H., Edrada-Ebel, R.A., Wray, V., Müller, W.*E.G.*, Kozytska S, Hentschel, U., Proksch, P., and Ebel, R. (2008). Bioactive metabolites from the endophytic fungus *Ampelomyces* sp. isolated from the medicinal plant *Urospermum picroides. Phytochemistry*, **69**: 1716-1725.

Arnold, A.E., and Lutzoni, F. (2007). Diversity and host range of foliar fungal endophytes: are tropical leaves biodiversity hotspots. *Ecology*, **88**: 541-549.

Atmosukarto, I., Castillo, U., Hess, W.M., Sears, J., and Strobel, G. (2005). Isolation and characterization of *Muscodor albus* I-41.3s, a volatile antibiotic producing fungus. *Plant Science*, **169:** 854-861.

Brady, S.F., Wagenaar, M.M., Singh, M.P., Janso, J.E., and Clardy, J. (**2000**). The cytosporones, new octaketide antibiotics isolated from an endophytic fungus. *Organic Lett*ers, **14:** 4043-4046.

Dai, J., Krohn, K., Draeger, S., and Schulz, B. (2009). New naphthalene-chroman coupling products from the endophytic fungus, *Nodulisporium* sp. from *Erica arborea*. *European Journal of Organic Chem*istry, **10**: 1564-1569.

Dhankhar, S., Dhankhar, S., Kumar, S., and Yadav, J.P. (2012).Antioxidant activity of Fungal endophytes isolated from *Salvadora oleoides* Decne. *International Journal of Pharmacy and Pharmaceutical sciences*. **4:** 380-385.

Dhankhar, S., Dhankhar, S., and Yadav, J.P. (2013).Investigating antimicrobial properties of Endophytic fungi associated with *Salvadora oleoides* Decne. *Anti Infective Agents*, **11**: 48-58.

Fernandes, M.D.R.V., Silva, T.A.C., Pfenning, L.H., Aparecida, D.L.M., and Miguel, D.C.C. (2009). Biological activities of the fermentation extract of the endophytic fungus *Alternaria alternata* isolated from *Coffea arabica* L Brazilian. *Journal of Pharmaceutical Science*, **45**: 677-685.

Fill, T.P., Silva, B.F., and Rodrigues, F. (2010). Biosynthesis of phenylpropanoid amides by an endophytic *Penicillium brasilianum* found in root bark of *Melia azedarach*. *Journal of Microbiology and Biotechnology*, **20:** 622-629.

Gangadevi, V., and Muthumary, J. (2009). Taxol production by *Pestalotiopsis terminaliae*, an endophytic fungus of *Terminalia arjuna* (Arjun tree). *Biotechnology and Applied Biochemistry*, **52**: 9-15.

Gond, S.K., Mishra, A., Sharma, V.K., Verma, S.K., Kumar, J., Kharwar, R.N., and Kumar, A. (2011). Diversity and antimicrobial activity of endophytic fungi isolated from *Nyctanthes arbortristis*, a well known medicinal plant of India. *Mycoscience.* doi: 10.1007/s10267-011-0146-z.

Guo, B., Wang, Y., Sun, X., and Tang, K. (2008). Bioactive natural products from endophytes: a review. *Applied Biochemistry and Microbiology*, **44:** 136-142.

Guo, L.D., Huang, G.R., and Wang, Y. (2008). Seasonal and tissue age influences on endophytic fungi of *Pinus tabulaeformis* (Pinaceae) in the Dongling Mountains Beijing. *Journal of Integrative Plant Biology*, **50**: 997-1003.

Guo, L.D., Hyde, K.D., and Liew, E.C.Y. (2000). Identification of endophytic fungi from *Livistona chinensis* based on morphology and rDNA sequences. *New Phytology*, **147**: 617-630.

Guo, B., Dai-Jin-Rui, S., Ng, Y., Huang, C., Leong, W., Ong, B.K., and Carte. (2000) Cytonic acids A and B: novel tridepside inhibitors of hCMV protease from the endophytic fungus *Cytonaema* species *Journal of Natural Products*, **63:** 602-604.

Harper, J.K., Arif, A.M., Ford, E.J., Strobel, G.A., Porco, J.A., Tomer, D.P., Oneill, K.L., Heider, E.M., and Grant, D.M. (2003). Pestacin: a 1,3 –dihydroisobenzofuran from *Pestalotiopsis microspora* possessing antioxidant and antimycotic activities. *Tetrahedron*, **59**: 2471-2476.

Hoffman, A.M., Mayer, S.G., Strobel, G.A., Hess,W.M., Sovocool G.W., Grange, A.H., and Kelley-Swift, *E.G.* (2008). Purification, identification and activity of phomodione, afurandione from an endophytic *Phoma* species. *Phytochemistry*, **69:** 1049–1056.

Huang, Z., Cai, X., Shao, C., She, Z., Xia, X., and Chen, Y. (2008). Chemistry and weak antimicrobial activities of phomopsins produced by mangrove endophytic fungus *Phomopsis* sp. ZSU-H76. *Phytochemistry*, **69**: 1604-1608.

Hussain, H., Akhtar, N., Schulz, B., Draeger, S., Schulz, B., Pescitelli, G., Salvadori, P., Antus, S., Kurtan, T., and Krohn, K. (2009). New bioactive 2,3-epoxycyclohexenes and isocoumarins from the endophytic fungus *Phomopsis* sp. From *Laurus azorica*. *European Journal of Organic Chemistry*, **2009**: 749-756.

Isaka, M., Berkaew, P., Intereya, K., Komwijit, S., and Sathitkunanon, T. (2007). Antiplasmodial and antiviral cyclohexadepsipeptides from the endophytic fungus *Pullularia* sp. BCC 8613. *Tetrahedron*, **63:** 6855-6860.

Jadulco, R., Brauers, G., Edrada, R.A., Ebel, R., Wray, V., Sudarsono, V., and Proksch, P. (2002). New metabolites from sponge derived fungi *Curvularia lunata* and *Cladosporium herbarum*. *Journal of Natural Products*, **65**: 730-733.

Kardošová, A., and Machová, E. (2006). Antioxidant activity of medicinal plant polysaccharides. *Fitoterapia*, **77**: 367373.

Kakolyris, S., Agelidou, A., Androulakis, N., Geroyianni, A., and Georgoulias, V. (2006). Cisplatin plus etopside chemotherapy followed by thoracic irradiation and paclitaxel plus cisplatin consolidation therapy for patients with limited stage small cell lung carcinoma. *Lung Cancer*, **53**: 59-65.

Kusari, S., Zühlke, S., and Spiteller, M. (2009). An endophytic fungus from *Camptotheca acuminata* that produces camptothecin and analogues. *Journal of Natural Products,* **72**: 2-7.

Li, J.Y., and Strobel, G.A. (2001). Jesterone and hydroxy-jesterone antioomycete cyclohexenenone epoxides from the endophytic fungus–*Pestalotiopsis jesteri. Phytochemistry,* **57:** 261-265.

Leuchtman, A. (2003). Ed. by White, J.F., Bacon, J., Hywel –Jones, N.L., and Spatafora, J.W. Marcel-Decker, New York, USA, pp. 341-175.

Li, E., Tian, R., Liu, S., Chen, X., Guo, L., and Che, Y. (2008). Pestalotheols A-D, bioactive metabolites from the plant endophytic fungus *Pestalotiopsis theae. Journal of Natural Products,* **71**: 664-668.

Liu, J.Y., Harper, J.K., Grant, D.M., Tombe, B.O., Bashyal, B., Hess, W.M., and Strobel, G.A. (2001). Ambuic acid, a highly functionalized cyclohexenone with antifungal activity from *Pestalotiopsis* sp. and *Monochaetia* sp. *Phytochemistry,* **56**: 463–468.

Liu, J.Y., Song, Y.C., Zhang, Z., Wang, L., Guo, Z.J., Zou, W.X., and Tan, R.X. (2004). *Aspergillus fumigatus* CY018, an endophytic fungus in *Cynodon dactylon* as a versatile producer of new and bioactive metabolites. *Journal of Biotechnology,* **114:** 279-287.

Liu, X., Dong, M., Chen, X., Jiang, M., Lu, X., and Yan, G. (2007). Antioxidant activity and phenolics of an endophytic *Xylaria* sp. from *Ginkgo biloba. Food Chemistry,* **105**: 548-554.

Liu, M., Dong, X., Chen, M., Jiang, X., and Zhou, J. (2008). Antimicrobial activity of an endophytic *Xylaria* sp.YX-28 and identification of its antimicrobial compound 7-amino- 4-methylcoumarin. *Applied Microbiology and Biotechnology,* **78**: 241-247.

Madki, M.A., Manzoor, A.S., Powar, S.P., and Patil, S.P. (2010). Isolation and Biological Activity of endophytic fungi from *Withania somnifera. International Journal of Pharmaceutical Sciences,* **2**: 848-858.

Owen, N.L., and Hundley, N. (2004). Endophytes- The chemical synthesizers inside plants. *Science Progress,* **87**: 79-99.

Petrini, O. (1991). Fungal endophytes of tree leaves. *In:* Microbial Ecology of Leaves, Ed. By Andrews, J.H. and Hirano, S.S. Springer, Verlag, New York. pp. 179-197.

Priti,V., Ramesha, B.T., Singh, S., Ravikanth, G., Ganeshaiah, K.N., Suryanarayana, T.S., and Uma Shanker, R. (2009). Opinion: How promising are endophytic fungi as alternative sources of plant secondary metabolites? *Current Science,* **97:** 477-478.

Qin, S., Krohn, K., Hussain, H., Schulz, B., and Draeger, S. (2011). Pestalotheols E-H: Antimicrobial metabolites from an endophytic fungus isolated from tree *Arbutus unedo. European Journal of Organic chemistry,* **26**: 5163-5166.

Rodriguez, R.J., White, J.F., Arnold, A.E., and Redman, R.S. (2009). Fungal endophytes: diversity and functional roles. *New Phytologists,* **182**: 314-330.

Sadananda, T.S., Nirupama, R., Chaithra, K., Govindappa, M., Chandrappa, C.P., and Raghavendra, B. (2011). Antimicrobial and antioxidant activities of endophytes from *Tabebuia argentea* and identification of anticancer agent (lapachol). *Journal of Medicinal Plant Research,* **16**: 3643-3652.

Saikkonen, K. (2007). Forest structure and fungal endophytes. *Fungal Biology Review,* **21:** 67-74.

Sappapan, R., Sommit, D., Ngamrojanavanchi, N., Pengpreecha, S., Wiyakrutta, S., Sriubolamas, N., and Pudhom, K. (2008). *Journal of Natural products,* **71**: 1657-1659.

Schulz, B., and Boyle, C.(2005). The endophytic continuum: *Review Mycology Research,* **109** : 661-698.

Silva, G.H., Teles, H.L., and Trevisan, H.C. (2005). New bioactive metabolites produced by *Phomopsis cassiae*, an endophytic fungus in *Cassia spectabilis*. *Journal of Brazilian Chemical Society,* **16**: 1463-1466.

Singh, S.B., Ondeyka, J.G., Tsipouras, N., Ruby, C., Sardana, V., Schulman, M., Sanchez, M., Pelaez, F., Stahlhut, M.W., Munshi, S., Olsen, D.B., and Lingham, R.B. (2004). Hinnuliquinone, a C2-symmetric dimeric non-peptide fungal metabolite inhibitor of HIV-1 protease. *Biochemistry Biophysics Research Communication,* **324**: 108-113.

Song, Y.C., Huang, W.Y., Sun, C., Wang, F.W., and Tan, R.X. (2005). Characterization of graphislactone A as the antioxidant and free radical-scavenging substance from the culture of *Cephalosporium* sp. IFB-E001, an endophytic fungus in *Trachelospermum jasminoides*. *Biol. Pharma. Bul.,* **28**: 506-509.

Song, Y.C., Li, H., Ye, Y.H., Shan, C.Y., Yang, Y.M., and Tan, R.X. (2004). Endophytic naphthopyrone metabolites are co-inhibitors of xanthine oxidase, SW1116 cell and some microbial growths. *FEMS Microbiology Letters,* **241:** 67-72.

Srinivasan, L., Jagadish, K., Shenbhagaraman, R., and Muthumary, J.(2010). Antioxidant activity of endophytic fungus *Phyllosticta* sp. isolated from *Guazuma tomentosa*. *Journal of Phytology and Phytochemistry,* **2** : 37-41.

Stierle, A., Strobel, G., and Stierle, D. (1993). Taxol and taxane production by *Taxomyces andreanae*, an endophytic fungus of Pacific yew. *Science,* **260:** 214-216.

Strobel, G.A., and Long, D.M. (1998). Endophytic microbes embody pharmaceutical potential. *Amererican Society of Mycology News,* **64:** 263-268.

Strobel, G.A. (2003). Endophytes are sources of bioactive products. *Microbial Infections,* **5:** 535-544.

Strobel, G., and Daisy. (2003). Bioprospecting for microbial endophytes and their natural products. *Microbiology and Molecular biology Review,* **67:** 491-502.

Strobel, G.A., Ford, E., Worapong, J., Harper, J.K., Arif, A.M., Grant, D.M., Fung, P., and Chau, R.M.W. (2002). Isopestacin, a unique isobenzofuranone from

Pestalotiopsis microspora possessing antifungal and antioxidant properties. *Phytochemistry*, **60**: 179-183.

Sun, C., Wang, J.W., Fang, L., Gao, X.D., and Tan, R.X. (2004). Free radical scavenging and antioxidant activities of EPS2, an exopolysaccharide produced by a marine filamentous fungus *Keissleriella* sp. YS 4108. *Life Science*, **75**: 1063-1073.

Tan, R.X., and Zou, W.X. (2001). Endophytes: a rich source of functional metabolites. *Natural Products Report*, **18**: 448-459.

Tejesvi, M.V., Kini, K.R., Prakash, H.S., and Ven Subbiah, Shetty, H.S. (2007). Genetic diversity and antifungal activity of species of Pestalotiopsis isolated as endophytes from medicinal plants. *Fungal diversity*, **24**: 37-54.

Verma, V.C., Gond, S.K., Kumara, A., Kharwar, R.N., and Strobel, G.A. (2007). The endophytic mycoflora of bark, leaf and stem tissues of *Azadirachta indica* A. Juss (neem) from Varanasi (India).*Microbial Ecology*, **54:** 119-125.

Wagenaar, M.M., Corwin, J., Strobel, G., and Clardy, J. (2000). Three new cytochalasins produced by an endophytic fungus in the genus *Rhinocladiella*. *Journal of Natural Products*, **63**: 1692-1695.

Wang, F.W., Jiao, R.H., Cheng, A.B., Tan, S.H., and Song, Y.C. (2007). Antimicrobial potentials of endophytic fungi residing in *Quercus variabilis* and brefeldin A obtained from *Cladosporium* sp. *World Journal of Microbiology and Biotechnology*, **23**: 79-83.

Wicklow, D.T., Roth, S., Deyrup, S.T., and Gloer, J.B. (2005). A protective endophyte of Maize: *Acremonium zeae* antibiotics inhibitory to *Aspergillus flavus* and *Fusarium verticilloides. Mycology Research*, **109**: 610-618.

You, F., Han, T., Wu, J.Z., Huang, B.K., and Qin, L.P. (2009. Antifungal secondary metabolites from endophytic *Verticillium* sp. *Biochemistry System Ecology*, **37**: 162-165.

Yu, R., Yang, W., Song, L., Yan, C., Zhang, Z., and Zhao,Y. (2007). Structural characterization and antioxidant activity of a polysaccharide from the fruiting bodies of cultured *Cordyceps militaris*. *Carbohydrate Polysaccharides*, **70**: 430-436.

Zhang, B., Salituro, G., Szalkowski, D., Li, Z., Zhang, Y., and Royol. (1999). Discovery of small molecule insulin mimetic with antidiabetic activity in mice. *Science*, **284**: 974-981.

Bioactive Phytochemicals: Perspectives for Modern Medicine Vol. 2 (2014) *Pages* **327–354**
Editor: **V.K. Gupta**
Published by: **DAYA PUBLISHING HOUSE, NEW DELHI**

15

Importance of *Cichorium* Species: Insight into the Wide Range of Biological Activities and Phytochemical Ingredients

Ipek Suntar[1]* and Esra Kupeli Akkol[1]

ABSTRACT

For the improvement of the efficiency of the discovery of novel drugs, exploring the plant resources is an alternative way. Chicorium genus is one of the prominent resources due to its nutritional quality as well as its biological active components. The whole plant contains a number of medicinally important compounds such as inulin, esculin, volatile compounds (monoterpenes and sesquiterpenes), coumarins, flavonoids and vitamins. This research work overviewed the published researches on the phytochemical analysis for the presence of various secondary metabolites and biological activities on the genus Cichorium in the light of the potential use in discovering new effective medicine.

Keywords: Asteraceae, Biological activity, *Cichorium*, Folk medicine, Phytochemistry.

Introduction

Biologically active compounds from natural sources have always been a great interest for the scientists. *Cichorium* L. is a medicinally important genus, from the family Asteraceae, is commonly known as chicory or endive and native to Europe, tropical Asia and North Africa. Common chicory (*Cichorium intybus*) is a small aromatic biennial or perennial herb having blue, lavender or white flowers (Nandagopal and Ranjitha Kumari, 2007).

1 Department of Pharmacognosy, Faculty of Pharmacy, Gazi University, Etiler 06330, Ankara, Turkey

* *Corresponding author*: E-mail: ipesin@gazi.edu.tr

According to German Commission E monographs and Physicians' Desk Reference for Herbal Medicines, *C. intybus* was reported to be used against dyspepsia and loss of appetite, for negative chronotropic and inotropic effects due to the sequiterpene lactones, cinnamic acid type compounds and flavonoids (Blumenthal, 1998; Fleming, 2000).

According to Turkish ethnobotanical researches, the ahes of roots are mixed with butter to prepare a wound healing remedy (Sezik *et al.*, 1991); decoction prepared from the roots is consumed against kidney stones and cancer; the leaves are mixed with *Anchusa* roots and butter and this remedy is applied for rapid healing of wounds (Sezik *et al.*, 2001); the decoction from the aerial parts is used against haemorrhoids and eczema (Yesilada *et al.*, 1999); and the latex is used for warts (Tuzlaci *et al.*, 2010). In India, the seeds of *C. intybus* are used for the preparation of coffee; the aqueous extract of the seeds are used against liver diseases and diarrhea (Gadgoli and Mishra, 1997); In Pakistan, the fresh shoots are consumed as food and used for the treatment of urinary infections and stomachache; the poultice prepared from the roots is used for pain relief (Shah *et al.*, 2006). The aqueous root extract is used against malaria in Afghanistan, (Bischoff *et al.*, 2004); for the treatment of warts in Iran (Syed *et al.*, 2008); against liver diseases and digestion problems in Poland (Kisiel and Michalska, 2002); as laxative and diuretic in Italy and Serbia (Petrovic *et al.*, 2004; Pushparaj *et al.*, 2007; Mulinacci *et al.*, 2001). The dried roots of *C. intybus* are used to prepare coffee like beverage and used as stomachic in Belgium, France and USA (Mulinacci *et al.*, 2001).

Phytochemical Studies on *Cichorium* Species

According to the published literatures, *Cichorium* species are mainly rich in flavonoids, coumarins, caffeic acid derivatives, sesquiterpene lactones, triterpenoids, phytosterols and inulins. The compounds of these species were presented in the tables below with regard to their chemical groups.

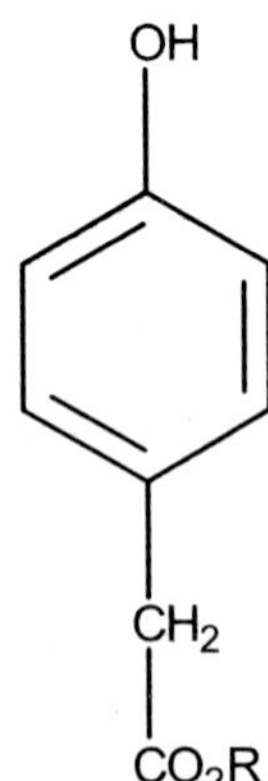

Table 15.1: Phenolic compounds isolated from *Cichorium* species.

Compound	R	Species	Reference
Methyl-*p*-hydroxyphenylacetate	Me	*endivia* var. *crispum*	Kisiel ve Michalska, 2006
Ethyl-*p*-hydroxyphenylacetate	Et		

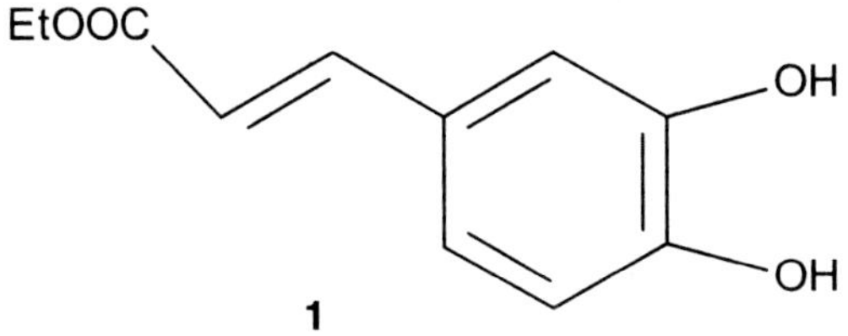

EtOOC
OH
OH

1

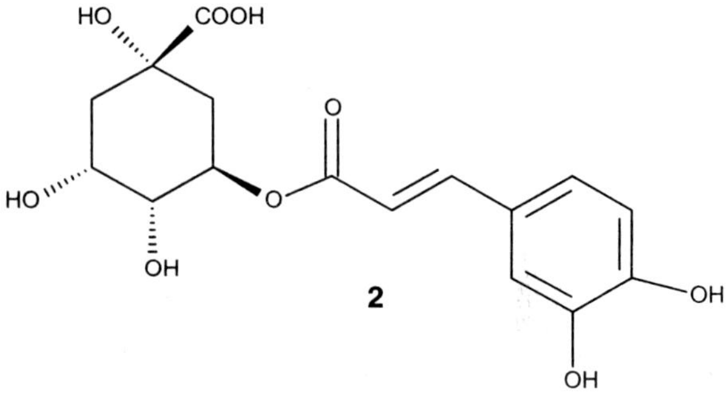

HO COOH
O
HO
O
OH

2

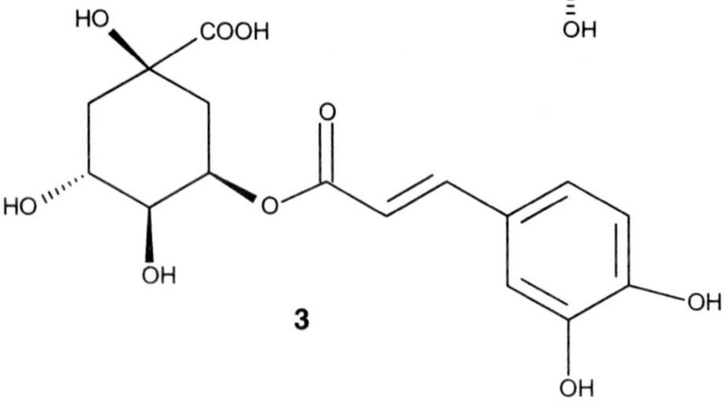

OH
OH

HO COOH
O
HO
O
OH

3

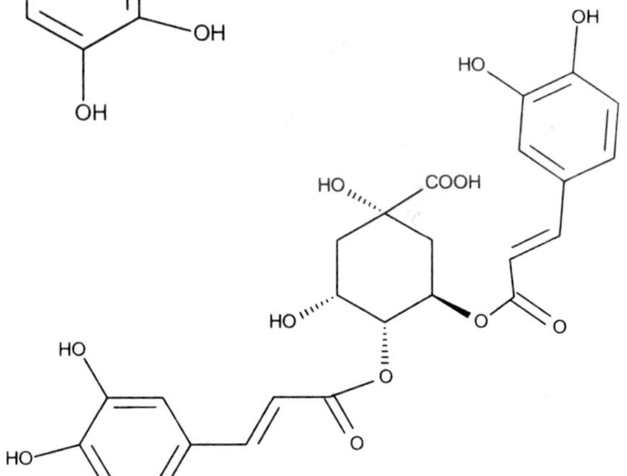

OH
OH

OH
HO
HO COOH
HO
O
O
HO
O
HO
O

4

HO COOH
O
O
O
O
OH
HO
OH
OH
OH

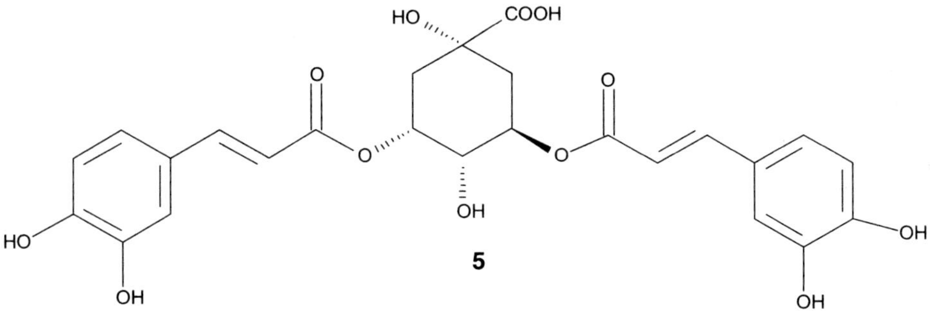

5

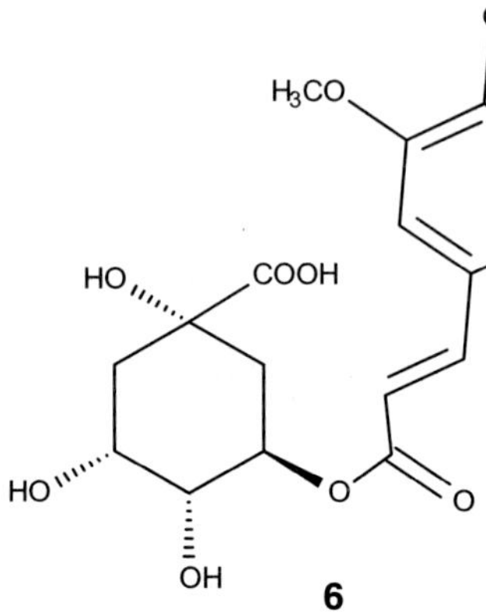

Table 15.2: Phenolic compounds isolated from *Cichorium* species.

Compound	*Species*	*Reference*
Ethyl-*trans*-caffeate **(1)**	*endivia* var. *crispum*	Kisiel and Michalska, 2006
Chlorogenic acid **(2)**	*intybus*	Mulinacci *et al.*, 2001
5-*O*-Caffeoylquinic acid **(3)**		
3,4-Di-*O*-Caffeoylquinic acid (**4**)	*endivia* var. *crispum*	Papetti *et al.*, 2008
3,5-Di-*O*-Caffeoylquinic acid (**5**)		
5-*O*-Feruloylquinic acid (**6**)		

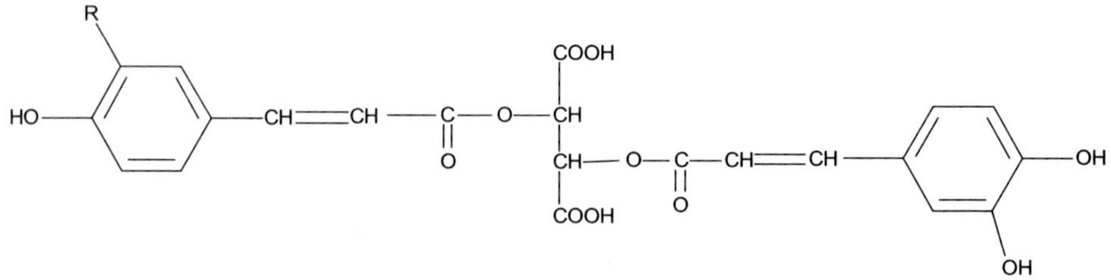

Table 15.3: Phenolic compounds isolated from *Cichorium* species.

Compound	*R*	*Species*	*Reference*
Chicoric acid	OH		
Caffeoyl feruloyl tartaric acid	OCH_3	*intybus*	Mulinacci *et al.*, 2001
Caffeoyl *p*-hydroxycinnamoyl tartaric acid	H		

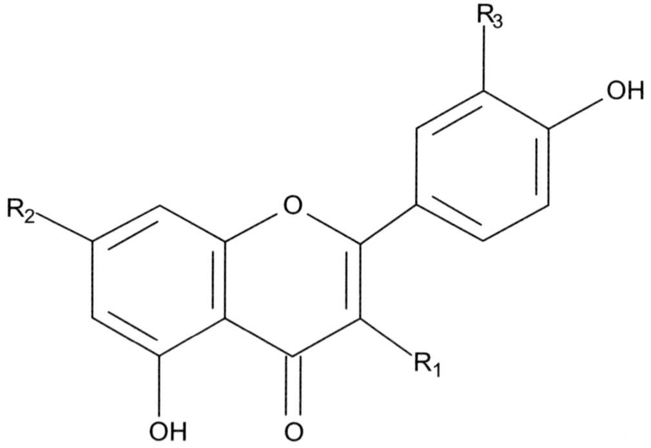

Table 15.4: Flavonoids isolated from *Cichorium* species.

Compound	R_1	R_2	R_3	*Species*	*Reference*
Luteolin-7-*O*-glucuronide	H	OGlucu	OH	*intybus*	Mulinacci *et al.*, 2001
Quercetin-3-O-glucuronide	OGlucu	OH	OH		
Quercetin-3-O-glucoside	OGlc	OH	OH	*intybus* *pumilum*	Mulinacci *et al.*, 2001 Kisiel *et al.*, 2004
Kaemferol	OH	OH	H	*endivia*	Chen *et al.*, 2011
Kaempferol-3-*O*-glucoside	OGlc	OH	H		

Glc: Glucose; Glucu: Glucuronic acid.

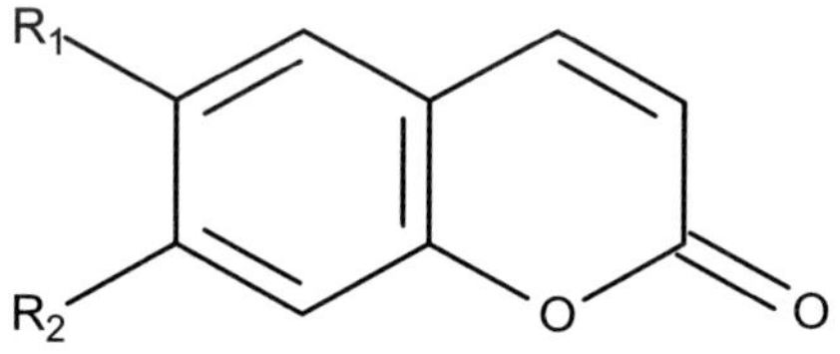

Table 15.5: Coumarins isolated from *Cichorium* species.

Compound	R_1	R_2	*Species*	*Reference*
Scopoletin	OMe	OH	*spinosum*	Michalska and Kisiel, 2007
Umbelliferone	H	OH		
Cichoriin	OH	OGlc	*spinosum* *intybus*	Michalska and Kisiel, 2007 Kisiel and Michalska, 2002
Aesculetin	OH	OH	*intybus*	Hazra *et al.*, 2002
Cichoriin-6-*p*-hidroksifenilasetat	OH	OGlc-A		Kisiel and Michalska, 2002

Glc: Glucose

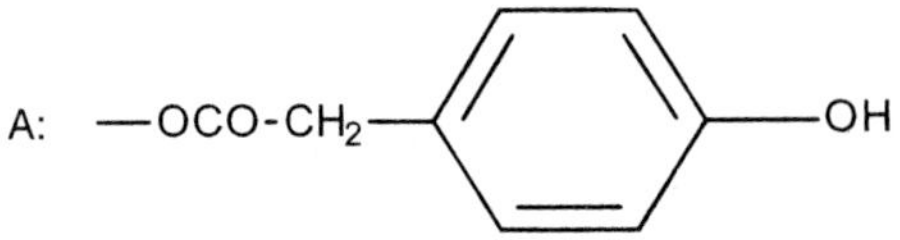

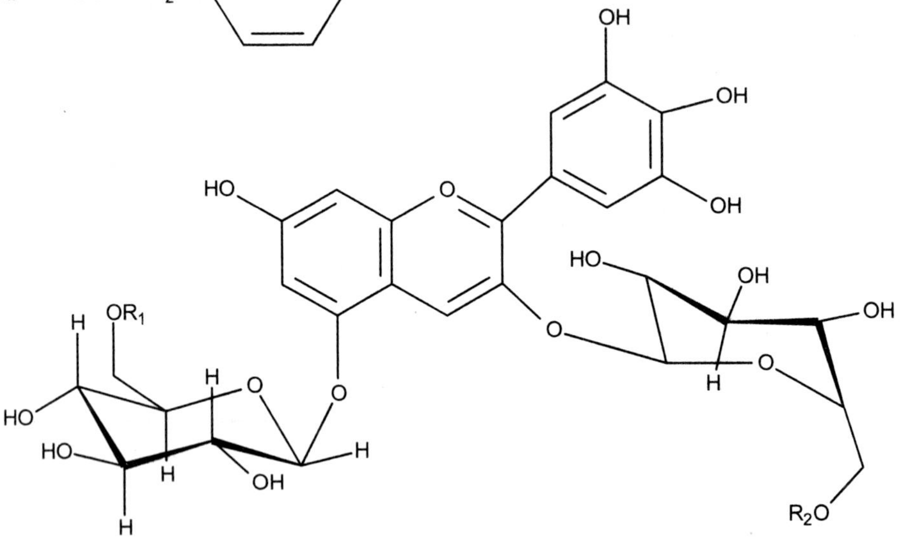

A: —OCO-CH2—C6H4—OH

Table 15.6: Anthocyanins isolated from *Cichorium* species.

Compound	R_1	R_2	*Species*	*Reference*
Delphinidin3,5-di-O-(6-O-malonyl-β-D-glucoside)	Malonyl	Malonyl		
Delphinidin 3-O-(6-O-malonyl-β-D-glucoside)-5-O-β-D-glucoside	Malonyl	H	*intybus*	Norbaek *et al.*, 2002
Delphinidin 3-O-β-D-glucoside-5-O-(6-O-malonyl-β-D-glucoside)	H	Malonyl		
Delphinidin 3,5-di-O-β-D-glucoside	H	H		

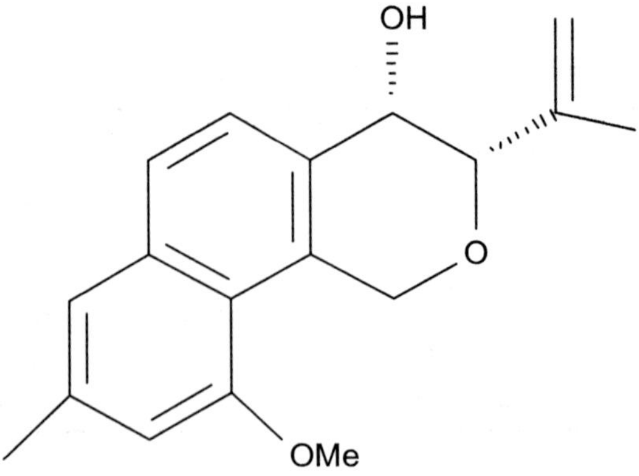

Table 15.7: Benzo-isochromene type compound isolated from *Cichorium* species.

Compound	*Species*	*Reference*
Cicohrin A	*intybus*	Hussain *et al.*, 2011

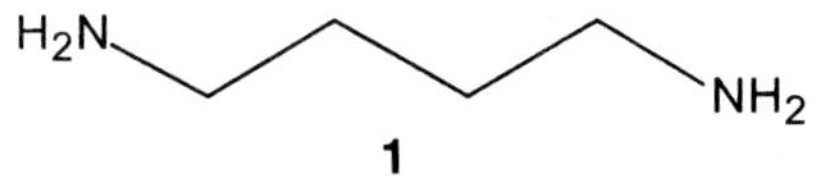

H_2N NH_2

1

H_2N NH NH_2

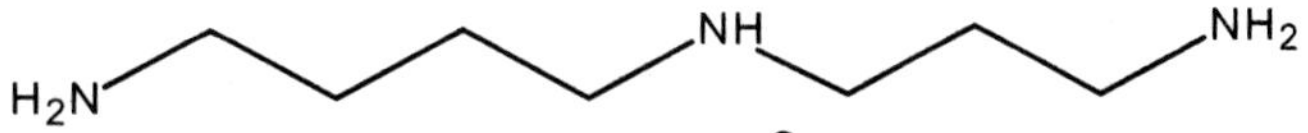

2

H N O N

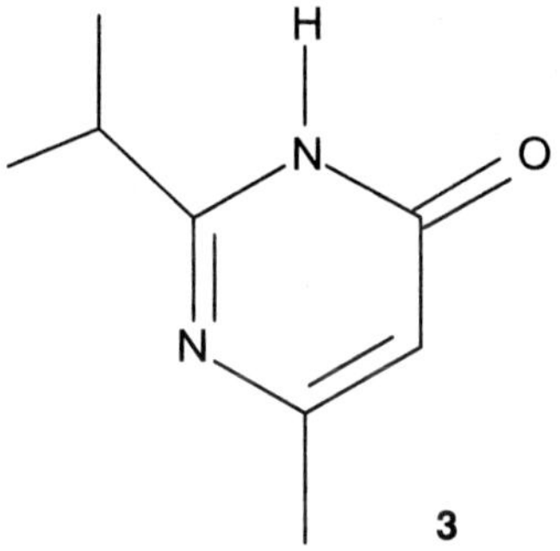

3

Table 15.8: N-containing compounds isolated from *Cichorium* species.

Compound	*Species*	*Reference*
Putresin **(1)** Spermidin **(2)**	*intybus*	Krebsky *et al.*, 1999
2-Isopropyl-6-methylpyrimidin-4(3*H*)-one **(3)**	*glandulosum*	Wu *et al.*, 2011

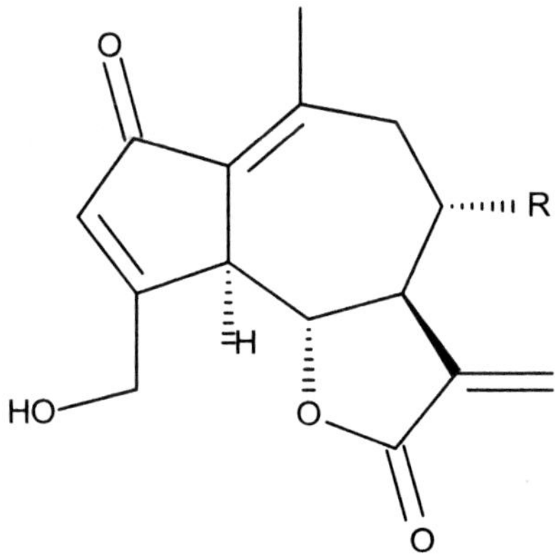

O R H HO O O

Table 15.9: Terpenic compounds isolated from *Cichorium* species.

Compound	*R*	*Species*	*Reference*
8-Deoxylactucin	H	*endivia* var. *crispum* *intybus*	Kisiel and Michalska, 2006; Kisiel and Zielinska, 2001;
Lactucopicrin	A		Malarz *et al.*, 2002
Lactucin	OH	*endivia* var. *crispum* *glandulosum* *intybus* *spinosum*	Kisiel and Michalska, 2006 Wu *et al.*, 2007; Kisiel and Zielinska, 2001; Michalska and Kisiel, 2007

A: —OCO-CH_2— —OH

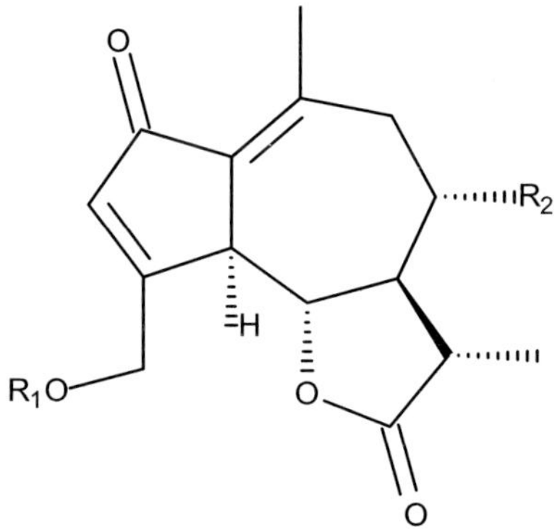

O
R_2
H
R_1O
O
O

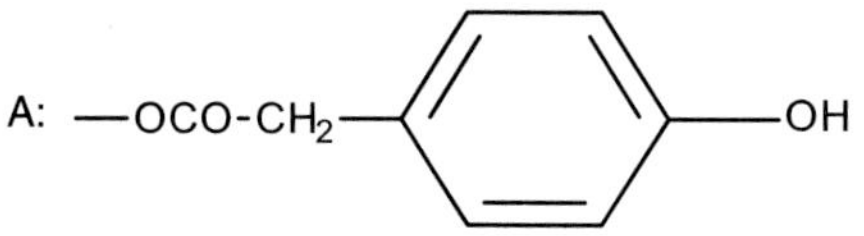

Table 15.10: Terpenic compounds isolated from *Cichorium* species.

Compound	*R_1*	*R_2*	*Species*	*Reference*
11β,13-dihydrolactucin	H	OH	*endivia* var. *crispum* *glandulosum* *intybus* *spinosum*	Kisiel and Michalska, 2006; Wu *et al.*, 2007; Kisiel and Zielinska, 2001; Michalska and Kisiel, 2007
11β,13-dihydrolactucopicrin	H	A	*endivia* var. *crispum* *intybus*	Kisiel and Michalska, 2006; Kisiel and Zielinska, 2001; Malarz *et al.*, 2002
Crepidiaside	Glc	H		
Cichorioside B	Glc	OH	*intybus*	Kisiel and Zielinska, 2001
Jacquinelin	H	H	*intybus*	Kisiel and Zielinska, 2001

Glc: Glucose

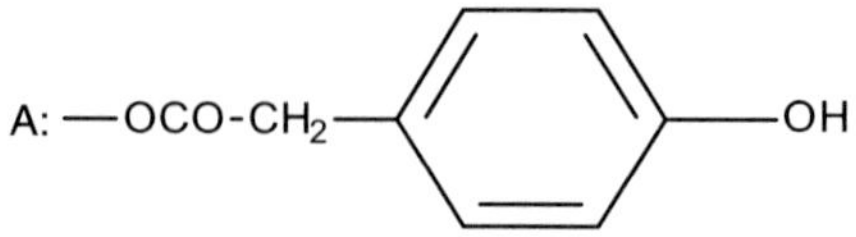

A: —OCO-CH_2— —OH

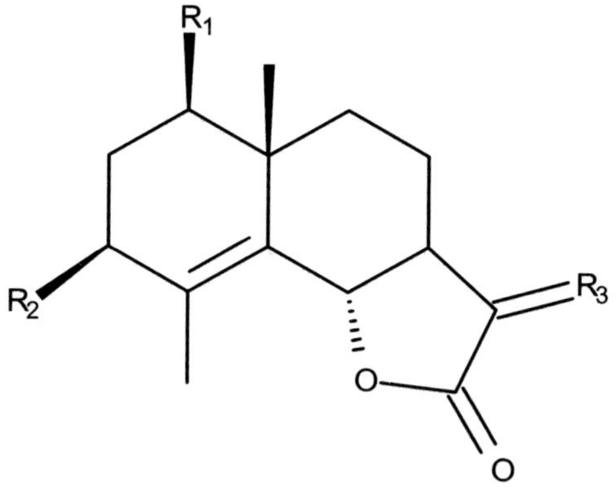

Table 15.11: Terpenic compounds isolated from *Cichorium* species.

Compound	R_1	R_2	R_3	*Species*	*Reference*
Artesin	OH	H	H		
Cichoriolide	H	OH	CH_2	*intybus*	Kisiel and Zielinska, 2001
Cichorioside	H	OGlc	CH_2		
Sonchuside C	H	OGlc	H		
Magnolialide	OH	H	CH_2	*endivia* var. *crispum* *intybus*	Kisiel and Michalska, 2006; Kisiel and Zielinska, 2001

Glc: Glucose

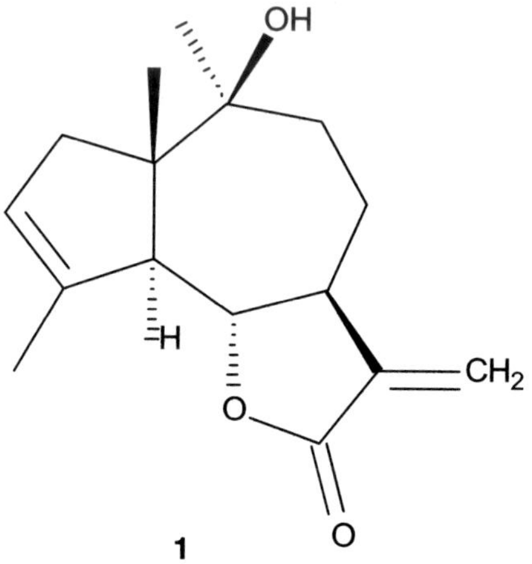

1

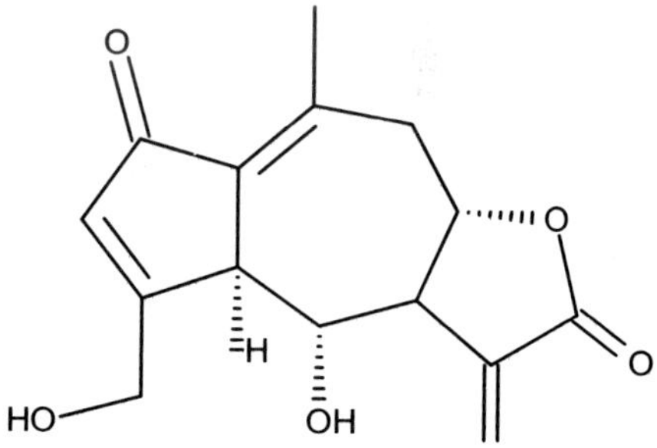

2

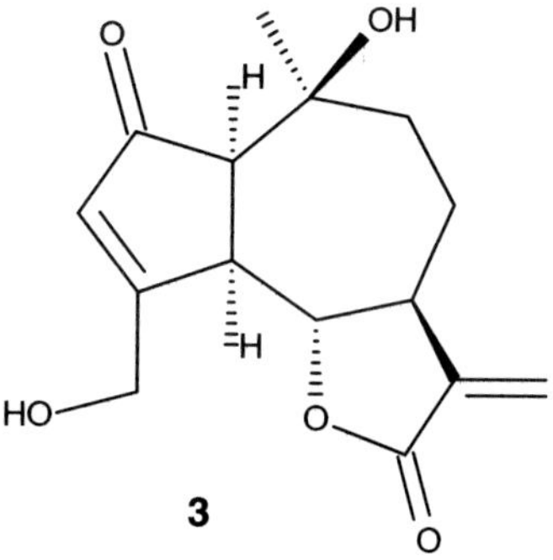

3

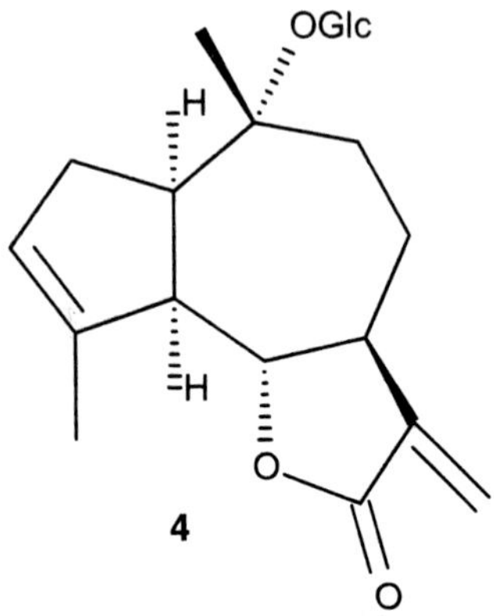

OGlc H H O O 4

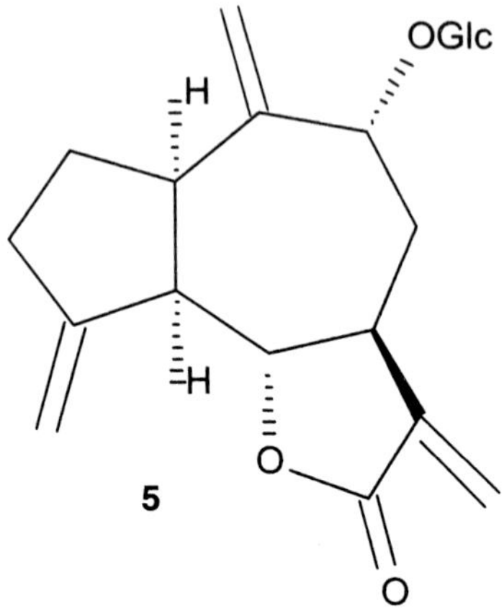

OGlc H H O O 5

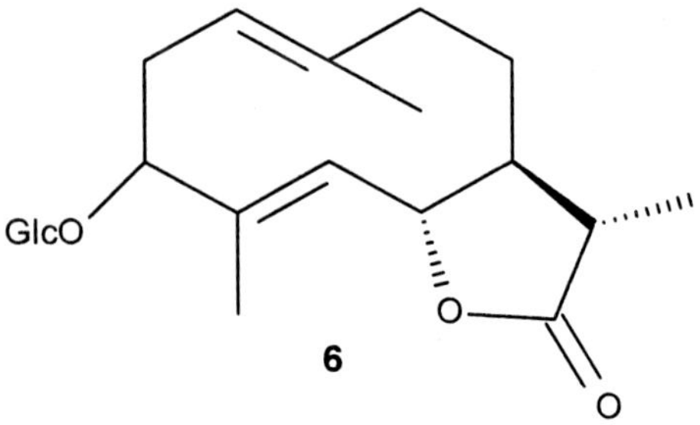

GlcO O O 6

Table 15.12: Terpenic compounds isolated from *Cichorium* species.

Compound	*Species*	*Reference*
Cichopumilide **(1)**	*intybus*	Kisiel and Zielinska, 2001
Intybulide **(2)**		
Hieracin II **(3)**	*endivia* var. *crispum*	Kisiel and Michalska, 2006
Macrocliniside G (**4**)		
Ixerisoside D (**5**)	*endivia* var. *crispumintybus*	Kisiel and Michalska, 2006; Malarz *et al.*, 2002
Sonchuside A **(6)**	*endivia* var. *crispumintybus*	Kisiel and Michalska, 2006; Kisiel and Zielinska, 2001; Malarz *et al.*, 2002

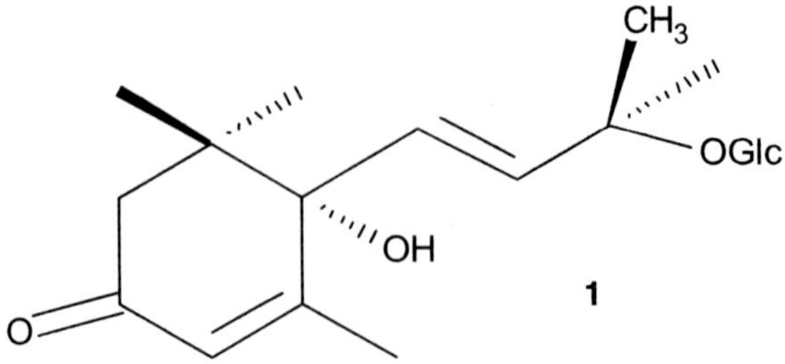

CH_3 OGlc OH O 1

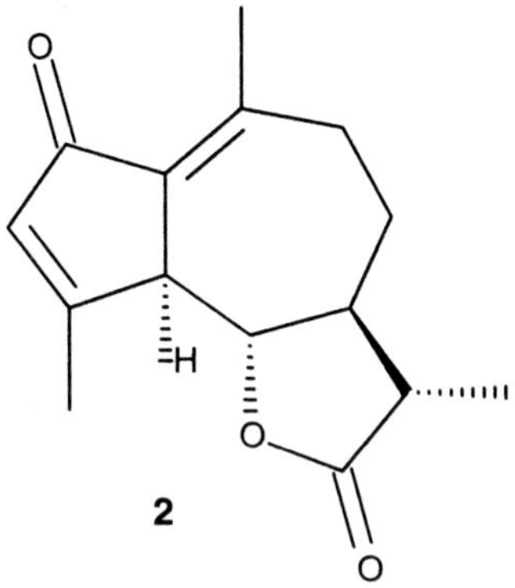

Table 15.13: Terpenic compounds isolated from *Cichorium* species.

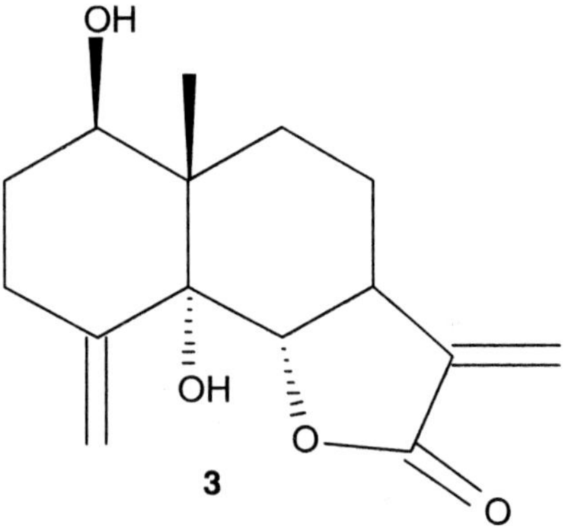

Compound	*Species*	*Reference*
Roseoside **(1)**	*pumilum*	Kisiel *et al.*, 2004
Leucodin **(2)**	*spinosum*	Michalska and Kisiel, 2007
Tanacetin **(3)**		

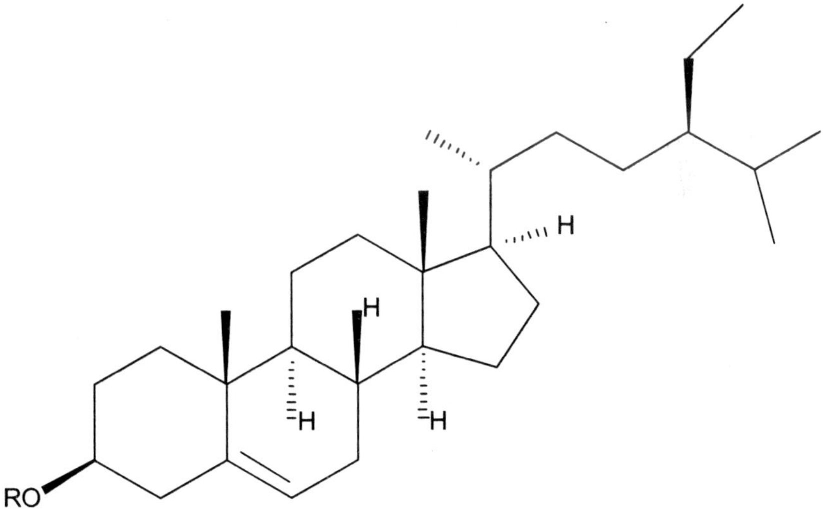

Table 15.14: Steroids isolated from *Cichorium* species.

Compound	*R*	*Species*	*Reference*
β-Sitosterol	H	*intybus*	Krebsky *et al.*, 1999
β-Sitosterol glucopyranoside	Glc		Hussain *et al.*, 2011

Glc: Glucose

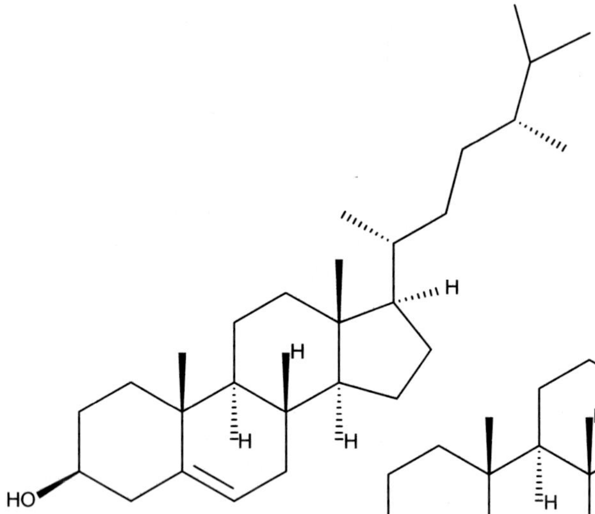

Table 15.15: Triterpenes and steroids isolated from *Cichorium* species.

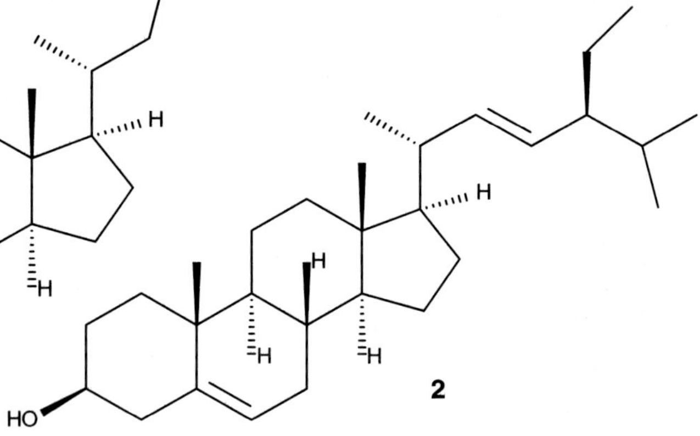

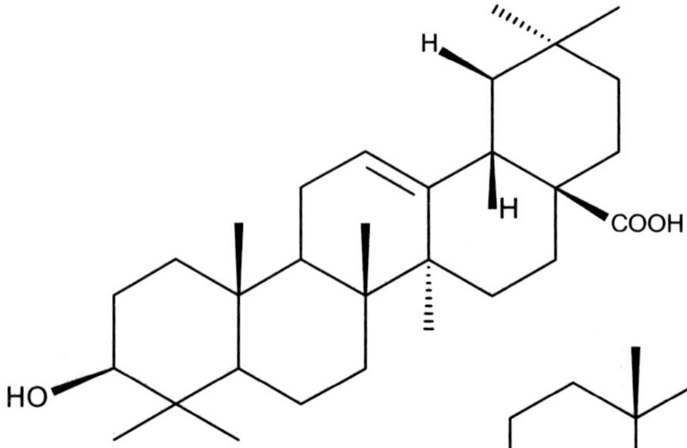

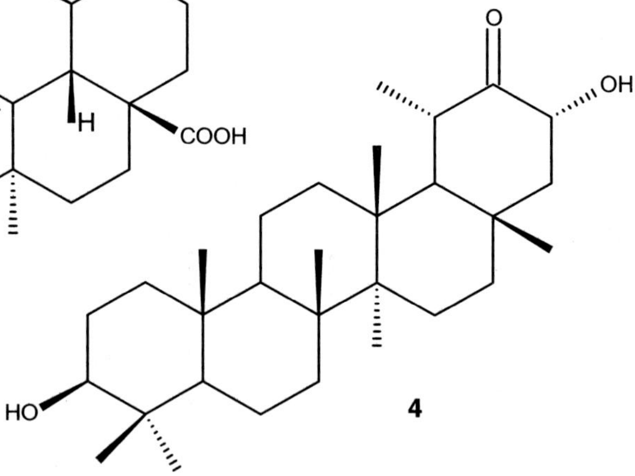

Compound	*Species*	*Reference*
Campesterol **(1)**		Krebsky *et al.,* 1999
Stigmasterol **(2)**	*intybus*	
Oleanolic acid **(3)**		Hussain *et al.,* 2011
21α-hydroxy-taraxasterol **(4)**	*glandulosum*	Wu *et al.,* 2011

Biological Activity Studies on *Cichorium* species

Antibacterial Effects

Antibacterial activity of the water, ethanol and ethyl acetate extracts of *Cichorium intybus* was investigated by using disc diffusion method. All the tested extracts showed antibacterial activity against *Bacillus subtilis, Sarcina lutea, Staphylococcus aureus, Agrobacterium tumefaciens, Erwinia carotovora, Escherichia coli, Pseudomonas aeruginosa* and *Pseudomonas fluorescens*. The ethyl acetate extract was found to be the most active. Root extracts were reported to possess more intensive antibacterial activity than extracts from whole plant (Petrovic *et al.*, 2004).

Antimalarial Effects

In Afghanistan the aqueous extract prepared from the roots of *Cichorium intybus* (L.) is used against malaria. Through biological activity guided fractionation assay, sesquiterpene lactones lactucin and lactucopicrin were isolated and found to be active against HB3 clone of strain *Honduras-1* of *Plasmodium falciparum*. The results reported in this study displayed that lactucin, lactucopicrin, or other structurally related sesquiterpene lactones could be active components of the aqueous extract of *Cichorium*. Previously isolated compounds, 8-deoxylactucin, jacquilenin, 11,13-dihydrolactucin, 11,13-dihydrolactucopicrin, crepediaside B and cichorioside B present in the fresh aqueous extract could also exert varying degrees of antimalarial activity and that the combined activity may provide a cheap, readily available alternative therapy for the treatment of malaria (Bischoff *et al.*, 2004).

Anthelmintic Effects

Use of chicory pastures for gastrointestinal parasite control is encouraged (Scales *et al.*, 1995; Barry, 1998). Foster *et al.*, demonstrated the anthelmintic activity of the methanolic extract of the leaves of cultivated *C. intybus* rich in sesquiterpene lactones mainly lactucin, 8-deoxylactucin and lactucopicrin. Results from the study suggested that cultivars having a high concentration of sesquiterpene lactones dominated by deoxylactucin could be useful in controlling *Haemonchus contortus* (Foster *et al.*, 2011).

Antioxidant Effects

Pro- and antioxidant activities of the water soluble compounds of *C. intybus* var. *silvestre* based on the coupled oxidation of linoleic acid and β-carotene, were assessed. The obtained juice at 2°C possessed either anti- or pro-oxidant activity, while the boiled juice showed only strong antioxidant activity, proving that the vegetable prooxidant components were thermally instable. Various compounds with different moleculer weight and different polarity were found to have antioxidant effect after the fractionation of the aqueous extract. The results showed that *C. intybus* var. *silvestre* contained several water soluble components which able to strongly contrast lipid per-oxidation. Their activity could be observed only after the inactivation or elimination of pro-oxidant components. The antioxidant compounds were found to have different polar characteristics (Papetti *et al.*, 2002). The same researchers also investigated the antiradical activity of the water soluble components of *C. intybus* var. *silvestre*, *C. intybus* var. *foliosum*, *C. endivia* var. *latifolium* and *C. endivia* var. *crispum*by using two biological systems *i.e.* microsome membrane rat hepatocyties in which oxidative damage was induced by CCl_4 and gram-positive bacterium, *Staphylococcus aureus* cultures, subjected to damage with cumene hydroperoxide. The results showed that in both systems *C. intybus* var. *silvestre* has the strongest antioxidant properties and contains different antioxidant compounds. All water soluble components of the vegetables were also found to be active in the biological systems, nevertheless, they were reported to have different properties depending on the assay used for the analysis, and within the same system, on the type of vegetable and on the molecular weight of its components (Papetti *et al.*, 2006).

The antioxidant properties of the aqueous extract of *C. intybus* var. *silvestre* were evaluated *in vitro* in β-carotene-linoleic acid model system and *ex vivo* against rat liver cell microsome lipid peroxidation measured as 2-thiobarbituric acid-reactive substances (TBA-RS) generated by peroxide degradation. Electrophoretic analysis revealed that the most active fraction was a complex mixture of brown components at MW\300 000 (Gazzani *et al.*, 2000).

Lyophilized extract of *Cichorium intybus* var. *silvestre*, was examined for its phenolic profile and its oxidative stability as compared to butylated hydroxytoluene. The extract was shown to reduce lipid peroxidation of different oils in the Rancimat test. In addition, the antioxidant property of the chicory extract was studied in a model system by evaluating the *Saccharomyces cerevisiae* response to oxidative stress by means of gene expression. *Cichorium intybus* var. *silvestre* did not affect the expression of any target gene in nonstressed cells, which means that the tested dose did not exert any pro-oxidative effect. Extract, added to the yeast culture prior to oxidative stress induction, was found to exhibit a pleiotropic protective effect on stress responsive genes. The obtained results suggested the possibility of using the extract of *Cichorium intybus* var. *silvestre* without a pro-oxidative effect. On the other hand, the extract might be used as a cheap, natural, and abundant food additive for replacing synthetic antioxidants (Lante *et al.*, 2011).

Analgesic and Anti-inflammatory Effects

The potential anti-inflammatory activity of the ethyl acetate extract of *C. intybus* due to the inhibition of prostaglandin E2 (PGE2) production in human colon carcinoma HT29 cells was determined by Cavin *et al.*, a major sesquiterpene lactone of the roots of *C. intybus*, the guaianolide 8-deoxylactucin, was identified as the compound responsible from the activity (Cavin *et al.*, 2005). In previous researches, the guaianolides lactucin, 8-deoxylactucin, and lactucopicrin were identified as major sesquiterpene lactones of chicory (Van Beek *et al.*, 1990; Malarz *et al.*, 2002). Due to the different numbers of alkylating structural elements or other possible factors such as lipophilicity, important differences in activity may occur between individual sesquiterpene lactones (Kupchan *et al.*, 1971; Lee *et al.*, 1977; Rüngeler *et al.*, 1999; Snyder *et al.*, 1981). A high selectivity in the anti-inflammatory properties was observed for lactucin, 8-deoxylactucin, and lactucopicrin. 8-Deoxylactucin was found to be responsible for the inhibitory effect on COX-2 expression produced by chicory extract, therefore, confirming a role for specific sesquiterpene lactones in the anti-inflammatory properties of chicory. The data obtained from the study suggest that chicory components have anti-inflammatory properties (Cavin *et al.*, 2005).

Characteristic bitter sesquiterpene lactones of *Cichorium intybus,* lactucin and its derivatives lactucopicrin and 11β,13-dihydrolactucin, were evaluated for analgesic and sedative properties in mice. In the hot plate test, the compounds showed analgesic effects at doses of 15 and 30 mg/kg. In the tail-flick test, the analgesic activities of the compounds at a dose of 30 mg/kg were comparable to that of ibuprofen. Lactucopicrin appeared to be the most potent analgetic of the three tested compounds (Wesolovska *et al.*, 2006). It was reported that sesquiterpene lactones, including parthenolide and helenalin, exert anti-inflammatory activities by inhibiting the central transcription factor, nuclear factor-κB (NF-κB). According to Siedle *et al.*, 8-deoxylactucin from *Lactuca virosa* and *Cichorium intybus*, act as an inhibitor of NF-κB activation (Siedle *et al.*, 2004). More recently, an inhibitory effects of chicory enriched in sesquiterpene lactones on prostaglandin E2 (PGE2) production and COX-2 protein expression were reported by Cavin *et al.* (Cavin *et al.*, 2005). The data confirmed the ancient use of *Lactuca virosa* as an analgesic and sedative herbal remedy, and suggested that *Cichorium intybus* may also possess analgesic and sedative properties due to the presence of lactucin and lactucin-like guaianolides (Wesolovska *et al.*, 2006).

Antidiabetic Effects

C. intybus which is widely used in India as a traditional treatment for diabetes mellitus, was investigated for the hypoglycemic and hypolipidemic properties. Male Sprague Dawley rats were administered with streptozotocin (STZ) intraperitoneally to induce diabetes. Hypoglycaemic effects of the ethanolic extract of *C. intybus* were observed in an oral glucose tolerance test in which, a dose of 125 mg of plant extract/ kg body weight exhibited the most potent hypoglycemic effect. Moreover, daily administration of the ethanolic extract of *C. intybus* at 125 mg/kg for 14 days to diabetic rats attenuated serum glucose by 20 per cent, triglycerides by 91 per cent and total cholesterol by 16 per cent. Moreover, hepatic glucose-6-phosphatase activity (Glc-6-Pase) was markedly reduced by the ethanolic extract of *C. intybus* when compared to the control group. Administration of *C. intybus* produced a significant reduction in serum glucose, triglycerides and total cholesterol in STZ-induced diabetic rats without any effects on insulin secretion. Previous study demonstrated that the water-soluble extract of chicory was found to reduce glucose uptake from the perfused jejunum in normal rats (Kim and Shin, 1998). The potent hypercholesterolemic and hypotriglyceridemic effects of the extract of *C. intybus* was attributed to the presence of inulin (Pushparaj *et al.*, 2007).

By using *in vitro* studies, the effects of a dicaffeoyl ester, chicoric acid (CRA) purified from *C. intybus* on glucose uptake and insulin secretion were reported. Caffeic acid and chlorogenic acid (CGA), a mono-caffeoyl ester, have been described as potential antidiabetic agents. The results showed that chicoric acid and chlorogenic acid increased glucose uptake in L6 muscular cells, an effect only observed in the presence of stimulating concentrations of insulin. Moreover, both chicoric acid and chlorogenic acid were able to stimulate insulin secretion from the INS-1E insulin-secreting cell line and rat islets of Langerhans. Chicoric acid was found to be a new potential antidiabetic agent carrying both insulin sensitizing and insulin-secreting properties. In previous studies, CGA was identified as an inhibitor of glucose-6-

phosphatase (G6P) in rat liver microsomes (Wang *et al.*, 1989), a property inducing the inhibition of hepatic glucose production and consequently the reduction of hyperglycemia. Since glucose-6-phosphatase in β-cells is associated with regulation of insulin secretion (Hambrock *et al.*, 2007), it was hypothesized that this effect could be linked to inhibition of glucose-6-phosphatase activity (Tousch *et al.*, 2008).

The methanolic extract of *C. intybus* exhibited significant glucose uptake in 3T3-L1 adipocytes with a dose-dependent response. The activity continued after detannification process of the methanolic extract (Muthusamy *et al.*, 2008). In a previous report by Liu *et al.*, the tannins were shown to possess capacity in the treatment of metabolic syndromes like diabetes (Liu *et al.*, 2001). However, in data in this study suggested that, significant glucose uptake was observed even after removing tannins from *C. intybus*. Therefore, it was hypothesized that the tannins' role might be insignificant in the biological activity of the methanolic extract of *C. intybus*.

Kamel *et al.*, investigated the effect of aqueous suspension of *C. endivia* leaves powder on some biochemical parameters in streptozotocin-induced diabetic rats. Experimental group animals received aqueous suspension of *C. endivia* leaves powder, 500 mg/kg b. wt. Glibenclamide was useed as the reference drug. *C. endivia* leaves powder was reported to have ability to produce significant hepatoprotective effects by decreasing the activities of serum aminotransferases (AST and ALT), alkaline phosphatase (ALP), lactate dehydrogenase (LDH) and liver malondialdehyde (MDA) level as well as liver superoxide dismutase (SOD) and catalase (CAT) activities and increasing the liver glutathione peroxidase (GPx) and glutathione-S-transferase (GST) activities and reduced glutathione (GSH) level. In diabetes, the increase in oxygen free radicals could be primarily due to the increase in the blood glucose level and secondarily due to the effect of the diabetogenic agent (streptozotocin or alloxan) (Szkudelski, 2001). Malondialdehyde (MDA) is one of the lipid peroxidation products frequently used to determine the oxidant/antioxidant balance in diabetic patient (Cheeseman and Slate, 1993). In this research, after intraperitoneal injection of STZ, the hepatic MDA level increased significantly. Similarly, Punitha *et al.* (2006), Selvan *et al.* (2008) and Chakraborty and Das (2010) reported that the administration of STZ in rats elevated hepatic MDA level, which may be due to the high concentration of lipid (Patel *et al.*, 2009). In conclusion, the results obtained clearly indicated the protective effects of *C. endivia* leaves powder aqueous suspension (Kamel *et al.*, 2011).

Antiulcerogenic Effects

Five plant remedies (*Asphodelus aestivus, Cichorium intybus, Equisetum palustre, Viscum album* ssp. *album* and *Laurus nobilis*) used to treat stomachache were selected to test for their anti-ulcerogenic potency. Either decoction or methanol extracts were prepared from these plants and tested on ethanol-induced gastric ulcer model in rats. Pharmacological experiments clearly demonstrated that the relevant extracts of all the plants given orally showed significant stomach protection against this model of ulcerogenesis (Gürbüz *et al.*, 2002).

Anticarcinogenic and Immunomodulatory Effects

C. intybus was evaluated for *in vitro* antiproliferative properties, using the sulforodamine B (SRB) assay, on four human cancer cell lines: breast cancer MCF-7, prostate cancer LNCaP, amelanotic melanoma C32 and renal adenocarcinoma ACHN. Significant antiproliferative activity was observed on melanoma (Conforti *et al.*, 2008). Previous studies revealed that phenolic compounds exert a direct antiproliferative action and their antiproliferative effect is in correlation with the structural oxidation state and the position, number, and nature of substituents (Yáñez *et al.*, 2004). The effect of phenolics on the cell cycle could probably contribute to the tumor cell killing (Indap and Barkume, 2003). In conclusion, this study demonstrated that *C. intybus* could be an interesting source of antiproliferative principles such as phenolic compounds and phytosterols (Conforti *et al.*, 2008).

In another study, the tumour-inhibitory effect of an ethanolic extract of the roots of *C. intybus* was studied against Ehrlich ascites carcinoma in mice. Significant results were obtained at doses of 500 mg/kg (Hazra *et al.*, 2002).

70 per cent ethanolic extracts were prepared with a group of Iranian medicinal plants, *Silybum marianum*, *Matricaria chamomilla*, *Calendula officinalis*, *Cichorium intybus* and *Dracocephalum kotschyi* to assess their mitogenic activity on human peripheral blood lymphocytes and thymocytes. Among the plants studied, *C. intybus* and *C. officinalis* showed a complete inhibitory effect on the proliferation of lymphocytes. The extracts of *C. officinalis* and *C. intybus* were demonstrated to have anti-inflammatory activities (Patel and Venkatakrishna-Bhat, 1983; Wagner *et al.*, 1984). Therefore, the study suggeted that these herbs could affect the immune reactions through their anti inflammatory actions (Amirghofran *et al.*, 2000).

Effects of the ethanolic extract of *C. intybus* (CIEE) on the immunotoxicity of ethanol (EtOH) were investigated in mice. 300 mg/kg CIEE was orally administered to mice daily for 28 consecutive days. The combination of CIEE and EtOH showed significant increases in the circulating leukocytes and the relative weights of liver, spleen and thymus, as compared with those in mice treated with EtOH alone. Splenic plaque forming cells and hemagglutination titers to sheep red blood cells, and the secondary IgG antibody response to bovine serum albumin were markedly enhanced by CIEE plus EtOH treatment. There were also significant increases in delayed-type hypersensitivity reaction, phagocytic activity, natural killer (NK) cell activity and cell proliferation as well as interferon G secretion. It was reported that NK cell activity might play the role of host defense against various infectious diseases and cancer (Martinez *et al.*, 1992). Meadows *et al.*, reported abnormality of NK cell function associated with EtOH (Meadows *et al.*, 1989). The results of the experimental study similarly showed that NK cell activity significantly decreased in mice receiving EtOH alone and normal mice. However, this decrease in NK cell activity was significantly restored by the combined administration of CIEE and EtOH. Furthermore, EtOH alone treatment markedly suppressed the production of interferon G and IL-4 as well as mitogen-stimulated cell proliferation. The combination of CIEE and EtOH significantly diminished the suppression of interferon G production and mitogen-stimulated cell proliferation but not IL-4 secretion by EtOH alone. These findings revealed that the immunotoxicity induced by EtOH is significantly restored or prevented by CIEE treatment (Kim *et al.*, 2002).

Contraceptive Effects

The aqueous extract of roots of *C. intybus* was reported to have anti-spermatogenic activity and has been used as male anti-fertility remedy (Roy-Choudhury and Venkatakrishna-Bhatt,1983).

Crude ethanolic extract of the seeds of *C. intybus* was evaluated for postcoital contraceptive efficacy in adult female Sprague-Dawley rats. The results showed that ethanolic extract of *C. intybus* and its chloroform fraction possess contraceptive effect through biological activity guided fractionation assay (Keshri *et al.*, 1998).

Hepatoprotective Effects

Hepatoprotective properties of the seeds of *Cichorium intybus* were investigated in a rat model of liver injury induced by carbon tetrachloride (CCl_4) and paracetamol. Decoction prepared from powdered roots was fractionated with aceton and methanol. Therefore, 3 fractions namely aceton, methanol and remaining water fractions were obtained. Total water extract and fractions were administered to the Wistar albino rats at 50 mg/kg dose. The most active fractions were found to be methanol and remaining water fractions. From these active fractions, 8,15-dihydroxy-2-oxo-l-(10),3,11(13)-guaiatrien-12,6-olid and flavonoid type compounds were isolated. The hepatotoxicity induced by CCl_4 is due to its metabolite CCl, a free radical that binds to lipoprotein and leads to peroxidation of lipids of endoplasmic reticulum (Recknagael, 1967). The isolated compounds could be protecting the liver by being oxidized and therefore, preventing peroxidation of lipids of the endoplasmic reticulum (Gadgoli and Mishra, 1997).

Protective effects of the aqueous extracts prepared from the roots ans root callus of *C. intybus* were determined against CCl_4-induced liver damage in Wistar albino rats. Callus extract was found to have better activity than natural root extract (Zafar and Ali, 1998). Ethanolic extract prepared by cold percolation method and petroleum ether, ethyl acetate and methanol fractions of the seeds screened for antihepatotoxic activity on CCl_4-induced liver damage in albino rats. The degree of protection was measured using biochemical parameters like aspartate transaminase (AST), alanine transaminase (ALT), alkaline phosphatase (ALKP), and total protein (TP). The phenolic compounds of the methanolic fraction were found to possess a potent antihepatotoxic activity comparable to the standard drug Silymarin (Ahmed *et al.*, 2003). In another research, esculetin, a phenolic compound from *C. intybus* was investigated for its possible protective effect against paracetamol and CCl_4-induced hepatic damage. Administration of 6 mg/kg esculetin was found to reduce the death rate from 100 per cent to 40 per cent. In the same study, esculetin was shown to prevent CCl_4-induced prolongation in pentobarbital sleeping time confirming hepatoprotectivity. These results indicated that esculetin possesses anti-hepatotoxic activity (Gilani *et al.*, 1998). Leukotriene synthesis inhibitors were reported to provide partial protection against acute liver damage induced by different hepatotoxins (Perez-Alvarez *et al.*, 1993; Mourelle *et al.*, 1987; Reyes *et al.*, 1994). Therefore, the reported anti-inflammatory (Tubaro *et al.*, 1988; Loggia *et al.*, 1988) and 5-lipoxygenase inhibitory (Neichi *et al.*, 1983; Panossian, 1984; Leung and Ip., 1986; Morito *et al.*, 1983) activities of esculetin may also be partly involved in the protective effect of esculetin against paracetamol and CCl_4-induced hepatotoxicity. The results of this study indicated that esculetin displays hepatoprotective activity and the presence of this compound in the plants may explain their traditional usage in hepatic damage (Gilani *et al.*, 1998).

Hassan and Yousef studied the effect of *C. intybus*-supplemented diet against nitrosamnine-induced oxidative stress and hepatotoxicity in male rats. Nitrosamine precursors caused a significant increase in liver TBARS and total lipids, total cholesterol, bilirubin, and enzymes activity (AST, ALT, ALP and c-GT) in both serum and liver, while chicory-supplemented diet succeeded to modulate these observed abnormalities (Hassan and Yousef, 2010). The hypocholesterolemic effect of chicory could be attributed to presence of isoflavones suggested by Chen *et al.* (2005). On the other hand, various studies have reveled that plant fibers induce alterations of intestinal absorption, intestinal or pancreatic hormone secretion, lipoprotein metabolism, bile acid metabolism or fermentation of byproducts and their effects on hepatic cholesterol synthesis (Chen *et al.*, 1984; Ozturk, 2009). Therefore, the potent hypercholesterolemic and hypotri-glyceridemic effects of chicory extract could be attributed the presence of inulin which behaves like a soluble fiber and possesses hypolipidemic effect (Lairon, 1996). Moreover the administration of chicory-supplemented diet prevented protein oxidation and improved liver and other organs functions which synthesized plasma protein (Jamshidzadeha *et al.*, 2006).

Aktay *et al.*, demonstrated the antilipid peroxidase effect of the ethanolic extracts prepared from the seeds, leaves and roots of *C. intybus* (Aktay *et al.*, 2000). In spite of previous reports on the potent hepatoprotective effect of *C. intybus* seeds (Gadgoli and Mishra, 1997) and roots (Zafar and Ali, 1998), in this study the leaves of the plant were demonstrated to have significant activity only on lipid peroxidation, but did not improve AST and ALT activities. This finding suggested that the extract was ineffective at single dose administration in preventing the structural integrity of hepatocytes (Aktay *et al.*, 2000).

In vitro and *in vivo* hepatoprotective properties of *C. endivia* L. extract was investigated to identify its chemical constituents. *C. endivia* significantly blocked the oxidative stress and cytotoxicity induced by *tert*-butyl hydroperoxide (*t*-BHP) in HepG2 cells. Moreover, oral administration of CEE before the treatment of *t*-BHP displayed a significant protective effect by lowering serum levels of ALT and AST, inhibiting the changes in MDA, SOD, GSH and GST, as well as ameliorating the liver injuries. Kaempferol and kaempferol-3-*O*-β-D-glucoside isolated from *C. endivia* were found to have antioxidant potency. Thus, it was concluded that *C. endivia* protects hepatic tissue from oxidative damage *in vitro* and *in vivo*, potentially due to its phenolic substances and could be a safe remedy to cure liver disease (Chen *et al.*, 2011).

Insecticide Effects

The methanol and aceton ectracts prepared from the aerial parts of *C. intybus* were reported to give 70-100 per cent of mortality when applied topically at a dose of 3 mg: larva against the stored grain pest *Tribolium castaneum* Herbst (Pascual-Villalobos and Robledo, 1998).

Sedative Effects

Sesquiterpene lactones of *Cichorium intybus*, lactucin and its derivatives lactucopicrin and 11β,13-dihydrolactucin, were evaluated for their sedative effects. Lactucin and lactucopicrin showed sedative properties in the spontaneous locomotor activity test while 11β,13-dihydrolactucin did not (Wesolovska *et al.*, 2006).

Wound Healing Effects

The ethanolic extract of *C. intybus* was evaluated for wound healing activity in rats. The extract was tested in 50mg/mL and 100mg/mL concentrations on three groups of male albino rats. 4cm^2 wounds were created in the posterior neck area of each animal. As placebo control, normal saline was applied topically. Complete wound epithelization was observed in the animals treated with the 100mg/mL ethanolic extract of *C. intybus*. The study permits the conclusion that ethanolic extract of *C. intybus* has wound healing potential (Syed *et al.*, 2008). In the previous studies, ethanolic extract of *C. intybus* displayed strong antimicrobial activity against a wide range of human pathogens (Petrovic *et al.*, 2007; Nandagopal *et al.*, 2004). This antimicrobial activity could be a key factor in rapid wound healing process against infectious microorganisms.

In a recent research by Suntar *et al.* the wound healing effect of the aerial parts and roots of *C. intybus* were experimentally evaluated to define the components responsible from the activity by bioassay-guided procedures. Methanol extract of *C. intybus* roots was found to possess potent wound healing activity by *in vivo* linear incision and circular excision would models. The hydroxyproline content of the tissues treated with test ointments were also assessed for the activity evaluation. Moreover, the test samples were investigated for antioxidant and anti-inflammatory activity potentials. According to biological activity-guided fractionation and isolation assays β-sitosterol was isolated as the compound being responsible from the wound healing activity (Suntar *et al.*, 2012).

Other Effects

Rasmussen *et al.*, demonstrated the effect of the roots of *C. intybus* on the liver cytochrome enzymes. Oral administration of chicory root for 16 days to pigs increased the mRNA expression of CYP1A2, 2A and 2E1; and the protein expression of CYP1A2 and 2A. The activities of CYP1A2 and 2A were increased. In a previous research, the sesquiterpene lactone artemisinin was shown to regulate CYP expression (Simonsson *et al.*, 2006). In this study, the induction was partially attributed to the sesquiterpene lactones of chicory (Rasmussen *et al.*, 2011).

The aqueous extract of *C. intybus* was reported to inhibit systemic anaphylactic reaction induced by compound 48/80 in mice dose-dependently. Particularly, *C. intybus* inhibited the anaphylactic reaction 100 per cent and local anaphylactic reaction activated by anti-dinitrophenyl. The results indicated that *C. intybus* inhibits mast cell-mediated immediate-type allergic reactions *in vivo* and *in vitro* The study suggested that *C. intybus* may also be active in the inhibition of human mast cell degranulation and, therefore, can be used in the treatment of human allergic disorders (Kim *et al.*, 1999).

Toxicity

C. intybus should be used carefully by the patients who have gall stone and can be rarely caused allergic skin reactions according to the German Commission E monographs (Blumenthal, 1998).

Conclusion

The genus *Cichorium* contains several important components *i.e.* inulin, bitter sesquiterpene lactones, coumarins, flavonoids and vitamins and used as antihepatotoxic, antioxidant, analgesic, anti-inflammatory, antidiabetic, antiulcerogenic, anticarcinogenic and immunomodulator, sedtive, contraceptive, antimicrobial, appetizer, digestive, stomachic, diuretic and also as tonic and for wound healing. Inulin was reported to find usage in replacement of fat or sugar and reduction of the calories of food, which is suitable for consumption by diabetics (Niness, 1999). The root is rich in alkaloids, which forms an ingredient or adulterant in coffee (Nandagopal and Kumari, 2007). Therefore, it can be concluded that the active chemical ingredients present in chicory indicate that this herb should be studied more extensively to explore its potential in the several other biological activity researches, and should certainly find place in the treatment of various diseases.

References

Ahmed, B., Al-Howiriny, T.A., and Siddiqui, A.B. (2003). Antihepatotoxic activity of seeds of *Cichorium intybus*. *Journal of Ethnopharmacology*, **87:** 237-240.

Aktay, G., Deliorman, D., Ergun, E., Ergun, F., Yesilada, E., and Cevik, C. (2000). Hepatoprotective effects of Turkish folk remedies on experimental liver injury. *Journal of Ethnopharmacology*, **73:** 121–129.

Amirghofran, Z., Azadbakht, M., and Karimi, M.H. (2000). Evaluation of the immunomodulatory effects of five herbal plants. *Journal of Ethnopharmacology*, **72:** 167-172.

Barry, T.N. (1998). The feeding value of chicory (*Cichorium intybus*) for ruminant livestock. *Journal of Agricultural Science*, **131:** 251-257.

Bischoff, A.T., Kelley, C.J., Karchesy, Y., Laurantos, M., Nguyen-Dinh, P., and Arefi, A.G. (2004). Antimalarial activity of Lactucin and Lactucopicrin: sesquiterpene lactones isolated from *Cichorium intybus* L. *Journal of Ethnopharmacology*, **95:** 455-457.

Blumenthal, M. (1998). The Complete German Commission E Monographs. 1st edition. Boston: Integrative Medicine Communications.

Cavin, C., Delannoy, M., Malnoe, A., Debefve, E., Touche, A., Courtois, D., and Schilter, B. (2005). Inhibition of the expression and activity of cyclooxygenase-2 by chicory extract. *Biochemical and Biophysical Research Communications*, **327:** 742-749.

Chakraborty, U. and Das, H. (2010). Antidiabetic and antioxidant activities of *Cinnamomum tamala* leaf extracts in STZ-treated diabetic rats. *Global Journal of Biotechnolnology and Biochemistry*, **5(1):** 12-18.

Cheeseman, K.H., and Slate, T.F. (1993). An introduction to free radical biochemistry. *British Medical Bulletin*, **49:** 481-493.

Chen, W.J., Anderson, J.W., and Jennings, D. (1984). Propionate may mediate the hypocholesterolemic effects of certain soluble plant fibers in cholesterol-fed rats. *Proc. Soc. Exp. Biol. Med.*, **175:** 215–218.

Chen, Y.C., Nakthong, C., and Chen, T.C. (2005). Effects of chicory fructans on egg cholesterol in commercial laying hen. *International Journal of Poultry Science*, **4:** 109–114.

Chen, C.J. Deng, A.J., Liu, C., Shi, R., Qin, H.L., and Wang, A.P. (2011). Hepatoprotective activity of *Cichorium endivia* L. extract and its chemical constituents. *Molecules*, **16:** 9049-9066.

Conforti, F., Ioele, G., Statti, G.A., Marrelli, M., Ragno, G., and Menichini, F. (2008). Antiproliferative activity against human tumor cell lines and toxicity test on Mediterranean dietary plants. *Food and Chemical Toxicology*, **46:** 3325–3332.

Fleming, T. 2000. PDR for Herbal Medicines. 3rd edition. Montvale: Medical Economics Company.

Foster, J.C., Cassida, K.A., and Turner, K.E. (2011). *In vitro* analysis of the anthelmintic activity of forage chicory (*Cichorium intybus* L.) sesquiterpene lactones against a predominantly Haemonchus contortus egg population. *Veterinary Parasitology*, **180(3-4):** 298-306.

Gadgoli, C., and Mishra, S.H. (1997). Antihepatotoxic activity of *Cichorium intybus*. *Journal of Ethnopharmacology*, **58:** 131-134.

Gazzani, G., Daglia, M., Papetti, A., and Gregotti, C. (2000). *In vitro* and *ex vivo* anti- and prooxidant components of *Cichorium intybus*. *Journal of Pharmaceutical and Biomedical Analysis*, **23(223):** 127–133.

Gilani, A.H., Janbaz, K.H., and Shah, B.H. (1998). Esculetin prevents liver damage induced by Paracetamol and CCl_4. *Pharmacological Research*, **37(1):** 31-35.

Gürbüz, I., Üstün, O., Yesilada, E., Sezik, E., and Akyürek, N. (2002). *In vivo* gastroprotective effects of five Turkish folk remedies against ethanol-induced lesions. *Journal of Ethnopharmacology*, **83:** 241-244.

Hambrock, A., de Oliveira Franz, C.B., Hiller, S., Grenz, A., Ackermann, S., Schulze, D.U., Drews, G., and Osswald, H. (2007). Resveratrol binds to the sulfonylurea receptor (SUR) and induces apoptosis in a SUR subtype-specific manner. *Journal of Biological Chemistry*, **282(5):** 3347–3356.

Hassan, H.A., and Yousef, M.I. (2010). Ameliorating effect of chicory (*Cichorium intybus* L.)-supplemented diet against nitrosamine precursors-induced liver injury and oxidative stress in male rats. *Food and Chemical Toxicology*, **48:** 2163-2169.

Hazra, B., Sarkar, R., Bhattacharyya, S., and Roy, P. (2002). Tumour inhibitory activity of chicory root extract against Ehrlich ascites carcinoma in mice. *Fitoterapia*, **73:** 730-733.

Hussain, H., Hussain, J., Saleem, M., Miana, G.A., Riaz, M., Krohn, K., and Anwar, S. 2011. Cichorin A: a new benzo-isochromene from *Cichorium intybus*. *Journal of Asian Natural Products Research*, **13(6):** 566–569.

Indap, M.A., and Barkume, M.S. (2003). Efficacies of plant phenolic compounds on sodium butyrate induced anti-tumour activity. *Indian Journal of Experimental Biology*, **41:** 861–864.

Jamshidzadeha, A., Khoshnooda, J.M., Dehghanib, Z., and Niknaha, H. (2006). Hepatoprotective activity of *Cichorium intybus* L. leaves extract against carbon tetrachloride induced toxicity. *Iranian Journal of Pharmaceutical Research*, **1:** 41-46.

Kamel, Z.H., Daw, I., and Marzouk, M. (2011). Effect of *Cichorium endivia* leaves on some biochemical parameters in Streptozotocin-induced diabetic rats. *Australian Journal of Basic and Applied Sciences*, **5(7):** 387-396,

Keshri, G., Lakshmi, V., and Singh, M.M. (1998). Postcoital contraceptive activity of some indigenous plants in rats. *Contraception*, **57:** 357-360.

Kim, M., and Shin, H.K. (1998). Thewater-soluble extract of chicory influences serum and liver lipid concentrations, cecal short-chain fatty acid concentrations and fecal lipid excretion in rats. *Journal of Nutrition*, **128:** 1732–1736.

Kim, H.M., Kim, H.W., Lyu, Y.S., Won, J.H., Kim, D.K., Milee, Y., Morii, E., Jippo, T., Kitamura, Y., and An, N.H. (1999). Inhibitory effect of mast cell-medýated ýmmediate-type allergic reactions by *Cichorium intybus. Pharmacological Research*, **40(1):** 61-65.

Kim, J.H., Mun, Y.J., Woo, W.H., Jeon, K.S., An, N.H., and Park, J.S. (2002). Effects of the ethanol extract of *Cichorium intybus* on the immunotoxicity by ethanol in mice. *International Immunopharmacology*, **2:** 733-744.

Kisiel, W., and Zielinska, K. (2001). Guaianolides from *Cichorium intybus* and structure revision of *Cichorium* sesquiterpene lactones. *Phytochemistry*, 57: 523-527.

Kisiel, W., and Michalska, K. (2002). A new coumarin glucoside ester from *Cichorium intybus*. *Fitoterapia*, **73:** 544-546.

Kisiel, W., Michalska, K., and Szneler, E. (2004). Norisoprenoids from aerial parts of *Cichorium pumilum. Biochemical Systematics and Ecology*, **32:** 343–346.

Kisiel, W., and Michalska, K. (2006). Sesquiterpenoids and phenolics from roots of *Cichorium endivia* var. *crispum. Fitoterapia*, **77:** 354-357.

Krebsky, E.O., Geuns, J.M.C., and De Proft, M. (1999). Polyamines and sterols in *Cichorium* heads. *Phytochemistry*, **50:** 549-553.

Kupchan, S.M., Eakin, M.A., and Thomas, A.M. (1971). Tumor inhibitors. Structure-cytotoxicity relationships am the sesquiterpene lactones. *Journal of Medicinal Chemistry*, **14:** 1147-1152.

Lairon, D. (1996). Dietary fibers: effects on lipid metabolism and mechanisms of action. *European Journal of Clinical Nutrition*, **50:** 125–133.

Lante, A., Nardi, T., Zocca, F. Giacomini, A., and Corich, V. (2011). Evaluation of red chicory extract as a natural antioxidant by pure lipid oxidation and yeast oxidative stress response as model systems. *Journal of Agricultural and Food Chemistry*, **59:** 5318–5324.

Lee, K.H., Hall, I.H., Mar, E.C., Starnes, C.O., ElGebaly, S.A., Waddell, T.G., Hadgraft, R.I., Ruffner, C.G., and Weidner, I. (1977). Sesquiterpene antitumor agents: inhibitors of cellular metabolism. *Science*, **196:** 533-536.

Leung, K.H., and Ip M.M. (1986). Regulation of rat natural killing. II. Inhibition of cytolysis and activation of inhibitors of lipoxygenase: possible role of leukotrienes. *Cell Immunology*, **100:** 474-484.

Liu, F., Kim, J., Li, Y., Liu, X., Li, J., and Chen, X. (2001). An extract of *Lagerstroemia speciosa* L. has insulin-like glucose uptake–stimulatory and adipocyte differentiation–inhibitory activities in 3T3-L1 cells. *Journal of Nutrition*, **131:** 2242–2247.

Loggia, R.D., Raggazi, E., Tubaro, A., Fassina, G., and Vertua, R. (1988). Anti-inflammatory activity of benzopyrone that are inhibitors of cyclo- and lipoxygenase. *Pharmacological Research Communications*, **20:** 91-94.

Malarz, J., Stojakowska, A., and Kisiel, W. (2002). Sesquiterpene lactones in a hairy root culture of *Cichorium intybus. Zeitschrift für Naturforschung*, **57c:** 994-997.

Martinez, F., Abril, E.R., Earnest, D.L., and Watson, R.R. (1992). Ethanol and cytokine secretions. *Alcohol*, **9:** 455–458.

Meadows, G.G., Back, S.E., and Duncan, D.D. (1989). Influence of ethanol consumption on natural killer cell activity in mice. *Alcoholism: Clinical and Experimental Research*, **13:** 476-479.

Michalska, K., and Kisiel, W. (2007). Further sesquiterpene lactones and phenolics from *Cichorium spinosum. Biochemical Systematics and Ecology*, **35:** 714-716.

Morita, I., Takahashi, R., Saito, Y., and Murota, S. (1983). Stimulation of eicosapentaenoic acid metabolism in washed human platelets by 12-hydroperoxyeicosatetra-enoic acid. *Journal of Biological* Chemistry, **258:** 10197-10199.

Mourelle, M., Amezcua, J.L., and Perez-Alvarez, V. (1987). Reduction of apparent indicators of liver cirrhosis in rats by the arachidonate lipoxygenase inhibitor BW755C. *European Journal of Pharmacology*, **134:** 175-180.

Mulinacci, N., Innocenti, M., Gallori, S., Romani, A., La Marca, G., and Vincieri, F.F. (2001). Optimization of the chromatographic determination of polyphenols in the aerial parts of *Cichorium intybus* L. *Chromatographia*, **54:** 455-461.

Muthusamy, V.S., Anand, S., Sangeetha, K.N., Sujatha, S., Arun, B., and Lakshmi, B.S. (2008). Tannins present in *Cichorium intybus* enhance glucose uptake and inhibit adipogenesis in 3T3-L1 adipocytes through PTP1B inhibition. *Chemico Biological Interactions*, **174:** 69–78.

Nandagopal, S. and Ranjitha Kumari, B.D. (2007). Phytochemical and antibacterial studies of chicory (*Cichorium intybus* L.)–A multipurpose medicinal plant. *Advances in Biological Research*, **1 (1-2):** 17-21.

Neichi, T., Koshihara, Y., and Murota, S. (1983). Inhibitory effect of esculetin on 5-lipoxygenase and leukotriene biosynthesis. *Biochimica et Biophysica Acta*, **753:** 130-132.

Niness, K.R. (1999). Inulin and Oligofructose: What are they? *Journal of Nutrition*, **129:** 1402-1406.

Norbaek, R., Nielsen, K., and Kondo, T. (2002). Anthocyanins from flowers of *Cichorium intybus*. *Phytochemistry*, **60:** 357-359.

Ozturk, H. (2009). Effects of chicory inulin on ruminal fermentation *in vitro*. *Veterinary Journal of Ankara University*, **56:** 171–175.

Panossian, A.G. (1984). Inhibition of arachidonic acid 5-lipoxygenase of human polymorphonuclear leukocytes by esculetin. *Biomedica Biochimica Acta*, **43:** 351-355.

Papetti, A., Daglia, M., and Gazzani, G. (2002). Anti- and pro-oxidant activity of water soluble compounds in *Cichorium intybus* var. *silvestre* (Treviso red chicory). *Journal of Pharmaceutical and Biomedical Analysis*, **30:** 939-945.

Papetti, A., Daglia, M., Grisoli, P., Dacarro, C., Gregotti, C., and Gazzani, G. (2006). Anti- and pro-oxidant activity of *Cichorium* genus vegetables and effect of thermal treatment in biological systems. *Food Chemistry*, **97:** 157-165.

Papetti, A., Daglia, M., Aceti, C., Sordelli, B., Spini, V., Carazzone, C., and Gazzani, G. (2008). Hydroxycinnamic acid derivatives occurring in *Cichorium endivia* vegetables. *Journal of Pharmaceutical and Biomedical Analysis*, **48:** 472-476.

Pascual-Villalobos, M.J., and Robledo, A. (1998). Screening for anti-insect activity in Mediterranean plants. *Industrial Crops and Products*, **8:** 183-194.

Patel, V.K., and Venkatakrishna-Bhat, H. (1983). *Cichorium intybus*, a novel herbal preparation as a gum massage, dentifrice, anti inflammatory and anti plaque agent (review and applied study). *Therapie*, **38:** 405–415.

Perez-Alvarez, V., Bobadilla-Lugo, R.A., Muriel, P., Favari, L., and Villanueva-Lopez, C. (1993). Effects of leukotiene synthesis inhibition on acute liver damage induced by CCl_4. *Pharmacology*, **47:** 330-336.

Petrovic, J., Stanojkovic, A., Comic, L.J., and Curcic, S. (2004). Antibacterial activity of *Cichorium intybus*. *Fitoterapia*, **75:** 737-739.

Punitha, I.S.R., Shirwaikar, A., and Shirwaikar, A. (2006). Antidiabetic activity of benzyl tetra isoquinoline alkaloid berberine in streptozotocin-nicotinamide induced type 2 diabetic rats. *Diabetologia Croatica*, **34:** 117-128.

Pushparaj, P.N., Low, H.K., Manikandan, J., Tan, B.K.H., and Tan, C.H. (2007). Anti-diabetic effects of *Cichorium intybus* in streptozotocin-induced diabetic rats. *Journal of Ethnopharmacology*, **111:** 430-434.

Rasmussen, M.K., Zamaratskaia, G., and Ekstrand, B. (2011). *In vivo* effect of dried chicory root (*Cichorium intybus* L.) on xenobiotica metabolising cytochrome P450 enzymes in porcine liver. *Toxicology Letters*, **200:** 88–91.

Recknagael, R. (1967). Carbontetrachloride hepatotoxicity. *Pharmacological Review*, **19(2):** 145-196.

Reyes, M.T., Mourelle, M., Hong, E., and Muriel, P. (1994). Caffeic acid prevents liver damage and ameliorates liver fibrosis induced by CCl_4 in the rat. *Drug Development Research*, **36:** 125-128.

Roy-Choudhury, A., and Venkatakrishna-Bhatt, H. (1983). Spermatogenic inhibition by *Cichorium intybus* L. aqueous root suspension in mice. *Naturwissenschaften*, **70:** 365-366.

Rüngeler, O., Castro, V., Mora, G., Gören, N., Vichnewski, W., Pahl, H.L., Merfort, I., and Schmidt, T.J. (1999). Inhibition of transcription factor NF-κB by sesquiterpene lactones: a proposed molecular mechanism of action. *Bioorganic and Medicinal Chemistry*, **7**: 2343-2352.

Scales, G.H., Knight, T.L., and Saville, D.J. (1995). Effect of herbage species and feding level on internal parasites and production performance of grazing lambs. *New Zealand Journal of Agricultural Research*, **38:** 237-247.

Selvan, V.T., Manikandan, L., Senthil Kumar, G.P., Suresh, R., Kakoti B.B., Gomathi, P., Kumar, D.A., Saha, P., Gupta, M., and Mazumder, U.K. (2008). Antidiabetic and antioxidant effect of methanol extract of *Artanema sesamoides* in streptozotocin-induced diabetic rats. *International Journal Applied Research in Natural Products*, **1(1):** 25-33.

Sezik, E., Tabata, M., Yesilada, E., Honda, G., Goto, K., and Ikeshiro, Y. (1991). Traditional medicine in Turkey I. Folk medicine in Northeast Anatolia. *Journal of Ethnopharmacology*, **35:** 191-196.

Sezik, E., Yesilada, E., Honda, G., Takaishi, Y., Takeda, Y. and Tanaka, T. (2001). Traditional medicine in Turkey X. Folk medicine in Central Anatolia. *Journal of Ethnopharmacology*, **75:** 95-115.

Shah, S.R.U., Hassan, G., Rehman, A., and Ahmed, I. (2006). Ethnobotanical studies of the flora of district Musakhel and Barkhan in Balochistan, Pakistan. *Pakistan Journal of Weed Sciences Research*, **12(3):** 199-211.

Siedle, B., Garcia-Pineres, A.J., Murillo, R., Schulte-Mönting, J., Castro, V., Rüngeler, P., Klaas, Ch.A., Da Costa, F.B., Kisiel, W., and Merfort, I. (2004). Quantitative structure–activity relationship of sesquiterpene lactones as inhibitors of the transcription factor NF-κB. *Journal of Medicinal Chemistry*, **47:** 6042-6054.

Simonsson, U.S.H., Lindell, M., Raffalli-Mathieu, F., Lannerbro, A., Honkakoski, P., and Lang, M.A. (2006). *In vivo* and mechanistic evidence of nuclear receptor CAR induction by artemisinin. *Eur. J. Clin. Invest.*, **36:** 647–653.

Snyder, G.H., Cenerazzo, M.J., Karalis, A.J., and Field, D. (1981). Electrostatic influence of local cysteine environments on disulfide Exchange kinetics. *Biochemistry*, **20:** 6509–6519.

Suntar, I., Küpeli Akkol, E., Keles, H., Yesilada, E., Sarker, S.D., Baykal, T. (2012). Comparative evaluation of traditional prescriptions from *Cichorium intybus* L. for wound healing: Stepwise isolation of an active component by *in vivo* bioassay and its mode of activity. *Journal of Ethnopharmacology*, **143(1):** 299-309.

Syed, N.A., Hasan, T.N., and Aalam, S.M.M. (2008). Evaluation of wound healing potential of *Chicorium ntybus*. L (Asteraceae) in rats. *Iranian Journal of Pharmacology and Therapeutics*, **7:** 181-184.

Szkudelski, T. (2001). The mechanism of alloxan and streptozotocin action in β-cells of the rat pancreas. *Physiological Research*, **50:** 536-546.

Tousch, D., Lajoix, A.D., Hosy, E., Azay-Milhau, J., Ferrare, K., Jahannault, C., Cros, G., and Petit, P. (2008). Chicoric acid, a new compound able to enhance insulin release and glucose uptake. *Biochemical and Biophysical Research Communications*, 377: 131–135.

Tubaro, A., Negro, P.D., Ragazzi, E., Zampiron, S., and Loggia, RD. (1988). Anti-inflammatory and peripheral analgesic activity of esculetin *in vivo*. *Pharmacological Research Communications*, **20:** 83-85.

Tuzlaci, E., Alparslan Ýþbilen, D.P., and Bulut, G. (2010). Turkish folk medicinal plants. VIII: Lalapaºa (Edirne). *Marmara Pharmaceutical Journal*, **14:** 47-52.

Wu, H., Su, Z., Yang, Y., Ba, H., and Aisa, H.A. (2007). Isolation of three sesquiterpene lactones from the roots of *Cichorium glandulosum* Boiss. et Huet. by high-speed counter-current chromatography. *Journal of Chromatography A*, **1176:** 217–222.

Wu, H.K., Xin, X.L., Su, Z., and Aisa, H.A. (2011). 2-Isopropyl-6-methylpyrimidin-4(3*H*)-one and taraxasterol from the stems of *Cichorium glandulosum*. *Chemistry of Natural Compounds*, **47(4):** 664-666.

Van Beek, T.A., Maas, P., King, B.M., Leclercq, E., Voragen, A.G.J., and De Groot, A. (1990). Bitter sesquiterpene lactones from chicory root. *Journal of Agricultural and Food Chemistry*, **38:** 1035–1038.

Wagner, H., Proksch, A., Riess-Maurer, I., Vollmer, A., Odenthal, S., Stupper, H., Jurcic, K., Le Turdu, M., and Heur, Y.H. (1984). Immunostimulant action of polysaccharide (heteroglycans) from higher plants. *Preliminary communication. Arzneimittelforschung*, **34:** 659–661.

Wang, P.H., Moller, D., Flier, J.S., Nayak, R.C., and Smith, R.J. (1989). Coordinate regulation of glucose transporter function, number, and gene expression by insulin and sulfonylureas in L6 rat skeletal muscle cells. *Journal of Clinical Investigation*, **84(1):** 62–67.

Wesolowska, A., Nikiforuk, A., Michalska, K., Kisiel, W., and Chojnacka-Wojcik, E. (2006). Analgesic and sedative activities of lactucin and some lactucin-like guaianolides in mice. *Journal of Ethnopharmacology*, **107:** 254-258.

Yáñez, J., Vicente, V., Alcaraz, M., Castillo, J., Benavente-García, O., Canteras, M., and Teruel, J.A. (2004). Cytotoxicity and antiproliferative activities of several phenolic compounds against three melanocytes cell lines: relationship between structure and activity. *Nutrition and Cancer*, **49:** 191–199.

Yesilada, E., Sezik, E., Honda, G., Takaishi, Y., Takeda, Y., and Tanaka, T. (1999). Traditional medicine in Turkey IX. Folk medicine in North-west Anatolia. *Journal of Ethnopharmacology*, **64:** 195-210.

Zafar, R., and Ali, S.M. (1998). Anti-hepatotoxic effects of root and root callus extracts of *Cichorium intybus* L. *Journal of Ethnopharmacology*, **63:** 227-231.

Bioactive Phytochemicals: Perspectives for Modern Medicine Vol. 2 (2014) *Pages* **355–382**
Editor: **V.K. Gupta**
Published by: **DAYA PUBLISHING HOUSE, NEW DELHI**

16

Phytochemical Characterization and Pharmacological Profiling of Picroliv: An Overview

Papiya Bigoniya[1*] and C.S. Singh[1]

ABSTRACT

The consumption of medicinal plants and herbal products has somewhat surprisingly increased in recent years, constituting a promising pharmaceutical segment. This may be primarily due to easy access, low cost, popularity and cultural compatibility. Medicinal plants are an important source of bioactive substances with desired therapeutic effect. Many of them have become prototypes for the isolation/synthesis of a large number of bioactive synthetic compounds. Furthermore a single bioactive compound or drug from a plant in general does not reflect reality as only in a few cases drug activity depend upon single component. Generally the pharmacological activity of herb is the result of concerted activity of several active compounds as noted in case of picroliv or kutkin. Picroliv is a stable, mixed crystal of two iridoid glycosides, picroside-I and kutkoside (1: 1.5 w/w) isolated from the roots and rhizomes of Picrorhiza kurroa Royal ex Benth. (Family- Scrophulariaceae). The mixture of picroside I and kutkoside accounts for approximately 60 per cent of the constituents of picroliv. A number of analytical techniques including HPLC and HPTLC have been reported for the quantification of picroliv. Picroliv (6-12.5 mg/kg) has shown efficacy comparable to silymarin (10-20 mg/kg) in rodent models against cadmium, aflatoxin B1, oxytetracycline, galactosamine, paracetamol, thioacetamide and $CC1_4$ *induced hepatic damage. It has also shown anti-cholestatic effect in rats, guinea pigs and cats treated with* CCl_4*. Picroliv has reported anti-viral, immuno-modulatory, anti-allergic and anti-anaphylactic activities and is devoid of any significant CNS and CVS, autonomic and other systemic activity. The recent findings*

1 Radharaman College of Pharmacy, Fatehpur Dobra, Ratibad, Bhopal – 462 002, M.P., India

* *Corresponding author*: E-mail: p_bigoniya2@hotmail.com

suggest that picroliv in combination with fluconazole and miltefosine would be a new option for antileishmanial chemotherapy at lower cost and toxicity. The present review discusses the chemistry, analytical methodologies, including the extraction, isolation and characterization of active ingredient picroliv from P. kurroa as well as its therapeutic potential against various diseases.

Keywords: Bioactive compound, Picroliv, Picroside I, Kutkoside, Hepatoprotective, Immunomodulatory, Anti-leishmanial.

Introduction

Medicinal plants have been known for millennia and are highly esteemed all over the world as a rich source of therapeutic agents, health products and pharmaceuticals for the prevention of diseases and ailments. Countries with ancient civilizations such as China, India, South America, Egypt, etc. are still using several plant remedies for various conditions. Nature has bestowed our country with an enormous wealth of medicinal plants therefore India has often been referred to as the Medicinal Garden of the world.

Despite the major advances in the modern medicine, the development of new drugs from natural products is still considered important. This seems to be even more relevant for the developing countries, where the costs to develop a drug are prohibitive. Since 1980, the World Health Organization has been encouraging countries to identify and exploit traditional medicine and phytotherapy. In India, phytotherapy is a promising alternative due to its biodiversity and many studies devoted to the description of plant chemistry and their biological activities. However the development of phytotherapeutical products is dependent on their standardization, to guarantee not only the authenticity of the plant extract, but also to enable a dose related use. Chromatographic methods have been recommended to the quality control of phytotherapeutical products with many advantages over other methods.

Picrorhiza kurroa Royle ex Benth. (Fam: Scrophulariceae) is a small reputed alpine herb, endemic to Himalayan region of India, Pakistan, Nepal and China. In India, *P. kurroa* is distributed from Kashmir to Sikkim at altitude ranging from 2700-4500 m. Picroliv is the standard preparation containing mainly a mixture of two iridoid glycoside, picroside I and kutkoside (1: 1.5 w/w) obtained from the ethanolic extract of the 3-4 years old roots and rhizomes of *P. kurroa.*

Picroliv has been reported to be a potent hepatoprotective agent against various hepatotoxins including hepatitis B virus (Girish *et al.*, 2009; Yadav and Khandelwal, 2006; Singh *et al.*, 2005; Yadav *et al.*, 2005; Rastogi *et al.*, 2000). Picroliv enhances immunity in combination with piperine, curcumin and chloroquine (Pathak and Khandelwal, 2008; Dwivedi *et al.*, 2008; Puri *et al.*, 1992). Picoliv was found to be an effective inhibitor of hepaticarcinogenesis as evaluated by Anand *et al.* (2008) and Rajeshkumar and Kuttan (2000, 2001, 2003). Pharmacological studies revealed that picroliv possess testicular protective (Yadav and Khandelwal, 2008, Yadav *et al.*, 2005), anti-hyperlipidemic activity (Vivekanand *et al.*, 2007; Singh *et al.*, 1992a) and

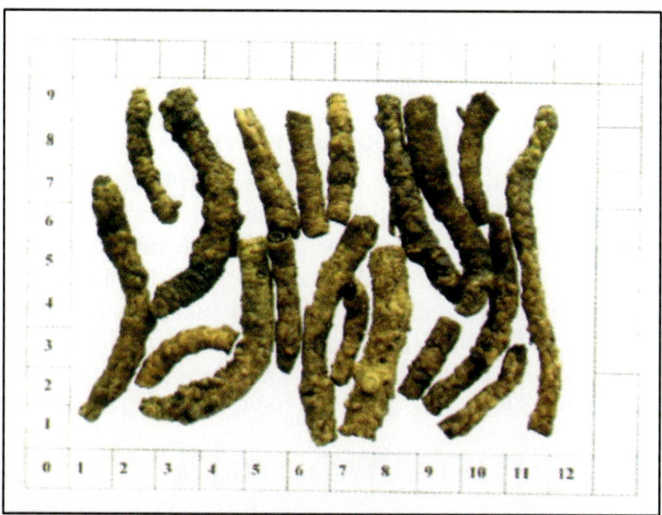

Figure 16.1a: Root and rhizome of *P. kurroa*.

Figure 16.1b: *P. kurroa* plant.

anti-oxidant activities (Rastogi *et al.*, 2001a; Chander *et al.*, 1992a; 1998). Picroliv restores choleretic parameters against carbon tetrachloride, paracetamol and ethynylestradiol induced hepatic damage in experimental animals by increasing biosynthesis and secretion of bile acids (Saraswat *et al.*, 1993; Shukla *et al.*, 1991). Recent findings revealed that picroliv in combination with antifungal and anti-leishmanial drug enhances the anti-leishmenial activity by enhancing cell mediated immunity (Shakya *et al.*, 2011a; Sane *et al.*, 2011; Gupta *et al.*, 2005; Mittal *et al.*, 1998).

There has been a tremendous interest in this bioactive compound, picroliv as evidenced by the voluminous work. Therefore, we aimed to compile an up date and comprehensive review of picroliv that covers its phytochemistry and pharmacology.

Phytochemistry

Medicinal values mentioned above have been attributed due to the presence of the iridoid glycosides, which mainly includes picroside I and picroside II. The picroliv contains two major iridoid glycosides (50-70 per cent) and mixture of cucurbitacins (4-5 per cent) with some uncharacterized minor compounds (Halliwel and Gutteridge, 1999). Among the iridoid glycosides present in the picroliv, one iridoid glycoside is identified as picroside I, while the identity of second iridoid glycoside is still controversial. However, it was treated as single compound and mentioned as kutkoside (Rastogi *et al.*, 2000; Baruah *et al.*, 1998; Kumar *et al.*, 2004) or picroside II

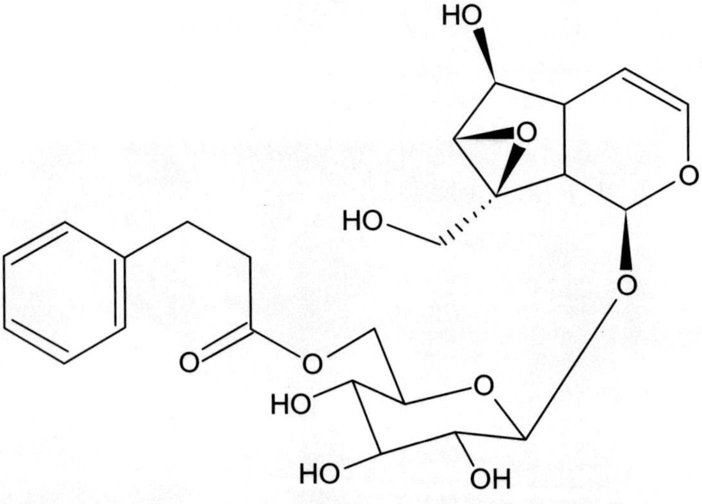

Picroside I

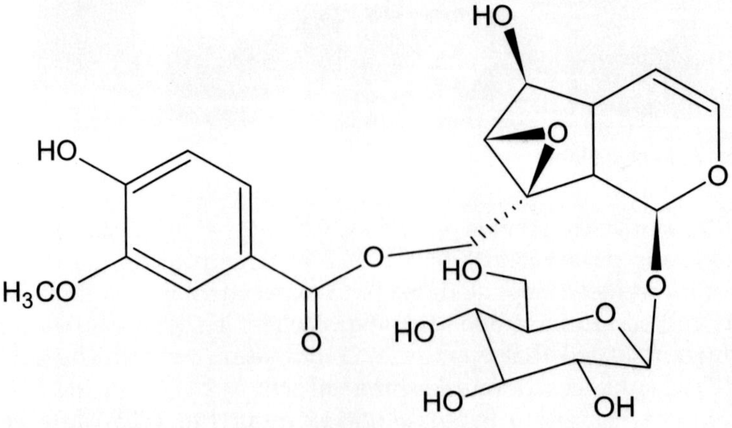

Kutkoside

(Stuppner and Wagner, 1992; Vaidya *et al.*, 1996; Stuppner and Wagner, 1989) in earlier reports. Picroside I as reported by Kitagawa *et al.* (1971) is an amorphous powder where as Singh and Rastogi (1972), have obtained it in crystalline form, having m.p. 128-130° and exhibited a red-purple colour with Godin's reagent. Kutkoside was obtained as an amorphous powder and gave same colour as picroside I with Godin's reagent. Picroliv, on repeated fractional crystallization affords a constant melting product, m.p. 210°. In TLC of picroliv, picroside I (6′-*O*-cinnamoylcatalpol) moves faster than kutkoside on Keiselgel G plates.

Extraction of Picroliv

Soak coarse powder of root and rhizome of *P. kurroa* in 95 per cent ethanol for four days with intermittent shaking, filter through muslin cloth and concentrate. Remove residual solvent under vacuum in a rotatory vacuum evaporator below 40° at 650 mm vacuum pressure resulting a solid blackish-brown mass.

Pulverized the solid mass to fine powder, dissolve in acetone and warm at 50°. Pour the mixture in 80 per cent acetone with continuous stirring for 1 h and set aside for 2 h. Filter the clear supernatant liquid and repeat the process for six times. Mix the different fractions of acetone and evaporate under reduced pressure at 50°, the dried residue thus obtained is picroliv (Bigoniya *et al.*, 2010).

Identification of Picroliv by TLC Method (Rajpal, 2002)

Sample preparation: 5 gm of powdered root and rhizome of *P. kurroa* is soaked overnight in 90 per cent methanol. The methanolic extract is decanted and concentrated.

Stationary phase: Silica gel G

Solvent system: Ethyl acetate (water saturated): methanol (47: 3)

Detecting reagent: Vannilin-sulphuric acid

Rf: Kutkoside at 0.26 and picroside I at 0.42

Determination of Picroliv by HPLC Method (IP, 2007)

Test solution: Dissolve 100 mg of the *P. kurroa* in 25 ml of methanol and filter.

Reference solution: A 0.1 per cent w/v solution of picroliv Reference Standard in methanol and filter.

Column: A stainless steel column 15 cm × 4.6 mm packed with octadecylsilane bonded to porous silica (5 μm)

Mobile phase: 83 volumes of 1 per cent v/v orthophosphoric acid in water and 17 volumes of acetonitrile

Flow rate: 1 ml/min

Absorbance: At 280 nm

Injector loop: 20 μl

Retention time: 13.80 min (approximate)

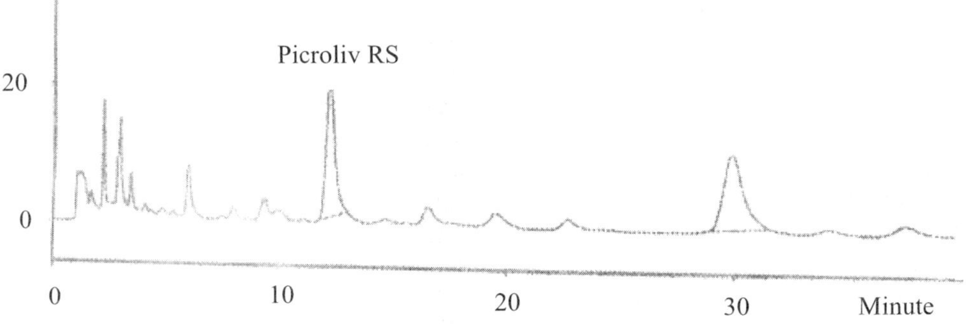

Figure 16.2: HPLC chromatogram picroliv reference standard (IP, 2007).

Pharmacological Activity

Hepatoprotective

Bigoniya *et al.* (2010) evaluated the functional integrity of hepatic microsomal enzyme on picroliv pretreatment against CCl_4 induced hepatotoxicity. They evaluated the hepatoprotective potential of picroliv at different dose ranges to establish its ED_{50} against CCl_4 induced hepatotoxicity. The CCl_4 was given orally (1 ml/kg) in liquid paraffin (1: 1). Picroliv at different doses and silymarin (20 mg/kg) standard were administered orally for 14 days. Body weight, biochemical parameters like SGPT, SGOT and ALP were estimated. Hepatic microsomal drug metabolizing enzyme functional integrity was accessed by *in vitro* bromosulphalein (BSP) uptake test of liver slices and thiopental (40 mg/kg) induced sleeping time. The results showed dose related hepatoprotective efficacy of picroliv. The ED_{50} values for effectiveness of liver enzyme metabolizing activity on BSP uptake test and sleeping time were 11.94 and 17.49 mg/kg, respectively.

Girish *et al.* (2009) evaluated the hepatoprotective activity of picroliv, curcumin and ellagic acid in comparison to silymarin using paracetamol induced acute liver damage. Hepatotoxicity was induced by administering a single oral dose of paracetamol (500 mg/kg) and was assessed by quantifying the serum enzyme activities; phenobarbitone induced sleeping time and histopathological analysis of liver tissues. The antioxidant parameters, malondialdehyde, reduced glutathione and catalase of the liver tissue were also assessed. The herbal drugs were administered for 7 days by oral route at 50 and 100 mg/kg. Paracetamol induced hepatic damage was manifested by a significant increase in the activities of marker enzymes (alanine transaminase, aspartate transaminase and ALP) in serum and malondialdehyde level in liver. There was also a significant decrease in activity of glutathione and catalase levels. The histopathological examination on toxic models revealed centrizonal necrosis and fatty changes. Pretreatment of mice with picroliv, curcumin and ellagic acid reversed these altered parameters towards normal values, which were compared with silymarin. The normalization of phenobarbitone induced sleeping time suggests the restoration of liver cytochrome P_{450} enzymes.

Yadav and Khandelwal (2006), evaluated the therapeutic efficacy of picroliv in male rats exposed to $CdCl_2$ (0.5 mg/kg, sc), 5 days/week for 18 weeks. Picroliv at two doses (6 and 12 mg/kg, po) was given to the Cd-administered group for the last 4 weeks. The Cd altered oxidative stress indices, such as increased lipid peroxidation and membrane fluidity, reduced levels of non-protein sulphydryls and Na^+K^+ATPase activity in the liver and kidney were found close to the control values by picroliv treatment, suggesting its antioxidant potential. The hepatoprotective action of picroliv was evident by its ability to lower the Cd-induced liver function parameters such as ALP, alanine aminotransferase, aspartate aminotransferase, gamma-glutamyl transpeptidase and lactate dehydrogenase. Bile flow and biliary Cd also increased as a result of picroliv's choleretic property. The Cd-induced serum urea and urinary excretion of proteins, calcium, Cd and enzymes, such as N-acetyl-beta-D-glucosaminidase and LDH, were less marked on picroliv treatment, indicating recovery from nephrotoxicity. Organ uptake of Cd and essential metals by Cd exposure was reduced on picroliv treatment. Cd-induced hepatic metallothionein (MT) was lowered by picroliv, whereas renal MT was unaltered. Cd-induced hepatic damage was also minimized. However, the renal morphological changes were marginally protected by picroliv. The result revealed picroliv at dose of 12 mg was more effective than the 6 mg/kg dose in causing amelioration of the above parameters.

Singh *et al.* (2005) investigated the hepatoprotective activity of picroliv against the hepatotoxic changes induced by CCl_4 and *E. histolytica* infection in three animal models- mastomys, gerbils and albino druckray rats. A total of 30 animals were used for each model and divided into five groups of six animals each. Group I consisted of normal animals. The rest received six doses of CCl_4 intraperitoneally. Group II served as hepatotoxic control. The remaining animals were infected intraperitoneally with *E. histolytica* trophozoites, of which group III was the hepatotoxic plus amoeba infected control. The remaining animals were divided into two groups, one received hepatoprotective agent picroliv and the other silymarin. All animals were sacrificed seven days post amoeba infection. Result showed that significant recovery obtained in serum enzyme levels in all animal models and against amoebic liver abscess in gerbils on treatment with picroliv indicated that it possess therapeutic activity against *E. histolytica* induced hepatic damage.

The potential of picroliv against acute Cd intoxication was evaluated in male rats by Yadav *et al.* (2005). Biochemical and histopathological profile in rats pretreated with picroliv (12 mg/kg, po) followed by a single dose of Cd as cadmium chloride (3 mg/kg, ip) revealed marked suppression of oxidative stress in liver. The Cd-induced enhanced levels of lipid peroxidation, membrane fluidity and reduced levels of nonprotein sulphydryls and Na^+K^+ATPase were significantly restored to near normal by picroliv pretreatment. In addition, the Cd-induced serum levels of SGOT, SGPT, gamma glutamyl transpeptidase and lactate dehydrogenase were restored to near basal levels. Hepatic histopathological damage was also minimized.

Single doses of aflatoxin B1 (2 mg/kg, ip) caused significant increases in the activities of tau-glutamyl transpeptidase, 5'-nucleotidase, acid phosphatase and acid ribonuclease, and decreases in the activities of succinate dehydrogenase and glucose-6-phosphatase in liver, after 8 weeks. The level of lipid peroxides, DNA, RNA, and cholesterol increased while glycogen decreased. It also increased the serum level of transaminases, sorbitol dehydrogenase, glutamate dehydrogenase, lactate dehydrogenase, acid phosphatase, ALP, and bilirubin. Oral administration of picroliv (25 mg/kg/day × 15 days), 6 weeks after aflatoxin B_1 toxication, significantly prevented the biochemical changes induced in liver and serum of aflatoxin B1 treated rats (Rastogi *et al.*, 2000).

Singh *et al.* (2000) evaluated the efficacy of picroliv against hepatic ischemia-reperfusion injury *in vivo*. Picroliv was fed to male Sprague Dawley rats in a dose of 12 mg/kg once daily by oral gavage for 7 days prior to hepatic ischemia. Ischemia was induced by occluding the hepatic pedicel with a microaneurysm clip for 30 min and reperfusion was allowed thereafter for varying period (15-120 min) by releasing the microaneurysm clip. Picroliv pretreatment resulted in better hepatocyte glycogen preservation and reduced apoptosis. Reduction in apoptosis was associated with decreased mRNA expression of caspase-3. Oxidant induced cellular damage as measured by tissue malondialdehyde levels was significantly less following picroliv pretreatment. Both a reduction in neutrophil infiltration and an increased level of intracellular antioxidant enzyme superoxide dismutase possibly contributed to the reduction in tissue lipid peroxidation. Tissue inflammatory cytokines level of interleukin-1α and interleukin-1β was also lower in picroliv group. Furthermore, picroliv pretreatment resulted in enhanced proliferating cell nuclear antigen immunoreactivity.

Saraswat *et al.* (1999) investigated the hepatoprotectve activity of picroliv against ethanol-induced hepatic injury in rats. Alcohol feeding (3.75 g/kg × 45 days) produced 20-114 per cent alteration in selected serum (AST, ALT and ALP) and liver markers (lipid, glycogen and protein). Further, it reduced the viability (44-48 per cent) of isolated hepatocytes (*ex vivo*) as assessed by Trypan blue exclusion and rate of oxygen uptake. Its effect was also seen on specific alcohol metabolizing enzymes (aldehyde dehydrogenase, 41 per cent; acetaldehyde dehydrogenase, 52 per cent) in rat hepatocytes. The levels of these enzymes were found to be reduced in the cells following alcohol intoxication. Ethyl alcohol also produced cholestasis (41-53 per cent), as indicated by reduction in bile volume, bile salts and bile acids. Picroliv treatment (3-12 mg/kg, po × 45 days) restored the altered parameters in a dose-dependent manner (36-100 per cent).

Picroliv, showed significant curative activity *in vitro* in primary cultured rat hepatocytes against toxicity induced by thioacetamide (200 µg/mL), galactosamine (400 µg/mL), and CCl_4 (3 µl/mL). Activity was assessed by determining the change in hepatocyte viability and rate of oxygen uptake and other biochemical parameters (SGOT, SGPT, and ALP). The toxic agents alone produced a 40-62 per cent inhibition of cell viability and a reduction of biochemical parameters after 24 h of incubation at 37° which (on removal of the toxic agents) was reversed after further incubation for 48 h. Incubation of damaged hepatocytes with picroliv exhibited a concentration (1-100 µg/mL) dependent curative effect in restoring altered viability parameters (Visen *et al.*, 1998).

Picroliv showed a dose dependent (1.5-12 mg/kg, po × 7 days) hepatoprotective activity against oxytetracycline induced hepatic damage in rat. Result showed that it increased the number of viable hepatocytes (*ex vivo*) significantly. Increase in bile volume and its contents in conscious rat suggest potent anticholestatic property. Picroliv also antagonised alterations in enzyme levels (SGOT, SGPT, and ALP) in isolated hepatocytes and serum, induced by oxytetracycline (200 mg/kg, ip) feeding. Picroliv was found to be more potent than silymarin (Saraswat *et al.*, 1997).

Rastogi *et al.* (1996) investigated the protective effect of picroliv against alcohol-induced chronic hepatotoxicity in rats. Daily administration of ethyl alcohol (3.76 g/kg, po × 45 days) resulted in significant changes in several biochemical parameters of the liver and serum of albino rats. After exposure to alcohol for 30 days when picroliv (12 mg/kg, po), was administered for 15 days along with alcohol, the degree of change in most of the parameters was reduced.

Perfusion of liver of rats toxicated with galactosamine or thioacetamide with a 0.02 per cent solution of picroliv for 30 min (1 ml/min; 6 mg/rat), significantly reversed toxicant-induced changes in the activities of several enzymes. Galactosamine-induced increase in the activities of ALP, γ-glutamyl transpeptidase, acid ribonuclease, acid phosphatase, succinate dehydrogenase and decreases in the activities of Na^+K^+ATPase and glucose-6-phosphatase (reversed by 40-87 per cent). Similarly, thioacetamide-induced inhibitions of the activities of Na^+K^+ATPase, Ca^{++}-ATPase, Mg^{++}-ATPase, succinate dehydrogenase and elevations in the activities of ALP, γ-glutamyl transpeptidase, and acid ribonuclease were also significantly reversed. A significant reversal of the toxicants-induced decrease in [14C]-leucine incorporation was also observed (Dwivedi *et al.*, 1993a).

Dwivedi *et al.* (1993b) investigated the hepatoprotecitve effect of picroliv by studying the protection of biochemical and histological changes induced in livers of rats given single oral doses (7 mg/kg) of aflatoxin B1. Administration of aflatoxin B1 resulted in a significant increase in 5'-nucleotidase, γ-glutamyl transpeptidase, acid ribonuclease, total lipids, cholesterol and lipid peroxides in liver and transaminases, ALP and bilirubin in serum. However, the activity of glucose-6-phosphatase and levels of cytochrome P_{450}, cytochrome B_5, DNA, RNA, proteins and glycogen in liver and total proteins in serum decreased. The liver histology showed ballooned hepatocytes, degeneration, microvesicular fat, focal necrosis, bile duct hyperplasia and proliferation of oval and spindle cells in portal tracts. When picroliv (25 mg/kg × 7 days) was given to aflatoxin B_1 toxicated rats, the majority of the biochemical and histological changes were significantly protected.

Picroliv showed a significant dose dependent (3-12 mg/kg, po × 7 days) protective activity against galactosamine-induced hepatic damage in rat as evaluated on the isolated hepatocytes (*ex vivo*) preparation. It markedly increased the percentage of viability of hepatocytes. It also restored the galactosamine-induced changes in the levels of enzymes (SGOT, SGPT and ALP) both in isolated hepatic cells as well as in serum. In addition, picroliv possessed a marked anti-cholestatic effect (Visen *et al.*, 1993).

Singh *et al.* (1992b) studied the effect of picroliv on protein and nucleic acid synthesis. Oral administration of picroliv, a standardised fraction of roots and rhizomes of *P. kurroa*, showed stimulation of nucleic acid and protein synthesis in rat liver.

Dwivedi *et al.* (1992a) studied the hepatoprotective effect of picroliv and its components kutkoside and picroside I against galactosamine-induced damage in rats. Oral administration of picroliv (12 mg/kg/day × 7 days), significantly prevented the biochemical changes in liver and serum of galactosamine-toxicated rats.Whereas kutkoside (12 mg/kg/day × 7 days) also protected against changes in most of the hepatic and serum constituents studied. Picroside I, at the same dose level could only prevent toxicant-induced changes in acid phosphatase, phospholipids and lipid peroxides in liver and ALP in serum. Mixture of picroside I and kutkoside in the ratio of 1: 1.5 at 12 mg/kg dose elicited lesser response than picroliv.

Chander *et al.* (1992b) studied the effect of picroliv on glutathione metabolism in liver and brain of *M. natalensis* infected with *P. berghei*. Administration of picroliv at a dose of 6 mg/kg, orally for two weeks showed significant protection against changes in liver and brain glutathione metabolism of *P. berghei* infected *M. natalensis*. The depletion of reduced glutathione level and inhibition of glutathione-*S*-transferase, glutathione reductase and glutathione peroxidase activities due to *P. berghei* infection was markedly recovered by picroliv. The increased levels of lipid peroxidation products in damaged tissues were also reduced along with the recovery of glutathione metabolism.

The efficacy of picroliv was studied by Dwivedi *et al.* (1992b) against the *A. phalloides*-induced biochemical changes in rat liver. *A. phalloides* (50 mg/kg) caused significant increases in the activities of hepatic 5'-nucleotidase, γ-glutamyl transpeptidase, acid ribonuclease, and succinate dehydrogenase, but a decrease in glucose-6-phosphatase. The level of cytochrome P_{450} in microsomal fraction and content of glycogen in liver showed significant depletions. Picroliv (25 mg/kg/day × 10 day) provided significant restorations of all the biochemical changes poisoned by *A. phalloides* except cytochrome P_{450} and glycogen.

Ansari *et al.* (1991) determined the hepatoprotective activity of picroliv in adult male albino rats against the hepatotoxic effects of paracetamol and galactosamine as evidenced by various biochemical and histopathological observations. Maximum hepatoprotective effect was observed with daily oral doses of 6 and 12 mg/kg for 7 or 8 days. The antihepatotoxic action of picroliv seems likely due to an alteration in the biotransformation of the toxic substances resulting in decreased formation of reactive metabolites.

Monocrotaline caused changes in most of the biochemical parameters in rats 12 days after a single dose of 120 mg/kg. This resulted in significantly increased activities of hepatic succinate dehydrogenase, acid ribonuclease, acid phosphatase, γ-glutamyl transpeptidase and 5'-nucleotidase and decreased in the activities of glucose-6-phosphatase and cytochrome P_{450}. The levels of DNA, RNA and glycogen in liver and albumin and protein in serum decreased while serum bilirubin increased. The histopathological changes in liver were characterized by diffused hepatocyte alterations in the form of ballooning, granular cytoplasm, indistinct cell outlines, nuclear changes, focal necrosis, and vascular damage. When picroliv was administered orally in a dose of 25 mg/kg simultaneously with monocrotaline, alterations in most of the biochemical parameters along with the histopathological changes in liver caused by monocrotaline were prevented (Dwivedi *et al.*, 1991a).

Thioacetamide (100 mg/kg), when administered to normal rats, caused a significant increase in the activities of 5'-nucleotidase and γ-glutamyl transpeptidase and a decrease in the activities of glucose-6-phosphatase and succinate dehydrogenase enzymes in the liver. DNA, RNA, and proteins were increased while the cytochrome P_{450} in the microsomal fraction and the glycogen content in the liver were decreased significantly. Elevations in the activities of SGOT, SGPT, ALP and bilirubin content in serum were also observed. Picroliv at doses of 12.5 and 25 mg/kg prevented most of the biochemical changes induced by thioacetamide in liver and serum (Dwivedi *et al.*, 1991b).

Floersheim *et al.* (1990) reported that survival of mice after lethal doses of lyophilizate from *A. phalloides* was markedly increased by pretreatment with single doses of picroliv, a mixture of iridoid glycosides picroside I and kutkoside isolated from the roots of *Picrorhiza kurroa*. The protective effect of picroliv was comparable to that of silibinin. The curative efficacy of picroliv appeared to be slightly superior.

Dwivedi *et al.* (1990) reported that administration of carbon tetrachloride to normal rats increased activities of hepatic 5(1)-nucleotidase, acid phosphatase, acid ribonuclease while the activities of succinate dehydrogenase, glucose-6-phosphatase, superoxide dismutase and cytochrome P_{450} were decreased. Levels of lipid peroxides, total lipids and cholesterol of liver were also increased. The activities of SGOT, SGPT and ALP were increased. Other serum parameters like bilirubin, proteins, cholesterol, triglycerides and lipoprotein-X showing changes after CCl_4 administration. Picroliv in doses of 6 and 12 mg/kg provided a significant protection against most of the biochemical alterations produced by CCl_4.

Chander *et al.* (1990) reported that administration of picroliv (3-12 mg/kg/day × 2 weeks) simultaneously with *P. berghei* infection showed significant protection against hepatic damage in *M. natalensis*. The increased levels of SGOT, SGPT, ALP, lipoprotein-X and bilirubin in the infected animals were marked reduced by different doses of picroliv. In the liver, picroliv decreased the levels of lipid peroxides and hydroperoxides and facilitated the recovery of superoxide dismutase and glycogen. Picroliv had no effect on the degree of parasitaemia.

Antioxidant

Rastogi *et al.* (2001a) investigated the comparative antioxidant effect of picroliv and silymarin on aflatoxin B_1 induced lipid-peroxidation in rat liver and kidney. Marked increases in lipid peroxide levels and a concomitant decrease in enzymic antioxidant levels were observed in aflatoxin B_1 (2 mg/kg, ip) -toxicated rats, while drug (picroliv and silymarin both) treatment reversed the condition to near normal levels. The effects of picroliv and silymarin were comparable.

Chander *et al.* (1998) reported that picroliv at the dose of 6 mg/kg, po, for two weeks provided significant protection against the generation of lipid peroxidation products in serum β-lipoproteins of *P. berghei* infected *M. coucha*. Incubation of normal rat hepatocytes with VLDL or LDL isolated from infected animals caused significant generation of lipid peroxides followed by a decrease in the viability of these cells, however these effects were partially reversed with the lipoproteins from infected and picroliv treated groups. HDL from infected animals was not toxic to hepatocytes *in vitro*.

Chander *et al.* (1994) investigated that picroliv at dose of 6 mg/kg, po for two weeks provided significant protection against depletion of reduced glutathione levels in liver and brain of *P. berghei* infected *M. natalensis*. The activation of γ-glutamyl transpeptidase enzyme and decreased levels of cysteine, sulphydryl groups as well as glutathione synthesis in both tissues due to *P. berghei* infection were reversed by picroliv. Enzymatic and non enzymatic lipid peroxidation in microsomes *in vitro* was significantly reduced by picroliv along with the recovery of reduced glutathione.

Picroliv and its main components which are a mixture of the iridoid glycosides, picroside-I and kutkoside, were studied *in vitro* as potential scavengers of oxygen free radicals (Chander *et al.*, 1992a). The superoxide (O^{2-}) anions generated in a xanthine-xanthine oxidase system, as measured in terms of uric acid formed and the reduction of nitroblue tetrazolium were shown to be suppressed by picroliv, picroside-I and kutkoside. Picroliv as well as both glycosides inhibited the non-enzymic generation of O^{2-} anions in a phenazine methosulphate NADH system. Malonaldehyde generation in rat liver microsomes as stimulated by both the ascorbate-Fe^{2+} and NADPH-ADP-Fe^{2+} systems was shown to be inhibited by the picroliv glycosides. Known antioxidants tocopherol (vitamin E) and butylated hydroxyanisole (BHA) were also compared with regard to their antioxidant actions in the above system. It was found that BHA afforded protection against ascorbate-$Fe^{(2+)}$-induced MDA formation in microsomes but did not interfere with enzymic or non-enzymic O^{2-} anion generation; and tocopherol inhibited lipid peroxidation in microsomes by both prooxidant systems and the generation of O^{2-} anions in the non-enzymic system but did not interfere with xanthine oxidase activity.

Anti-leishmeniasis

Shakya *et al.* (2011b) studied the leishmanicidal efficacy of ketoconazole in combination miltefosine and picroliv in *L. donovani*/hamsters model. Animals treated with combination of ketoconazole (50 mg/kg, po × 5 days) and miltefosine (5 mg/kg, po × 5 days) showed augmentation in efficacy against leishmania parasite (72 per cent) in comparison to those treated with ketoconazole (54.67 per cent) and miltefosine (54.77 per cent) separately. Co-administration of picroliv (10 mg/kg, po × 12 days) has further enhanced antileishmanial efficacy from 72 per cent to 82 per cent. Significant generation of reactive oxygen species reactive nitrogen species and hydrogen peroxide and increased phagocytosis was observed in animals treated with ketoconazole and miltefosine; however, addition of picroliv to this combination did not alter the level of metabolites and phagocytosis due to its antioxidative and non- leishmanicidal characteristics, respectively. Significant rise in cell mediated immunity witnessed in this group reveals the role played by the immunomodulator, picroliv and justifies the significance of enhanced cell mediated immunity in the therapy.

Leishmanicidal efficacy of fluconazole in combination with miltefosine, and picroliv, was evaluated by Shakya *et al.* (2011a) in hamsters infected with *L. donovani*. Animals treated with fluconazole (50 mg/kg, po × 5 days) and miltefosine (5 mg/kg, po × 5 days) showed enhancement in antileishmanial efficacy (77 per cent), reactive nitrogen species, reactive oxygen species, hydrogen peroxide, and phagocytosis index as compared to those treated with individual drugs. Addition of picroliv to this combination further increased the antileishmanial efficacy from 77 per cent to 88 per cent. Up regulation of cell-mediated immunity was also observed in animals of this group which strengthens the immunomodulatory role of picroliv.

Sane *et al.* (2011) investigated the immunomodulatory effect of picroliv on a combination of paromomycin and miltefosine using *L. donovani*/hamster model. Picroliv has significantly enhanced antileishmanial efficacy and lymphocyte proliferation when given in combination with paromomycin and miltefosine. Increased toxic oxygen metabolite generation and phagocytosis were also witnessed.

Gupta *et al.* (2005) tried to examine the possibility of minimizing long course treatment, long half life and side effects of miltefosine by applying miltefosine in lower doses in combination with picroliv, against *L. donovani* in hamsters (*M. auratus*). The picroliv showed no antileishmanaial potential. However, when given with suboptimal dose of miltefosine, it enhanced efficacy of the latter from 45 to 86 per cent on day 7 post treatments and from 32 to 64 per cent on day 28 post treatment.

Mittal *et al.* (1998) studied the anti-leishmanial efficacy of picroliv (12.5 mg/kg, po × 7 days) alone and in combination with sodium stibogluconate on parasitemia, lipid peroxidation and hepatic marker enzymes of golden hamsters during *L. donovani* infection. The results indicated that picroliv in terms of biochemical markers; have a significant antileishmanial activity implying that it can be utilized as an adjunct to chemotherapy or in combination therapy of kala azar along with sodium stibogluconate.

Antiviral

Picroliv and its major components picroside I, catalpol, kutkoside I, kutkoside, andrographolide, silymarin and *Phyllanthus niruri* extract were tested for the presence of anti hepatitis B virus surface antigen (anti HBs) like activity (Mehrotra *et al.*, 1990). HBsAg positive serum samples obtained from hepatitis B virus (HBV) associated acute and chronic liver diseases and healthy HBsAg carriers were used to evaluate the anti-HBs like activity of compounds/extract. The latter were mixed with serum samples and incubated at 37° overnight followed by HBsAg screening in the Elisa system. A promising anti-HBsAg like activity was noted in picroliv and catalpol, *P. niruri* which differed from the classical viral neutralization. Picroliv also inhibited purified HBV antigens (HBsAg and HBsAg) prepared from healthy HBsAg carriers.

Anticancer

Harikumar *et al.* (2008) evaluated the inhibitory effect of three non-toxic antiviral natural compounds namely berberine, curcumin or picroliv against FMuLv induced erythroleukemia in BALB/c mice. To understand the effect of these compounds in the initiation and progression of leukemia they performed a series of analysis, which include hematological and biochemical parameters, histopathological evaluations of the liver and the spleen and expression analysis of selected genes such as *Bcl-2*, p53, p45NFE2, Raf-1, Erk-1, IFNγ receptor and erythropoietin in spleen. The treatment with berberine, curcumin or picroliv were found to elevate the life span of leukemia harboring animals by more than 60 days, decreased the anemic condition which was highly prevalent in FMuLv alone treated group. Histopathological evaluations showed that the compounds tested here inhibited the massive leukemic cell infiltrations to sinusoidal spaces in spleen, decrease the expression of Bcl-2, Raf-1, Erk-1 IFNγ receptor and erythropoietin induce the expression of p53.

Anand *et al.* (2008) postulated that picroliv interfere with nuclear factor-κB (NF-κB) activation by inhibiting the activation of NF-κB-mediated signal cascade. Electrophoretic mobility shift assay showed that pretreatment with picroliv abrogated tumor necrosis factor (TNF)-induced activation of NF-κB. The glycoside also inhibited NF-κB activated by carcinogenic and inflammatory agents, such as cigarette smoke condensate, phorbol-12-myristate-13-acetate, okadaic acid, hydrogen peroxide, lipopolysaccharide, and epidermal growth factor. Picroliv inhibited activation of IκBalpha kinase, leading to inhibition of phosphorylation and degradation of IκBalpha. It also inhibited phosphorylation and nuclear translocation of p65. Studies revealed that picroliv directly inhibits the binding of p65 to DNA, which was reversed by the treatment with reducing agents, suggesting a role for a cysteine residue in interaction with picroliv. Mutation of Cys(38) in p65 to serine abolished this effect of picroliv. NF-κB inhibition by picroliv leads to suppression of NF-κB-regulated proteins, including those linked with cell survival (inhibitor of apoptosis protein 1, Bcl-2, Bcl-xL, survivin, and TNF receptor-associated factor 2), proliferation (cyclin D_1 and cyclooxygenase-2), angiogenesis (vascular endothelial growth factor), and invasion (intercellular adhesion molecule-1 and matrix metalloproteinase-9). Suppression of these proteins enhanced apoptosis induced by TNF.

The effect of picroliv treatment on the carcinogenic response and, hepatic and renal antioxidant enzymes of rats administered with 1, 2-dimethylhydrazine hydrochloride (DMH) was studied by Rajeshkumar and Kuttan, (2003) in male Sprague-Dawley rats. DMH-induced hepatic carcinogenic response and necrosis were inhibited by oral administration of picroliv (40 and 200 mg/kg). Liver gamma-glutamyl transpeptidase, elevated by DMH administration, was found to be reduced by picroliv treatment at 40 and 200 mg/kg, dose. Elevated number of Argyrophilic Nucleolar Organizer Region dots and clusters, an index of proliferation, of DMH treated rat liver was reduced by picroliv treatment. DMH-induced depletion of hepatic and renal antioxidant enzymes such as catalase and superoxide dismutase levels was also restored to normal by picroliv treatment. Picroliv treatment reduced the DMH-induced elevation of lipidperoxidation in liver, kidney and serum. Elevated levels of serum total bilirubin by DMH administration was reduced by picroliv treatment. Depleted renal glutathione-*S*-transferase and hepatic glutathione levels after DMH administered rats were found to be significantly increased by picroliv treatment. Histological analysis of the DMH administered rat liver showed hepatic cell necrosis, coalescent nodular areas and cystic hyperplasia of the bile ducts with inflammation. Picroliv treated liver resembled normal liver except the presence of a few degenerating cells. Renal anatomy was not altered by DMH administration.

Cancer chemoprevention of chemically induced tumours by picroliv was studied by Rajeshkumar and Kuttan, (2001) on 20-methylcholanthrene (20-MC)-induced sarcoma model and 7, 12-dimethylbenz[a]anthracene (DMBA)-initiated papilloma formation in BALB/c mice. Administration of picroliv (100 and 200 mg/kg, po) inhibited the sarcoma development by 47 and 53 per cent as estimated on day 200 after 20-MC administration 60 and 66 per cent of the animals survived in the picroliv treated group, 100 and 200 mg/kg, respectively. Picroliv exhibited anti-tumour promoting activity on a two stage carcinogenesis test on mouse skin using DMBA as an initiator and croton oil as a promoter. Topical application of picroliv (1 and 5 mg/mouse) 30 min prior to that of croton oil application resulted in a 50 per cent and 60 per cent reduction in the number of animals that developed papilloma's, and 48 per cent and 64 per cent reduction in the number of papilloma's per mouse. There was also a delay in the onset of first skin tumour in the group of animals treated with picroliv. Oral administration of picroliv (150 mg/kg, po) prior to DMBA application delayed the onset of papilloma and the percent of mice (60 per cent) with tumours indicates that picroliv inhibited the tumour initiation induced by DMBA. Picroliv administration was also found to increase the life span of transplanted Dalton's Lymphoma Ascites and Ehrlich Ascites Carcinoma harboring mice and reduced the volume of transplanted solid tumours.

Picroliv was found to be an effective inhibitor of hepatocarcinogenesis induced by N-nitrosodiethylamine (NDEA) in rats. Animals administered with NDEA had large hepatic nodules and the liver weight was increased. Picroliv administration (200 mg/kg) reduced the liver weight. Oral administration of picroliv reduced NDEA-induced elevation of γ-glutamyltranspeptidase in serum and liver to that of normal rats. Moreover, elevated levels of bilirubin, ALP, SGPT and serum peroxides were also found to be significantly reduced by picroliv administration. Similar observations were noticed in glutathione and glutathione-*S*-transferase levels. Histopathological analysis of the picroliv treated rat liver resembles that of a normal liver except for a few alterations such as hepatocytomegalia and karyomegalia in some foci (Rajeshkumar and Kuttan, 2000).

Anti-cholestatic

The effect of picroliv on hepatic microsomal mixed-function oxidases (MFO) and glutathione conjugating enzyme system in cholestatic rats was studied by Dhuley (2005). Dhuley (2005) found that bile duct ligation in male rats for one weeks caused significant increase in both serum sorbitol dehydrogenase activity and serum bile acid concentration indicating cholestatic liver injury. Furthermore, a rise in the hepatic hydroxyproline level indicating collagen accumulation was observed, as a result of these alterations, the hepatic microsomal MFO system was imparied as evidenced by a decrease in cytochrome P_{450} system content and in the activities of NADPH-cytochrome C reductase and aminopyrine demethylase. While the hepatic glutathione content remained unaffected, the cytosolic glutathione-*S*-transferase activity was clearly suppressed due to subchronic cholestasis. Oral administration of picroliv (25 mg/kg/day × 21 days) in bile ligation induced cholestatic rats; significantly prevented the biochemical changes induced in liver and serum of cholestatic rats.

Picroliv showed a dose (3-12 mg/kg, po × 7 days) dependent choleretic activity as evidenced by increase in bile flow and its contents (bile salts and bile acids). Significant anticholestatic activity was also observed against CCl_4 induced cholestasis in conscious rat, anaesthetized guinea pig and cat (Saraswat *et al.*, 1993).

Shukla *et al.*, 1991 studied the choleretic effect of picroliv showed a dose-dependent (1.5-12 mg/kg × 7 days) choleretic effect in conscious rats and anaesthetised Guinea pigs. It also possessed a marked anticholestatic effect against paracetamol and ethynylestradiol-induced cholestasis. It antagonized the changes in bile volume as well as the contents (bile salts and bile acids).

Immunomodulatory

Cadmium induces apoptosis in murine lymphocytes and alters the immune functions. Thus, for the amelioration of its effect, Pathak and Khandelwal (2008), selected three structurally different bioactive herbal extracts such as piperine-alkaloid, picroliv-glycosides and curcumin-polyphenols for ascertaining their immunomodulatory role various biochemical indices of cell damage such as cytotoxicity, oxidative stress (ROS, GSH), apoptosis (mitochondrial membrane potential, caspase-3 activity, phosphatidylserine externalization, apoptotic DNA) along with lymphocyte phenotyping, blastogenesis and cytokine secretion were assessed in thymic and splenic cell suspensions. Picroliv was found to be moderately active in increasing cell viability.

Dwivedi *et al.* (2008) evaluated the potential of combination therapy comprising of immunomodulator picroliv and antimalarial chloroquine against drug resistant *P. yoelii* infection in BALB/c mice. The immunomodulatory potential of picroliv was established by immunizing animals with model antigen along with picroliv. Immune response was assessed using T cell proliferation assay and also by determining the antibody isotype-profile induced in the immunized mice. In the next set of experiment, prophylactic potential of picroliv to strengthen antimalarial properties of chloroquine against *P. yoelii* (MDR) infection in BALB/c mice was assessed. T cell proliferation as well as antibody production study reveals that picroliv helps in evoking strong immuno potentiating response against model antigen in the immunized mice. Co-administration of picroliv enhances efficacy of chloroquine against experimental murine malaria.

Sinha *et al.* (1998) studied the effect of picroliv on proliferative T cell response to the mycobacterial purified protein derivative (PPD) antigen in subjects infected with or exposed to mycobacteria. Co-culture of their peripheral blood mononuclear cells with the optimal concentration of picroliv (0.5 μg/ml) significantly enhanced the proliferative response to 1/10 optimal PPD dose, as determined by [3H] thymidine incorporation, in the group of low responders. The response to PPD of cells from high responders and to PHA (phytohaemagglutinin, a non-specific T cell mitogen) remained unaffected by picroliv which did also not induce cell proliferation on its own.

Puri *et al.* (1992) investigated immunostimulant activity of picroliv, and its protective action against *L. donovani* infection in hamsters-1. Oral administration of picroliv (10 mg/kg × 7 days) in mice prior to immunization with sheep red blood cells (SRBC) resulted in a significant increase in haemagglutinating antibody titre, plaque forming cells, and delayed hypersensitivity response to SRBC. Picroliv enhanced the non-specific immune response characterized by an increase in macrophage migration index, [(14) C]-glucosamine uptake, phagocytosis of [(14) C]-leucine labelled *E. coli*, chemiluminescence of peritoneal macrophages, and higher uptake of [(3) H]-thymidine in the lymphocytes of treated mice. It also induced a high degree of protection in golden hamsters against challenge infection with *L. donovani* promastigotes.

Testicular Protective

Ameliorative potential of picroliv on Cd induced early and advanced testicular damage was investigated by Yadav and Khandelwal (2008) in male rats. In the first protocol, the rats were administered Cd as $CdCl_2$ (0.5mg/kg, sc) 5days/week for 18 weeks and picroliv at two doses (6 and 12 mg/kg, po) was given for the last 4 weeks to the Cd administered group. In the second protocol, the Cd administration continued for 24 weeks and picroliv was given from week 21 to 24. At 18 weeks, Cd caused alterations in oxidative stress indices like increased lipid peroxidation and reduced levels of non protein sulphydryls. They were found close to the control values by picroliv treatment, suggesting its antioxidant potential. The increased levels of Zn and Ca were reduced by picroliv, the Cd levels remained unaltered. The Cd induced testicular damage was also mitigated by picroliv. The higher dose (12 mg/kg) being more effective than the lower dose. However, at 24 weeks of Cd exposure, the oxidative stress indicators in testis were more pronounced along with the morphological alterations. On comparative evaluation of the two studies, 18 weeks Cd exposure caused moderate testicular damage, which could be reversed significantly by picroliv administration and correlated well with oxidative stress markers.

Yadav *et al.* (2005) evaluated the potential of picroliv against acute Cd intoxication in male rats. Biochemical and histopathological profile in rats pretreated with picroliv (12 mg/kg, po) followed by a single dose of cadmium chloride (3 mg/kg, ip) revealed marked suppression of oxidative stress in testes. The Cd-induced enhanced levels of lipid peroxidation, membrane fluidity and reduced levels of nonprotein sulphydryls and Na^+K^+ATPase were significantly restored to near normal by picroliv pretreatment. In addition, the Cd-induced serum levels of SGOT, SGPT, γ-glutamyl transpeptidase and lactate dehydrogenase were restored to near basal levels. Testicular histopathological damage was also minimized.

Anti-hyperlipidemic

Vivekanandan *et al.* (2007) studied the protective effect of picroliv against hydrazine (Hz)-induced hyperlipidemia in rats. Hz administration (50 mg/kg, i.p.) caused an increase in triglyceride (TG), cholesterol, free fatty acids, and total lipids (TL) in both the plasma and liver tissue of rats accompanied by a fall in phospholipids (PL) in the liver tissue 24 h after its administration, indicating its hyperlipidemic property. These abnormalities can prevent by simultaneous treatment of picroliv (50 mg/kg, po). Hz treatment caused an increase in the mobility of TG and TL from adipose tissue, and these results indicate that Hz administration could cause hepatic steatosis by nonhepatocellular factors (such as mobilization of depot fats). This effect was also prevented by simultaneous treatment of picroliv with Hz. Picroliv alone treatment, however, did not produce any change in the status of all the lipid parameters evaluated in plasma, liver, and adipose tissues. These results indicate that increased mobilization of depot fats from adipose tissue may contribute to the development of hepatic steatosis in addition to decreased lipoprotein secretion, increased hepatic TG biosynthesis, and increased hepatic uptake of free fatty acid.

Singh *et al.* (1992a) studied the effect of picroliv on low density lipoprotein receptor binding of rat hepatocytes in hepatic damage induced by paracetamol. Study showed reversal of low density lipoprotein (LDL) binding to paracetamol-induced damaged hepatocytes of rats. Changes in levels of glutamic oxaloacetic transaminase, glutamic pyruvic transaminase, ALP, conjugated dienes and lipids of hepatocytes were significantly prevented by picroliv at different doses. The effect of picroliv on enzyme levels, LDL receptor binding and lipids in damaged hepatocytes was found to be comparable to silymarin, a known hepatoprotective agent.

Wound Healing

Singh *et al.* (2007) investigated the effect of picroliv in an *ex vivo* rat aorta ring model of angiogenesis. Picroliv enhanced the sprouting and migration of endothelial cells. They also investigated punch wound healing on days 4 and 7 after wounding by histology, morphometry and collagenization. The data showed improved re-epithelialization, neovascularization and migration of various cells such as endothelial, dermal myofibroblasts and fibroblasts into the wound bed after picroliv treatment. Immunohistochemical localization showed increased *vascular endothelial* growth factor and alpha smooth muscle actin staining consistent with an increased number of microvessels in granulation tissue.

Anti-stress

Seth *et al.* (2003) studied the effect of picroliv on hepatocellular injury and redox status, as well as its possible mechanism of action in a hemorrhage-resuscitation (H/R) model in adult rats. Anesthetized rats were subjected to hemorrhagic shock by bleeding 30 mL/kg body weight. After 60 min of shock, rats were resuscitated with twice the shed blood volume of lactated Ringer's solution and were sacrificed 2 h after resuscitation. They observed that picroliv (12 mg/kg) pretreatment, given orally for 7 days, resulted in a significant decrease in serum aspartate transaminase and γ-glutamyl transpeptidase levels. Picroliv also inhibited the lipid peroxidation and nitric oxide release that occurred after H/R and altered the activity of glutathione reductase in a favorable manner, thereby suggesting better antioxidant status. Picroliv significantly down-regulated the stress sensitive transcription factor AP1 and decreased the level of *c-fos* mRNA as well as *c-jun* and *c-fos* proteins in liver tissue, indicating that its actions could be mediated through AP1 and associated signal transduction pathways.

Gaddipati *et al.* (1999) studied the modulation of IGF-I, IGF-II and IGF-IR in hypoxia and the ability of picroliv to modify their expression *in vivo*. Male Sprague Dawley rats, placed in 10 per cent oxygen for 4 days, were sacrificed, and the expression of IGF-I, IGF-II and IGF-IR was determined by immune histochemistry, *in situ* hybridization and reverse transcriptase polymerase chain reaction (RT-PCR) in brain, liver and lung. One group of animals was pretreated with picroliv and the other served as control. IGF-I and IGF-IR were expressed in distinct regions of the brain but not in liver or lung. IGF-I was mainly expressed in the hippocampus and cerebellum, whereas IGF-IR expression was also observed in the cortex. A significant reduction in the messenger RNA (mRNA) level of these genes was observed in response to hypoxia. They found that picroliv pretreatment not only prevented such down regulation but more importantly resulted in increased levels of IGF-I and IGF-IR. These observations correlated with reduced neuronal cell death observed in these animals. The mRNA of IGF-II was constitutively expressed and was not altered by hypoxia.

Seth *et al.* (2000) investigated the efficacy of picroliv in an *in vivo* model of renal ischemia-reperfusion injury in rats at a dose of 12 mg/kg orally for 7 days. The animals were killed at various times after reperfusion. Increased lipid peroxidation and apoptotic cell number reflected the oxidative damage following renal ischemia-reperfusion injury. Picroliv pretreated rats exhibited lower lipid peroxidation, improved antioxidant status, and reduced apoptosis, indicating better viability of renal cells. Immunohistochemical studies revealed that picroliv pretreatment attenuated the expression of intercellular adhesion molecule-1 in the glomerular region.

Gaddipati *et al.* (1999b) evaluated the protective effect of picroliv against hypoxic injury by examining lactate dehydrogenase (LDH) release in Hep 3B and glioma cells. The expression of hypoxia regulated genes, vascular endothelial growth factor (VEGF) and hypoxia inducible factor (HIF-1) was studied in human umbilical vein endothelial cells (HUVEC), Hep 3B and glioma cells. Picroliv reduced the cellular damage caused by hypoxia as revealed by a significant reduction in LDH release compared to untreated control. The expression of VEGF and HIF-1 subunits (HIF-1α and HIF-1β) was enhanced by treatment with picroliv during normoxia and hypoxia in HUVEC and Hep 3B cells and on reoxygenation the expression of these genes was significantly reduced as revealed by mRNA analysis using RT-PCR. Simultaneous treatment with picroliv during hypoxia inhibited VEGF and HIF-1 expression in glioma cells whereas the expression was not reduced by picroliv treatment during reoxygenation as evidenced by both RT-PCR and northern hybridization. VEGF expression as revealed by immunofluorescence studies correlates well with the regulations observed in the mRNA expression. They have also examined the kinase activity of tyrosine phosphorylated proteins and protein kinase C (PKC) in glioma cells treated with picroliv during hypoxia/reoxygenation. A selective inhibition of protein tyrosine kinase activity leading to tyrosine dephosphorylation of several proteins including 80 kd protein, and a reduction in PKC was seen in cells treated with picroliv and hypoxia. These findings suggest that picroliv may act as a protective agent against hypoxia/reoxygenation induced injuries, and the underlying mechanism may involve a novel signal transduction pathway.

Anti-allergic

Picroliv at a dose of 25 mg/kg, po inhibited passive cutaneous anaphylaxis in mice (82 per cent) and rats (50-85 per cent) and protected mast cells from degranulation (60-80 per cent) in a concentration-dependent manner. Its effect was also studied in sensitized Guinea pig ileum preparation *in vitro* (Schultz-Dale study) and in normal guinea pigs *in vivo* (Konzett-Rossler, in preparation). There was inhibition of the Schultz-Dale response in sensitized Guinea pig ileum, but the bronchospasm induced by histamine could not be antagonized or prevented by picroliv, indicating the absence of a direct post-synaptic histamine receptor blocking activity (Baruah *et al.*, 1998).

Discussion

Liver disorder may result from a wide variety of insults including infections, drugs, toxins, ischemia and autoimmune disorders. Picroliv has been shown to protect the liver cells from a wide variety of insults *viz.*, amantina poisoning, CCl_4, galactosamine, ethanol, aflatoxin B1, acetoaminophen, thiactamide, oxytetracycline and monocrotaline in both *in vivo* and *in vitro* experiments. The injury caused by toxicants disrupts the hepatocellular plasma membrane and release the enzymes normally present in the cytosol into blood stream. The liver injury due to toxins could result in defective excretion of marker enzyme (alanine transaminase, aspartate transaminase, ALP, γ-glutamyl transpeptidase, lactate dehydrogenase and bile) reflected by their increase levels in serum (Singh *et al.*, 2005; Visen *et al.*, 1998; Rastogi *et al.*, 1996; Dwivedi *et al.*, 1991a; 1992a; 1993b). Picroliv restored the serum marker enzyme values to normal level in different literature reports showing potential hepatoprotective activity. The protective effects may be the result of stabilization of plasma membrane there by preserving the structural integrity of cells as well as due to the repair of hepatic tissue damage caused by toxicants. Histopathological profile of the toxicant treated animals liver shows ballooned hepatocytes, granular cytoplasm, zonal necrosis and fatty degeneration with infiltration of inflammatory mediators. The animals treated with picroliv showed significant improvement in liver histopathology against the injury caused by inducing agent (Dwivedi *et al.*, 1991a; 1993b). Pretreatment with picroliv have added advantage as a hepatoprotective, as it promotes the assembly of ribosomes on endoplasmic reticulum to facilitate the protein synthesis and regenerative activity (Singh *et al.*, 1992b).

Another factor responsible for hepatoprotective potential of picroliv may be its antioxidant property, which may contribute by reducing lipid peroxidation and free radical damage. Chander *et al.* (1992b) found that picroliv inhibits the non-enzymatic generation of O^{2-} anions in phenazine methosulphate-NADH system, inhibited oxidative malonaldehyde generation by both the ascorbate- Fe^{2+} and NADPH-ADP-Fe^{2+} systems and scavenged superoxide (O^{2-}) anions generated in a xanthine-xanthine oxidase system. Glutathione is vital in maintaining a variety of intacellular function, including detoxification, oxidation, tertiary protein configuration and redox balance. Picroliv was found to restore depleted glutathione levels including glutathione-*S*-transferase, glutathione reductase and glutathione peroxidase (Rastogi *et al.*, 2001b). The structure activity relationship of picroliv revealed that the antioxidant potential of picroliv could be related to electophilic free radical scavenging nature of its iridoid glycoside (Girish *et al.*, 2009).

According to the World Health Organization, the population of nearly 88 countries is threatened by leishmaniasis and about 350 million people are at risk of the disease. The prevalence of leishmaniasis is 12 million with a rate of 2 million cases annually (Ashfordet *et al.*, 1992). The control of leishmaniasis is problematic because no effective vaccine is available. Moreover, the antileishmanial drugs, which include pentavalent antimonials such as sodium stibogluconate, N-methylglucamine or amphotericin B, have varying level of effectiveness, with their own side effects profile and also have associated high costs. The development of resistance to the medications by the leishmania parasite is also a matter of concern (Berman *et al.*, 1997; Modabber *et al.*, 1993). Treatment failure especially in patients with kala-azar and mucosal leishmaniasis is a common problem in many areas of endemicity (Sereno *et al.*, 1997). There is therefore, a need to develop newer drug therapies that are more efficacious and safer. Here, *P. kurroa* is noted to be a good source of bioactive agents which are useful against leishmaniasis. Recent findings revealed that picroliv in combination with antifungal and antileishmanial drug enhances the anti-leishmanial activity by enhancing cell mediated immunity (Shakya *et al.*, 2011a,b).

It is well known that one of the most important problems faced with anticancer drugs is their severe toxicity. This is where herbal medicine has a definite advantage over conventional chemotherapy. Picroliv inhibits carcinogenic action of 1, 2-dimethylhydrazine hydrochloride, 20 methylcholanthrene and N-nitrosodiethylamine. Picroliv signicantly decreased elevated number of Argyrophilic Nucleolar Organizer Region dots and clusters, an index of proliferation, gamma-glutamyl transpeptidase, ALP, SGPT and serum peroxidase. Picroliv also normalize hepatic cell necrosis, coalescent nodular areas and cystic hyperplasia of the bile ducts caused by the carcinogenic agents (Anand *et al.*, 2008; Rajeshkumar and Kuttan, 2000, 2001, 2003).

The host's immune system plays an important role for the prophylactic effect and control of various diseases. A number of agents have been explored with promising immunostimulatory activity, but relatively few studies have been conducted on the clinical applications, indicating the bioavailability to eliminate toxic manifestation of the potent compounds to make them clinically acceptable. In recent years, there has been an increase in the clinical use of indigenous drugs. Among the various bioactive identified, picroliv has been reported as one of the promising immunostimulant agent. Studied revealed that picroliv helps in evoking strong immuno potentiating response against model antigen in the immunized mice. Co-administration of picroliv with chloroquine enhances efficacy of chloroquine against experimental murine malaria (Dwivedi *et al.*, 2008). Puri *et al.* (1992) investigated that picroliv enhanced the non-specific immune response characterized by an increase in macrophage migration index, [(14) C]-glucosamine uptake, phagocytosis of [(14) C]-leucine labelled *E. coli*, chemiluminescence of peritoneal macrophages, and higher uptake of [(3) H]-thymidine in the lymphocytes of treated mice.

Cd is one of the main environmental pollutants of metallic origin. Exposure to Cd reduces male fertility in both human beings and rodents and as low as 0.5 to 3 mg/kg body weight exposure can cause testicular damage without pathological changes to other organs. Most of the toxic effects of Cd are related to its ability to generate free radicals at a rate high enough to overwhelm the natural antioxidant defense system of the body. Picroliv was reported to exhibit antioxidant and testicular protective activity that was attributed to increased androgen production. The significant protective effect picroliv can be attributed to inhibition of oxygen free radicals generation, thereby preventing testicular tissue peroxidation. Picroliv administration to Cd-treated rats showed a remarkable protective effect in all aspects as evident from significant recovery from histological alterations, tissue peroxidation, and Cd accumulation along with elevated antioxidant capacity and functional enzymes activity (Yadav and Khandelwal, 2008; Yadav *et al.,* 2005).

Cardio vascular diseases are leading cause of death in both industrialized and developing nations. Disorders of lipid metabolism, following oxidative stress are the prime risk factors for initiation and progression of these diseases (Sharma and Garg, 2009). The current anti-hyperlipidemic therapy includes principally statins and fibrates, the former corrects the altered blood lipid profile by inhibiting the bio-synthesis of cholesterol and the latter acts by enhancing the clearance of triglyceride rich lipo-proteins. The investigation of picroliv will be a useful strategy in the discovery of new lead molecule as anti-hyperlipidemic which eliciting improved activity by increasing the mobilization of depot fats from adipose tissue may contribute to the development of hepatic steatosis in addition to decreased lipoprotein secretion, increased hepatic TG biosynthesis, and increased hepatic uptake of free fatty acid (Vivekanandan *et al.,* 2007; Singh *et al.,* 1992a).

Conclusion

Pharmacological activity of the presumed herbal bioactive in general does not reflect reality. Only in a few cases drug activity depend upon single component. Generally, it is the result of concerted activity of several active compounds as in picroliv containing picroside I, kutkoside and mixture of cucurbitacins. Their may be possibility that the pharmacological activity exerted by picroliv may be due to concerted activity of these bioactive compounds. Considering the above facts, there is still need to elucidate, the structure of picroliv and come to the point that which bioactive is actually responsible for which activity individually as well as in combination. The extensive literature survey revealed that picroliv, is a potential bioactive compounds with diverse promising pharmacological activity spectrum. Picroliv is excellent hepatoprotective with reported anti-hepatotoxic effect against a wide variety of liver insults. Its noteworthy anti-oxidant, anti-leishmanial and immunostimulant activity should be explored in details to unfold the mechanism underlined and also to serve the community with an effective phytochemical. Picroliv can be used as a lead molecule in search of new bioactive synthetic compound by structural modifications or substitutions. The pharmacological action of picroliv can be improved by preparing its derivatives or using it in combination which may show efficacy in various experimental pharmacological models. Picroliv in combination with other bioactive molecules may be developed as future drug for use in clinical practice for liver diseases, immuno-stimulant and leishmanial infections. The recent research trends showed that the scientists are exploring therapeutic potential of picroliv in combination with other phytochemicals and synthetic drugs against liver damage and leishmaniasis.

References

Anand, P., Kunnumakkara, A.B., Harikumar, K.B., Ahn, K.S., Badmaev, V., and Aggarwal, B.B. (2008). Modification of cysteine residue in p65 subunit of nuclear factor-κB (NF-κB) by picroliv suppresses NF-κB-regulated gene products and potentiates apoptosis. *Cancer Research*, **68** (21): 8861-8870.

Ansari, R.A., Tripathi, S.C., Patnaik, G.K., and Dhawan, B.N. (1991). Antihepatotoxic properties of picroliv: an active fraction from rhizomes of *Picrorhiza kurrooa. Journal of Ethnopharmacology*, **34** (1): 61-68.

Ashford, R.W., Desjour, P., and De Raadt, P. (1992). Estimation of population at risk of infection and number of cases of leishmaniasis. *Parasitology Today*, **8**: 104-105.

Baruah, C.C., Gupta, P.P., Nath, A., Patnaik, L.G., and Ramesh, B.N. (1998). Anti-allergic and anti-anaphylactic activity of picroliv-α standardized iridoid glycoside fraction of *Pichrorhiza kurroa. Pharmacology Research*, **38** (6): 487-492.

Berman, J.D. (1997). Human leishmaniasis: clinical, diagnostic, and chemotherapeutic developments in the last 10 years. *Clinical Infectious. Diseases*, **24**: 684-703.

Bigoniya, P., Shukla A., and Singh, C.S. (2010). Evaluation of hepatic microsomal enzyme functional integrity on picroliv pretreatment against CCl_4 induced hepatotoxicity. *International Journal of Pharmacology*, **6**: 200-207.

Chander, R., Dwivedi, Y., Rastogi, R., Sharma, S.K., Garg, N.K., Kapoor, N.K., and Dhawan, B.N. (1990). Evaluation of hepatoprotective activity of picroliv (from *Picrorhiza kurroa*) in *Mastomys natalensis* infected with *Plasmodium berghei. Indian Journal of Medical Research*. **92**: 34-37.

Chander, R., Kapoor, N.K., and Dhawan, B.N. (1992a). Effect of picroliv on glutathione metabolism in liver and brain of *Mastomys natalensis* infected with *Plasmodium berghei. Indian Journal of Experimental Biology*, **30** (8): 711-714.

Chander, R., Kapoor, N.K., and Dhawan, B.N. (1992b). Picroliv, picroside-I and kutkoside from *Picrorhiza kurroa* are scavengers of superoxide anions. *Biochemical Pharmacology*, **44** (1): 180-183.

Chander, R., Kapoor, N.K., and Dhawan, B.N. (1994). Picroliv affects gamma-glutamyl cycle in liver and brain of *Mastomys natalensis* infected with *Plasmodium berghei. Indian Journal of Experimental Biology*, **32** (5): 324-327.

Chander, R., Singh, K., Visen, P.K., Kapoor, N.K., and Dhawan, B.N. (1998). Picroliv prevents oxidation in serum lipoprotein lipids of *Mastomys coucha* infected with *Plasmodium berghei. Indian Journal of Experimental Biology*, **36** (4): 371-374.

Dhuley, J.N. (2005). Effect of picroliv administration on hepatic microsomal mixed function oxidases and glutathione-conjugating enzyme system in cholestatic rats. *Hindustan Antibiotic Bulletin*, **47-48** (1-4): 13-19.

Dwivedi, V., Khan, A., Vasco, A., Fatima, N., Soni, V.K., Dangi, A., Misra-Bhattacharya, S., and Owais, M. (2008). Immunomodulator effect of picroliv and its potential in treatment against resistant *Plasmodium yoelii* (MDR) infection in mice. *Pharmaceutical Research*, **25** (10): 2312-2319.

Dwivedi, Y., Rastogi, R., Chander, R., Sharma, S.K., Kapoor, N.K., Garg, N.K., and Dhawan, B.N. (1990). Hepatoprotective activity of picroliv against carbon tetrachloride-induced liver damage in rats. *Indian Journal Medical Research*, **92**: 195-200.

Dwivedi, Y., Rastogi, R., Garg, N.K., and Dhawan, B.N. (1992a). Picroliv and its components kutkoside and picroside I protect liver against galactosamine-induced damage in rats. *Pharmacology and Toxicology*, **71** (5): 383-387.

Dwivedi, Y., Rastogi, R., Garg, N.K., and Dhawan, B.N. (1992b). Effects of picroliv, the active principle of *Picrorhiza kurroa*, on biochemical changes in rat liver poisoned by *Amanita phalloides*. *Zhongguo Yao Li Xue Bao*, **13** (3): 197-200.

Dwivedi, Y., Rastogi, R., Garg, N.K., and Dhawan, B.N. (1993a). Perfusion with picroliv reverses biochemical changes induced in livers of rats toxicated with galactosamine or thioacetamide. *Planta Medica*, **59** (5): 418-420.

Dwivedi, Y., Rastogi, R., Mehrotra, R., Garg, N.K., and Dhawan, B.N. (1993b). Picroliv protects against aflatoxin B1 acute hepatotoxicity in rats. *Pharmacological Research*, **27** (2): 189-199.

Dwivedi, Y., Rastogi, R., Sharma, S.K., Mehrotra, R., Garg, N.K., and Dhawan, B.N. (1991a). Picroliv protects against monocrotaline-induced hepatic damage in rats. *Pharmacological Research*, **23** (4): 399-407.

Dwivedi, Y., Rastogi, R., Sharma, S.K., Garg, N.K., and Dhawan, B.N. (1991b). Picroliv affords protection against thioacetamide-induced hepatic damage in rats. *Planta Medica*, **57** (1): 25-28.

Floersheim, G.L., Bieri, A., Koenig, R., and Pletscher, A. (1990). Protection against *Amanita phalloides* by the iridoid glycoside mixture of *Picrorhiza kurroa* (kutkin). *Agents Actions*, **29** (3-4): 386-387.

Gaddipati, J.P., Madhavan, S., Sidhu, G.S., Singh, A.K., Seth, P., and Maheshwari, R.K. (1999b). Picroliv–α natural product protects cells and regulates the gene expression during hypoxia/reoxygenation. *Molecular and Cellular Biochemistry*, **194** (1-2): 271-81.

Gaddipati, J.P., Mani, H., Banaudha, K.K., Sharma, S.K., Kulshreshtha, D.K., and Maheshwari, R.K. (1999a). Picroliv modulates the expression of insulin-like growth factor (IGF)-I, IGF-II and IGF-I receptor during hypoxia in rats. *Cellular and Molecular Life Sciences*, **56** (3-4): 348-355.

Girish, C., Koner, B.C., Jayanthi, S., Ramachandra, R.K., Rajesh, B., and Pradhan, S.C. (2009). Hepatoprotective activity of picroliv, curcumin and ellagic acid compared to silymarin on paracetamol induced liver toxicity in mice. *Fundamental and Clinical Pharmacology*, **23** (6): 735-745.

Gupta, S., Ramesh, S.C., and Srivastava, V.M. (2005). Efficacy of picroliv in combination with miltefosine, an orally effective antileishmanial drug against experimental visceral leishmaniasis. *Acta Tropica*, **94** (1): 41-47.

Halliwel, B., and Gutteridge, M.C. (1999). Free radical in biology and medicine. *In:* The Study of Generalized Light Emission, Edn. 3rd, University Press, Oxford, pp. 387.

Harikumar, K.B., Kuttan, G., and Kuttan, R. (2008). Inhibition of progression of erythroleukemia induced by Friend virus in BALB/c mice by natural products-berberine, curcumin and picroliv. *Journal of Experimental Therapeutics and Oncology*, **7** (4): 275-84.

IP. (2007). Indian Pharmacopoeia, Government of India. Ministry of health and family welfare, India.

Kitagawa, I., Hino, K., Nishimura, T., Iwata, E., and Yosioka, I. (1971). On the constituents of *Picrorhiza kurroa*. (1) The structure of picroside I, a bitter principle of the subterranean part. *Chemical and Pharmaceutical Bulletin,* **19** (12): 2534-2544.

Kumar, V., Mehrotra, N., Lal, J., and Gupta R.C. (2004). Pattern profiling of the herbal preparation picroliv using liquid chromatography –tandon mass spectrometry. *Journal of Chromatography*, **1045** (1-2): 145-152.

Mehrotra, R., Rawat, S., Kulshreshtha, D.K., Patnaik, G.K., and Dhawan, B.N. (1990). *In vitro* studies on the effect of certain natural products against hepatitis B virus. *Indian Journal of Medicinal Research*, **92**: 133-138.

Mittal, N., Gupta, N., Saksena, S., Goyal, N., Roy, U., and Rastogi, A.K. (1998). Protective effect of picroliv from *Picrorhiza kurroa* against *Leishmania donovani* infections in *Mesocricetus auratus*. *Life Sciences*, **63** (20): 1823-1834.

Modabber, F. (1993). Chagas'disease, In: Tropical Disease Research, World Health Organisation, Geneva, Ninth Programme, Report of the UNDP/World Bank/WHO.

Pathak, N., and Khandelwal, S. (2008). Comparative efficacy of piperine, curcumin and picroliv against Cd immunotoxicity in mice. *Biometals*, **21** (6): 649-661.

Puri, A., Saxena, R.P., Guru, P.Y., Kulshreshtha, D.K., Saxena, K.C., and Dhawan, B.N. (1992). Immunostimulant activity of picroliv, the iridoid glycoside fraction of *Picrorhiza kurroa*, and its protective action against *Leishmania donovani* infection in Hamsters1. *Planta Medica*, **58** (6): 528-532.

Rajeshkumar, N.V., and Kuttan, R. (2001). Protective effect of picroliv, the active constituent of *Picrorhiza kurroa*, against chemical carcinogenesis in mice. *Teratog Carcinog Mutagen*, **21** (4): 303-313.

Rajeshkumar, N.V., and Kuttan, R. (2003). Modulation of carcinogenic response and antioxidant enzymes of rats administered with 1, 2-dimethylhydrazine by picroliv. *Cancer Letters*, **191** (2): 137-143.

Rajeshkumar, N.V., and Kuttan, R. (2000). Inhibition of N-nitrosodiethylamine-induced hepatocarcinogenesis by picroliv. *Journal of Experimental and Clinical Cancer Research,* **19** (4): 459-465.

Rajpal, V. (2002). Standardization of Botanicals Testing and Extraction Methods of Medicinal Herbs. Eastern Publishers, New Delhi.

Rastogi, R., Saksena, S., Garg, N.K., Kapoor, N.K., Agarwal, D.P., and Dhawan B.N. (1996). Picroliv protects against alcohol-induced chronic hepatotoxicity in rats. *Planta Medica,* **62** (3): 283-285.

Rastogi, R., Srivastava, A.K., and Rastogi, A.K. (2001a). Long term effect of aflatoxin B_1 on lipid peroxidation in rat liver and kidney: effect of picroliv and silymarin. *Phytotherapy Research,* **15** (4): 307-310.

Rastogi, R., Srivastava, A.K., and Rastogi, A.K. (2001b). Biochemical changes induced in liver and serum of aflatoxin B1-treated male Wistar rats: preventive effect of picroliv. *Pharmacology and Toxicology,* **88** (2): 53-58.

Rastogi, R., Srivastava, A.K., Srivastava, M., and Rastogi, A.K. (2000). Hepatocurative effect of picroliv and silymarin against aflatoxin B1 induced hepatotoxicity in rats. *Planta Medica,* **66** (8): 709-713.

Sane, S.A., Shakya, N., and Gupta, S. (2011). Immunomodulatory effect of picroliv on the efficacy of paromomycin and miltefosine in combination in experimental visceral leishmaniasis. *Experimental Parasitology,* **127** (2): 376-381.

Saraswat, B., Visen, P.K., Patnaik, G.K., and Dhawan, B.N. (1997). Protective effect of picroliv, active constituent of *Picrorhiza kurroa,* against oxytetracycline induced hepatic damage. *Indian Journal of Experimental Biology,* **35** (12): 1302-1305.

Saraswat, B., Visen, P.K., Patnaik, G.K., and Dhawan, B.N. (1993). Anticholestatic effect of picroliv, active hepatoprotective principle of *Picrorhiza kurroa,* against carbon tetrachloride induced cholestasis. *Indian Journal of Experimental Biology.* **31** (4): 316-318.

Saraswat, B., Visen, P.K., Patnaik, G.K., and Dhawan, B.N. (1999). *Ex vivo* and *in vivo* investigations of picroliv from *Picrorhiza kurroa* in an alcohol intoxication model in rats. *Journal of Ethnopharmacology,* **66** (3): 263-269.

Sereno, D., and Lemesre, J.P. (1997). Axenically cultured amastigote forms as an *in vitro* model for investigation of antileishmanial agents. *Antimicrobial Agent Chemotherapy,* **41** (5): 972-976.

Seth, P., Kumari, R., Madhavan, S., Singh, A.K., Mani, H., Banaudha, K.K., Sharma, S.C., Kulshreshtha, D.K., and Maheshwari, R.K. (2000). Prevention of renal ischemia-reperfusion-induced injury in rats by picroliv. *Biochemical and Pharmacology,* **59** (10): 1315-1322.

Seth, P., Sundar, S.V., Seth, R.K., Sidhu, G.S., Sharma, S.C., Kulshreshtha, D.K., and Maheshwari, R.K. (2003). Picroliv modulates antioxidant status and down-regulates AP1 transcription factor after hemorrhage and resuscitation. *Shock,* **19** (2): 169-175.

Shakya, N., Sane, S.A., and Gupta, S. (2011a). Antileishmanial efficacy of fluconazole and miltefosine in combination with an immunomodulator—picroliv. *Parasitology Research*, **108** (4): 793-800.

Shakya, N., Sane, S.A., Vishwakarma, P., Bajpai, P., and Gupta, S. (2011b). Improved treatment of visceral leishmaniasis (kala-azar) by using combination of ketoconazole, miltefosine with an immunomodulator-Picroliv. *Acta Tropica*, **119** (2-3): 188-193.

Sharma, N., and Garg, V. (2009). Antidiabetic and antioxidant potential of ethanolic extract of *Butea monosperma* leaves in alloxan-induced diabetic mice. *Indian J Biochemical and Biophysics*, **46** (1): 99-105.

Shukla, B., Visen, P.K., Patnaik, G.K., and Dhawan, B.N. (1991). Choleretic effect of picroliv, the hepatoprotective principle of *Picrorhiza kurroa*. *Planta Medica*. **57** (1): 29-33.

Singh, A.K., Mani, H., Seth, P., Gaddipati, J.P., Kumari, R., Banuadha, K.K., Sharma, S.C., Kulshreshtha, D.K., and Maheshwari, R.K. (2000). Picroliv preconditioning protects the rat liver against ischemia-reperfusion injury. *European Journal of Pharmacology*, **395** (3): 229-239.

Singh, A.K., Sharma, A., Warren, J., Madhavan, S., Steele, K., RajeshKumar, N.V., Thangapazham, R.L., Sharma, S.C., Kulshreshtha, D.K., Gaddipati, J., and Maheshwari, RK. (2007). Picroliv accelerates epithelialization and angiogenesis in rat wounds. *Planta Medica*, **73** (3): 251-256.

Singh, B., and Rastogi, R.P. (1972). Chemical examination of *Picrorhiza kurroa* Benth. Part VI. Reinvestigation of kutkin. *Indian Journal of Chemistry*, **10**: 29-31.

Singh, M., Tiwari, V., Jain, A., and Ghoshal, S. (2005). Protective activity of picroliv on hepatic amoebiasis associated with carbon tetrachloride toxicity. *Indian Journal of Medical Research*, **121** (5): 676-682.

Singh, V., Kapoor, N.K., and Dhawan, B.N. (1992b). Effect of picroliv on protein and nucleic acid synthesis. *Indian Journal of Experimental Biology*, **30** (1): 68-69.

Singh, V., Visen, P.K., Patnaik, G.K., Kapoor, N.K., and Dhawan, B.N. (1992a). Effect of picroliv on low density lipoprotein receptor binding of rat hepatocytes in hepatic damage induced by paracetamol. *Indian Journal of Biochemistry and Biophysics*. **29** (5): 428-32

Sinha, S., Mehrotra, J., Bala, L., Jaiswal, A.K., and Dhawan, B.N. (1998). Picroliv, the iridoid glycoside fraction of *Picrorhiza kurroa*, selectively augments human T cell response to mycobacterial protein antigens. *Immunopharmacology and Immunotoxicology*, **20** (4): 579-588.

Stuppner, H., and Wagner, H. (1989). New cucurbitacin glycosides from *Picrorhiza kurroa*. *Planta Medica*, **55**: 559-563.

Stuppner, H., and Wagner, H. (1992). HPLC and DC-untersuchungenvon iridoids, cucurbitacin, and phenol-giykosiden aus *Pichrorhiza kurroa*. *Scientia pharmaceutica*, **60**: 73-85.

Vaidya, A.B., Anantakar, D.S., Doshi, J.C., Bhatt, A.B., Ramesh, V., Bora, P.V., Pasissond, D., Baxi, A., and Kale, P.M. (1996). *Pichrorhiza kurroa* (kutaki) Royle ex Benth as hepatoprotective agent-experimental and clinical studies. *Journal of Postgraduate Medicine*, **42** (4): 105-108.

Visen, P.K., Saraswat, B., and Dhawan, B.N. (1998). Curative effect of picroliv on primary cultured rat hepatocytes against different hepatotoxins: an *in vitro* study. *Journal of Pharmacological and Toxicological Methods*, **40** (3): 173-179.

Visen, P.K., Shukla, B., Patnaik, G.K., and Dhawan, B.N. (1993). Prevention of galactosamine-induced hepatic damage by picroliv: study on bile flow and isolated hepatocytes (*ex vivo*). *Planta Medica*, **59** (1): 37-41.

Vivekanandan, P., Gobianand, K., Priya, S., Vijayalakshmi, P., and Karthikeyan, S. (2007). Protective effect of picroliv against hydrazine-induced hyperlipidemia and hepatic steatosis in rats. *Drug and Chemical Toxicology*, **30** (3): 241-52.

Yadav, N., and Khandelwal, S. (2006). Effect of picroliv on cadmium-induced hepatic and renal damage in the rat. *Human and Experimental Toxicology*, **25** (10): 581-591.

Yadav, N., and Khandelwal, S. (2008). Effect of picroliv on cadmium induced testicular damage in rat. *Food and Chemical Toxicology*, **46** (2): 494-501.

Yadav, N., Dogra, R.K., Khan, M.Y., and Khandelwal, S. (2005). Prevention of acute cadmium toxicity by picroliv. *Human and Experimental Toxicology*, **24** (10): 529-536.

Bioactive Phytochemicals: Perspectives for Modern Medicine Vol. 2 (2014) *Pages* **383–409**
Editor: **V.K. Gupta**
Published by: **DAYA PUBLISHING HOUSE, NEW DELHI**

17

Pharmacognosy of *Aspilia africana* Plant: Phytochemistry and Activities

O.O.K Oko[1*], E.A. Agiang[1] and E.E. Osim[2]

ABSTRACT

Aspilia africana (Pers.) C. D. Adams (Bush marigold) is commonly referred to as "iodine or haemorrhage plant" due to its ability to stop blood flow in fresh wounds. It is a semi-woody herb naturally distributed throughout the subtropical region of Africa. A. africana is an important herb of the African system of herbal medicine, where it is recommended for the treatment of several ailments including arthritis, rheumatism, abscesses, ulcer, cirrhosis, hepatitis and digestive disorders. In Nigeria, concoction from its leaf, stem and flower are extensively employed to prevent colic disorders, suppress cough and heal injuries as well as to treat liver diseases. Its root is chewed to cure toothache and aphthae. While extracts of the whole plant and leaf are used for treating malaria, syphilis and psoriasis. It is a preference browse for ruminant and wild animals. Recent ethno-pharmacological studies have proved that A. africana plant possesses several physiological properties and activities such as anti-inflammatory, anti-microbial, anti-ulcerative, anti-oxidant and growth promoting activities. The leaves are rich in proteins, minerals, polyunsaturated fatty acids, essential oils and numerous bioactive compounds. Crude extracts of A. africana leaf also demonstrated varying physiological effects in vivo. To harness the potentials of A. africana further studies are required that could lead to the discovery and synthesis of more potent drugs useful for pharmaceutical and agricultural purposes. This review summaries recent research and biotechnology on A. africana in relation to the bioactive compounds present in its leaf and extracts.

Keywords: *Aspilia africana*, Biological properties, Chemical composition, Tropical forbs, Medicinal values.

1 Animal Physiology Unit, Department of Animal Science, University of Calabar, Nigeria

2 Department of Physiology, University of Calabar, Nigeria

* *Corresponding author*: E-mail: oluwatosin.kennedyoko@gmail.com

Introduction

Scientific investigation of indigenous knowledge of many communities for the development of new pharmaceutical drugs from natural sources is the current trend in ethnopharmacological studies (Oko and Agiang, 2011a). Ethnoveterinary practices are being developed for the management of health and animal productivity using local remedies and surgical manipulating technologies (Allen *et al.*, 1997; Oko *et al.*, 2010). According to Gupta *et al.* (2005), ethnobotanical studies conducted in African and Asian communities' revealed indigenous plants and their products as the major constituents of traditional herbal systems. Therefore the significance of medicinal plants in health care delivery systems cannot be overemphasized since about 70 per cent of the communities rely solely on traditional medicines as their primary sources of health care delivery services (Oyesola *et al.*, 2010; Oko and Agiang, 2011b). More so, the World Health Organization is recommending the incorporation of medicinal plants into national drug policies and their commercial pharmaceutical exploration (Oyesola *et al.*, 2010). Forbs commonly known as herbs constitute one of the primary pharmacological interventions attempted by rural communities for primary health care services (Nalubega *et al.*, 2011).

Active components of these plants are now being investigated, extracted and developed into drugs with little or no negative effects or contra-indication (Oluyemi *et al.*, 2007). Over 50 per cent of all modern chemical drugs originated from natural plant sources and their derivatives are the major source of drug development in pharmaceutical industry (Burton *et al.*, 1983). In order to standardize natural plant products and their derivatives, determine their efficacies and possible health risks, several scientific investigations are being conducted.

In Nigeria, research on indigenous plants including weeds used in herbal medicine is under investigation (Oko, 2010). One of such plants is *Aspilia africana* (Asteraceae). It is a semi-woody herb from a perennial woody root stock up to 2 meters high. It is very polymorphic and occurs throughout the region on wasteland of the savannah forest. It is also widely distributed across tropical Africa (Dalziel, 1973). In African herbal medicine, A*spilia africana* enjoys folk reputation due to its ability to stop bleeding even from severed artery, its rapid wound and sore healing properties as well as in the management of cardiovascular-related diseases. Some reports indicated that *Aspilia africana* plant had been incorporated in the feeding of cattle, sheep, goats, rabbits and hares (Carew *et al.*, 1980; Boyejo, 1994).

Definition

Ethnobotany is the science that study how modern and indigenous societies use plants. According to Muthu *et al.* (2006), the World Health Organization (WHO) defines traditional medicine as practices, knowledge and belief systems which use minerals, plants and animal-based remedies, spiritual therapies and exercises to prevent, treat and maintain health. The African community has a long history with the use of plants with up to 90 per cent of the population rely on tropical medicinal plants as source of food, shelter and drugs (Hostettmann *et al.*, 2000).

Medicinal plants are aromatic plants which contain certain growth-promoting, prophylactic and therapeutic substances or substances that act as precursors for the synthesis of some (useful) drugs (Sofowora, 1993; Tabata *et al.*, 1994). Indigenous knowledge on medicinal plants as gained as the early man monitored animals feeding behavior in the wild (Huffmann, 2005) and communicated this information to their progenies. Traditional healers claimed that these medicines are available, affordable and more effective than orthodox medicines.

Phytochemicals, also known as Plant secondary metabolites (PSM), represent a diverse group of natural products (Wink, 2004) and are speculated to possess both nutritional and anti-nutritional properties (Acamovic and Brooker, 2005). Therefore, the biological activities of phytochemicals have been the focus of many studies (Ushie and Adamu, 2010). Several countries have programmes screening indigenous medicinal plants for their antimicrobial activities (Atai *et al.*, 2009; Kumar *et al.*, 2010; Thakare *et al.*, 2010; Oko and Agiang, 2011a). Results of Phytochemical analysis reveal that *Aspilia africana* leaf has high crude protein content (Oko and Agiang, 2011a and b) and is rich in saponins, tannins, flavonoids and alkaloids (Adeniyi and Odufowora, 2000; Abi and Onuoha, 2010). The presence of four essential oils; sesquiterpenes, monoterpenes, germacrene-d and alpha-pienene have also been reported (Kuiate *et al.*, 1999). Assays on the growth promoting effects of leaf meal and crude extracts of *Aspilia africana* leaf in animal production are on-going (Oko, 2010).

Morphological Characteristics

The morphological features of *A. africana* has been fully described (Akobundu, 1987). It is a semi-woody herb from a perennial root stock that grows up to two meters high, depending on rainfall and soil fertility (Dalziel, 1973). *A. africana* plant reproduces from seeds. The leaves are rough, opposite, ovate-lanceolate, creased accordion-style covered with trichomes (short flexible hairs). They are 6–25cm long and 3–7cm wide (Figure 17.1), rounded at the base with petioles about 1cm long with 3 prominent veins (Obute and Adubor, 2005). The stem is stiff at the base, with many branches and rough to touch. The inflorescence consists of solitary terminal flower head with hairy stalk about 4-10cm long. The flowers have showy-yellow (heads with several) florets. The fruits are 4-angled achenes about 5mm long, bristly and minutely hairy (Akobundu, 1987; Adjanohoun *et al.*, 1996; Obute and Adubor, 2005).

It is an indigenous plant of East Africa and grows throughout tropical Africa. A weed of field crops, it is abundant along roadsides and farm lands in most ecological forest zones where, they compete and cause heavy crop losses to farmers (Akinola *et al.*, 2000). *Aspilia africana* is widely known and used in many Nigerian communities. It has many different names across tropical Africa. For instance, in Nigeria, it is known as "bush marigold", "headband", "friend of pepper" or "to draw out mucus". Geo-politically, it is also known as; *Edemeron* in Efik, *Uranjila* (Igbo), *Tozalin/Kalankwa* (Hausa), *Yun Yun* (Yoruba), *Ndinuene* (Ibibio), *Uwhoridhoatu* (Ekpeye) and *Ottabi* (Eutung). Figure 17.2 presents the distribution of *Aspilia africana* plant in Africa. *Aspilia africana* belongs to the *genus – Aspilia* comprising of approximately 140 species. Some reports also indicated that *Aspilia africana* plant had been incorporated in the feeding of cattle, sheep, goats, rabbits and hares (Carew *et al.*, 1980; Boyejo, 1994).

Adeniyi and Odufowora (2000) and Oko and Agiang (2011a) described the macroscopic characteristics of crude extracts of *Aspilia africana* plant and leaves.

Figure 17.1: Photograph of *Aspilia africana* leaf.

Their results showed that a higher per cent extraction yield was obtained using alcohol-based solvents (methanol and ethanol). Compared to the cold extraction method, higher per cent yield was obtained using the soxhlet method. The colour of *Aspilia africana* leaf products vary from yellowish green to dark brown based on the processing and extraction methods.

The physico-chemical report of Obuzor and Ntui (2010) compared the oil of *Aspilia africana* leaf with some commonly used oils (palm, peanut, olive and coconut oils) and found out that, it is edible, non-drying, of low susceptibility to oxidative rancidity and has a long shelf life (Table 17.1).

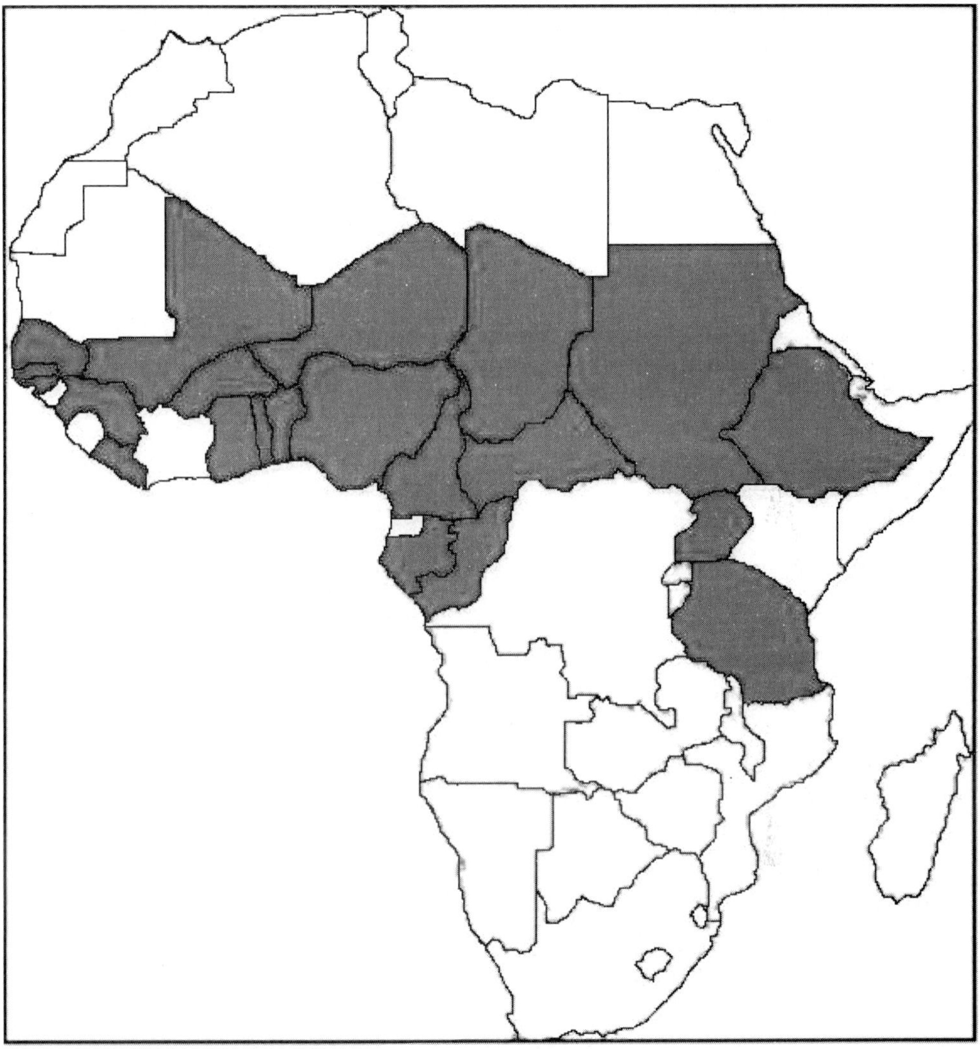

Source: http: //www.zipcodezoo.com/Animals/A/Aspilia_africana/Default.asp

Countries highlighted in the map include; Benin • Burkina Faso • Cameroon • Central African Republic • Chad • Congo • Ethiopia • Gabon • Ghana • Guinea • Liberia • Mali • Niger • Nigeria • Senegal • Sudan • Tanzania • Togo • Uganda

Figure 17.2: Distribution of *Aspilia africana* in Africa.

Processing of *Aspilia africana* Plant

Aspilia africana can be used as whole plant, roots, leaves, stems and extract. It can be administered in fresh, dried or in extract form. Different solvents and extraction (cold or hot) methods have been used in the extraction of bioactive principles from *Aspilia africana* plant. Solvents used; methanol, distilled water, ethanol, chloroform (Oko and Agiang, 2011a) and petroleum ether, while essential oils are obtained using hexane and methane solutions (Adeniji and Odufowora, 2000).

Table 17.1: Physico-chemical properties of *Aspilia africana* leaf.

Sl.No.	*Parameter*	*Leaf Oil Value*
1.	Free fatty acid (oleic)	8.84 per cent
2.	Saponification	201.21mg KOH/g of oil
3.	Peroxide value	16.17 MEq of active Oxygen/kg oil
4.	Iodine value	79.00 (g of 1/100g of oil)
5.	Acid value	0.86mg KOH/g of oil
6.	K (Potassium)	13,800.37 ppm
7.	Ca (Calcium)	11,550.56 ppm
8.	Mg (Magnesium)	3,202.53 ppm
9.	Na (Sodium)	262.09 ppm
10.	Cu (Copper)	15.46 ppm

Source: Obuzor and Ntui, 2010.

Phytochemistry of *Aspilia africana* Plant

Several studies have demonstrated the chemical composition and bioactive constituents of *A. africana* plant. Details are given below:

Proximate Composition

Proximate analysis of the whole plant of *A. africana* plant revealed that it has over 17 per cent crude protein with appreciable amount of carbohydrate, amino-acids, minerals and vitamins. The leaves are rich in crude protein with over 14–26 per cent crude protein in its dried leaf and extracts (Oko and Agiang, 2011a). Table 17.2 presents some results on the chemical composition of *Aspilia africana* plant. Reports of Isah *et al.* (2008) and Taiwo *et al.* (2008) gave crude protein value of 14.26 per cent and 24.86 per cent, respectively. The high protein value of *Aspilia africana* leaf supports its use as a feed ingredient in animal production (Oko, 2010).

Vitamin and Mineral contents

Table 17.3 shows that *A. africana* leaf and whole plant contains ascorbic acid, riboflavin, thiamine and niacin. Studies indicated that *A. africana* leaf has high concentration of ascorbic acid (vitamin C), known for its anti-inflammatory and wound healing properties (Okwu and Josiah, 2006; Okoli *et al.*, 2007; Abi and Onouha, 2011). Riboflavin is implicated in proper functioning of the skin, eye as well as in growth enhancement.

Reports of Abi and Onouha (2011) and Oko and Agiang (2011a and b) also indicated that *Aspilia africana* leaf is a good source of minerals such as calcium, potassium, phosphorus, magnesium, sodium, iron and zinc. Calcium is the most predominant mineral and zinc the highest trace element. Calcium is known for its potency in blood coagulation and this could explain the potentials of *A. africana* in wound healing therapy. Zinc provides a protective mechanism against virus and its deficiency could lead to the weakening of the immune system, retard wound healing

Table 17.2: Proximate composition of *Aspilia africana* plant.

Herb	*DM*	*CP*	*EE*	*CF*	*Ash*	*NFE*	*Energy*	*ADF*	*NDF*	*LDF*	*Reference*
Aspilia africana	88.66	24.89	4.64	19.94	17.00	22.21	4.21	34.79	—	—	Taiwo *et al.*, 2008
	83.16	15.09	19.00	23.00	20.50	—	—	42.00	45.00	16.67	
	90.96	18.00	15.17	20.89	13.62	23.28	—	35.14	43.34	—	Ayo-Enwerem *et al.*, 2008
	92.50	14.26	8.53	6.75	11.99	51.00	1.42	25.88	5.47	18.49	Oko and Agiang, 2011a

and cause mental depression (Strausel and Saltman, 2000). The presence of selenium and copper in the plant reinforces its wound healing ability, since selenium acts an antioxidant that stimulates the immune system and contributes to the formation of antibodies against infectious agents (Abi and Onouha, 2011). While copper has the ability to kill a broad spectrum of pathogens. Magnesium, calcium and phosphorus in *Aspilia africana* plant support the claim that it is a remedy for osteoporosis and rheumatism. The presence of these vitamins and minerals in the plant could account for its use in wound healing, treatment of skin and eye diseases (Oko, 2010).

Table 17.3: Mineral composition of *Aspilia africana* leaf.

Mineral (mg/100g)		*Leaf*	
Ascorbic acid	–	11.00	26.42
Niacin	–	3.05	0.09
Thiamine	–	1.94	0.11
Riboflavin	–	0.13	0.20
Sodium	203.56	–	0.02
Potassium	345.10	57.50	0.03
Calcium	667.94	246.00	1.04
Magnesium	248.09	88.10	0.12
Zinc	512.88	34.65	5.68
Iron	11.93	–	3.78
Phosphorous	474.51	–	0.32
Chromium	0.24	–	–
Manganese	0.08	–	–
Copper	ND	–	–
Selenium	–	10.97	–
Nitrogen	–	213.00	–
Lead	ND	7.43	–
References	Oko and Agiang, 2011a	Abi and Onouha, 2011	Okwu and Josiah, 2006

ND: Not Detected; –: Not analyzed.

Phytonutrients

Phytochemical screening (Table 17.4) showed the presence of Alkaloids, tannins, saponins, flavonoids, phenols and steroids in the leaf and extracts of *Aspilia africana*. Results of the phytochemical composition indicated that saponins and tannins were the most abundant compounds (Obadoni and Ochuko 1998), while other studies showed that alkaloids are the highest concentration, followed by saponins, flavonoids and tannins (Adeniyi and Odufowora, 2000; Abi and Onouha, 2011; Oko and Agiang, 2011b). Some studies indicated the absence of glycosides in the ethanolic extract of *Aspilia africana* leaf. Alkaloids and saponins are known to have antimicrobial activities, flavonoids act as super antioxidants, anti-inflammatory, anti-ulcerative and anti-allergic agents (Harborne, 1984). Phenols act as bactericidal agents while tannins have astringent properties which hasten wound healing process (Oko and Agiang, 2011a and b). These metabolites are usually responsible for the pharmacological activities of medicinal plants. Results from these studies indicated variations in the concentrations of phytonutrients suggesting that the method of processing, type of plant materials used and the location of the plant could affect the concentrations of phytonutrients in *A. africana* plant. Masato and Wu (1994) reported the presence of thiarubrine A, derivates of 1, 2-dithiocyclohexa-3, 5-diene in the leaves and root of *A. africana*. Thiarubrine has generated considerable interest among synthetic and theoretical chemists (Freeman *et al.*, 1992) because of its unique structures and bioactivities.

Table 17.4: Phytochemical screening and composition of *Aspilia africana*.

Constituent	*Activity*	*Whole Plant*	*Methanolic Extract*	*Concentration g/100g dry wt.*	*Concentration per cent*
Saponins	++	++	+	–	2.26 ± 0.15
Tannins	++	++	+	0.04 ± 0.10	0.88 ± 0.04
Phenolics	++		+	1.46 ± 0.02	0.11 ± 0.02
Alkaloids	+	+++	+	1.24 ± 0.11	6.35 ± 0.84
Glycoside	–	+	–	–	–
Terpernoids	+/-		+	–	–
Steroids	+/-		+	–	–
Flavonoids				1.48 ± 0.20	2.01 ± 0.11
References	Anibijuwon *et al.*, 2010	Adeniyi and Odufowora, 2000	Okoli *et al.*, 2007	Okwu and Josiah, 2006	Abi and Onuoha, 2011

+++: Strongly positive; +: Positive; +/-: Weakly positive; –: Not detected.

Adapted from Oko, 2010.

Some quantitative studies showed that *Aspilia africana* leaves contain higher amount of flavonoid – gossypetin (Obute and Adubor, 2005; Okwu and Josiah, 2006), compared to phenols and alkaloids but with lower amounts of tannins. The presence of these phytochemicals explains why the leaf of *Aspilia africana* is an important herb in African herbal medicine.

Fatty Acid Contents

The relative fatty acid composition and concentration of *Aspilia africana* is presented in Table 17.5. Report of Obuzor and Ntui (2010) identified 16 fatty acids in *Aspilia africana* (nine in quantifiable amounts and seven in negligible quantities). Fatty acid composition of *Aspilia africana* was in the order of polyunsaturated (50.56 per cent) > saturated (30.36 per cent) > monounsaturated (19.07 per cent).

Gamma-linolenic acid (GLA- omega-6 fatty acid), an essential fatty acid found in plant-based oil are considered necessary for the maintenance of human health. GLA lowers cholesterol and maintain proper levels of various blood fats (Obuzor and Ntui, 2010). The presence of GLA's in *Aspilia africana* leaf with its anti-inflammatory properties have led to its use in the treatment of eczema and psoriasis, rheumatoid arthritis, nerve pain caused by diabetes and chronic fatigue syndrome. In epidemiologic and clinical studies, stearic acid (C18: 0) is associated with lowered low-density lipoprotein cholesterol in comparison with other saturated fatty acids. Oleic acid (monounsaturated omega-9 fatty acid) is the only monounsaturated fatty acid found in *Aspilia africana* at a concentration of 18.84 per cent.

Table 17.5: Relative fatty acid concentration (per cent) of *Aspilia africana*.

Sl.No.	*Name*	*Concentration (per cent)*	*Ret. Time (min)*
1.	Butyric Acid (C5: 0)	Trace	9.04
2.	Caproic Acid (C6: 0)	Trace	10.28
3.	Caprylic Acid (C8: 0)	Trace	11.12
4.	Capric Acid (C10: 0)	Trace	11.25
5.	Lauric Acid (C12: 0)	Trace	13.04
6.	Myristic Acid (C14: 0)	0.001	14.70
7.	Palmitic Acid (C16: 0)	11.77	16.46
8.	Palmitolaic Acid (C16: 1)	Trace	18.16
9.	Stearic Acid (C18: 0)	15.10	19.69
10.	Oleic Acid (C18: 1)	18.85	21.11
11.	Linoleic Acid (C18: 2)	50.28	22.61
12.	Linolenic Acid (C18: 3)	0.28	23.97
13.	Arachidic Acid (C20: 0)	0.54	25.11
14.	Behenic Acid (C22: 0)	2.95	26.36
15.	Erucic Acid (C22: 1)	0.22	27.48
16.	Lignoceric Acid (24: 0)	Trace	28.53
TOTAL		**99.99**	

Source: Obuzor and Ntui, 2010.

Essential Oil Contents

Essential oils are highly concentrated liquid that is generally distilled from stems, roots, barks, leaves, flowers, or other elements of a natural plant (Obuzor and Ntui, 2010). Usman *et al.* (2010) analyzed the essential oils in *Aspilia africana* leaves growing in Port Harcourt, Nigeria and found out that they contain two types of terpenoids; monoterpenes and sesquiterpenes (Table 17.6). Obuzor and Ntui (2011) also identifies 35 monoterpenes (2-Carene and Thymol the predominant), 20 sesquiterpenes (Germacrene D, Germacrene B and β-Caryophyllene were the major constituents) and 3 essential amino acids (histidine, phenylalanine and arginine) in the essential oil of *Aspilia africana* leaves (Table 17.7). Earlier study by Kuiate *et al.* (1999) identified Germacrene D (a sesquiterpene) as the most predominant compound in the oils of the Bafoussam grown *A. africana* while α-pinene (a monoterpene) is the most abundant compound in the oils of Yaounde grown *Aspilia africana*. The presence of diterpenes, kaurenoic and grandiflorenic acids was identified in the leaf oil of *Aspilia africana* (Page *et al.*, 1992). The oil obtained from Bafoussam grown *A. africana* is of germacrene D chemotype, the oil of Yaounde grown *A. africana* is of α-pinene chemotype while the Jos, Nigerian grown *A. africana* is of α-cubebene chemotype. Variations in the quantitative compositions of the oils may be due to their agroclimatic and geographical conditions.

Table 17.6: Chemical composition of the leaf oil of *Aspilia africana*.

Compound[a]	*RI*[b]	*Percentage Composition*	*Mass Spectra Data*
α-thujene	933	5.1	136, 121, 105, 91, 65
α-pinene	941	6.7	136, 121, 93, 79, 67
α-myrcene	990	1.7	136, 121, 115, 107, 93
Car-3-ene	1010	5.0	136, 121, 105, 93, 41
α-phellandrene	1030	0.2	136, 105, 93, 77, 65
Limonene	1031	2.5	136, 93, 79, 67, 53
β-ocimene	1038	1.6	136, 121, 105, 93, 79
α-cubebene	1351	31.1	204, 161, 145, 133, 119
α-copaene	1376	7.0	204, 189, 161, 145, 133
Patchoulene	1380	2.7	204, 189, 175, 161, 139
β-elemene	1391	2.1	189, 175, 161, 121, 105
α-cedrene	1410	2.0	204, 189, 161, 147, 119
Isocaryophyllene	1411	2.1	189, 175, 161, 147, 91
β-caryophyllene	1423	4.8	189, 175, 161, 147, 91
α-guaiene	1442	1.4	204, 189, 161, 147, 105
Germacrene D	1554	1.2	204, 161, 147, 133, 119
Germacrene B	1556	3.1	209, 204, 161, 121, 93
β-santalol	1713	0.2	202, 173, 161, 122, 94
Total		80.00	

[a]Compounds are listed in order of elution from silica capillary column coated on CP-Sil 5; [b]retention indices on fused silica capillary column coated with CP-Sil 5.

Source: Usmani *et al.* (2010).

Germacrene D (1Z, 6Z)-1-methyl-5-methylidene-8-propan-2-ylcyclodeca-1,6-diene is a subset of volatile organic hydrocarbons (sesquiterpenes) produced in some plant species for their antimicrobial and insecticidal properties. They also serve as insect pheromones. The presence of this compound in *Aspilia africana* leaf could be responsible for its antiparasitic and insecticidal activities. β-Caryophyllene (βCP) with a concentration of 13.22 per cent is the third highest constituent in *Aspilia africana*, a natural bicyclic sesquiterpene usually found as a mixture with isocaryophyllene (the *cis* double bond isomer). βCP is the first dietary cannabinoid approved by FDA probably due to its anti-inflammatory activity. Okoli *et al.* (2007)

and Usman *et al.* (2010) noted that this attribute of β-Caryophyllene must have contributed to the use of *A. africana* in arresting bleeding from fresh wounds, inhibition of microbial growth and acceleration of wound healing process.

Table 17.7: Chemical composition of the essential oil of the leaves of *Aspilia africana*.

Sl.No.	Name	Per cent	Ret. Time (min.)	Type of terpene
1.	Isoamyl acetate	0.04	2.97	Monoterpenes
2.	Cymen	0.04	3.63	Monoterpenes
3.	Sabine	0.20	4.14	Monoterpenes
4.	Camphene	0.39	4.79	Monoterpenes
5.	α-Phellandrene	3.11	6.95	Cyclic monoterpenes
6.	2-Carene	5.85	7.80	Cyclic monoterpenes
7.	α-Pinene	0.60	9.83	Monoterpenes
8.	Limonene	0.59	11.45	Monoterpenes
9.	*Trans*-Ocimene	0.08	12.15	Monoterpenes
10.	*Cis*-Ocimene	3.14	12.91	Monoterpenes
11.	Germacrene d	37.98	18.78	Sesquiterpene
12.	Histidine	0.03	19.49	Essential amino acid
13.	Phenyl alanine	0.05	21.01	Essential amino acid
14.	Arginine	0.07	21.31	Non essential amino acid
15.	γ-Cadinene	0.13	22.87	Bicyclic sesquiterpene
16.	β-Caryophyllene	13.22	22.43	Sesquiterpene
17.	Caryophyliene oxide	0.10	28.03	Sesquiterpene oxide
18.	Germacrene b	21.52	29.31	Sesquiterpene
19.	Bulnesol	0.01	29.55	Sesquiterpene alcohol
20.	Isobornyl isobutyrate <8-isobutyryloxy>	0.06	29.79	Sesquiterpene
21.	Bisabolone oxide	0.01	32.00	Sesquiterpene oxide

Extracted from Obuzor and Ntui (2011).

Traditional Uses

Medicinal Values

Man's inquisitiveness and study of animal behavior have proven useful tools in effective management of health in most rural communities (Huffman, 2005). Different communities named their herbal system according to their experience, wisdom and knowledge. Reports indicated that Plants from Nigerian biomass are utilize in indigenous medicines in the treatments of malaria (Ajaiyeoba *et al.*, 2006), fungal and bacterial infections (Edeoga *et al.*, 2005; Okoli *et al.*, 2007) and in livestock management (Oguike and Udeh, 2009; Oko, 2010). Extraction of bioactive compounds from medicinal plants permits demonstration of their physiological activity. It also facilitates pharmacology studies leading to discovery of synthesis of more potent drugs.

According to Rodriguez and Wrangham, (1992) and Huffman (2005), plants from the *Aspilia* genus gained attention within the research community following publications on its use by chimpanzee in Gombia National Park and Mahale Mountains National Park of Tanzania. Chimps swallow the leaves whole through the "Velcro effect" and defeacate the leaves whole without any visible evidence of passing through the digestive tract (Huffman, 2005).

In African folk medicine, *Aspilia africana* has been established as an effective medicinal plant due to its wide applications. Akujobi *et al.* (2004) noted that several parts of the plant; root, leaf, stem or whole plant can be used for herbal remedy either in dried form or as crude extracts. It is generally used to stop bleeding even from severed artery, promotes rapid healing of wounds and sores as well as for the management of cardiovascular-related diseases. It also remove corneal opacities, induce delivery and in the treatment of anaemia and various stomach complaints (Iwu, 1993; Adjanohoun *et al.*, 1996). They are used in the treatment of rheumatism pain (Oliver, 1960), bee and scorpion stings (Sofowora, 1993).

Among the diseases that have been managed successfully with *Aspilia africana* by traditional (herbal) medicine include malaria, epilepsy, infertility, convulsion, diarrhoea, dysentery, gonorrhoea, flatulence, tonsillitis, bacterial and fungal infections, mental illness and worm infections (Sofowora, 1996), gonorrhea, tuberculosis, cough, rheumatic pains, stomach trouble, corneal opacity, wounds and insect bites (Agonihotri *et al.*, 2010).

In Kenya, it is used in eliminating intestinal worms from the gastrointestinal tracts. In Uganda, it is used in the treatment of gonorrhea (Page *et al.*, 1992). In Ghana, the root decoction is taken for tuberculosis while the leaves are used as cough medicine for children. Decoctions from the leaves are used to treat eye problems and serve as lotion for the face to relieve febrile headache. They are used to cure ringworm and dysentery (Abii and Onuoha, 2011). In Cameroon ethnomedicine, it is used in the treatment of stomach ailments. In Nigeria, the leaves of *Aspilia africana* are used in wound healing, treatment of skin diseases and eye infections (Okoli *et al.*, 2007). The decoction of the root is a remedy for lumbago and sciatica neuralgia (Macfoy and Cline, 1990) and membrane stabilization (Oyedapo *et al.*, 1998). The infusion of the leaves is administered to pregnant women as enema before and after delivery. A pinch of salt and lime juices are mixed with the leaf juice and applied to the eyes for the removal of corneal opacities and foreign bodies from the eyes. Native chalk (Kaolinite) is also mixed with the leaf decoction to cure stomach troubles. The leaves are also used to remove the slimy substance from snail in food preparation. Decoction of the leaves also serve as tonic (Obute and Adubor, 2005).

Nutritional Values

In most villages in Africa, *Aspilia africana* is one of the most browsed plants by cattle, sheep, goats, rabbits and hare (Boyejo, 1994). Animal preference for *Aspilia africana* is partially due to its nutritional qualities, succulent of the plant as well as its availability all year round. Its utilization is however hampered by the presence of some anti-nutrients. In monogastric diets, 5 per cent inclusion level is recommended since its bitter taste at high concentration can reduce its palatability.

Biological Activities

Aspilia africana is one of the plants that have been reported to contain a wide range of biological activities such as: antiviral, fungicidal and anti-bacterial activities due to the presence of phytochemicals especially thiarubrine A (Masato and Wu, 1994). Below are some of its activities:

Antimicrobial Activity

The wide range of *in vitro* antimicrobial activities of *Aspilia africana* have been published (Adeniji and Odufowora, 2000; Okoli *et al.*, 2007b; Anibijuwon *et al.*, 2010; Nalubega *et al.*, 2011; Agiang and Oko, 2010; Oko and Agiang, 2009; 2011a). Most studies suggested that the main components of *A. africana* leaf that exhibit *in vitro* anti-microbial activity are responsible for its activity. The different extracts of *Aspilia africana* leaf displaying anti-microbial activity are shown in Table 17.8. Results showed that the minimum inhibitory concentrations (MIC) of *A. africana* leaf differ and vary between experiments. Findings of Nalubega *et al.* (2011) showed that ether extracts exhibited greater antibacterial activity compared to the ethanol extracts. These studies

Table 17.8: Antimicrobial activity of the meal and extracts of *Aspilia africana* leaf.

Microorganism	*Leaf Meal*	*MIC 95 per cent Ethanolic Extract, mm*	*MIC Hexane Extract, mg/ml*	*Extract, mm*
Ruminococcus flavefaciens	+			
Sarcina ventrivculi	–			
Bacillus subtilis	–			12.00
Esherichia coli	+	8.5		
Streptococci bovis	–			
Staphylococcus aureus	–	10.5	0.50	12.00
S. faecalis		10.3		
K. pneumoniae		11.5		
P. aeruginosa		21.0	0.125	14.00
P. fluorescens			0.25	
Aspergillus flavus	+			17.00
Fumigatus spp.	+			–
Penicillum notatum	+			–
Candida albican				20.00
References	Agiang and Oko, 2010	Anibijuwon *et al.*, 2005	Okoli *et al.*, 2007	Adeniji and Odunfowora, 2000

+: Microorganism is sensitive to the extract.

–: Microorganism is insensitive to the extract.

demonstrated that *Aspilia africana* leaf and extracts reacted to most pathogenic bacterial species, indicative of the presence of compounds with broad spectrum activity. Some gram positive *Bacillus subtilis* and *Micrococcus lutea* are sensitive to compounds present in *Aspilia africana*. Gram-negative bacteria appeared less susceptible to *Aspilia africana* leaf extracts probably due to their unique structures.

Anibijuwon *et al.* (2010) compared the antibacterial activity of the ethanolic and aqueous extracts of *Aspilia africana* leaf to those of chloramphenicol, tetracycline and ampicillin and reported similar activity at concentrations of 12.5–100mg/ml. The hot aqueous extract however did not exert any activity. Their results further revealed that the ethanolic extract of *Aspilia africana* showed the highest antibacterial activity against *Pseudomonas aeruginosa* (31.5m) and lowest activity (17.5mm) against *Streptococcus faecalis* and *Escherichia coli* at 100mg/ml concentration. The Minimum Inhibitory concentration (MIC) of the ethanolic extracts was 12.5mg/ml for all the organisms. The methanol extract of *Aspilia africana* leaf also had higher anti-microbial activities compared to its fractions (Okoli *et al.*, 2007b). Reports of Agiang and Oko (2010) also demonstrated the susceptibility of *Escherichia coli, Staphylococcus aureus, Streptococcus bovis,* and *Bacillus subtilis* to extracts of *Aspilia africana* leaf.

The antibacterial activity of crude extracts of *Aspilia africana* leaf on tested organisms justifies the active principle or ingredient observed in herbal physician in their preference for the local gin "ogogoro" as extract in the preparation of crude drugs from medicinal plant materials. Local gin obtained from fermented palm wine distillation is known to contain a high concentration of alcohol (Anibijuwon *et al.*, 2010).

The exact anti-microbial mode of action of *A. africana* as other plant materials is still poorly understood (Lee *et al.*, 2004). However, it has been suggested that their chemical structure and lipophilic property could be responsible for these activities. Alkaloids have been implicated for this antimicrobial activity (Adeniyi and Odufuwora, 2000). Okoli *et al.* (2007b) investigated how phenols and phenylpropanoid exert their antimicrobial effects on *Escherichia coli* O157 and *Salmonella typhimurium*. They disintegrated the membrane of bacteria, leading to the release of membrane-associated materials from the cells to the external medium. The lipophilicity, aromaticity and presence of the functional groups in terpenoids and phenylpropanoid compounds of plant materials are implicated for their antibacterial activities.

Shapiro and Guggenheim (1995) indicated that membrane perforation is the main principal mode of action of antibacterial agents. These cell membrane perforation leads to an increase of permeability and leakage of vital intracellular constituents, thus impairment of bacterial enzyme systems. The selective inhibition by *Aspilia africana* leaf to pathogenic, intestinal bacteria may have pharmacological role in balancing the intestinal microbiota. Some of these bacteria have been implicated in diseases as diarrhoea, oral, dental and wound infections.

The antifungal activity of *A. africana* plant has also been documented (Adeniyi and Odufuwora, 2000; Oko and Agiang, 2011a). These studies also indicated that *Aspergillus flavus, Aspergillus parasiticus, Penicillum notatum* and *Candida albican* were susceptible to *Aspilia africana* leaf extracts indicating that *Aspilia africana* leaf possesses strong anti-fungal properties. The mechanism of antifungal action of plants is proposed to occur through the inhibition of fungal-cell-wall synthesizing enzymes and complex formation with sulfhydryl groups required for fungal growth. The formation of charge transfer complexes with electron donors in the fungus cells lead to cell division, thus interfering with cell metabolism. Thiarubrine A found in *Aspilia africana* is reported to exhibit a wide spectrum of potent antiviral (Hudson *et al.*, 1993), anti-tumor and anticancer activities. Pure alkaloids isolated from *A. africana* leaves are known to used as basic medicinal agents due to their analgesic, antispasmodic and bactericidal effects (Okwu and Okwu, 2004).

These vast anti-microbial activities of *Aspilia africana* have attracted attention and its subsequent application as prophylactic and therapeutic agents in animal production. Dietary supplementation of *Aspilia africana* is expected to affect the gastrointestinal microflora composition and population. In some studies (Oko, 2011; Agiang *et al.*, 2011; Oko and Agiang, 2011; Oko *et al.*, 2011a; Oko and Offiong, 2012; Oko *et al.*, 2012a and b) we investigated *Aspilia africana* leaf as potential antibiotics in poultry. Some growth promoting effects of *Aspilia africana* was observed in quails and broilers.

Anti-inflammatory Activity

Inflammation according to Agnihorti *et al.* (2010) is a natural response of the mammalian body to a variety of hostile agents including parasites, pathogenic micro organisms, toxic chemical substances and physical damage to tissue. Grover *et al.* (2006) further noted that rheumatoid arthritis (RA) is a common chronic and systemic autoimmune disorder characterized by inflammation of the synovial joints and concomitant destruction of cartilage and bones. Anti-inflammatory activity of hexane leaf extract of *Aspilia africana* has also been reported (Okoli *et al.*, 2007a; Agonihotri *et al.*, 2010). The hexane extract of *Aspilia africana* leaf is reported to relieve rheumatism as well as suppresses inflammatory response initiated by the tissue injury (Agnihorti *et al.*, 2010). They explained that the extract inhibited the release of actions chemical mediators such as histamine, 5-HT, kinnins and prostanoids known to mediate acute inflammation. Sterols (β- sitosterol) and terpenoids present in the plant implicated for this anti-inflammatory activity (Okoli *et al.*, 2007a). Findings by Changa *et al.* (2008) revealed that terpenoids has several mechanisms useful in inhibiting the development of chronic joint swelling.

Anti-oxidant Activity

The relationship between the antioxidant properties and the chemical composition of the essential oil of plant materials has been established. The higher the amount of phenols in the plant the greater anti-oxidative activity. The phenolic OH group acts as hydrogen donors to the peroxyl radicals produced during the initial lipid oxidation process, thus inhibiting the formation of hydrogen peroxide. Preethi *et al.* (2010) evaluated the antioxidative efficacy of some medicinal plants against food borne pathogens.

Studies indicated that *A. africana* leaf could act as effective free radical scavengers and influence the *in vivo* antioxidant defense systems such as superoxide, dismustase, gluthathione peroxidases and vitamin E. Bostsoglou *et al.* (1997) noted that the antioxidant components in plants can be transferred to egg yolk based on a lower concentration of malonaldehyde in the yolk. Oko *et al.* (2012b) and Oko and Offiong (2012) showed that dietary *Aspilia africana* leaf meal and extracts could acts as antioxidant in egg and meat of quails and chickens when supplemented into their diets. Significant reductions in abdominal fat contents in the meat of poultry were also reported (Oko *et al.*, 2012b). The high amount of essential oils as well as phenolic contents in the leaves of *Aspilia africana* leaf is responsible for the antioxidant properties (Usman *et al.*, 2010; Oko and Agiang, 2011a). However, further studies on its effects on meat stability are required.

Antiseptic Activity

The extract and fractions arrested bleeding from fresh wounds by reducing bleeding/clotting and whole blood coagulation time which are important indices of haemostatic activity (Achonye, 1976). It is also used to clean the surfaces of sores, in the treatment of rheumatic pains, bee and scorpion stings and for removal of opacities and foreign bodies from the eyes (Okoli *et al.*, 2007b). These studies indicate that leaf extracts of *A. africana* have good potentials for use in wound care and further provide a rationale for the use of the leaves of this plant in wound management in traditional medicine practice.

Anti-ulcerative Activity

Nguelefack *et al.* (2005) assessed the methanol extract of the leaves of *A. africana* against gastric ulcerations induced by HCl/ethanol and pylorus-ligation. They discovered that in both methods, the extract exerted a dose- related inhibition on gastric ulcerations. The anti-ulcer effects of the methanol extract of the leaves of *Aspilia africana* may be due to the presence of triterpenes in the extract or the enhancement of the mucosal barrier through the increase production of prostaglandins (Nguelefack *et al.*, 2005). Some triterpernic compounds such as nimbidin have been shown to possess anti-ulcer activity (Pillai and Santhakumari, 1984) while prostaglandins are known to provide gastric cyto-protection against strong necrotizing irritants without reducing gastric acid secretion in rats. Ubaka *et al.* (2010) also assessed the antiulcer activity of leaves of *Aspilia africana* leaf in rats and observed that the aqueous extract exhibited dose dependent anti-ulcer activity comparable to those of standard drugs. These findings clearly show that aqueous and methanol extract of the leaves of *A. africana* possess potent anti-secretory and antiulcer properties.

Anti-Plasmodic and Anti-Parasitic Activities

Rodriguez and Wrangham (1992) noted that *Aspilia africana* leaves possess some anti-parasitic compounds. Species of *Aspilia* genus have also been found to possess anti-malaria activities against *P. falciparum* and galactogogne activity (Page *et al.*, 1992). Waako *et al.* (2005) and Chepkwony *et al.* (2007) assessed the anti-malaria and anti-viral effect of *Aspilia africana* leaf while Okokon *et al.* (2006) investigated the antiplasmodic activity of ethanolic extract of *A. africana* leaf. Results indicated that *Aspilia africana* leaf exhibited significant blood schizonticidal activity as evident from the chemosuppression reaction obtained during the early stage of infection. This implies that the extract of *A. africana* leaf possesses strong anti-malaria and anti-plasmodial activities which can be explored in malaria therapy. Christensen and Kharazmi (2001) suggested that the presence of alkaloids, terpenes and flavonoids in the leaf may be responsible for this anti-malaria and anti-plasmodic activities. These results agree with *in vitro* studies reported by Waako *et al.* (2005) and affirm its usage in folk medicine as malaria anecdote. *A. africana* is not only used as an anti-parasitic but also as an effective insecticide due to the presence of precocene–1 (Kuaite *et al.*, 1999). Enikuomehin (2005) also reported the efficacy of aqueous extract of *A. africana* to control cercospora leaf spot – acting as a pesticides and fungicides.

Growth Promoting Activity

Several medicinal plants have also been used as dietary adjuncts especially in poultry production (Hernandez *et al.*, 2004; Oko *et al.*, 2012a and b). There are evidences to suggest that medicinal plants such as herbs, spices and their derivates have appetite and digestion stimulating properties (Kamel, 2001; Hernandez *et al.*, 2004).

Few studies have been conducted on the effects of *Aspilia africana* plants in animals. Our previous studies (Agiang *et al.*, 2011a; Oko *et al.*, 2011a) in laying quails demonstrate that *Aspilia africana* leaf has the potential to boost growth performance, egg production and improve internal egg quality with positive impacts on egg composition and yolk colour (Tables 17.9 and 17.10). A slight improvement in the digestibility of nutrients has also been reported (Oko *et al.*, 2012a).

Aspilia africana leaf products demonstrated non-significant improvement in the growth performance of quail chickens. Similarly, results from our ongoing studies (supported by the International Foundation for Science – B/4719) is indicating same pattern of growth in broilers and laying hens when *Aspilia africana* leaf meal was included either in mash or pelleted form. Compared to the mash form, the pelleted form appears to exert more phytogenic effect in both broiler and laying hens. The aqueous extract demonstrated higher phytogenic activity compared to the leaf meal or ethanolic extract. This finding strongly supported submissions by Alicicek *et al.* (2004) that plant extracts improved performances of birds.

Nevertheless at high concentrations of 7.5–10 per cent *Aspilia africana* leaf inclusions, feed utilization, growth performances and egg production parameters were impaired. It is hypothesized that the presence of phytochemicals such as phytates, tannins, saponins and alkaloids could have negative effects on either the taste of the feeds thus affecting feed intake or diminish nutrient bioavailability therefore limiting the feed quality. From our studies in quails and chickens, optimum performance is attained at 5 per cent *Aspilia africana* leaf supplementations.

Toxicological Effects of *Aspilia africana*

Generally, *Aspilia africana* plant is classified among substances with low to moderate toxicity, with an LD_{50} of 1.41–6.6g kg^{-1} in rat (Okokon *et al.*, 2006; Taziebou *et al.*, 2007). In one of our studies (Oko *et al.*, 2011b) we investigated the toxicological activities of *Aspilia africana* leaf and extracts in mice and found out that the degree of

Table 17.9: Influence of Aqueous extract of *Aspilia africana* Leaf (AeAaL) on quail performance.

Mean Performance	*0 per cent AeAaL*	*2.5 per cent AeAaL*	*5 per cent AeAaL*	*7.5 per cent AeAaL*	*10 per cent AeAaL*	*Sem*	*LSD*	*Level of Sig.*
Initial body weight (g)	16.52	15.52	14.45	14.00	17.11	0.53	4.22	NS
Final body weight (g)	107.32	126.87	120.65	116.21	116.12	3.65	26.60	NS
Average body weight (g)	84.75^{a}	82.19^{b}	84.71^{a}	80.41^{b}	82.22^{b}	6.13	2.27	**
Daily feed intake (g/day)	15.70^{a}	13.97^{b}	14.31^{b}	14.26^{b}	14.25^{b}	0.18	1.18	*
Daily weight gain (g/day)	18.87	18.79	19.69	18.85	18.07	0.62	2.91	NS
Feed conversion Ratio (per cent)	0.83	0.74	0.73	0.77	0.79	0.04	0.26	NS
Mortality (per cent)	4.00^{a}	1.33^{b}	1.33^{b}	1.33^{b}	1.33^{b}	0.02	0.10	*

AeAaL: Aqueous extract of *Aspilia africana* Leaf.

Source: Agiang *et al.*, 2011.

toxicity of *Aspilia africana* leaf is route of exposure, extractant and dose-specific. Signs of behavioural toxicity include; nervous and respiratory disorders as well as piloerection in mice. Oral LD_{50} of *Aspilia africana* leaf was indicated at 12,589.25mg/kg, 1,995.26mg/kg and 10,000mg/kg in the aqueous, cholorform and ethanolic extract, respectively.

Table 17.10: Influence of *Aspilia africana* leaf products on the egg quality in laying quails.

Morphometric Trait	*Control*	*5 per cent AeAaL*	*5 per cent AaLM*	*5 per cent EeAaL*	*Oxytetra-cycline*
Egg Size (g)	9.61	9.54	9.15	9.86	9.65
Hen Day Production (per cent)	61.38	86.24	72.55	77.42	69.20
Albumen Weight (per cent EW)	47.80	52.07	50.27	52.04	51.95
Yolk Weight (per cent EW)	30.63	32.64	34.64	32.77	31.03
Shell Weight (per cent EW)	21.57	15.29	15.09	15.13	17.03
Per cent Crack	0.87	0.68	0.78	0.84	0.85
Shell Thickness (mm)	0.29	0.33	0.30	0.35	0.30
Yolk Colour	3.03	5.28	5.75	5.30	4.31
Daily Weight Gain (g/day)	0.30	0.17	0.19	0.33	0.20
Dressed Weight (per cent)	65.53	72.64	68.94	70.44	67.10
Crude protein (per cent)	48.83	48.75	46.72	43.05	45.68
Crude fat (per cent)	38.14	38.22	36.77	38.43	39.47

AeAal: Aqueous extract of *Aspilia africana* leaf; EeAaL: Ethanolic extract of *Aspilia africana* leaf; AaLM: *Aspilia africana* leaf meal.

Adapted from Agiang *et al.*, 2011 and Oko, 2010.

Other studies have also indicated concerns on the health risk of prolong use of *Aspilia africana* plant. Findings of Mukherjee *et al.* (2006), Okwuonu *et al.* (2007), Oluyemi *et al.* (2007), Taziebou *et al.* (2007) and Eweka (2009) demonstrated the toxic effects of *Aspilia africana* leaf on animal physiology to include; reductions in uterine tissue, alterations in the duration of estrus cycle, anti-ovulatory effects in female rats and anti-fertility effects in male animals. Oluyemi *et al.* (2007) showed that intraperitoneal administration of the methanolic extract of *Aspilia africana* leaf caused significant delay in estrus cycle and severe damages to the uterine tissues in wistar rats. Eweka (2009) similarly revealed that an oral administration of the aqueous extract of *Aspilia africana* leaf could distort the histology of the ovaries in the female wistar rats, thus impairing fertilization. These studies suggest that *Aspilia africana* plant could be toxic as well as pose some health and reproductive risks to animals.

The presence of high levels of pytoestrogens such as saponins and essential oils has been implicated. Phytoestrogenic plants have both estrogenic and antiestrogenic effects on animal reproductive systems. Following prolong administration, *Aspilia africana* plant products have been found to reduce fertility in animals. They prevent implantations and other estrogen-dependent activities in the reproductive system by causing hormonal imbalances in the systems.

Dimo *et al.* (2002) also investigated the effect of methylene chloride/methanol extract of *Aspilia africana* in rats. They observed increases *in vitro* vascular smooth muscle contraction in rats' aortic ring preparations. The extract acted in a dose – dependant pattern on the smooth muscles of the blood vessels and myometrium leading to their atrophy. Oyesola *et al.* (2010) investigated the effects of aqueous extract of *Aspilia africana* (500 – 1000 $mgkg^{-1}$ BW) on rats and reported a significant decrease in weight, growth rate and a dose-dependent reduction in the ovulation rate in wistar rat. The extract caused significant inflammation of the fallopian tube, degeneration in the ovarian cortex and disruption of the endometrium of the uterus. These observations suggest that prolonged administration of aqueous extract of *A. africana* could have adverse effects on the estrus cycle and ovulation processes in animals. They suggested that the presence of flavonoids in the leaf could be responsible for the inhibition of cyclooxygenase activity and prostaglandin synthesis in the ovary. This supports its use as an anti-fertility agent in herbal medicine. Previous study (Okwuonu *et al.*, 2008) showed no effect of methanolic extract of *Aspilia africana* on weight loss in rats.

Results further revealed that oral exposure and the aqueous extract are safer than the intraperitoneal exposure and ethanolic or the cholorm extract. It is therefore recommended that oral administration of up to 10,000 mg/kg body weight of aqueous and ethanolic extracts of *Aspilia africana* leaf are safe for human and animal use (Oko *et al.*, 2011b). These findings support the common practice of oral administration of either aqueous or ethanolic extract of *Aspilia africana* as a medicinal decoction in herbal medicine.

Summary and Conclusion

Plant-based antimicrobials represent a vast untapped source of medicines with enormous potentials for therapeutic purposes. The phytochemical screening of *Aspilia africana* leaf and extracts revealed that it is a rich source of nutrients and bioactive components. The leaf of *Aspilia africana* could sever as both macro and micro-nutrients required for normal body functioning of the body system. It can also serve as source of potassium to balance the high sodium intake resulting from food additives, calcium for strong teeth and bones, copper as an antioxidant and regulation of gene expression. The fatty acid results indicate that *Aspilia africana* is an edible leaf. Result from the biological assays of *Aspilia africana* leaf meal and extracts support some of the ethno-pharmacological claims about the plant in the treatment of diseases. Processing of the leaf could give valuable and essential products for various industrial applications.

Further studies are required in order to maximize the potentials of *Aspilia africana* leaf as cheap, available feed ingredient and alternative antibiotic growth promoter. Also, researches tailored at standardizing the proper extraction procedures for its discovery as potent drugs for health care management should be considered. For sustainable use of medicinal plants, indigenous and scientific knowledge should be explored and documented. Therefore, documentation of the traditional therapeutic know-how could lead to the discovery of new drugs as well as contribute to the conservation and sustainable management and use of plant resources.

Acknowledgments

This research was supported by the International Foundation for Science, Stockholm, Sweden and the Organisation for the prohibition of Chemical Weapons, The Hague, The Netherlands, through a grant number "IFS Gt B/4719 Olawoyin" to evaluate the potentials of *Aspilia africana* leaf as growth promoter effects in poultry. We are also grateful to other colleagues/authors who contributed to the knowledge on *Aspilia africana*.

References

Abi, T. A., and Onuoha, E. N. (2011). The chemical constituents of the leaf of *Aspilia africana* as a scientific backing to its tradomedical potentials. *Agricultural Journal*, **6(1)**: 28 – 30.

Acamovic, T., and Brooker, J. D. (2005). Biochemistry of plants secondary metabolites and their effects in animals. *Proceedings of the Nutrition Society*, **64**: 403–412.

Achonye, E. L. (1976). A Pharmacological Investigation of the haemostatic action of pressed leaf extract of *Aspilia latifolia* (Compositae) B. Pharm. Thesis, Department of Pharmacology and Toxicology, University of Nigeria, Nsukka.

Adeniyi, B. A., and Odufowora, R. O. (2000). *In-vitro* anti-microbial properties of *Aspilia africana* (Compositae). *African Journal of Biomedical Research*, **3**: 167–170.

Adjanohoun, J. E., Aboubakar, N., Dramane, K., Ebot, M. E., Ekpere, J.A., Enow-Orock, E.G., Focho, D., Gbile, Z.O., Kamanyi, A., Kamsu kom, J., Keita, A., Mbenkum, T., Mbi, C.N., Mbielle, A.L., Mbome, I.L., Mubiri, N.K., Nancy, W.L., Nkongmeneck, B., Satabie, B., Sofowa, A., Tamze, V., and Wirmum, C. K. (1996). Traditional Medicine and Pharmacopeia-contribution to ethnobotanical and floristic studies in Cameroon, CNPMS, Porto-novo, Benin.

Agiang, E. A., and Oko, O. O. K (2010). Antimicrobial activities of crude extracts from *Aspilia africana* leaves. *Proceedings of the Animal Science Association of Nigeria*, **15:** 123 – 125.

Agiang, E. A., Oko, O. O. K., and Essien, G. E. (2011). Quails response to aqueous extract of bush marigold *Aspilia africana* leaf. *American Journal of Animal and Veterinary Sciences*, **6(4):** 130 – 134.

Agnihortri, S., Wakode, S., and Agnihortri, A. (2010). An overview on anti-inflammatory properties and chemo-profiles of plants used in traditional medicines. *Indian Journal of Natural Products and Resources*, **1(2):** 150 – 167.

Ajaiyeoba, E. O., Abiodun, O. O., Falade, M. O., Ogbole, N. O., Ashidi, J. S., Happi, C. T., and Akinboye, D. O. (2006). *In vitro* cytotoxicity studies of 20 plants used in Nigerian antimalaria ethnomedicine. *Phytomedicine* **13**: 295–298.

Akinola, J. O., Larbi, A., Farinu, G. O., and Odunsi, A. A. (2000). Economic analysis of Nigerian poultry industry. A study commissioned by the Federal Livestock Department, Lagos State.

Akobundu, I. O. (1987). Weed science in the Tropics: Principles and practices. John Wiley and Sons, Chichester.

Akujobi, C.O., Ogbulie, J.N., and Okorondu, T. (2004). Antibacterial and nutrient potentials of *Gongronema latitolium* and *Piper guineenses* used in herbal remedies and as species. *Nigerian Journal of Microbiology,* **18(1-2)**: 241- 246.

Allen, P. C., Lyzon, J., and Danforth, H. D. (1997). Effects of components of *Artemisia annua* on coccidia infections in chickens. *Poultry Science,* **76**: 1156–1163.

Anibijuwon, I. I., Duyilemi, O.P., and Onifade A. K. (2010). Antimicrobial activity of leaf of *Aspilia africana* on some pathogenic organisms of clinical origin. *Nigerian Journal of Microbiology,* **24(1)**: 2048 – 2055.

Atai, Z., Matapour, M., and Mohseni, M. (2009). Inhibitory effects of ginger extract on *Candida albicans. American Journal of Applied Science,* **6(6)**: 1067 – 1069.

Botsoglou, N. A., Athanassios, L. Y., Dimitrios, J. F., Angela, S. T., and Daschalis, D. F. (1997). Effect of dietary thyme on the oxidative stability of egg yolk. *Journal of Agricultural and Food Chemistry,* **45**: 3711–3716.

Boyejo, O. O. (1994). Siam weed (*Chromolaena odorata*) in rabbit ration: Effect on performance, haematological and biochemical indices. M.Sc. Thesis, Department of Animal Science, University of Ibadan, Nigeria.

Burton, G. W., Joyce, D., and Ingold, K. U. (1983). Is vitamin E the only soluble chain leaving anti-oxidant in human blood plasma and erythrocyte membrane. *Archives of Biochemistry and Biophysics,* **222**: 228–290.

Carew, B. A. R., Mosi, A. K., Mba, A. U., and Egbunike, G. N. (1980). The potential of browse plants in the nutrition of small ruminants in the humid forest and derived savanna zones of Nigeria. *International Livestock Centre for Africa,* Ibadan. pp. 307 – 311.

Changa, C., Wena, Z., Wangc, S., and Duha, C. (2008). New inflammatory steroids from the Formosan soft coral *Clavularia viridis. Steroids,* **73**: 562 – 567.

Dalziel, J. M. (1973). The useful plants of West Africa. Crown Agents, London.

Edeoga, H. O., Okwu, D. E., and Mbaebie, B. O. (2005). Phytochemical constituents of some Nigerian medicinal plants. *African Journal of Biotechnology,* **4(7)**: 685–688.

Eweka, A. O. (2009). Histological studies of the effects of oral administration of A*spilia africana* (*Asteraceae*) leaf extract on the ovaries of female Wistar rats. *African Journal of Traditional, Complementary and Alternative Medicines,* **6 (1):** 57 – 61.

Freeman, F., Kim, D. S. H. L., and Rodriguez, E. J. (1992). Biochemistry of medicinal plants. *Organic Chemistry,* **57**: 1722 – 1727.

Grover, S. Tandon, R., Misra, R., and Aggrawal, A. (2006). Interlukin-I receptor antagonist gene polymorphism in patients with rheumatoid arthritis in India. *Indian Journal of Medical Research,* **123(6)**: 815–820.

Gupta, M. P, Solis, P. N., Calderon, A. I., Guinneau-Sinclair, F., and Correa, M. (2005). Medical ethnobotany of the tribes of bocas del toco, Panama. *Journal of Ethnopharmacology,* **96:** 389 – 390.

Harborne, J. B. (2001). Twenty-five years of chemical ecology. *Natural Product Reports,* **18**: 361–379.

Hostettmann K, Marston A, Ndjoko K., and Wolfender J-L. (2000). The potential of African medicinal plants as a source of drugs. *Current Organic Chemistry*, **4**: 973-1010.

Hudson, J. B., Balza, F., Harris, L., and Towers, G. H. N. (1993). Light-mediated activities of Thiarubrines against human immunodeficiency virus (HIV). *Photochemistry and Photobiology*, **57**: 491–496.

Huffman, M. (2005). Animal instinct for finding treatment. Primate Research Institute of Kyoto University, Japan.

Isah, O. A., Omoniyi, L. A., Olaniyan, O. M., Adefowope, L. M., and Tanimonure, O. F. (2008a). Identification and chemical composition of feed resources used in ruminant production in three local government areas of Ogun State. *Proceedings of the Animal Science Association of Nigeria*, **13**: 242–244.

Iwu, M. M. (1993). Handbook of African Medicinal Plants. CRP Press, Florida.

Kuiate, J. R., Amvam-Zollo, P. H., Lamaty, G., Menut, C., and Bessiere, J. M. (1999). Composition of the essential oils from the leaves of two varieties of *Aspilia africana* (Pers). C.D. Adams from Cameroon. *Flavour and Fragrance Journal*, **14 (3)**: 167-169.

Kumar, A. Alzahal, O., Singh, K. K., and McBride, B. W. (2010). *In vitro* nutritional evaluation of chaya (*Cnidoscolus aconitifolius*) for use as livestock feed. *Indian Journal of Animal Sciences*, **80 (10)**: 1008–1010.

Lee, K. W., Everts, H., and Beynen, A. C. (2004). Essential oils in broiler nutrition. *International Journal of Poultry Science*, **3**: 738–752.

Macfoy, C. A., and Cline, E. I. (1990). *In vitro* antibacterial activities of three plants used in traditional medicine in Sierra Leone. *Journal of Ethnopharmacology*, **28:** 323–327.

Masato, K., and Wu, Y. (1994). Chemistry of 1, 2 – Dithinns synthesis of the potent antibiotic Thiarubrine A. *Journal of American Chemist Society*, **116:** 10793–10794.

Mukherjee, S., Banerjee, R., Upadhyay, S. N., Hajra, J., Podder, K. N., Murkherjee, A., and Sana, A. (2006). Reproductive effects of ethnomedicinal formulation of Tapevine leaves in female rats. *Biology and Pharmacology Bulletin*, **29(9)**: 1916–1922.

Muthu, C., Ayyanar, M., Raja, N., and Ignacimuthu, S. (2006). Medicinal plants used by traditional healers in Kancheepuram District of Tamil Nadu, India. *Journal of Ethnobiology and Ethnomedicine*, **2**: 43.

Nalubega, R., Kabasa, J. D., Olila, D., and Kateregga, J. (2011). Evaluation of antibacterial activity of selected ethnomedicinal plants for poultry in Masaka district, Uganda. *Research Journal of Pharmacology*, **5(2)**: 18 – 21.

Nash, R. J. (2004). Remedies from nature. *Chemistry World*, **7**: 20 – 23.

Nguelefack, T. B., Watcho, P., Wansi, S. L., Mbonuh, N. M., Ngamga, D., Tane, P., and Kamanyi, A. (2005). The antiulcer effects of the methanol extract of the leaves of

Aspilia africana (*Asteraceae*) in rats. *African Journal of Traditional and. Complementary Alternative Medicine,* **2 (3):** 233 – 237.

Obadoni, B. O., and Ochuko, P.O. (1998). Phytochemical studies and comparative efficacy of crude extracts of some haematostatic plants in Edo and Delta states in Nigeria. *http: /www.life.vivc.edubio.324/vert_. plant.html.* Retrived July 10, 2009.

Obute, G. C., and Adubor, G. O. (2005). Chemicals detected in plants used for folk medicine in Southeastern Nigeria. *http./www.siu.edu/hebl/leaflets/obute.html.* Retrieved October 23, 2007.

Obuzor, G. U., and Ntui, J. N. (2010). Fatty acid composition of *Aspilia africana* (Nigerian). *International Archive of Applied Sciences and Technology,* **1 (2)**: 13-19.

Obuzor, G.U., and Ntui, J.N. (2011). Essential oil composition of *Aspilia africana* (pers.) C.D. Adams of Port Harcourt, Nigeria. I*nternational Journal of Academic Research,* **3(2):** 140 – 143.

Oguike, M. A., and Udeh, M. E. (2009). Effects of *Spondia mombin* L. on milk composition of lactating West African Dwaf ewes. *Nigerian Journal of Animal Production,* **36 (2):** 335–343.

Oko, O. O. K., and Offiong (2012). Performance indices of broiler chickens fed *Aspilia africana* leaf meals. Under review.

Oko, O. O. K. (2010). Efficacies of *Aspilia africana* leaf meal and extracts as alternative antibiotic growth promoters in quails. Ph. D. Dissertation, University of Calabar, Nigeria.

Oko, O. O. K. (2011). Comparative efficacy of meal and extracts of *Aspilia africana* leaf in laying quails. *Journal of Animal Science,* **89(E-Suppl. 1)**: 724.

Oko, O. O. K., and Agiang, E. A. (2009). Phytochemical evaluation of 3 extractions of *Aspilia africana* leaves. A paper presented at the International conference on global food crisis. Federal University of Technology, Owerri, Nigeria. Pp 87 – 92.

Oko, O. O. K., and Agiang, E. A. (2011a). Phytochemical activities of *Aspilia africana* leaf using different extractants. *Indian Journal of Animal Sciences,* **81 (8)**: 814 – 818.

Oko, O. O. K., and Agiang, E. A. (2011b). Phytochemical properties of *Aspilia africana* leaf. *Planta Medica,* **77:** 1307.

Oko, O. O. K., Agiang, E. A., and Afolabi, A. F. (2011a). *Aspilia africana* leaf as egg booster and yolk colourant in quails. *International Journal of Agriculture,* **3(1)**: 103 – 108.

Oko, O. O. K., Agiang, E. A., and Eneji, C. A. (2012b). Alterations in lipid profile of quails following *Aspilia africana* leaf extracts. *World Poultry Science Journal,* **68(Suppl. 1):** 136.

Oko, O. O. K., Agiang, E. A., and Iso, I. E. (2012a). Dietary *Aspilia africana* leaf on nutrient digestibility and physic-chemical properties of intestinal segments in quails. *Journal of Animal Science,* **90(suppl.3)**: 136.

Oko, O. O. K., Agiang, E. A., and Mbre, J. I. (2010). Plantain (*Musa paradisicia*) root meal as a coccidiostat in growing cockerels. *International Journal of Animal Science*, **2(1)**: 34 – 38.

Oko, O. O. K., Agiang, E. A., Osim, E. E., and Asuquo, O. R. (2011b). Toxicological evaluation of crude extracts of *Aspilia africana* leaf. *American Journal of Pharmacology and Toxicology*, **6(3):** 96 – 101.

Okokon, J. E., Nwidu, L. L., and Essiet, G. A. (2006). Evaluation of *in vivo* antiplasmodial activity of *Aspilia africana. International Journal of Pharmacology*, **2(3):** 348 – 351.

Okoli,C.O., Akah, P.A., Nwafor,S.V., Anisiobi, A.J., Ibegbunam I.N., and Erojikwe, O. (2007a). Antiinflammatory activity of hexane leaf extract of *Aspilia africana. Journal of Ethnopharmacology*, **109(2)**: 219-225.

Okoli, C. O., Akah, P. A., and Okoli A. S. (2007b). Potentials of leaves of *Aspilia africana* (Compositae) in wound care: an experimental evaluation. *Biomedcentral (BMC)*, 7: 24.

Okwu, D. E., and Josiah, C. (2006). Evaluation of the chemical compostion of two Nigerian medicinal plants. *African Journal of Biotechnology*, **5**: 357 – 361.

Okwu, D.E., and Ekeke, O. (2003): Phytochemical screening and mineral composition of chewing sticks in South-Eastern Nigeria. *Global Journal of Applied Sciences*, **9:** 235-238.

Okwuonu, C. U., Oluyemi, K. A., Baxter, G. D., Adesanya, O. A., Ukwenya, V. O., Odion, B. I., and Ofusori, D. A. (2008). Effects of methanolic extract of *Aspilia africana* leaf on the ovarian tissue and weights of Wistar rats. *International Journal of Alternative Medicine*, **5(1):** 12 – 16.

Oliver, B. E. P. (1960). Medicinal plants in Nigeria. Nigerian College of Arts, Science and Technology, Lagos Nigeria.

Oluyemi, K. A., Okwuonu, U. C., Baxter, P. G., and Oyesola, T. O. (2007). Toxic effects of methanolic extract of *Aspilia africana* leaf meal on the estrus cycle and uterine tissue of wistar rats. *International Journal of Morphology*, **25(3):** 609–614.

Oyedapo, O.O, Akindele, V. R., and Okunfolami, O. K. (1998). Effects of extracts of Olax *subscorpioides* and *Aspilia africana* leaf on bovine red blood cells. *Phytotherapy Research*, **19**: 633 – 642.

Oyesola, T. O., Oyesola, O. A., and Okoye, C.S. (2010). Effects of aqueous extract of *Aspilia africana* on reproductiove functions of female wistar rats. *Pakistan Journal of Biological Sciences*, **13 (3):** 126 – 131.

Page, J. E., Baiza, F., Nishida, T., and Towers, G. H. M. (1992). Biologically active diterpenes from *Aspilia mossambicensis*- a Chimpanzee medicinal plant. *Phytochemistry*, **31(10):** 3437-3439.

Preethi, R., Devanathan, V. V., and Loganathan, M. (2010). Antimicrobial and antioxidant efficacy of some medicinal plants against food borne pathogens. *Advances in Biological Research*, **4 (2):** 122– 125.

Rodriguez, E., and Wrangham, R. W. (1992). Zoopharmacology: Medicinal plant used by wild apes and monkeys'. *Paper presented at the 1992 Meeting of American Association for the Advancement of Science (AAAS),* Chicago. pp. 295.

Singh, R., and Sawhney, S. K. (1988). Advances in frontier areas of plant biochemistry. Prentice-Hall of Indian Private Ltd, New Delhi. Pp 489.

Soforowa, E.A (1996). Research on medicinal plants and traditional medicine in Africa. *J. Alternative and Complementary Med.,* **2(3):** 365-372.

Sofowora A. (1993). Medicinal plants and traditional medicine in Africa. Spectrum books limited: Ibadan.

Strausel, L., and Saltman, P. (2000). Trace mineral interactions. Department of Biology, University of California, San Diego, USA.

Tabata, M., Seziek, E., Honda, G., Yesilada, E., Fukui, H., Goto, K., and Ikeshiro, Y. (1994). Traditional medicine in Turkey. III. Folk medicine in East Anatolia, Van and Bitlis, India. *International Journal of Pharmacology,* **32:** 3–12.

Taiwo, B. B. A., Sekoni, A. O., and Fayemi, A. O. (2008). Flora composition of rangeland of Egbado North, Ogun State, Nigeria. *Proceedings of the 13th Annual Conference of Animal Science Association of Nigeria (ASAN),* September 15–19, Ahmadu Bello University, Zaria. pp. 629 – 634.

Taziebou, L. C., Etoa, F. X., Nkeogum, B., Pieme, C.C.A., and Dzeufiet, D. P. D. (2007). Acute and sub-acute toxicity of *Aspilia africana* leaves. *African Journal of Traditional and Complementatry Medicine,* 4: 127 – 134.

Ubaka, M.C., Ukwe, V.C., Okoye, C.T., and Adibe, O.M. (2010). Investigation into the anti-ulcer activity of the aqueous leaf extract of *Aspilia africana* C.D. Adams. *Asian Journal of Medical Sciences,* **2(2)**: 40-43.

Usman, L. A., Oladosu, I. A., Olawore, N. O., Fakunle, C. O., Hamid, A.A., and Lawal, A. (2010). Chemical composition of leaf oil of Nigerian grown *Aspilia africana* C. D. Adams. *Journal of Medicinal Plants Research,* **4(4):** 346-348.

Waako, P. J., Smith, P., and Folb, P. T. (2005). *In vitro* interactions of *Aspilia africana* (Pers.) C.D. Adams, a traditional antimalaria medicinal plant, with arternisinin against *Plasmodium falciparum. Journal of Ethnopharmacology,* **102**: 262 – 268.

WHO, (1990). Evaluation of certain veterinary drug residues in Food. 36th Report of the Joint FAO/WHO expert committee on food additives. World Health Organization. *Technical. Report Series,* **799:** 45–56.

Wink, M. (2004). Evolution of toxins and antinutritional factors in plants with special emphasis on *Leguminosae.* In: T. Acamovic, C. S. Stewart and T. W. Pennycott (Eds.). *Poisonous plants and related toxins.* CABI Publishing, Wallingford. pp. 1 – 25.

Bioactive Phytochemicals: Perspectives for Modern Medicine Vol. 2 (2014) *Pages* **411–420**
Editor: **V.K. Gupta**
Published by: **DAYA PUBLISHING HOUSE, NEW DELHI**

18

Phytochemical Composition and *In vitro* Anti-oxidant Property of *Elsholtzia communis*

C.C Barua[1*], J.D. Roy[1], D.C. Pathak[2], M.J. Bordoloi[3] and K. Khate[4]

ABSTRACT

Elsholtzia communis, a herb belongs to the family- Lamiaceae, is one of the most endemic species of North East India. Traditionally the herb is used as tonic, condiment, carminative, astringent and anti-septic. Various extracts used in this study are Elsholtzia communis ethanol extract (GAP-253/1/1), Elsholtzia communis ethanol-water extract (GAP-253/1/2) and Elsholtzia communis water extract (GAP-253/1/3). The extract were subjected to phytochemical screening along with quantitative analysis of the important phytoconstituents responsible for antioxidant property which was screened by 2, 2-Diphenyl-1-picryl hydrazyl (DPPH) radical scavenging activity, reductive ability and Superoxide radical scavenging activity method. In this study, GAP/253/1/1 showed presence of diterpenes, phenolics, steroids, triterpenes, flavonoids, tannin, GAP/253/1/2 showed presence of diterpenes, phenolics, steroids, triterpenes, flavonoids, tannin and GAP/253/1/3 showed presence of phenolics, steroids, flavonoids, tannin and glycosides. GAP/253/1/1 showed maximum phenolic content with 0.157 ± 0.001

1 Department of Pharmacology and Toxicology, College of Veterinary Science, Khanapara, Guwahati – 2, Assam, India

2 Department of Pathology, College of Veterinary Science, Khanapara, Guwahati – 2, Assam, India

3 Department of Natural Product Chemistry, NEIST, Jorhat – 22, Assam, India

4 Department of Livestock Production and Management, College of Veterinary Science, Khanapara, Guwahati – 2, Assam, India

* *Corresponding author*: E-mail: chanacin@gmail.com; chanacin@satyam.net.in

mg tannic acid/g of dry plant material and GAP/253/1/3 showed maximum flavonoid content with 0.043 ± 0.020 mg quercetin/g of dry plant material. Among the three extracts of Elsholtzia communis, the aqueous extracts i.e., GAP/253/1/3 showed best in-vitro antioxidant activity in DPPH method as compared to the ethanol and hydro-ethanol extracts of the plants.

Keywords: Anti-oxidant, 2,2-Diphenyl-1-picryl hydrazyl (DPPH), *Elsholtzia communis*, Flavonoids phenolics.

Introduction

Plants and plant-based medicaments are the basis of many of the modern pharmaceuticals we use today for our various ailments (Abraham, 1981). Knowledge of the chemical constituents of plants is desirable, not only for the discovery of therapeutic agents, but also because such information may be of value in disclosing new sources of such economic materials as tannins, oils, gums, precursors for the synthesis of complex chemical substances. Phytochemicals are increasingly purported to exert potent beneficial actions to support health and may play a role in reducing synthetic drug use for the treatment of metabolic complications. To this effect, research has focused on the identification and isolation of compounds from natural products with high antioxidant capacities (Okamura *et al.*, 2003; Parasakthy *et al.*, 1996; Katsuzaki *et al.*, 1993). Recent studies have shown that a number of plant products including polyphenols, terpenes and various plant extracts exerted an antioxidant action. Phenolic acids, flavonoids are the most commonly found polyphenolic compounds in plant extracts (Wolfe *et al.*, 2003; Naik *et al.*, 2006). Flavonoids are 15-carbon compounds generally distributed throughout the plant kingdom (Harborne, 1988). Flavonoids and many other phenolic compounds of plant origin have been reported as scavengers of reactive oxygen species (ROS), and are viewed as promising therapeutic drugs for free radical pathologies (Parshad *et al.*, 1998; Chang *et al.*, 2007). The implication of oxidative and free radical–mediated reactions in degenerative processes related to aging and chronic disease conditions is important. Free radicals in the form of reactive oxygen species (ROS) and reactive nitrogen species are implicated in numerous diseases such as inflammation, metabolic disorders, reperfusion damage, atherosclerosis, and carcinogenesis (Robak *et al.*, 1988; Ames *et al.*, 1993). There is increasing evidence that indigenous antioxidants may be useful in preventing the deleterious consequences of oxidative stress and there is increasing interest in the protective biochemical functions of natural antioxidants contained in spices, herbs and medicinal plants.

Elsholtzia communis, locally known as "Lomba" in Manipur and "Nimhu" in Nagaland belongs to the family Lamiaceae is used as tonic, astringent, carminative and antiseptic. Traditionally the herb is used as tonic, condiment, carminative, astringent and anti-septic. Decoction of leaves and flowers are used in tonsilitis, fever, cough, high blood pressure, nose bleeding, menstrual disorder, treatment of body itching, headache, indigestion, as salad and for flavouring food preparation for its characteristic aroma. It is also believed to be stimulant and also aphrodisiac (Singh *et al.*, 2003; Singh, 2007). Literature survey has shown that *E. communis* var. purple flower and *E. communis* var. white flower were screened for polyphenolic compounds and free radical scavenging activity (Khomdram and Singh, 2011); the plant was also screened for trace elements (Singh *et al.*, 2010). Although it is used as spice by many tribes of North East India for its aromatic, tonic and digestant property but scientific study on its antioxidant or other medicinal property are still lacking. Hence, the present study is undertaken to determine its anti-oxidant property *in vitro* along with phytochemical analysis.

Materials and Methods

Plant Material

Leaves of *Elsholtzia communis* were collected during the month of August – September, 2011 from Jharnapani, Nagaland. The plant specimens were identified by Dr. Iswar Barua, Principal Scientist, Department of Agronomy, AAU, Jorhat, Assam and voucher specimens were deposited in the herbarium.

Preparation of Various Extracts

With 95 per cent Alcohol

About half Kg of the dried/fresh plant material was submerged with sufficient quantity of ethanol; the material was filtered and was further extracted this way three/four times. The filtrate obtained was distilled under reduced pressure at 50°C. Per cent recovery for each extract was noted (GAP/253/1/1).

Aqueous Ethanol Extract

About half Kg of the dried/fresh plant material was submerged with sufficient quantity of ethanol and water (1: 1), the material was filtered and was further extracted this way three/four times. The filtrate obtained was distilled under reduced pressure at 50°C. Per cent recovery for each extract was noted (GAP/253/1/2).

With Water

The same procedure was followed as described above for the extraction of the plant material with cold distilled water. The extract was then distilled below 50°C under vacuum. Then it was dried using a lyophilizer per cent recovery for each extract was noted (GAP/253/1/3).

Phytochemical Screening

The extracts were subjected to phytochemical screening as per the method of Harborne, 1991 for presence of different phytoconstituents.

Quantitative Study of Phytoconstituents

Determination of Total Phenolics

Using modified Folin Ciocalteu method, total phenol content in the extracts was determined. An aliquot of the extract was mixed with 5 ml Folin Ciocalteu reagent (previously diluted with water 1: 10 v/v) and 4 ml (75 gm/l) of sodium carbonate. The tubes were vortexed for 15 seconds and allowed to stand for 30 minutes at 40°C for colour development. Absorbance was then measured at 765 nm using spectrophotometer (UV 2100, Chemito). Samples of extracts were evaluated at a final concentration of 0.1 mg/ml. Total phenolics were expressed as mg/g tannic acid equivalent using the following equation based on the calibration curve: y=0.1216x where, x is the absorbance, y is the tannic acid equivalent (mg/g).

Determination of Total Flavonoids

Estimation of the total flavonoids in the plant extracts was carried out using the method of Ordon Ez *et al.*, 2006. To 0.5 ml of sample, 0.5 ml of 2 per cent $AlCl_3$ solution in ethanol was added. After 1 hour incubation at room temperature, the absorbance was measured at 420 nm (UV 2100, Chemito). A yellow colour indicated the presence of flavonoids. Extract samples were evaluated at a final concentration of 0.1 mg/ml. Total flavonoids were expressed as quercetin (mg/g) equivalent using the following equation based on the calibration curve: y=0.0255x where, x is the absorbance, y is the quercetin equivalent (mg/gm).

Antioxidant Activity

2,2-Diphenyl-1-picryl Hydrazyl (DPPH) Radical Scavenging Activity

The free radical scavenging activities of the plant extracts against 2, 2-Diphenyl-1-picryl hydrazyl (DPPH) radical (Sigma-Aldrich) were determined by using UV spectrophotometer (UV 2100, Chemito) at 517 nm (Ayoola *et al.*, 2006). Vitamin C was used as antioxidant standard at different concentrations. The free radical scavenging activity was calculated using the following formula:

Per cent inhibition = $\{[A_b - A_a]/Ab\}$ x,

where,

A_b is the absorption of the blank sample, A_a is the absorption of the extract.

Reductive Ability

Reducing power of plant extract is based on the ability of anti-oxidants to form coloured complex with potassium ferricyanide, TCA and $FeCl_3$. One ml of different concentrations of plant extract (to produce final concentration 100-600 μg/ml) were mixed with 2.5 ml potassium ferricyanide (1 per cent) and 2.5 ml of phosphate buffer (pH 6.6). This mixture was incubated at 50° C for 20 minutes. Thereafter, 2.5 ml of TCA (10 per cent) was added to it and centrifuged at 3000 rpm for 10 minutes, followed by addition of 2.5 ml of supernatant, 2.5 ml of water and 0.5 ml of $FeCl_3$ (0.1 per cent) (Jayprakesh *et al.*, 2001). Absorbance was measured at 700nm by using UV spectrophotometry (UV 2100, Chemito).

Superoxide Radical Scavenging Activity

Superoxide anion scavenging activity of plant extract was measured according to the method of Robak *et al.*, 1988 with some modifications. All the solutions were prepared in 100mM phosphate buffer (pH 7.5). One ml of nitroblue tetrazolium chloride (NBT) (156 μM), 1 ml of NADH (468 μM) and 3 ml of plant extracts were mixed. The reaction was started by adding 100 μl of phenazine methosulphate (PMS) (60 μM) and the mixture was then incubated at 25° C for 5 minutes followed by measurement at 560 nm (U V 2100, Chemito). The percentage inhibition was calculated from the formula:

Per cent inhibition = $\{[A_b - A_a]/Ab\} \times 100$

A_b is the absorption of the blank sample, A_a is the absorption of the extract.

Results and Discussion

Over the past few years, investigations for phenolics compounds in medicinal herbs have gained importance due to their high antioxidative properties (Zhu *et al.*, 2004). The present study involves the determination of flavonoids and phenolic content as well as the antioxidant activity of few fractions of *Elsholtzia communis* which is known for such activities.

The phytochemical screening of GAP/253/1/1 showed presence of diterpenes, phenolics, steroids, triterpenes, flavonoids, tannin; GAP/253/1/2 showed presence of diterpenes, phenolics, steroids, triterpenes, flavonoids, tannin and GAP/253/1/3 showed presence of phenolics, steroids, flavonoids, tannin and glycosides (Table 18.1).

Phenolic compounds are the principal antioxidant constituents of natural products and are composed of phenolic acids and flavonoids, which are potent radical terminators (Kahkonen *et al.*, 1999; Shahidi *et al.*, 1997). They donate hydrogen to radicals and break the reaction of lipid oxidation at the initiation step (Gilicin *et al.*, 2004). Caffeic acid, ferulic acid, and vanillic acid are examples of phenolic acids isolated as natural antioxidants in fruits, vegetables, and other plants (Javanmardi *et al.*, 2003; Larson, 1988). The high potential of polyphenols to scavenge free radicals may be because of their many phenolic hydroxyl groups (Sawa *et al.*, 1999). In this study, among the three extracts, GAP/253/1/3 (*Elsholtzia communis* ethanol extract) showed maximum phenolic content with 0.035 ± 0.006 mg tannic acid/g of dry plant material (Table 18.1). Flavonoids exhibit inhibition of mutagenicity induced by chemical mutagens and have anticarcinogenic, antioxidant and anti-inflammatory activities (Miyazawa *et al.*, 2000). In this study, GAP/253/1/3 (*Elsholtzia communis* water extract) exhibited maximum flavonoid content with 0.043 ± 0.020 mg quercetin/g of dry plant material (Table 18.1).

DPPH forms a stable molecule on accepting an electron or a hydrogen atom and thus has applications in the determination of radical scavenging activity of natural products (Jun *et al.*, 2004). *In situ*, free radicals like polyaromatic hydrocarbon cations have been linked with carcinogenesis (Yen and Chen, 1995). Thus, products that will scavenge DPPH *in vitro* may also scavenge polyaromatic hydrocarbon cations *in vivo*. GAP/253/1/3 exhibited the best *in vitro* anti-oxidant activity among the three extracts at three concentrations of 50 µg/ml, 100 µg/ml and 500 µg/ml with percent inhibition of 95.50, 95.70 and 98.50 respectively. The standard Vitamin –C exhibited the maximum antioxidant activity at 2000 µg/ml (Table 18.2).

A substance may act as an antioxidant due to the ability to reduce ROS by donation of hydroxyl atom. The reducing property of various extracts of the plant implies that it is capable of donating hydrogen. The high content of phenol compound in the extract may be a contributing factor towards anti-oxidant activity because the

Table 18.1: Physical and phytochemical characteristics of various extracts of *Elsholtzia communis.*

Extract Code	*Name of the Extract*	*Colour*	*Consistency*	*Recovery (per cent)*	*Possible Phyto-constituents Present*	*Phenolic Content (mg tannic acid/ g of dry Plant material)*	*Total Flavonoid Content (mg quercetin/g of dry plant material)*
GAP-253/1/1	*Elsholtzia communis* ethanol extract	Dark brown	Semi solid	4.8	Diterpenes, Phenolics, steroids, triterpenes, flavonoids, tannin	0.025 ± 0.004*	0.022 ± 0.007*
GAP-253/1/2	*Elsholtzia communis* ethanol-water extract	Dark brown	Semi solid	5.29	Diterpenes, phenolics, steroids, triterpenes, flavonoids, tannin	0.018 ± 0.003*	0.020 ± 0.002*
GAP-253/1/3	*Elsholtzia communis* water extract	Dark brown	Solid	4.00	Phenolics, steroids, flavonoids, tannin and glycosides	0.035 ± 0.006	0.042 ± 0.080

Values are Mean ± SE. N=6; *P < 0.05.

Table 18.2: *In-vitro* antioxidant activity of various extracts of *Elsholtzia communis* by DPPH method.

Plants Extract Code	*Concentration (µg/ml)*				
	50	*100*	*500*	*1000*	*2000*
			Per cent inhibition		
GAP/253/1/1	82.73 ± 0.850**	84.20 ± 1.590**	94.45 ± 1.520**	56.20 ± 3.000**	67.40 ± 1.000**
GAP/253/1/2	82.11 ± 1.090**	79.50 ± 4.000**	67.90 ± 3.000	52.60 ± 2.000**	65.30 ± 3.000**
GAP/253/1/3	95.50 ± 3.000	95.70 ± 3.000	98.70 ± 3.000**	78.80 ± 3.000	71.40 ± 1.000
Vitamin- C 81.92 ± 0.600		50.17 ± 1.450	52.36 ± 0.480	56.78 ± 0.240	59.75 ± 0.240

Values are Mean ± SE; N=6; **: P< 0.01, highly significant.

Table 18.3: Reductive ability of various extracts of *Elsholtzia communis.*

Plants Extract Code	*Concentration (μg/ml)*					
	100	*200*	*300*	*400*	*500*	*600*
	Absorbance at 700 nm					
GAP/253/1/1	0.913 ± 0.020**	0.520 ± 0.006**	0.340 ± 0.014**	0.150 ± 0.002**	0.250 ± 0.011	0.040 ± 0.003**
GAP/253/1/2	0.912 ± 0.019**	0.522 ± 0.007**	0.350 ± 0.011**	0.152 ± 0.003**	0.280 ± 0.002	0.041 ± 0.004**
GAP/253/1/3	1.508 ± 0.036	1.910 ± 0.003	1.890 ± 0.020	2.290 ± 0.061	2.550 ± 0.014**	1.450 ± 0.012

Values are Mean ± SE. N=6; **P< 0.01, highly significant.

Table 18.4: Superoxide radical scavenging activity of various extracts of *Elsholtzia communis.*

Plants Extract Code	*Concentration (μg/ml)*					
	20	*40*	*60*	*80*	*100*	*110*
	Per cent inhibition					
GAP/253/1/1	92.88 ± 1.140**	85.32 ± 0.710	93.25 ± 1.120**	54.03 ± 1.360	63.46** ± 1.170	71.12 ± 1.095
GAP/253/1/2	91.30 ± 0.330**	92.18 ± 0.500	93.10 ± 1.250**	92.58 ± 0.780	92.69 ± 1.20	83.50 ± 1.18
GAP/253/1/3	96.99 ± 0.820**	73.17 ± 1.28	63.31 ± 0.67	61.09 ± 0.280	62.79** ± 1.020	59.83 ± 0.970

Values are Mean ± SE. N=6; **P< 0.01, highly significant.

phenol compounds are known to have direct antioxidant property due to the presence of hydroxyl group which functions as hydrogen donor (Dud *et al.*, 1999; Dreosti, 2000). In this study, it was observed that in comparison to other two extracts, GAP/253/1/3 showed maximum reducing ability (2.539 ± 0.100) at 500 μg/ml correlating with the findings of DPPH radical scavenging activity in which the extract was found to possess potent anti-oxidant activity. The extract was found to be rich in their phenolics and flavonoid content which might contribute to its superior reducing activity (Table 18.3).

Superoxides are produced from molecular oxygen due to oxidative enzymes (Sainani *et al.*, 1997) of the body as well as via non enzymatic reaction such as auto oxidation by catecholamine (Hemmani and Parihar, 1998). In this study, superoxide radical reduces NBT to a blue coloured formazan that is measured at 560 nm (Khanam *et al.*, 2004). The result is shown in Table 18.4. Among the three extracts, GAP/253/1/3 exhibited maximum superoxide radical scavenging activity at the concentration of 20μg/ml. The probable mechanism of scavenging of the superoxide anions may be due to the inhibitory effect of extracts towards generation of superoxide ion in the *in vitro* reaction mixtures.

The potent antioxidant activity can be explained by the presence of various phyto constituents in the extract. It has been reported that the antioxidant activity of many compounds of botanical origin is proportional to antioxidant content suggesting a correlation between total phenolics and antioxidant activity (Rice-Evans *et al.*, 1997; Veglioglu *et al.*, 1998). As a result, extracts with high phenolic and flavonoids contents, exhibit better anti-oxidant activity. The plant *Elsholtzia communis* specially its aqueous extract *i.e.* GAP/253/1/3 has exhibited maximum antioxidant and reducing ability in comparison to the other two extracts. As such the plant is rich in phenolic and flavonoid content and exerts its beneficial anti oxidant activity due to these phytoconstituents.

Acknowledgements

The authors are grateful to Department of Biotechnology, Govt. of India, New Delhi for providing financial assistance to carry out this work. Physical facility provided by the Director of Research (Vety), F.V.Sc is also gratefully acknowledged.

References

Abraham, Z., 1981. Glimpses of Indian Ethnobotany. Oxford and Publishing Co., New Delhi, pp. 308-320.

Ames, B.N., Shigenaga, M. K., and Hagen, T.M. (1993). Oxidants, antioxidants and the degenerative diseases of aging. *Proceeding of National Academy Sciences*, USA, **90:** 7915-7922.

Ayoola, G.A., Sofidiya, T., Odukoya, O., and Coker, H.A.B. (2006). Phytochemical screening and free radical scavenging activity of some Nigerian medicinal plants. *Journal of Pharmacy Science and Pharmaceutical Practice*, **8:** 133-136.

Chang, H., Huang, G., Agrawal, D.C., Kuo, C., Wu, C., and Tsay, H. (2007). Antioxidant activities and polyphenol contents of six folk medicinal ferns used as "Gusuibu". *Botanical Studies*, **48:** 397-406.

Deosti, *I.E.* (2000). Antioxidant polyphenols in tea, cocoa and wine. *Nutrition*, **16:** 269.

Dud, P. H., Tu, Y.Y., and Yen, G. C. (1999) Antioxidant activity of aqueous extract of han jyut (*Chrysanthemum morifolium* Ramat). *Lebesmittel-Wissenchaft und Technologie*, **32:** 269.

Gqlcin, I., Beydemir, S., Alici, H.A., Elmasta,Y.M., and Bqyqkokurovlu, M.E. (2004). *In vitro* antioxidant properties of morphine. *Pharmacological Research*, **49:** 59-66.

Harborne, J. B. (1991). Phytochemical Screening –Guide to modern techniques of plant analysis. 2nd edition, Chapman and Hall, pp. 653.

Hemmani, T., and Parihar, M. S. (1998). Reactive oxygen species and oxidative DNA damage. *Indian Journal of Physiology and Pharmacology*, **42:** 440.

Javanmardi, J., Stushnoff, C., Locke, E., and Vivanco, J.M. (2003). Antioxidant activity and total phenolic content of Iranian *Ocimum* accessions. *Food Chemistry*, **83:** 547-550.

Jayprakash, G. K, Singh, R.P., and Sakariah, K.K. (2001). Anti-oxidant activity of grape seed extracts on peroxidation models *in vitro*. *Journal of Agricultural and Food Chemistry*, **55:** 1018.

Jun, M., Tohru, U., Jianzhang, L., and Takeshi, F. (2004). Identification and evaluation of antioxidant activities of bamboo extracts. *Study in China*, **6:** 1-5.

Katsuzaki, H., Kawashiki, S., and Osawa, T. (1993). Structure of novel antioxidative lignan triglucoside isolated from sesame seed. *Heterocycles*, **36:** 933- 936.

Khanam, S., Shivprasad, H. N., and Kshama, D. (2004). *In vitro* antioxidant screening models: a review. *Indian Journal of Pharmaceutical Education and Research*, **38:** 180

Khkfnen, M. P., Hopia, A.I., Vuorela, H. J., Rauha, J. P., Pihlaja, K., and Kujala, T.S. (1999). Antioxidant activity of plant extracts containing phenolic compounds. *Journal of Agricultural and Food Chemistry*, **47:** 3954- 3962.

Larson, R. A. (1988) The antioxidants of higher plants. *Phytochemistry*, **27:** 969- 978.

Naik, G.H., Priyadarsini, K.I., and Mohan, H. (2006). Free radical scavenging reactions and phytochemical analysis of triphala, an ayurvedic formulation. *Current Science*, **90:** 1100-1105.

Noda, Y., Anzai-Kmori, A., Kohono, M., Shimnei, M., and Packer, L. (1997). Hydroxyl and superoxide anion radical scavenging activities of natural source antioxidants using the computerized JES-FR30 ESR spectromoter system. *Biochemistry and Molecular Biology International*, **42:** 35-44.

Okamura, H., Mimura, A., Yakou, Y., Niwano, M., and Takahara, Y. (1993). Antioxidant activity of tannins and flavonoids in *Eucalyptus rostrata*. *Phytochemistry*, **33:** 557- 561.

Ordon Ez, A. A. L., Gomez, J. D., Vattione, M. A., and Isla, M.L. (2006). Anti-oxidant activities of *Sechium edule* (Jacq.) Swarts. *Food Chemistry*, **97:** 452-458.

Osawa., T. (1994). Post harvest biochemistry. In: Uritani I, Garcia VV, Mendoza, EM, editors. Novel neutral antioxidant for utilization in food and biological systems. Japan Scientific Societies Press, Japan, pp. 241-251.

Parasakthy, K., Shanthi, S., Deepalakshmi, P., and Niranjali, S. D. (1996). The antioxidant effect of eugenol on carbon tetrachloride induced erythrocyte damage in rats. The *Journal of Nutritional Biochemistry*, **7:** 23-28.

Parshad, R., Sanford, K.K., Price, F.M., Steele, V.E., Tarone, R.E., Kelloff, G.J., and Boone, C. W. (1998). Protective action of plant polyphenols on radiation-induced chromatid breaks in cultured human cells. *Anticancer Research*, **18:** 3263-3266.

Rice-Evans, C.A., Miller, N.J., and Paganga, G. (1997). Antioxidant properties of phenolic compounds. *Trends in Plant Science*, **2:** 152-159.

Robak, J., Shridi, F., Wolbis, M., and Krolikowska, M. (1988). Screening of the influence of flavonoids on lipoxygenase and cyclooxy-genase activity, as well as on nonenzymatic lipid oxidation. *Polish Journal of Pharmacology and Pharmacy*, **40:** 451- 458.

Sainani, G. S., Manika, J. S., and Sainani, R. G. (1997). Oxidative stress: a key factor in pathogenesis of chronic diseases. *Medical Update*, **1:** 1.

Sawa, T., Nakao, M., Akaike, T., Ono, K., and Maeda, H. (1999). Alkylperoxyl radical scavenging activity of various flavonoids and other phenolic compounds: implications for the antitumor promoter effect of vegetables. *Journal of Agricultural and Food Chemistry*, **47:** 397- 402.

Shahidi, F., and Wanasundara, P.K.J.P.D. (1992). Phenolic antioxidants. *Critical Review in Food Science and Nutrition*, **32:** 67 -103.

Veglioglu, Y. S., Mazza, G., Gao, I., and Oomah, B.D. (1998). Antioxidant activity and total phenolics in selected fruits, vegetables and grain products. *Journal of Agricultural and Food Chemistry*, **46:** 4113-4117.

Wolfe, K., Wu, X., and Liu, R. H. (2003). Antioxidant activity of apple peels. *Journal of Agricultural and Food Chemistry*, **51:** 609-614.

Yen, G. C., and Chen, H. Y. (1995). Antioxidant activity of various tea extracts in relation to their antimutagenicity. *Journal of Agricultural and Food Chemistry*, **46:** 849-854.

Zhu, Y. Z., Huang, S. H., Tan, B. K. H., Sun, J., Whiteman, M., and Zhu, Y. C. (2004). Antioxidants in Chinese herbal medicines: a biochemical perspective. *Natural Product Report*, **21:** 478-489.

Bioactive Phytochemicals: Perspectives for Modern Medicine Vol. 2 (2014)
Editor: **V.K. Gupta**
Published by: **DAYA PUBLISHING HOUSE, NEW DELHI**

Pages **421–438**

19

Lignoids of Burseraceae

Igor M.de Assis[1], Juliana S. Pereira[1] and Valdir F. da Veiga-Junior[1*]

ABSTRACT

The Burseraceae family is a well-known source of oleoresin exudates, rich in terpenoids, that are used for many purposes, for example, in perfumery. Lignoids constitute an important group of plant phenolic compounds that are structurally characterized by the coupling of two phenylpropanoid units. It is a group with wide structural variation resulting in the wide variety of classes found in nature, such as lignans and neolignans. Substances belonging to this class have shown important pharmacological effects, such as the antiviral activity of (-)-podophyllotoxin and activity against leishmaniasis of the neolignans virolinsurinamensin. The occurrence of lignans in the family Burseraceae is described in only five genera, Bursera, Commiphora, Protium Crepidospermum and Tetragastris. The genus Bursera is the one with more studies and greater diversity of classes of lignans, as aryltetralin classes, dibenzylbutane, dibenzylbutyrolactone and furanics. The presence of four unusual neolignans was also described in this genus, and the reported biological activities, such as anti-inflamatory, and inhibitory effects on the fibrosarcoma cells, and the cytotoxic effects against cell lines of the human colon, have also led to increased interest in these substances. In the genus Commiphora lignans were identified belonging to the classes and aryltetralin and furofuran, some of them with cytotoxic effects on tumor cells and inhibitory activity against the enzyme α-glucosidase. In Crepidospermum and Tetragastris genus only dibenzylbutyrolactone and furan lignans classes were identified, respectively. In the genus Protium, the most studied the classes of lignans arylnaftalene, coumarinolignoid, and dibenzylbutyrolactol dibenzylbutyrolactones have

1 Departamento de Química, Instituto de Ciências Exatas, Universidade Federal do Amazonas, Av. Gal. Rodrigo Otávio Jordão Ramos, 3000, 69077-040 Manaus–AM, Brazil

* *Corresponding author*: E-mail: valdirveiga@ufam.edu.br

already been described. In the genus, the presence of two neolignans also described. In these last three genera, Crepidospermum, Tetragastris and Protium, no biological studies with isolated lignans have been found in the scientific literature.

Keywords: Lignans, Neolignans, *Bursera, Commiphora, Crepisdospermum, Tetragastris, Protium.*

Burseraceae Family

Burseraceae comprises approximately 700 species in 19 genera (Thulin *et al.*, 2008). In the region of the Neotropics, there are 230 species distributed in eight genera and in the Brazilian Amazon, species of the genera *Bursera, Crepidospermum, Dacryodes, Protium* and *Tetragastris trattinnickia*, can be found (Weeks *et al.*, 2005).

The species of the Burseraceae family are known as "breus" and their resins have been used for many years as insect repellent, and in the paints, varnishes and cosmetics industries. They have also long been used, folk medicine, in the form teas and in the tropical use, due to their important medicinal properties. Their use a tonics, stimulants (Costa, 1975; PioCorrea, 1984), hemostatics, antiblennorrhagics, and antirheumatics, for the treatment of lung and stomach diseases, and in dentistry, have been described (Pio Correa, 1984), as well as their laxative (Quignard *et al.*, 2004), healing, analgesic, and contraceptive effects among other applications (Costa, 1975; Maia *et al.*, 2000; Vieira Junior *et al.*, 2005).

Besides the resins, other parts of the Burseraceae species trees have been studied, such as the leaves, fruits and woo. They are described as having aterpenicchemical composition, with large quantities of mono, sesqui and triperpenes (Figure 19.1) as described by Rüdiger, Siani and Veiga Junior in a review article (2007).

Lignoids: Chemical and Biological Properties

Lignoids are small molecules whose skeleton is formed exclusively, or in conjunction with other groups, the phenylpropanoic group $(C_6.C_3)_n$, being n restricted to a few units, 1 to 3 (Gottlieb and Yoshida, 1984). It is a group with wide structural variation, with 500 lignoids having been described, resulting in a wide variety of classes found in nature, including lignans, neolignans, alolignans, norlignan as, oligolignoids, and heterolignoids. Of these lignans and neolignans the most numerous (Simões, 2004).

Lignans belong to a large class of special metabolites that form an important group of plant phenolic compounds. This term was introduced by Haworth in the mid-1940s to refer to carbon skeletons in which two n-propylbenzenics units (C_6C_3) are connected by β carbons of their propyl side chains (C_3). When these couplings occur between other carbons of these units, these constituents are called neolignans (Figure 19.2). (Umezawa, 2003; Gottlieb and Yoshida, 1984; Moss, 2000).

Lignans and neolignans have attracted great interest for many years, due to their widespread occurrence in nature, and their wide range of biological activities (Agrawal and Thakur, 1985). The podophyllotoxin, for example, has been used against a variety of cancers including testicular cancer and acute lymphocytic leukemia. It

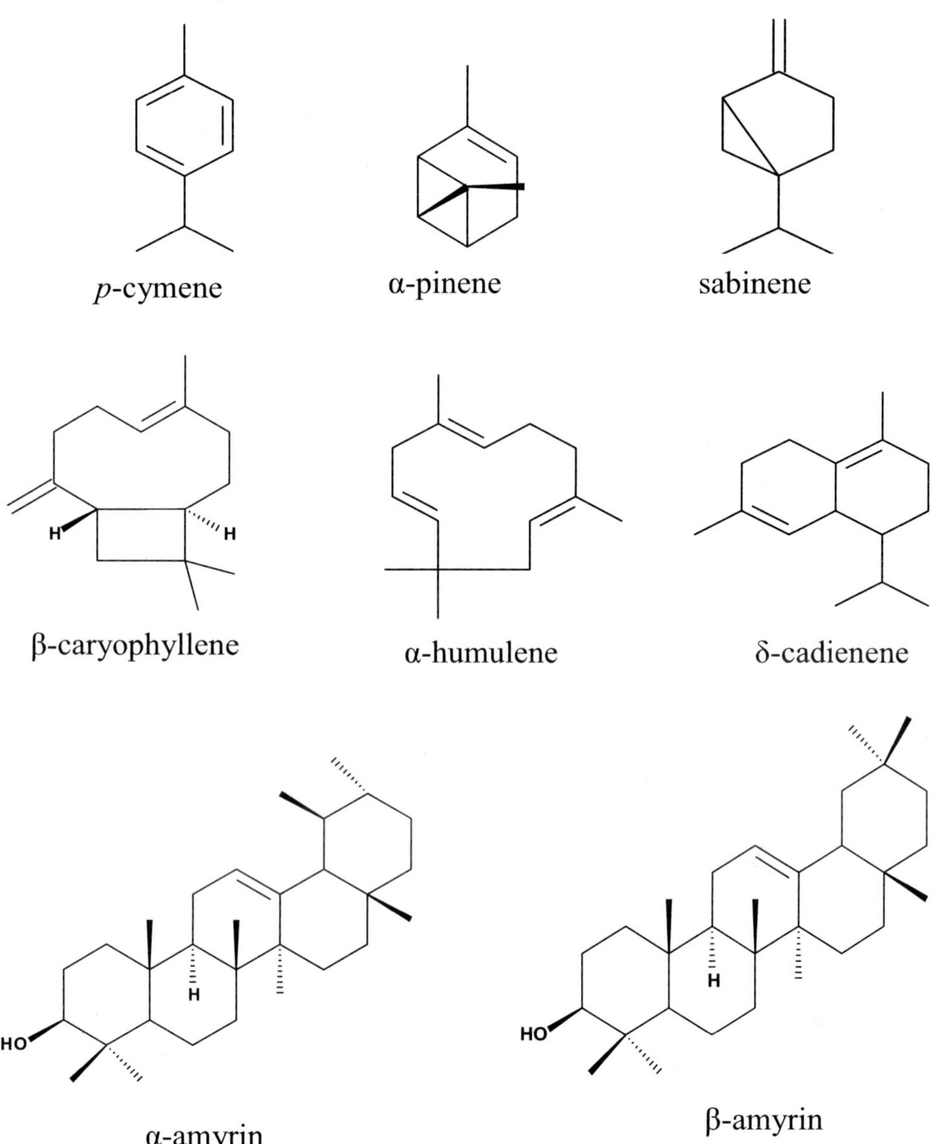

Figure 19.1: Main terpenoids found in Burseraceae.

also presents remarkable antiviral activity, as demonstratedin a study of the alcoholic extract of *Podophyllum peltatum* plants, in which (-)-podophyllotoxin (Figure 19.3) is the main component and has been reported as an antiviral agent for treatment of venereal warts (Ayres, 1990).

The viriolin and susarinamensinneolignans isolated from leaves of *Virola surinamensis* showed activity against leishmaniasis, a tropical disease transmitted by mosquitoes (Costa and Takahata, 1996).

Neolignan

Lignan

Figure 19.2: Basic structures of lignans and neolignans.

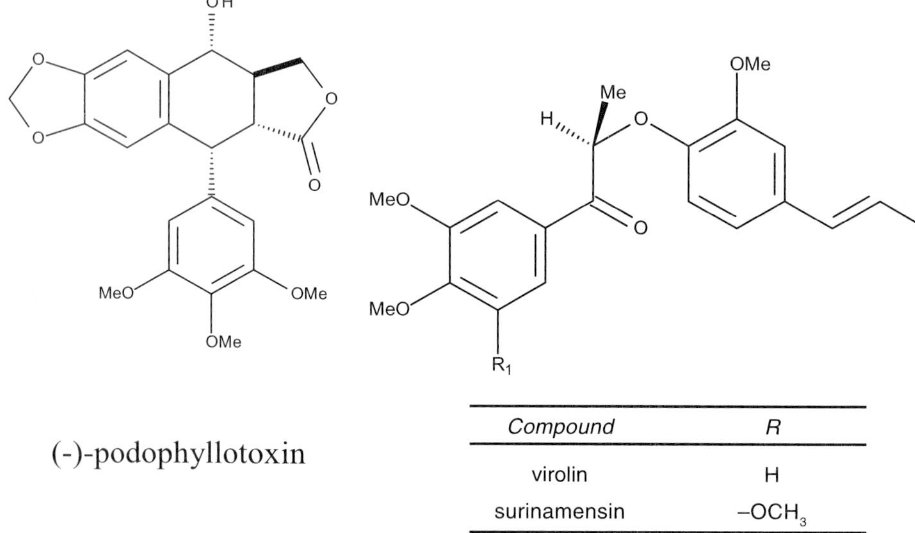

Compound	*R*
virolin	H
surinamensin	$-OCH_3$

Figure 19.3: Aryltetralinlignan (-)-podophyllotoxin and the neolignansvirolin and surinamensin.

Lignans and Neolignans of Burseraceae

Lignans and neolignans were found in both oleoresins, and in leaves, wood and roots. 37 lignans and 6 neolignans are described in the Burseraceae family (Peraza-Sanchez *et al.*, 1992, Siqueira *et al.*, 1995; Jutiviboonsuk *et al.*, 2005). The occurrence of these lignoids is described in only five genera of this family, *Bursera, Commiphora, Protium, Crepidospermum* and *Tetragastris* (Table 19.1).

Lignoids isolated from species of Burseraceae have several biological activities, such as antitumor, anticancer, anti-viral and anti-inflammatory (Wickramaratne *et al.*, 1995; Noguera *et al.*, 1992; Canel *et al.*, 2000; Jutiviboonsuk *et al.*, 2005).

Table 19.1: Lignans and neolignans of Burseraceae.

Sl.No.	Lignans and Neolignans	Species	References
1.	(7'R,8R,8'R)-(-)-deoxypodophyllotoxin	*Bursera permollis,Bursera morelensis, Bursera fagaroides, Bursera microphylla*	Jolad *et al.*, 1977 Velázquez-Jiménez *et al.*, 2011 Bianche *et al.*, 1968
2.	3-(3,4-methylenedioxybenzyl)-4-(3',4',5'-trimethoxybenzyl) tetrahydrofuran burseran	*Bursera microphylla*	Cole *et al.*, 1969
3.	methyl-β-peltatin A	*Bursera simaruba*	Noguera *et al.*, 2004
4.	picropolygamain	*Bursera graveolens, Bursera simaruba*	Nakanishi *et al.*, 2005 Peraza-Sánchez and Peña Rodríguez, 1992
5.	β-peltatin-*O*-β-D-glucopyranoside	*Bursera simaruba*	Maldini *et al.*, 2009
6.	(8R,8'R)-(-)-yatein	*Bursera fagaroides, Bursera simaruba*	Velázquez-Jiménez *et al.*, 2011 Maldini *et al.*, 2009
7.	hinokinin	*Bursera simaruba*	Maldini *et al.*, 2009
8.	bursehernin	*Bursera simaruba*	Maldini *et al.*, 2009
9.	(7R,7'R,8R,8'R)-(-)-podophyllotoxin	*Bursera fagaroides*	Velázquez-Jiménez *et al.*, 2011
10.	(7R,7'R,8R,8'R)-(-)-acetylpodophyllotoxin	*Bursera fagaroides*	Velázquez-Jiménez *et al.*, 2011 Rojas-Sepulveda *et al.*, 2009
11.	(7'R,8R,8'R)-(-)-morelensin	*Bursera fagaroides*	Velázquez-Jiménez *et al.*, 2011
12.	(8R,8'R)-(-)-5'-desmethoxyyatein	*Bursera fagaroides*	Velázquez-Jiménez *et al.*, 2011
13..	desoxypodorhizol	*Bursera fagaroides*	Rojas-Sepulveda *et al.*, 2009
14.	burseranin	*Bursera graveolens*	Nakanishi *et al.*, 2005
15.	(-)- *trans*-2-(3'',4'',5''-trimethoxybenzyl)-3-(3',4'-methylenedioxybenzyl)butyrolactone	*Bursera schlechtendalii*	McDoniel and Cole, 1972
16.	(-)- *trans*-2-(3'',4''-dimethoxybenzyl)-3-(3',4'-methylenedioxybenzyl)butyrolactone	*Bursera schlechtendalii*	McDoniel and Cole, 1972

Contd...

Table 19.1–*Contd...*

Sl.No.	*Lignans and Neolignans*	*Species*	*References*
17.	(2R,3R)-bis-2,3-(3,4-methylenedioxybenzyl-)-1,4-butanediol diacetateariensin	*Bursera ariensis*	Hernández *et al.*, 1983
18.	β-peltatinmethylether	*Bursera permollis*	Wickramaratne *et al.*, 1995
19.	nemerosin	*Bursera permollis*	Wickramaratne *et al.*, 1995
20.	5'-desmethoxydeoxypodophyllotoxin	*Bursera morelensis*	Jolad and Cole, 1977
21.	burselignan	*Bursera tonkinensis* Guillaum	Jutiviboonsuk *et al.*, 2005
22.	(+)-isolariciresinol	*Bursera tonkinensis* Guillaum	Jutiviboonsuk *et al.*, 2005
23.	5-methoxy-(+)-isolariciresinol	*Bursera tonkinensis* Guillaum	Jutiviboonsuk *et al.*, 2005
24.	4'-demethyldesoxypodophyllotoxin	*Bursera tonkinensis* Guillaum	Jutiviboonsuk *et al.*, 2005
25.	4'-demethyldesoxypodophyllotoxin-4-*O*-β-D-glucoside	*Bursera tonkinensis* Guillaum	Jutiviboonsuk *et al.*, 2005
26.	1-(4'-hydroxy-3'-methoxyphenyl)-2-[4"-(3-hydroxypropyl)-2",6"-dimethoxyphenoxy]propane-1,3-diol	*Bursera tonkinensis* Guillaum	Jutiviboonsuk *et al.*, 2005
27.	Dihydrodehydrodiconiferylalcohol	*Bursera tonkinensis* Guillaum	Jutiviboonsuk *et al.*, 2005
28.	5-methoxy-*trans*-dihydrodehydrodiconiferyl alcohol	*Bursera tonkinensis* Guillaum	Jutiviboonsuk *et al.*, 2005
29.	burseneolignan	*Bursera tonkinensis* Guillaum	Jutiviboonsuk*et al.*, 2005
30.	2β,3β-dhidroxy-6-methoxy-4β-angeloyloxyisopicropolygamatin	*Commiphora erlangeriana*	Habtemarian,2003 Dekebo *et al.*, 2002
31.	2α-acetoxy-6-methoxipicropolygamatin	*Commiphora erlangeriana*	Habtemarian,2003 Dekebo *et al.*, 2002
32.	2α-acetoxydeoxypicropodophyllin	*Commiphora erlangeriana*	Habtemarian,2003 Dekebo *et al.*, 2002
33.	2α-acetoxy-3β-hydroxyisodeoxypodophyllotoxin	*Commiphora erlangeriana*	Habtemarian,2003 Dekebo *et al.*, 2002
34.	epiexcelsin	*Commiphora mukul*	Abbasi*et al.*, 2005, 2009

Contd...

Table 19.1–*Contd...*

Sl.No.	*Lignans and Neolignans*	*Species*	*References*
35.	5'-demethoxyepiexcelsin	*Commiphora mukul*	Abbasi *et al.*, 2005, 2009
36.	5,5'-tetrahydro-1H,3H-furo[3,4-c]furan-1,4-diylbis [7-(methoxy)-1,3-benzodioxole]	*Commiphora mukul*	Francis *et al.*, 2004
37.	Parabenzolactone	*Crepidospermum rhoifolium*	Lima *et al.*, 2001
38.	(-)-savinin	*Tetragastris altíssima*	Lima *et al.*, 2001
39.	(-)-cubebin	*Protium tenuifolium*	Siqueira *et al.*, 1995
40.	(+)-(2S,3S)-2-(3",4"-methylenedioxybenzyl)-3-(3',4'-methylenedioxyacetophenone)-butyrolactone	*Protium tenuifolium*	Siqueira *et al.*, 1995
41.	5-methoxyjusticidin A	*Protium unifoliolatum*	Siani *et al.*, 1998
42.	Propacin	*Protium opacum*	Zoghbi *et al.*, 1981; Marques and Yoshida, 1990
43.	5-methoxypropacin	*Protium unifoliolatum*	Magalhães *et al.*, 2006

Genus *Bursera*

Bursera is a genus that has more studies and the highest diversity of classes of lignans. The class esaryltetralin, dibenzylbutane, dibenzylbutyrolactone and furan have been identified, and the presence of neolignanshas also been described (Figure 19.4).

In studies of Bianchi *et al.* (1968) and Cole *et al.* (1969), on the wood of *Bursera microphylla*, the lignans deoxypodophyllotoxin [1] and burseran [2] were identified, belonging to the aryltetralin and furan classes, respectively, and presenting potential antitumor activity.

Figure 19.4: Lignans and neolignans of the genus *Bursera*.

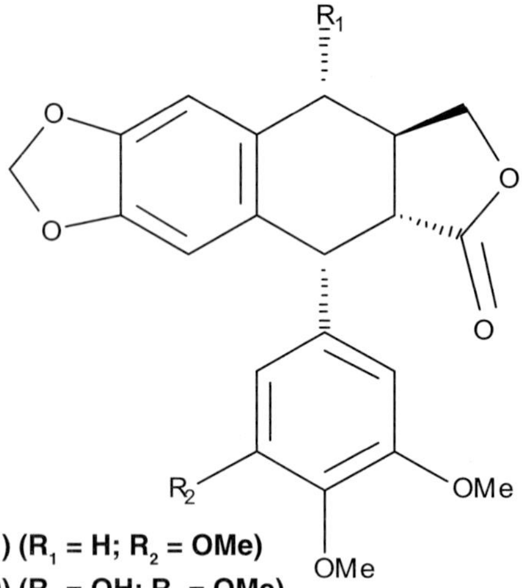

(1) (R_1 = H; R_2 = OMe)
(9) (R_1 = OH; R_2 = OMe)
(10) (R_1 = OAc; R_2 = OMe)
(11) (R_1 = H; R_2 = H)

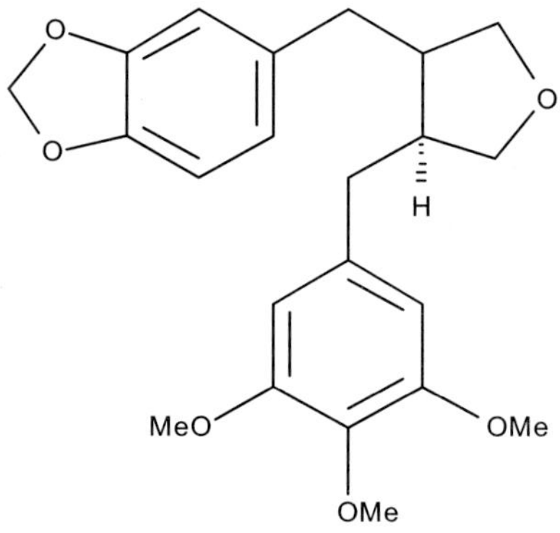

(2) burseran

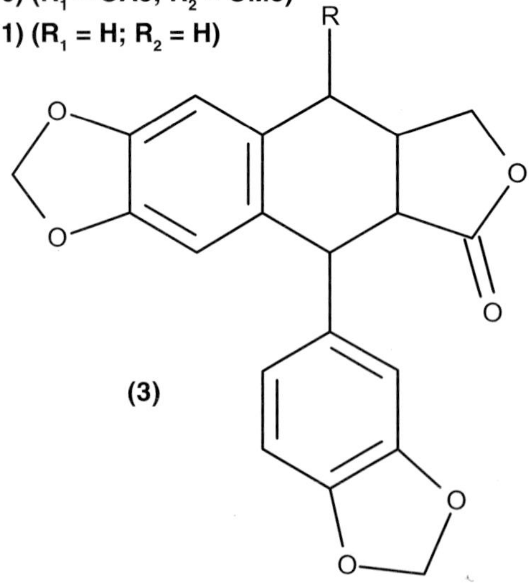

(3)

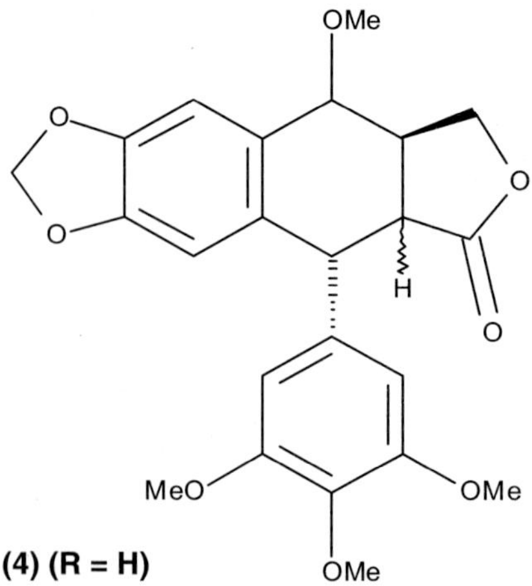

(4) (R = H)
(14) (R = OCH_3)

Contd...

Figure 19.4–*Contd...*

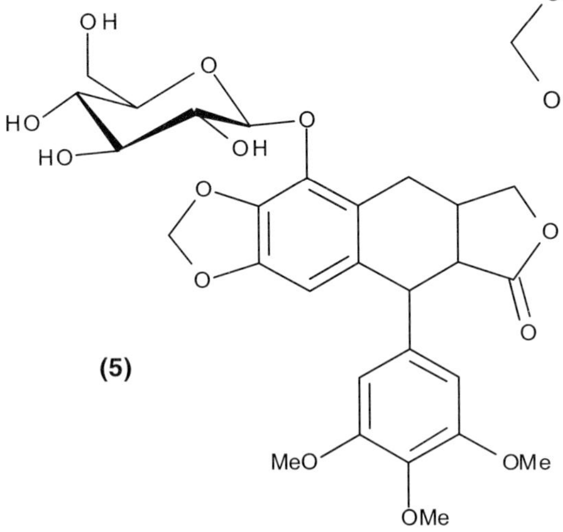

(5)

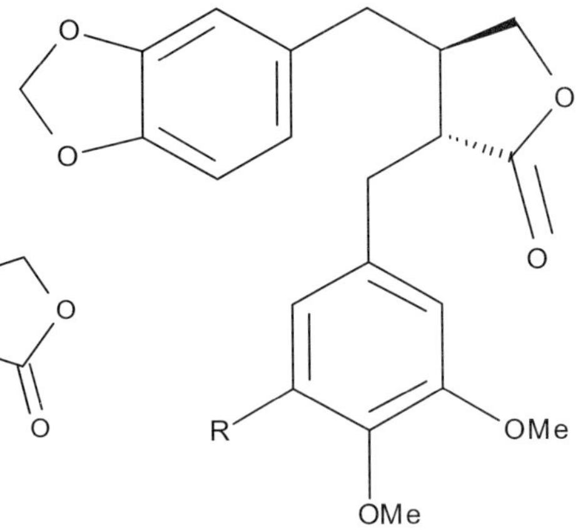

(6) (R = OMe)
(12) (R = H)

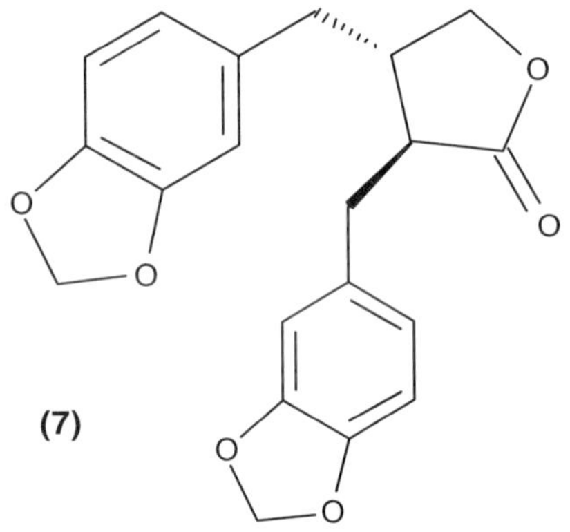

(7)

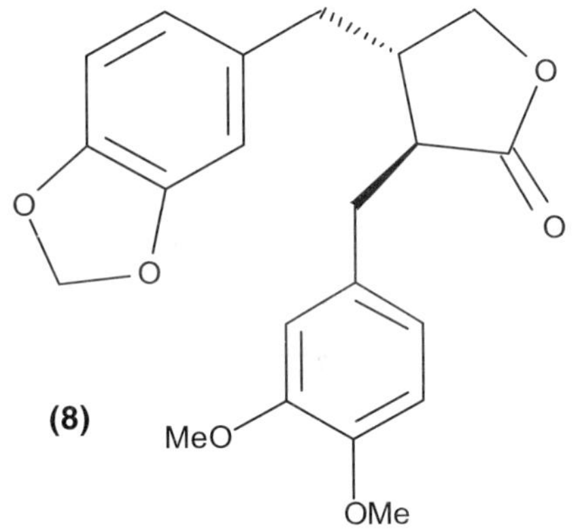

(8)

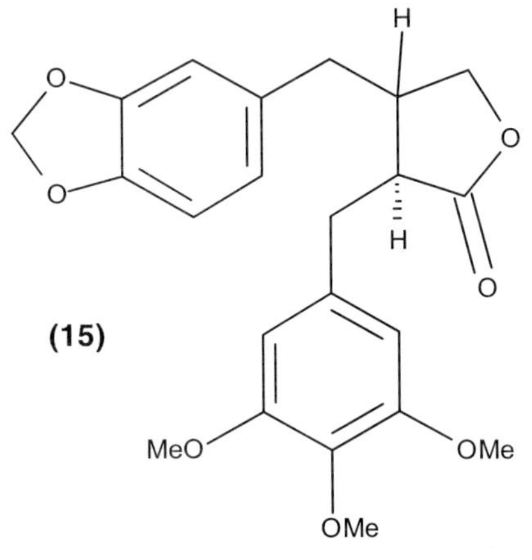

(15)

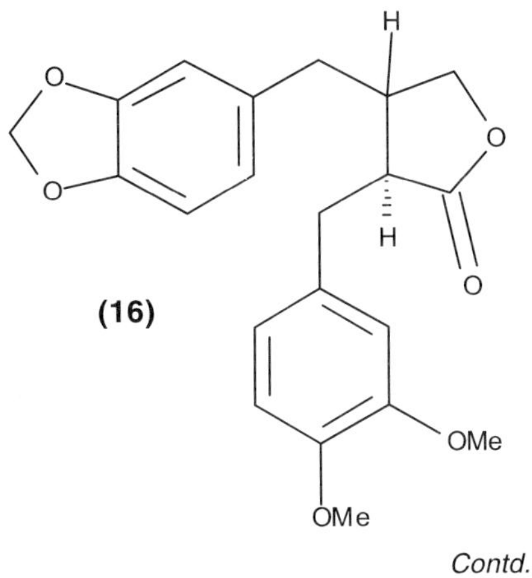

(16)

Contd...

Figure 19.4–*Contd...*

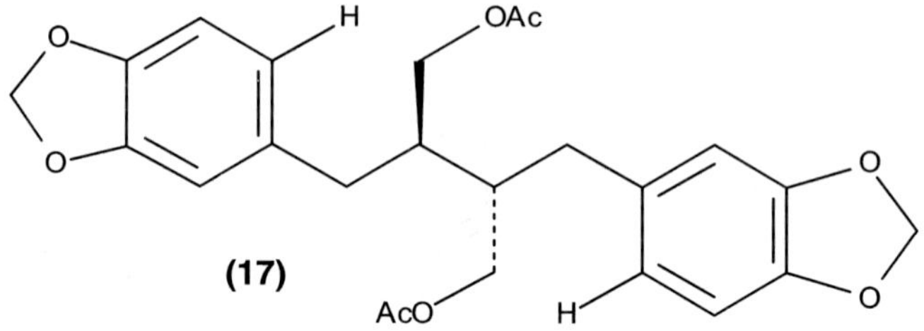

(17)

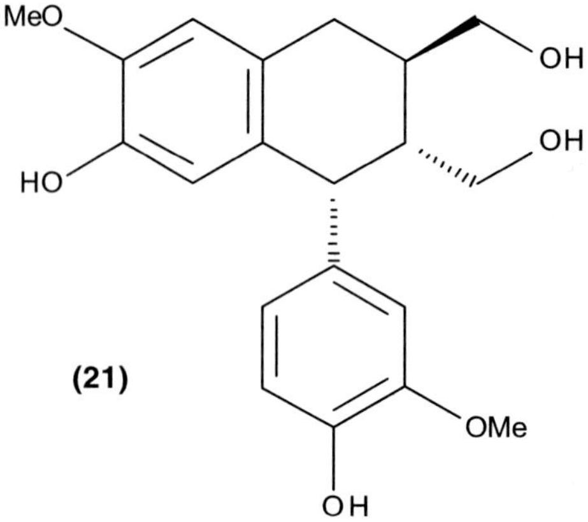

(21)

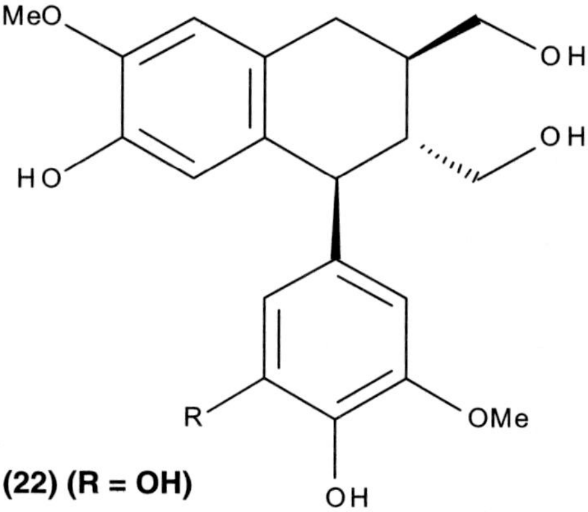

(22) (R = OH)
(23) (R = OMe)

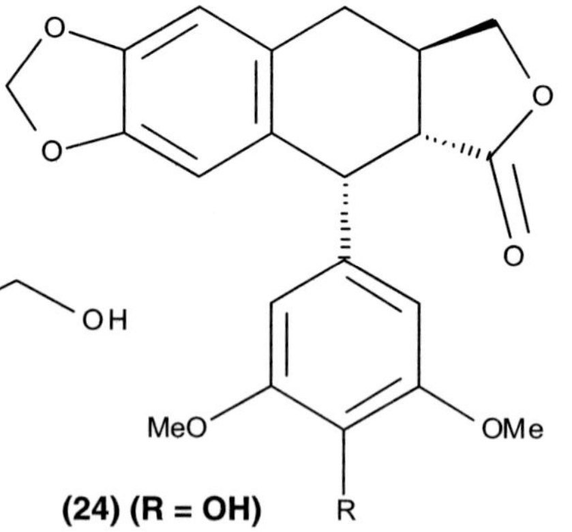

(24) (R = OH)
(25) (R =Glc)

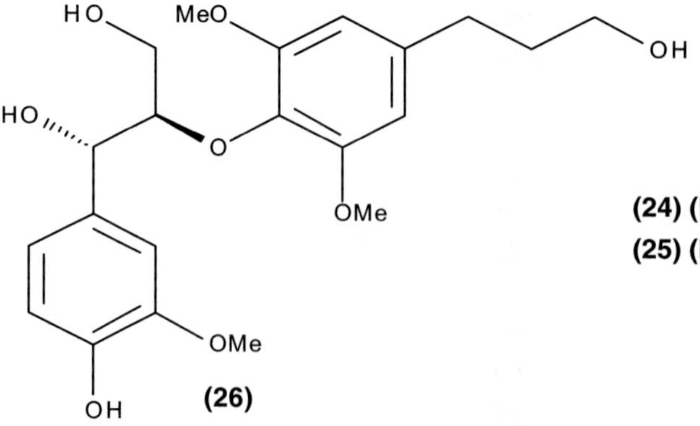

(26)

Contd...

Figure 19.4–*Contd...*

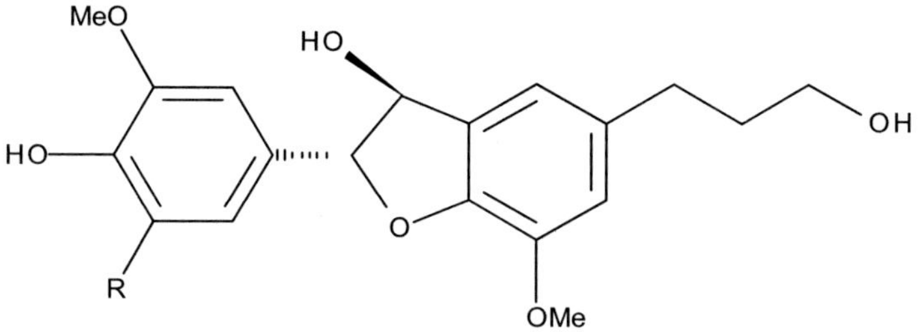

(27) (R = H)
(28) (R = OMe)

In a recent study of the leaves of the species *B. simaruba*, the aryltetralin-type lignansβ-methyl-petaltin A [3] was identified, and it has been suggested that this lignan may be the active principle involved in the anti-inflammatory activity of leaves of this species (Noguera *et al.*, 2004). From the resin of the same species, picropolygamain [4] was isolated (Peraza-Sánchez and Rodríguez-Peña, 1992).

Studying phenolic compounds from the bark of *B. simaruba*, Maldini *et al.* (2009) identified the aryltetralin-type lignanβ-peltatin-O-β-D-glucopyranoside [5] and the dibenzylbutyrolactone-types yatein [6], hinokinin [7] and bursehernin [8]. Quantitative analysis by mass spectrometry confirmed that these lignans are the main compounds of the shells, and that β-O-peltatin-β-D-glucopyranoside is the most abundant.

In the resin of *B. fagaroides*, the aryltetralin-type lignans (-)-podophyllotoxin [9], (-)-acetyl podophyllotoxin [10], (-)-deoxypodophyllotoxin [1] and morelesin [11] were isolated, and the dibenzylbutyrolactone-type lignans (-) -yatein [6] and (-) -5 ′-desmethoxyyatein [12] (Velázquez-Jiménez *et al.*, 2011). Another study with the same species, but this time with the branch, revealed the presence of the aryltetralin-type lignans acetyl podophyllotoxin [10] and desoxypodorhizol [13], (Rojas-Sepulveda *et al.*, 2009).

In *B. graveolens*, the aryltetralin -type lignansburseranin [14] and picropolygamain [4] were identified, isolated from the branches. These lignans showed inhibitory effect on fibrosarcoma cells (Nakanshi *et al.*, 2005).

In the species *B. schlechtendalii*, two lignans belonging to the dibenzybutirolactone class were identified: (-)-trans-2-(3 ′, 4 ″, 5″-trimethoxybenzyl) -3–(3′, 4′-methylene-dioxybenzyl) butyrolactone [15] and (-) -trans) -2–(3 ″,″ 4-dimethoxybenzyl) -3–(3′, 4′-methylenedioxybenzyl) butyrolactone [16], isolated from the leaves and branches (McDoniel, 1992).

From the resin of *B. ariensis* a lignan belonging to the dibenzylbutane class ariesin [17] was identified (Hernandez *et al.*, 1983). In the *B. permollis*, the aryltetralin-type lignans deoxypodophyllotoxin-type [1], β-petaltin methyl ether [18] and nemerosin [19] were identified, isolated from the bark, and showed significant antimitotic potential againsta strocyte cell lines (Wickramaratne *et al.*, 1995). Jolad *et al.* (1977) isolated the aryltetralin-type lignans deoxypodophyllotoxin-type and 5 ′-desmethoxideoxypodophyllotoxin [20] from the resin of *B. morelensis.*

In *B.tonkinensis*, the aryltetralin-type lignansburselignan [21], (+)-isolariciresinol [22], 5-methoxy (+)-isolariciresinol [23], 4 ′dimethyldesoxypodophyllotoxin [24] and 4′- dimethyldesoxypodophyllotoxin-4-*O*-β-D-glucoside [25] have been described, isolated in the roots. Also in this species, the presence of the neolignans 1-(4′-hydroxy-3′-methoxyphenyl)-2-[4"-(3-hydroxypropyl)-2",6"-dimethoxyphenoxy]pronane-1,3-diol [26], dihydrodehydroconiferyl alcohol [27] and 5-methoxy-*trans*-dihydrodehydrodiconiferyl alcohol [28] have been reported (Jutiviboonsuk *et al.*, 2005).

Genus *Commiphora*

Studies performed with resins of *C. erlangeriana* have identified lignans belonging to the aryltetralin classes 2β-dihydroxy-6-methoxy-4β-angeloyloxyisopicro-polygamatin [30], 2α-Acetoxy-6-methoxypicropolygamatin [31], acetodeoxypicro-podophyllin-2α [32] and 2α-Acetoxy-3β-hydroxyisodeoxypodophyllotoxin [33] (Dekebo *et al.*, 2002). Habtemariam (2003) also isolated these same constituents from the same species, and evaluated their cytotoxic effect on human and murine cells, observing potential activity consistent with cytotoxic mode of action in the cell lines described above.

Abbasi *et al.* (2005) isolated the lignansepiexcelsin [34] and 5 ′-demethoxye-piexcelsin [35] (Figure 19.5), both belonging to the furofuran class, from the leaves of the species *C. mukul*. These lignans presented inhibitory effect against α-glucosidase enzyme and cytotoxic against murine macrophage cells (Abbasi *et al.*, 2005; Abbasi *et al.*, 2009).

Genera *Crepisdospermum* and *Tetragastris*

In the type *C. rhoifolium*, the lignin parabenzolactone [37] was identified, belonging to the class dibenzylbutyrolactone and in *T. altissima*, the lignan (-)-savinin [38] was identified, also belonging to the dibenzylbutyrolactone class, as shown in Figure 19.6 (Lima *et al.*, 2001).

Genus *Protium*

In a study with the wood of the species *P. tenuifolium*, Smith *et al.* (1995) identified epimeric mixtures of (-)-cubebin [39] and the lignin oxo-parabenzolactone [40] (Figure 19.7), belonging to the dibenzylbutyrolactone and furan classes, respectively. Siani *et al.* (1998) isolated from the wood of this species the arylnaphthalene-type 5-methoxyjusticidin A [41].

In the genus *Protium* neolignans are also described. Studies with the wood of the species *P. opacum* showed the presence of the neolignanpropacin [42] (Zoghbi *et al.*, 1981; Marques and Yoshida, 1990), and Adams *et al.* (2002) isolated the neolignanpropacin from the barks of P. *heptaphyllum*. In other species of this genus, *P. unifoliolatum*, Magalhães *et al.* (2006) isolated the neolignan 5-methoxypropacin [43] from the bark.

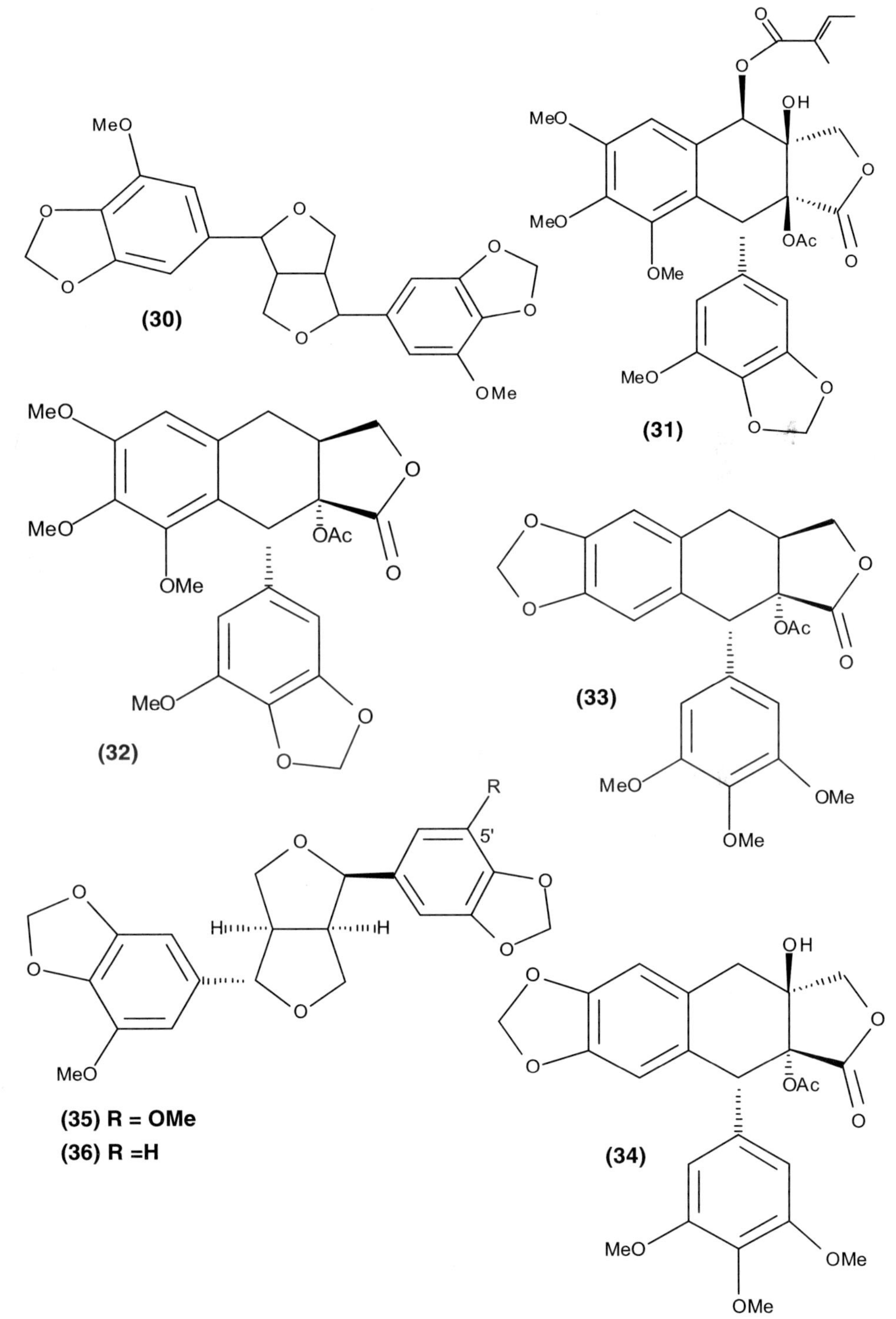

(35) R = OMe
(36) R = H

Figure 19.5: Lignans of the genera *Commiphora*.

Figure 19.6: Lignans of the genera *Crepisdospermum* and *Tetragastris.*

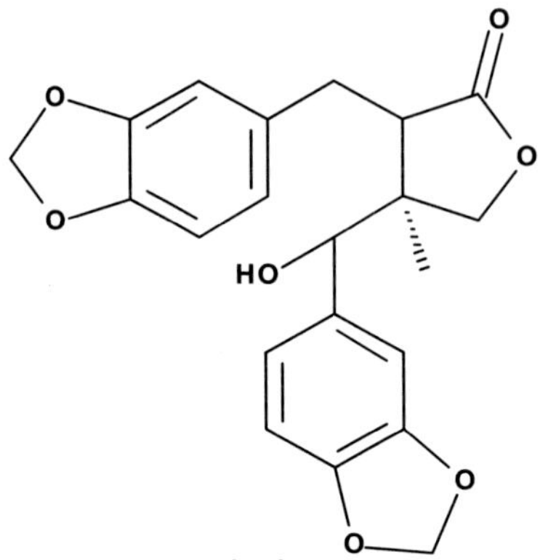

(37)

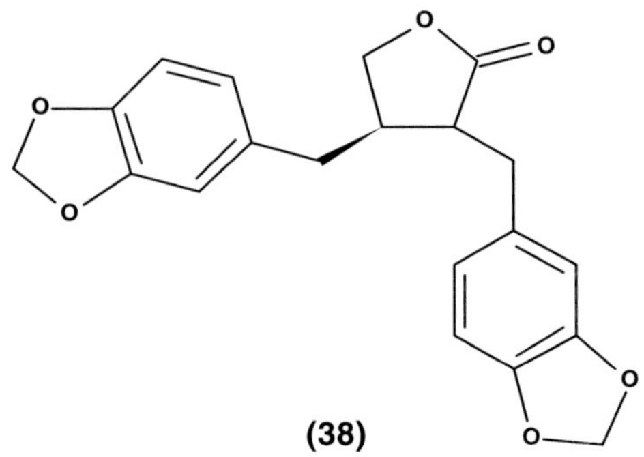

(38)

Figure 19.7: Lignans and neolignans of the genus *Protium.*

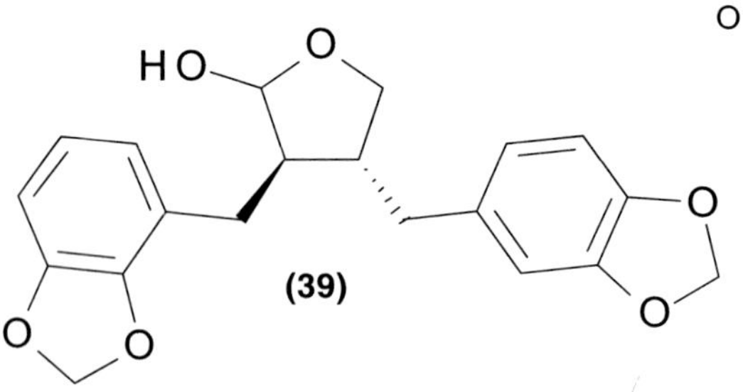

(39)

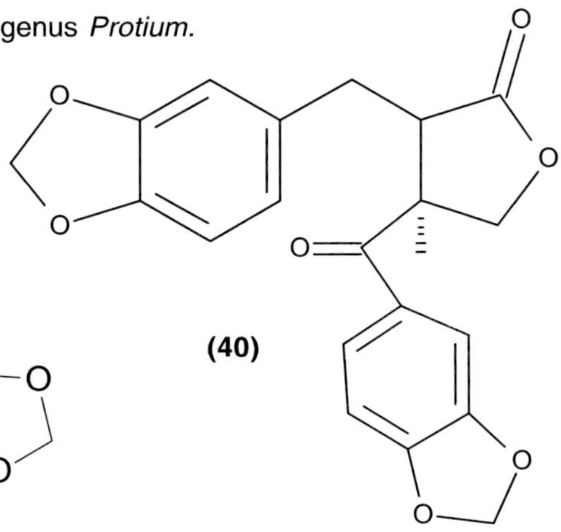

(40)

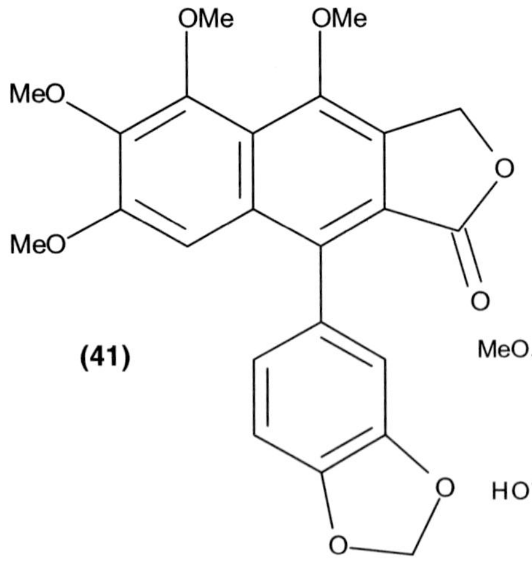

(41)

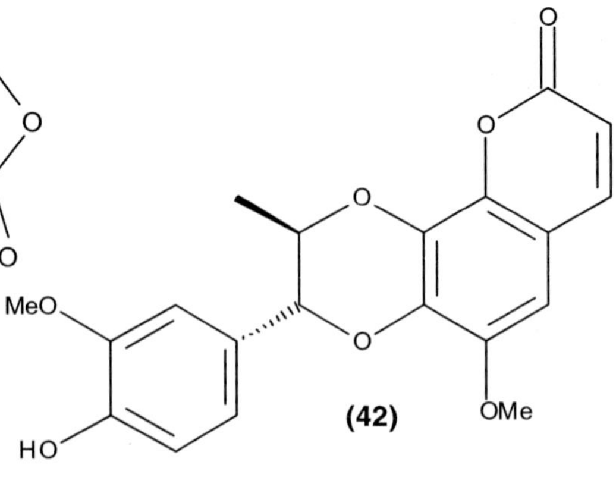

(42)

Contd...

Figure 19.7–*Contd...*

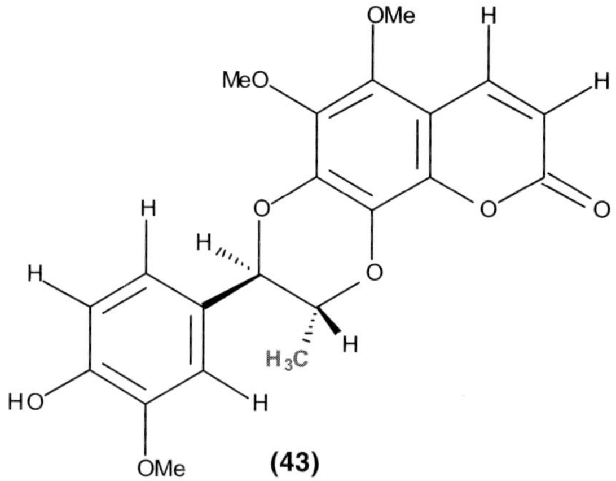

(43)

Conclusion

Burseraceae is an important botanical family with several medicinal and commercial applications around the world. It has a predominantly terpenic chemical composition, but phenolic constituents have also been described, in which 37 lignans and 6 neolignans have been isolated and identified, belonging to different classes and presenting important pharmacological activities. Recently published studies show that the phenolic composition of Burseraceae needs to be further studied, since several highly active lignans have been to be discovered.

References

Abbasi, M. A., Ahmad, V. U., Zubair, M., Khan, S. N., Lodhi, M. A., and Choudhary, M. I. (2005). α-glucosidase and chymotrypsin inhibiting lignans from *commiphora mukul*. *Procedings of Pakistan Academy Sciences*, **42:** 23-26.

Abbasi, M. A., Lodhi, M. A., and Choudhary, M. I. (2009). Kinetics studies on the lignan class of natural compounds that inhibits α-chymotrypsin. *Journal of Asian Natural Products Research*, **11:** 933–939.

Agrawal, P. K., and Thakur, R. S. (1985). ^{13}C NMR spectroscopy of lignan and neolignan derivatives. *Magnetic Resonance in Chemistry*, **23:** 389-418.

Almeida, E. X., Conserva L. M., and Lemos R. P. L. (2002). Coumarins, coumarinolignoids and terpenes from *Protium heptaphyllum*. *Biochemical Systematics and Ecology*, **30:** 685–687.

Ayres, D. C., and Loike, J. D. (1990). Lignans: Chemical, Biological and Clinical Properties. Cambridge University Press.

Bianchi, E., Caldwell, M E., and Cole, J R. (1968). Antitumor agents from *Bursera microphylla* (Burseraceae) I. Isolation and characterization of deoxypodophyllotoxin. *Journal of Pharmaceutical Sciences*, **57:** 696-697.

Canel, C., Moraes, R. M., Dayan, F. E., and Ferreira, D. (2000). Podophyllotoxin. *Phytochemistry*, **54:** 115–120.

Cole, J R., Bianchi, E., and Trumbull, E R. (1969). Antitumor agents from *Bursera microphylla* (Burseraceae). II. Isolation of a new lignan-burseran. *Journal of Pharmaceutical Sciences*, **58:** 175-176.

Costa, A. F. (1975). Farmacognosia, V. 1. Lisboa: Fundação CalousteGulbeukian, .

Costa, M. C. and Takahata Y. (1997). Conformational analysis of synthetic neolignans active against leishmaniasis, using the molecular mechanics method (MM2). *Journal of computational chemistry*, **18:** 5.

Dekebo, A., Lang, M., Polborn, K., Dagne. E., and Steglich, W. (2002). Four lignans *Commiphora erlangeriana*. *Journal Natural Products*, **65:** 1252-1257.

Habtemariam S. (2003). Cytotoxic and cytostatic activity of erlangerins from *Commiphora erlangeriana*. *Toxicon*, **41:** 723–727.

Hernández, J. D., Román, L. U., Espiñeira, J., and Joseph-Nathan, P. (1983). Ariensin, a new lignan from *Bursera ariensis*. *Planta Medica*, **47:** 215-217.

Jolad, S.D., Wiedhopf, R.M., and Cole, J.R. (1977). Cytotoxic agents from *Bursera klugii* (Burseraceae). 1. Isolation of sapelins-A and B. *Journal of Pharmaceutical Sciences*, **66:** 889-890.

Jutiviboonsuk, A., Zhang, H., Tan, G. T., Cuiying, M.A., Hung, N. V., Cuong, N. M., Bunyapraphatsara, N. D., Soejarto, D., and Fong H.H.S. (2005). Bioactive constituents from roots of *Bursera tonkinensis*. *Phytochemistry*, **66:** 2745–2751.

Lima, M.P., Castro, F. B. G., Silva, M. F. G. F., Ferreira, A. G., Fo, E. R., Fernandes, J. B., and Vieira, P. C. (2001). Phytochemistry of *Crepidospermum rhoifolium, Tetragastris altissima, Protium icicariba* and *P. apiculatum*: Chemosystematic implications. *Revista Latinoamericana de Quimica*, **29:** 135-144.

Magalhães, A., Zoghbi, M. G. B., and Siani, A. C. (2006). 5-Methoxypropacin, a novel coumarinolignoid from *Protium unifoliolatum*. *Natural Product Research*, **20:** 43–46.

Maia, R.M., Barbosa, P.R., Cruz, F.G., Roque, N.F., and Fascio, M. (2000). Triterpenos da resina de *Protium heptaphyllum* March (Burseraceae): caracterização em misturas binárias. *Química Nova*, **23:** 623-626.

Maldini, M., Montoro, P., Piacente, S., and Pizza, C. (2009). Phenolic compounds from *Bursera simaruba* Sarg. bark: Phytochemical investigation and quantitative analysis by tandem mass spectrometry. *Phytochemistry*, **70:** 641-649.

Marques, O. M., and Yoshida. (1990). Estrutura de propacina. *Química Nova*, **13:** 245-246.

McDoniel, P.B., and Cole, J.R. (1972). Antitumor activity of *Bursera schlechtendalii* (Burseraceae): isolation and structures determination of two new lignans. *Journal of Pharmaceutical Sciences*, **61:** 1992–1994.

Moss, G. P. (2000). Nomenclature of lignans and neolignans. *Pure and Applied Chemistry*, **72:** 1493–1523.

Nakanishi, T., Inatomi, Y., Murata, H., Shigeta, K., Iida Naoki, A., Inada Jin, Murata, Farrera, M.L A.L P., Iinuma, M., Tanaka, T., Tajima, S., and Oku, N. (2005). A new and known cytotoxic Aryltetralin-type lignans from stems of *Bursera graveolens*. *Chemical Pharmaceutical Bulletin*, **53:** 229–231.

Noguera, B., Díaz, E., García, M.V., Feliciano, A. S., López-Perez, J.L., and Israel, A. (2004). Anti-inûammatory activity of leaf extract and fractions of *Bursera simaruba* (L.) Sarg (Burseraceae). *Journal of Ethnopharmacology*, **92:** 129–133.

Otto, R., Gottlieb, and Massayosh Yoshida. (1984). Com atenção especial à química das neolignanas. *Química Nova*, **7:** 250-273.

Peraza-Sánchez, S. R., and Peña-Rodríguez L. M. (1992). Isolation of picropolygamain from the resin of *Bursera simaruba*. *Journalof Natural Products*, **55:** 1768-1771.

Pio Correa, M. (1984). Dicionários de plantas úteis do Brasil e das exóticas cultivadas. Brasília: Ministério da Agricultura, 1.

Quignard, E.L.J., Nunomura, S.M., Pohlit, A.M., Alecrim, A.M., Pinto, A.C.S., Portela, C.N., Oliveira, L.C.P., Don, L.C., Silva, L.F.R., Henrique, M.C., Santos, M., Pinto, P.S., and Silva, S.G. (2004). Median lethal concentrations of Amazonian plant extracts in the brine shrimp assay. *Pharmaceutical Biology*, **42:** 253-257.

Rojas-Sepulveda, A., Antúnez, M., Pérez, A. M. P., Marquina, S., and Alvarez, L. (2009). Aryltetralinlignans from the stem bark of *Bursera fagaroides* var *fagaroides* (Burseraceae). Resumo Southeastern Regional Meeting of the American Chemical Society, 104-A.

Rüdiger, A. L., Siani, A. C. and Veiga Junior, V. F. (2007). The chemistry and pharmacology of the South America genera *Protium burm.* f. (Burseraceae); *Pharmacognosy Reviews*, **1:** Jan- May.

Siani, A. C., Zoghbi, M. G. B., Wolter E. L. A., and Vencato I. (1998). 5-Methoxyjusticidin A, a New Arylnaphthalene Lignan from *Protium unifoliolatum*. *Journal of Natural Products*, **61:** 796-797.

Simões, C.M.O., Schenkel, E.P., Gosmann, G., Mello, J.C.P., Mentz, L.A., and Petrovick, P.R., Farmacognosia da Planta ao Medicamento, 5ª ed., Editora da UFRGS/ UFSC, 2004.

Siqueira, J. B. G.; Zoghbi, M. G. B.; Cabral, J. A.; Wolter, F. W.J. (1995). Lignans from *Protium tenuifollum*. *Journal of Natural Products*, **58:** 730-732.

Thulin, M., Beier, Björn-Axel, Sylvain G. Razafimandimbison, and Hannah I. Banks. (2008). Ambilobea, a new genera from Madagascar, the position of Aucoumea, and comments on the tribal classification of the frankincense and myrrh family (Burseraceae). *Nordic Journal of Botany*, **26:** 218-229.

Umezawa, T. (2003). Diversity in lignan biosynthesis. *Phytochemistry Reviews*, **2:** 371–390.

Velázquez-Jiménez, R., Torres-Valencia, J. M., Cerda-García-Rojas, C. M., Hernández-Hernández, J. D., Román-Martín, L. U., Manríquez-Torres, J. J., Gómez-Hurtado, M. A., Valdez-Calderón, A., Motilva, V., García-Mautiño, S., Talero, E., Ávila, J., and Joseph-Nathan. P. (2011). Absolute configuration of podophyllotoxin related lignans from *Bursera fagaroides* using vibrational circular dichroism. *Phytochemistry*, **72:** 2237-2243.

Vieira Junior, G.M., Souza, C.M.L., and Chaves, M.H. (2005). Resina de *Protium heptaphyllum*: isolamento, caracterização e avaliação das propriedades térmicas. *Química Nova*, **28:** 183-187.

Weeks, A., Daly, D.C., and Sympson, B.B. (2005). The phylogenetic history and biogeography of the frankincense and myrrh family (Burseraceae) based on nuclear and chloroplast sequence data. *Molecular Phylogenetics and Evolution*, **35:** 85-101.

Wickramaratne, D.B.M., Mar, W., Chai, H., Castillo, J.J., Farnsworth, N.R., Soejarto, D.D., Cordell, G.A., Pezzuto, J.M., and Kinghorn, A.D. (1995). Cytotoxic constituents of *Bursera permollis*. *Planta Medica*, **61:** 80–81.

Zoghbi, M. G. B., Roque, N. F., and Gottlieb, O. R. (1981). Propacin, A coumarinolignoid from *Protium opacum*. *Phytochemistry*, **20:** 180.

Bioactive Phytochemicals: Perspectives for Modern Medicine Vol. 2 (2014) *Pages* **439–451**
Editor: **V.K. Gupta**
Published by: **DAYA PUBLISHING HOUSE, NEW DELHI**

20

Antioxidant Activity of Amazon Fruits

Klenicy K.L. Yamaguchi[1], Carlos V. Lamarão[1], Emerson S. Lima[2] and Valdir F. Veiga Junior[1*]

ABSTRACT

Plant foods such as fruits play an important role in human consumption, due to their nutritional and sensory attributes.In this context the Amazon region has numerous fruits species with exotic flavors, combined with various chemical substances that are beneficial to human health.The chemical composition of this fruits has been linked to beneficial activities for the human body, being associated with a lower incidence of chronic and degenerative diseases. This chapter presents the available literature data concerning the antioxidant activity of Amazonian fruits and substances that are related to this activity.The protective effect of these foods has been attributed to the presence of antioxidant compounds such as vitamins A, C and E, carotenoids and phenolic compounds such as flavonoides, anthocyanins and phenolic acids.Amazonian acai and other fruits, such as acerola, buriti, camu camu, caja and piquia, present the highest evidence of compounds antioxidants. The antioxidant activity related to the polyphenols is due to its reducing properties and chemical structure.This allows the scavenging of free radicals or chelation of transition metal.The resonance of the aromatic ring present in these compounds is primarily responsible for the formation of stable intermediate after the reduction process of free radicals.

Keywords: Amazonia, Polyphenols, Antioxidant.

1 Chemistry Department – Amazonas Federal University, Av. Gal. Rodrigo Octávio, 3000, ICE, Blerk 10 Japiim, 69077-000, Manaus – AM, Brazil

2 Faculty of Pharmaceutical Sciences – Amazonas Federal University, Alexandre Amorin street, 330, Aparecida, 69010-300, Manaus – AM, Brazil

* *Corresponding author*: E-mail: valdirveiga@ufam.edu.br

Introduction

Plant foods, such as fruits, play an important role in human consumption due to their nutritional and therapeutic effects. In this context, the Amazon region is a hotbed of exotic fruits that are rich in beneficial substances to the human body. These are being highlighted each year in the national and international market (Rabelo, 2012).

The use of Amazonian fruits, due to their therapeutic properties, have been described in the literature for many decades. Guarana (*Paullinia cupana*) was used by the indigenous peoples as an energy booster. Nowadays, It is used among contemporary populations for its fruits (seeds), which are sold in the form of powder or extract (Rodrigues *et al.*, 2007).

As an example of the use of fruits in the Amazon, the Amazonian Indians have used Brazil nuts for centuries as the basis of their diet. The seeds are commonly sold, both in the domestic market and abroad (Rodrigues *et al.*, 2007). Brazil nut is eaten raw or processed. According to Souza (2006), the species is also used as a source of raw material in the production of handicrafts (lamps, ashtrays, necklaces, rattles, etc.), cosmetics (soap, shampoo, moisturizing cream, oil and lipstick) and drugs (repellent healing, etc.).

The flavor and the chemical properties of substances in the pulp of these fruit, which contains different compounds, has attracted the interest of many research groups. This is due to the biological properties that may be related to activities which are beneficial to the human body, such as preventing and minimizing the incidence of chronic and degenerative diseases due to the stabilizing activity of free radicals (Velioglu *et al.*, 1998; Zibade *et al.*, 2007; Melo *et al.*, 2008; Vieira *et al.*, 2011; Silva *et al.*, 2012; Poulose *et al.*, 2012; Kang *et al.*, 2012).

A characteristic of free radicals is that they exhibit the presence of one unpaired electron in an outer orbit. These chemicals play essential role in cell metabolism, but when in excess, they are involved in molecular damage to the cell structures, relating to functional alterations and loss of vital functions in various tissues and organs, such as the muscle, liver, adipose tissue, vascular tissue and brain (Duarte *et al.*, 1993; Halliwell, 1994, 1996; Anderson, 1996; Guaratini *et al.*, 2007; Dröge, 2002, Nordberg and Arner, 2001; Rover Junior *et al.*, 2001).

The balance between the formation and removal of reactive free radical species can be reestablished by antioxidant systems. These are substances that when present at low concentrations compared to oxidizable substrate, significantly regenerate or prevent oxidation (Barreiros and David, 2006; Halliwell *et al.*, 2000).

Foods not only provide essential nutrients necessary for life, but also bioactive compounds with biological properties that are said to promote health. Thus, the intake of antioxidant capable of neutralizing the free radicals may play a role in disease prevention (Murugananthan and Pabbthi, 2012).

In this context Amazonian fruits appear as promising sources for improving the quality of life of people. This is due to its antioxidant capacity, related to its chemical composition with high levels of phenolics, flavonoids, carotenoids and vitamins that prevent the harmful action of free radicals (Chisté *et al.*, 2012; Gonçalves *et al.*, 2010; Canuto *et al.*, 2010).

Species like cupuaçu (*Theobroma grandiflorum*), bacuri (*Platonia insignis*), uxi (*Endopleura uxi*) piquia (*Caryocar villosum*), buriti (*Mauritia flexuosa*) and guarana (*Paullinia cupana*) have been highlighted (Canuto *et al.*, 2010; Couto and Canniatti-Brazaca, 2010; Gonçalves, 2010; Rosso and Mercadante, 2012; Dalonso and Petkowicz, 2012).

In light of the potential problems caused by the consumption of synthetic antioxidants, such as problems of toxicity, carcinogenesis and many other evils: increased liver weight and significant proliferation of the endoplasmic reticulum, the research has looked for natural products with antioxidant activity that will replace synthetic products, or make associations between them (Melo and Guerra, 2002; Simão, 1985; Guaratini *et al.*, 2007; Souza *et al.*, 2007; Tang *et al.*, 2008; Augustyniak *et al.*, 2010).

This chapter presents a survey of the available scientific literature on the antioxidant activity of Amazonian fruits and their phenolic substances that are related to the benefits of this activity.

Material and Methods

A literature search was performed using the databases SCOPUS, ISI, PubMed, SCIELO, and search tools on the Internet. Search keywords such as "Amazon fruit", "antioxidant" and "phenolic compounds" were used, plus the names of the main fruits in various combinations.

Results and Discussion

In Brazil as a whole, and recently in the Amazon region, studies have shown the fruits as promising sources of phenolic compounds with antioxidant activity. They are thought to have action as kidnappers of free radicals, with protective action against the start and/or development of degenerative processes associated with various diseases (Poulose *et al.*, 2012; Rosso and Mercadante, 2012; Dalonso and Petkowicz, 2012; Kang *et al.*, 2012; Vieira, 2011;Gonçalves *et al.*, 2010; Sousa *et al.*, 2007; Silva *et al.*, 2005).

Recent studies have revealed a positive correlation between the antioxidant activity and the amount of phenolic compounds present in fruits.

Structurally, phenolic compounds have at least one hydroxyl group attached directly to a benzene ring. They are widely found in nature, and are metabolized by both acetate and shikimic acid. They are the main antioxidants found in plants, and fit into several categories, such as simple phenols, phenolic acids (derivatives of benzoic acid and cinnamic acid), coumarins, flavonoids, stilbenes, tannins, lignins and lignans (Dewick, 2002). Some examples of antioxidants are shown in Figure 20.1.

The antioxidant capacity of the Amazonian fruits araca (*Psidium guineensis*), araca-boi (*Eugenia stipitata*), camucamu (*Myrciaria dubia*), bacuri (*Scheelea phalerata*), cupuassu (*Theobroma grandiflorum*), manna-cubiu (*Solanum sessiliflorum*) tucuma

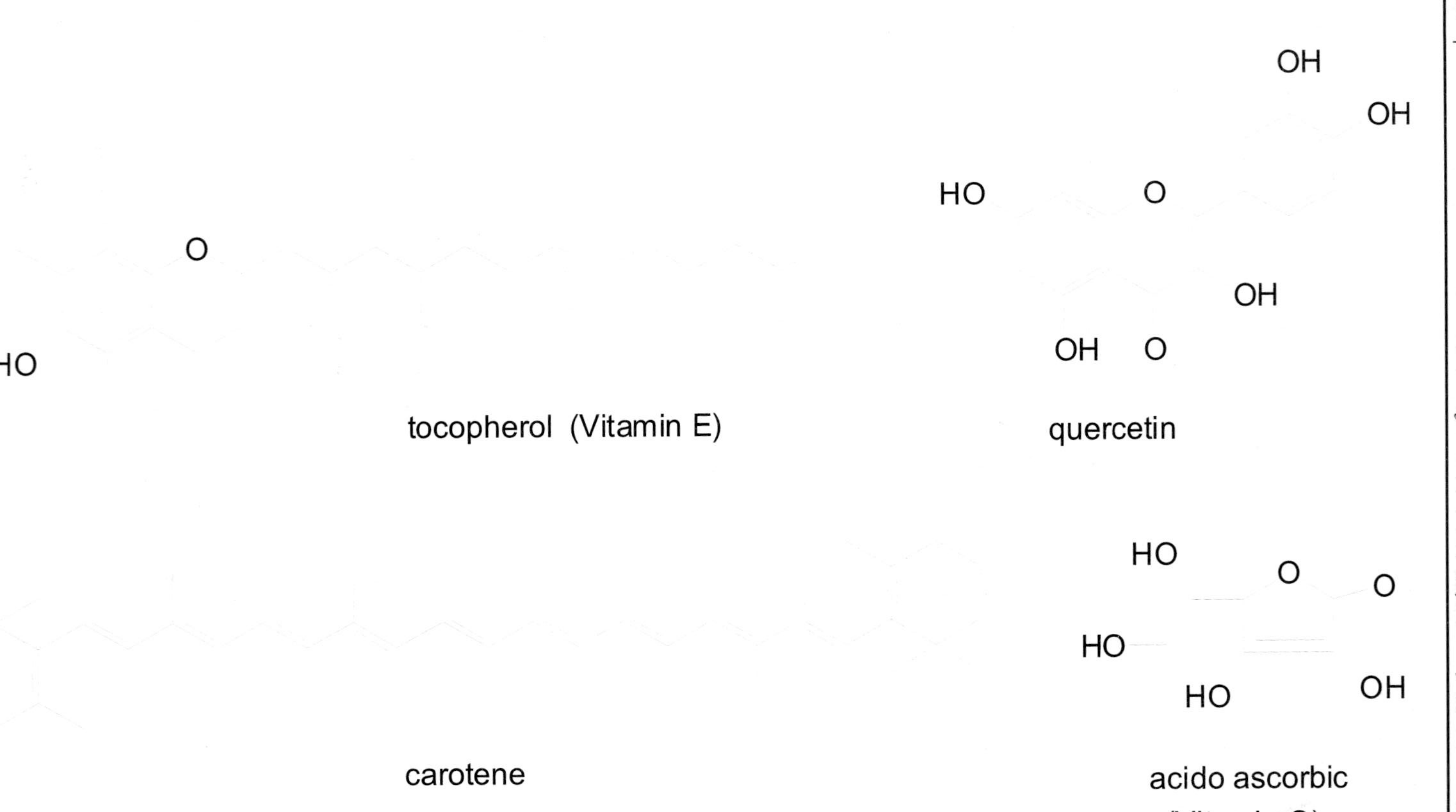

Figure 20.1: Chemical structures of some antioxidants obtained in the diet.

(*Astrocaryum aculeatum*) uxi (*Endopleura uchi*) and abiu (*Pouteria caimito*) show a good correlation between the amount of phenol by the determination of Folin-Ciocalteu reducing capacity and scavenging ability of free radical DPPH (2,2-diphenyl-1-picrylhydrazyl), oxygen radical absorption capacity (ORAC), and bleaching of β-carotene (Gonçalves, 2008). Among the above mentioned fruits, camu camu, uxi and tucuma are the ones that show the best antioxidant capacity, with a significant correlation between the values of the tests analyzed.

Cupuassu, soursop and araza-boi fruits show the high levels of ascorbic acid, 110, 82 and 78mg per 100g of material. Among the flavanols, the tucumã and soursop showed catechin levels of 79 and 15.9 mg per 100 g of sample, respectively. Quercetin and its derivatives were detected in araca-boi (14.4 mg), araca (40mg), tucuma (2.96), buriti(0.6) and camucamu, with high content of 42mg/100g. This shows a high content of anthocyanins, 306mg/100g sample (Gonçalves *et al.*, 2010). The chemical structure of the main falvonoides is illustrated in Figure 20.2.

The pulp of camu camu has proven to be an excellent source of vitamin C in the diet, with high levels of ascorbic acid (Gonçalves *et al.*, 2010; Gonçalves *et al.*, 2008; Maeda *et al.*, 2006; Alves *et al.*, 2000; Andrade *et al.*, 1995). These fruits, which grow to

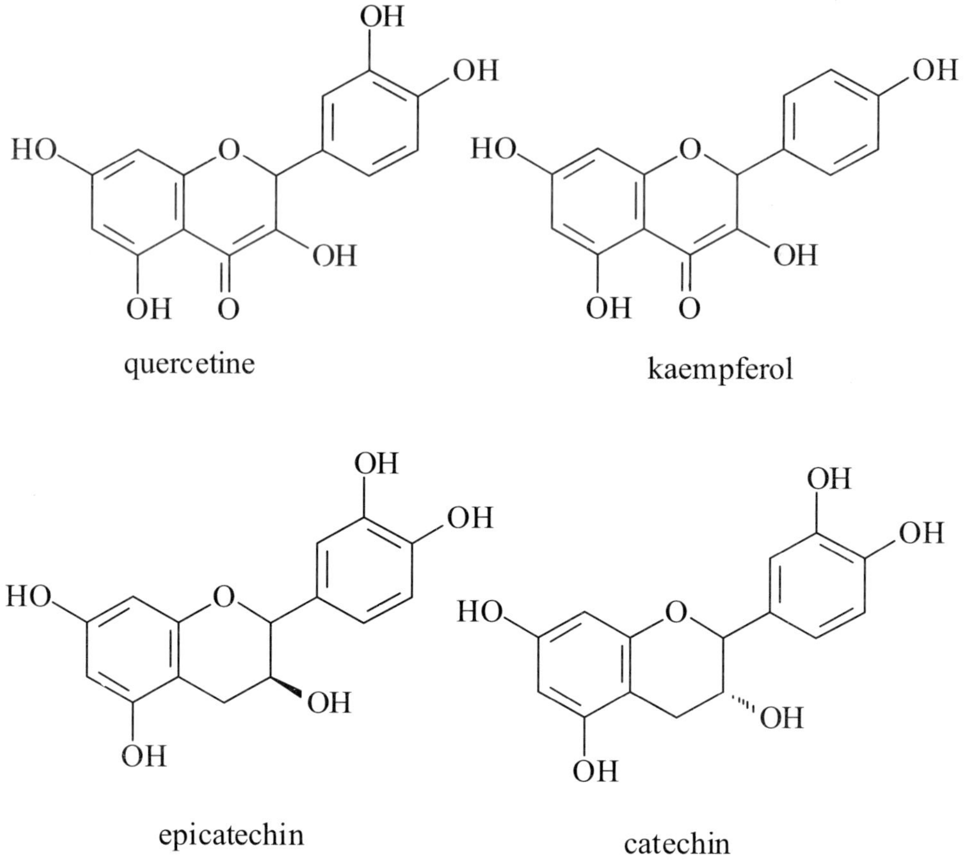

Figure 20.2: Structure of the main flavonoids described in Amazonian fruits.

the east of the State of Roraima (RR), show high concentrations of ascorbic acid, ranging from 3.6 to 6.1 g/100g, making this one of the richest fruits in vitamin C in the world (Yuyama *et al.*, 2002). The action of ascorbic acid as an antioxidant, with formation of a stable radical, is shown in Figure 20.3.

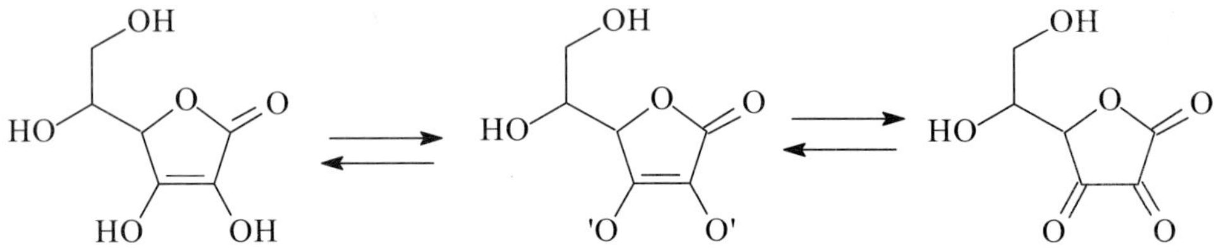

Figure 20.3: Scheme of the oxidation of ascorbic acid to dehydroascorbic acid with the formation of the ascorbyl radical as intermediate compound (*Source*: Schafer and Buettner, 1997).

The content of chemical substances can vary between different regions, depending on seasonality, temperature, intensity of light, nutrients, moisture, and diversity in the maturation stage of the fruit used in the preparation of the pulps (Canuto *et al.*, 2010; Gobbo-Neto and Lopes, 2007).

The antioxidant activity of fifteen samples of fruit pulp (abiu, cherry, acai, guava-ox bacaba, bacuri, buriti, caja, cajarana, cashew, cupuassu, soursop, muruci, noni, and tamarind) presented activity by the Trolox Equivalent Antioxidant Capacity (TEAC) method, evaluating the capacity of the sample to absorb 2,2 '-azino-bis (3-ethylbenzothiazoline-6-sulfonic acid ($ABTS^{\bullet+}$). The pulps with the highest values were cherry, acai and noni, with results for total phenol of 3.5, 2.4 and 1.2, expressed as mmol/L of gallic acid, and for free antirradical activity (TEAC) of 12.1, 10 and 3.9 mol/L equivalent to trolox, respectively (Canuto *et al.*, 2010; Barreto *et al.*, 2009; Alves *et al.*, 2009).

However, in frozen fruit pulp (cherry, caja and bacuri), the high amount of polyphenols was confirmed only for the cherry pulp (835 and 449 mg/100g in hydroalcoholic and aqueous extracts, respectively). It was found that the hydroalcoholic extracts had higher sequestration capacity (concentration equivalent EC_{50}) for DPPH than those found in aqueous extract of the fruits analyzed. Also, the cherry extract was the one that showed the highest antioxidant activity (EC_{50} 1.74 mg/mL) (Vieira *et al.*, 2011).

Piquia (*Caryocar villosum*) and pequi (*Caryocar brasilense*), species found in different regions (central Amazon and the border between the Amazon and the Brazilian cerrado (tropical savanna) respectively), have antioxidant capacity and the presence of gallic acid in common (Ascari *et al.*, 2010; Roesler *et al.*, 2007).

Piquia (*Caryocar villosum*), a native fruit of the Amazon, was regarded as an interesting source of bioactive compounds. Chiste *et al.* (2012) demonstrated that antioxidant sequestration capacity *in vitro* of the extracts is closely related to their content of phenolic compounds. Extracts of different polarities were tested for their sequestration capacity *in vitro* of reactive oxygen species (ROS) and reactive nitrogen species (RNS). The hydroalcoholic and aqueous extracts showed the highest phenolic levels (5163 and 1745 mg/g extract, respectively). The phenolic compounds detected by high performance liquid chromatography coupled with a diode array detector system and a mass spectrometer (HPLC-DAD-MS/MS) were ellagic acid and gallic acid. The antioxidant activity (EC_{50}) of gallic acid ranged from 0.4 to 226 ug/mL for ROS and from 0.04 to 0.12 g/ml to RNS), while that of the ellagic acid ranged from 1.7 to 108 ug/mL for ROS and from 0.05 to 0.59 ug/mL for RNS.

In the quantification of phenolic substances in piquia by high performance liquid chromatography (HPLC-DAD-ESI-MS/MS) gallic acid (182.4 ug/g pulp), followed by ellagic acid rhamnosides (107 ug/g pulp) and ellagic acid (104 ug/g pulp) were the phenolic compounds present in higher quantities. The behaviour of this fruit as kidnapper in response to the formation of a peroxyl radical (ORAC assay) presented satisfactory results (3.7 mmolTrolox/100 g pulp), confirming this fruit as a food antioxidant potential (Joke and Mercadante, 2012).

Another much appreciated fruit in the Amazon region that is being distributed internationally is acai. The juice is dark purple in colour due to its high concentration of anthocyanins (Coisson*et al.*, 2005). Several tests *in vivo* and *in vitro* have been performed to prove and determine the broad range of biological activities of these fruits. Its antioxidant activity is the most widely studied activity in the research carried out to date.

According with Rogez (2000), the approximate content of anthocyanins in acai is 440 mg/kg of fruit. Expressing this content per 100g of dry extract gives an average value of 1.02 g, in other words, about 1 per cent of the dry matter is composed of antioxidants.

Anthocyanins are known for their potential health benefits due to their antioxidant activity (Kuskoski *et al.*, 2006; Lima *et al.*, 2011; Gouvea *et al.*, 2012). The basic structure of anthocyanins found in fruits can be represented by the ion phenyl-2-benzopyrone (Figure 20.4).

In twelve samples of commercially marketed assai pulp in Ceara (Brazil) the anthocyanins present ranged from 13.93 to 54.18 mg/100g (Santos *et al.*, 2008). These levels are consistent with those found in the work of Rogez (2000), Kalt *et al.* (1999) and Kuskoski *et al.* (2005) who obtained similar results to those found in this study, and attributed the high levels of antioxidant activity to phenolic compounds and anthocyanins.

The antioxidant capacity of fruits may be related to the prevention of degenerative diseases, by improving cognitive and motor functions and reducing inflammation. Alcohol extracts (ethanol and methanol) of acai (*E. oleracea* Mart) have been shown to have a positive effect both in protecting microglial cells BV-2 mice and in reducing inflammation through the mechanisms of cyclooxygenase-2 (COX-2), protein kinase (p38-MAPK), tumor necrosis factor-α (TNF α) and nuclear factor-kB (NF-κB). Anthocyanins detected in these fruits were cyanidin, delphinidin, malvidin, peonidin and pelargonidin (Poulose *et al.*, 2012).

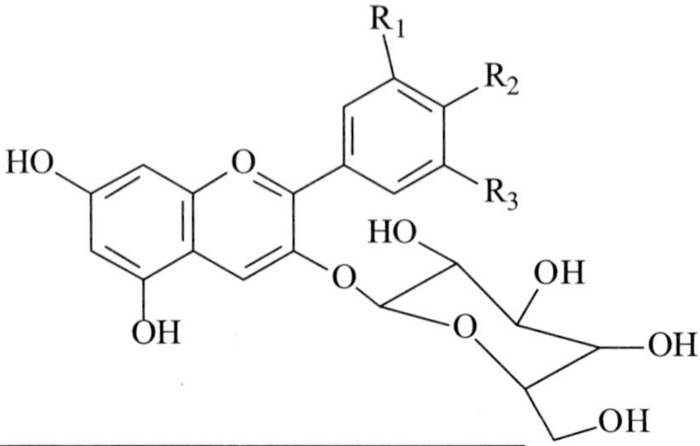

Anthocyanidins	substitution pattern		
	R_1	R_2	R_3
Pelargonidin (plg)	H	OH	H
Cyanidin (cyd)	OH	OH	H
Delphinidin (plg)	OH	OH	H
Peonidin (pnd)	OCH_3	OH	H
Petunidin (ptd)	OCH_3	OH	OH
Malvidin (mvd)	OCH_3	OH	OCH_3

Figure 20.4: Basic structure representing the major anthocyanins found in fruits (*Source*: Kuskoski *et al.*, 2004).

Anthocyanins isolated from 12 genotypes of cherry fruits grown in the Federal Rural University of Pernambuco showed ability to sequester free radicals by stabilizing DPPH. The initial concentration was 16.7 mg/mL, and after 5 minutes and 2 hours of reaction, the percentage varied from 36.97 to 63.92 per cent and from 73.27 to 94.54 per cent, respectively (Lima *et al.*, 2011).

Conclusion

Amazonian fruits have been shown to be efficient sources of phenolic substances, such as flavonoids and anthocyanins, with contents similar or higher than other fruits traditionally known as sources of this nutrient in the diet.

They are also good sources of bioactive compounds, with potential beneficial effects on human health, due to their free radical scavenging activity, particularly acai, camu camu and cherry.

This study therefore shows that Amazonian fruits have potential beneficial effects on human health, justifying further studies that may contribute to the production of biotechnological activities, both in the food industry and in the production of natural medicines with possible therapeutic actions.

References

Alves, R.E., Filgueiras, H.A.C., and Moura, C.F.H. (2000).Caracterização de frutas nativasda América Latina. *SérieFrutasNativas*, **9**. Jaboticabal: Funep, 66 pp.

Anderson, D. (1996). Antioxidant defences against reactive oxygen species causing genetic and other damage. *Mutation Research,* **350**: 103-108.

Andrade, J. S., Aragão, C. G., Galeazzi, M. A. M., and Ferreira, S. A. N. (1995).Changes in the concentration of total vitamin C during maturation and ripening of camu-camu (Myrciariadúbia (H.B.K.) McVaugh) fruits cultivated in the upland of Brazilian Central Amazon. *Acta Horticultura,* **370**: 177-180.

Ascari, J., Takahahi, A., and Boaventura, M. A. D. (2010). Phytochemical and biological investigations of Caryocarbrasiliense Camb. *Boletin Latinoamericano y del Caribe de Plantas Medicinales y Aromáticas,* **9**: 20-28.

Augustyniak, A., Bartosz, G., Cipak, A., Duburs, G., Horáková, L., Luczaj, W., Majekova, M., Odysseos, A. D., Rackova, L., Skrzydlewska, E., Stefek, M., Strosová, M., Tirzitis, G., Venskutonis, P.R., Viskupicova, J., Vraka, P. S., and Zarkovic, N.(2010). Natural and synthetic antioxidants: an updated overview. *Free Radical Research,* **44**: 1216-1262.

Barreiros, A. L. B. S., and David, J. M. (2006). Estresse oxidativo: relação entre geração de espécies reativas e defesa do organismo. *Química Nova,* **29:** 113-123.

Barreto, G.P.M., Benassi, M.T., and Mercadante, A.Z. (2009). Bioactive compounds from several tropical fruits and correlation by multivariate analysis to free radical scavenger activity. *Journal of Brazilian Chemical Society,* **20**: 1856-1861.

Bianchi, M. L. P., and Antunes, L. M. G. (1999). Radicais livres e os principais antioxidantes da dieta. *Revista de Nutrição,* **12**: 123-130.

Boots, A. W., Haenen, G. R., and Bast, A. (2008). Health effects of quercetin: From antioxidant to nutraceutical. *Eur J Pharmacol,* **13**: 325-37.

Buettner, G. R., and Schafer, F. Q. (1997). Ascorbate (Vitamin C) its antioxidant chemistry. *Oxygen Society of Education Program*: 1-20.

Canuto, G. A. B., Xavier, A. A. O., Neves, L. C., and Benassi, M. T. (2010). Caracterização físico-química de polpas de frutos da Amazônia e sua correlação com a atividade anti-radical livre. *Revista Brasileira de Fruticultura,* **32**: 1196-1205.

Chisté, R. C., Freitas, M., Mercadante, A. Z., and Fernandes, E. (2012). The potential of extracts of *Caryocar villosum* pulp to scavenge reactive oxygen and nitrogen species. *Food Chemistry,* **27**.

Chisté, R. C., and Mercadante, A. Z. (2012). Identification and quantification by HPLC-DAD-MS/MS, of carotenoids and phenolic compounds from the Amazonian fruit *Caryocar villosum*. *Journal of Agricultural and Food Chemistry,* **60**: 5884-5892.

Coisson, J.D., Travaglia, F., Piana, G., Capasso, M., and Arlorio, M. (2005). *Euterpe oleracea* juice as a functional pigment for yogurt. *Food Technology,* **38**: 893-897.

Couto, M. A. L., and Canniatti-Brazaca, S. G. (2010). Quantificação de vitamina C e capacidade antioxidante de variedades cítricas. *Ciência e Tecnologia de Alimentos,* **30**: 15-19.

Cunha, A. P., Salgueiro, L. and Roque, O. R. (2005). Farmacognosia e fitoquímica, 22, Fundação Calouste Gulenkian.

Dalonso, N., and Petkowicz, C. L. O. (2012). Guarana powder polysaccharides: Characterisation and evaluation of the antioxidant activity of a pectic fraction. *Food Chemistry*, **134**: 1804-1812.

Dewick, P.M. (2002). Medicinal Natural Products: a biosynthetic approach. 2nd ed. England: John Wiley and Sons Ltd.

Dröge, W. (2002). Free radicals in the physiological control of cell function. *Physiological Reviews*, **852**: 47-95.

Duarte, J. A., Appell, H. J., Carvalho, F., Bastos, M. L., and Soares, J. M. (1993). Endothelium-derived oxidative stress may contribute to exercise-induced muscle damage. *International Journal of Sports Medicine*, **14**: 440-443.

Elsayed, N. M. (2001). Antioxidant mobilization in response to oxidative stress: a dynamic environmental–nutritional interaction. *Nutrition*, **17**: 828-834.

Gobbo-Neto, L., and Lopes, N. P. (2007). Plantas medicinais: fatores de influência no conteúdo de metabólitos secundários. *Química Nova*, **30**: 374-381.

Gonçalves, A. E. S. S. (2008). Avaliação da capacidade antioxidante de frutas e polpas de frutas nativas e determinação dos teores de flavonoides e vitamina C. Dissertação de Mestrado,Universidade de São Paulo, São Paulo. Brazil.

Gonçalves, A. E. S. S., Lajolo, F. M., and Genovese, M. I. (2010). Chemical composition and antioxidant/antidiabetic potential of Brazilian native fruits and commercial frozen pulps. *Journal of Agricultural and Food Chemistry*, **58**: 4666-4674.

Gouvêa, A. C. M. S., Araújo, M. C. P., Schulz, D. F., Pacheco, S., Godoy, R. L. O., and Cabral, L. M. C. C. (2012). Anthocyanins standards (cyanidin-3-O-glucoside and cyanidin-3-O-rutinoside) isolation from freeze-dried açaí (*Euterpe oleraceae* Mart.) by HPLC.*Ciência e Tecnologia de alimentos*, **24**: 43-46.

Guaratini, T., Medeiros, M. H. G., and Colepicolo, P. (2007). Antioxidantes na manutenção do equilíbrio redox cutâneo: uso e avaliação de suaeficácia. *Química Nova*, **30**: 206-213.

Halliwell, B. (1996). Antioxidants in human health and disease. *Annual Review of Nutrition*, **16**: 33-50.

Halliwell, B. (1994). Free radicals and antioxidants: a personal view. *Nutrition Reviews*, **52**: 253-265.

Halliwell, B. (2000). The antioxidant paradox.*The Lancet*, **355**: 1179-1180.

Kalt, W., Forney, C. F., Martin, A., and Prior, R. L. (1999). Antioxidant capacity, vitamin C, phenolics and anthocyanins after fresh strorage of small fruits. *Journal Agriculture Food Chemistry*, **47**: 4638-4644.

Kang, J., Thakali, K. M., Xie, C., Kondo, M., Tong, Y., Ou, B., Jensen, G., Medina, M. B., Schauss, A. G., and Wu, X. (2012). Bioactivities of açaí (*Euterpe precatoria* Mart.) fruit pulp, superior antioxidant and anti-inflammatory properties to *Euterpe oleracea* Mart. *Food Chemistry*, **133**: 671-677.

Kuskoski, E. A., Asuero, A. G., Garcia-Parilla, M. C., Troncoso, A. M., and Fett, R. (2004). Actividad antioxidante de pigmentos antociánicos.*Ciências e Tecnologia de Alimentos*, **24**: 691-693.

Kuskoski, E. A., Asuero, A. G, Morales, M. T., and Fett, R. (2006). Frutos tropicais silvestres e polpas de frutas congeladas: atividade antioxidante, polifenóis e antocianinas. *Ciência Rural*, **36**: 1285-1286.

Kuskoski, E. M., Asuero, A. G., Troncoso, A. M., Mancini-Filho, J., and Fett, R. (2005). Aplicación de diversos métodos químicos para determinar actividad antioxidante enpulpa de frutos. *Ciências e tecnologia de alimentos,* **25**: 726-732.

Lima, A. de J. B., Côrrea, A. D., Saczk, A. A., Martins, M. P., and Castilho, R. O. (2011). Anthocyanins, pigment stability and antioxidant activity in jabuticaba [*Myrciariacauliflora* (Mart.) O. Berg]. *Revista Brasileira de Fruticultura,* **33**: 877-887.

Lima, V. L. A. G., Melo, E. A., Pinheiro, I. O., and Guerra, N. B. (2011). Antioxidant capacity of anthocyanins from acerola genotypes. *Ciências e Tecnologia de Alimentos*, **31**: 86-92.

Maeda, R. N., Pantoja, L., Yuyama, L. K., and Chaar, J. M. (2006). Determinação da formulação e caracterização de néctar de camu-camu (Myrciaria dúbia McVaugh). *Ciencias e Tecnologia de Alimentos*, **26**: 70-74.

Melo, E. A., Maciel, M. I. S., Lima, V. L. A. G., and Nascimento, R. J. (2008). Capacidade antioxidante de frutas. *Revista Brasileira de Ciências Farmacêuticas,* **44**: 193-201.

Melo, E.A., and Guerra, N.B. (2002). Ação antioxidante de compostos fenólicos naturalmente presentes em alimentos. *Bol. SBCTA*.**36**: 1-11.

Murugananthan, G., and Pabbthi, S. C. (2012). Medicinal plants with potent antioxidant constituents. *International Journal of Pharmaceutical Sciences and Research*, **3**: 1268-1273.

Nordberg, J., and Arnér, S.J. (2001). Reactive oxygen species, antioxidants, and the mammalian thioredoxin system. *Free Radical Biology and Medicine,* **31**: 1287-1312.

Poulose, S. M., Fisher, D. R., Larson, J., Bielinskit, D. F., Rimando, A. M., Carey, A. N., Schauss, A. G. and Shukitt-Hale, B. (2012). Anthocyanin-rich açai (*Euterpe oleracea* Mart.) fruit pulp fractions attenuate inflammatory stress signaling in mouse brain BV-2 microglial cells. *Journal Agriculture Food Chemistry*, **60**: 1084-1093.

Rabelo, A. (2012). Frutos nativos da Amazônia: comercialização nas feiras de Manaus-AM. INPA.

Rodrigues, S., Caetano, D. G., and Caetano, C. M. (2007). Especies frutíferas do centro-sul do Estado de Rondônia, Amazônia brasileira. *Acta Agronomica*,**56**: 69-74.

Roesler, R., Malta, L. G., Carrasco, L. C., Holanda, R. B., Souza, C. A. S., and Pastore, G. M. (2007). Atividade antioxidante de frutas do cerrado. *Ciência e Tecnologia de Alimentos*, **27**: 53-60.

Rogez, H. (2000). Açaí: Preparo, Composição e Melhoramento da Conservação. Belém: EDUFPA.

Rosso, V. V., and Mercadante, A. Z. (2012). Identification and quantification of carotenoids, by HPLC-PDA,MS/MS, from Amazon fruits. *Journal of agricultural and food chemistry*, **55**: 5062-5072.

Rover Júnior, L., Höehr, N. F., and Vellasco, A. P. (2001).Sistema antioxidante envolvendo o ciclo metabólico da glutationa associado a métodos eletroanalíticos na avaliação do estresse oxidativo. *Quimica Nova*, **24**: 112-119.

Santos, G. M. (2008). Correlação entre atividade antioxidante e compostos bioativos de polpas comerciais de açaí (*Euterpe oleracea* Mart). *Archivos Latinoamericanosde Nutrición*, **58**: 187-192.

Santos, R. I. (2004). Metabolismo básico e origem dos metabólitos secundários. In: C.M.O. Simões *et al.* (eds.). Farmacognosia da planta ao medicamento. Universidade UFRGS/Ed. da UFSC, 5° edição rev., Porto Alegre, capítulo 16, Florianópolis.

Silva, C.G., Herdeiro, R.S., Mathias, C.J., Panek, A.D., Silveira, C.S., Rodrigues, V.P., Renno, M.N., Falcao, D.Q., Cerqueira, D.M., Minto, A.B.M., Nogueira, F. L. P., Quaresma, C.H., Silva, J.F.M., Menezes, F.S., and Eleutherio, E.C.A. (2005). Evaluation of antioxidant activity of Brazilian plants. *Pharmacological Research*, **52**: 229-233.

Silva, F. C., Arruda, A., Ledel, A., Dauth, C., Romão, N. F., Viana, R. N., Ferraz, A. B. F., Picada, J. N., and Pereira, P. (2012). Antigenotoxic effect of acute, subacute and chronic treatments with Amazonian camu-camu (Myrciariadúbia) juice on mice blood cells. *FoodandChemistryToxicology*, **50**: 2275-2281.

Simão, A.M. (1985). Aditivos para alimentos sob o aspecto toxicológico. São Paulo: Nobel.

Soares, S. E. (2002). Ácidos fenólicos como antioxidantes. *Revista de Nutrição*,**15**: 71-81.

Sousa, C.M.M., Silva, H.R., Vieira-Jr, G.M., Ayres, M.C.C., Costa, C.L.S., Araújo, D.S., Cavalcante, L.C.D., Barros, E.D.S., Araújo, P.B.M., Brandão, M.S., and Chaves, M.H. (2007). Fenóis totais e atividade antioxidante de cinco plantas medicinais. *Química Nova*, **30**: 351-355.

Souza, I.F. (2006). Cadeia de produção de castanha do Brasil (Bertholletia excelsa) no estado de Mato Grosso. Dissertação de Mestrado, Universidade Federal de Mato Grosso do Sul, Campo Grande, Brazil.

Tang, H., Wang, A., Salley, S. O., and Simon Ng, K. Y. (2008). The effect of natural and synthetic antioxidants on the oxidative stability of biodiesel. *Journal of the American Oil Chemists Society*, **85**: 373–382.

Velioglu, Y.S., Mazza, G., Gao, L., and Oomah, B.D. (1998). Antioxidant activity and total phenolics in selected fruits, vegetables, and grain products. *Journal of Agricultural and Food Chemistry*, **46**: 4113-4117.

Vieira, L. M., Sousa, M. S. B., Mancini-Filho, J., and Lima, A. (2011). Fenólicos totais e capacidade antioxidante *in vitro* de polpas de frutos tropicais. *Revista Brasileira de Fruticultura*, **33**: 888-897.

Yuyama, K., Aguiar, J.P.L., and Yuyama, L.K.O. (2002). Camu-camu: Um fruto fantástico como fonte de vitamina C. *Acta Amazonica*, **32**: 169-174.

Zibade, S., Farid, R., Moriguchi, S., Lu, Y., Foo, L., Tehrani, P., Ulreich, J., and Watson, R. (2007). Oral administration of purple passion fruit peel extract attenuates blood pressure in female spontaneously hypertensive rats and humans. *Nutrition Research*, **27**: 408-416.

Bioactive Phytochemicals: Perspectives for Modern Medicine Vol. 2 (2014)
Editor: **V.K. Gupta**
Published by: **DAYA PUBLISHING HOUSE, NEW DELHI**

Pages **453–470**

21

Phytochemical Analysis of Bioactive Constituents in Coriander (*Coriandrum sativum* L.) Seeds

C.U. Rajeshwari[1*] and B. Andallu[1]

ABSTRACT

Plants now occupy an important position in allopathic medicine, herbal medicine, homoeopathy and aromatherapy in the modern world. Many plants are cheaper and more accessible to most people especially in the developing countries than the orthodox medicine, and there is lower incidence of adverse effects after use of plant products. Many spices have been recognized to have medicinal properties and possess many beneficial effects on health, such as antioxidant activity, digestive stimulant action, anti-inflammatory, antimicrobial, hypolipidemic, antimutagenic, anticarcinogenic potential. The medicinal value of these plants lies in some chemical substances that produce definite physiological actions in the human body. The most important of these bioactive constituents of plants are polyphenols, tannins, flavonoids, steroids, terpenoids etc. Phenolics and polyphenolics form the most desirable food bioactive compounds because of their antioxidant activity, brought about by the capacity to scavenge free radicals that cause oxidative damage to lipids, proteins and nucleic acids. Among the various spices, coriander (Coriandrum sativum L.) seeds have been popular from ancient times for their use in folk medicine and they are used as a traditional spice in many foods in India. Hence, present investigation was aimed to identify the presence of phytochemical compounds viz. carbohydrates, amino acids, organic acids, tannins, glycosides, saponins, triterpenoids, flavonoids and flavones in various fractions viz. hexane, benzene, ethyl acetate, n-butanol and water obtained by the sequential fractionation of methanolic extracts of coriander

1 Sri Sathya Sai Institute of Higher Learning, Anantapur campus, Anantapur- 515 001, A.P., India

* *Corresponding author*: E-mail: sairajic@gmail.com

seeds. All the fractions responded positively for qualitative tests for various bioactive compounds viz. carbohydrates, tannins, triterpenoids, flavonoids and flavones, while only few fractions responded positively for amino acids, glycosides and organic acids. However, ethyl acetate fraction possessed maximum content of carbohydrates, amino acids, glycosides, saponins, triterpenoids, flavonoids and n-butanol fraction contained higher tannins due to the difference in their solubility in various solvents. This study revealed coriander seeds to be a storehouse of a number of bioactive compounds providing the chemical basis for the wide use of coriander as therapeutic agent for preventing and treating various ailments.

Keywords: Coriander (*Coriandrum sativum* L.) seeds, Methanolic extract, Ethyl acetate, n-Butanol fraction, Bioactive compounds.

Introduction

There has been growing interest in exploiting the biological activities of different ayurvedic medicinal herbs, owing to their natural origin, cost effectiveness and lesser side effects (Chopra *et al.*, 2009). Medicinal plants are nature's priceless gift to human beings. The development in the field of modern medicine temporarily subdued the traditional herbal medicine. But it has now staged a "herbal renaissance" and is blooming across the world (Sumathi and Parvathi, 2010). Medicinal plants are of great importance to the health of individuals and communities. The most important of these bioactive constituents of plants are phenolics, tannins, flavonoids, steroids, saponins and glycosides. Many of these indigenous medicinal plants are used as spices and food plants (Okwu, 2001). Many attempts of modern scientists are directed towards finding evidence of identification of various bioactivities of responsible active compounds and their molecular targets (Tapsell *et al.*, 2006).

Medicinal principles are present in different parts of the plant. These medicinal principles are separated by different processes; the most common being extraction which is the separation of the required constituents from plant materials using a solvent. In order to study the therapeutic effects of various phytochemical compounds present in herbs and spices, it is necessary to extract them from the source prior to the analysis. To facilitate the extraction, the solvent should diffuse inside the cell and the substance must be sufficiently soluble in the solvent. The ideal solvent for complete extraction is one that is most selective, has the best capacity for extraction and is compatible with the properties of the material to be extracted (Sivarajan and Balachandran, 1994). Extraction of phenolic compounds in plant materials is influenced by their chemical nature, the extraction method employed, sample particle size, storage time and conditions, as well as presence of interfering substances. Phenolic extracts of plant materials are always a mixture of different classes of phenolics that are soluble in the solvent system used. Additional steps may be required for the removal of unwanted phenolics and non-phenolic substances such as waxes, fats, terpenes and chlorophylls (Naczk and Shahidi, 2004). An attempt to obtain bioactive agents from plants is a worthwhile exercise since only 10 per cent of all plants have been investigated in detail. Among the plant species, only a small percentage has been investigated phytochemically and the fraction submitted to biological or pharmacological screening is even smaller. Thus, any phytochemical investigation of a given plant will reveal the spectrum of its constituents. (Herborne, 1973). *Coriandrum sativum* L. is one such plant which is widely used in indigenous system of medicine and also as a spice (Ayyanar and Ignacimuthu, 2008). No much research has been reported on phytochemicals screening of coriander seeds.

Hence, the present study was aimed to sequentially fractionate methanolic extracts of coriander seeds using solvents of increasing polarity *viz.* hexane, benzene, ethyl acetate, n-butanol and water and to examine the fractions for the presence of bioactive compounds using standard qualitative tests for carbohydrates, amino acids, organic acids, tannins, glycosides, saponins, triterpenoids, flavanoids and flavones.

Materials and Methods

Procurement of Sample

Coriander seeds purchased from the local departmental stores of Anantapur were cleaned; shade dried and powdered finely using a blender.

Extraction

The powdered coriander seeds were subjected to extraction of various bioactive compounds using 80 per cent methanol for three successive days. The combined extract was then concentrated by evaporation and the residue was dissolved in equal amount of water. This was then fractionated sequentially with hexane, benzene, ethyl acetate, n-butanol and water (Hashim *et al.,* 2005). Each extract was rota-evaporated to dryness and the weight of each residue was noted and used for the qualitative analysis to know the presence of various phytochemicals in the various fractions of methanolic extract.

Qualitative Analysis

Detection of Carbohydrate Derivatives

The qualitative analysis of various extracts of coriander seeds was carried out for carbohydrates by the methods of Ramakrishnan *et al.* (1994). Extract (10mg) was dissolved in 10 ml of water and filtered and the filtrate was subjected to tests for carbohydrates.

Molish Test

To 2 ml of the test solution, two drops of Molish reagent (5 per cent α-naphtol in alcohol) were added, shaken well, 1ml of conc. sulphuric acid was added slowly along the sides of the test tube, purple ring indicated the presence of carbohydrates.

Barfoed's Test

To 1ml of filtrate, 1ml of Barfoed's reagent (6.5 per cent cupric acetate in distilled water, filtered and added 0.9 ml of glacial acetic acid) was added and heated in a boiling water bath for 2min. Red precipitate indicated the presence of reducing monosaccharides.

Benedict's Test

To 0.5ml of filtrate, 0.5ml of Benedict's reagent (173g sodium citrate and 100g sodium carbonate were dissolved in 800 ml of distilled water and boiled to make it clear. 17.3g cupric sulphate dissolved in 100 ml distilled water was added to it) was added and heated in a boiling water bath for 2 min. Orange-red precipitate indicated the presence of reducing sugars.

Fehling's Test

One ml filtrate was boiled in a water bath with 1ml each of Fehling's solution A (7 per cent cupric sulphate in distilled water) and B (35g potassium sodium tartarate and 10g sodium hydroxide in 100 ml water). A red precipitate indicated the presence of reducing sugars.

Detection of Amino Acids

The qualitative analysis of amino acids in various extracts of coriander seeds was carried out by the method of Ruthmann (1970). The extract (10mg) was dissolved in 10 ml of distilled water, filtered through Whatman no.1 filter paper and the filtrate was subjected to qualitative tests for amino acids.

Million's Reaction

To 1ml of the test solution, few drops of Million's reagent (1g mercury in 9ml of fuming nitric acid) were added; red colour indicated the presence of aromatic amino acid-tyrosine.

Hopkin Cole Reaction

To 2ml of the test solution, 2ml of glacial acetic acid and 2ml of conc. H_2SO_4 were added through the sides of the test tube, appearance of purple ring indicated the presence of aromatic amino acid-tryptophan.

Ehrlich Reaction

To1ml of the test solution, 2ml of Ehrlich reagent (10 per cent p-dimethyl amino benzaldehyde in conc. HCl) were added. The appearance of dark pink colour indicated the presence of aromatic amino acid–tryptophan.

Pauly's Reaction

To 2ml of test solution, 1ml of 1 per cent sulphanilic acid was added followed by 1ml of 5 per cent sodium nitrate solution. After 3min, solution was made alkaline by adding 2ml of 1 per cent Na_2CO_3 solution. The appearance of yellow, orange and red colour indicated the presence of tyrosine, tryptophan and histidine respectively.

Nitroprusside Test

To 1ml of the test solution, 1ml of aqueous solution of freshly prepared 2 per cent sodium nitroprusside and 40 per cent NaOH solutions were added. Appearance of blood red/dark pink colour indicated the presence of sulphur containing amino acids *i.e.*, cysteine.

Detection of Flavonoids

The qualitative analysis of flavonoids in various extracts of coriander seeds was carried out by the method of Edeoga *et al.* (2005).

Sodium Hydroxide Test

The extract was treated with 20 per cent sodium hydroxide; formation of yellow colour indicated the presence of flavonoids.

Sulphuric Acid Test

The extract was treated with conc. sulphuric acid; formation of yellow or orange colour indicated the presence of flavones.

Sodium Acetate Test

Little quantity of extract was treated with amyl alcohol, sodium acetate and ferric chloride. Disappearance of yellow colour formed on the addition of an acid indicated the presence of flavonoids.

Detection of Sterols and Terpenoids

The qualtitative analysis of various samples for sterols and triterpenoids was carried out (Finar, 1986).

Libermann-Buchard Test

To 1mg of test substance dissolved in few drops of chloroform, 3ml of acetic anhydride, 3ml of glacial acetic acid were added, warmed and cooled under the tap and few drops of conc. H_2SO_4 were added along the sides of the test tube. Appearance of bluish-green colour showed the presence of sterols.

Salkowski Test

To 1mg of test substance dissolved in few drops of chloroform, equal volume of conc. H_2SO_4 was added. Formation of bluish-red to cherry colour in chloroform layer and green fluorescence in the acid layer indicated the presence of sterol components in the extract.

Detection of Tannins and Phenol Compounds

The qualitative analysis of various extracts for tannins and phenolic compounds was carried out by standard methods (Herborne, 1973).

Ferric Chloride Test

To the test solution, few drops of neutral 5 per cent ferric chloride solution were added; dark green colour indicated the presence of phenolic compounds.

Potassium Ferricyanide Test

A small quantity of the test solution was treated with 2 per cent potassium ferricyanide and ammonia solution. A deep red colour indicated the presence of tannins.

Potassium Dichromate Test

To the test solution, 2 per cent potassium dichromate solution was added. A yellow precipitate indicated the presence of tannins.

Detection of Glycosides

The qualitative analysis of various samples for glycosides was carried out by the methods given by Sofowara (1993). The test solution was prepared by dissolving the extract in alcohol or boiling with hydro-alcoholic solution.

Baljet's Test

The test solution was treated with 2 per cent sodium picrate. Appearance of yellow to orange colour indicated the presence of glycosides.

Raymond's Test

The test solution was treated with dinitrobenzene in hot methanolic alkali. Appearance of violet colour indicated the presence of glycosides.

Bromine Water Test

Test solution when mixed with bromine water gave yellow precipitate indicating the presence of glycosides.

Keller-Killani Test for Digitoxose

The test solution was treated with few drops of 5 per cent $FeCl_3$ solution, mixed, and then H_2SO_4 containing $FeCl_3$ solution was added, formation of two layers wherein lower layer was reddish brown and upper layer turned bluish green indicated positive test.

Legal's Test

Test solution was treated with pyridine and made alkaline, addition of 2 per cent sodium nitroprusside gave pink to red colour indicating the presence of glycosides.

Detection of Anthroquinone Glycosides

Borntrager's Test

Test solution was treated with 5 ml of 10 per cent sulphuric acid for 5 min., filtered while hot, cooled, and the filtrate was shaken gently with equal volume of benzene. Benzene layer was separated and treated with half of its volume with a solution of 10 per cent ammonia and allowed to separate it. The rose pink colour in ammonial layer indicated the presence of anthroquinones (Herborne, 1973).

Detection of Organic Acids

Aqueous extract was neutralised with dilute NH_4OH solution and then the following tests were performed (Jaiswal *et al.*, 2012).

Calcium Chloride Test

To 2 ml of the test solution, few drops of 5 per cent $CaCl_2$ solution were added. Formation of white precipitate immediately, indicated the presence of oxalic acid.

Precipitate observed on shaking indicated the presence of tartaric acid, precipitate observed on boiling and cooling indicated the presence of citric acid.

Test for Oxalic Acid

To 2 ml of the test solution, few drops of 5 per cent lead acetate solution were added; white precipitate confirmed the presence of oxalic acid.

Test for Citric Acid

To 2 ml the test solution, one drop of dilute NH_4OH and excess $AgNO_3$ solution were added, boiled for 15 min. Blackish silver mirror formed, indicated the presence of citric acid.

Detection of Saponins

Foam Test

Test solution when shaken, the formation of foam, stable for at least 5 minutes, confirmed the presence of saponins (Kumar *et al.*, 2009).

Results and Discussions

The curative properties of medicinal plants are perhaps due to the presence of various secondary metabolites such as flavonoids, tannins and phenolic compounds, sterols and terpenoids, glycosides etc. Polarity of the solvent used for extraction may influence the group of bioactive compounds obtained from the plant material. Therefore, in the present study, different solvents used for the extraction of various compounds from coriander seeds in the increasing order of polarity responded differently to the qualitative tests. The results of phytochemical screening in terms of qualitative tests carried out to identify carbohydrates, amino acids, flavonoids, tannins and phenolic compounds, sterols and terpenoids and glycosides are presented in Tables 21.1 and 21.2.

All the fractions responded positively for qualitative tests for various phytochemical constituents' *viz.* carbohydrates, tannins, triterpenoids, flavanoids and flavones, while only a few fractions responded positively for amino acids, glycosides and organic acids. However, ethyl acetate fraction possessed maximum of carbohydrates, aminoacids, flavonoids, triterpenoids, glycosides and saponins; n-butanol fraction had higher amounts of tannins. Among all the fractions, ethyl acetate and n-butanol fractions showed maximum content of various bioactive compounds.

Carbohydrates

Carbohydrates or saccharides are the most abundant among the four major classes of biomolecules. All carbohydrates have C=O and –OH functional groups. Polysaccharides are ubiquitous can be either homopolysaccharides or hetero polysaccharide and occur in algae (*e.g.* alginates), plants (*e.g.* pectin, guar gum, mannan), microbes (*e.g.* dextran, xanthan gum) and animals (*e.g.* chitosan, chondroitin) and they can also be produced by means of recombinant DNA techniques (Sinha and Kumaria, 2001).

They are highly stable, safe, non toxic, hydrophilic and biodegradable. Polysaccharides having large number of reactive groups, a wide range of molecular weight, varying chemical composition etc. contribute to their diversity in structure and in property. Hydrophilic groups in polysaccharide form non-covalent bonds with biological tissues causes bioadhesion (Lee *et al.*, 2000). Complex carbohydrates obtained from natural sources such as plants have shown diverse biological activities such as wound healing, enhancement of the reticulo-endothelial system, stimulation of the immune system, treatment of tumours and effects on the hematopoietic system (Tamalde *et al.*, 2004).

Table 21.1: Carbohydrates, amino acids and organic acids in methanolic extract and various fractions of coriander seeds

Sl.No.	*Plant Constituents*	*Me*	*He*	*Be*	*Ea*	*nBu*	*Aq*
1.	*Carbohydrates derivatives*						
	a. Molisch's test	++	+	++	+++	++	++
	b. Barfoed's test	+	+	+	+++	++	++
	c. Benedict's test	++	traces	+	+	++	traces
	d. Fehling's test	++	++	++	+++	++	++
2.	*Amino acids*						
	a. Million's test	+	traces	+	++	++	traces
	b. Hopkin-cole test	+	+	+	++	++	traces
	c. Ehrlich test	+	traces	+	++	+	–
	c. Pauly's test	+++	+	+	++	+	++
	c. Nitroprusside test	+	traces	+	++	++	traces
3.	*Organic acids*						
	a. Calcium chloride test	+	+	+	+	+	+
	b. Oxalic acid test	traces	traces	++	+++	+++	traces
	c. Tartaric acid test	++	++	+++	++++	+++	traces
	d. Citric acid test	+	+++	+	++	+++	++

++++: Abundant; ++: Average; +: Meagre; –: Absence of constituent.

In the present investigation, various tests (Molisch, Benedict's, Fehling's, Barfoed's) carried out for carbohydrate derivatives for all the fractions *viz.* hexane, benzene, ethyl acetate, n-butanol and aqueous fractions showed positive response for reducing sugars as well as reducing compounds. However, maximum response for the tests was given by ethyl acetate fraction followed by n-butanol fraction, methanolic extract and aqueous fraction while least response was shown by benzene and hexane fractions in proportion to the solubility of different carbohydrate derivatives in various solvents (Table 21.1) indicating water insolubility of the tested.

Amino Acids

Amino acids are the chemical substances that make up protein and are found in our bodies and also in plants; they have a role in metabolism, and are important in nutrition. They form short polymer chains called peptides, as well as longer chains that are called polypeptides or proteins. About 75 per cent of the human body is made up of chains of amino acids. All the chemical reactions that occur in the body depend on amino acids and the proteins they build. The protein component of the foods we consume is broken down into individual amino acids by the digestive action of enzymes in the stomach and small intestine. The free amino acids can then be used to form any of the millions of specific proteins required for running the human body. However plant sources of protein are often deficient in one or more essential amino acids, these deficiencies can be overcome by consuming a wide variety of plant foods (http: //www.vaxa.com/aminoacids-main.cfm).

In the present study, Millions test indicative of the presence of phenol containing amino acids (tyrosine), Hopkin-cole test, positive for indole group containing amino acids like tryptophan, Ehrlich test indicative of tryptophan, Pauly's test positive for imidazole containing amino acid like histidine, nitroprusside test that indicates the presence of sulphur containing amino acid–cysteine were responded by ethyl acetate fraction in a higher degree followed by n-butanol, methanolic extract, benzene and hexane fractions while least response was seen with aqueous fraction (Table 21.1) indicating water insolubility of the amino acids tested.

Organic Acids

Among a variety of candidates for the replacement of antibiotics, organic acids have been broadly applied worldwide with reasonable success (Mroz, 2005) reported that the supplementation of organic acids in drinking water not only improved the growth performance but reduced the number of *E. coli* in the gut of post-weaning pigs. Ever since then, many findings have shown that organic acids can provide beneficial effects similar to those of feeding antibiotics (Sciopioni *et al.*, 1978; Mathew *et al.*, 1991). The probable mode of actions of organic acids includes reducing the pH in the gastrointestinal tract (GIT) (Ravindran and Kornegay, 1993), regulating the balance of microbial populations in the gut, stimulating the secretion of digestive enzymes (Harada *et al.*, 1988; Thaela *et al.*, 1998) and promoting the growth and recovery of the intestinal morphology (Galfi and Bokori, 1990).

Citric acid is an alpha hydroxy acid, used as an active ingredient in chemical peels. Citric acid is widely used as a pH adjusting agent in creams and gels of all kinds. In this role, it is classified in most jurisdictions as a processing aid and so does not need to be listed on ingredient lists. Citric acid is commonly used as a buffer to increase the solubility of brown heroin. Single-use citric acid sachets have been used as an inducement to get heroin users to exchange their dirty needles for clean needles in an attempt to decrease the spread of AIDS and hepatitis (Garden *et al.*, 2003). Other acidifiers used for brown heroin are ascorbic acid, acetic acid and lactic acid; in their absence, a drug user will often substitute lemon juice or vinegar. Citric acid is used as one of the active ingredients in the production of antiviral tissues (http: // www.google.com/html).

In the present investigation, calcium chloride test conducted for the identification of organic acids, gave less precipitate with all the fractions. Confirmatory test for oxalic acid gave maximum precipitate for ethyl acetate and n-butanol fractions, followed by benzene fractions and the least by the methanolic extract, hexane and aqueous fraction. Confirmatory test for tartaric acid was responded well by ethyl acetate and n-butanol fractions followed by benzene fraction, methanolic extract and hexane fraction while traces are seen with aqueous fraction (Table 21.1).

Flavonoids

Flavonoids are phenolic substances isolated from a wide range of vascular plants, with over 8000 individual compounds known. Dietary flavonoids represent an important source of antioxidants, since their intake may reach 800 mg/day. They act in plants as antioxidants, antimicrobials, photoreceptors, visual attractors, feeding repellants, and for light screening. Many studies have suggested that flavonoids exhibit biological activities, including antiallergic, antiviral, anti-inflammatory, and vasodilating actions. However, most interest has been devoted to the antioxidant activity of flavonoids, which is due to their ability to reduce free radical formation and to scavenge free radicals. Both the absorbed flavonoids and their metabolites may display an *in vivo* antioxidant activity, which seems to involve differently the physiological antioxidants, resulting in a sparing effect on α-tocopherol and β-carotene. They have antioxidant potentials hence could offer protection against heart disease and cancer probably by enhancing the body defense against pathology induced by free radical generation (Pier-Giorgio, 2000).

In the present study, methanolic extract and its fractions showed very good response to the qualitative tests (sodium hydroxide test, sodium acetate test and sulphuric acid test) for flavonoids. Ethyl acetate fraction of coriander seeds possessed highest amount of flavonoids followed by methanolic extract, n-butanol and aqueous fractions, while benzene and hexane fractions had minimal amount of flavonoids in proportion to their solubility and polarity (Table 21.2).

Typically, when discussing phenolics in plants foods, flavonoids are the predominant class, because they account for approximately two-thirds of the dietary phenols. However, phenolic acids account for almost all of the remaining one-third, and there is an increasing awareness and interest in the antioxidant behavior and potential health benefits associated with these simple phenolic acids (Robbins, 2007). Flavonoids are hydrogen-donating radical scavengers (antioxidants) by complexing with iron ions, flavonoids suppress the superoxide-driven Fenton reaction. By reducing the β-tocopheroxyl radical, flavonoids regenerate β-tocopherol and also quench singlet oxygen (Rice-Evans *et al.*, 1996). Copper complexation is also an important activity of certain flavonoids, especially those with the catechol structure in the β-ring (Brown *et al.*, 1998).

Tannins and Polyphenols

Polyphenolics is a highly inclusive term that covers many different subgroups of phenolic acids and flavonoids. More than 5000 polyphenolics, including over 2000 flavonoids have been identified and the number is still growing (Merken and Beecher, 2000). Polyphenolics vary in structures: hydroxybenzoic acids and hydroxycinnamic acids have a single-ring structure, while flavonoids can be further classified into anthocyanins, flavan-3-ols, flavones, flavanones and flavonols.

Polyphenols are especially important antioxidants, because of their high redox potentials, which allow them to act as reducing agents, hydrogen donors and singlet oxygen quenchers. In addition, they have a metal chelating potential. Among them, the antioxidant activity of the flavonoids is considered to be much greater than that of the essential vitamins, therefore contributing significantly to the health benefits of fruits (Delgado-Vargas *et al.*, 2000).

Table 21.2: Phytochemical constituents in methanolic extract and various fractions of coriander seeds.

Sl.No.	*Plant Constituents*	*Me*	*He*	*Be*	*Ea*	*nBu*	*Aq*
1.	*Flavonoids*						
	a. Sodium hydroxide test	+++	++	++	++++	+++	+++
	b. Sodium acetate test	+++	++	++	+++	++	++
	c. Sulphuric acid test	+++	++	++	+++	++	++
2	*Tannins and Polyphenols*						
	a. Ferric chloride test	++	traces	+	++	+++	+
	b. Potassium dichromate test	++	traces	+	++	++	+
	c. Potassium ferricyanide test	++	traces	+	+++	+++	traces
3	*Glycosides*						
	a. Legal's test	+	traces	traces	–	++	++
	b. Balget's test	–	traces	traces	–	+	–
	c. Keller-Killani test	+	traces	–	–	+	traces
	d. Borntrager's test	+	traces	–	–	++	traces
	e. Raymond's test	–	–	–	+	+	–
	f. Bromine water test	+	–	+	++	++	++
4	*Saponins*						
	a. Foam test	++	+	++	+++	++	++
5	*Steroids and terpenoids*						
	a. Leibermann buchard's test	+	+++	+++	++++	+++	traces
	b. Salkowski test	+++	++	++	+++	+++	traces

++++: Abundant; ++: Average; +: Meagre; –: Absence of constituent.

Tannins, also called as proanthocyanidins detoxify carcinogens and scavenge harmful free radicals. Tannins can be an effective antiviral, antibacterial and antiparasitic compounds. Foods rich in tannins can be used in the treatment of hemochromatism, a hereditary disease characterized by excessive absorption of dietary iron resulting in a pathological increase in total body iron stores (Bele *et al.*, 2010).

In the present investigation, methanolic extract of coriander seeds and all the other fractions gave a very good response for the qualitative tests (ferric chloride test, potassium ferricyanide test and potassium dichromate test) conducted for tannins and polyphenols except benzene and hexane fractions. n-butanol fraction exhibited maximum content of tannins and polyphenols followed by ethyl acetate fraction and methanolic extract, while aqueous, benzene, and hexane fractions had the least. Better response was seen with potassium ferricyanide test than potassium dichromate test (Table 21.2).

Glycosides

Glycosides are unique molecules used by the plants to store potentially dangerous compounds and by animals to eliminate toxins. They consist of sugar group called a glycone, chemically attached to non carbohydrate compound (aglycone). A wide variety of sugars are attached to natural cardiac glycosides and the most common are glucose, galactose, mannose, rhamnose and digitalose. Although sugars themselves have no activity, the addition of sugars affects the pharmacodynamic and pharmacokinetic profile of each glycoside. For example, free aglycones are absorbed more rapidly and are metabolized more easily than their glycosylated counter parts, moreover, the type of the attached sugar influences the potency of the compound. New findings indicate that cardiac glycosides can regulate several cellular processes, which are beyond their well-established role in ion homeostasis. Overall, it is now clear that the ultimate response to cardiac glycoside treatment is dependent on the tissue, exposure time and dose (Mijatovic, 2007). Notably, responses of cancer cells to cardiac glycoside treatment seem to be different, which has stimulated interest in their potential as anticancer drugs (Winnicka *et al.*, 2006).

In the present investigation, qualitative tests (Legal, Balget's, Keller-killani, Borntrager's, Raymond's, Bromine water) for glycosides gave meagre response with all the fractions of methanolic extract of coriander seeds indicating lesser amount of glycosides while n-butanol fraction showed response to Legal, Bromine water and Borntrager's tests indicating presence of glycosides in n-butanol fraction of coriander seeds (Table 21.2).

Saponins

Saponins, used in traditional medicine preparations are reported to possess antitumor and antimutagenic activities and can lower the risk of human cancers, by preventing cancer cells from growing. Saponins can bind cholesterol-type compound present in the cancer cell membranes and thus interfere with cell growth and division. Saponins have been found to be useful in the treatment for hypercholestrolemia. Saponins bind with cholesterol and cause a depletion of body cholesterol by preventing its reabsorption (Asl *et al.*, 2008). Saponins function as natural antibiotics as they fight against viruses, combat microbenzenes and fungal infections (Xu *et al.*, 1996).

In the present investigation, foam test carried out for saponins was responded positively by methanolic extract and its various fractions. Maximum response was shown by ethyl acetate fraction, followed by n-butanol fraction, methanolic extract, aqueous fraction, benzene and hexane fractions (Table 21.2).

Steroids and Terpenoids

Sterols may be found either as free sterols, acylated, alkylated and sulphated. They are lipids resistant to saponification and are found in good quantity in all animal and vegetable tissues. Studies showed that phytosterols have more efficacy in lowering LDL cholesterol and they also have anti-inflammatory (Bouic, 2001), antioxidative (Homma *et al.*, 2003) and anti-atherosclerotic activities. Phytosterols were also found to be protective to ulcers (Jayaraj *et al.*, 2003) and to have antifungal activity (Li *et al.*, 2007). The intake of plant sterols is beneficial to prevent or treat many different types of cancers *viz.* breast (Awad *et al.*, 2003), esophagus (De stefani *et al.*, 2000), endometrial, ovary (Mc Cann *et al.*, 2003) etc.

Isoprenoids, more commonly called terpenoids are the most functionally and structurally varied group of plant metabolites and plant produced terpenoids are essential components of the human diet (Ikeda *et al.*, 2008). Triterpenoids are ubiquitously distributed throughout the plant kingdom in the form of free acid or aglycones for saponins, types of triterpenoids (Wang and Jiang, 1992) and have long been considered to have a number of biological effects. Some are increasingly being used for medicinal purposes for a variety of clinical diseases in many Asian countries as antioxidant, anticancer and anti-inflammatory agents (Manez *et al.*, 1997)

Terpenoids, the largest and the most widespread class of secondary metabolites, are found in higher plants, mosses, liverworts, algae and lichens as well as in lower invertebrates *i.e.* insects, microbes or marine organisms. Terpenoids are derived from a common biosynthetic pathway based on mevalonate as parent, and are named terpenoids, terpenes or isoprenoids with the subgroup of steroids among them as a class (Bohlmann and Keeling, 2008). Diterpenoids containing four isoprene units (C-20) constitute a large group of compounds found in higher plants, fungi, insects and marine organisms. Most of the diterpenoids found in recent years have been isolated from Asteraceae and Lamiaceae, and this class of terpenoids has one of the widest ranges of biological activities. Diterpenoids are classified according to their chemical structure (De las Heras *et al.*, 2008). Several biological actions have been reported for diterpenes including antibacterial, antifungal, anti-inflammatory, antileishmanial, cytotoxic and antitumour activities (Singh *et al.*, 1999).

In the present investigation, methanolic extract of coriander seeds and various fractions gave a very good response to Libermann-Buchard and Salkowski tests indicating the presence of sterols in all sample extracts. However, ethyl acetate fraction gave maximum response, followed by n-butanol, benzene and hexane fractions and methanolic extract while aqueous extract showed minimum response due to insolubility of steroids and terpenoids in water (Table 21.2). Steroidal compounds are of importance and interest in pharmacy due to their relationship with compounds such as sex hormones (Edeoga *et al.*, 2005).

Thus the preliminary screening tests may be useful in the detection of the bioactive principles and subsequently may lead to the drug discovery and development. Further, these tests facilitate their quantitative estimation and qualitative separation of pharmacologically active chemical compounds. Also, the data generated from these preliminary investigations could provide the chemical basis for the wide use of this plant as therapeutic agent for treating various ailments. However, there is need to further carry out advanced hyphenated spectroscopic studies in order to elucidate the quantity and structure of these compounds. Furthermore, this data may be handy in probing of biochemistry of this plant in the future.

Conclusions

The findings confirm the presence of phytochemicals in various fractions of the methanolic extract of the seeds of *Coriandrum sativum* L. which are dependent upon various solvents used for extraction. Ethyl acetate fraction had abundance of carbohydrates, amino acids, organic acids, flavonoids, saponins, steroids and terpenoids and n-butanol fraction had max amounts of tannins and polyphenols and glycosides. The methanolic extract proved to be multi-component by having all the compounds but in lesser concentrations. The phytochemicals in coriander seeds are of the great importance in development of pharmaceutical, cosmetic, and food products and also to improve the health status. Further investigations on the ethyl acetate fraction of *Coriandrum sativum* L. seeds are being conducted in order to isolate, identify, characterize and elucidate the structure of the bioactive compounds.

Acknowledgements

The authors gratefully acknowledge Sri Sathya Sai Institute of Higher Learning for the facilities provided to conduct the research.

References

Asl, M.N., and Hossein, H. (2008). Effects of *Glycyrrhiza* sp. and its bioactive compounds. *Phytotherapy Research,* **22(6):** 709-724.

Awad, A.B., Roy, R., and Fink, C.S. (2003). β sitosterol, a plant sterol, induces apoptosis and activates key casposes in MDA-MB- 231 human breast cancer cells. *Oncology Reports,* **10:** 497-500.

Ayyanar, M., and Ignacimuthu, S. (2008). Medicinal uses and pharmacological actions of five commonly used Indian medicinal plants: A mini review. *Iranian Journal of Pharmacology and Therapeutics,* **7(1):** 107-114.

Bele, A. A., Jadhav, V. M., and Kadam, V. J. (2010). Potential of tannins: a review. *Asia Journal of Plant Sciences,* **9(4):** 209-214.

Bohlmann, J., and Keeling, C.I. (2008). Terpenoid biomaterials. *Plant Journal.,* **54:** 656-669.

Bouic, P.J. (2001). The role of phytosterols and phytosterolins in immune modulation: a review of the past 10 years. *Current Opinions in Clinical Nutrition Metabolism and Care,* **4:** 471-475.

Brown, J.E., Khodr, H., Hider, H., and Rice-Evans, C.A. (1998). Structural dependence of flavonoid interactions with Cu^{2+} ions: implications for their antioxidant properties. *Biochemistry Journal,* **330:** 1173-1178.

Chopra, R.N., Nayar, S.L., and Chopra, I.C. (2009). Glossary of Indian Medicinal Plants. CSIR, New Delhi, pp.185-210.

De las Heras, B., Rodriguez, B., Bosca, L., and Villar, A.M. (2008). Terpenoids: Sources, structure elucidation and therapeutic potential in inflammation. *Current Topics in Medicinal Chemistry,* **3:** 171-185.

Delgado-Vargas, F., Jimenez, A.R., and Paredes-Lopez, O. (2000). Natural pigments: Carotenoids. *Critical Reviews in Food Science and Nutrition.* **40**: 173.

Edeoga, H.O., Okwu, D.E., and Mbaebie, B.O. (2005). Phytochemical constituents of some Nigerian medicinal plants. *African Journal of Biotechnology,* **4:** 685-688.

Finar, I.L. (1986). Stereo chemistry and the chemistry of natural products, Vol 2. Longman, Singapore.

Funatogawa, K., Hayashi, S., Shimomura, H., Yoshida, T., Hatano, T., Ito, H., and Hirai, Y. (2004). Antibacterial activity of hydrolysable tannins derived from medicinal plants against *Helicobactor pylori. Microbiology and Immunology,* **48(4):** 251-261.

Galfi, P., and Bokori, J. (1990). Feeding trial in pigs with a diet containing sodium n-butyrate. *Acta Veterinaria Hungarica,* **38:** 3-17.

Garden, J., Roberts, K., Taylor, A., and Robinson, D. (2003). Evaluation of the provision of single use citric acid sachets to injecting drug users, Scottish Center for Infection and Environmental Health. NHS Greater Glasglow, Lanarkshire, pp. 1-18.

Harada, E., Hiroko, K., Kobayashi, E., and Tsuchita, H. (1988). Postnatal development of biliary and pancreatic exocrine secretion in piglets. *Composite Biochemical Physiology,* **91A:** 43-51.

Hashim, M.S., Lincy, S., Remya,V., Teena, M., and Anila, L. (2005). Effect of polyphenolic compounds from *Coriandrum sativum* on H_2O_2-induced oxidative stress in human lymphocytes. *Food Chemistry,* **92:** 653-660.

Herborne, J.B. (1973). Phytochemical methods, 3rd edn. Chapman and Hall Ltd. London, pp.135-203.

Homma, Y., Keda, I., Shikawa, T., Tateno, M., Sugano, M., and Nakamura, H. (2003). A randomized, placebo. controlled trial: Decrease in plasma low-density lipoprotein cholesterol, apolipoprotein B, cholesteryl ester transfer protein, and oxidized Low-density lipoprotein by plant ptanol sterol- containing spread. *Nutrition,* **19(4):** 369-374.

http: / /www.google.com/html (accessed in June, 2012).

http: / /www.vaxa.com/aminoacids-main.cfm/html (accessed in July, 2012).

Ikeda, Y., Murakami, A., and Ohigashi, H. (2008). Ursolic acid: An-anti and pro-inflammatory triterpenoid. *Molecular Nutrition Food Research,* **52**: 26-42.

Jaiswal, Y., Naik, V., Tatke, P., Gabhe, S., and Vaidya, A. (2012). Pharmacognostic and preliminary phytochemical investigations of *Anacardium occidentale* (Linn.) leaves. *Academic Sciences.* **4(3):** 625-631.

Jayaraj, P.A., Tovey, F.L., and Hobsley, M. (2003). Duodenal nuclear prevalence research into the nature of possible protective. *Dietary Lipids Phytotherapy Research,* **17:** 391-398.

Katan, M.B., Grundy, S.M., Jones, P., Law, M., Miettinen, T., and Paoletti, R. (2003). Efficacy and safety of plant stanols and sterols in the management of blood cholesterol levels. *Mayo Clinic Proceedings*, 78(8): 965–978.

Khan, H. A., Sadiq, U. M., Asif, M.S., Md Meraj, A., Umar, H., Wasim, A., Shadab, A. S., and Sayeed, A. (2012). Anti-inflammatory and antioxidant activity of *Coriandrum sativum* seeds in collagen-induced arthritis in rats. *International Journal of Drug and Developing Research*, **4(1):** 210-219.

Kumar, A., Ilavarasan, R., Jayachandran, T., Decaraman, M., Aravindan, P., Padmanaban, N., and Krishnan, M. (2009). Phytochemical investigation on a tropical plant. *Pakistan Journal of Nutrition*, **8(1):** 83-85.

Lee, J. W., Park, J. H., and Robinson, J. R. (2000). Bioadhesive-based dosage forms: The next generation. *Journal of Pharmaceutical Sciences*, **89**: 850-866.

Li, H-LIV., Zhang, J.X., and Ji, G. (2007). Antioxidiabetic activity of Lipophilic (-). Epigallocatechin – 3- gallati derivative and its role of *Alpha glucosidase* inhibition. *Biomedicine Pharmaco Therapy*, **61(1):** 91-96.

Manez, S., Recio, M.C., Giner, R.M., and Rios, J.L. (1997). Effect of selected triterpenoids on chronic dermal inflammation. *European Journal of Pharmacology*, **334**: 103-105.

Mathew, A.G., Franklin, M.A., Upchurch, W.G., and Chattin, S.E. (1991). Influence of weaning age on ileal microflora and fermentation acids in young pigs. *Nutrition Research*, **16:** 817-27.

Mc Cann, S.E., Freudenheim, J.L., Marshal, J.R., and Graham, S. (2003). Risk of human ovarian cancer is related to dietary intake of selected nutrients, phytochemicals food groups. *Journal of Nutrition*, **133:** 1937-1942.

Mendilaharou, M. (2000). Plant sterols and risk of stomach cancer: A case-control study in Urguay. *Nutrition Cancer*, **37:** 140–144.

Merken, H.M., and Beecher, G.R. (2000). Measurement of food flavonoids by high-performance liquid chromatography: A review. *Journal of Agriculture Food Chemistry*, **48**: 577.

Mijatovic, T., Van Quaquebeke, E., Delest, B., *et al.* (2007). Cardiotonic steroids on the road to anti-cancer therapy. *Biochimica Biophysica Acta*, **1776**: 32–57.

Mroz, Z. (2005). Organic acids as alternatives to antibiotic growth promoters for pigs. In: advances in pork production, G. Foxcroft, University of Alberta Press, Edmonton, Alberta, pp.169-182.

Naczk, M., and Shahidi, F. (2004). Extraction and analysis of phenolics in food. *Journal of Chromatography A*, **1054:** 95-111.

Okwu, D.E. (2001). Evaluation of the chemical composition of indigenous spices and flavouring agents. *Global Journal of Pure and Applied Sciences*, **7(3):** 455- 459.

Pier-Giorgio, P. (2000). Flavonoids as antioxidants. *Journal of Natural Products*, **63:** 1035-1042.

Ramakrishnan, S., Prasannan, K.G., and Rajan, R. (1994). Text book of Medical Biochemistry, Orient Longman, New Delhi, India.

Ravindran, V., and Kornegay, E.T. (1993). Acidification of weaner pig diets: A review. *Journal of Science and Food Agriculture*, **62**: 313-322.

Rice-Evans, C.A., Miller, N.J., and Paganga, G. (1996). Structure antioxidant activity relationships of flavonoids and phenolic acids. *Free Radical Biology and Medicine*, **20**: 933-956.

Robbins, R.J. (2007). Phenolic acids in foods: An overview of analytical methodology. *Journal of Agriculture Food Chemistry*, **51**: 2866-2887.

Ruthmann, A. C. (1970). In: Phytochemical Techniques, Raaman N, New India publishing agency, New Delhi, pp.19-24.

Schiavone, A., Guo, K., and Tassone, S. (2008). Effects of a natural extract of chestnut wood on digestibility, performance traits, and nitrogen balance of broiler chicks. *Poultry Sciences*, **87 (3):** 521–527.

Sciopioni, R.G., Zaghini, and Biavati, B. (1978). Researches on the use of acidified diets for early weaning of piglets. *Zootechnology, Nutrition of Animals*, **4**: 201-218.

Singh, M., Pal, M., and Sharma, R.P. (1999). Biological activity of the labdane diterpenes. *Planta Medicine*, **65**: 2-8.

Sinha, V.R., and Kumria, R. (2001). Polysaccharides in colon-specific drug delivery. *International Journal of Pharmacy*, **224**: 19-38.

Sivarajan, V.V., and Balachandran, I. (1994). Ayurvedic Drugs and their Plant Sources. Oxford and IBH Publishing Co. Pvt. Ltd., New Delhi, p. 570.

Sofowora, A. (1993). Screening plants for bioactive agents. In: medicinal plants and traditional medicine in Africa, 2nd edn, Spectrum Books Ltd. Sunshine house, Ibadan, Nigeria, pp. 134-156.

Sumathi, P., and Parvathi, A. (2010). Antimicrobial activity of some traditional medicinal plants. *Journal of Medicinal Plant Research*, **4**: 316-321.

Tabacco, E., Borreani, G., Crovetto, G.M., Galassi, G., Colombo, D., and Cavallarin, L. (2006). Effect of chestnut tannin on fermentation quality, proteolysis, and protein rumen degradability of alfalfa silage. *Journal of Dairy Science*, **89 (12):** 4736–4746.

Talmadge, J., Chavez, J., Jacobs, L., Munger, C., Chinnah, T., Chow, J.T., Williamson, D., and Yates, K. (2004). Fractionation of *Aloe vera* L. inner gel, purification and molecular profiling of activity. *International Journal of Immunopharmacology*, **4:** 1757-1773.

Tapsell, L., Hemphill, I., Cobiac, L., Patch, C., Sullivan, D., and Fenech, M. (2006). Health benefits of herbs and spices: The past, the present, the future. *Medical Journal of Australia.*, **185(4):** S4–S24.

Thaela, M.J., Jensen, M.S., Pierzynowski, S.G., Jakob, S. and Jensen, B.B. (1998). Effect of lactic acid supplementation in pigs after weaning. *Journal of Animal Feed Science,* **7:** 181-183.

Wang, B., and Jiang, Z. (1992). Study of Oleanolic acid. *Chinese Pharmocology Journal,* **27**: 393-397.

Winnicka, K., Bielawski, K., and Bielawska. (2006). Cardiac glycosides in cancer research and cancer therapy. *Acta Poloniae Pharmaceutica -Drug Research,* **63(2):** 109-115.

Xu, R., Zhao, W., Xj shao, B., and Qin, G. (1996). Studies on bioactive saponins from Chinese medicinal plants. *Advances in Experimental Medicine and Biology,* **404:** 371-82.

Bioactive Phytochemicals: Perspectives for Modern Medicine Vol. 2 (2014) *Pages* **471–498**
Editor: **V.K. Gupta**
Published by: **DAYA PUBLISHING HOUSE, NEW DELHI**

22

Utilization of Cyclodextrin and its Derivatives in Novel Delivery of Herbal Constituents and Bioactives

Amber Vyas[1*], Manju Singh[1], Deependra Singh[1], Vishal Jain[1], Bina Gidwani[1] and Nagendra Singh Chauhan[2]

ABSTRACT

Herbal medicine is the most popular form of medicine with a long history. The traditional medicine involves the use of different plant extracts or the bioactive constituents. Global sales of herbal products, including herbal medicines, are already over USD 10^{11} and expected to exceed USD 10^{12} in the next 20 years at the present growth rate. Not only in under developed countries but also in developed countries a large proportion of population relies on the herbal remedies. The use of herbal medicines has increased due to their miraculous therapeutic effects and fewer adverse effects as compared to the modern medicines. However, their use is still limited due to various drawbacks like the delivery system available for herbal medicines are out-dated, resulting in reduced efficacy of the drug. Also, the herbal medicines were not considered for development of novel formulations due to the lack of scientific justification and processing difficulties, such as standardization, extraction and identification of individual drug components in complex polyherbal systems. In order to overcome most of these limitations, cyclodextrin complexation came up as a novel approach, although the utilization of cyclodextrin complexation is not new and is being used from last 10-15 years in different fields and lead to promising and successful results. Thus, can be explored for herbals also.

1 University Institute of Pharmacy, Pt. Ravishankar Shukla University, Raipur – 492 010, Chhattisgarh, India

2 Drugs Testing Laboratory Avam Anusandhan Kendra, Government Ayurvedic College Campus, Raipur – 492 010, Chhattisgarh, India

* *Corresponding author*: E-mail: ambervyas@gmail.com

These cyclodextrins serve as multi-functional drug carriers, through the formation of inclusion complex and potentially serve as novel drug carriers. This article outlines the applications and comparative benefits of cyclodextrins (CDs) and their derivatives with herbal constituents in the designing of novel delivery systems like liposomes, microspheres, nanoparticles, nanosponges etc. Apart from these, cyclodextrins also help to achieve controlled/sustained release, drug safety, and stability. This article covers all the aspects related to herbal active constituents, their advantages and drawbacks, especially highlighting the benefits of cyclodextrin complexation with herbal constituents. From last two years, the use of cyclodextrins and their derivatives with respect to herbals is tremendously increasing with enhanced therapeutic effect, efficacy and reduced toxicity of herbal constituents. Development of novel delivery system based on cyclodextrin complexation will serve as novel approach for herbal constituents in the coming future.

Keywords: Herbal constituents, Bioactives, Cyclodextrin, Inclusion complex, Phase solubility.

Introduction

"Health for all" is a dream and a goal which humanity at large shares and strives for. Alternative and traditional medicines, largely herbal in nature, are now regarded as important but underutilized tools against disease. The World Health Organization (WHO) recognized this fact in the early 1970s that knowledge of herbal medicines for disease prevention and health promotion is essential (Mosihuzzaman *et al.*, 2008).

Although recently developed tools such as computer aided molecular drug design or high-throughput screening techniques, have increased the success rate in the discovery of novel bioactive molecules, traditional remedies are still among the most reliable sources in the discovery and development of new medicines (Mukherjee, 2009).

'Traditional medicine' is a basically the sum total of knowledge, skills and practices based on the theories, beliefs, and personal experiences indigenous to different cultures, that are used to maintain health, prevent disease spread, minimize disease or treat it completely both physical and mental wellness (Acharya, 2008; Fabricant, 2001). Not only in underdeveloped countries but also in developed countries a large proportion of population relies on the herbal remedies. High treatment cost and side effects along with drug resistance are major problem associated with modern medicines and synthetic drugs. Thus, the use of herbal medicine has increased due to its miraculous therapeutic effects and fewer side effects as compared to the modern medicine. Nearly 80 per cent of the world's population relies on herbal medicine for primary health care, most of which involve the use of plant extracts (Sandhya *et al.*, 2006). In India, almost 95 per cent of the prescriptions were plant based in the traditional systems of Unani, Ayurveda, Homeopathy and Siddha (Satyavati, 1987). Conventional delivery systems used for herbals are outdated with less potency and reduced efficacy of drugs. This accounts for the development of novel delivery system for herbal drugs with increased efficacy and potency. Drugs obtained from natural origin play a major significant role in the public health care system of any nation. In general, drug substance from natural herbal source offer four vital and appreciable roles in the modern system of medicine thereby adequately justifying their legitimate presence in the prevailing therapeutic response, namely:

1. Serve as extremely useful natural drugs.
2. Provide basic compounds affording less toxic and more effective drug molecules.
3. Explore the biologically active prototypes towards newer and better synthetic drugs.
4. Modify the inactive natural products by suitable biological/chemical means into potent drugs.

On a recent survey conducted by the World Health Organization (WHO) globally, around 20,000 medicinal plants are being used profusely either in pharmaceutical industry or in folk medicines. Interestingly, about 1.4 per cent do possess well established, widely-proven and broadly accepted active constituents (Ashutoshkar, 2007). Herbal drugs/constituents have some distinct advantages, which accounts for their maximum utilization. Figure 22.1 highlights the four evergreen advantages of using herbal drugs.

Advantages of Herbal Drugs

1. Low Risk of Side Effects

Mostly all the herbal drugs are well tolerated by the patients, having fewer unintended consequences and fewer side effects and may be safer to use.

2. More Effectiveness

Herbal drugs are more effective for long-standing health complaints that do not respond well to other medicines.

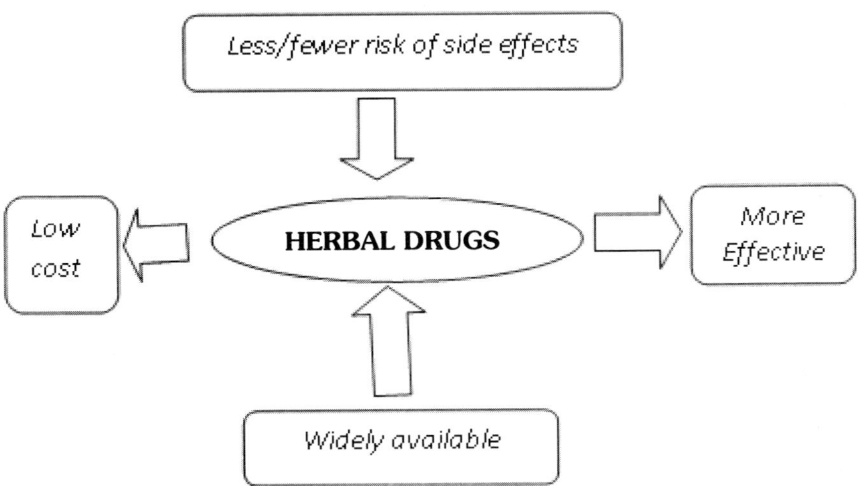

Figure 22.1: Advantage of herbal drugs.

3. Lower Cost

Cost of herbal drugs/bioactives is much less than prescription medications. Research, testing and marketing add considerably to the cost of prescription medicines. Herbs tend to be inexpensive compared to drugs.

4. Wide Spread Availability

Herbs are available without a prescription. Moreover, they can be cultivated also.

Herbal/Plant Products

Plant products have been part of phytomedicines since time immemorial. These can be derived from any part of the plant like bark, leaves, flowers, seeds, etc. (Cragg *et al.*, 2001) *i.e.*, any part of the plant may contain active components. Knowledge of the chemical constituents of plants is desirable because such information will be of value for the synthesis of complex chemical substances (Siddiqui *et al.*, 2009). The constituents of an extract can be divided into active and accompanying substances or co-effectors. Co-effectors may have an influence on the physicochemical properties such as stability, solubility, or the half-life of the active compounds. For example, Saponins increase the absorption of corticosteroids, some antibiotics, flavones, phytosterols, and silicic acid. The concentration of kavain and yangonin in mouse brain samples is higher after administration of a *Piper methysticum* extract than after administration of the purified single compounds in the same amount.

Active Constituents

The pharmacological and therapeutic activity of drug mainly depends on the active constituent present in it. Herbs and spices are highly rich in volatile oils, which give pleasurable aromas. In addition, all the herbs contain major and minor constituents in them, which are responsible for their activity. These active constituents are of greater interest to the pharmacologists. Some of the main active constituents in herbals are as described in Table 22.1 (Brown, 1995; De Guzman and Sienonsma, 1999).

All these active constituents are responsible for the pharmacological action and therapeutic activity. However, most of them are bitter in taste, irritant, poorly soluble in water.

Formulations Based on Herbals

Herbal drugs, either in raw form or in processed form have been used in treatment of ailments/diseases from time immemorial. These formulations are generally regarded as safe (GRAS). "Herbal formulation mean a dosage form consisting of one or more herbs or processed herb(s) in specified quantities to provide specific nutritional, cosmetic benefits, and/or other benefits meant for use to diagnose treat, mitigate diseases of human beings or animals and/or to alter the structure or physiology of human beings or animals".

Herbal preparations are obtained by subjecting the whole plant, fragmented or cut plants, plants parts to treatments such as extraction, distillation, expression, fractionation, purification, concentration or fermentation. These include comminuted or powdered herbal substances, tinctures, extracts, essential oils, expressed juices, and processed exudates (Cott, 1995).

Table 22.1: Properties exhibited by plant constituents.

Sl.No.	*Active Constituent*	*Features/Properties*
1.	**Alkaloids**	Bitter, often based on alkaline nitrogenous compounds. They affect the central nervous system.
2.	**Anthraquinones**	Bitter, irritant and laxative, acting also as dyes.
3.	**Iridoides and sesquiterpenes**	Bitter, increases and improves digestion.
4.	**Coumarins**	Antibacterial, anticoagulant with a smell of new-mown hay.
5.	**Flavones/ flavonoids**	Bitter or sweet, often diuretic, antiseptic, antispasmodic and anti-inflammatory.
6.	**Resins (oleoresins/ oleo-gum resins)**	Acrid, astringent, antiseptic, healing.
7.	**Saponins**	Sweet, stimulant hormonal, anti-inflammatory, or diuretic, soapy in water.
8.	**Tannins**	Astringent, antiseptic, checks bleeding and discharges.
9.	**Volatile oils**	Aromatic, antiseptic, fungicidal, irritant and stimulant.

Limitations/Shortcomings with the Herbals

Herbal medicines suffer from a range of limitations/shortcomings. These include insufficient and unacceptable evidences of safety, efficacy, standardization, and inconsistent production practices (Mosihuzzaman *et al.*, 2008). High treatment cost and side effects along with drug resistance are major problem associated with synthetic drugs. Since traditional medicine is not easily accessible but also affordable, there is an increased emphasis on the use of plants to treat human diseases. Therefore, the global markets are turning to plants as potential and realistic source of ingredients for health care products (Tiwari *et al.*, 2011). Global sales of herbal products, including herbal medicines, are already over USD 10^{11} and are expected to exceed USD 10^{12} in the next 20 years at the present growth rate. Along with herbal medicines, other herbal products such as cosmetics, fragrances, teas, health foods, and nutraceuticals are equally popular and constitute a large proportion of global herbal business.

Why Novel Drug Delivery Systems for Herbal Drugs?

Herbal drugs are becoming more popular in the modern world for their application to cure variety of diseases with less toxic effects and better therapeutic effects (Atmakuri *et al.*, 2010).

The conventional available dosage forms of herbal constituents/bioactives suffer from following limitations, which can be overcome by novel carrier based delivery of herbal products.

Drawbacks of Conventional Dosage Forms

1. Poor patient compliance, increased chances of missing the dose of a drug with short half-life for which frequent administration is necessary.
2. The unavoidable fluctuations of drug concentration may lead to under medication or over medication.
3. A typical peak-valley plasma concentration time profile is obtained which make attainment of steady-state condition difficult.
4. The fluctuations in drug levels may lead to precipitation of adverse effects especially of a drug with small therapeutic index whenever over medication occur (Jain, 2005).

Previously, the herbal drugs could not attract scientists towards the development of novel drug delivery systems due to processing, standardizing, extracting, and identification difficulties (Goyal *et al.*, 2011). Now-a-days with the advancement in the technology, novel drug delivery systems (NDDS) opens the door towards the development of herbal drug delivery systems. Novel drug delivery technologies have gained the importance to achieve modified delivery of herbal drugs thereby increasing the therapeutic value as well as reducing toxicity. For last one decade, many novel carriers such as liposomes, nanoparticles, phytosomes, and implants have been reported for successful modified delivery of various herbal drugs.

Benefits of Novel Herbal Drug Carriers

1. Novel herbal drug carriers cure particular disease by targeting exactly the affected zone inside a patient's body and transporting the drug to that area.
2. Novel drug delivery system is advantageous in delivering the herbal drug at predetermined rate and delivery of drug at the site of action, which minimizes the toxic effects with the increase in bioavailability of the drugs.
3. In novel drug delivery technology, control of the distribution of drug is achieved by incorporating the drug in carrier system or by changing the structure of the drug at molecular level (Biju *et al.*, 2006).

The novel carriers should ideally fulfill two prerequisites. Firstly, it should deliver the drug at a rate directed by the needs of the body, over the period of treatment. Secondly, it should channel the active entity of herbal drug to the site of action. Conventional dosage forms including prolonged-release dosage forms are unable to meet none of these. Thus, there is a need for novel delivery system of herbal to achieve better effect (Ajazuddin, 2010).

1. There is a great possibility for herbal drugs that many compounds will be destroyed in the highly acidic pH of the stomach. The liver might metabolize other components before reaching the blood. As a result, the actual amount of the drug may not reach the blood. If the drug does not reach the blood at a minimum level, which is known as 'minimum effective level', then there will be no therapeutic effect.

2. Drug delivery system used for administering the herbal medicine to the patient is traditional and out-of-date, resulting in reduced efficacy of the drug. Also for a long time herbal medicines were not considered for development as novel formulations owing to lack of scientific justification and processing difficulties, such as standardization, extraction and identification of individual drug components in complex poly herbal systems.
3. If the advanced drug delivery technology is applied in herbal medicine, it may help in increasing the efficacy and reducing the side effects of various herbs and herbal compounds (Musthaba, 2009).

Modern phyto-pharmaceutical research can solve the scientific needs (such as determination of pharmacokinetics, mechanism of action, site of action, accurate dose required etc.) of herbal medicines to be incorporated in novel drug delivery system such as nanoparticles, microemulsions, matrix systems, solid dispersions, liposomes, solid lipid nanoparticles and so on. Thus, the delivery of herbal drugs also requires modifications with the purpose to achieve (Kapil, 2012):

- ☆ Desired controlled/sustained release
- ☆ To increase patient compliance etc.
- ☆ To minimize drug degradation and loss
- ☆ To prevent harmful side effects and
- ☆ To increase drug bioavailability

These modifications of herbal actives, extractives can be done by various techniques like physical modifications, chemical modifications, complexation etc. of

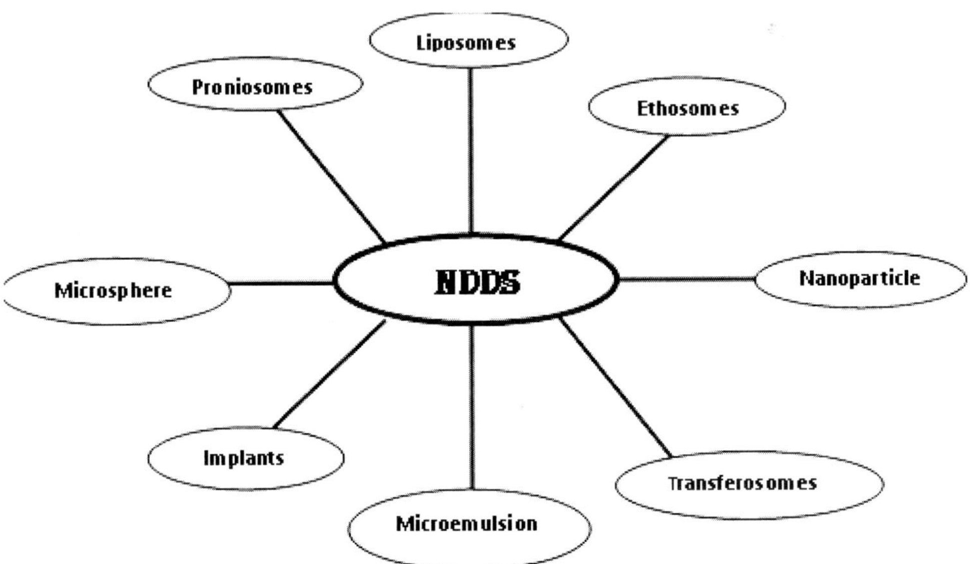

Figure 22.2: Highlights the different novel carriers and drug delivery systems for herbals.

all these complexation with cyclodextrins and their derivatives is found to be most potent.

Advantages of Cyclodextrin Complexation with Herbals

The novel cyclodextrin based formulations of herbals are reported to have remarkable advantages over conventional formulations of plant actives and extracts which include enhancement of solubility, bioavailability, protection from toxicity, enhancement of pharmacological activity, enhancement of stability, improved tissue macrophages distribution, sustained delivery, and protection from physical and chemical degradation. Thus, the nano sized cyclodextrin based carriers/novel drug delivery systems of herbal drugs have a potential future for enhancing the activity and overcoming problems associated with plant medicines. Table 22.2 enlist the comparative analysis of novel drug delivery carriers of herbal constituents and bioactive before and after cyclodextrin complexion.

Table 22.2: Comparison of micro/nanocarriers prepared with/without cyclodextrin complexation.

Herbals Carriers Before Cyclodextrin Complexation	*Herbals Carriers After Cyclodextrin Complexation*
Microcarriers/Microparticles	
Low–Moderate entrapment efficiency	Higher entrapment efficiency
Variable drug release	Provide efficient release
Nanocarriers/Nanoparticles	
Limited Safety and efficacy	More efficacy
Drug loading and entrapment efficiency is limited	Increased loading capacity

For herbal products/constituents to be complexed with cyclodextrins, are firstly subjected to drying and successive extraction, followed by phytochemical screening to find out the active constituent responsible for therapeutic activity and then the isolation of biological active constituent. These constituents are then complexed with cyclodextrin and its derivatives.

In case of herbal active constituents, cyclodextrins are considered better than the organic solvents due to following reasons:

- ☆ Frequently less irritating after i.v. and i.m. injection
- ☆ Frequently less toxic
- ☆ The drug does not precipitate after i.v. injection

Cyclodextrins

Cyclodextrins, a class of macrocyclic compounds composed of α-1,4-coupled six (α-cyclodextrin), seven (β-cyclodextrin), or eight (γ-cyclodextrin), nine (δ-CD), ten (α-CD) or more α-D-glucopyranose units that form cyclic oligomers. These cyclodextrins act as host, the caves of which have a diameter of 5-8 and exhibit a hydrophobic internal cavity and a hydrophilic outer layer. They have hydroxyl groups on the outer surface and a void cavity in the center and are generally represented as truncated cones with conical cavities. This structure and the hydrophobic cavity of CDs can act as host for various suitable guest molecules. This unique property/ special feature allows them to form inclusion complexes with various drugs of proper size and polarity by non-covalent interaction forces (hydrophobic bonds, hydrogen bonds, vander Waals forces). They are also known as cycloamyloses, cyclomaltoses and Schardinger dextrins (Villers, 1891), (Eastburn, 1994). They are produced as a result of intra-molecular transglycosylation reaction from degradation of starch by cyclodextrin glucanotransferase (CGTase) enzyme (Szetjili, 1998). They possess a cage-like supramolecular structure, which is the same as the structures formed from cryptands, calixarenes, cyclophanes, spherands and crown ethers. Figure 22.3 shows the structures and features of natural cyclodextrims.

Table 22.3: Advantages of cyclodextrin based NDDS.

Advantages of Combination approach for cyclodextrin based novel delivery of herbal plant extractives/actives	
Cyclodextrin Complexation	*Novel Delivery System/Nanocarriers*
Increase in solubility	Drug targeting to specific sites
Increase in stability	Sustained/controlled release
Fixation of volatile substances	Improved pharmacokinetic properties
Convenient for liquid extracts/semisolid components	Improved patient compliance
Protection from microbial growth	Increase permeability of drugs
Helpful in masking taste/smell	Decrease dose and dosing frequency
Enhance drug delivery	Reduce side effects
Prevent interactions	
Reduce local irritation	

Properties of Cyclodextrins

The three natural cyclodextrins (α, β and γ) are referred to as first generation or parent cyclodextrins. Cyclodextrins crystallize in two main types of crystal packing, channel structures and cage structures, depending on the type of cyclodextrin and guest compound. Among the natural CDs, the cavity size of α-cyclodextrin is insufficient for many drugs and γ -cyclodextrin is expensive. β-Cyclodextrin is the most accessible, the lowest-priced, and generally the most useful. It is mainly used for the inclusion process because of the perfect cavity size and ability to adopt a bucket-like arrangement with a hydrophilic exterior and hydrophobic interior. Recently, various hydrophilic, hydrophobic and ionic cyclodextrin derivatives have been utilized to extend the physicochemical properties and inclusion capacity of natural cyclodextrin (Hiramaya, 1999). Hydroxypropyl-β-cyclodextrin is a water-soluble derivative of β- cyclodextrin, which has been widely studied as a complexation agent for many pharmaceuticals (Misiuk, 2009). Table 22.4 enlists the features of three natural parent cyclodextrins.

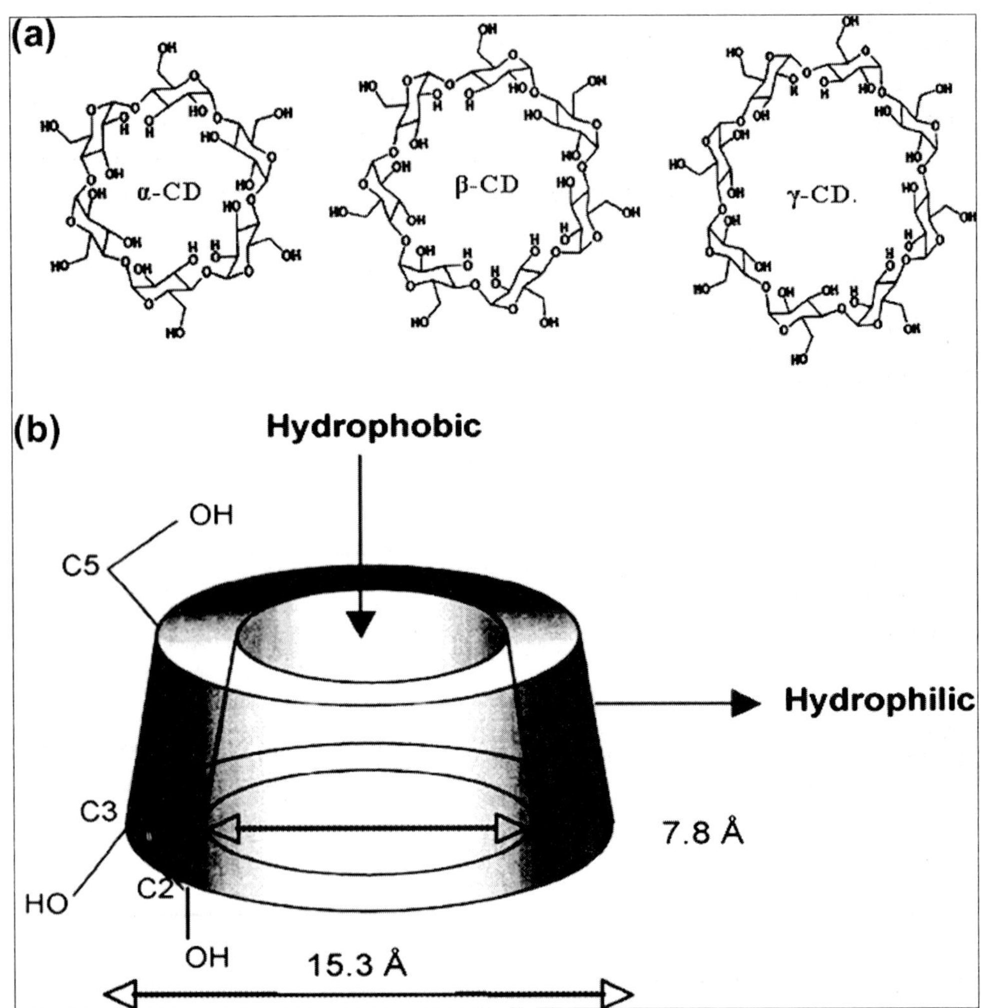

Figure 22.3: (a) Chemical structures of α-CD, β-CD and γ-CD. (b) The internal and external widths for β-cyclodextrin are noted in the figure the height of the cavity is presumed to be 8 Å and width size is 15.3 Å.

Table 22.4: Properties of natural cyclodextrins.

Property	*α-Cyclodextrin*	*β-Cyclodextrin*	*γ-Cyclodextrin*
Number of glucopyranose units	6	7	8
Molecular weight (g/mol)	972	1135	1297
Solubility in water at 25ºC (per cent, w/v)	14.5	1.85	23.2
Outer diameter (Å)	14.6	15.4	17.5
Cavity diameter (Å)	4.7–5.3	6.0–6.5	7.5–8.3
Height of torus (Å)	7.9	7.9	7.9
Cavity volume (Å)	174	262	427

Criteria for Cyclodextrin Complexation

All the drugs/constituents are not suitable substrates for CD complexation. Drug molecule to be complexed with CD should have certain characteristics:

1. More than five atoms (C, P, S, N) form the skeleton of the drug molecule.
2. Melting point temperature of the substance is below 250 °C.
3. Solubility in water is less than 10 mg/mL.
4. The guest molecule consists of less than five condensed rings.
5. Molecular weight between 100 and 400.

Toxicological Aspect/Regulatory Status of Cyclodextrins and their Derivatives

Cyclodextrins are listed in a number of pharmacopoeia sources including the US Pharmacopoeia/National Formulary (USP/NF), European Pharmacopoeia (Ph.Eur.) and Japanese Pharmaceutical Codex (JPC).

In general, the natural cyclodextrins and their hydrophilic derivatives are only able to permeate lipophilic biological membranes, such as the eye cornea, with considerable difficulty. Even the somewhat lipophilic randomly methylated β-cyclodextrin does not readily permeate lipophilic membranes, although it interacts more readily with membranes than the hydrophilic cyclodextrin derivatives (Satyavati, 1987). All toxicity studies have demonstrated that orally administered cyclodextrins are practically non-toxic, due to lack of absorption from the gastrointestinal tract (Irie, 1997). Furthermore, a number of safety evaluations have shown that γ-cyclodextrin, 2-hydroxypropyl-β-cyclodextrin, sulphobutylether-β-cyclodextrin, sulphated-β-cyclodextrin and maltosyl-β-cyclodextrin appear to be safe even when administered parenterally. However, toxicological studies have also shown that the parent α and β cyclodextrin and the methylated β cyclodextrins are not suitable for parenteral administration.

The negligible cytotoxic effects of CDs are an important attribute in applications such as rug carrier, food and flavors, cosmetics, packing, textiles, separation processes, environment protection, bioconversion, packaging and textile, fermentation and catalysis (Mammucari, 2008).

The three parent cyclodextrins, (α-CD, β-CD and γ-CD) were introduced into the generally regarded as safe (GRAS) list of the FDA in 2004, 2001 and 2000, respectively. Also, the derivative of β-CD, HP-β-CD is cited in the FDA's list of Inactive Pharmaceutical Ingredients (IPI) and SBE-β-CD (sulfobutylated beta cyclodextrin) is listed in the FDA's compilation of inactive pharmaceutical ingredients. This exploits the official use of cyclodextrins in herbals.

Formation of Inclusion Complex

The most notable feature of cyclodextrin and their derivatives is the ability to form solid inclusion complexes (host–guest complexes) with a very wide range of solid, liquid and gaseous compounds by a molecular complexation (Villers, 1981). In these complexes, a guest molecule is held within the cavity of the cyclodextrin host molecule. Complex formation is a dimensional fit between host cavity and guest molecule (Munoz-botella, 1995). The lipophilic cavity of cyclodextrin molecules provides a microenvironment into which appropriately sized non-polar moieties can enter to form inclusion complexes (Loftsson, 1996). No covalent bonds are broken or formed during formation of the inclusion complex (Schneiderman, 2000). The main driving force of complex formation is the release of enthalpy-rich water molecules from the cavity. Water molecules are displaced by more hydrophobic guest molecules present in the solution to attain an apolar–apolar association and decrease of cyclodextrin ring strain resulting in a more stable lower energy state (Szejtli, 1998). The binding of guest molecules within the host cyclodextrin is not fixed or permanent but rather is a dynamic equilibrium. Complexes can be formed either in solution or in the crystalline state and water is typically the solvent of choice. Inclusion complexation can be accomplished in a co-solvent system and in the presence of any non-aqueous solvent.

Formation of inclusion complex with a guest molecule is a function of two key factors. The first is steric and depends on the relative size of the cyclodextrin to the size of the guest molecule or certain key functional groups within the guest. The second critical factor is the thermodynamic interactions between the different components of the system (cyclodextrin, guest, solvent). Table 22.3 covers the advantages of combination approach of cyclodextrin complexation and novel drug delivery synstem.

In general, therefore, four energetically favourable interactions help shift the equilibrium to form the inclusion complex:

- ☆ The displacement of polar water molecules from the apolar cyclodextrin cavity.
- ☆ The increased number of hydrogen bonds formed as the displaced water returns to the larger pool.
- ☆ A reduction of the repulsive interactions between the hydrophobic guest and the aqueous environment.
- ☆ An increase in the hydrophobic interactions as the guest inserts itself into the apolar cyclodextrin cavity.

The ability of CDs to form inclusion complexes is affected by size, shape, hydrophobicity, and the form of the guest molecule. Such a complexation leads to changes in their physicochemical and biopharmaceutical properties; therefore, complexation with CDs enhances the drug's solubility, bioavailability, dissolution rate, and chemical stability, and reduces their side effects, toxicity, and unpleasant odor and taste (Loftsson, 1996). One or two guest molecules can be entrapped by one, two or three cyclodextrins. The complexation phenomenon often results in a significant variation in the photo physical and photochemical properties of guest molecules due to the micro environmental difference inside the CD and the aqueous medium (Szejtli, 1998).

Complexation Techniques

The preparation technique of inclusion complexes between drugs and CD can influence the characteristics of the drug/CD complex products (Blanco, 1991). Both solid and liquid drugs can be complexed with cyclodextrins. The methods used for preparation of non covalent inclusion complexes are Physical mixing, Kneading, Solvent evaporation, Freeze drying, Spray drying, Co-precipitation, Co-grinding, Microwave irradiation, Neutralization, Melting in solution, Sealed heated products and Suspension method (Aleem, 2010; Mural *et al.*, 1997). Each method has its own advantages and disadvantages. Some of them are discussed as under:

1. *Grinding* – This method involves simple grinding of equimolar amount of constituent/drug with the cyclodextrin.
2. *Solid dispersion/co-evaporation* – The drug and cyclodextrin are dissolved in ethanol and water separately, mixed and stirred to attain equilibrium.
3. *Kneading* – In this method paste of cyclodextrin is prepared with small amount of water to which the drug is added.
4. *Spray- drying*–Monophasic solution of drug and cyclodextrin is prepared using suitable solvent, and then stirred to attain equilibrium and the solvent is removed by spray drying.

Of all these methods, for making solid inclusion complexes between CDs and many drugs, kneading, co-evaporation, sealed-heating, co-grinding, spray-drying and freeze-drying are mostly preferred (Mura, 2005). However, these methods have drawbacks such as lengthy multistage processing techniques, high toxic residual solvent content, thermal degradation of the drug, and high-energy consumption (Sauceau, 2008). Table 22.5 shows the advantages and disadvantages of different methods of complexation.

Table 22.5: Comparative study of different methods of inclusion complexes.

Method	*Advantage*	*Disadvantage*
Spray drying	Common technique, continuous process, scaleable, works well	limited commercial capacity
Freeze drying	Works well	Expensive, low throughput, limited commercial capacity.
Kneading (Paste complexation)	Scalable, equipment commonly available	Not as effective as spray drying or freeze-drying.
Co- precipitation	–	Low yield, difficult to scale, expensive, requires about 48 hrs
Solid phase complexation/grinding	Simple method	Less effective

Characterization of Inclusion Complexes

Molecular encapsulation of drugs in cyclodextrin cavity occurs both in solid and in solution state. In solid state, guest molecules (drug/herbal constituent) can be enclosed within the cavity or may be aggregated to the outside of the cyclodextrin molecule and in solution state; there is equilibrium between complexed and non-complexed guest molecules. A guest molecule experiences changes in the physicochemical properties when it gets incorporated within the cyclodextrin cavity. Changes in the physicochemical properties provide methods to characterize whether guest molecules are included in the cyclodextrin cavity (Ramnik, 2010).

Characterization of Inclusion Complexation in the Solid State

Inclusion complexation in the solid state are characterized by different thermo-analytical methods like Scanning Electron Microscopy (SEM), X-ray diffractiometry, single crystal X-ray structure analysis, wettability, dissolution tests, Infra-Red (IR) spectroscopy, Thin Layer Chromatography (TLC) etc.

Characterization of Inclusion Complexation in Solution State

Inclusion complexation in solution state are characterized by Polarography, Conductivity, Polarimetry, Solubility studies, Spectroscopy methods like Nuclear Magnetic Resonance (NMR) spectroscopy, Electron Spin Resonance (ESR), Ultraviolet/Visible (UV/VIS), Fluorescence spectroscopy, Circular Dichroism (CD), pH-Potentiometric Titration and Microcalorimetry.

Phase-Solubility Analysis

Phase-solubility analysis involves an examination of the effect of a solubilizer, *i.e.* cyclodextrin or ligand (L) on the drug being solubilized. Experimentally, the drug is added to several vials always in excess. The presence of solid drug in these systems in necessary to maximize the thermodynamic activity of the dissolved substrate (Prameela, 2009). To the drug or substrate (S) a constant volume of water, containing successively larger concentrations of the cyclodextrin or ligand (L) is added. The vials are mixed at constant temperature until equilibrium is established (which frequently takes about 1 week). The solid drug is then removed and the solution

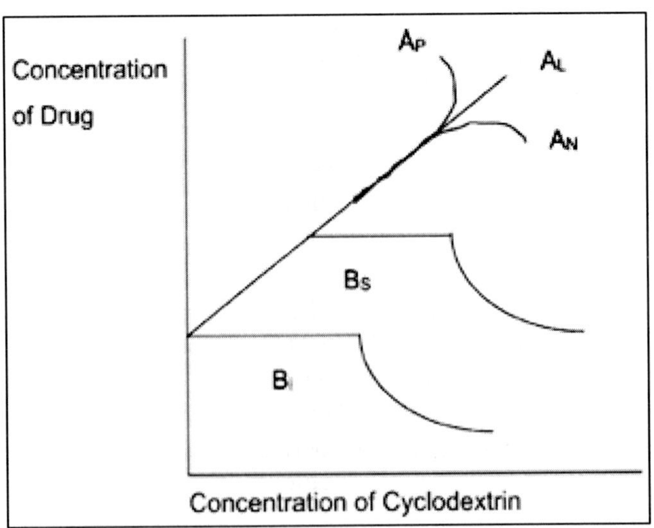

Figure 22.4: Phase solubility curve.

assayed for the total concentration of S. A Phase-solubility diagram is constructed by plotting the total molar concentration of S on the *y*-axis and the total molar concentration of L added on the *x*-axis (Figure 22.4).

Phase-solubility diagrams fall into two main categories, A- and B-types. A-type curves are indicative for the formation of soluble inclusion complexes while B-type behaviors are suggestive of the formation of inclusion complexes of poor solubility (Nicolescu, 2010). A B_S-type response denotes complexes of limited solubility and a B_I-curve are indicative of insoluble complexes. The A-curves are sub-divided into A_L (linear increases of drug solubility as a function of cyclodextrin concentration), A_P (positively deviating isotherm) and A_N (negatively deviating isotherms) subtypes. While β-cyclodextrin often gives rise to B-type curves due to the poor water solubility of the ligand itself, the chemically modified CDs including HP β-CD and SBE β-CD usually produce soluble complexes (*i.e.*, A-type systems).

Cyclodextrin Complexation with Herbal Constituents

Cyclodextrins and their derivatives have been used in complexation of drugs obtained from natural and synthetic origin. This section covers the examples of cyclodextrin complexation of herbal drugs for the enhancement of physical, chemical and pharmacological activity. Figure 22.5 shows the basic structure of some active herbal constituents. Results have proved that complexation of herbal constituents with cyclodextrins and their derivatives have led to increase in solubility, stability, pharmacological activity etc to several folds. Table 22.6 enlists the examples of cyclodextrin complexation with herbal constituents.

Enhancement of Solubility and Stability

Solubility is an essential criteria for a drug to exhibit pharmacological action. Poor solubility of drug leads to poor bioavailability. Most of the herbal constituents isolated from plants also exhibit poor solubility which can be enhanced by cyclodextrin complexation. The solubility enhancement of different categories of active constitiuents like carotenoids, terpenoids, thiocynates, flavonoids, resins etc. is discussed in detail.

Carotenoids

Carotenoids have been found to provide several common biological functions such as photo protection, antioxidant effects including singlet oxygen quenching, immunomodulatory and anticancer activity in both humans and rodents (Krinsky and Johnson, 2005).

Astaxanthin is the main carotenoid pigment naturally synthesized in some plants. It can easily be decomposed by light and oxygen, which can cause the loss of antioxidant properties in processing and store. Furthermore, the human body because of its poor bioavailability cannot readily absorb it (Anarjan *et al.*, 2010).

The inclusion complex was formed between astaxanthin and hydroxypropyl-β-cyclodextrin (HP β-CD). The water solubility and stability of astaxanthin was increased. Furthermore, the release of astaxanthin from the inclusion complex was controlled (Yuan *et al.*, 2008).

Table 22.6: Complexation of herbal constituents with cyclodextrins and its derivative.

Herbal Actives	*Source*	*Limitations*	*CD Used*	*Method of Complexation*	*Outcome*	*Reference*
Astaxanthin	Carotenoid pigment found in marine animals and plants	Poor solubility, bio-availability, heat unstable	HP-β-CD	Freeze drying	Enhancement of solubility and stability, molecular modeling of structure	Yuan, 2012
Curcumin	Curcuminoid (natural phenols) found in plant *Curcuma longa* Linn.	Poor solubility, bio-availability, rapid hydrolysis, photo-degradation etc.	HP-γ-CD	Freeze drying	Greater cellular uptake and longer half-lifein the cells, superior anti-proliferative and anti-inflammatory activities	Yadav, 2012
Betulin	Triterpene, found in the white bark of the birch tree	Poor solubility	HP-γ-CD	Physical mixing and kneading	Enhancement of solubility, anti-cancer activity	Falamas, 2011
Isothiocynates	Occur naturally in cruciferous vegetables	Heat unstable	RM-β-CD	Spray drying	Heat stabilization	Neoh, 2012
Asiaticoside	Trisaccharide triter-penoid, obtained from *Centella asiatica* Linn.	Poor solubility, bio-availability	HP-β-CD	Freeze drying	Enhancement of solubility, increase in release of AC from films.	Panichpakdee, 2011
Chlorogenic acid		Poor pharmacological activity	HP-β-CD	Co-precipitation	Increase in anti-oxidant activity	Chao, 2012
9-Nitro-camp-tothecin	Alkaloid obtained from *Camptotheca acuminata*	Poor physico-chemical property	HP-β-CD	Co-lyophilization	Increase in solubility, stability and anti tumor activity	Jiang, 2010
Garlic oil	Obtained from *Allium sativum* Linn.		β-CD	Co-precipitation	Improvement in physico-chemical properties	Wang, 2011
β-lacophene	Natural napthoquinone obtained from bark of *Tabebuia avellanedae*		HP-β-CD	Freeze drying	Enhancement of solubility, stability and anti-cancer activity	Cavalcanti, 2011
Taxifolin	Flavonoids, exist as seco-ndary metabolite in plants		β-CD, HP-β-CD	–	Enhancement of solubility and bioavailability	Yang, 2011
Vanillin	Food additive	Poor solubility, light sensitive, oxidative, poor stability	β-CD	–	Enhancement of solubility, stability, prolonged shelf life	Kayaci, 2012

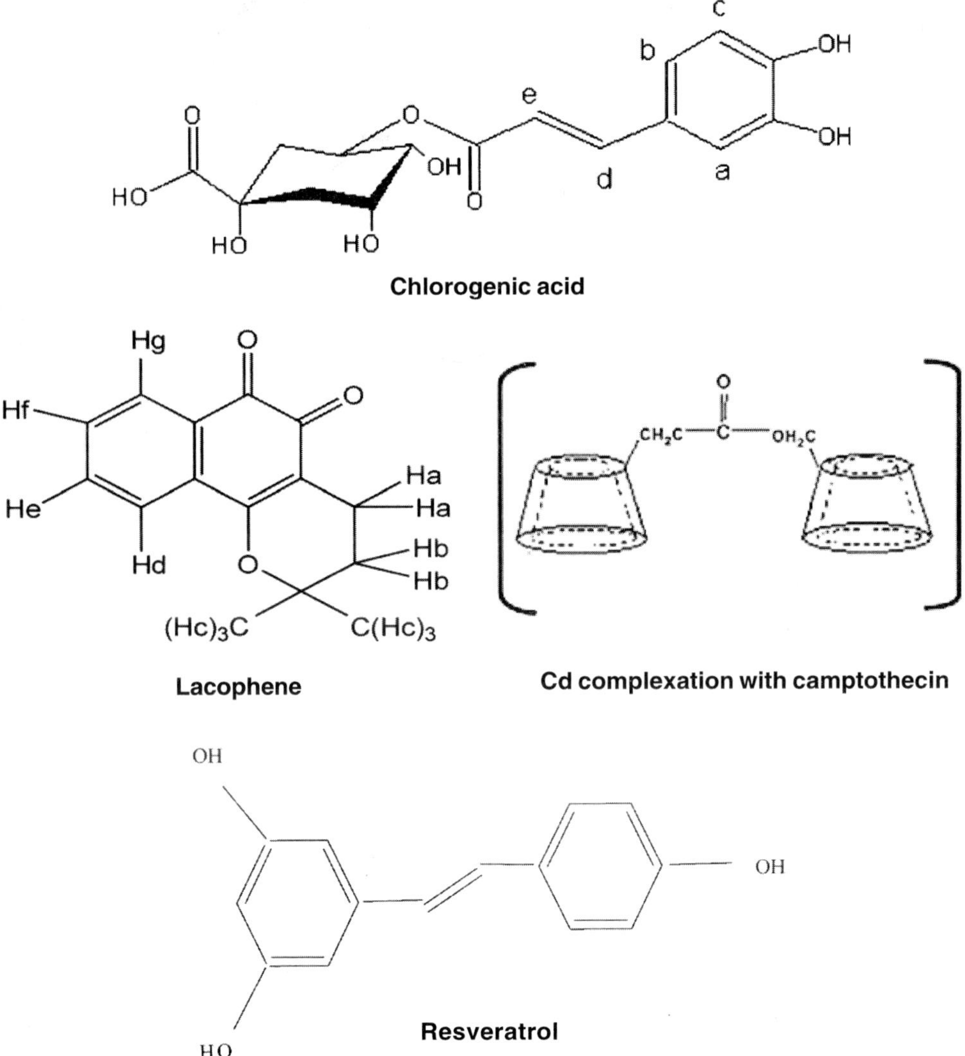

Figure 22.5: Some herbal active constituents.

Isothiocynates

Isothiocyanates (ITCs) are hydrolysis products of sulphur-containing compounds called glucosinolates, which occur naturally in cruciferous vegetables, such as broccoli and cabbage. The application of ITCs as flavors in food systems is restrained due to their high volatility, poor aqueous solubility, and pungency. In addition, the incompatibility of the ITCs' appearance in the form of oil with major food packaging materials also limited their use as antimicrobial and antifungal agents (Li *et al.*, 2007).

Inclusion complexation reactions between isothiocyanates (ITCs), namely, allyl isothiocyanate (AITC) and phenyl isothiocyanate (PITC), and randomly methylated β-cyclodextrin (RM-β-CD) displayed AP type solubility isotherms, indicating continuous increases in the stoichiometry of the inclusion complexes. The complex formed led to suppression in volatility and physical stabilisation by inclusion complexation and are essentially useful for the development of an active food packaging material functionalized by the antimicrobial and antifungal ITCs.

Terpenoids

Asiaticoside (AC), one of the four major trisaccharide triterpenoid components (*i.e.*, asiatic acid, asiaticoside, madecassic acid and madecassoside) of the extract from the medicinal plant *Centella asiatica* (L.) AC has been regarded as one of the most active compounds associated with the healing of wounds, as evidenced from the observed increase in antioxidant levels at an early stage of recovery of excision-type cutaneous wounds in rats (Shukla *et al.*, 1999).

Inclusion complex of AC with HP-β-CD was prepared and investigated for its potential for use in solvent cast cellulose acetate films as carrier for topical and transdermal delivery. Various properties (*i.e.*, morphology, size of self-assembling aggregates, water retention and weight loss) of both the AC- and the AC/HP β CD complex-loaded CA films and the release characteristics of AC from these materials in a phosphate buffer saline solution that contained methanol was investigated.

The phase solubility profile of inclusion complex was AL-type and indicated a 1: 1 molar ratio of the AC/HP-β-CD complex. Unlike the crystalline nature of the native AC, the stoichiometric AC/HP β CD complex powder was found to be amorphous in nature. The maximal cumulative released amount of AC, a sparingly water-soluble substance, from the CA films containing a mixture of AC/HP-β-CD was seen.

Lacophene

β-lacophone (β -lap) is a semi synthetic compound obtained from lapachol, a natural ortho-naphthoquinone extracted from the bark of *Tabebuia avellanedae*, a native tree originating in South America and found particularly in Brazil. It possess anti-tumor and anticancer activity (Coelho *et al.*, 2010).

The therapeutic use of β-lap remains limited due to its poor water solubility and physico-chemical properties. β–lap and HP-β-CD complexes were prepared using the freeze-drying technique.

Results indicated that incorporation of β–lap complexed HP-β-CD into liposomes led to encapsulation efficiency above 90 per cent. The inclusion complex provided an alternative means of overcoming the low aqueous solubility of β-lapachone and the encapsulation of inclusion complex could modify drug pharmacokinetic and lead eventually to β-lapachone uses in cancer researches.

Flavonoids

Flavonoids are a large group of phenolic compounds and constitute one of the largest groups of secondary metabolites in plants (Xiao *et al.*, 2008).

The flavonoid taxifolin or dihydroquercetin, is widely distributed in barks of the genus *Pinus* or *Larix* and in seeds of the genus *Silybum* (Daniel *et al.*, 2006). It has been widely used in the treatment of cerebral infarction and sequela, cerebral thrombus, coronary heart disease and angina pectoris (Shiko *et al.*, 2009).

The potential use of taxifolin as a natural antioxidant or herbal medicine is greatly limited by its low water solubility and bioavailability. Although much effort has been made to improve its water solubility and stability by introducing some nanodispersion or enzymatic condensation techniques (Daniel *et al.*, 2006; Shiko *et al.*, 2009), it is still not possible to sufficiently dissolve taxifolin in water, which prevents its usage for some therapeutic and cosmetic applications.

The inclusion complexation behavior, characterization and binding ability of taxifolin with native cyclodextrin and its derivative hydroxypropyl β cyclodextrin (HP-β-CD) enhanced the water solubility and thermal stability of taxifolin and led to an important step in the design of a novel formulation of taxifolin for herbal medicine or healthcare products.

Polyphenols

Resveratrol, a polyphenol, has been found in a wide variety of about 70 plant species, including grapes, mulberries, and peanuts, which could protect against a range of illnesses and diseases, *e.g.*, cardiovascular diseases (Niles, 2003), cancer, viral attacks, neurodegenerative diseases and ageing (Delmas, 2005). Furthermore, it could increase lifespan and physical activity of diverse species. Various studies have also indicated that resveratrol shows several interesting biological effects such as antioxidant, anti-inflammatory and anti-proliferative activities (Olas, 2005). However, due to its low water solubility, resveratrol cannot be used for oral administration.

The stability and bioavailability of resveratrol was enhanced by complexation with β cyclodextrin making it suitable for oral administration.

Enhancement of Anti-oxidant Activity

A large number of active constituents isolated from plants possess prominent anti-oxidant activity but their use is limited due to one or more properties. Cyclodextrin complexation can overcome these limitations leading to enhancement in activity.

Polyphenols

5-O-Caffeoylquinic acid (chlorogenic acid, CGA) is a kind of polyphenol derivative that occurs widely in fruits, vegetables, black teas, soybeans, wheat and some traditional Chinese medicines. Due to its ability to directly interact with reactive oxygen species (Kono, 1997), CGA has many beneficial properties such as antioxidant effect and anxiolytic activity (Shi, 2007). Through its antioxidant function, CGA has been shown to inhibit hypertension and stimulate the flowering in plants. It also affects enzyme activities (trypsin, amylase and other enzymes) (Beninca, 2009).

However, the presence of an ortho-dihydroxyphenol in the CGA molecule makes it more likely to undergo rapid oxidation by gamma irradiation or other oxidative stress to highly reactive electrophilic chlorogenoquinone. Additionally, trans-esterification reaction can occur during storage or processing (Xiang, 2008).

Solid complex of CGA with HP β CD was prepared at a 1: 1 molar ratio. HP β CD formed stable inclusion complex with CGA and therefore can be used for the encapsulation/release of this potent antioxidant. The DPPH radical scavenging ability of CGA on complexation with HP β CD was increased as compared to uncomplexed CGA.

Enhancement of Anti-cancer Activity

The anti-cancer activity exhibited by herbal drugs as compared to synthetic drugs is safer and more effective. An increase in the activity of herbal constituents can be achieved by formation of inclusion complex with different cyclodextrins and its derivative.

Curcumin

Curcumin, a yellow pigment and an active component present in turmeric (*Curcuma longa*), has been used as a spice and as an Ayurvedic medicine for centuries in the Indian subcontinent. This nutraceutical has been shown to suppress carcinogenesis in the skin, liver, lung, colon, stomach, and breast. Extensive research within the last two decades has shown that this agent exhibits antioxidant, anti-inflammatory, anti-survival, antiproliferative, anti-invasive, and antiangiogenic activity (Kunnumakkara *et al.*, 2008).

Despite of its multiple beneficial activities, the optimum potential is still limited due to its poor bioavailability and lack of solubility in aqueous solvents. Poor oral absorption of curcumin due to its extremely low aqueous solubility or extensive presystemic metabolism may be responsible for the unfavourable pharmacokinetics of this molecule (Holder *et al.*, 1978).

Various methods have been tried to enhance curcumin delivery, including its incorporation into liposomes (Li *et al.*, 2005), (Takahashi *et al.*, 2008), phospholipid vesicles (Sou, 2008), and nanoparticles (Anand *et al.*, 2009).

Inclusion complex was prepared by a pH shift method using a highly alkaline solution for curcumin dissolution and contacting with hydroxypropyl-γ -cyclodextrin (HP-γ-CD).

Using the electrophoretic mobility shift assay, it was found that cyclodextrin complexed curcumin (CDC) was more active than free curcumin in inhibiting TNF-induced activation of the inflammatory transcription factor NF-κB and in suppressing gene products regulated by NF-κB, including those involved in cell proliferation (cyclin D1), invasion (MMP-9), and angiogenesis (VEGF). CDC was also found to be more active than free curcumin in inducing the death receptors DR4 and DR5.

Betulin

Betulin (lup-20(29)-ene-3β,28-diol), a triterpene exists abundantly in the white bark of the birch tree (Lugemwa, 1990). Betulin and its metabolic successor betulinic acid, exhibits anti-cancer activities (Zalesinska, 2009) used in skin malignancies, including melanoma and skin cancer. The major problem in formulating pharmaceutical compounds from this triterpene is its low hydro-solubility.

Two preparation methods (physical mixing and kneading) and 1: 1 and 1: 2 M ratios were used in to obtain the inclusion complexes based on betulin and cyclodextrins. Cyclodextrin complexation leads to enhancement of poor aqueous solubility of betulin, making it potential for anti-cancer activity.

Camptothecin (CPT)

Camptothecin is an inhibitor of the DNA-replicating enzyme topoisomerase-I, which is believed to act by stabilizing a topoisomerase-I induced single strand break in the phosphodiester backbone of DNA, thereby preventing relegation.

9-Nitro-camptothecin (9-NC) is a potent topoisomerase-I inhibitor. Pharmacological studies indicated that the anti-tumor activity of 9-NC was superior to that of camptothecin (CPT) in human tumors xenografted in nude mice (Gao *et al.*, 2008). However, the clinical application of 9-NC is hampered because of its poor aqueous solubility and stability under physiologic conditions (Verschraegen *et al.*, 1998).

The inclusion complex of 9-NC with HP β CD was prepared by the co-lyophilization technique. The 9-NC complexed in HP β CD, instead of dissolved in the organic solvent, presented significant anti-tumor activity and low toxicity for the treatment of cancer. Taken together, the complexation of HP β CD provides a promising carrier for 9-NC, and suitable for parenteral delivery.

Enhancement of Physico-chemical Properties

Garlic (Allium sativum L.)

It is a widely distributed plant and is used throughout the world not only as a spice and a food, but also as a folk-medicine, and many of the beneficial health-related biological effects have been attributed to its characteristic organosulphur compounds (Rybak, *et al.*, 2004).

The use of garlic oil in the food industry is limited due to its volatility, strong odor, insolubility in water, and low physico-chemical stability (Corzo-Martínez *et al.*, 2007). GO was complexed with β -CD using co-precipitation method in a molar ratio of 1: 1. The results of UV–visible spectroscopy, FT-IR, DSC and XRD demonstrated that GO/β-CD complex has different physico-chemical characteristics from free GO. The aqueous solubility and stability of GO were significantly increased by inclusion in β -CD. The GO release rate from the GO/β -CD complex was controlled.

Conclusions

History records the fact that almost every culture around the world has noted the contribution of herbs and their use as foods and medicines. It is somewhat ironic that this 'return to nature', as far as medicinal substances are concerned, has occurred at a time when medicine has become increasingly technologically sophisticated, both in the equipment and products used for diagnosis and treatment, and also in the design and research into the mechanisms underlying disease.

India is a vast repository of medicinal plants that can be developed as safe and economical system of medicine for treatment of various diseases and ailments. The contribution of cyclodextrin in development of novel delivery system for herbals has led to enhanced therapeutic effects. Through these novel systems, it is possible to achieve drug targeting, modify pharmacokinetic parameters and reduce the drug toxicity.

Cyclodextrins are safe, biocompatible, water soluble and versatile excipients that can incorporate appropriately sized lipophilic compounds into their hydrophobic cavities, forming non-covalent inclusion complexes. They comprise a family of three well-known industrially produced major, and several rare, minor cyclic oligosaccharides. The three major cyclodextrins α, β and γ are crystalline, homogeneous and non-hygroscopic substances, having torus-like macro-rings built up from glucopyranose units. As they are obtained from enzymatic degradation of starch, are considered as natural products. The use of cyclodextrins (CDs) is a practical and economical way to improve the physicochemical and pharmaceutical properties of administered drug molecules. Studies in both humans and animals have shown that cyclodextrin can be used to improve the drug delivery from almost any type of drug formulations. Not all the drugs can be complexed to cyclodextrins, for complexation, certain criteria are to be fulfilled by guest molecule like molecular weight between 100- 400, solubility below 10 mg/ml, melting point below 250 etc.

The drawbacks/limitations associated with herbal drugs and their delivery system can be successfully alleviated by cyclodextrins. The herbal active constituent isolated and complexed with cyclodextrin are more beneficial compared to uncomplexed drugs and constituents. The enhancement in physico-chemical and pharmacological property of drugs after complexation is approx 50- 100 folds. The purpose of this review is to outline how cyclodextrin and its derivatives are suitable and beneficial as carrier in development of novel delivery systems of herbal actives.

Herbal medicine is being continuously used all over the world since ancient times and has been recognized by physicians and patients for their better therapeutic value and fewer adverse effects as compared with modern medicines. Novel drug delivery systems based on cyclodextrin system not only reduce the repeated administration to overcome non-compliance, but also help to increase the therapeutic value by reducing toxicity and increasing the bioavailability etc. Recently, the scientists have shifted their focus in designing a delivery system for herbal medicines using a scientific approach. For a long time herbal medicines were not considered for development as novel formulations owing to lack of scientific justification and processing difficulties, such as standardization, extraction and identification of individual drug components in complex polyherbal systems. However, modern phyto-pharmaceutical research solves the scientific needs for herbal medicines as in modern medicine, which gives way for developing novel formulations such as nanoparticles, microemulsions, matrix systems, solid dispersions, liposomes, solid lipid nanoparticles, and so on. An extensive research is going on in the area of novel drug delivery and targeting for plant actives and extracts. However, research in the field of herbals with cyclodextrins and their delivery system is still at the exploratory stage. Herbal drugs have enormous therapeutic potential, which should be explored through some value added drug delivery systems.

References

Acharya, D., and Shrivastava, A. (2008). Indigenous Herbal Medicines: Tribal Formulations and Traditional Herbal Practices. New Delhi.

Ajazuddin, Saraf, S. (2010). Applications of novel drug delivery system for herbal formulations. *Fitoterapia,* **81:** 680–689.

Anand, P., Nair, H.B., Sung, B., Kunnumakkara, A.B., Yadav, V.R., and Tekmal, R.R. (2009). Design of curcumin-loaded PLGA nanoparticles formulation with enhanced cellular uptake and increased bioactivity *in vitro* and superior bioavailability *in vivo. Biochemical Pharmacology,* **79**: 330–8.

Anarjan, N., Mirhosseini, H., Baharin, B.S., and Tan, C.P. (2010). Effect of processing conditions on physicochemical properties of astaxanthin nanodispersions. *Food Chemistry,* **123(2):** 477–483.

Arun, R. (2008). Cyclodextrins as drug carrier molecule: A review. *Scientia Pharmaceutica,* **76**: 567–598

Ashutosh, K. (2007). Pharmacognosy and pharmacobiotechnology.

Atmakuri, L.R., and Dathi, S. (2010). Current trends in herbal medicines. *Journal of Pharmaceutical Research,* **3(1):** 109-113.

Beninca, C., Kaskantzis, G., and Zanoelo, E.F. (2009). Experimental study of chemical kinetics and simulation of 5-O-caffeoylquinic acid oxidation during manufacturing of mate. *Biosystems Engineering,* **104**: 503–509.

Biju, S.S., Talegaonkar, S., Mishra, P.R., and Khar, R.K. (2006). Vesicular system: an overview. *Indian Journal of Pharmaceutical Science,* **68(2):** 141-153.

Blanco, J., Vila-Jato, J.L., Otero, F., and Anguiano, S. (1991). Influence of method of preparation on inclusion complexes of naproxen with different cyclodextrins. *Drug Development and Industrial Pharmacy,* **17**: 943–957.

Cavalcanti, I., and Mendonça, A.M. (2011). The encapsulation of β-lapachone in 2-hydroxypropyl-β-cyclodextrin inclusion complex into liposomes: A physicochemical evaluation and molecular modeling approach. *European Journal of Pharmaceutical Sciences,* **44:** 332–340.

Chao, J., and Wang, H. (2012). Investigation of the inclusion behavior of chlorogenic acid with hydroxypropyl-β-cyclodextrin. *International Journal of Biological Macromolecules,* **50**: 277– 282.

Coelho, T.S., Silva, R.S.F., Pinto, A.V., Pinto, M.C.F.R., Scaini, C.J., Moura, K.C.G., and Silva, P.A. (2010). Activity of β-lapachone derivatives against rifampicin-susceptible and -resistant strains of *Mycobacterium tuberculosis. Tuberculosis,* **90:** 293– 297.

Connors, K.A. (1995). Population characteristics of cyclodextrin complex stabilities in aqueous solution. *Journal of Pharmaceutical Science,* **84:** 843–848.

Corzo-Martinez, M., Corzo, N., and Villamiel, M. (2007). Biological properties of onions and garlic. *Trends in Food Science and Technology,* **18**: 609–625.

Cott, J. (1995). Natural product formulations available in Europe for psychotropic indications. *Psychopharmacology Bulletin*, **31**: 745.

Cragg, G.M., and David, J.N. (2001). Natural product drug discovery in the next millennium. *Pharmaceutical Biology*, **39**: 08 – 17.

Crupi, V., Ficarra, R., Guardo, M., and Majolino, D. (2007). UV–Vis and FTIR–ATR spectroscopic techniques to study the inclusion complexes of genistein with β-cyclodextrin. *Journal of Pharmaceutical and Biomedical Analysis*, 44: 110–117.

Daniel, A., Renaud, N., Patrick, R., and Fabrice, L. (2006). Water-soluble phenolics derivatives with dermocosmetic and therapeutic applications. European Patent. EP1867729.

Delmas, D., Jannin, B., and Latruffe, N. (2005). Resveratrol: preventing properties against vascular alterations and ageing. *Molecular Nutrition and Food Research*, **49:** 377–395.

Eastburn, S.D., and Tao, B.Y. (1994). Applications of modified cyclodextrins. *Biotechnology Advances*, **12**: 325–39.

Fabricant, D.S., and Farnsworth, N.R. (2001). The value of plants used in traditional medicine for drug discovery. *Environmental Health Perspectives*, **109**: 69-75.

Falamas, A., and Pinzaru, C.S. (2011). Spectroscopic investigations of newly formed betulin-cyclodextrin guest–host type complexes as potential anti skin cancer candidates. *Journal of Molecular Structure*, **993**: 297–301.

Gao, J.M., Ming, J., He, B., Gu, Z.W., and Zhang, X.D. (2008). Controlled release of 9-nitro-20(S)-camptothecin from methoxy poly (ethylene glycol)-poly (δ-lactide) micelles. *Biomedical. Materials*, **3:** 13–20.

Goyal, A., Kumar, S., Nagpal, M., Singh, I., and Arora, S. (2011). Potential of novel drug delivery systems for herbal drugs. *Indian Journal of Pharmaceutical Education and Research*, **45(3):** 225- 235.

Hirayama, F., and Uekama, K. (1999). Cyclodextrin-based controlled drug release system. *Advance Drug Delivery Review*, **36:** 125–141.

Holder, G.M., Plummer, J.L., and Ryan, A.J. (1978). The metabolism and excretion of curcumin (1, 7-bis-(4-hydroxy-3-methoxyphenyl)-1, 6-heptadiene-3, 5-Dione) in the rat. *Xenobiotica*, **8**: 761–768.

Hougton, P.J., and Mukherjee, P.K.(2009). Evaluation of Herbal Medicinal Products.

Irie, T., and Uekama, K. (1997). Pharmaceutical applications of cyclodextrins. III. Toxicological issues and safety evaluation. *Journal of Pharmaceutical Science*, **86**: 147– 62.

Jain, S.K., Awasthi, A.M., Jain, N.K., and Agrawal, G.P. (2005). Calcium silicate based microspheres of ripaglinide for gastroretentive floating drug delivery system of ripaglinide: Preparation and *in vitro* characterization. *Journal of Controlled Release*, **107**: 300-309.

Jiang, Y., and Sha, X. (2010). Complex of 9-nitro-camptothecin in hydroxypropyl-β-cyclodextrin: *In vitro* and *in vivo* evaluation. *International Journal of Pharmaceutics*, **397**: 116–121.

Kapil, K., and Rai, A.K. (2012). Miraculous therapeutic effects of herbal drugs using novel drug delivery systems. *International Research Journal of Pharmacy*, **3 (2):** 27-30.

Kayaci, F., and Uyar, T. (2012). Encapsulation of vanillin/cyclodextrin inclusion complex in electro spun polyvinyl alcohol (PVA) nanowebs: Prolonged shelf life and high temperature stability of vanillin. *Food Chemistry*, **133**: 641–649.

Kono, Y., Kobayashi, K., Tagawa, S., Adachi, K., and Ueda, A. (1997). Antioxidant activity of polyphenolics in diets: Rate constants of reactions of chlorogenic acid and caffeic acid with reactive species of oxygen and nitrogen. *Biochimica etbiophysica Acta*, **1335:** 335–342.

Krinsky, N.I., and Johnson, E.J. (2005). Carotenoid actions and their relation to health and disease. *Molecular Aspects of Medicine*, **26 (6):** 459–516.

Kunnumakkara, A.B., Anand, P., and Agarwal, B.B. (2008). Curcumin inhibits proliferation, invasion, angiogenesis and metastasis of different cancers through interaction with multiple cell signaling proteins. *Cancer Letter*, **269:** 199–225.

Li, L., Braiteh, F.S., and Kurzrock, R. (2005). Liposome-encapsulated curcumin: *In vitro* and *in vivo* effects on proliferation, apoptosis, signaling, and angiogenesis. *Cancer Letter*, **104**: 1322–31.

Li, X., Jin, Z., and Wang, J. (2007). Complexation of allyl isothiocyanate by α and β-cyclodextrin and its controlled release characteristics. *Food Chemistry*, **103**: 461–466.

Loftsson, T., and Brewster, M.E. (1996). Pharmaceutical applications of cyclodextrins: 1. Drug solubilization and stabilization. *Journal of Pharmaceutical Science*, **85:** 1017–25.

Lugemwa, F.N., Huang, F.Y., and Bentley, M.D. (1990). Birch bark processing and the isolation of natural products from birch bark. *Journal of Agriculture and Food Chemistry*, **38**: 493– 496.

Drag-Zalesinska, M.J., Kulbacka, J., Saczko, T., Wysocka, M., Zabel, P., and Surowiak, M. (2009). Esters of betulin and betulinic acid with amino acids have improved water solubility and are selectively cytotoxic toward cancer cells. *Bioorganic and Medicinal Chemistry Letters*, **19:** 4814–4817.

Mammucari, R., and Foster, N.R. (2008). Dense gas technology and cyclodextrins: state of the art and potential. *Current Organic Chemistry*, **12**: 476–491.

Maria, C.B., Anna, R.B., and Lorenzo, B. (2007). Studies on the interactions between some flavonols and cyclodextrins. *Bioorganic and Medicinal Chemistry Letters*, **17**: 5744–5748.

Misiuk, W., and Zalewska, M. (2009). Preparation and characterization of the emamectin benzoate/hydroxypropyl-β-cyclodextrin inclusion complex. *Carbohydrate Polymer*, **77:** 482–488.

Mosihuzzaman, M., and Iqbal, C.M. (2008). Protocols on safety, efficacy, standardization, and documentation of herbal medicine. *Pure and Applied Chemistry,* **80 (10):** 2195–2230.

Munoz-Botella, S., Del-Castillo, B., and Martyn, M.A. (1995). Cyclodetrin properties and applications of inclusion complex formation. *American Association of Pharmaceutical Sciences,* **36**: 187–98.

Mura, P., Furlanetto, S., Cirri, M., Maestrelli, F., Corti, G., and Pinzauti, S. (2005). Interaction of naproxen with ionic cyclodextrins in aqueous solution and in the solid state. *Journal of Pharmaceutical and Biomedical Analysis,* **37**: 987–994.

Musthaba, S.M., Baboota, S., and Ahmed, S.J. (2009). Status of novel drug delivery technology for phyto therapeutics. *Expert Opinion on Drug Delivery,* **6:** 625-37.

Neoh, T.L., and Yamamoto, C. (2012). Heat stability of allyl isothiocyanate and phenyl isothiocyanate complexed with randomly methylated β–cyclodextrin. *Food Chemistry,* **131**: 1123–1131.

Nicolescu, C., and Arama, C. (2010). Phase solubility studies of the inclusion complexes of repaglinide with β-cyclodextrin and β-cyclodextrin derivatives. *Farmacia,* **58: (5)** 620-628.

Niles, R.M., McFarland, M., Weimer, M.B., and Redkar, A. (2003). Resveratrol is a potent inducer of apoptosis in human melanoma cells. *Cancer Letter,* **190**: 157–163.

Olas, B., and Wachowicz, B. (2005). Resveratrol, a phenolic antioxidant with effects on blood platelet functions. *Platelets,* **16:** 251–260.

Panichpakdee, J., and Supaphol, P. (2011). Use of 2-hydroxypropyl β-cyclodextrin as adjuvant for enhancing encapsulation and release characteristics of asiaticoside within and from cellulose acetate films. *Carbohydrate Polymers,* **85**: 251–260.

Prameela, R.A., Siva, T.P., and Archana, N. (2009). Phase solubility studies on oral antidiabetic drugs with β-cyclodextrin and Hydroxypropyl-β cyclodextrin. *International Journal of Pharmaceutical Technology and Research,* **1 (4):** 1632-1637.

Ramanik, S. (2010). Characterization of cyclodextrin inclusion complexes: A review. *Journal of Pharmaceutical Science and Technology,* **2 (3):** 171-183.

Rybak, M.E., Calvey, E.M., and Harnly, J.M. (2004). Quantitative determination of allicin in garlic: Supercritical fluid extraction and standard addition of alliin. *Journal of Agricultural and Food Chemistry,* **52**: 582–687.

Sandhya, B.S., Thomas, W., and Shenbagarathai, R. (2006). *Complementary and alternative medicines.*

Satyavati, G.V., Gupta, A.K., and Tandon, N. (1987). Medicinal plants of India. *Indian Council of Medical Research.*

Sauceau, M., Rodier, E., and Fages, J. (2008). Preparation of inclusion complex of piroxicam with cyclodextrin by using supercritical carbon dioxide. *Journal of Supercritical Fluids,* **47:** 326–332.

Schneiderman, E., and Stalcup, A.M. (2000). Cyclodextrins: A versatile tool in separation science. *Journal of Chromatography*, **745**: 83–102.

Shi, G.R., Rao, L.Q., and Yu, H.Z. (2007). Yeast-cell-based microencapsulation of chlorogenic acid. *Journal of Food Engineering*, **80**: 1060–1067.

Shiko, A.N., Pozharitskaya, O.N., Miroshnyk, I., Mirzab, S., Urakova, I.N., and Hirsjavi, S. (2009). Nanodispersions of taxifolin: Impact of solid-state properties on dissolution behavior. *International Journal of Pharmaceutics*, **377**: 148–152.

Shukla, A., Rasik, A.M., and Dhawan, B.N. (1999). Asiaticoside-induced elevation of antioxidant levels in healing wounds. *Phytotherapy Research*, **13:** 50–54.

Siddiqui, S., Verma, A., Rather, A.A., and Jabeen, F. (2009). Preliminary phytochemical analysis of some important medicinal and aromatic plants. *Advances in Biological Research*, **3 (5-6):** 188-195.

Sou, K., Inenaga, S., Takeoka, S., and Tsuchida, E. (2008). Loading of curcumin into macrophages using lipid-based nanoparticles. *International Journal of Pharmaceutics*, **352:** 287–93.

Szejtli, J. (1998). Introduction and general overview of cyclodextrin chemistry. *Chemical Reviews*, **98**: 1743-54.

Takahashi, M., Kitamoto, D., Imura, T., Oku, H., Takara, K., and Wada, K. (2008). Characterization and bioavailability of liposomes containing an ukon extract. *Bioscience, Biotechnology and Biochemistry*, **72**: 1199–205.

Tavornvipas, S., Arima, H., Hirayama, F., Uekama, K., and Ishiguro, T. (2002). *Journal of Inclusion Phenomenon and Macrocyclic Chemistry*, **44**: 391–394.

Tiwari, K.L., Jadhav, S.K., and Joshi, V. (2011). An updated review on medicinal herb genus *Spilanthes*. *Journal of Chinese integrative medicine*, **9 (11):** 1170-1178.

Totterman, A.M., Schipper, N.G., Thompson, D.O., and Mannermaa, J.P. (1997). Intestinal safety of water-soluble β-cyclodextrins in paediatric oral solutions of spironolactone: effects on human intestinal epithelial Caco-2 cells. *Journal of Pharmacy and Pharmacology*, **49:** 43–48.

Verschraegen, C.F., Natelson, E.A., Giovanella, B.C., Kavanagh, J.J., Kudelka, A.P., Freedman, R.S., Edward, C.L., and Stehlin, J.S. (1998). A phase-I clinical and pharmacological study of oral 9-nitrocamptothecin, a novel water-insoluble topoisomerase I inhibitor. *Anticancer Drug*, **9:** 36–44.

Villiers, A. (1891). Cyclodextrin based catalysts and molecular reactors. *Compendus Rendus*, **112:** 536.

Wang. J., and Cao, Y. (2011). Physicochemical and release characterization of garlic oil-β-cyclodextrin inclusion complexes. *Food Chemistry*, **127:** 1680–1685.

Wenz, G. (1994). Aptamer Bioconjugate Drug Delivery Device. *Angewantde Chemie*, **106**: 851-70.

Xiang, Z.N., and Ning, Z.X. (2008). Scavenging and antioxidant properties of compound derived from chlorogenic acid in South-China honeysuckle LWT. *Food Science and Technology*, 1189–1203.

Xiao, J.B., Suzuki, M., Jiang, X.Y., Chen, X.Q., Yamamoto, K., and Xu, M. (2008). Influence of B-ring hydroxylation on interactions of flavonols with bovine serum albumin. *Journal of Agricultural and Food Chemistry*, **56**: 2350–2356.

Yadav, V.R., Prasad, S., and Kannappan, R. (2010). Cyclodextrin-complexed curcumin exhibits anti-inflammatory and antiproliferative activities superior to those of curcumin through higher cellular uptake. *Biochemical Pharmacology*, 1021–1032.

Yang, J.L., and Chen, W. (2011). Host–guest system of taxifolin and native cyclodextrin or its derivative: Preparation, characterization, inclusion mode and solubilization. *Carbohydrate Polymers*, **85**: 629–637.

Yuan, C., Jin, Z., Xu, X., Zhuang, H., and Shen, W. (2008). Preparation and stability of the inclusion complex of astaxanthin with hydroxypropyl-β-cyclodextrin. *Food Chemistry*, **109(2):** 264–268.

Yuan, C., Jin, Z., and Xu, X. (2012). Inclusion complex of astaxanthin with hydroxypropyl-β -cyclodextrin: UV, FTIR, ^{1}H NMR and molecular modeling studies. *Carbohydrate Polymers*, 1-5.

Bioactive Phytochemicals: Perspectives for Modern Medicine Vol. 2 (2014) *Pages* **499–530**
Editor: **V.K. Gupta**
Published by: **DAYA PUBLISHING HOUSE, NEW DELHI**

23

Bauhinia variegata (Caesalpiniaceae): A Review of its Traditional Uses, Pharmacognosy, Phytochemistry, Pharmacology and Analysis

Yogesh A. Kulkarni[1*], Sona Warrier[1] and Vikash S. Makadiya[1]

ABSTRACT

Bauhinia variegata (Caesalpiniaceae) is an important medicinal plant in India used in Ayurveda. This systematic review provides details of Bauhinia variegata, with respect to its traditional uses, pharmacognosy, phytochemistry related to various parts, pharmacological activities and analytical methods for identification. A number of important phytoconstituents belonging to class of flavonoids, glycosides, terpenes, and steroids are reported from various parts of the plant. Different extracts of this plant showed significant pharmacological activities like anticancer, antidiabetic, antioxidant, anti-inflammatory, etc. However, there is need to study this plant and its constituents for mechanistic approach in pharmacology and also in clinical trials in order to validate traditional use in the light of a rational phytotherapy.

Keywords: Analysis, *Bauhinia variegata*, Kanchnar, Pharmacognosy, Pharmacology, Phytochemistry.

Introduction

The use of different medicinal plants in the India can be traced back to the Vedic period. Ancient literatures like Charaka Samhita, Sushruta Samhita, Rigveda and Atharveveda describe uses of different medicinal plants for treating different

1 SPP-School of Pharmacy and Technology Management, SVKM's NMIMS University, V.L. Mehta Road, Vile Parle (W), Mumbai – 400 056, Mh., India

* *Corresponding author*: E-mail: yogeshkulkarni101@yahoo.com

disorders. *Bauhinia variegata* is an important medicinal plant mentioned in Ayurveda-an important traditional system of medicine. Remarkable attention has been given by scientific community to *Bauhinia variegata*, as evidenced by various studies carried out in recent few years. The present review assembles the scattered information on the traditional uses, pharmacognosy, phytochemistry, pharmacology and analysis of this medicinal plant. The present information will emphasize the importance of *Bauhinia variegata* and may offer a new path for researchers in the future.

Morphology

It is a deciduous tree. The leaves are 10-15 cm long as broad as or rather broader than long, cleft 1/4 to 1/3 of the way down into two obtuse lobes. They are pubescent beneath when young. The base of the leaves is deeply cordate (Kirtikat and Basu, 1999). The bark is grey in colour with longitudinal cracks and pale pink inside. The flowers are large, fragrant white or purplish in colour. The flowers appear in short axillary or terminal, few flowered, grey pubscent racemes. The petals are 5-6.3 cm long, obovate with long rather broad claws. The pods of the plant are long, hard, flat, glabrous, dehiscent. It shows presence of 10-15 seeds (Kirtikat and Basu, 1999; Anonymous, 2003).

Distribution and Propagation

The tree is found in Sub-Himalayan tract and outer Himalaya of the Punjab, ascending to 4000 ft. from the Indus eastwards, Kumaon between 1000-6000 ft., Sikkim, through India proper to Burma (Kirtikat and Basu, 1999). The tree is cultivated in the garden for its showy flowers. It is propagated from seeds sown in line; the soil is well loosened and is followed by regular watering and weeding. Stump planting also gives good results (Anonymous, 2003).

Traditional Uses

The parts of plant used for traditional medicines are roots, bark, buds and flowers.

Roots

The decoction of roots is used in dyspepsia and flatulence; it is also an anti-fat remedy (Nadkarni, 2000; Anonymous, 2003). The juice of root is given in snake-bite (Kirtikar and Basu, 1999).

Bark

Bark is used as an alterative, tonic and astringent. The addition of extract of acacia pods and pomegranate flowers to the bark extract, forms a gargle and it is a remedy in salvation and sore throat. An emulsion formed with rice water and bark with the addition of ginger is used in the scrofulous enlargement of glands in the neck. Decoction of the bark is a useful wash in ulcers and skin diseases and a remedy in diarrhea. A paste made up of ginger and bark is applied to tumors (Nadkarni, 2000). The bark is tonic to liver. It is used for treating biliousness, asthma, leprosy and leucoderma. It is also useful in dysmenorrhoea, menorrhagia and ulcers. Fresh juice of bark and fresh juice of flowers of *Strobilanthes citrata* is given as expectorant in Konkan region of India (Kirtikar and Basu, 1999). It is useful as anthelmenthic and used for scrofula and skin diseases. A decoction of the bark is taken for dysentery by the Nagas in India (Anonymous, 2003).

Buds and Flowers

The decoction of buds is used in cough, bleeding piles, haematuria and menorrhagia. Dried buds are also useful in diarrhea, worms, piles and dysentery. The flowers are laxative (Nadkarni, 2000; Kirtikar and Basu, 1999; Anonymous, 2003).

Pharmacognostical Studies

Bark

The outer surface bark was greyish-brown and inner surface dark reddish brown in colour. It showed quilled shape, rough surface, fibrous fracture; prominent lenticels. Microscopy of bark indicated presence of 25-30 cells thick periderm, distinct into 3 zones–cork cells are empty, thick walled; phellogen 3-5 cells thick; phelloderm, arranged like bricks, few thick-walled cells have irregular groupings associated with idioblasts containing prismatic crystals; crystal idioblasts also found scattered throughout cortex in irregular groups; cells of phloem mostly contain druses along with few prismatic crystals, medullary rays indistinct and discontinuous; usually broader towards phloem (3-5 cells thick) and narrow towards xylem (1 cell thick) in between medullary rays lie irregular patches of phloem fibres associated with idioblasts containing prismatic crystals, some of the parenchymatous cells are filled with light-brown tannin; cortex and phloem cells contain scanty starch grains (Srivastava *et al.*, 1988).

Leaves

The leaves showed the presence of thin walled epidermis, parenchymatous cells, abundant solitary calcium oxalate crystals, vascular bundles and multicellular covering trichomes (Gupta *et al.*, 2009a).

Phytochemical Studies

Roots

Isolation of a flavonol glycoside 5,7,32,42-tetrahydroxy-3-methoxy-7-O-α-L-rhamnopyranosyl (1-3)-O-β-D-galactopyranoside (I) is reported from the roots of *Bauhinia Variegata* (Yadava, 2003).

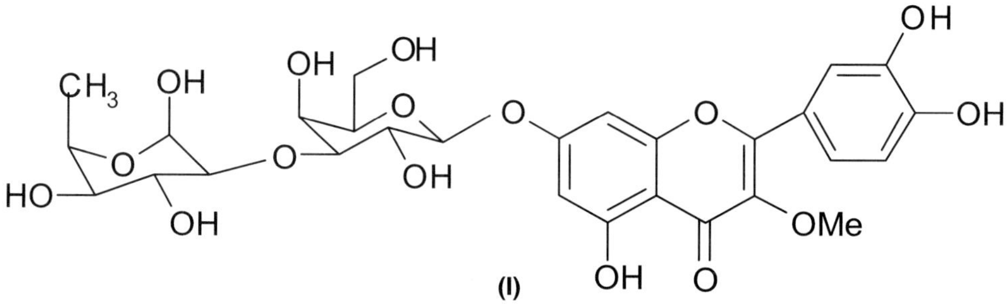

CH_3 OH OH OH O O O O OH HO OH OH OH OMe OH O

(I)

The root bark of *Bauhinia variegata* showed the presence of a new flavanone, (2*S*)-5,7-dimethoxy-3′,4′-methylenedioxyflavanone (II) and a new dihydrodibenzoxepin, 5,6-dihydro-1,7-dihydroxy-3,4-dimethoxy-2-methyldibenz [b,f]oxepin (III) together with three known flavonoids (Reddy *et al.*, 2003).

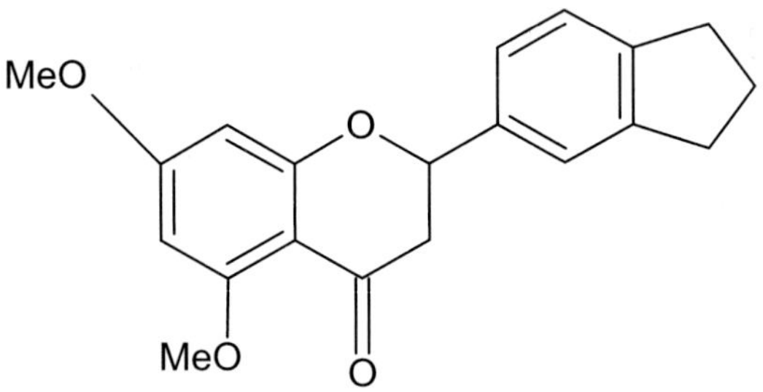

(II)

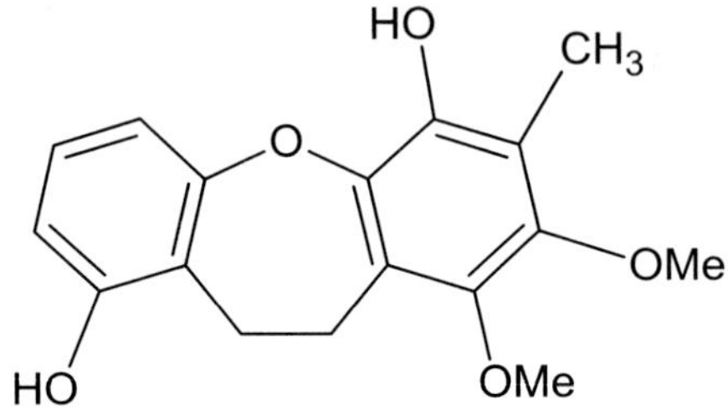

(III)

Flowers

Flowers of *Bauhinia variegata* showed the presence of 3-hydroxy-5, 7, 4′-trimethoxy flavanol. The glycoside is identified as kaempferol- 3-galactoside (Rahman and Begum, 1966).

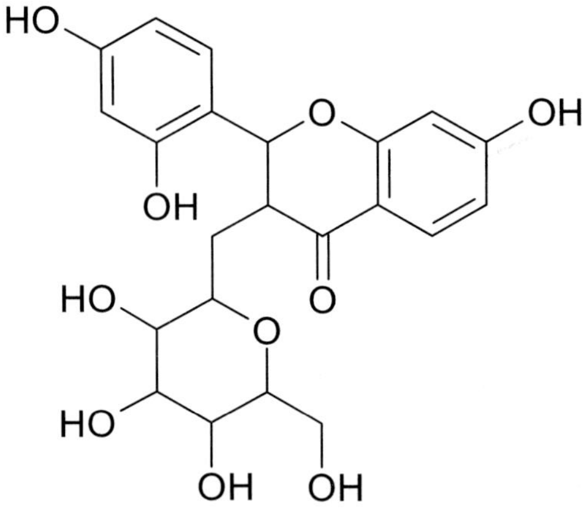

Kaempferol-3-galactoside

Kaempferol glycosides were reported in *B. variegata* flowers by Duret and Paris (1977).

The pale violet flowers contain cyanidin-3-glucoside, malvidin-3-glucoside, malvidin-3-diglucoside, peonidin-3-glucoside, peonidin-3-diglucoside. The white flowers contain kaempferol-3-galactoside and kaempferol-3-rhamnoglucoside (Anonymous, 2003).

The volatile fraction of flowers of *Bauhinia variegata* was found to be 0.04 per cent (el-Hamouly and Salama, 2004). The important nutrients in the flower of *B. variegata* var. *candida* were tested and compared with those in the flower of *Hemerocallis citrine*.

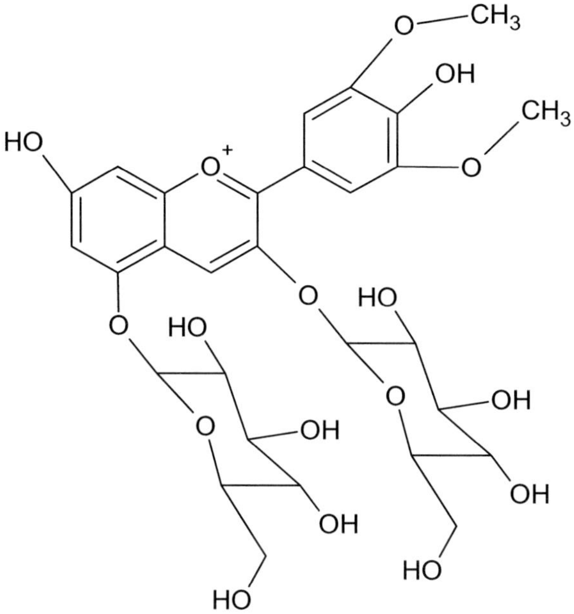

Malvidin 3,5-diglucoside

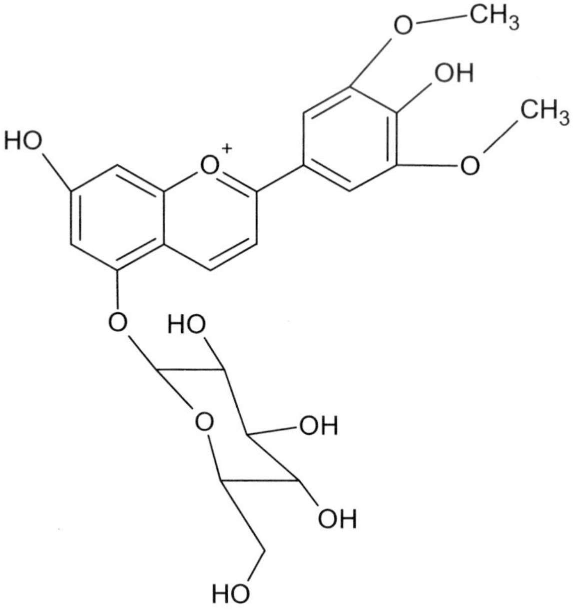

Malvidin-3-glucoside

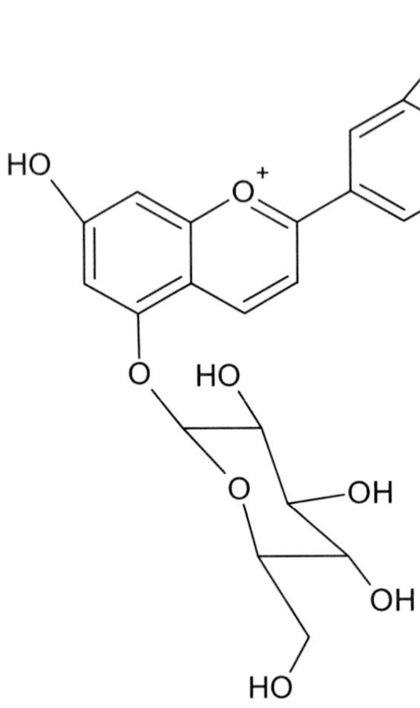

Peonidin-3-glucoside

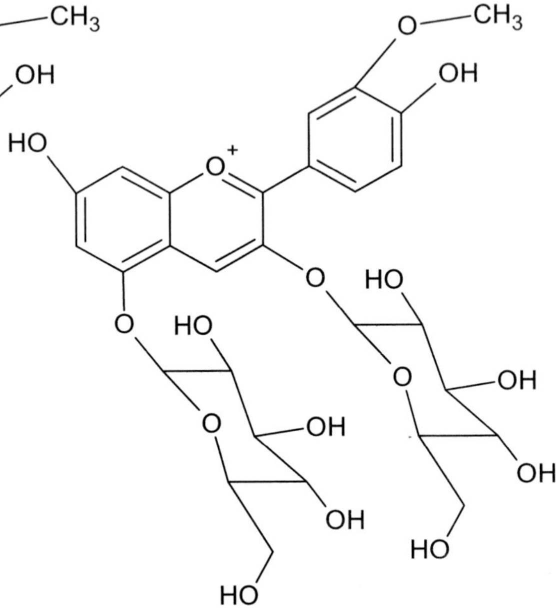

Peonidin-3,5-diglucoside

The contents of P, K, Na, Ca, Mg, Fe and Cu in *B. variegata* var. *candida* were higher than in *H. citrina,* but the content of Zn was lower. The total protein content of *H. citrina* and *B. variegata* var. *candida* was 19.19 per cent and 15.79 per cent, but the content of total amino acids contents in protein was 9.99 per cent and 12.74 resp. The close degree of essential amino acids contents in *H. citrina* and *B. variegata* var. *candida* to that in egg protein (the std.) were 0.73 and 0.85 (You-kai *et al.,* 2004). Keto acids of flowering buds of *B. variegata* were analyzed for keto acids during bud development and correlated with the free amino acids. Initially 4 amino acids appeared in early stages which increased to 9. The amino acids present were alanine, aspartic acid, glycine, serine, and glutamic acid (Mukherjee and Laloraya, 1975). The oils obtained from flowers of *B. variegata, B. variegata* var. *candida* were investigated by TLC, gas chromatography, combined gas chromatography-mass spectroscopy, and IR spectroscopy. The oils contained 24 and 16 terpene constituents, respectively (Wassel *et al.,* 1986).

Stem

From ethanolic extract of the stem of *Bauhinia variegata,* a flavanone glycoside called as 5,7-dihydroxyflavanone-4′-*O*-α-L-rhamnopyranosyl-β-D-glucopyranoside is deduced (Gupta *et al.,* 1979c).

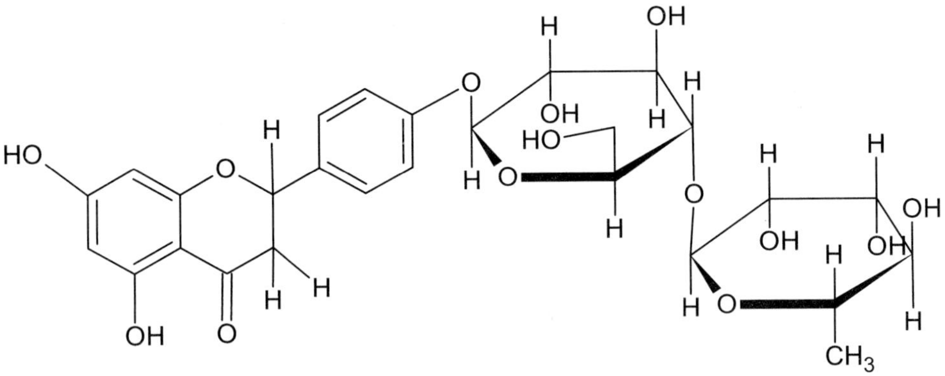

5,7-dihydroxyflavanone-4'-O-α-L-rhamnopyranosyl-β-D-glucopyranoside

5,7-dimethyoxyflavanone 4'-O-α-L-rhamnopyranosyl-β-D-glucopyranoside was obtained from the benzene extract of stems of *B. variegata*.

A phenanthraquinone named Bauhinione was isolated from 95 per cent ethanol extract of stems of *Bauhinia variegata* (Zhao, 2005a).

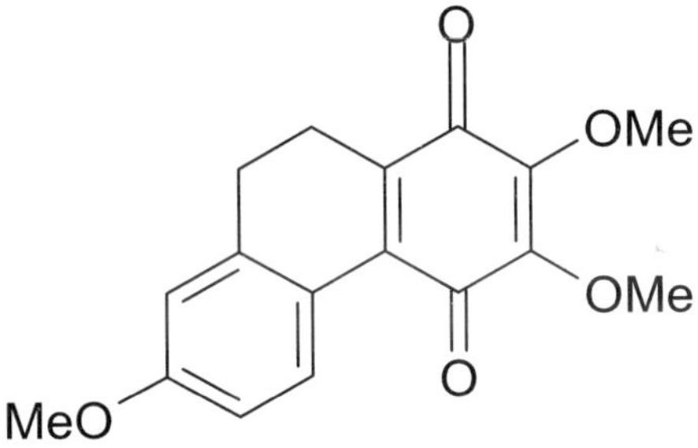

Bauhinione

A 5,7,4′-Trimethoxy flavavone has been isolated from ethyl acetate fraction of 95 per cent ethanol extract of stem of *Bauhinia variegata*. The concentrated ethyl acetate fraction (15 g) of the alcohol extract of the plant was subjected over a silica gel column

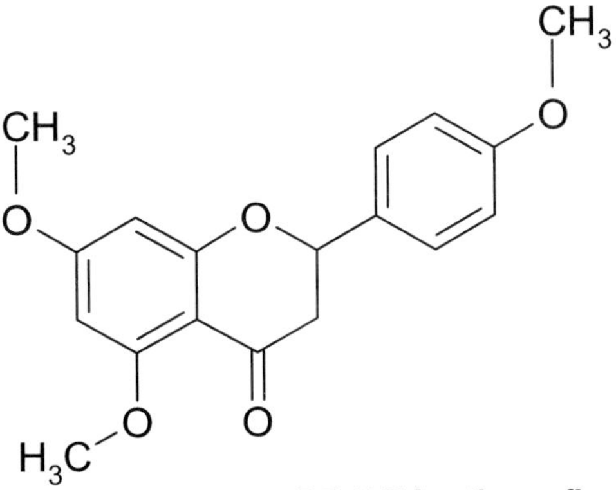

5,7,4′-Trimethoxy flavavone

using a hexane–EtOAc step gradient. The hexane : EtOAc (35 : 65) eluates yielded a yellow residue (50 mg).

Stem bark of *Bauhinia variegata* Linn also showed the presence of a polyphenolic aromatic glycoside characterized as 2,4,8,9,10-pentahydroxy-3,7-dimethoxyanthracene-6-O-α-L-rhamnopyranoside (Bodakhe *et al.*, 2010)· Chemical evaluation of the stem bark of *Bauhinia variegata* showed the presence of 30-norhopane, etcerotate, friedelin, palmitic acid, β-sitosterol, aplysterol, taraxeran-3-α-ol, naringenin, kaempferol, quercetin, n-octacosane, β-sitosterol palmitate and docosanoic acid (Jain *et al.*, 2004). Ethanolic extract of the stem of *B. variegata* showed the presence of a flavanol glycoside which was characterized as kaempferol-3-glucoside. The sugars and amino acids present in the stem were isolated and identified (Gupta and Chauhan, 1984).

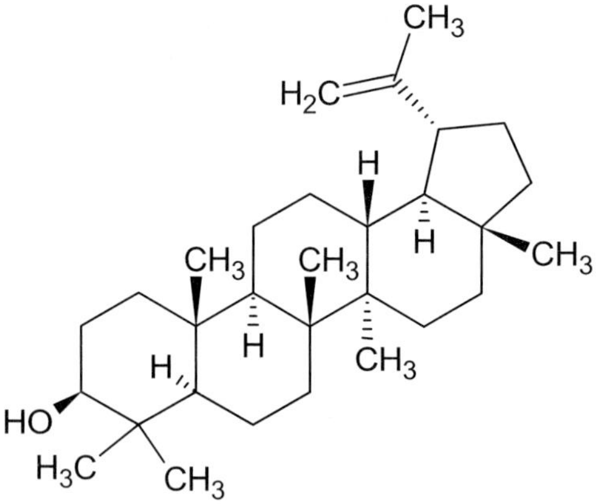

Lupeol

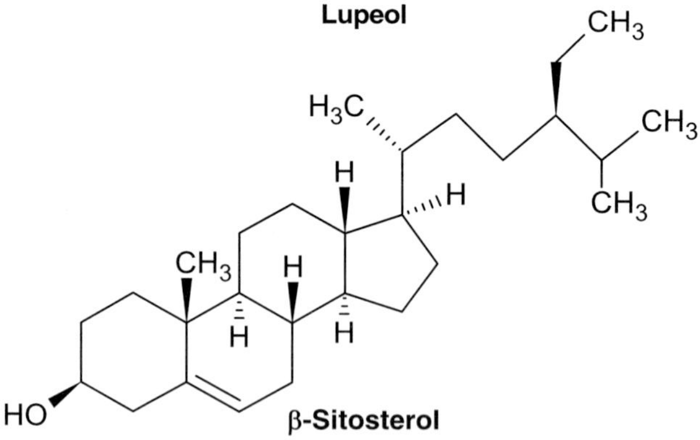

β-Sitosterol

The benzene extract of *Bauhinia variegata* stem showed the presence of 5,7-dimethyoxyflavanone 4'-O-α-L-rhamnopyranosyl-β-D-glucopyranoside. Further, β-sitosterol and lupeol was obtained from the petroleum ether extract of the stem (Gupta *et al.*, 1980).

Seeds

Myricetol glycosides were reported in *B. variegata* seeds (Duret and Paris, 1977). The seeds *Bauhinia variegata* showed presence of crude proteins (31.6 per cent). The essential amino acids detected were: lysine, threonine, valine, methionine, isoleucine, and phenylalanine. Other amino acids found in adequate amount were histidine, arginine, aspartic acid, serine, glutamic acid, proline, glycine, alanine, and tyrosine (Wassel *et al.*, 1989).

Seeds of *Bauhinia* were high in aspartic and glutamic acids, and had a good balance of essential amino acids (Dhan and Misra, 1983). The seed oil of the plant was analyzed by many researchers for its composition. The seed oil showed presence of glycerides of myristic acid (1 per cent), palmitic acid (17 per cent), stearic acid (13.4 per cent), lignoceric acid (1 per cent), oleic acid (31.8 per cent) and linoleic acid (35.9 per cent). Sitosterol was also present (Puntambekar and Krishna, 1940). The palmitic, stearic, oleic, and linoleic acid contents were 20.5, 12.0, 12.5, and 35.4 per cent, respectively, for *Bauhinia variegata* seed oils (Badami and Shanbhag, 1973). Fatty Acid composition as determined by GLC for *Bauhinia variegata* seeds revealed oils like linolenic (0.55 per cent), Linoleic (36.84 per cent) oleic (26.14 per cent), stearic (16.95 per cent), palmitic (19.52 per cent) acids (Zaka *et al.*, 1983). Seeds were examined for oil content, oil properties, and triglyceride composition. *Bauhinia variegata* seeds also indicated presence of triglycerides (Tyagi and Vasishtha, 1984). Carbohydrates content of seeds of *B. variegata* was found to be 19.92 per cent (Niranjan *et al.*, 1985). Trypsin inhibitors of two varieties of *Bauhinia variegata* seeds have been isolated and characterized. *Bauhinia variegata* candida trypsin inhibitor (BvcTI) and *B. variegata* lilac trypsin inhibitor (BvlTI) are proteins with molecular weight of about 20,000 without free sulfhydryl groups. Amino acid analysis showed a high content of aspartic acid, glutamic acid, serine, and glycine, and a low content of histidine, tyrosine, methionine, and lysine in both inhibitors. The trypsin inhibitors show Ki values of 6.9 and 1.2 nM for BvcTI and BvlTI, respectively (Di Ciero *et al.*, 1998). The protein

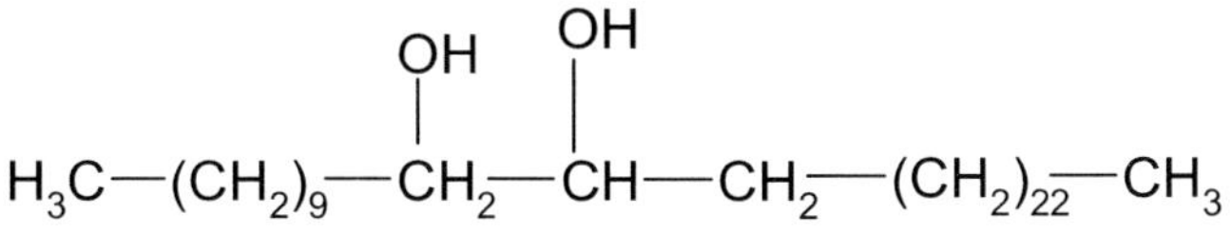

$$H_3C-(CH_2)_9-\underset{\substack{|\\OH}}{CH_2}-\underset{\substack{|\\OH}}{CH}-CH_2-(CH_2)_{22}-CH_3$$

(I)

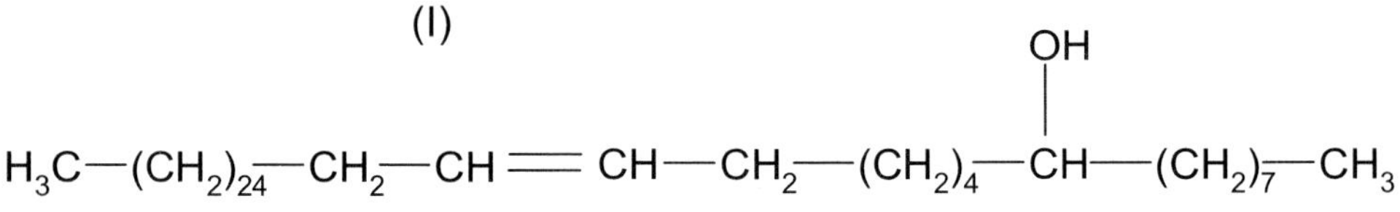

$$H_3C-(CH_2)_{24}-CH_2-CH=CH-CH_2-(CH_2)_4-\underset{\substack{|\\OH}}{CH}-(CH_2)_7-CH_3$$

(II)

$$H_3C-(CH_2)_{24}-CH_2-CH=CH-(CH_2)_4-CH_2-\underset{\substack{|\\OH}}{CH}-(CH_2)_7-CH_3$$

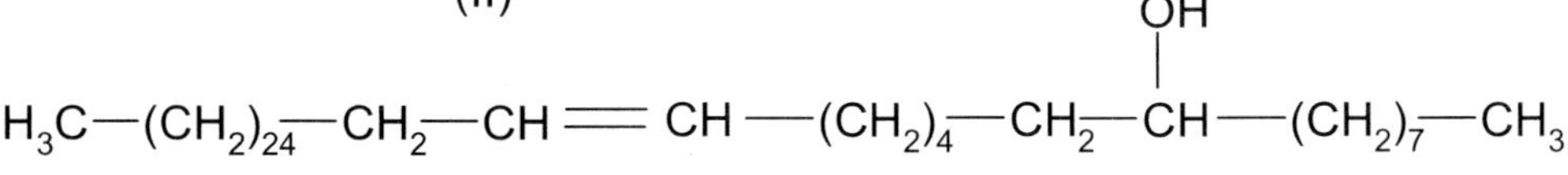

(IIa)

content and oil content of seeds of *B. variegata* was found to be 46.5 per cent and 17.3 per cent respectively (Dhan *et al.*, 2001).

A flavone glycoside namely 5-hydroxy 7,3′,4′,5′-tetra- methoxyflavone 5-O-β-D-xylopyranosyl-(1→2)-α-L-rhamnopyranoside was isolated from the 95 per cent ethanolic extract of the seeds of *Bauhinia variegata* Linn (Yadava and Reddy, 2001).

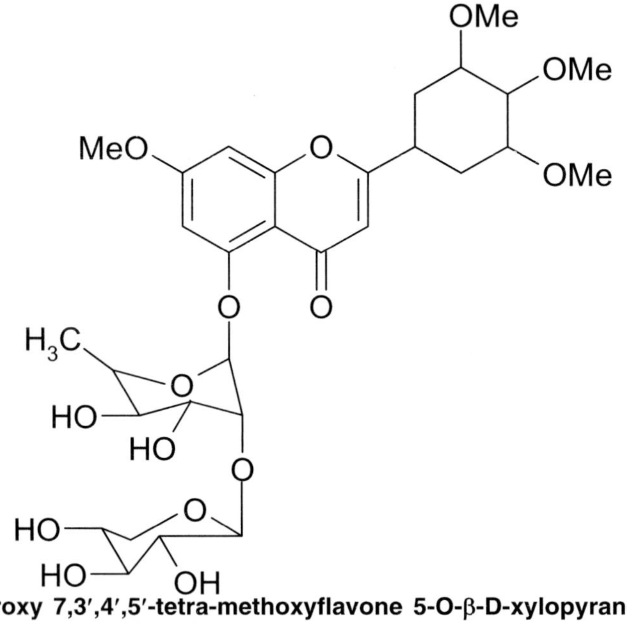

5-hydroxy 7,3′,4′,5′-tetra-methoxyflavone 5-O-β-D-xylopyranosyl-(1→2)-α-L-rhamnopyranoside

A lectin (BvcL) from seeds of *Bauhinia variegata* candida was isolated by gel filtration chromatography on Sephadex G75 and affinity chromatography on immobilized D: -lactose column. Amino acid analysis revealed a large content of Ser, Gly, Thr, Asp and Glu and low concentrations of Met, Cys and His (Silva Jose *et al.*, 2007). A dimeric 64-kDa melibiose-binding lectin was isolated from the seeds of *Bauhinia variegata* by affinity chromatography on Affi-gel blue gel, ion exchange chromatography on Mono Q, and gel filtration on Superdex 75 (Lin Peng and Ng Tzi Bun, 2008). A galactose-specific lectin was purified from seeds of *Bauhinia variegata* by affinity chromatography on lactose-agarose. The molecular mass of BVL subunit is 32 871 Da, determined by MALDI-TOF spectrometry (Pinto Luciano, 2008). A new Kunitz-type trypsin inhibitor (termed BvvTI) was isolated from seeds of *Bauhinia variegata* (Fang *et al.*, 2010).

Leaves

Isolation of naringenin- a flavonoid aglycone is reported from leaves of *Bauhinia variegata* (Arafa, 2001). Chemical investigations of *Bauhinia variegata* leaves showed presence of six compounds *i.e.* quercetin 3-Me ether, luteolin, rulin, narcissin, isoquercitrin and daucosterol (El-Dondaity *et al.*, 2005). The leaves of *Bauhinia variegata* also showed the presence of a new chalcone, 2'-hydroxy-4',6'-dimethoxy-3,4-methylenedioxy chalcone together with 2 flavonol glycosides, kaempferol-3-O-D-glucopyranoside and kaempferol-3-O-β-L rhamnoside (Maheswara *et al.*, 2005). Leaves of species of *Bauhinia variegata* were submitted to volatile oil extraction by steam distillation. Oils from leaves showed the presence of germacrene D, spathulenol, α- and β-cadinene (Duarte-Almeida *et al.*, 2004). Two long chain compounds, heptatriacontane-12,13-diol and dotetracont-15-en-9-ol (I, II, IIa) were isolated from the leaves of *B. variegata* (Singh *et al.*, 2006).

Six flavonoid compounds–quercetin, 3,3′-dimethoxy quercetin,3,3′,6-trimethoxy quercetin, quercetin 3-*O*-β-D-4C1-glucopyranoside, quercetin 3-*O*-β-D-4C1-galacturonopyranoside, quercetin 3-*O*-α-L-1C4-rhamnopyranosyl- (1‴→2″)-*O*-β-D-4C1-glucopyranoside and two cinnamic acid derivatives- caffeic acid, ferulic acid were isolated from ethyl acetate fraction of methanol extract of leaves.

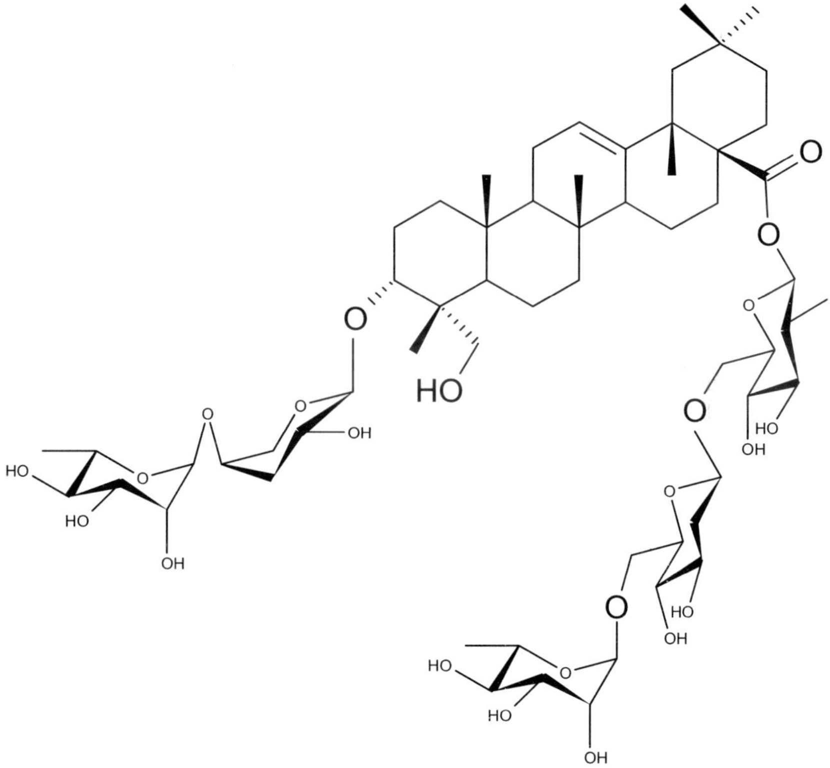

Triterpene saponin

N-butanol fraction of methanol extract yielded a triterpene saponin, named as 23-hydroxy-3α-[O-α-L-1C4-rhamnopyranosyl-(1''→4')-O-α-L-4C1-arabinopyranosyl-oxy]olean-12-en-28-oic acid O-α-L-1C4-rhamnopyranosyl-(1''''→4''')-O-β-D-4C1-glucopyranosyl-(1'''→6'')-O-β-D-4C1 glucopyranosyl ester (Mohamed *et al.*, 2009).

A chloroplast protein with a molecular mass similar to that of bovine insulin was extracted from leaves by 2-mm thick 15 per cent SDS-PAGE gel electrophoresis and fractionated with a 2 × 24 cm Sephadex G-50 column (Azevedo *et al.*, 2006).

Aerial Parts

(-)-Epi-catechin, protocatechuic acid, protocatechuic acid ethyl ester, methyl gallate, ethyl gallate, p-hydroxybenzoic acid, schizandriside were isolated from plant (Zhao *et al.*, 2005b).

Flavonoids namely kaempferol, ombuin, kaempferol 7,4'-dimethyl ether 3-O-β-D-glucopyranoside, kaempferol 3-O- β -D-glucopyranoside, isorhamnetin 3-O-β-D-glucopyranoside and hesperidin, together with one triterpene caffeate, 3 β -trans-(3,4-dihydroxycinnamoyloxy)olean-12-en-28-oic acid were isolated from the non-woody aerial parts of *Bauhinia variegata* (Rao *et al.*, 2008). Friedelin, 24(R)-9,19-cyclolanost-3-one-24,25-diol and 24(S)-9,19-cyclolanost-3-one-24,25-diol, stigmast-3-one, stigmast-4-en-3-one, stigmast-3-ol-6-one, stigmast-5-en-3-ol-7-one, β-sitosterol and physcion were isolated from plant (Zhao *et al.*, 2004a). Alkaloids namely N-naphthyl-2-aniline, bicycloamhanimbine, mahainbine, girinimbine were isolated from plant of *Bauhinia variegata* (Zhao *et al.*, 2004b). Five flavonoids were isolated from the different organs of *B. variegata* var. *candida* cultivated in Egypt. They were identified as quercetin, rutin, quercitrin, apigenin, and apigenin 7-O-glucoside.

Pharmacological Studies

Antitumor Activity

The antitumour activity of ethanol extract of stem of *Bauhinia variegata* has been evaluated against Dalton's ascetic lymphoma in Swiss albino mice. A significant enhancement of mean survival time of extract treated tumour bearing mice was found with respect to control group. Extract treatment was found to enhance peritoneal cell counts. After 14 days of inoculation, extract was able to reverse the changes in the haemotological parameters, protein and PCV consequent to tumour inoculation (Rajkapoor *et al.*, 2003a). The ethanol extract of *Bauhinia variegata* has been evaluated for antitumour activity against Ehrlich ascites carcinoma in Swiss albino mice. The mean survival time of extract treated tumour bearing mice was found to be significantly increased with respect to the control group. The peritoneal cell counts were found to be enhanced after treatment with extract. The changes in the haemotological parameters, protein and PCV were reversed by extract after 14 days of inoculation. Oral administration of ethanol extract showed significant reduction in solid tumour mass development induced by Ehrlich ascites carcinoma cells (Rajkapoor *et al.*, 2003b). Four isolated alkaloids- N-naphthyl-2-aniline, bicycloamhanimbine, mahainbine, girinimbine form *Bauhinia variegata* were studied for their anticancer activity. The cell proliferation inhibitory and apoptosis inducing activities were assayed using K562 cells by the SRB method and flow cytometry respectively. The apoptosis inducing activities of N-naphthyl-2-aniline was reported first time from this plant (Zhao *et al.*, 2004b). The anticarcinogenicity and antimutagenicity activity of Kachanar extract were assessed using a skin carcinogenesis and melanoma tumour model along with micronucleus and chromosomal aberration tests, in Swiss albino mice. In the skin papilloma model, significant prevention, with delayed appearance and reduction in the cumulative no. of papillomas was observed in the DMBA + Kachanar + Croton oil treated group as compared to the DMBA + Croton Oil group. C57 Bl mice which received a 50 per cent methanolic extract of Kachanar extract at the doses of 500 and 1,000 mg/kg body weight for 30 days showed increase in life span and tumour size was significantly reduced as compared to controls. In antimutagenicity studies, a single application of Kachanar extract at doses of 300, 600 and 900 mg/kg dry weight, 24 hours prior the *i.p.* administration of cyclophosphamide (at 50 mg/kg) significantly prevented micronucleus formation and chromosomal aberrations in bone marrow cells of mice, in a dose dependent manner (Agrawal *et al.*, 2009). Anticarcinogenic activity of methanol extract of *Bauhinia variegata* was tested using two stage protocol in skin papillioma model in Swiss albino mice. The significant prevention of papillomas in DMBA + extract (500 and 1000 mg/kg body weight) + croton oil treated group was found to be effective in decreasing the rate of tumor incidence in comparison to the control. The cumulative number of papillomas, tumor yield and tumor burden were also reduced. The depleted levels of glutathione were restored in extract treated groups (Pandey and Agrawal, 2009). The main constituent of *Bauhinia variegata*, bauhinione, was assessed for the inhibitory effect on the proliferation of human chronic myeloid leukemia K562 cells. The IC_{50} values of bauhinione on K562 cells and mouse tsFT210 cells were detected by sulforhodamine B (SRB) assay and the cell cycle inhibitory and apoptosis inducing effects of bauhinione were analyzed by flow cytometry (FCM) accompanied with morphol. The minimal inhibitory concentration (MIC) was assayed with the multiple dilution method. The IC_{50} values of bauhinione on K562 cells and mouse tsFT210 cells were 24.9 ± 2.1 mg/L and 111.36 ± 3.5 mg/L, resp. The MIC value of bauhinione on K562 cells was 6.25 mg/L (Zhao *et al.*, 2007). Chemopreventive potential of flower extracts of *Bauhinia variegata* is reported. Two-stage skin carcinogenesis was induced by a single topical application of 7, 12-dimethylbenz(a)anthracene (DMBA) (104 μg/100 μl acetone), and one weeks later, promoted by repeated application of croton oil (1 per cent in acetone/twice in a week) till the end of the expt. (16 weeks) in Swiss albino mice. Single topical application of flower extract at a dose of 500 and 1000 mg/kg body weight along with DMBA + Croton oil was found to be effective in decreasing the cumulative number of papilloma, tumor incidence, tumor yield and tumor burden as compared to control (DMBA + Croton oil) group (Pandey and Agrawal, 2010a). Flavanone isolated form stem of *Bauhinia variegata* was tested for *in vitro* cytotoxicity using the sulphorhodamine B (SRB) assay method. The compound was tested at five concentrations against 57 human tumor lines, representing leukemia, non-small cell lung, colon, central nervous system, melanoma, ovarian, renal, prostate and breast cancers. The flavanone had cytotoxic activity against human tumor cell lines. The flavanone showed significant cytotoxic activity with average 50 per cent growth inhibition in the range of 4.06–4.81 mM. The phytoconstituents was more selective against OVCAR-5 and SKOV-3 cell lines of ovarian cancer (Rajkapoor *et al.*, 2009).

A dimeric 64-kDa melibiose-binding lectin isolated from the seeds inhibited proliferation in hepatoma HepG2 cells and breast cancer MCF7 cells with an IC_{50} of 1.4 μM and 0.18 μM, respectively (Lin Peng and Ng Tzi Bun, 2008).

Ethanolic extract of *Bauhinia variegata* showed chemopreventive and cytotoxic effect on N-nitrosodiethylamine (DEN) induced experimental liver tumor in rats and human cancer cell lines. Oral administration of ethanol extract of *Bauhinia variegata* (250 mg/kg) effectively suppressed liver tumor induced by DEN as revealed by decrease in DEN induced elevated levels of SGPT, SGOT, ALP, total bilirubin, GGTP, LPO, glutathione peroxidase (GPx) and GST. The extract produced an increase in enzymatic antioxidant (superoxide dismutase and catalase) levels and total proteins when compared to those in liver tumor bearing rats. The histopathological changes of liver samples were compared with respective controls. Ethanol extract was found to be cytotoxic against human epithelial larynx cancer (HEp2) and human breast cancer (HBL-100) cells (Rajkapoor *et al.*, 2006). Kunitz-type trypsin inhibitor (BvvTI) from seeds of *Bauhinia variegata* significantly inhibited the proliferation of nasopharyngeal cancer CNE-1 cells in a selective way. This may partially be contributed by its induction of cytokines and apoptotic bodies. BvvTI also showed anti-HIV-1 reverse transcriptase activity (Fang *et al.*, 2010).

Nephroprotective Effect

Both ethanolic and aqueous root extracts of *Bauhinia variegata* Linn. produced significant nephroprotective activity in Gentamicin induced nephrotoxicity model as evident by decrease in elevated serum creatinine, serum urea, urine creatinine and BUN levels, which was further confirmed by histopathological study. Gentamicin induced glomerular congestion, blood vessel congestion, and epithelial desquamation, accumulation of inflammatory cells and necrosis of the kidneycells were found to be reduced in the groups receiving the root extract of *Bauhinia variegata* along with gentamicin (Sharma *et al.*, 2010). The nephroprotective activity of the ethanolic extract of *Bauhinia variegata* (Linn.) whole stem against cisplatin-induced nephropathy was investigated by an *in vivo* method in rats. Acute nephrotoxicity was induced by *i.p.* injection of cisplatin (7 mg/kg of body weight). Ethanol extract at 400 mg/kg decreased the serum level of creatinine (0.65 ± 0.09; $P<0.001$) and urea (32.86 ± 5.88; $P<0.001$) associated with a significant increase in body weight (7.16 ± 1.10; $P<0.001$) and urine volume output (11.95 ± 0.79; $P<0.05$) as compared to the toxic control group. The ethanol extract of *Bauhinia variegata* at 400 mg/kg exhibited significant and comparable nephroprotective potential to that of the standard polyherbal drug cystone (Pani *et al.*, 2011).

Hepatoprotective Effect

The hepatoprotective activity alcoholic extract of stembark of *Bauhinia variegata* was investigated in carbon tetrachloride (CCl_4) intoxicated Sprague-Dawley rats. *Bauhinia variegata* at different doses (100 and 200 mg/kg) were administered orally to male Sprague-Dawley rats weighing between 100-120 g. The effect of SBE on the serum marker enzymes, *viz.*, AST, ALT, ALP and GGT and liver protein and lipids were assessed. The extract exhibited significant hepatoprotective activity (Bodakhe and Alpana, 2007).

Hepatoprotective activity of ethanolic extract of *Bauhinia variegata* leaves was tested in paracetamol induced hepaptotoxicity in rats. Extract was administered orally at different doses (200 mg/kg and 400 mg/kg). The extract showed a significant reduction in level of SGOT (53.26 per cent), SGPT (41.64 per cent), SALP (72.30 per cent) and bilirubin (68.18 per cent). It also preventing liver histopathological changes in rats induced with PCM hepatotoxicity (Sahu and Jain, 2011).

Anti-hyperlipidaemic Activity

Extracts of *Bauhinia variegata* Linn. were studied for its anti-hyperlipidaemic effects and anti-diabetic effects by measuring total cholesterol, triglycerides, phospholipids, LDL, VDL and HDL in diabetic and nondiabetic rats (Thiruvenkatasubramaniam *et al.*, 2010).

The blood cholesterol level and triglycerides levels were measured in normal and Triton WR-1339 induced rats. There was significant increase in cholesterol and triglyceride level in triton induced hyperlipidemic rats from 60 mg/dl in normal rats to 190 mg/dl. Administration of methanolic extract of *Bauhinia variegata* (Linn.) at a dose of 100 mg/kg and standard drug 65 mg/kg the cholesterol level is 99.65 mg/dl and triglyceride level is 142.0 mg/dl as compare to standard drug fenofibrate where decrease of cholesterol level is by 65.24 mg/dl and triglyceride level is by 77.96 mg/dl in triton induced hyperlipidemic rats (Kumar *et al.*, 2011).

The extracts were also evaluated for antihyperlipidemic activity in Triton WR-1339 (iso-octylpolyoxyethylene phenol)-induced hyperlipidemic albino rats by estimating serum triglyceride, very low density lipids (VLDL), cholesterol, low-density lipids (LDL), and high-density lipid (HDL) levels (Rajani *et al.*, 2009).

Antiobesity Activity

Antiobesity effect of methanol extract of stem and root barks of *Bauhinia variegata* was tested in female rats. The extract was administered at doses of 200 and 400 mg/kg to rats on hypercaloric diet for 40 days. Administration of extract at selected doses significantly reduced cholesterol, VLDL, LDL level and increased HDL level. Treatment of obese animals with extract showed significant ($p<0.01$) increase in brain serotonin level (Balamurugan and Muralidharan, 2010).

Anti-diabetic Activity

The antidiabetic activity of ethanol (95 per cent) bark extract of *Bauhinia variegata* Linn. was evaluated in alloxan induced diabetic rats. The various parameters like blood glucose, body weight, total cholesterol, triglycerides, high density lipoprotein, insulin, oral glucose tolerance test and histopathology of pancreas were studied. The results were found to be significant antidiabetic by reducing the blood glucose level, improving body weight, attenuating the altered lipid profile toward the normal and by regenerating the Islets of langerhans (Koti *et al.*, 2009).

The insulin-secreting cell line INS-1 was used to examine the effects of the crude ethanolic extract of leaves of *Bauhinia variegata* Linn. var. *candida* Voidt and its major metabolite (6S,7E,9R)-9-hydroxymegastigma-4,7-dien-3-one-9-β-glycopyraroside (roseoside) on insulinotropic activity The crude extracts and the major metabolite were shown to increase insulin secretion in a dose-dependent manner (Frankish *et al.*, 2010).

A chloroplast protein with a molecular mass similar to that of bovine insulin was extracted from 2-mm thick 15 per cent SDS-PAGE gels and fractionated with a 2 × 24 cm Sephadex G-50 column. The effect of the insulin-like protein on serum glucose levels of diabetic mice was determined in alloxan treated mice. The leaf extract and the insulin-like protein isolated from leaves of *B. variegata* reduced the glucose levels of the diabetic mice. The hypoglycemic activity of the insulin-like protein isolated from leaves was similar to that of commercial swine insulin used as control (Azevedo *et al.*, 2006). The total flavonoid content proved to be effective as hypoglycemic agent (Abd-El-Wahab *et al.*, 1987).

In vitro testing of the ethanol extract of *Bauhinia variegata* on insulin secretagogue activity was carried out. The extracts were tested at various concentrations (between 1 and 40 microg/mL) for insulin release from INS-1 cells in the presence of 5.5 mM glucose. *Bauhinia variegata* showed significant insulin secretagogue activity ($p < 0.05$) at 20 µg/mL (Hussain, 2004).

Ethanol extract of *Bauhinia variegata* stem bark was assessed for antidiabetic activity in alloxan induced diabetes in rats. The extract at doses of 200 and 400 mg/kg showed significant reduction in blood glucose levels after 7 days oral administration. The extract also showed significant reduction of glucose levels in oral glucose tolerance test (Kumar *et al.*, 2012).

Anti-oxidant Activity

The alcohol and water extract of the plant were tested for antioxidant activity by using a model system consisting of β-carotene and linoleic acid. Ethanol and water extract showed 91.5 and 90.3 per cent of antioxidant activity (Kaur and Kapoor, 2002).

In vitro antioxidant activity of ethanolic and aqueous extracts of the stem bark and root of *Bauhinia variegata* Linn. was assessed by various methods namely total reducing power, scavenging of various free radicals such as 1,2-diphenyl-2-picrylhydrazyl (DPPH), super oxide, nitric oxide, and hydrogen peroxide (Rajani *et al.*, 2009).

In vitro antioxidant activity of petroleum ether, benzene, chloroform, ethyl acetate, acetone, ethanol and water extracts of leaves of *B. variegata* was tested by DPPH radical scavenging assay and phosphomolybdate method. In DPPH assay, ethanol and acetone extracts (1000 µg/ml) exhibited about 88-89 per cent free radical scavenging activity while petroleum ether and aqueous extracts showed below 25 per cent inhibition. In phosphomolybdate method, the antioxidant capacity of ethanol, acetone and chloroform extracts (333 ppm) was 257.5, 163.6 and 251.7 µg propyl gallate equivalent per gram of sample of extract respectively. Aqueous extract showed lowest activity (5.7-55.5 µg PGE/g) at all test concentrations (Mishra *et al.*, 2011).

Antioxidant activity of 80 per cent methanol extract of bark and its fractions like hexane, chloroform n-butanol, ethyl acetate and remaining extract was tested against 2,2-diphenyl-1-picrylhydrazyl (DPPH) free radicals, reducing power assays and H_2O_2-induced oxidative damage to pBR322 DNA. The methanol extract, ethyl acetate fraction, n-butanol fraction and remaining extract showed greater antioxidant activity (EC_{50} 44.07, 58.09, 69.68 and 51.81 µg/ml, respectively) in comparison to non-polar fractions. All the fractions effectively protected pBR322 plasmid DNA from the H_2O_2 induced damage (Sharma *et al.*, 2011).

Antioxidant activity of methanol extract of leaf, stem bark and floral buds was estimated by Deoxyribose assay for OH^- radical scavenging activity. IC_{50} value of *Bauhinia variegata* leaf, stem bark and floral buds was reported 17.9, 19.5 and 17.2 ug/ml respectively.

Anti-inflammatory Activity

A flavanol glycoside, 5, 7, 3′, 4′-tetrahydroxy-3-methoxy-7-O-α-L-rhamnopyranosyl (1→3)-O-β-galactopyranoside, isolated from roots showed anti-inflammatory activity in carrageenin induced hind paw oedema. Administration of the test compound showed 62 per cent reduction in paw oedema (Yadava *et al.*, 2003).

Alcohol (70 per cent) extract of *Bauhinia variegata* L. leaves in carrageenan-induced paw oedema showed significant inhibition of inflammation (El-Dondaity *et al.*, 2005).

Kaempferol, ombuin, kaempferol 7,42-dimethyl ether 3-*O*-β -D-glucopyranoside, kaempferol 3-*O*-β -D-glucopyranoside, isorhamnetin 3-*O*-β-D-glucopyranoside, hesperidin and with one triterpene caffeate, 3β-trans-(3,4-dihydroxycinnamoyloxy) olean-12-en-28-oic acid isolated from the non- woody aerial parts of *Bauhinia variegata* were tested for their anti inflammatory effects. All compounds significantly and dose dependently inhibited lipopolysaccharide (LPS) and interferon (IFN)-γ induced nitric oxide (NO), TNF-α and IL-12 (Rao *et al.*, 2008).

A triterpene saponin, named as 23-hydroxy-3α-[O-alpha-L-1C4-rhamnopyranosyl-(1"→4')-O-α-L-4C1-arabinopyranosyl-oxy]olean-12-en-28-oic acid O-α-L-1C4-rhamnopyranosyl-(1""′→4"″)-O-β-D-4C1-glucopyranosyl-(1""→6")-O-β-D-4C1 glucopyranosyl ester, isolated from the leaves of *Bauhinia variegata* was tested for its anti-inflammatory. The compound at 40, 80, 100 and 200mg/kg dose indicated reduction in oedema by 21.3 per cent, 26.2 per cent, 39.6 per cent, and 28.3 per cent when compared to the control group (Mohamed *et al.*, 2009).

Anti-pyretic Activity

Ethanolic extract (70 per cent) of aerial parts of *Bauhinia variegata* at dose of 200 mg/kg exhibited significant anti-pyretic activities in Brewer's yeast induced pyrexia in rats. The antipyretic activity of extracts *of Bauhinia variegata* is probably by inhibition of prostaglandin synthesis in hypothalamus (Mandal *et al.*, 2011).

Analgesic Activity

Analgesic activity of aqueous and ethanolic extract (90 per cent) of roots was evaluated by using Eddy's hot plate method and acetic acid induced writhing. Both aqueous and ethanolic extract at dose levels of 200 and 400 mg/kg showed dose depended increase in latency of response in hot plate method. In acetic acid induced writhings, aqueous and ethanolic extract at 400 mg/kg showed maximum inhibition 81 per cent and 77 per cent respectively. Both extracts at selected dose levels showed significant analgesic activity ($p< 0.01$) in hot plate test (Kumar *et al.*, 2011).

A triterpene saponin, named as 23-hydroxy-3α-[O-α-L-1C4-rhamnopyranosyl-(1''→4')-O-α-L-4C1-arabinopyranosyl-oxy]olean-12-en-28-oic acid O-α-L-1C4-rhamnopyranosyl-(1'''''→4'''')-O-β-D-4C1-glucopyranosyl-(1''''→6''')-O-β-D-4C1 glucopyranosyl ester, isolated from the leaves of *Bauhinia variegata* showed significant anti-antinociceptive activity. The compound was tested for its antinociceptive activity in hot plate and acetic acid induced writhings test. The compound at a dose of 200 mg/kg showed 54 per cent inhibition in acetic acid induced writhings test and at the same dose it showed 47 per cent potency in hot plate test (Mohamed *et al.*, 2009).

Anti-microbial Activity

The anti-microbial activity of ethanol extract, total seed proteins, globulins, albumins, fractions of albumins and lectins of *Bauhinia variegata* seeds were investigated. The activity was tested against *Mycobacterium rhodochrous, Bacillus circus, Bacillus megaterium, Bacillus sphaericus, Escherichia coli, Serratia marcescens, Cornybacterium xerosis* and *Staphylococcus aureus.* The lectins and one of the albumin fractions showed antimicrobial activity. Ethanol extract and globulins showed no antimicrobial activity (Sammour and El-Shanshoury, 1992).

The methanol extract of leaves of *Bauhinia variegata* has been tested for antimicrobial activity against *Bacillus anthracis, Proteus vulgaris, Escherichia coli* and *Streptococcus agalacties* and antifungal activity against various fungi, *Aspergillus fumigatus* and *Aspergillus niger* (Sharma and Saxena, 1996).

Antibacterial and antifungal activities of hexane and methanol extracts (400 µg/ml) of air-dried plant material were determined by photometric microtiter broth dilution protocol. Methanol extract showed significant activity against *Salmonella typhi, Pseudomonas aeruginosa, Shigella sonii* and *Staphylococcus pyogenes* with percentage inhibition of 91.76, 98.83, 98.22 and 87.60 respectively. The hexane extract showed no activity except against *Corynebacterium diptheriae* (33.09 per cent), *Pseudomonas aeruginosa* (48.44 per cent) and *Staphylococcus aureus* (2.51 per cent). In antifungal studies methanol extract showed maximum activity against *Trichophyton schoenleinii* with percentage inhibition of 99.84. The hexane extract failed to show any antifungal activity against *Trichophyton mentagrophytes, Trichophyton simii, Fusarium solani, Macrophomina phaseolina, Rhizoctonia solani* and *Trichophyton longifuses* (Ali *et al.*, 1999).

The alcoholic extract of bark of *Bauhinia variegata* was reported for its antimicrobial activity against *Bacillus subtilis, Pseudomonas aeruginosa, Salmonella typhi, Shigella dysenteriae, Staphylococcus aureus* and *Vibrio cholera.* The maximum zone of inhibition was 18 and 15 mm against *Bacillus subtilis* and *Staphylococcus aureus* respectively. The minimum bactericidal concentration was found to be 0.39 mg/ml against *Bacillus subtilis, Shigella dysenteriae* and *Staphylococcus aureus* (Pokhrel *et al.*, 2002).

The alcohol extract of the plant exhibited significant activity against *Staphylococcus aureus* (Jain *et al.*, 2004).

Antimicrobial screening showed that volatile fraction of the studied plant organs has promising antimicrobial activity against a no. of microorganisms (el-Hamouly and Salama, 2004).

2'-hydroxy-4',6'-dimethoxy-3,4-methylenedioxy chalcone, kaempferol-3-O-D glucopyranoside and kaempferol-3-O-L-rhamnoside isolated from the leaves *Bauhinia variegata.* Showed significant antifungal activity against *Aspergillus niger* and *Candida albacans* (Maheswara *et al.*, 2005).

The antibacterial activity of petroleum ether, 1,4-dioxan, acetone, acetone, methanol, methanol, DMF, distilled water extracts of *Bauhinia variegata* L. bark was determined by agar well diffusion method at three different concentrations *i.e.*, 10 mg/ml, 5 mg/ml and 2.5 mg/ml. The antibacterial activity of defatted extracts of *Bauhinia variegata* L. was more than those without defatting. Maximum activity was observed at highest concentration *i.e.*, 10 mg/ml. Defatted acetone and methanol extracts of *Bauhinia variegata* L. were most active as compared to other extracts against *P. pseudoalcaligenes, E. coli* and *B. cereus* organisms. Petroleum ether extracts of *Bauhinia variegata* L. was inactive against *P. pseudoalcaligenes, E. coli* and *B. cereus* organisms (Parekh *et al.*, 2006).

The methanol extract of the bark of *Bauhinia variegata* has antibacterial activity against *B. cereus, S. epidarmidis* and *P. vulgaris* but not against *E. aerogenes* and *S. typhimurium.* Aqueous extract has not shown antibacterial activity against *B. cereus, S. epidarmidis, P. vulgaris, E. aerogenes* and *S. typhimurium* (Parekh *et al.*, 2007).

Dimeric 64-kDa melibiose-binding lectin isolated from the seeds of *Bauhinia variegata* was tested for its antifungal activity against fungal species *Botrytis cinerea, Mycosphaerella archidicola* and *Fusarium oxysporum.* The lectin showed no activity against selected fungal species (Lin *et al.*, 2008).

Petroleum ether, chloroform, methanol and aqueous extracts of leaves of *Bauhinia variegata* were tested for their anti-bacterial activity against gram positive and gram negative bacteria. The aqueous extract has antibacterial activity against *Escherichia coli, Bacillus subtilis, Pseudomonas aeruginosa* and *Staphylococcus aureus* while methanol extract has this activity against *Pseudomonas aeruginosa* and *Staphylococcus aureus.* The MIC for aquous extract against *Staphylococcus aureus, Escherichia coli, Bacillus subtilis,* and *Pseudomonas aeruginosa* was 10 µg/mL, 10 µg/mL, 50 µg/mL and 50 µg/mL, respectively. The MIC for methanol ext. was 500 µg/mL (*Staphylococcus aureus*) and 50 µg/mL (*Pseudomonas aeruginosa*), for petroleum ether ext. was 100 µg/mL (*Staphylococcus aureus*) and 250 µg/mL (*Escherichia coli*), for chloroform ext. was 250 µg/mL (*Escherichia coli*) and 500 µg/mL (*Staphylococcus aureus*) (Gupta *et al.*, 2009b).

The antibacterial activity of aquous, ethanol and acetone extracts of *Bauhinia variegata* was determined against 33 UTI isolates *i.e., Proteus mirabilis* (10), *Escherichia coli* (6), *Proteus vulgaris* (6), *Klebsiella pneumoniae* (5), *Enterobacter cloacae* (2), *Providencia pseudomallei* (2), *Pseudomonas aeruginosa* (1) and *Klebsiella oxytoca* (1) by disk diffusion method. The acetone and ethanol extracts exhibited significant activity against selected UTI pathogens (Sharma *et al.*, 2009).

Antimicrobial activity of the ethanol extract of *Bauhinia variegata* leaves was tested at 100, 250, 500 1000 µg/ml against gram positive and gram negative bacteria. The extract showed maximum inhibitory activity against *Salmonella typhi* (27mm) followed by *Vibrio cholera, K. pneumonia, E. coli* and *S. aureus.* The least activity was observed for *S. aureus* (18mm). Antifungal activity of the extract was evaluated at 100 μg/ml against *Trychophyton mentagrophytes,Trychophyton rubrum, Aspergillus niger, Fusarium oxysporum* and *Mucor hiemalis.* Ethanol extract was very effective against *T. mentagrophytes* and *Mucor hiemalis* (Gunalan *et al.,* 2011).

Anti-goitrogenic Activity

The effects of *Bauhinia variegata* were studied on rats with goiter induced by neomercazole and found to be effective in bringing the goitrogenic thyroid to normal level at a dose of 200 mg/day (Veena *et al.,* 1975).

Immunomodulatory Activity

The ethanol extract of stem bark of variegata was tested for its immunomodulatory activity in carbon clearance test and neutrophil adhesion test. The extract at 250 and 500 mg/kg showed significant increase in primary and secondary humoral antibody response. The phagocytic index and percentage neutrophil adhesion was significantly increased after administration of ethanol extract of bark (Ghaisas *et al.,* 2009). The acetone-water, aqueous extracts and isolated compound (tannin) of *Bauhinia variegata* stem bark were tested for their possible immunomodulatory activity by using nitroblue tetrazolium test, phagocytosis of killed *Candida albicans,* candidacidal assay, neutrophil locomotion and chemotaxis. The acetone: water and isolated compound of *Bauhinia variegata* Linn. stem bark showed significant activity on *in vitro* human neutrophils in all parameters (Patil *et al.,* 2010).

Anthelmintic Activity

The effect of aqueous and chloroform extracts of bark of *Bauhinia variegata* were tested for their anthelmintic activity against *Pheretima posthuma* and *Ascardia galli.* The selected extracts exhibited a dose dependent (25, 50 and 100 mg/ml) inhibition of spontaneous motility (paralysis) and time of death of the worms. The extracts showed maximum effect at the concentration of 100 mg/ml (Bairagi *et al.,* 2012).

Anti-eosinophilic Activity

Aqueous and ethanolic extract of *Bauhinia variegata* were active against milk induced leucocytosis and eosinophilia in albino mice. The result of the study revealed that pretreatment with both the extracts caused significant reduction in total leucocyte and eosinophil counts in animals in dose dependent manner. From this result, it can be concluded that the plant *Bauhinia variegata* having antieosinophilic activity (Mali and Dhake, 2011).

Mast Cell Stabilizing Activity

The effect of *Bauhinia variegata* (Caesalpiniaceae) stem bark extracts on mast cell stabilizing activity was investigated to explore the scientific basis for the use of the decoction of its stem bark in the treatment of asthma and other respiratory complications in traditional Ayurvedic system of medicine. The authenticated plant material was defatted using petroleum ether and extracted with ethanol and distilled water to obtain ethanol and aqueous extract, respectively. The stabilization potential was studied on mouse mast cells and anti-anaphylactic activity was evaluated by determining the mortality rate of mice treated with toxicant (compound 48/80). The ethanol extract at dose of 400 mg.kg^{-1} significantly protected mast cell from degranulation (71.18 per cent) and reduced the mortality (50 per cent) when compared to standard drug disodium cromoglycate (DSCG) used in the study. Ethanol extract possessed potent mast cell stabilizing activity, suggesting potential for use in the treatment of asthma (Mali and Dhake, 2011).

Anti-Ulcer Activity

Alcohol extract of stem of *Bauhinia variegata* was tested in ulcers induced by pyloric ligation and aspirin treatment in rats. In pylorus ligation model, the alcohol extract (250 mg/kg) showed significant reduction ($p<0.01$) in the total volume of gastric juice, free and total acidity of gastric secretion in rats. The extract treatment also reduced the ulcer score (1.6 ± 0.02) when compared with control rats (3.1 ± 0.33), in aspirin induced ulcer model (Rajkapoor *et al.*, 2003c). The anti-ulcer activity of aqueous and ethanol extract (90 per cent) of roots was evaluated by using pylorus ligation, ethanol and aspirin induced ulcer models. Both extracts at dose of 200 and 400 mg/kg showed 99 per cent protection in ethanol induced ulcer model. In aspirin induced ulcer model, aqueous and ethanol extract at dose of 400 mg/kg showed 83 and 99.9 per cent protection. Aqueous and ethanol extract at the dose of 400 mg/kg showed 99 per cent protection in pylorus ligation ulcer model (Kumar *et al.*, 2011).

Anticataract Activity

Rhamnocitrin isolated from stem bark of *Bauhinia variegata* was tested for anticataract activity in ovine and chick embryo lens model. Rhamnocitrin showed significant anticataract activity by protection against cloudiness in lenses induced by H_2O_2 and hydrocortisone (Bodakhe *et al.*, 2012).

Anti-N Phytagglutinin Activity

Bauhinia variegata L. was found to have a promising anti-N-agglutinin activity (Schertz *et al.*, 1958). Seeds of *Bauhinia variegata* was homogenized and the seeds provided anti-N-agglutinin activity. Addition of glucose or melibiose inhibited nonspecific agglutination without altering the specific activity, but human serum and poly(vinylpyrrolidinone) enhanced both specific and nonspecific agglutination (Fletcher, 1959).

Antimutagenicity Activity

Methanol extract of stem bark of *Bauhinia variegata* was evaluated for its antimutagenicity effect in Swiss mice. Preventive effect of extract was tested against cyclophosphamide induced micronucleus formation in the mouse bone marrow cells. The extract at the dose of 125, 250 and 375 mg/kg body weight, 24 hours prior the administration of cyclophosphamide (50 mg/kg) have significantly prevented the micronucleus formations in dose dependent manner in bone marrow cells of mice as compared to only cyclophosphamide treated group. However, methanol extract alone has not induced micronucleus formations in bone marrow cells as compared to control group. Thus methanol extract have a preventive potential against cyclophosphamide -induced micronucleus formation in *Swiss* mouse bone marrow cells (Pandey and Agrawal, 2010b).

Haemagglutinating Activity

The saline extract of seed exhibited haemagglutination activity against erythrocytes of man, monkey, rabbit, rat, goat, sheep, cow, buffalo, horse, mule and Fowl (Roy *et al.*, 1981). The *Bauhinia variegata* showed significant haemagluttinating activity (Wassel *et al.*, 1989).

Dimeric 64-kDa melibiose-binding lectin isolated from the seeds of *Bauhinia variegata* showed hemagglutinating activity which was stable after 30-min exposure to temperatures up to 70°C (Lin *et al.*, 2008).

HIV-1 Reverse Transcriptase Activity

Kunitz-type trypsin inhibitor (termed BvvTI) from seeds showed significant anti HIV-1 reverse transcriptase activity (Fang *et al.*, 2010). HIV-1 reverse transcriptase activity was inhibited by dimeric 64-kDa melibiose-binding lectin isolated from the seeds of *Bauhinia variegata* with an IC_{50} of 1.02 μM (Lin *et al.*, 2008).

Wound Healing Activity

Effect lectin isolated from *Bauhinia variegata* was studied in incision wound model in mice. Various parameters like wound size, contraction rate, epithelialization rate and histopathological findings were assessed during the study. Topical administration of lectin for 12 days showed faster wound closure in animals as compared to control mice. The lectin showed effect possibly by acting on cells of the immune system, enhancing proinflammatory response, collagen synthesis by fibroblasts and angiogenesis by modulating the release of inflammatory cytokines and growth factors (Neto *et al.*, 2011).

Analysis

Analysis of Leaves

The leaves of *Bauhinia variegata* were analysed for organic matter, crude proteins, ash, calcium, phosphorus. The cell wall and tannin content was also determined. The leaves showed 89.8 and 16.1 per cent organic matter and crude proteins respectively. The ash content was found to be 10.3 per cent. The percentage of leaves calcium, phosphorus was found to be 2 and 0.4 respectively. The cell wall showed 19.6 per cent of hemicelluloses, 18.5 per cent of cellulose and 12.7 per cent of lignin. The leaves showed 7.5 per cent of tannins (Sharma and Sharma, 2006).

Two statistical mixtures designs were used to optimize the proportions of solvents used in both the extraction medium and the reversed liquid chromatography mobile phase to improve the quality of chromatography fingerprints of *Bauhinia variegata* L extracts. For modeling, the number of peaks was used as a measure of fingerprint information. Three mobile phases, each with chromatography strength of 2, gave good results. A methanol/water (77: 23 vol./vol.) mixture resulted in 17 peaks in the chromatography fingerprint whereas acetonitrile/water (64.5: 35.5 vol./vol.) and methanol/acetonitrile/water (35: 35: 30 vol./vol./v) mixts. resulted in 18 and 20 peaks, resp. The corresponding optimum solvent compns. to ext. chem. substances for these 3 mobile phases were ethanol/acetone (25: 75 vol./vol./v) and dichloromethane/acetone (70: 30 vol./vol.) mixtures, and pure dichloromethane, respectively (Delaroza and Scarminio, 2008).

Analysis of Seeds

Seeds of *Bauhinia* were studied by scanning electron microscopy. Presence of stomata was observed which are mostly normal in appearance but lack subsidiary cells (Rugenstein and Lersten, 1981).

Determination of Metal Elements

Stem bark of *Bauhinia variegata* was powdered and analyzed by inductively coupled plasma for five common heavy metals As, Cr, Mg, Pb and Zn. The concentration of magnesium was highest (30800 ppm) as compared to other ions (Bhandarkar *et al.*, 2008).

An easy method for the determination of metal elements in *Bauhinia variegata* L. is reported. The samples were digested by microwave digestion and subjected to atomic absorption spectrometry for determination of Ca, Mg, Cu, Mn, Fe, and Zn. The detection limits were found to be below 0.0081 μg/mL and the relative standard deviations were less than 4.19 per cent (Zhang *et al.*, 2010).

Determination of Total Phenolics and Total Flavonoids

The total phenolics in the plant of *Bauhinia variegata* was determined by Folin- Ciocalteau method. The ethanol extract (80 per cent) showed 275.0 ± 10.1 [mg (100 g^{-1})] of total phenolics (Kaur and Kapoor, 2002).

Total phenolics and total flavonoids of 80 per cent methanol extract of bark and its fractions like hexane, chloroform n-butanol, ethyl acetate and remaining extract were determined. Methanol extract showed highest content of total phenolics (440.0 ± 3.02 mg gallic acid equivalent/g of extract) and total flavonoids (937.5 ± 5.37 mg rutin equivalent/g of extract). The ethyl acetate fraction of methanol extract showed maximum concentration of total phenolics and total flavonoids as compared to other fractions (Sharma *et al.*, 2011).

High Performance Thin Layer Chromatography (HPTLC) Method

A simple, sensitive, selective, precise, and robust HPTLC method was developed and validated for determination of flavonoids in *Bauhinia variegata*. The HPTLC of flavonoids was carried out by using RP-18 F254 TLC plates with dual run, water (5 per cent formic acid)/methanol (70: 30) and water (5 per cent formic acid)/methanol (50: 50) as mobile phases. Densitometric scanning of flavonoids was performed at λ_{max} 280 nm in reflectance/absorbance mode. The the concentration range of 150-800 ng/spot for apigenin and rutin and 200-1000 ng/spot for quercetin, luteolin, and quercitrin showed good results. The average recovery for apigenin, quercetin, rutin, luteolin, and quercitrin was 97-99.8 per cent indicating the excellent reproducibility (Bhandari *et al.*, 2007).

A HPTLC method is reported to standardize the bark of *Bauhinia variegata* Linn. Lupeol was used as marker in this method. The methanol extract of bark was used for quantification. The HPTLC of the extract was performed by using solvent system toluene-acetone-acetonitrile, 9: 0.8: 0.2 v/v. The R_f of lupeol was 0.58. The densitometric scanning was performed at λ_{max} of 520 nm in normal mode for quantitation. The response to lupeol was a linear function of concentration over the range 5 to 35µg mL^{-1}. The concentration of lupeol in bark was found to be 0.10 mg g^{-1}. The method was validated for system suitability, linearity, precision and accuracy (Kekare, 2008).

Total RNA Extraction Method

An extraction method for the total RNA of *Bauhinia variegata* was developed that could efficiently eliminate the interference of polyphenols and polysaccharides. The results of 1 per cent agarose gel electrophoresis and UV spectrophotometer analysis showed that the total RNA of *Bauhinia variegata* extended through this method had a good purity without obvious degradation (Haiwu *et al.*, 2006).

Composition and Processing

Freshly harvested samples of *Bauhinia* were analyzed for moisture, protein, ascorbic acid and ash before and after dehydration and during 270-day storage of the dehydrated product prepared by steam blanching, dipping in 1 per cent Na_2SO_3, and drying. The white variety had a higher drying rate, lower final moisture content and higher rehydration ratio than purple Kachnar. Subjective evaluation of colour, flavor and appearance confirmed the superiority of the white over the purple variety. Though the ascorbic acid and SO_2 contents and rehydration ratio gradually decreased during 270 days of storage of the dehydrated samples, the product remained acceptable until the end of the experiment (Shah *et al.*, 1977).

Conclusion

The available data indicated that different parts of *Bauhinia variegata* like bark, leaves, roots and flowers are being used in traditional system of medicine. The survey revealed that there is no systematic information on identification of all parts and standardization by using well accepted techniques like HPLC and HPTLC. The leaves and seeds of the plant has been exhaustively worked for its phytochemistry, but there are few reports on phytochemical studies of other parts of this important plant. The important constituents found in the plant are flavonoids and flavonoid glycosides. The pharmacological studies conducted on *Bauhinia variegata* indicate the enormous potential of this plant. Various studies related to its anticancer activity has been reported which give scientific basis for its use as per traditional system of medicine. The plant has shown promising effects in conditions such as diabetes, wounds, ulcer and hypelipidemia. It has shown antioxidant and anti-inflammatory effects as oxidative injury underlies many of various diseases. However, the varied pharmacological effects of the extracts and few isolated phytoconstituents have only been assayed by *in vivo* tests. Thus there is unmet need for clinical studies, as the results in the preclinical studies may not necessarily be transferable to the situation in humans. The mechanistic approach for these reported pharmacological activities is lacking and need to be addressed with systematic studies. Further, meticulous research, in relation with formulation development can only strengthen *Bauhinia variegata* to sustain its importance current therapeutic practice.

References

Abd-El-Wahab, S.M., Wassel, G.M., Ammar, N.M., and Hanna, T. (1987). Flavonoid constituents in the different organs of selected *Bauhinia* species and their effect on blood glucose. *Herba Hungarica,* **26(1):** 27-39.

Agrawal, R.C., and Pandey, S. (2009). Evaluation of anticarcinogenic and antimutagenic potential of *Bauhinia variegata* extract in Swiss albino mice. *Asian Pacific Journal of Cancer Prevention,* **10(5):** 913-916.

Ali, M.S., Azhar, I., Amtul, Z., Ahmad, V.U., and Usmanghani, K. (1999). Antimicrobial screening of some caesalpiniaceae. *Fitoterapia,* **70:** 299-304.

Anonymous. (1988). The Wealth of India. A Dictionary of Indian Raw Materials and Industrial Products. CSIR, New Delhi.

Arafa, R. (2001). Naringenin as an antidermatophyte flavonoid aglycone from *Bouhinia variegata. Al-Azhar Journal of Microbiology,* **51**: 17-23.

Azevedo, C.R., Maciel, F.M., Silva, L.B., Ferreira, A.T., da Cunha, M., Machado, O.L., Fernandes, K.V., Oliveir, A.E., and Xavier-Filho, J. (2006). Isolation and intracellular localization of insulin-like proteins from leaves of *Bauhinia variegata. Brazilian Journal of Medical and Biological Research,* **39(11):** 1435-1444.

Badami, R.C., and Shanbhag, M.R. (1973). Component acids of six seed oils. *Journal of the Oil Technologists' Association of India,* **5(4):** 68-70.

Bairagi, S.M., Aher, A.A., and Nimase, P.K. (2012). *In vitro* anthelmintic activity of *Bauhinia variegata* bark (leguminosae). *International Journal of Pharmacy and Pharmaceutical Sciences,* **4(3):** 672-674.

Balamurugan, G., and Muralidharan, P. (2010). Antiobesity effect of *Bauhinia variegata* bark extract on female rats fed on hypercaloric diet. *Bangladesh Journal of Pharmacology,* **5:** 8-12.

Bhandari, P., Kumar, N., Gupta, A.P., Singh, B., and Kaul, V.K. (2007). A rapid RP-HPTLC densitometry method for simultaneous determination of major flavonoids in important medicinal plants. *Journal of Separation Science,* **30(13):** 2092-2096.

Bhandarkar, N. K., Kekare, M. B., Champanerkar, P., and Vaidya, V.V. (2008). Determination of heavy metals from *Bauhinia variegata* using inductively coupled plasma technique. *Nature, Environment and Pollution Technology,* **7(3):** 569-570.

Bodakhe, S.H., Alpana, Ram., and Verma, S. (2012). Anticataract activity rhamnocitrin isolated from *Bauhinia variegata* stem bark. *Oriental Pharmacy and Experimental Medicine,* **12(3):** 227-232

Bodakhe, S.H., Alpana, Ram, Bodakhe, K.S., Pandey, and Devi P. (2010). New polyphenolic aromatic glycoside from *Bauhinia variegata* L. stem bark. *Asian Journal of Chemistry,* **22(5):** 3549-3553.

Bodakhe, S.H., and Alpana, Ram. (2007). Hepatoprotective Properties of *Bauhinia variegata* bark extract. *Yakugaku Zasshi,* **127(9):** 1503- 1507.

Delaroza, F., and Scarminio, I.S. (2008). Mixture design optimization of extraction and mobile phase media for fingerprint analysis of *Bauhinia variegata* L. *Journal of Separation Science*, **31(6-7):** 1034-1041.

Dhan, P., and Misra, P.S. (1983). Amino acid composition of *Bauhinia* and *Cassia* seeds. *Fitoterapia*, **54(6):** 257-60.

Dhan, P., Niranjan, A., Tewari, S.K., and Pushpangadan, P. (2001). Underutilised legumes: Potential sources for low-cost protein. *International Journal of Food Sciences and Nutrition*, **52(4):** 337-341.

Di Ciero, L., Oliva, M.L., Torquato, R., Kohler, P., Weder, J.K., Camillo Novello J., Sampaio, C.A., Oliveira, B., and Marangoni, S. (1998). The complete amino acid sequence of a trypsin inhibitor from *Bauhinia variegata* var. *candida* seeds. *Journal of Protein Chemistry*, **17(8):** 827-834.

Duarte-Almeida, Joaquim M., Negri, Giuseppina and Salatino, Antonio. (2004). Volatile oils in leaves of *Bauhinia* (Fabaceae Caesalpinioideae). *Biochemical Systematics and Ecology*, **32(8):** 747-753.

Duret, S., and Paris, R.R. (1977). Plants of Nepal. V. The flavonoids of various *Bauhinia: Bauhinia valhii* Wight et Arn., *B. variegata* Lindl. and *B. malabarica* Roxb., Luguminosae-Caesalpinia. *Plantes Medicinales et Phytotherapie*, **11(3):** 213-221.

El-Dondaity, S.E., Mahdy, M., El-Hamouly, M., and Ammar, H. (2005). Chemical and/or biological studies of *Bauhinia variegata* L. and *Cleome droserifolia* (Forssk.) Del. growing in Egypt. *Egyptian Journal of Biomedical Sciences*, **19:** 181-203.

El-Hamouly, M.M.A., and Salama, A.T. (2004). Anatomical and phytochemical investigation of flowers and/or fruits of some ornamental plants growing in Egypt. *Egyptian Journal of Biotechnology*, **16:** 162-176.

Fang, Evandro F., Wong, Jack H., Bah, Clara Shui Fern., Lin, Peng., Tsao, Sai Wah., and Ng, Tzi Bun. (2010). *Bauhinia variegata* var. *variegata* trypsin inhibitor: From isolation to potential medicinal applications. *Biochemical and Biophysical Research Communications*, **396(4):** 806-811.

Fletcher, G. (1959). The anti-N phytagglutinin of *Bauhinia variegata*. *Australian Journal of Science*, **22**: 167-168.

Frankish, N., de Sousa Menezes, F., Mills, C., and Sheridan, H. (2010). Enhancement of insulin release from the β-cell line INS-1 by an ethanolic extract of *Bauhinia variegata* and its major constituent roseoside. *Planta Medica*, **76(10):** 995-997.

Ghaisas, M.M., Shaikh, S.A., and Deshpande A.D. (2009). Evaluation of immunomodulatory activity of ethanolic extract of the stem bark of *Bauhinia variegata* Linn. *International Journal of Green Pharmacy*, 70-74.

Gunalan, G., Saraswathy, A., and Krishnamurthy, V. (2011). Antimicrobial activity of medicinal plant *Bauhinia variegata* Linn. *International Journal of Pharmacy and Biological Sciences*, **1(4):** 400-408.

Gupta A.K, Vidyapati, T.J., and Chauhan J.S. (1980a). Chemical examination of the stem of *Bauhinia variegata*. *Planta Medica*, **38(2):** 174-176.

Gupta, A.K., and Chauhan, J.S. (1984). Constituents from the stem of *Bauhinia variegata. National Academy Science Letters (India)*, **7(1):** 15-16.

Gupta, A.K., Vidyapati, T.J., and Chauhan, J.S. (1979). 5, 7-dihydroxyflavanone-4'O-α-L-rhamnopyranosyl-β-D-glucopyranoside from the stem of *Bauhinia variegata. Indian Journal of Chemistry*, **18B(1):** 85-86.

Gupta, R., and Paarakh, P.M. (2009a). Antibacterial activity of different extracts of the plant *Bauhinia variegata. International Journal of Chemical Sciences*, **7(4):** 2377-2383.

Gupta, R., Paarakh, P.M., and Gavani, U. (2009). Pharmacognostical and phytochemical screening of *Bauhinia variegata* Linn leaves. *Journal of Pharmacy Research*, **2(7):** 1196-1198.

Haiwu, X., Liuxin L., and Guixin C. (2006). New method for total RNA extraction in legume of *Bauhinia variegata. Fenzi Zhiwu Yuzhong*, **4(1):** 147-149.

Hussain, Z., Waheed, A., Aleem, Q.R., Burdi, D.K., Verspohl, E.J., Khan, N., and Hasan, M. (2004). The effect of medicinal plants of Islamabad and Murree region of Pakistan on insulin secretion from INS-1 Cells. *Phytotherapy Research*, **18:** 73-77.

Jain, R., Nagpal, S., Jain, S., and Jain, S.C. (2004). Chemical and biological evaluation of *Bauhinia* species. *Journal of Medicinal and Aromatic Plant Sciences*, **26(1):** 48-50.

Kaur, C., and Kapoor, H.C. (2002). Anti-oxidant activity and total phenolic content of some Asian vegetables. *International Journal of Food Science and Technology*, **37**: 153-161.

Kekare, M.B., Bhandarkar, N., Vaidya, V. V., and Charegaonkar, G. (2008). High performance thin layer chromatographic method for the quantification of lupeol in bark powder of *Bauhinia variegata* linn. *Analytical Chemistry: An Indian Journal*, **7(8):** 606-608.

Kirtikar, K.R., and Basu, B.D. (1999). Indian Medicinal Plants. International book distributors, Dehradun.

Koti, B.C., Biradar S.M., Karadi R.V., Taranalli A.D., and Benade V.S. (2009). Effect of *Bauhinia variegata* bark extract on blood glucose level in normal and alloxanised diabetic rats. *Journal of Natural Remedies*, **9(1):** 27-34.

Kumar, D., Parch, V., Dhulia, I., and Maithani, A. (2011). Evaluation of Anti-hyperlipidemic activity of methanol extract of *Bauhinia variegata* (Linn.) leaves in triton WR-1339 (tyloxapol) induced hyperlipidemic rats. *Journal of Pharmacy Research*, **4(1):** 58-60.

Kumar, P., Baraiya S., Gaidhani, S.N., Gupta, M.D., and Wanjari, M.M. (2012). Antidiabetic activity of stem bark of *Bauhinia variegata* in alloxan-induced hyperglycemic rats. *Journal of Pharmacology and Pharmacotherapeutics*, **3(1):** 64-66.

Kumar, Y.R., and Rajani, G.P. (2011). Analgesic and anti-ulcer activities of ethanol and aqueous extracts of root of *Bauhinia variegata* Linn. *International Journal of Pharmacology*, **7(5):** 616-622.

Lin, Peng., and Ng Tzi, Bun. (2008). Preparation and biological properties of a melibiose binding lectin from *Bauhinia variegata* seeds. *Journal of Agricultural and Food Chemistry*, **56(22):** 10481-10486.

Maheswara, M., Rao, Y.K., Siddaiah, V., and Rao, C.V. (2005). Isolation of new chalcone from the leaves of *Bauhinia variegata*. *Asian Journal of Chemistry*, **18(1):** 419-422.

Mali, R.G., and Dhake, A.S. (2011a). Evaluation of effects of *Bauhinia variegate* stem bark extracts against milk-induced eosinophilia in mice. *Journal of Advanced Phamaceutical Technology and Research*, **2(2):** 132-134.

Mali, R.G., and Dhake, A.S. (2011b). Mast cell stabilizing activity of *Bauhinia variegata*. *Journal of Herbs, Spices and Medicinal Plants*, **17(3):** 268-274.

Mandal, S., Upadhyay, N., Sharma, I., Rohit, S., and Mandloi, A. (2011). A comparative antipyretic activity of the crude extracts of the ariel parts of *Glycosmis pentaphylla* and *Bauhinia variegata*. *Recent Research in Science and Technology*, **3(7):** 16-18.

Mishra, A., Kumar, S., Bhargava, A., Sharma, B., and Pandey, A.K. (2011). Studies on *in vitro* antioxidant and antistaphylococcal activities of some important medicinal plants. *Cellular and Molecular Biology (Noisy-le-grand)*, **57(1):** 16-25.

Mohamed, M.A., Mammoud, M.R., and Hayen, H. (2009). Evaluation of antinociceptive and anti-inflammatory activities of a new triterpene saponin from *Bauhinia variegata* leaves. *Zeitschrift fuer Naturforschung, C: Journal of Biosciences*, **64(11-12):** 798-808.

Mukherjee, D., and Laloraya, M.M. (1975). Keto acids and amino acids in the floral buds of *Bauhinia variegata*. *Journal of the Indian Botanical Society*, **54(3-4):** 207-10.

Nadkarni, K.M. (2000). Indian Materia Medica. Popular prakashan, Mumbai.

Neto, L.G.N., de Souza, P.L., Bastos, R.M., Evaristo, F.F.V., Vasconcelos, M.A., Carneiro, V.A., Arruda, F.V.S., Porto, A.L.F., Leal, R.B., Junior, V.A.dS., Cavada, B.S., and Teixeira, E.H. (2011). Effect of the lectin of *Bauhinia variegata* and its recombinant isoform on surgically induced skin wounds in a murine model. *Molecule*, **16**: 9298-9315.

Niranjan, G.S., Agrawal, R., and Srivastava, A. (1985). Chemical examination of some wild leguminous seeds. *Journal of the Indian Chemical Society*, **62(9):** 693-694.

Pandey, S., Agrawal, R.C., and Maheshwari S. (2012). *In vitro* antioxidant and free radical scavenging activity of *Bauhinia variegata* Linn. *International Journal of Scientific and Research Publications*, **2(2):** 1-5.

Pandey, S., and Agrawal, R.C. (2009). Effect of *Bauhinia variegata* bark extract on DMBA-induced mouse skin carcinogenesis: A preliminary study. *Global Journal of Pharmacology*, **3 (3):** 158-162.

Pandey, S., and Agrawal, R.C. (2010a). Chemopreventive potential of *Bauhinia variegata* flower extract against DMBA-induced skin papillomagenesis in mice. *Pharmacologyonline*, **1:** 39-46.

Pandey, S., and Agrawal, R.C. (2010b). Clastogenic analysis of *Bauhinia variegata* bark extract using micronucleus assay in mouse bone marrow cells. *American-Eurasian Journal of Toxicological Sciences*, **2(2):** 112-114.

Pani, S.R., Mishra, S., Sahoo, S., and Panda, P.K. (2011). Nephroprotective effect of *Bauhinia variegata* (Linn.) whole stem extract against cisplatin-induced nephropathy in rats. *Indian Journal of Pharmacology*, **43(2):** 200-202.

Parekh, J., and Chanda, S. (2007). Antibacterial and phytochemical studies on twelve species of Indian medicinal plants. *African Journal of Biomedical Research*, **10:** 175-181.

Parekh, J., Karathia, N., and Chanda, S. (2006). Evaluation of antibacterial activity and phytochemical analysis of *Bauhinia variegata* L. bark. *African Journal of Biomedical Research*, **9(1):** 53-56.

Patil, J. K., Jalalpure, S. S., Hamid, S., and Ahirrao, R. A. (2010). *In-vitro* immunomodulatory activity of extracts of *Bauhinia vareigata* linn stem bark on human neutrophils. *Iranian Journal of Pharmacology and Therapeutics*, **9(2):** 41-46.

Pinto, L.S., Nagano, C.S., Oliveira, T.M., Moura, T.R., Sampaio, A.H., Debray, H., Pinto, V.P., Dellagostin, O.A., and Cavada, B.S. (2008). Purification and molecular cloning of a new galactose-specific lectin from *Bauhinia variegata* seeds. *Journal of Biosciences*,**33(3):** 355-363.

Pokhrel, N.R., Adhikari, R.P., and Baral, M.P. (2002). *In-vitro* evaluation of the antimicrobial activity of *Bauhinia variegata* locally known as koiralo. *World Journal of Microbiology and Biotechnology*, **18(1):** 69-71.

Puntambekar, S.V., and Krishna, S. (1940). Fatty oil from the seeds of *Bauhinia variegata* Linn. *Journal of the Indian Chemical Society*, **17:** 96-100.

Rahman, W., and Begum, Sardar Jehan. (1966). Flower pigments: flavonoids from the white flowers of *Bauhinia variegata*. *Naturwissenschaften*, **53(15):** 385.

Rajani, G.P., and Ashok, P. (2009). *In vitro* antioxidant and antihyperlipidemic activities of *Bauhinia variegata* Linn. *Indian Journal of Pharmacology*, **41(5):** 227-232.

Rajkapoor, B., Jayakar, B., and Murugesh, N. (2003a). Antitumour activity of *Bauhinia variegata* on Dalton's ascetic lymphoma. *Journal of Ethnopharmacology*, **89**: 107-109.

Rajkapoor, B., Jayakar, B., and Murugesh, N. (2003b). Antitumour activity of *Bauhinia variegata* against ehrlich ascites carcinoma induced mice. *Pharmaceutical Biology*, **41(8):** 604-607.

Rajkapoor, B., Jayakar, B. and Murugesh, N., and Sakthisekaran, D. (2006). Chemoprevention and cytotoxic effect of *Bauhinia variegata* against N-nitrosodiethylamine induced liver tumors and human cancer cell lines. *Journal of Ethnopharmacology*, **104(3):** 407-409.

Rajkapoor, B., Jayakar, B., Anandam, R., and Kavimani, S. (2003c). Anti-ulcer effect of *Bauhinia variegata* Linn in rats. *Journal of Natural Remedies*, **3:** 215-216.

Rajkapoor, B., Murugesh, N., and Krishna, D.R. (2009). Cytotoxic activity of a flavanone from the stem of *Bauhinia variegata* Linn. *Natural Product Research*, **23(15):** 1384-1389.

Rao, Y.K., Fang, S.H., and Tzeng, Y.M. (2008). Anti-inflammatory activities of flavonoids and a triterpene caffeate isolated from *Bauhinia variegata*. *Phytotherapy Research*, **22(7):** 957- 962.

Reddy, M.V., Reddy, M.K., Gunasekar, D., Caux, C., and Bodo, B. (2003). A flavanone and a dihydrodibenzoxepin from *Bauhinia variegata*. *Phytochemistry*, **64(4):** 879-882.

Roy, S., and Bhalla, V. (1981). Haemagglutinins and lysins in plants and the application in characterizing human and animal red cells. *Australian Journal of Experimental Biology and Medical Science*, **59:** 195-201.

Rugenstein, S., and Lersten, N. (1981). Stomata on seeds and fruits of *Bauhinia* (Leguminosae: Caesalpinioideae). *American Journal of Botany*, **68(6):** 873- 876.

Sahu, G., Jain, S.K., and Pathak, S. (2011). Hepatoprotective activity of ethanolic extract of *Bauhinia variegata* Linn. leaves. *Pharmacologyonline*, **3:** 721-728.

Sammour, Reda Helmy; El-Shanshoury, and Abd El Raheem R. (1992). Antimicrobial activity of legume seed proteins. *Botanical Bulletin of Academia Sinica*, **33(2):** 185-190.

Schertz, K.F., Boyd, W.C., Jurgelsky, W., and Cabanillas E. (1958). Seed extracts with agglutinating activity for human blood. *Economic Botany*, **14(3):** 232-240.

Shah, B.N., Shah, M.R., Modi, D.C., and Sahu, D. (2010). Phyto-pharmacological profile of *Bauhinia variegata*. *Pharmacologyonline*, 829-837.

Shah, W.H., Jabbar, A., and Kausar, Saida. (1977). Studies on the composition and processing of *Bauhinia* species (Kachnar). *Pakistan Journal of Scientific and Industrial Research*, **20(6):** 384-388.

Sharma, A., Verma, R., and Ramteke, P. (2009). Antibacterial activity of some medicinal plants used by tribals against UTI causing pathogens. *World Applied Sciences Journal*, **7(3):** 332-339.

Sharma, K., and Sharma, S. D. (2006). Chemical composition and in sacco degradability of some fodder tree leaves, shrub leaves and herbaceous plants. *Indian Journal of Animal Sciences*, **76(7):** 538-541.

Sharma, N., Bhardwaj, R., Kumar, S., and Kaur, S. (2011). Evaluation of *Bauhinia variegata* L. bark fractions for *in vitro* antioxidant potential and protective effect against H_2O_2-induced oxidative damage to pBR322 DNA. *African Journal of Pharmacy and Pharmacology*, **5(12):** 1494-1500.

Sharma, R., Rajani, G.P., Sharma, V., and Komala, N. (2010). Effect of ethanolic and aqueous extracts of *Bauhinia variegata* Linn. on gentamicin-induced nephrotoxicity in rats. *Indian Journal of Pharmaceutical Education and Research*, **45(2):** 192-198.

Sharma, R.N., and Saxena, V.K. (1996). *In vitro* antimicrobial efficacy of leaves extracts of *Bauhinia variegata*. *Asian Journal of Chemistry*, **8(4):** 811-812.

Silva Jose, A., Damico Daniela, C.S., Baldasso Paulo, A., Mattioli Marcelo, A., Winck Flavia, V., Fraceto Leonardo, F., Novello Jose, C., and Marangoni S. (2007).

Isolation and biochemical characterization of a galactoside binding lectin from *Bauhinia variegata candida* (BvcL) seeds. *The Protein Journal,* **26(3):** 193-201.

Singh, R.S., Pandey, H.S., and Ghanshyam. (2006). Two new long chain compounds from *Bauhinia variegata* Linn. *Indian Journal of Chemistry,* **45B(9):** 2151-2153.

Srivastava, G.N., Bagchi, G.D. and Srivastava, A.K. (1988). Pharmacognosy of Ashoka stem bark and its adulterants. *International Journal of Crude Drug Research,* **26(2):** 65-72.

Thiruvenkatasubramaniam, R., and Jayakar, B. (2010). Anti-hyperglycemic and anti-hyperlipidemic activities of *Bauhinia variegata* L. on streptozotocin induced diabetic rats. *Der Pharmacia Lettre,* **2(5):** 330-334.

Tyagi, P., and Vasishtha, A.K. (1984). Component glycerides of seed oils. *Journal of the Oil Technologists' Association of India,* **15(2):** 84-86.

Veena, L., Prasad, G.C., Singh, K.P., and Udupa, K.N. (1975). Preventive effect of some indigenous drugs on experimental goitre in rats. *Journal* of *Research* in *Indian Medicine,* **10:** 12-18.

Wassel, G.M., Abdel Wahab, S.M., and Ammar, N.M. (1986). Constituents of the essential oils from *Bauhinia variegata* L. and *Bauhinia purpurea* L. flowers. *Scientia Pharmaceutica,* **54(4):** 357-361.

Wassel, G.M., El-Wahab, S.A., and Ammar, N.M. (1989). Seed proteins of selected *Bauhinia* species and their hemagglutinating effect. *Herba Hungarica,* **28(1-2):** 123-125.

Yadava, R.N., and Reddy, V.M. (2001). A new flavone glycoside, 5-hydroxy 7,3',4',5'-tetra-methoxyflavone 5-O-beta-D-xylopyranosyl-(1—>2)-alpha-L-rhamnopyranoside from *Bauhinia variegata* Linn. *Journal of Asian Natural Product Research,* **3(4):** 341-346.

Yadava, R.N., and Reddy, V.M. (2003). Anti- inflammatory activity of a novel flavanol glycoside from *Bauhinia variegata* Linn. *Natural Products Research,* **17(3):** 165-169.

You-kai, Xu., Hong-mao, Liu., Xiang-sheng, Dao., Chun-fen, Xiao., Chun-tao, Cai., Guang-hong, Chen., and Zi-yan, Xu. (2004). Nutrients content of *Bauhinia variegata* var. *candida* and its value as an edible wild flower. *Yunnan Daxue Xuebao, Ziran Kexueban,* **26(1):** 88-92.

Zaka Sh., Saleem M., Shakir N., and Ahmad Khan Sh. (1983). Fatty acid composition of *Bauhinia variegata* and *Bauhinia malabarica* seed oils–Comparison of their physico-chemical properties. *European Journal of Lipid Science and Technology,* **85(4):** 169- 170.

Zhang, Q., Song, S., Guo, J., Yang, M., Bai, W., and Ye, Y. (2010). Determination of metal elements in *Bauhinia* by microwave digestion and atomic absorption spectrometry. *Yunnan Minzu Daxue Xuebao, Ziran Kexueban,* **19(2):** 119-121.

Zhao Y.Y. (2005a). A new phenanthraquinone from the stems of *Bauhinia variegata* L. *Journal of Asian Natural Products Research,* **7:** 835- 838.

Zhao, Y., Cui, C., and Sun, Q. (2007). Inhibitory effect on the proliferation of human chronic myeloid leukemia K562 cells of Bauhinione from *Bauhinia variegata*. *Shenyang Yaoke Daxue Xuebao,* **24(2):** 109-112.

Zhao, Y., Cui, C., Cai, B., and Sun, Q. (2004a). Isolation and identification of chemical constituents from *Bauhinia variegata* L. *Zhongguo Yaowu Huaxue Zazhi,* **14(5):** 294-297.

Zhao, Y., Cui, C., Cai, B., and Sun, Q. (2005b). Chemical constituents from *Bauhinia variegata* L. *Zhongguo Yaowu Huaxue Zazhi,* **15(5):** 302-304.

Zhao, Y., Cui, C., Cai, B., Han, B., and Sun, Q. (2004b). Study on the constituents with anticancer activities of alkaloids from *Bauhinia variegata* L. *Zhongguo Yaowu Huaxue Zazhi,* **14(3):** 169-171.

Bioactive Phytochemicals: Perspectives for Modern Medicine Vol. 2 (2014) *Pages* **531–545**
Editor: **V.K. Gupta**
Published by: **DAYA PUBLISHING HOUSE, NEW DELHI**

24

Embelin: A Lead Molecule for the Future

Zeena S. Pillai*[1] and Beena Joy[2]

ABSTRACT

Medicinal plants play a prominent role in health care from time immemorial. Majority of drugs or drug formulations used in Ayurveda are principally derived from herbs and plants. A study of the morphological and pharmacological characteristics of the plant products can provide a better understanding of their active principles and mode of action. The isolation, characterization and standardization of the bioactive principles of the plants are a burning topic in Ayurvedic drug industry today.

The herbal medicine, Embelin isolated from Embelia ribes, belonging to the family Myrsinaceae, is getting significance in recent years due to its natural abundance and multifaceted activities. A wide spectrum of biological and physiological activities such as antihelmintic, antifertility, antitumour, antimicrobial, analgesic, antiinflammatory etc. are exhibited by Embelin. This chapter is a compilation of the research work carried out on the synthesis as well as the antibacterial properties of the derivatives of Embelin.

Keywords: *Embelia ribes*, Lead molecule, Embelin derivatives, Embelin-metal complexes, Antimicrobial effects, Photophysical studies.

1 Department of Chemistry, Amrita School of Arts and Sciences, Amrita Vishwa Vidyapeetham, Amritapuri, Kollam – 690 525, Kerala, India

2 Agroprocessing and Natural Product Division, National Institute for Interdisciplinary Science and Technology,Thiruvananthapuram, Kerala, India

* *Corresponding author*: E-mail: zeenapillai@gmail.com

Introduction

Majority of the world's population use traditional medicine (those derived from plant extracts) for health care (Ambati *et al.*, 2010; Bhutani *et al.*, 2007; Krishna *et al.*, 2007). Some of the recent statistics show that the annual turnover of herbal medicinal products manufactured by pharmaceutical companies is more than 100 times of that of modern drugs. It is well known that Ayurveda–the age old system of medicine prescribes herbal preparations for the prevention and treatment of diseases. With increase in the number of diseases among the rapidly growing human population the need for identification of the chemical compounds in these herbal preparations and identification of molecular targets has become highly essential. Immense work based on quinones is being carried out these days as it is one among the abundantly occurring natural product (Pardasani *et al.*, 2011). The colour and biological activity of these compounds add to their attraction (Nohl *et al.*, 1986). Several quinone derivatives are in different stages of clinical and preclinical development. They are believed to be involved in numerous cellular functions, thus, participating in the electron transfer and hydrogen transfer processes (Pardasani *et al.*, 2011). This in turn leads to the highly reactive oxygen species. Reactive oxygen species modifies and degrades nucleic acids and proteins with in cells (Waris and Ahsan, 2006; Kuo *et al.*, 1999).

Embelin: A Naturally Occurring Quinone

Embelia ribes is a threatened woody shrub sparsely distributed in the moist deciduous forests of the Western Ghats in India, Sri Lanka, Malaysia and South China. This is one of several embelia species used in traditional Chinese medicine.

Figure 24.1: Dried berries of *Embelia ribes*.

In Indian system of medicine 'Ayurveda', the plant is popularly known as 'Vidanga'. Embelin (Figure 24.2), a naturally occurring quinonoid compound, forms one of the simplest 1, 4-benzoquinone isolated from berries of the plant *Embelia ribes* (Figure 24.2), a species in Myrsinaceae family (Fernandes *et al.*, 1992; Agarwal *et al.*, 1999; Githori *et al.*, 2002). Myrsinaceae family comprise 35 genera and over 1000 species and is a rich source of natural products including quinines. Embelin is also a popular member of traditional system of medicine, Ayurveda. In addition to embelin, the major compounds previously isolated from *Embelia ribes* are quercitol, christembine and vilangin (Shahid *et al.*, 2009; Tripathi *et al.*, 2008; Choudhury *et al.*, 2007).

Figure 24.2: Embelin (2, 5-dihydroxy-3-undecyl-1, 4-benzoquinone).

Uses of *Embelia ribes*

According to World health Organization 'a medicinal plant is any plant which in one or more of its organs contains substances that can be used for therapeutic purposes or which are precursors for chemopharmaceutical semisynthesis'.

Embelia ribes is a woody shrub (Figure 24.3). The root bark is acrid, astringent, antihelmintic, antioestrogenic, carminative, digestive, laxative, soothing, stimulant, stomachic, and thermogenic. Leaves and fruits are astringent, thermogenic, contraceptive, diuretic, febrifuge, laxative demulcent and depurative in nature. It is useful in skin diseases, leprosy, dyspepsia, flatulence, colic, tumors, asthma, fever etc. (Bhutani *et al.*, 2007; Krishna *et al.*, 2007).

Figure 24.3: *Embelia ribes* shrub.

It is effective against intestinal parasites and intestinal worms. It is used in abdominal disorders, skin fungal infections, flatulence, constipation, indigestion, headache, hemorrhoids, lung diseases, obesity, piles, pneumonia, mouth ulcers, toothache and sore throat. Its decoction is useful in treating mental disorders and heart disease (Bhutani *et al.*, 2007; Krishna *et al.*, 2007).

Chemical structure of Embelin (2,5-dihydroxy-3-undecyl-1, 4-benzoquinone) is having quite resemblance with that of natural Coenzyme Q10 (ubiquinones) and the role of this is well defined in various biochemical protective mechanisms (Osbrink *et al.*, 2008; Krishnaswamy and Purushotaman, 1980)

Biological Studies of Embelin

Embelin has a long alkyl chain (undecyl) as a substituent, which confers solubility in the non-polar phase and cell permeability. The adjacent quinone and hydroquinone groups on embelin form intramolecular hydrogen bonds. Embelin exhibits a wide spectrum of biological and physiological activities. It has been reported to possess anti-fertility and antiimplantation (Chitra *et al.*, 1994) properties. In earlier studies, the antitumour, antiinflammatory, analgesic and antioxidant activities (Swamy *et al.*, 2007) of embelin were investigated. In this study, the effect of embelin on lipid profile in cancer has been assessed as alterations in lipid levels might serve as indicators of tumour regression or progression. It has been shown to decrease testosterone levels (Githui *et al.*, 1991) induce apoptosis in human myeloid HL-60 cells by targeting microtubular proteins (Xu *et al.*, 2005) and induce cleavage of receptor-interacting protein (RIP). The beneficial effects of embelin against oxidative tissue damage have been reported earlier (Joy and Lakshmi, 2010) The antibacterial activity of embelin was reported by Chitra *et al.* (1994). In this study, among the analyzed bacteria, the inhibition was highly significant against *Staphylococcus aureus*, *Streptococcus pyogenes*, *Shigella flexneri*, *Shigella sonnei* and *Pseudomonas aeruginosa*. Embelin shows antitumor activity in methylcholanthrene induced fibrosarcoma in albino rats besides enhancing their survival time. The study of embelin by Joshi *et al.* (2007) suggests that it can act as an antioxidant in physiological conditions. Embelin has been found to scavenge hydroxyl, glutathiyl, peroxyl radical and the inhibition of lipid peroxidation. Reduction of Fe (III) and restoration of Mn-superoxide dismutase *in vitro* studies suggest that it can be an efficient antioxidant in physiological system (Joy and Lakshmi, 2010). The wound healing activity of embelin (Swamy *et al.*, 2007) and the ethanol extract of the leaves of *Embelia ribes* were screened by excision, incision and dead space wound models on Swiss albino rats. The report from Sreepriya and Bali (2005) reveals that embelin and curcumin provide a good support for the chemopreventive effect against N-nitrosodiethylamine/phenobarbital induced hepatocarcinogenesis in Wistar rats. Recent studies have shown that embelin is a fairly potent non-peptidic cell permeable inhibitor of X-linked inhibitor of Apoptosis (XIAP) (Apoptosis is a genetically regulated cell-death mechanism) and it represents a promising lead compound for designing an entirely new class of anticancer agents (Nikolovska *et al.*, 2004; Wang *et al.*, 2006).

Isolation of Embelin

About 100 g of the powdered berries of *E. ribes* was extracted with n-hexane in the solvent ratio 1: 15(w/v). After the complete extraction, the solvent was removed by evaporation under reduced pressure. The hexane extract was subjected to column chromatography over silica gel (100–200 mesh). Elution of the column with hexane/ethyl acetate (1 : 9) yielded an orange colour powder. The crude product was recrystallised from cold hexane yielding orange flakes of embelin. The characterization has been carried out by spectroscopic techniques IR, MS, and NMR. The purity of embelin has been confirmed by using HPLC analysis.

Melting point: 134°C

Molecular formula: $C_{17}H_{26}O_4$.

Molecular weight: 294.

Objective of the Present Investigation

In recent years, the herbal medicine such as embelin is gaining popularity mainly due to its natural abundance and multifaceted activities. It can control cell division there by inhibiting cancer. It can be used as antifertility reagent there by controlling reproduction. Antitumour, antiinflammatory, analgesic and antioxidant properties (Chitra *et al.*, 1994) of embelin have been investigated earlier. In order to harvest these properties, and make it available for the welfare of mankind it is highly essential to prepare biocompatible, biologically active derivatives of embelin. A detailed study on the structure-activity correlation has to be carried out in order to understand the mechanistic scheme in detail.

Present Work

Embelin has a long alkyl chain (undecyl) as a substituent, with adjacent quinone and hydroquinone groups. Several research groups have synthesised various derivatives of embelin (Gree *et al.*, 2008; Braun *et al.*, 2008).

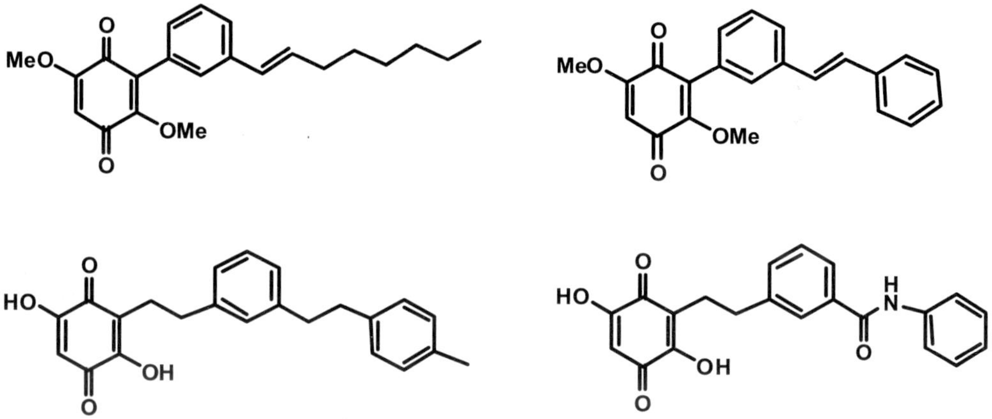

MeO O OMe O

MeO O OMe O

HO O OH O

HO O OH O H N O

Chart 1 (Gree *et al.*, 2011)

They have been prepared by Domino Knoevenagel hetero Diels-Alder reactions by Braun *et al.* (2008). Srinivas *et al.* (2010) has reported the anticancer, antimicrobial and antimitotic activity of a few derivatives of embelin. Several others are involved in the studies based on antifungal, antibacterial activities of embelin (Tripathi *et al.*, 2008; Choudhury *et al.*, 2007). As per some of the earlier reports, increasing the number of aromatic substituents and increasing the chain length of the alkyl group improves biological activity (Srinivas, 2010). Embelin is expected to produce semiquinone radical in the oxidation and reduction reaction. Few attempts have been made to characterize the semiquinone radical by pulse radiolysis and to do the structural investigation by quantum chemical methods (Joshi *et al.*, 2007, 2009).

Biological Studies of Embelin Derivatives

Biological studies of only a very few derivatives of embelin has been carried out so far.

Anti-mitotic Activity (Srinivas, 2010)

The derivatives listed here are prepared by phase transfer catalysis and are reported to exhibit antimitotic activity.

a) 2, 5-dihydroxy-3-undecylcyclohexa-2, 5-diene-1, 4-dione
b) 2-hydroxy-5-methoxy-3-undecylcyclohexa-2, 5-diene-1, 4-dione
c) 5-ethoxy-2-hydroxy-3-undecylcyclohexa-2, 5-diene-1, 4- dione
d) 2-hydroxy-5-propoxy-3-undecylcyclohexa-2, 5-diene-1, 4-dione
e) 5-butoxy-2- hydroxy-3-undecyl benzo-1, 4-quinone
f) 5-allyloxy-2- hydroxy-3-undecyl benzo-1, 4-quinone
g) 5-benzyloxy-2- hydroxy-3-undecyl benzo-1, 4-quinone

Anticancer and Antimicrobial Activity (Srinivas, 2010)

h) 2,5-dihydroxy-3-undecylcyclohexa-2,5-diene-1,4-dione
i) 2-hydroxy-5-methoxy-3-undecylcyclohexa-2,5-diene-1,4- dione
j) 5-ethoxy-2-hydroxy-3-undecylcyclohexa-2,5-diene-1,4- dione
k) 2-hydroxy-5-propoxy-3-undecylcyclohexa-2,5-diene-1,4- dione
l) 5-butoxy-2- hydroxy-3-undecyl benzo-1,4-quinone
m) 5-allyloxy-2- hydroxy-3-undecyl benzo-1,4-quinone
n) 5-benzyloxy-2- hydroxy-3-undecyl benzo-1,4-quinone

Among these derivatives 5-benzyloxy-2- hydroxy-3-undecyl benzo-1, 4-quinone has showed maximum anti cancer properties and 2-hydroxy-5-methoxy-3-undecylcyclohexa-2, 5-diene-1, 4-dione has showed maximum anti bacterial properties. Even though few reports on the biological activity of the derivatives are present (Ambati *et al.*, 2010), a systematic study on the structure activity correlation is yet to be explored in detail.

Our Studies and Reports

Despite all the reports on embelin and its derivatives to ascertain the biological activity hardly any efforts have been made so far to elucidate the photophysical properties and thus have a structure-activity correlation. With these aspects in mind we prepared a few derivatives of embelin as per reported procedure (*Wie* and Du, *1963;* Christopher *et al.*, 2008; Brahmeshwari, 2009) (Chart 2). Our preliminary studies show that the derivatives **1-4** possess antibacterial activity when compared with streptomycin. We were interested in looking into the photo physical properties of the derivatives in order to better understand the mechanism involved in its activity.

Preliminary photo physical properties of the derivatives were carried out. Figure 24.4 shows the absorption spectra of embelin in solvents of varying polarity.

The solvents chosen for the study are DMSO, chloroform, acetonitrile and methanol. The absorption maxima of embelin in different solvents are listed in chart 1. A slight absorption centered around 450 nm is also observed in methanol. Table 24.1 consolidates the absorption characteristics of embelin in various solvents.

Chart 2

The extinction coefficient of embelin in acetonitrile was determined to be 11480 M^{-1} cm^{-1}. It exhibits a weak emission centered around 325 nm (quantum yield < 0.001). Figure 24.5 shows the absorption spectra of compounds **1, 2, 3** and **4** recorded in acetonitrile. Spectral profile of embelin and methoxy derivative looks similar. The

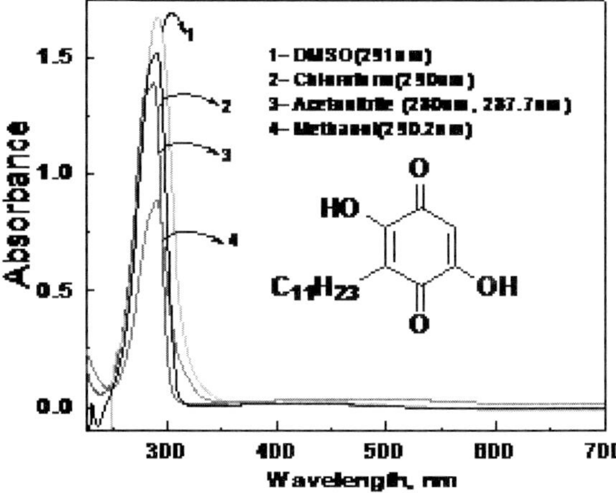

Figure 24.4: Absorption spectra of embelin in solvents of varying polarity.

benzhydrazide adduct (Compound 4, Chart 2) has a blue shifted band centered around 371 nm and a shoulder band centered around 235 nm.

Table 24.1: Absorption characteristics of embelin in various solvents.

Solvent	*Absorption Maxima*
DMSO	291 nm
Chloroform	290 nm
Acetonitrile	280 nm, 287.7 nm
Methanol	290.2 nm

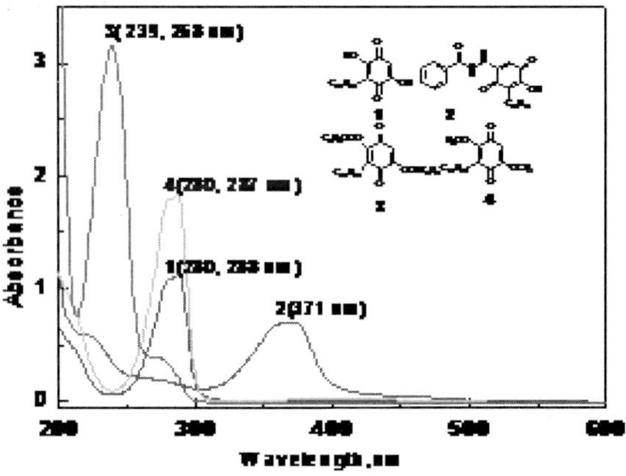

Figure 24.5: Absorption spectra of **1, 2, 3 and 4** in acetonitrile.

It is worth noting that the emission quantum yields of all the derivatives (**1-4**) in acetonitrile reported here is very low. It is <0.001. Thus the derivatives under study may be possessing a triplet state with a reasonably good quantum yield. More over some of the preliminary results from our research group suggest that embelin could be used for photodynamic therapy. Photodynamic Therapy (PDT) is used for treating cancer (McCaughan, 1999; Hoebeke *et al.*, 2001). After intravenous injection of a PDT drug as a photo sensitizer, the drug is selectively retained by the tumor cells. After a certain period of time the tumor becomes more sensitive than the adjacent normal tissues.

When such PDT drug is activated by laser, the tumor is more likely to be killed by cytotoxic agents. The main cytotoxic agent in PDT is widely believed to be singlet oxygen ($^1O_2$), a highly reactive oxygen species that oxidizes various biological substrates. $^1O_2$ is believed to be a major mediator of photodynamic damage in biological systems. In PDT, $^1O_2$ is produced via the following Type-II photochemical pathway (Patterson *et al.*, 2002),

$$S_0 + h\nu_0 \rightarrow S_1 \quad (1)$$

$$S_1 \rightarrow T_1 \quad (2)$$

$$T_1 + {}^3O_2 \rightarrow S_0 + {}^1O_2 \quad (3)$$

where, S_0, S_1, T_1 are the photosensitizers in ground state, the first excited singlet state and the first excited triplet state and $^3O_2$ and $^1O_2$, the ground-state triplet and the excited singlet states of oxygen respectively. Therefore, $^1O_2$ is strongly related to PDT, and monitoring $^1O_2$ generation is important in PDT drug development. Once produced, $^1O_2$ molecules can either undergo oxidization of surrounding biomolecules or undergo radiative decay at around 1270 nm (Patterson *et al.*, 2002).

Embelin-Metal Complexes

Immobilisation of transition metal complexes within the super cages of zeolites or mesoporous materials has been a topic of intense research in recent years (Yusuff and Abraham, 2003). The reduction of dioxygen by hydrogen in presence of transition metal complexes is an area of considerable interest with respect to gas separations. Though several transition metal complexes were found to be active in catalysing the reaction, their application on an industrial scale is restricted due to the decomposition of ligand by attack of oxygen due to dimerisation (Kiprono *et al.*, 2005) Zeolite encapsulation is known to safeguard the complexes.

The ability of complexes to function as catalyst for the reduction of oxygen was explored and the results indicate that, the complexes are catalytically active; and the thermal and mechanical stability increases as a result of encapsulation (Rao and Usharani, 2010). The optimization of the reaction conditions or modifications of the complex could lead to far reaching applications in gas purification. *Even though catalytic properties of the metal-complexes have been looked into hardly any studies have been carried out to see if they are biologically active.*

Synthesis of Embelin–Metal Complexes

Scheme 1

Metal salts were reacted with embelin in a 1: 2 ratio to yield different complexes (Vlietinck and Berghe, 1991). The complexes were characterized using IR spectroscopy. IR analysis of metal complexes showed that the metal ions were coordinated through the carbonyl and hydroxyl oxygen atoms. The ligands chosen for our present study are NO_2, SO_4, H_2O etc. and the metal ions include Zn, Fe, Co and Cu.

The complexes prepared for our studies are shown in Chart 3.

The C=O and O-H bands in embelin appear at 1610 and 3300 cm^{-1} respectively. The IR spectrum of the complex showed a shift of the carbonyl peak to 1520 cm^{-1}. This indicates a decrease in the stretching frequency C=O as a consequence of coordination through the carbon and oxygen atom. The absorption attributed to the hydroxyl group appeared as a broad peak in the region 3600 to 3050 cm^{-1} with a maximum at

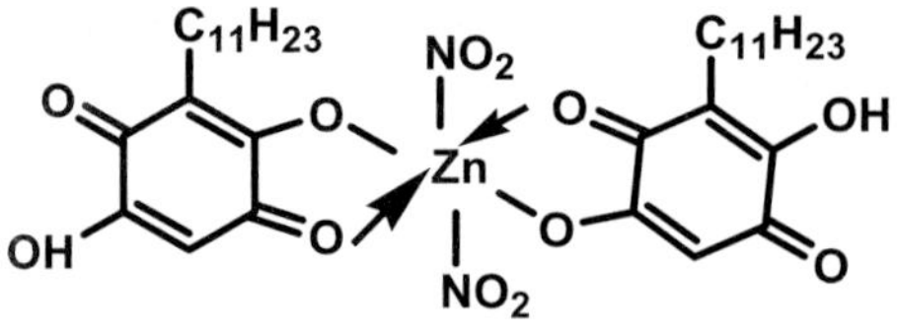

7a 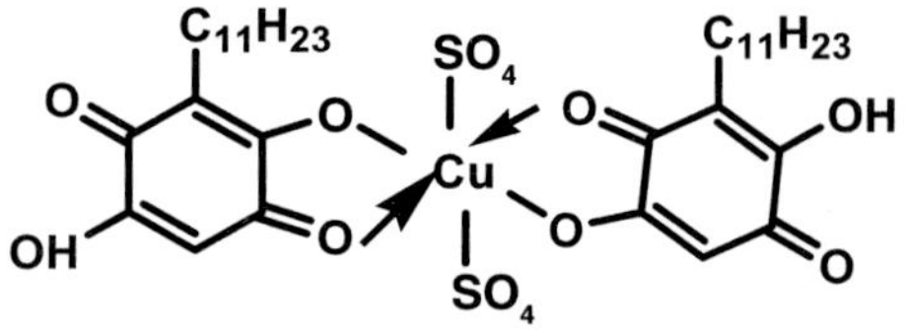 **7b**

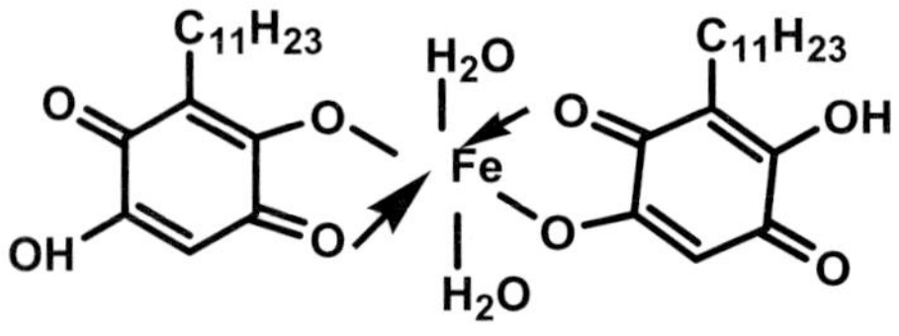

7c 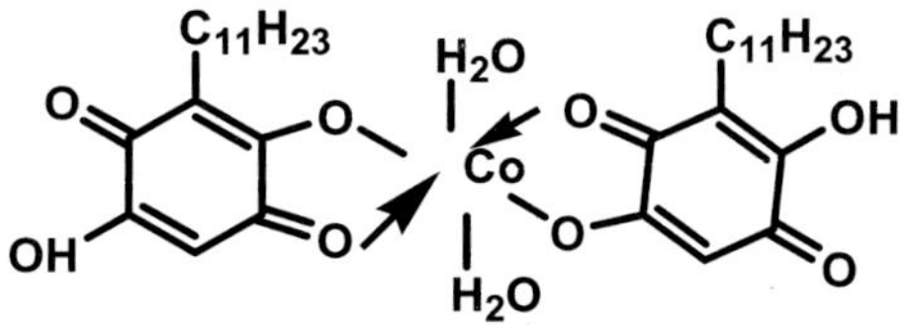 **7d**

Chart 3

3550 cm^{-1}. The peak at 1365 cm^{1} in complex **7a** is assigned to coordinated nitrate, suggesting that the nitrate group contribute to the coordination sphere in the case of zinc. In the spectrum of complex **7b,** there is a considerable shift of the carbonyl absorption from 1610 cm^{-1} to 1460 cm^{-1}. The broad peak at 1400 cm^{-1} may indicate the presence of coordinated sulphate group. The rest of the IR bands are similar to that observed in the case of embelin. The overall IR evidence suggests that the embelin residue acts as a bi-dentate molecule and coordinates through carbonyl and the hydroxyl oxygen atoms.

Antibacterial Screening of Embelin–Metal Complexes

Screening of compounds **5**, **6** and **7a-7d** for antibacterial activity was done by the Disc Diffusion method (Sherman and Cappucino, 1999). Standard strains [Microbial Type Culture Collection (MTCC)] of Gram-positive and Gram-negative bacteria obtained from the Institute of Microbial Technology, Chandigarh, India. The results are presented in Table 24.4. The derivatives show weak antimicrobial activity against the microorganisms tested.

Agar Diffusion Assay

The bacteria were grown in nutrient broth for 24h (pH 7.2–7.4) and were used as inoculums. The Mueller-Hinton agar medium were poured into the plates to uniform depth of 5 mm and allowed to solidify. Then the microbial suspensions in a concentration of 5×10^{-6} cfu/ml were spread over the surface of media using a sterile cotton swab to ensure the confluent growth of the organism. Aliquots of 10μL samples were impregnated on Whatman No.1 filter paper discs of 6 mm diameter. These discs were then aseptically applied to the surface of the agar plates at well-spaced intervals. The plates were incubated at 37 °C for 24 h and observed inhibition zones including the diameter of the discs were measured. Control discs impregnated with 10μL of the solvent DMSO and streptomycin (10μg/disc), reference for bacteria was used alongside the test discs in each experiment.

The 10 μl of solution from clear wells were spread on nutrient agar plates. The number of surviving organisms was determined after overnight aerobic incubation at 27°C. All experiments were performed in triplicate. Disc capacity is 10 μl/disc.

Table 24.4: Anti microbial effects of embelin complexes.

Organism	*5*	*6*	*7a*	*7b*	*7c*	*7d*	*S*	*D*
Antibacterial								
1. *Proteus vulgaris* [426]	–	–	–	–	–	–	14	
2. *Bacillus cereus* [430]	7	8	–	–	–	–	15	
3. *Escherichia coli* [443]	7	–	–	–	–	–	10	
4. *Serratia marcescens* [97]	–	7	–	–	–	–	16	
5. *Pseudomonas aeruginosa* [741]	8	–	–	–	–	–	14	
6. *Bacillus subtilis* [441]	–	7	–	–	–	–	11	
7. *Salmonella typhi* [733]	8	–	–	–	–	–	12	
8. *Staphylococcus aureus* [96]	–	–	–	–	–	–	21	
9. *Klebsiella pneumoniae* [109]	–	8	–	–	–	–	10	
Antifungal								**F**
1. *Candida albicans* [3017]	–	–	–	–	–	–	21	
2. *Candida albicans* [1637]	–	–	–	–	–	–	8	
3. *Candida albicans* [227]	–	–	–	–	–	–	10	
4. *Candida glabrata* [3019]	–	–	–	–	–	–	9	

S: Streptomycin; D: DMSO; F: Flucanazole.

Streptomyces and Flucanazole were used as the +ve control and DMSO was used as the negative control. Unlike the derivatives **1-4**, the metal complexes of embelin hardly exhibited any antibacterial or antifungal activity. Vilangin **5** and compound **6** showed very weak activity in comparison with Streptomycin.

Conclusions

There has been considerable interest in and support for the search of novel biologically active natural products. Terrestrial plants serve as unique renewable source of potential new drugs and biological entities. The use of bioactive natural products as cardioactive, hypolipidaemic, antiviral, antituberculosis, antitumor agents etc are a few examples of immense chemical diversity in plants. Proper derivatization of the lead molecule using simple synthetic steps is essential to improve the efficacy. In the present work, we have prepared a few derivatives of the natural product embelin. Preliminary photophysical studies as well as biological studies were carried out. Unlike the alkyl/aryl substituted derivatives, metal complexes of embelin were found to be biologically inactive. The dimer vilangin was observed to be weakly active. Preliminary studies carried out in our group show that embelin could be used for photodynamic therapeutic applications. The low fluorescence of the derivatives prompts us to expect reasonably good triplet yield as well as singlet oxygen generation efficiency for these derivatives. A detailed investigation into the photophysical properties is highly essential to draw a structure-activity correlation. The parent compound being an abundantly available natural product, this is an easy and economic method to prepare biologically active compounds. Our preliminary studies on the synthesis and biological studies showed that embelin, the lead molecule of *Embelia ribes* is a promising molecule for the future.

Acknowledgements

The authors thank Dr.Pradeep, Scientist, TBGRI, Palode. The infrastructural and financial support from Amrita Viswa Vidyapeetham, Amritapuri Campus, Karunagapally and NIIST, Trivandrum is gratefully acknowledged.

References

Agarwal, S, Chauhan, S.K., and Singh, B.P. (1999). Determination of embelin from *Embelia ribes* by high performance thin layer chromatography. *Indian Drugs*, **36**: 41-43.

Ambati, S., Jyothi, V., and Jyothi, A.V.(2010). Phramacological, pharmacognostic and phytochemical review of *Embelia ribes*. *International Journal of Pharmacy and Technology*, **2**: 525- 539.

Bhutani, K.K., Singh, I.P., Sandip, B., Bharati, A., and Singh (2007). Fate of embelin in pippalyadi yoga an ayurvedic oral contraceptive: Structure of embeli-borax complex and evaluation of anti-fertility activity. *Indian Journal of Chemistry*, **46(B)**: 320-325.

Brahmeshwari,G. (2009). Synthesis of 6,7-dihydroxy-3-aryl-5-undecyl-4,1,2-benzoxadiazines. *Ind. Journal of Chemistry*, **10**: 1457.

Braun, Estévez, A., Jimenez, S.A., Chávez, H., Ravelo, Á. G., and Feresin, A. Tapia. (2008). An efficient synthesis of embelin derivatives through domino Knoevenagel hetero Diels–Alder reactions under microwave irradiation. *Tetrahedron*, **64:** 8938-8942.

Chitra, M., Sukumar, E., Suja, V., and Devi, C.S. (1994). Antitumor, anti-inflammatory and analgesic property of embelin, a plant product. *Chemotherapy*, **40:** 109–113.

Choudhury, R.P., Md. Ibrahim, A., Bharathi, H.N., and Venkatasubramanian, P. (2007). Quantitative analysis of embelin in *Myrsine africana* L. (Myrsinaceae) using HPLC and HPTLC. *Electronic Journal of Food and Plants Chemistry*, **2(1):** 20-24.

Christopher, M. J., Loic, G., and David, T. (2008). Synthesis of the reported structure of the bisbenzoquinone lanciaquinone, isolated from *Maesa lanceolat*. *Org. lett.*, **10:** 4505-08.

Fernandes, J.B., Januaro, A.H., Fatima, D.M., De Silva, F., and Viera, P.C.(1992). Dammarane and cycloartane triterpenoids from three *Rapanea* species. *Phytochemistry*, **4:** 1251-1253.

Githori, J.B., Hoglund, J., Waller, P.J., and Baker, L.R. (2002). Anthelmintic activity of preparations derived from *Myrsine africana* and *Rapanea melanophloeos* against the nematode parasite, *Haemonchus contortus*, of sheep. *J.O.Ethnopharmacology,* **80:** 187-191.

Githui, E.K., Makawiti, D.W., and Midiwo, J.O. (1991). Changes in the concentrations of testosterone, luteinising hormone and progesterone associated with administration of embelin. *Contraception,* **44**: 311–317.

Gree, R., Viault, G., Gree, D., Das, S., and Singh, J.Y. (2011). Synthesis of a focused chemical library based on derivatives of embelin, a natural product with proapoptotic and anticancer properties. *Eur.J.Org.Chem.,* 1233-1241.

Hoebeke, M., Damoiseau,X., Schuitmaker, H.J., and Lagerberg, J.W.M. (2001). Increase of the photosensitizing efficiency of the bacteriochlorin *a* by liposome-incorporation. *J Photochem. Photobiol B-Biol.,* **60**: 50–60.

Joshi, R., Kamat, J.P., and Mukherjee,T. (2007). Free radical scavenging reactions and antioxidant activity of embelin: Biochemical and pulse radiolytic studies. *Chemico-Biological Interaction,* **167:** 125-134.

Joshi,R., Ghanty,T.K., and Mukherjee, T. (2009). Formation of semiquinone radical in the reaction of embelin (2,5-dihydroxy-3-undecyl-1,4-benzoquinone) with reductants as well as oxidants. Characterization by pulse radiolysis and structure investigation by quantum chemical study. *Journal of Molecular Structure,* **928(1–3):** 46–53.

Joy, B., and Lakshmi, S. (2010). Antiproliferative properties of *Embelia ribes. The Open Process Chemistry Journal,* **3:** 17-22.

Kiprono, C. P., Cherutoi, J. K., and Cheruiyot, L. L. (2005). Synthesis and characterisation of Zn(II) and Copper (II) complexes of embelin. *Bull. Chem. Soc. Ethiop.,* **19:** 295.

Krishna, V., Kumara Swamy, H.M., Shankarmurthy, K., Abdul Rahiman, B., Mankani, K.L., Mahadevan, K.M., Harish, B.G., and Raja Naika, H. (2007). Wound healing activity of embelin isolated from the ethanol extract of leaves of *Embelia ribes* Burm. *Journal of Ethnopharmacology,* **109**: 529–534.

Krishnaswamy, M., and Purushothaman, K.K. (1980). Antifertility properties of *Embelia ribes* (embelin). *Indian J Exper Biol.,* **18(11):** 1359-1360.

Kuo Min-Liang, Shine-Gwo Shiah, Chau-Jong Wang, and Shuang-En Chuang. (1999). Suppression of apoptosis by Bcl-2 to enhance benzene metabolites-induced oxidative DNA damage and mutagenesis: A possible mechanism of carcinogenesis. *Mol Pharmacol.,* **55:** 894-901.

McCaughan, J. S. Jr. (1999). Photodynamic therapy: a review. *Drugs Aging,* **15(1):** 49-68.

Nikolovska-Coleska, Z., Xu, L., Hu, Z., Tomita, Y., Li, P., Roller, P.P., Wang, R., Fang, X., Guo, R., Zhang, M., Lippman, M.E., Yang, D., and Wang, S.(2004). Discovery of embelin as a cell-permeable, small-molecular weight inhibitor of XIAP through structure-based computational screening of a traditional herbal medicine three-dimensional structure database. *J. Med. Chem.,* **47**: 2430-40.

Nohl, H., Jordan, W., and Youngman, R.J. (1986). Quinones in biology: Functions in electron transfer and oxygen activation. *Advances in Free Radical Biology and Medicine*, **2(1):** 211–279.

Osbrink, W.L.A., Kobaisy. M., Cantrell, C.L., Mims, A.B., Lax, A.R., and Tellez, M.R. (2008). Activity of 1, 4-Benzoquinones against Formosan Subterranean Termites (*Coptotermes formosanus*). *J Agric Food Chem.*, **56**: 4021-4026.

Pardasani, T.P., Abraham, I., Joshi, R., and Pardasani, P. (2011). Recent advances in 1,4-benzoquinone chemistry. *J. Braz. Chem. Soc.,* **22***:* 385-421.

Patterson, M.S, Niedre, M., and Wilson, B.C. (2002). Direct near-infrared luminescence detection of singlet oxygen generated by photodynamic therapy in cells *in vitro* and tissues *in vivo*. *Photochemistry and Photobiology,* **75**: 382-391.

Rao, J.V., and Usharani, G. N. (2010). Chemical speciation of binary complexes of embelin with some biologically important metal ions. *Acta Chim. Slov.,* **57**: 916.

Shahid, M., Shahzad, A., Sobia, F., Sahai, A., Tripathi, T., Singh, A., and Khan, H.M. (2009). Plant natural products as a potential source for antibacterial agents: Recent trends. *Anti-Infective Agents in Medicinal Chemistry,* **8(3):** 211-225.

Sherman, N., and Cappucino, J. G. (1999). *Microbiology–A Laboratory Manual* California, Benjamin Cumming Science Publishing,

Sreepriya, M. (2006). Effects of administration of embelin and curcumin on lipid peroxidation, hepatic glutathione antioxidant defense and hematopoietic system during N nitrosodiethylamine/Phenobarbital- induced hepatocarcinogenesis in Wistar rats. *Mol Cell Biochem.,* **284(1–2):** 49–55.

Sreepriya, M., and Bali, G. (2005).Chemopreventive effects of embelin and curcumin against *N*-nitrosodiethylamine/phenobarbital-induced hepatocarcinogenesis in Wistar rats. *Fitoterapia,* **76**: 549-555.

Srinivas, K. (2010). Anti-mitotic activity of embelin derivatives. *International Journal of Phytopharmacology,* **1(2):** 97-102.

Srinivas, K. (2010). Anticancer and antimicrobial activity of embelin derivatives. *International journal of preclinical and pharmaceutical research,* **1**: 35-38.

Swamy, H.M., Krishna, V., Shankarmurthy, K., Abdul, R.B., Mankani, K.L., and Mahadevan, K.M. (2007).Wound healing activity of embelin isolated from the ethanol extract of leaves of *Embelia ribes* Burm. *Journal of Ethnopharmacology,* **109**: 529-534.

Tripathi, C.D., Bhandari, U., Nazam Ansari, M., and Islam, F. (2008). Protective effect of aqueous extract of *Embelia ribes* Burm fruits in middle cerebral artery occlusion-induced focal cerebral ischemia in rats. *Indian Journal of Pharmacology,* **40(4):** 152-157.

Vlietinck, A. J., and Berghe, D. A.V. (1991). *Screening Methods for Antibacterial and Antiviral Agents from Higher Plants, Methods in Plant Biochemistry*. Dey, P. M.; Harborne, J. B. Editors London, Academic Press.

Waris, G., and Ahsan, H. (2006). Reactive oxygen species: Role in the development of cancer and various chronic conditions. *J. Carcinog.*, 5-14.

Wang, S., Chen, J., Nikolovska-Coleska, Z., Wang, G., and Qiu, S.(2006), Design, synthesis, and characterization of new embelin derivatives as potent inhibitors of X-linked inhibitor of apoptosis protein. *Bioorganic and Medicinal Chemistry Letters*, **16(22):** 5805-5808.

Wie, J.S., and Du, Y.C. (1963). Study of vermifuges. I. Isolation of embelin from the fruit of mugua-wha (*Embelia oblongifolia* Hemsl.). *Yao Xue Xue Bao.*, **10:** 578-580.

Xu, M., Cui, J., Fu, H., Proksch, P., Lin, W., and Li, M. (2005). Embelin derivatives and their anticancer activity through microtubule disassembly. *Planta Med.*, **71:** 944-948.

Yusuff, K. K. M., and Abraham, R. (2003) Copper(II) complexes of embelin and 2-aminobenzimidazole encapsulated in zeolite Y-potential as catalysts for reduction of dioxygen. *Journal of Molecular Catalysis A. Chemical*, **198:** 175-183.

Previous Volume Contents

Bioactive Phytochemicals: Perspectives for Modern Medicine Vol 1/*Gupta, V K et al eds*

2012, xi+516p., col. plts., figs., tabls., ind., 25 cm

ISBN 978-81-7035-779-7 Rs. 2800/US$ 122

Index

A

B

D

E

F

M

O

P

Q

U

V